The Palgrave Handbook of Literature and Mathematics

Robert Tubbs · Alice Jenkins ·
Nina Engelhardt
Editors

The Palgrave Handbook of Literature and Mathematics

palgrave
macmillan

Editors
Robert Tubbs
University of Colorado Boulder
Boulder, CO, USA

Alice Jenkins
University of Glasgow
Glasgow, UK

Nina Engelhardt
Department for English Literatures
University of Stuttgart
Stuttgart, Baden-Württemberg, Germany

ISBN 978-3-030-55477-4 ISBN 978-3-030-55478-1 (eBook)
https://doi.org/10.1007/978-3-030-55478-1

© The Editor(s) (if applicable) and The Author(s), under exclusive license to Springer Nature
Switzerland AG 2021, corrected publication 2022
Chapter 1 is licensed under the terms of the Creative Commons Attribution 4.0 International
License (http://creativecommons.org/licenses/by/4.0/). For further details see license
information in the chapter.
This work is subject to copyright. All rights are solely and exclusively licensed by the Publisher,
whether the whole or part of the material is concerned, specifically the rights of translation,
reprinting, reuse of illustrations, recitation, broadcasting, reproduction on microfilms or in any
other physical way, and transmission or information storage and retrieval, electronic adaptation,
computer software, or by similar or dissimilar methodology now known or hereafter developed.
The use of general descriptive names, registered names, trademarks, service marks, etc. in this
publication does not imply, even in the absence of a specific statement, that such names are
exempt from the relevant protective laws and regulations and therefore free for general use.
The publisher, the authors and the editors are safe to assume that the advice and information in
this book are believed to be true and accurate at the date of publication. Neither the publisher
nor the authors or the editors give a warranty, expressed or implied, with respect to the material
contained herein or for any errors or omissions that may have been made. The publisher remains
neutral with regard to jurisdictional claims in published maps and institutional affiliations.

Cover credit: Konstantin Kirillov/Alamy Stock Photo

This Palgrave Macmillan imprint is published by the registered company Springer Nature
Switzerland AG
The registered company address is: Gewerbestrasse 11, 6330 Cham, Switzerland

ACKNOWLEDGMENTS

We thank all of the scholars who agreed to contribute to this Handbook and then submitted such engaging, wide-ranging chapters.

We also thank our editor at Palgrave, Allie Troyanos, and production assistant Rachel Jacobe, for their guidance, tireless professionalism, and patience.

Without all of these people this Handbook would not have been possible.

Robert Tubbs
Alice Jenkins
Nina Engelhardt

CONTENTS

1   **Introduction: Relationships and Connections Between
Literature and Mathematics**   1
Nina Engelhardt and Robert Tubbs

Part I   Mathematics in Literature

2   **Numbered Possibilities: Chaucer and the Evolution
of Late-Medieval Mathematics**   23
David Baker

3   **Mercantile Arithmetic, Financial Profit, and Ben Jonson's
*The Devil Is an Ass***   41
Joe Jarrett

4   **Mathematics and Poetry in the Nineteenth Century**   61
Daniel Brown

5   **Non-Normative Euclideans: Victorian Literature
and the Untaught Geometer**   81
Alice Jenkins

6   **Mathematical Contrariness in George Eliot's Novels**   97
Derek Ball

7   **Mathematics in Russian Avant-Garde Literature**   113
Anke Niederbudde

viii CONTENTS

8   Uses of Chaos Theory and Fractal Geometry in Fiction   129
Alex Kasman

9   Mathematical Clinamen in the Encyclopedic Novel: Pynchon, DeLillo, Wallace   149
Stuart Taylor

10   Squaring the Circle: A Literary History   169
Robert Tubbs

Part II   Mathematics and Literary Forms

11   Mathematics and Poetic Meter   189
Jason David Hall

12   Randomizing Form: Stochastics and Combinatorics in Postwar Literature   207
Alison James

13   Oulipian Mathematics   227
Warren Motte

14   Mathematics and Dramaturgy in the Twentieth and Twenty-First Centuries   243
Liliane Campos

15   Nonlinearity, Writing, and Creative Process   263
Ira Livingston

Part III   Mathematics, Modernism, and Literature

16   Mathematics and Modernism   281
Nina Engelhardt

17   Mathematics in German Literature: Paradoxes of Infinity   299
Howard Pollack-Milgate

18   The Ghosts of Departed Quantities: Samuel Beckett and Gottfried Wilhelm Leibniz   319
Chris Ackerley

CONTENTS ix

19 "Numbers Have Such Pretty Names": Gertrude Stein's
Mathematical Poetics 339
Anne Brubaker

20 Modernist Literature and Modernist Mathematics I:
Mathematics and Composition, with Mallarmé,
Heisenberg, and Derrida 361
Arkady Plotnitsky

21 Modernist Literature and Modernist Mathematics II:
Mathematics and Event, with Mallarmé, Gödel,
and Badiou 385
Arkady Plotnitsky

Part IV Relations Between Literature and Mathematics

22 *King Lear*, Without the Mathematics: From Reading
Mathematics to Reading Mathematically 399
Travis D. Williams

23 Newton, Burns, and a Poetics of Figure: Toward
a Prehistory of Consilience 419
Matthew Wickman

24 The Mathematics of Associationism in Laurence Sterne's
Tristram Shandy 439
Aaron Ottinger

25 Romantic Parts and Wholes, Statistical and Literary 457
Margaret Kolb

26 "Colours of the Dying Dolphin": Nineteenth-Century
Defenses of Literature and Mathematics 473
Imogen Forbes-Macphail

27 Combinatorial Characters 493
Andrea Henderson

28 Datelines 513
Steven Connor

x   CONTENTS

**29   The Metaphor as an Equation: Ezra Pound
       and the Similitudes of Representation**   529
Jocelyn Rodal

Part V   Mathematics as Literature

**30   Rehearsing in the Margins: Mathematical Print and
       Mathematical Learning in the Early Modern Period**   553
Benjamin Wardhaugh

31   Mathematics, Narrative, and Temporality   569
Marcus Tomalin

**32   A Cognitive and Quantitative Approach to Mathematical
       Concretization**   589
Marc Alexander

**Correction to: Introduction: Relationships and Connections
Between Literature and Mathematics**   C1
Nina Engelhardt and Robert Tubbs

Index   609

A Chronological Organization
of the Chapters

Note: Some chapters appear in several sections

Middle Ages

2. Numbered Possibilities: Chaucer and the Evolution of Late-Medieval Mathematics
David Baker

11. Mathematics and Poetic Meter
Jason David Hall

10. Squaring the Circle: A Literary History
Robert Tubbs

17. Mathematics in German Literature: Paradoxes of Infinity
Howard Pollack-Milgate

Early Modern Period

30. Rehearsing in the Margins: Mathematical Print and Mathematical Learning in the Early Modern Period
Benjamin Wardhaugh

3. Mercantile Arithmetic, Financial Profit, and Ben Jonson's *The Devil Is an Ass*
Joe Jarrett

11. Mathematics and Poetic Meter
Jason David Hall

10. Squaring the Circle: A Literary History
Robert Tubbs

xii   A CHRONOLOGICAL ORGANIZATION OF THE CHAPTERS

17. Mathematics in German Literature: Paradoxes of Infinity
Howard Pollack-Milgate

22. *King Lear*, Without the Mathematics: From Reading Mathematics to
Reading Mathematically
Travis D. Williams

Eighteenth Century

11. Mathematics and Poetic Meter
Jason David Hall

23. Newton, Burns, and a Poetics of Figure: Toward a Prehistory of Consilience
Matthew Wickman

24. The Mathematics of Associationism in Laurence Sterne's *Tristram Shandy*
Aaron Ottinger

17. Mathematics in German Literature: Paradoxes of Infinity
Howard Pollack-Milgate

31. Mathematics, Narrative, and Temporality
Marcus Tomalin

Nineteenth Century

25. Romantic Parts and Wholes, Statistical and Literary
Margaret Kolb

11. Mathematics and Poetic Meter
Jason David Hall

5. Non-Normative Euclideans: Victorian Literature and the Untaught
Geometer
Alice Jenkins

27. Combinatorial Characters
Andrea Henderson

26. "Colours of the Dying Dolphin": Nineteenth-Century Defenses of
Literature and Mathematics
Imogen Forbes-Macphail

28. Datelines
Steven Connor

4. Mathematics and Poetry in the Nineteenth Century
Daniel Brown

6. Mathematical Contrariness in George Eliot's Novels
Derek Ball

A CHRONOLOGICAL ORGANIZATION OF THE CHAPTERS   xiii

Twentieth Century—First Half

11. Mathematics and Poetic Meter
Jason David Hall

29. The Metaphor as an Equation: Ezra Pound and the Similitudes of Representation
Jocelyn Rodal

20. Modernist Literature and Modernist Mathematics I: Mathematics and Composition, with Mallarmé, Heisenberg, and Derrida
Arkady Plotnitsky

21. Modernist Literature and Modernist Mathematics II: Mathematics and Event, with Mallarmé, Gödel, and Badiou
Arkady Plotnitsky

7. Mathematics in Russian Avant-Garde Literature
Anke Niederbudde

16. Mathematics and Modernism
Nina Engelhardt

10. Squaring the Circle: A Literary History
Robert Tubbs

17. Mathematics in German Literature: Paradoxes of Infinity
Howard Pollack-Milgate

Twentieth Century—Second Half, and Twenty-First Century

11. Mathematics and Poetic Meter
Jason David Hall

9. Mathematical Clinamen in the Encyclopedic Novel: Pynchon, DeLillo, Wallace
Stuart Taylor

13. Oulipian Mathematics
Warren Motte

12. Randomizing Form: Stochastics and Combinatorics in Postwar Literature
Alison James

14. Mathematics and Dramaturgy in the Twentieth and Twenty-First Centuries
Liliane Campos

19. "Numbers Have Such Pretty Names": Gertrude Stein's Mathematical Poetics
Anne Brubaker

18. The Ghosts of Departed Quantities: Samuel Beckett and Gottfried Wilhelm Leibniz
Chris Ackerley

31. Mathematics, Narrative, and Temporality
Marcus Tomalin

8. Uses of Chaos Theory and Fractal Geometry in Fiction
Alex Kasman

15. Nonlinearity, Writing, and Creative Process
Ira Livingston

32. A Cognitive and Quantitative Approach to Mathematical Concretization
Marc Alexander

Notes on Contributors

Chris Ackerley is Emeritus Professor of English at the University of Otago, New Zealand. His field is modernism, and his chief scholarly interest is annotation, particularly of the works of Samuel Beckett and Malcolm Lowry. With Stanley Gontarski he co-authored the Grove Press *Companion to Samuel Beckett* (2004). He has written book-length commentaries on *Murphy* and *Watt*, and he has annotated a range of Beckett's texts from the early fiction to the *Textes pour rien/Texts for Nothing* and the late plays. He is currently completing a study of *Samuel Beckett and Science*.

Marc Alexander is Professor of English Linguistics at the University of Glasgow and is Director of the *Historical Thesaurus of English*. His research primarily focuses on the study of words, meaning, and effect in English, including historical semantics and lexicology, parliamentary discourse from 1803 to the present, and the stylistics of manipulative and popularized texts.

David Baker works in secondary education, as Deputy Head at Denstone College, Staffordshire, England. He read English at Peterhouse, Cambridge, where he also studied for his M.Phil. in Medieval Literature. He completed his Ph.D. on "Literature, Logic and Mathematics in the Fourteenth Century" at Durham University in 2013. Previous relevant publications include "The Gödel in Gawain: Paradoxes of Self-Reference and the Problematics of Language in *Sir Gawain and the Green Knight*" (*The Cambridge Quarterly*, 2003) and "A Bradwardinian Benediction: The Ending of the *Nun's Priest's Tale*" (*Medium Aevum*, 2013).

Derek Ball obtained a First in Mathematics in Cambridge in 1964 and subsequently worked as a mathematics teacher and university lecturer. Much later he obtained a degree in English Literature from the Open University, and subsequently an M.A. in Victorian Studies from the University of Leicester. His Ph.D., *Mathematics in George Eliot's Novels*, also obtained from Leicester,

xvi NOTES ON CONTRIBUTORS

was awarded in 2016. Since then he has retained his abiding interest in how an understanding of mathematical ways of thinking illuminates widely diverse fields of knowledge.

Daniel Brown is a Professor of English literature at the University of Southampton. Most of his work focuses upon relations of literature to science and philosophy. His books include *Hopkins' Idealism: Philosophy, Physics, Poetry* (Oxford, 1997) and *The Poetry of Victorian Scientists: Style, Science and Nonsense* (Cambridge, 2013). He is currently completing a book on the place of women in Victorian professional science, as it is disclosed in poetry of the period.

Anne Brubaker is a Lecturer in the Writing Program at Wellesley College, where she teaches courses on modern American writing, literature and science, and science fiction. She earned her Ph.D. in English in 2011 from the University of Illinois at Urbana-Champaign. Her research on the intersections among literary criticism, modern American literature, and mathematics has appeared in *New Literary History* and *Genre: Forms of Discourse and Culture*.

Liliane Campos is a Lecturer in English and Theater studies at the Sorbonne Nouvelle University in Paris, and a fellow of the Institut Universitaire de France. She is the author of two books on science in theatre, *Sciences en scène dans le théâtre britannique contemporain* (PUR, 2012) and *The Dialogue of Art and Science in Tom Stoppard's* Arcadia (PUF, 2011), and has published articles on contemporary performance in *Contemporary Theatre Review, New Theatre Quarterly, Etudes Anglaises, Epistémocritique, Etudes Britanniques Contemporaines* and *Interdisciplinary Science Reviews*. She co-directs a Science and Literature seminar at the Sorbonne Nouvelle (http://litorg. hypotheses.org/).

Steven Connor is Grace 2 Professor of English at the University of Cambridge and Director of Cambridge's Center for Research in Arts, Social Sciences, and Humanities (CRASSH). He has published books on many topics, including Dickens, Beckett, Joyce, value, ventriloquism, skin, flies, and air. His most recent books are *Living by Numbers: In Defence of Quantity* (2016), *Dream Machines* (2017), *The Madness of Knowledge* (2019), and *Giving Way* (2019). His website at www.steven-connor.com includes lectures, broadcasts, unpublished work, and work in progress.

Nina Engelhardt is the author of the monograph *Modernism, Fiction and Mathematics* (Edinburgh University Press, 2018). She received her Ph.D. in English literature from the University of Edinburgh and, after holding positions at the Institute for Advanced Studies in the Humanities Edinburgh, and the University of Cologne, joined the Department of English Literatures and Cultures at the University of Stuttgart, Germany in 2019. Her research in the field of literature and mathematics has led to publications on the works of, among others, Thomas Pynchon, Virginia Woolf, and Robert Musil.

Imogen Forbes-Macphail is a Ph.D. student at the University of California, Berkeley. Her research focuses on the relationship between literature and mathematics in the nineteenth century, and has included work on Samuel Taylor Coleridge, Charles Babbage, Ada Lovelace, and Gerard Manley Hopkins.

Jason David Hall is Associate Professor of English at the University of Exeter. His books include *Seamus Heaney: Poet, Critic, Translator* (2007), *Seamus Heaney's Rhythmic Contract* (2009), *Meter Matters: Verse Cultures of the Long Nineteenth Century* (2011), *Decadent Poetics: Literature and Form at the British Fin de Siecle* (2013), and *Nineteenth-Century Verse and Technology: Machines of Meter* (2017).

Andrea Henderson is a Professor of English at the University of California, Irvine. She is the author of *Romantic Identities: Varieties of Subjectivity, 1774–1830* (Cambridge University Press, 1996) and *Romanticism and the Painful Pleasures of Modern Life* (Cambridge University Press, 2008). Her book, *Algebraic Art* (Oxford University Press, 2018), is a study of the influence of Victorian mathematical formalism on literature and photography.

Alison James is Associate Professor of French at the University of Chicago. Her research interests include the Oulipo group, the contemporary French novel, representations of everyday life, and questions of fact and fiction. She is the author of *Constraining Chance: Georges Perec and the Oulipo* (Northwestern UP, 2009) and *The Documentary Imagination in Twentieth-Century French Literature* (Oxford UP, 2020). She has edited journal issues on "Forms of Formalism" (*L'Esprit créateur* 48.2, 2008), "Valère Novarina" (with Olivier Dubouclez, *Littérature*, no. 176, 2014), and "Fieldwork Literatures" (with Dominique Viart, *Revue critique de fixxion française contemporaine*, no. 18, 2019). With Christophe Reig, she co-directed the volume *Frontières de la non-fiction: littérature, cinéma, arts* (Presses Universitaires de Rennes, 2013).

Joe Jarrett is a Junior Research Fellow at Magdalene College, Cambridge. He was previously a Postdoctoral Research Associate between the Faculty of English and the Center for Research in the Arts, Social Sciences, and Humanities, both at the University of Cambridge.

Alice Jenkins is Professor of Victorian Literature and Culture at the University of Glasgow. Her research centers on the emergence of the knowledge economy in the nineteenth century. Publications include *Space and the 'March of Mind': Literature and the Physical Sciences, 1815–1850* (Oxford University Press, 2007). She is a co-editor of the *Palgrave Studies in Literature, Science and Medicine* book series and is currently completing a

xviii   NOTES ON CONTRIBUTORS

cultural history of Euclidean geometry in the nineteenth century, as well as a study of Victorian ideas about the unity of knowledge.

Alex Kasman received his Ph.D. in mathematics from Boston University in 1995 and subsequently held prestigious postdoctoral positions in Athens, Montreal, and Berkeley. Since 1999 he has been a professor at the College of Charleston. He has published over 30 research papers in mathematics, physics, and biology journals. He also maintains a website that lists, reviews, and categorizes all works of "mathematical fiction." The American Mathematical Society published his textbook on soliton theory in 2010 and the Mathematical Association of America published a book of his short stories in 2005.

Margaret Kolb is a Lecturer in English at University of California, Berkeley, where she is completing a book manuscript analyzing the relationship between the rise of the novel and the mathematics of probability, entitled *The Romance of the Probable*. An essay drawn from the project has appeared in *Victorian Studies*, another is forthcoming in *Configurations*. "Romantic Parts and Wholes, Statistical and Literary" is drawn from her second book project, which investigates how nineteenth-century literary and mathematical texts conceptualized the relations between part and whole.

Ira Livingston is a Professor of Humanities and Media Studies at Pratt Institute in Brooklyn, New York, where he directs the transdisciplinary initiative Poetics Lab. He is the author of several books, most recently *Magic Science Religion* (Brill, 2018) and the digital/interactive *Poetics as a Theory of Everything* (Poetics Lab, 2015; available free on iBooks and Kindle).

Warren Motte is College Professor of Distinction at the University of Colorado. He specializes in contemporary French literature, with particular focus upon experimentalist works that put accepted notions of literary form into question. His most recent books include *Fables of the Novel: French Fiction since 1990* (2003), *Fiction Now: The French Novel in the Twenty-First Century* (2008), *Mirror Gazing* (2014), and *French Fiction Today* (2017).

Anke Niederbudde is Associate Professor (Akademische Rätin) of *Slavic Studies* at the Ludwig-Maximilians-University Munich (LMU). She has researched and published in the area of Russian Symbolism and Avant-Garde. She is the author of *Mathematische Konzeptionen in der russischen Moderne: Florenskij - Chlebnikov - Charms* (2006) and the editor (with Nora Scholz) of *Revolution und Avantgarde* (2018).

Aaron Ottinger is an Instructor in the Department of Digital Technology and Cultures at Seattle University. His other publications on literature and mathematics include an essay on geometry and Wordsworth (*Essays in Romanticism*) and an essay on mathematics and the gut in Lord Byron's *Don Juan* (*Romanticism and Speculative Realism*). His current book project, *Astral Romanticism: Poetics, Mathematics, Models, and Realism*, explores objectivity and the role of mathematics in literature from the mid-eighteenth century through the Romantic period.

Arkady Plotnitsky is a distinguished Professor in the English and Philosophy and Literature Program at Purdue University. He authored eight books and two hundred articles on continental philosophy, romanticism, modernism, and the philosophy of science. His last book was *The Principles of Quantum Theory, From Planck's Quanta to the Higgs Boson* (Springer/Nature, 2016). Recent articles include "'The Heisenberg Method': Geometry Algebra, and Probability in Quantum Theory" (*Entropy*, 2018), and "La réalité sans le réalisme: Penser (avec) l'impensable en physique and en philosophie," in *Choses en soi. Métaphysique du réalisme*, ed. E. Alloa and E. During, Presses Universitaires de France, 2018.

Howard Pollack-Milgate is Professor of German at DePauw University in Greencastle, Indiana. He has published a translation of Heinrich Heine's *Zur Geschichte der Religion und Philosophie in Deutschland* as well as several articles on topics such as Novalis and mathematics and the public–private distinction in E.T.A. Hoffmann. He is currently working on the Romantic logic of interiority and its resonances in the philosophical anthropology of Helmuth Plessner and the psychoanalysis of Melanie Klein.

Jocelyn Rodal is an Associate Research Scholar and Lecturer in English at Princeton University. She received her Ph.D. in English from U.C. Berkeley in 2016, and she has held postdoctoral appointments at Rutgers University and Penn State University. She is currently at work on a book manuscript titled *Modernism's Mathematics: From Form to Formalism*. That project reads literary modernism alongside a contemporaneous modernist movement in mathematics. Examining authors such as Virginia Woolf, Ezra Pound, T.S. Eliot, and James Joyce, the book argues that literary modernists used mathematics to reconceive form—form that, in turn, engendered formalism in literary studies. *Modernism's Mathematics* uses mathematical theories of syntax and semantics to understand form, arguing that literary formalism has structural and historical roots in mathematics.

Stuart Taylor teaches at the University of Glasgow, where he completed his doctoral dissertation entitled "Encyclopedic Architectures: Mathematical Structures in the Works of Don DeLillo, Thomas Pynchon, & David Foster Wallace" in 2019. He is a founding member of the David Foster Wallace Research Group at Glasgow and his writing features in *Lettera Matematica*, *Orbit*, and *Postmodern Culture*.

Marcus Tomalin is a Fellow at Trinity Hall, Cambridge, and a Senior Research Associate in the Cambridge University Machine Intelligence Laboratory. His academic work focuses on the literature and culture of the long eighteenth century, as well as the complex relationship between mathematics and linguistics. He is also actively involved in developing more ethical language-based machine learning systems. His many publications include the monographs *Linguistics and the Formal Sciences* (2006), *Romanticism and Linguistic Theory* (2011), *The French language and British Literature, 1756–1830* (2016), and *Telling the Time in British Literature, 1675–1830* (2020).

Robert Tubbs is a professor of mathematics at the University of Colorado, Boulder, Colorado U.S.A. His academic work has focused on number theory, mathematics education, and mathematics and the humanities. In addition to research articles in mathematics, he has published four books: two in number theory, *Making Transcendence Transparent*, with Edward Burger (2004) and *Hilbert's Seventh Problem: Solutions and Extensions* (2016), both with Springer; and two in mathematics and the humanities, *What is a Number? Mathematical Concepts and Their Origins* (2009), and *Mathematics in Twentieth-Century Literature and Art* (2014), both with the Johns Hopkins University Press.

Benjamin Wardhaugh holds a doctorate from the University of Oxford and has been engaged in research and teaching in the history of mathematics since 2006. His ten books include the textbook *How to Read Historical Mathematics* (Princeton, 2010) and *A Wealth of Numbers* (Princeton, 2012), an anthology of 500 years of popular mathematics writing; they also include five volumes of critical editions. His research interests span the sixteenth to the eighteenth century and include mathematical theories of music, the transmission of mathematical texts, and the history of mathematics teaching and numeracy in that period.

Matthew Wickman is Professor of English at Brigham Young University and Founding Director of the BYU Humanities Center. He is the author of *Literature After Euclid* (2016), *The Ruins of Experience* (2007), and dozens of articles on literary history, literary theory, and interdisciplinary subjects across and outside the humanities.

Travis D. Williams is Associate Professor of English at the University of Rhode Island. He specializes in Shakespeare and other early modern literature, and is completing a book on the shared writing cultures of literature and mathematics in early modern Europe. He is a winner of the *Rhetorica* Prize from the International Society for the History of Rhetoric and is co-editor of *Shakespeare Up Close: Reading Early Modern Texts* (Arden, 2012).

LIST OF FIGURES

Fig. 8.1 Repeatedly "zooming in" on the Mandelbrot Set confirms that arbitrarily small pieces of it are similar in shape and complexity to the entire set. This "self-similarity" is a hallmark of fractal geometry (*Source* Produced by author) — 136

Fig. 8.2 Plant leaves exhibit self-similarity in that the branching patterns of the smaller veins look like little copies of the branching of the larger veins (*Source* Produced by author) — 138

Fig. 8.3 Start with an equilateral triangle and then repeatedly add another equilateral triangle to the center third of each edge of the resulting polygon. The object that is the limit as the process is iterated infinitely many times is the fractal known as the Koch Snowflake (*Source* Produced by author) — 142

Fig. 11.1 Edward Manwaring, *Stichology: Or, a Recovery of the Latin, Greek, and Hebrew Numbers* (London: n. pub., 1737), n. p. — 193

Fig. 11.2 View of gears from John Clark's Eureka Latin Hexameter Machine (*Source* Photograph by the author, 2015) — 197

Fig. 25.1 Frederick Eden Morton, "Cumwhitton," *The State of the Poor,* 2.74 (*Source* Reproduced courtesy of HathiTrust) — 459

Fig. 25.2 Frederick Eden Morton, "Cumwhitton," *The State of the Poor,* 2.75 (*Source* Reproduced courtesy of HathiTrust) — 459

Fig. 25.3 Frederick Eden Morton, "The Parish of Ainstable," *The State of the Poor,* 2.46 (*Source* Reproduced courtesy of HathiTrust) — 461

Fig. 25.4 *Estimate of the Number of Inhabitants of Great Britain and Ireland,* p. 25 (*Source* Reproduced courtesy of HathiTrust) — 464

Fig. 27.1 Hamilton's "Icosian game" (Photo courtesy of the Puzzle Museum. Copyright 2017 Hordern-Dalgety Collection. https://puzzlemuseum.org) — 494

Fig. 27.2 Traveller's Dodecahedron, or A Voyage Round the World (Photo courtesy of the Puzzle Museum. Copyright 2017 Hordern-Dalgety Collection. https://puzzlemuseum.org) — 495

xxii LIST OF FIGURES

Fig. 27.3 Kirkman's solution to "Kirkman's Schoolgirls" problem (From *The Lady's and Gentleman's Diary* 148 [London: J. Greenhill, 1851], p. 48) 496

Fig. 27.4 Cartesian coordinate graph. 345Kai at the English language Wikipedia (Public domain, GFDL [http://www.gnu.org/copyleft/fdl.html] or CC-BY-SA-3.0 [http://creativecommons.org/licenses/by-sa/3.0/]), via Wikimedia Commons 497

Fig. 27.5 Map of the bridges of Königsberg. Leonhard Euler, "Solution of a Problem in the Geometry of Position," *Commentarii Academiae Scientarum Imperialis Petropolitanae* 8 ([1736] 1741), Plate VIII 498

Fig. 27.6 Graph of the Königsberg Bridge Problem 498

Fig. 27.7 Figure 58h from Franco Moretti, *Atlas of the European Novel, 1800–1900* (London: Verso, 1998) 508

Fig. 29.1 Three triangles, two of which are congruent 535

Fig. 32.1 A dispersion plot of key analogical categories in *MP*; the plot represents the book from its start on the left to its end on the right, and each small vertical line indicates a point in *MP* where either *HT* category is used 595

Fig. 32.2 Twenty random concordance lines of *way* and *far* from MP, output using the author's programs. Concordances such as these are read by running one's eye down the central column, in bold, and reading just enough context as is necessary to establish the sense of each word. Only a set amount of characters on either side of the central word are reproduced in a concordance, and concordances should not be read like normal sentences 596

Fig. 32.3 Frequency of key terms in *MP*, arranged by the 400-word 'chunks' the book was split into for analysis 597

Fig. 32.4 A 'map' of the input spaces in the analysed extract from *MP* 603

CHAPTER 1

Introduction: Relationships and Connections Between Literature and Mathematics

Nina Engelhardt and Robert Tubbs

"A good preface must be at once the square root and the square of its book" (Schlegel [1797] 2003, p. 239). This statement by the German poet Friedrich Schlegel (1772–1829) is an example of a literary writer drawing on mathematics to communicate the ideal aim of a written text—calling up associations of mathematics with truth, clarity, and rigidity as well as implying the impossible: the "quadrature of the circle" of simultaneously getting to the root or core of a book as well as going far beyond its range by "multiplying it with itself." Mathematically, Schlegel's condition has the number 1 as its nonzero solution: the root of 1 is 1, and the square of 1 is 1. Figuratively, Schlegel's "good preface" would thus be the book itself. We will follow this "mathematically deduced" conclusion and let the collection of essays speak for itself but also aim to use this introduction—knowing full well that we will inevitably fail to square this circle—to both address the fundamentals of relations between literature and mathematics and to give a broader context for the chapters to follow.

Literature and mathematics might seem to constitute entirely different domains of knowledge, practice, and meaning. Literature is often associated

---

The original version of this chapter was previously published non-open access. A correction to this chapter is available at https://doi.org/10.1007/978-3-030-55478-1_33

N. Engelhardt (✉)
Department of English Literatures and Cultures, University of Stuttgart, Stuttgart, Germany

R. Tubbs
University of Colorado Boulder, Boulder, CO, USA

© The Author(s) 2021, corrected publication 2022
R. Tubbs et al. (eds.), *The Palgrave Handbook of Literature and Mathematics*, https://doi.org/10.1007/978-3-030-55478-1_1

with subjective, individual experience, emotional depth, and the vagaries of human life, and as produced and read in particular historical, cultural, and social contexts. In contrast, mathematics is commonly seen as a system of eternal truths that are established by objective, rigorous methods employed in a steady accumulation of knowledge. Where literature is at least theoretically accessible to all literate readers and might develop its greatest power and appeal when giving rise to various interpretations, mathematics is celebrated for its certainty and precision and sometimes revered as the realm of geniuses. The chapters in this Handbook vividly demonstrate that these stereotypes and associations are at best half of the story. Neither literature nor mathematics lends itself to easy characterization, both fields experience remarkable changes, crises, and unresolved questions, and the relation between them is not one of clear-cut contrast but includes manifold connections, intricate parallels, and creative borrowings. This Handbook addresses interrelations of literature and mathematics in five categories, which work to organize and group together the chapters to follow. Like any categorization, the five groups can only delineate rough tendencies, cannot hope to cover all aspects in a broad field, and do not do justice to many of the chapters as these are wide-ranging and could be included in several or even all of the parts "Mathematics in Literature," "Mathematics and Literary Form," "Mathematics, Modernism, and Literature," "Relations between Literature and Mathematics," and "Mathematics as Literature."

Mathematics in Literature

The first part presents chapters that examine literary texts' employment of mathematics on the levels of plot and language, as topic, theme, and metaphor. This can include characters who practice mathematics as a profession, direct discussions of mathematical problems, and also the use of mathematical vocabulary and symbols. While some texts employ numbers to stand for the threat of inhuman rationalization, others introduce them in positive contexts as allowing for order or draw on the metaphorical potential of irrational numbers or imaginary numbers to suggest the "mathematically proven" existence of realms beyond reason and physical reality. Similarly, simple counting and quantification can have positive as well as negative implications: the successive reduction of King Lear's knights in Shakespeare's play suggests the unstoppable development of a mathematical series and the power that comes with commanding numbers (see Chapter 22 by Travis Williams). The protagonist in Samuel Beckett's *Molloy* experiences the sense of order and control that counting and quantification can provide when he collects sixteen stones and attempts to rotate them between his four pockets in such a way as to take them out in a specific order. At the same time, the calculation of his rate of farting as being one fart every 3.62 minutes suggests the absurdity of quantifying life (see Chapter 18 by Chris Ackerley). The lures and dangers of quantifying and calculating probabilities have a presence in literature at least since the Middle Ages (see Chapter 2 by David Baker) and show their sometimes sterile, inhuman aspects in financial speculation and profit-making projects

1 INTRODUCTION: RELATIONSHIPS AND CONNECTIONS ...   3

such as those of the character Merecraft in Ben Jonson's *The Devil Is an Ass* (see Chapter 3 by Joe Jarrett).

The incorporation of mathematical symbols in literary texts showcases, in an immediately visible way, the differences between these systems of notation. Charles Bernstein's poem "Erosion Control Area 2" is creatively typeset and includes symbols from mathematics throughout:

Clothe ≤ ma
oμ β wolμ iε
   Whicɸ t∩ ou ≥ (Bernstein 1996, p. 17).[1]

The focus here is on the visual impression of these mathematical symbols rather than their sound or meaning, and their strangeness draws attention to the materiality of the text and the fact that words in alphabetical letters similarly do not give immediate access to meaning but are printed symbols on paper. The Russian avant-garde writer Velimir Khlebnikov (1885–1922) employs the symbol for an imaginary number, $\sqrt{-1}$, in his short prose piece "We Climbed Aboard" (1916): "We climbed aboard our $\sqrt{-1}$ and took our places at the control panel" (Khlebnikov 1989, p. 82). The mathematical symbol stands out from ordinary printed letters and visually expresses the imaginary position above everyday reality that allows the speaker and the poem to leave reality behind and observe how "centuries of warfare passed before me" (p. 82) (see Chapter 7 by Anke Niederbudde). While $\sqrt{-1}$ is a well-known mathematical symbol and it easily lends itself to associations with imaginary and fictional domains, Thomas Pynchon's novel *Gravity's Rainbow* (1972) displays a partial differential equation that readers cannot be expected to understand but that visually communicates that complex mathematics is involved in the development of the V-2 rocket during the Second World War (see Chapter 9 by Stuart Taylor).

The term "imaginary number," which was introduced by René Descartes in *La Géométrie*, an appendix to *Discourse on Method* (1637), implies that this mathematical entity has no correspondence in reality while other numbers have a direct relation to the physical world. The idea of mathematics as the language of the book of nature (Galileo 1960, pp. 183–84) came under increasing pressure during the nineteenth century when mathematical concepts seemed to leave reality behind, for example, by formulating a four-dimensional space that goes beyond the three dimensions that can be physically experienced. Mathematically, the fourth, fifth, or sixth dimension does not differ from the first three, but literary texts, as well as occult and spiritual movements, interpreted further dimensions in mathematics as proof of a realm beyond material existence. In Joseph Conrad and Ford Madox Ford's *The Inheritors* (1901) the fourth dimension harbors a superhuman race, and in *The Time Machine* (1895) by H. G. Wells, it is understood as time and can be manipulated to travel into the future and the past. While the mathematics of higher dimensions was taken to point to realms beyond physical existence, other developments showed long-established methods of calculation to lead to inadequate descriptions of the world: while Euclidean geometry works well to calculate triangles and spheres, "[c]louds are not spheres and mountains are not cones," as Benoît Mandelbrot (1924–2010) put it ([1977] 1982, p. 1). Mandelbrot's fractal geometry, which he developed in *The Fractal Geometry of Nature* (1977), can be used to describe more complex natural systems. As Chapter 8 by Alex Kasman demonstrates, fractal geometry and chaos theory appear in literary fiction, often, but not always, metaphorically or to take advantage of nonmathematical properties of these areas.

While Mandelbrot proposed a geometry better suited to describe the physical world than the geometry formulated by Euclid in the third century BCE, the absolute truth of Euclidean geometry had already come under attack in the nineteenth century when Nikolai Lobachevsky (1792–1856) and János Bolyai (1802–1860) described an alternative geometry which does not rest on the so-called Parallel Postulate. Euclidean geometry was mainstay in mathematics education, particularly in the nineteenth century, and literary texts refer to it across the centuries (see Chapter 5 by Alice Jenkins). In the early fourteenth century, Dante appealed to the classical problem of squaring a circle with only using "Euclidean tools," a straightedge and a compass, and its presumed impossibility, as a metaphor for humans' inability to understand the Incarnation in Christianity, and this problem reappears in later literary texts (see Chapter 10 by Robert Tubbs). In the Romantic period, Euclid's *Elements* inform William Wordsworth's "Arab Dream," (see Chapter 4 by Dan Brown), as well as Samuel Taylor Coleridge's poem "A Mathematical Problem" (1840), which begins:

> On a given finite Line
> Which must no way incline;
> To describe an equi--
> --lateral Tri--
> --A, N, G, L, E. (Coleridge 1840, p. 24)

The poem goes on with the proof of Proposition 1 of Book I of the *Elements*, which describes how to construct an equilateral triangle on a given line segment, and, considering that many men encountered Euclid as a profound presence in their mathematics education, alludes to a commonly experienced type of mathematical problem in the nineteenth century. While Euclidean geometry thus works as a "language" that connects many Victorians, access to mathematical education for girls and children from the working class was very limited—a fact that George Eliot addresses in several of her novels (see Chapter 6 by Derek Ball). Tensions between understanding mathematics as universal language and knowledge, and considering it in specific historical, cultural, and social contexts grow more acute in the twentieth century (see below).

Mathematics and Literary Forms

The chapters in the second part address ways in which literary texts engage with numbers and other mathematical constructs through their forms. Literary form can appeal to readers through its regularity but also by breaking with order and allowing for creative fluidity, and formal restraint can be limiting as well as lead to unforeseen results and inspire new structures. Poetry, which often plays with establishing and breaking regularity in rhyme, rhythm, and stanza structure, is a particularly apt genre for considering mathematics and literary form. In Chapter 11, Jason Hall shows that poets and metrical theorists across the centuries use mathematical vocabulary and paradigms to explain the organization of poems and draw on mathematics for theories of ratio, harmony, and abstraction. Mathematical structures can also play a role in the production of literature even if these are no longer visible in the end result. The Oulipo, a group of writers and mathematicians founded in 1960, aimed to "propose new 'structures' to writers, mathematical in nature, or to invent new artificial or mechanical procedures that will contribute to literary activity: props for inspiration as it were, or rather, in a way, aids for creativity" (Queneau 1986, p. 51). This included imposing constraints on literary practice, for example, in Jacques Jouet's "metro poems," composed on the Parisian metro in the time between two stops. While the structure of the metro poems is not overtly mathematical, their number and lengths depends on "chance," determined by the time between stops and the number of stops on the line. *One Hundred Thousand Billion Poems* (*Cent mille milliards de poèmes*, 1961) by Raymond Queneau consists of ten sonnets, which are written so that if the first line of any sonnet is combined with the second line of any sonnet and so forth, you will obtain a new sonnet, so that the book contains 10^{14} possible poems (see Chapter 13 by Warren Motte on Oulipian mathematics, and Chapter 12 by Alison James on relationships between chance, numbers, and literary form). The appeal of unpredictability and open-endedness that mathematical models and metaphors can provide also shows on the theater stage, even if the abstract nature of mathematics might seem to contradict the presence and immediacy of the theater (see Chapter 14 by Liliane Campos). Moreover, and connecting to the Oulipo members' interest in the role of mathematics for writing and creativity, mathematics can provide models for and help understand forms of creative practice, including relationships between nonlinearity, writing and language, and creative process (see Chapter 15 by Ira Livingston).

Examining mathematics and literary form in his monograph *The English Renaissance Stage: Geometry, Poetics, and the Practical Spatial Arts 1580–1630* (2006) has led Henry S. Turner to the conclusion that what literary scholars usually understand as "form" is in need of reconsideration, not least by accounting for mathematical notions of form. According to Turner, traditional concepts of form can be grouped into four categories—namely, stylistic notions of form, such as verbal patterning or metrical language; structural notions of form that include plot and stanzaic structures; material notions such as the page size and layout; and social notions of form that include class organization, economic production, and political systems (Turner 2010, pp. 580–81). Turning to mathematics and mathematical form can, so Turner argues, help rethinking these traditional concepts that are overly focused on linguistic and textualist models. In *Unified Fields: Science and Literary Form* (2014) Janine Rogers discusses various sciences, yet she also considers the specific relation of mathematics and literary form and argues for paying greater attention to the fact that form is not merely a product of knowledge, but that it is a way of knowing and source of meaning in its own right. This epistemological "*function* of form ... is shared by both science and literature" (Rogers 2014, p. xvii), and mathematical form in particular shares with its literary counterpart a focus on unity and beauty (Rogers 2014, pp. 48–65). Indeed, in the early twentieth century the popular science writer J. W. N. Sullivan expressed a common idea when identifying similarities between mathematics and art in their ability to develop outside of experience and with a focus on beauty: "Although the simple, primary mathematical ideas were doubtless originally suggested by experience, the mathematician's development of them has been very largely independent of the teachings of experience. He has been guided chiefly by considerations of form—a criterion which is probably, at bottom, aesthetic" (Sullivan 1933, pp. 243–44). The idea that mathematics is focused on form, rather than deriving from experience or representing the physical world, developed rapidly in the nineteenth century. For example, Augustus De Morgan (1806–1871) explained: "no word nor sign of arithmetic or algebra has one atom of meaning throughout this chapter, the object of which is *symbols, and their laws of combination*" (1849, p. 101; original emphasis). And Henri Poincaré (1854–1912) declared in 1902: "Mathematicians study not objects, but relations between objects; the replacement of these objects by others is therefore indifferent to them, provided the relations do not change. The matter is for them unimportant, the form alone interests them" (2015, p. 44). As Andrea Henderson shows in her monograph on mathematical formalism and Victorian culture, "[n]ineteenth-century mathematicians were very aware that in distinguishing content from form and privileging the latter they were fundamentally changing their discipline and its claims upon truth" (2018, p. 30). In the early twentieth century, a number of visual artists and literary writers drew on this formalist notion of mathematics when developing and experimenting with new conceptualizations of art that similarly prioritize form over content and meaning. In this way, mathematics plays a particularly important role for literary form in modernism.

Mathematics, Modernism, and Literature

The third part focuses on developments in literature and mathematics in the first half of the twentieth century. The rise of mathematical formalism in the nineteenth century (see above), that is, the notion that mathematics is not concerned with objects and meaning but is a self-contained formal system, led to concern about its foundations: if mathematics is not grounded in a relation to nature, then the foundations that guarantee its truth and consistency have to be found elsewhere. Three main mathematical schools—logicism, formalism, and intuitionism—attempted to establish secure foundations for mathematics, and their respective prioritizing of logical, formal, and intuitive viewpoints plays out broader tensions in a period in which celebrations of rapid advancements in science and technology coexist with attempts to escape threatening rationalization in realms outside reason and calculation. None of the approaches succeeded in providing mathematics with stable foundations however, and mathematicians as well as non-professionals increasingly became aware of unresolved questions, unsolvable paradoxes, and the need to revise what had seemed to be certain knowledge. The historian of mathematics Jeremy Gray summarizes: "the mathematics of the nineteenth century is marked by a growing appreciation of error leading to a note of anxiety, hesitant at first but persistent by 1900" (Gray 2004, p. 23). The period from the 1880s to around 1930 is accordingly called the "foundational crisis" of mathematics, and the questions and anxieties surrounding mathematics also occupied non-professionals and appear in literary texts.

Chapter 17 by Howard Pollack-Milgate demonstrates that the concerns coming to the fore in the foundational crisis of mathematics are not unique to early twentieth-century modernism but have a long prehistory both in mathematics and in literature. Paradoxes of infinity crystallize the following assertions and contrasts from the fifteenth to the twentieth century: "the condition of the modern world, the utility of results versus the problems of a foundation, the notion of the mathematical as, on the one hand, certain and clear, on the other, perplexing and contradictory" (Pollack-Milgate by Chapter 17). Chapter 18 by Chris Ackerley also relates modernist concerns with earlier thinking when examining how Samuel Beckett draws on work by Gottfried Wilhelm Leibniz (1646–1716) "to gain perspective on a major concern of his times," namely, a paradox discovered by Bertrand Russell in 1901 that showed that the notion of a set motivated by trying to formalize Georg Cantor's work leads to a contradiction. If formalism is the inquiry into mathematical signs and their relations, a similar movement appeared in early twentieth-century linguistics: Ferdinand de Saussure argued that language is a self-referential system that should be examined without consideration of a sign's referent in the real world. The aim of establishing a system of signs that makes possible ordered, exact communication thus connects mathematics and language at the beginning of the twentieth century, and literary writers such as Gertrude Stein compare these projects and find in mathematics a model for the limits of representation in literary language (see Chapter 19 by Anne Brubaker).

When in the early twentieth century, "mathematicians fashioned for themselves a new image of the subject: autonomous, abstract, largely axiomatic, and unconstrained by applications even to physics" (Gray 2008, p. 305), mathematics exhibits characteristics more commonly associated with

8 N. ENGELHARDT AND R. TUBBS

modernist literature and art. Therefore, a number of historians of mathematics argue to view modern mathematics as part of the culture of modernism and to therefore speak of a "modernist mathematics" (Mehrtens 1990; Gray 2008). Chapters 20 and 21 by Arkady Plotnitsky examine relationships between modernist literature and modernist mathematics. Chapter 20 focuses on the movement toward independence and self-determination that characterizes both modernist literature and mathematics; Chapter 21 brings into view the question of mathematics and ontology. Even if ontological questions are more evident in the postmodernist literature of the second half of the twentieth century, modernist texts engaging with mathematics are both concerned with epistemological questions about the possibility and certainty of mathematical knowledge and with ontological considerations regarding the structures and "worlds" that mathematics creates and the ways in which these can be compared to literary fiction (see Chapter 16 by Nina Engelhardt).

RELATIONS BETWEEN LITERATURE AND MATHEMATICS

Part four collects chapters that directly address the question of how literature and mathematics connect to each other as areas of knowledge, education, and practice. While, as detailed above, literature and mathematics can be seen as opposites in a number of ways—regarding universality and individuality of knowledge, certainty and vagueness, accessibility, and so on—they also share characteristics that suggest a surprisingly close relationship between them, closer than between literature and science. Indeed, although mathematics and the natural sciences are often thought of together—for example, in discussions about the "STEM" subjects (science, technology, engineering, mathematics)—mathematics is not implicit in the S for science: it is not based on empirical research. Rather, it shares characteristics with the humanities when it "can be considered a creative cultural achievement since it is only accountable to human thinking" (Mühlhölzer et al. 2008). But counting mathematics as a discipline of the humanities is problematic too since it is not concerned with human beings or their cultural achievements (Mühlhölzer et al. 2008). The German physicist and philosopher Carl Friedrich von Weizsäcker held that asking whether mathematics is part of the natural sciences or the humanities is based on an incomplete classification, and that it belongs to a third category, namely, that of structural science (Weizsäcker [1971] 1980). Bernd-Olaf Küppers, like Weizsäcker a physicist and philosopher, explains:

> The distinguishing feature of this type of science is that—unlike the natural sciences and the humanities—it deals with the over-arching structures of reality, independent of the question of where these structures actually occur, whether they are found in natural or artificial, living or non-living systems. Owing to their high degree of abstraction, the structural sciences include a priori the entire realm of reality as the area of their applicability. (Küppers 2018, p. 176)

And since the structural sciences, for which mathematics is the "archetype," is abstract and has an "integrative function," it can, so Küppers argues, serve as a link between the natural sciences and the humanities (Küppers 2018, pp. 176, 178).

Chapter 26 by Imogen Forbes-Macphail locates mathematics in reference to the Huxley-Arnold-debate in the 1880s, a well-known negotiation of the respective educational, social, and cultural value of literary and scientific knowledge between Thomas Henry Huxley and Matthew Arnold. Shifting the focus to mathematics, Forbes-Macphail notes that both literature and mathematics had to defend themselves against the growing importance of science education. She demonstrates how the poet Matthew Arnold and mathematicians James Joseph Sylvester and William Spottiswoode characterized their fields in similar ways, as not immediately lending themselves to application but being pursued for the sake of knowledge and beauty. Moreover, this nineteenth-century discussion values mathematics, like literature, for its ability to connect ideas within mathematics and between disciplines—a characteristics of the structural sciences that Küppers similarly notes today: "They already link up large areas of natural science, economics and the humanities" (Küppers 2018, p. 178).

The then following chapters examine further concepts of the relation between literature and mathematics. For example, the idea of consilience, formulated by E. O. Wilson in 1998, describes the convergence between different areas of knowledge, particularly of the humanities and the sciences. Chapter 23 by Matthew Wickman examines the prehistory of consilience in Newton's fluxional calculus and discusses a consilient logic of figure in reference to Newton's formulation of the calculus and a poem by Robert Burns. Chapter 28 by Steven Connor notes the self-referentiality of both mathematics and language and argues that this constitutes not equality or identity but a convergence of congruences. The foreignness of actual dates in literary fiction, for example, in novels by Charles Dickens, shows that words and numbers might share similar structures but that ultimately, they remain external to each other. Chapter 29 by Jocelyn Rodal puts into focus the notion of equality itself, demonstrating how Ezra Pound takes equation to describe things that are different yet also show sameness and how this notion informs his use of metaphor, comparison, and juxtaposition. Again, a concept in mathematics—here an equivalence relation—offers abstract understanding and illuminates the way in which mathematics as a structural science can work to describe relationships and link different fields of knowledge.

Chapter 25 by Margaret Kolb and Chapter 27 by Andrea Henderson are not directly concerned with the relationship between literature and mathematics but examine both as engaging with and contributing to wider developments, in different yet comparable ways. In the nineteenth century, newly accessible data, for example collected during a census in Britain in 1801, opened up questions that, so Kolb asserts, reverberate in both mathematics and literature: "How should numbers be aggregated, arranged, and read? What are the limits of numerical representation? Can a part—what we now call a sample—explain a larger whole?" (Kolb, Chapter 25). Henderson argues that the logic of late-Victorian capitalism, placing value not in individuals but in their links to others, shows in late-Victorian characterization that privileges characters' relations to each other, and is epitomized in the mathematical field of combinatorics. Chapter 24 by Aaron Ottinger identifies similarities between geometry and propositional logic and between probability and associationist logic and examines how Laurence Sterne's *Tristram Shandy* (1759–1767) combines these to challenge readers' accidental associations and elicit moral feelings. Also examining mathematical thinking and reading response, Chapter 22 by Travis Williams begins by analyzing explicit references to number and calculation in William Shakespeare's *King Lear* to then develop a way of "reading mathematically" that is independent of such direct engagement with mathematics. He argues that imaginary numbers—which in Shakespeare's time were seen as purely mental creations that enjoyed a liminal, "imaginary" existence but served a practical purpose in calculation—depend on a similar logic as reading or viewing *King Lear* where the audience is led to imagine what is later exposed as a pretense and this process serves a function. In these chapters, engaging with reality by taking into account "merely" probable or imagined states emerges as a strategy that links mathematics and literature.

MATHEMATICS AS LITERATURE

Chapters in the last part address ways in which mathematical writing—in research, education, and popularization—exhibits literary qualities and can usefully be examined with the tools of literary analysis. Chapter 30 by Benjamin Wardhaugh shows that a distinctive early modern culture of mathematical reading and activity can be traced in the marginal annotations of printed texts, on waste paper, and slates: learning mathematics involved manually doing mathematics, using blank spaces to copy diagrams, supplement proofs, and correct printing errors. Chapter 31 by Marcus Tomalin argues that mathematical texts possess literary qualities and examines the relationship between mathematics, narrative, and temporality in classical proofs by Leonhard Euler, Carl Friedrich Gauss, and a late twentieth century proof of a lemma that appeared in Andrew Wiles's paper establishing Fermat's Last Theorem. Next to mathematical practice in research and education, literary texts and devices also play an important role in the communication of mathematical knowledge to non-professionals. Tom Stoppard's play *Arcadia* (1993) is a prime example of the way in which metaphors, dialogue, and performance on stage can convey complex mathematical ideas such as chaos theory, fractal geometry, and Fermat's Last Theorem (see Chapter 14 by Liliane Campos). The successful way in which *Arcadia* makes mathematical intricacies understandable, fun, and relevant to everyday life was acknowledged by the Royal Institution of Great Britain when it short-listed the play for an award in the category "best science book ever written." Chapter 32 by Marc Alexander examines Marcus du Sautoy's *The Music of the Primes* (2003) as a popular science book that communicates complex mathematical concepts, in particular regarding its use of analogy as a way to give non-experts a sense of understanding. Alexander's chapter also is an example of taking a mathematical approach to literature when it employs a quantitative methodology to analyze texts. As he argues, "[q]uantitative approaches cannot and should not replace an analyst's reading of a text, but they can supplement our existing methods for finding areas worth studying" (Alexander, Chapter 32). The chapters in this part demonstrate the fruitfulness both of taking a quantitative, mathematical approach to literature and of examining mathematical writing with the tools of literary analysis.

Literature and Mathematics Studies

The remainder of this introductory chapter charts the development of the study of literature and mathematics, with reference to the broader field of literature and science studies. Literature and mathematics is commonly understood as a subfield of literature and science, and while it shares key questions, concerns, and developments with the larger field, mathematics differs from the natural sciences in several important aspects. While the natural sciences rely heavily on observation and experiment, much of mathematics is done without specialized instruments or reference to nature. While this is far less the case today, as computers are indispensable parts of mathematical research and applications, historically and for many non-professionals who primarily encounter mathematics while in school, mathematics retains a stronger image of abstract, theoretical knowledge. At the same time, of course, mathematics plays a crucial role in the natural sciences, in scientific practice, and in gaining knowledge about the world. Indeed, a key development toward modern science was its mathematization: Isaac Newton (1643–1727) put natural philosophy—the precursor of modern science—on mathematical foundations in his *Philosophiae Naturalis Principia Mathematica* (1687). One consequence of Newton's immensely influential approach was to replace Cartesian vortex theory, an alternative, earlier seventeenth-century theory to explain planetary motion and gravitation that used verbal formulations and could not compete with the comprehensiveness and predictive power of Newton's explanations in mathematical form. The increasingly close connection between mathematics and natural philosophy implied a correspondingly larger distance to literature and the humanities, particularly with the professionalization and further mathematization of scientific disciplines in the nineteenth century: "In the mid-nineteenth century, scientists still shared a common language with other educated readers and writers of their time.... [Scientific writers] shared a literary, non-mathematical discourse which was readily available to readers without a scientific training" (Beer [1983] 2000, p. 4). While Gillian Beer here presents mathematization as an obstacle to shared discourse, mathematics itself served as a common language in the nineteenth century: mathematical education focused on studying geometry, in particular Euclid's *Elements*, and this ensured an educational experience shared by mathematicians and non-mathematicians alike: "mathematics ... was a shared experience and a 'common knowledge' for nineteenth-century readers and writers, and its impact on society and culture was immense" (see Jenkins 2017, p. 217). However, the mathematization of the sciences led to a growing distance between professionals and the larger public: by and large, as Beer points out in the 1980s, "[n]on-scientists do not expect to be able to follow the mathematical condensation of meaning in scientific journals, and major theories are more often presented as theorems than as discourse" (Beer [1983] 2000, p. 4). Since the mathematization of the sciences is a decisive factor in the historical development of the sciences and in the changing relations between literature and science, any study of literature and mathematics or literature and science necessitates reflection about the other.

The early academic study of relations between literature and science stresses the role of literary texts in interpreting science and its implications. For example, Marjorie Hope Nicolson gives her 1950 *The Breaking of the Circle* the subtitle: *Studies in the Effect of the "New Science" Upon Seventeenth-Century Poetry*. As Martin Willis summarizes in his overview of early literature and science criticism: "Such a one-way model of influence has fallen out of favor—in both literature and science and history of science scholarship—and is one of the key beliefs that the contemporary criticism has worked to overturn" (Willis 2015, p. 42; for the overview see pp. 32–51). Scholarship in the 1980s made a decisive development away from researching influence in favor of considering mutual interrelations and shared discourses, with pioneers Gillian Beer and George Levine both focusing on the work of Charles Darwin in particular. As Levine and Beer examine the narrative, metaphorical, and creative qualities of Darwin's writing, mathematics does not play a large role in their works. Levine even stresses the absence of mathematical thinking and notation that brings to the fore the more literary character of Darwin's writing: "Darwin demonstrates the regularity and comprehensibility of phenomena without reducing them to the strict form of logic and mathematics" (Levine [1988] 1991, p. 19). Alice Jenkins summarizes for scholarship on science in nineteenth-century Britain: "historicist studies in the tradition of Beer (1983), Levine (1988), Shuttleworth (1984), Dawson (2007), and O'Connor (2007) have given comparatively little attention to the mathematical sciences, and especially to mathematics itself" (Jenkins 2017, p. 219).

While mathematics plays only a very limited role in the quickly growing field of literature and science studies in the 1980s, Linda Dalrymple Henderson's (1983) *The Fourth Dimension and Non-Euclidean Geometry in Modern Art* has its main focus on mathematics, albeit primarily in its relation to the visual arts. Henderson explores ways in which modern art engages with and contributed to the new understanding of space that developed from non-Euclidean geometry and the mathematics of higher dimensions. The mathematical expansion of familiar notions of space feeds into modern movements' departures from representing visual reality and exploring possibilities of perception outside the restrictions of visible, three-dimensional reality, for example, in Cubism, Dadaism, or Surrealism. Henderson's research has been immensely influential, yet, its relevance for literature and mathematics studies long remained untapped. As recently as 2018, Mark Blacklock developed the implications for literary studies that Henderson's (1983) monograph touched upon: "Henderson's work demonstrates that the geometries developed in the nineteenth century and their popular and occultist elaborations informed Modernist production in the visual arts and outlines potential lines of inquiry in the literary arts. I have followed these leads" (Blacklock 2018, p. 6).

Yet, mathematics is not completely ignored even in the first decade of research on the "two-way traffic" between literature and science that Beer advocates in the 1980s ([1983] 2000, p. 5). Indeed, Beer employs mathematics as an extreme example to illustrate the fruitfulness of understanding literature and science as mutually affecting each other rather than only taking science to influence the literary. In her 1990 lecture "Translation or Transformation? The Relations of Literature and Science" at the Royal Society, she points out that mathematicians are no strangers to creative and figurative language. When Mandelbrot terms structures in his fractal geometry "Cross Lumped Curdling Monsters" and "Knotted Peano Monsters, Tamed," he allows nonmathematical readers to "glimpse the implications of the theorems that are interspersed between the sentences": "A verbal mimesis of his [Mandelbrot's] own theoretical work is implied, in which the random, the inordinate, the non-Euclidean is granted an appropriate language that bulges, miniaturizes and grows gargantuan, constantly shifting across registers of scale and distance to achieve its imaginative effects" (Beer 1990, p. 90). Thus, while mathematical symbols and formulas can result in excluding non-professionals, Beer highlights that literary scholars' examination of metaphors and language can illuminate even the field often seen as furthest removed from everyday language.

The notion that science is not an objective accumulation of truths about reality but subject to the possibilities of language, historical circumstances, and social conditions and participates in constructing reality is central to research in science studies, beginning in the 1960s and intersecting with the field of literature and science. Milestones in work on the historical, social, and cultural dimensions of science include *The Structure of Scientific Revolutions* (1962) in which Thomas Kuhn argues that science is not a linear accumulation of truths but characterized by paradigm shifts, sociologists Bruno Latour and Steve Woolgar's *Laboratory Life: The Construction of Scientific Facts* (1979) introducing the important role of writing and text in scientific practice, and *The Manufacture of Knowledge* (1981) by Karin Knorr-Cetina which advances the thesis that products of science are not disinterested uncoverings of truth but constructions dependent on the social and historical context in which they are produced. Many more names and works could be added here, but as a comprehensive survey cannot be our aim here, we focus on highlighting the role of mathematics in the surge of interest in the sociology and history of science from the 1960s onwards.

The sociologist David Bloor pointed out in 1973: "One of the central problems of the sociology of knowledge is that status of logic and mathematics. These branches of knowledge are so impersonal and objective that a sociological analysis scarcely seems applicable" (Bloor 1973, p. 173). Bloor argues that, while Karl Mannheim (1893–1947) "could not see how to think sociologically about how twice two equals four" (1973, p. 173), Ludwig Wittgenstein's (1889–1951) *Remarks on the Foundations of Mathematics* demonstrates the possibility and value of a sociology of mathematics. Bloor's *Knowledge and Social Imagery* (1976) develops at greater length that mathematics can be part of a so-called strong program in the sociology of scientific knowledge that sees scientific knowledge and epistemic standards as context-dependent, social phenomena. Bloor admits "that these 'constructive proofs' cannot be offered in abundance" (Bloor [1976] 1991, p. 84) since a long tradition has established mathematics as the epitome of objective and true knowledge. The development of quaternions, formulated by William Rowan Hamilton in 1843 as an extension of complex numbers to applications in three-dimensional space, is a main example in Bloor's project of showing "that there is nothing obvious, natural or compelling about seeing mathematics as a special case which will forever defy the scrutiny of the social scientist" (p. 84) and that an alternative mathematics is imaginable (see Bloor [1976] 1991, chapter six, "What Would an Alternative Mathematics Look Like?"). In the 1990s, Andrew Pickering also drew on Hamilton's work on quaternions to illustrate that mathematical concepts are not found in nature or a pre-existing Platonic realm but are "constructed" in specific historical and cultural circumstances (Pickering 1995, chapter 4 "Concepts: Constructing Quaternions," and Pickering and Stephanides 1992). More generally, Pickering used examples from mathematics to analyze how knowledge is produced not only in the better-explored experimental sciences but also in theoretical practice (Pickering 1995).

In a 2010 essay on the cultural strategies, resources, and conjunctures of mathematical practices, Moritz Epple concludes: "Detailed historical analyses of the practices of mathematisation and mathematical argument in science as *cultural* practices are still rare compared to the recent history of experiment" (Epple 2010, p. 219). The editors of *Perspectives on Mathematical Practice: Bringing Together Philosophy of Philosophy of Mathematics, Sociology of Mathematics, and Mathematics Education* (2007), Bart Van Kerkhove and Jean Paul Van Bendegem, arrive at a similar answer to the question:

> Is mathematics finally going through the Kuhnian revolution that the sciences or, more precisely, the philosophers, historians, sociologists, economists,

psychologists of science, ... have been able to deal with ever since the magical year of 1962? ... [O]ne cannot easily identify a book that has played the part that *The Structure* has played – of course, Lakatos' *Proofs and Refutations* comes pretty close, but it does not possess the generality of Kuhn's work. (Van Kerkhove and Van Bendegem 2007, p. vii)

This summarizing assessment in 2007 highlights, firstly, that mathematics still occupies a special position in the history, philosophy, and sociology of science in the twenty-first century, and, secondly, the exceptional role of Imre Lakatos's *Proofs and Refutations*. Published in 1976, *Proofs and Refutations*—an allusion to the famous paper "Conjectures and Refutations" by Karl Popper—argues that the development of mathematics is not a steady accumulation of truths but a dialectical process and that mathematics is fallible, for example, in the sense that theorems can be refuted by finding counterexamples that require adjusting the theorem. Lakatos's *Proofs and Refutations* marks an important development in the philosophy of mathematics, and it is also remarkable in the way in which its literary form—it is written as a fictional dialogue between teacher and students—is part of the argument. Lakatos explains that "[t]he dialogue form should reflect the dialectic of the story," namely, the story of the development of mathematics as the community decides which proofs are valid (Lakatos [1976] 2015, p. 5). As the dialogic form in *Proofs and Refutations* suggests, it is worth paying attention to the ways in which mathematics is practiced, negotiated, and communicated among scholars, learners, and non-professionals, and literary scholars are well-positioned to explore this aspect.

Work in the history, philosophy, and sociology of science that stresses the constructed and context-dependent nature of scientific knowledge initiated intense debate that culminated in the so-called Science Wars of the 1990s. On one side, scientists and scholars in the humanities insisted that scientific knowledge be valued as an objective description of reality and criticized what they denounced as relativist views of social constructivism and arbitrary postmodernist positions. The other side emphasized the need to recognize the role of historical, social, and cultural conditions on what is perceived as scientifically valid and true, and the fact that these agreements and the realities they construct undergo change. *Mathematics, Science, and Postclassical Theory* (1997), edited by Barbara Herrnstein Smith and Arkady Plotnitsky, participates in the Science Wars by giving exceptional prominence to mathematics and with an unusual calming note in the sometimes heated discussion of the 1990s. As the editors explain in their introduction, what they call "postclassical" theory—involving critical analyzes of concepts such as knowledge, objectivity, truth, and proof—also has implications for mathematics, yet, the chapters in the collection "dealing with mathematics suggest that the relations—both historical and conceptual—between mathematics and postclassical theory are on the whole quite cordial and that, even where those relations are complex, they do not involve any wholesale refutations or underminings in either direction" (Smith and Plotnitsky 1997, p. 3). In contrast, at the height of the Science Wars in 1996, in what has become known as the Sokal Hoax, physicist Alan Sokal published a fabricated paper in the journal *Social Text*: it argued that quantum gravity is a linguistic and social construct, and was aimed at exposing the lack of intellectual rigor in postmodern critical theory and, ultimately, the unfoundedness of constructivist arguments. The Sokal Hoax serves as a concrete, if maybe overly discussed, moment in the relations between the humanities and the sciences, and it has a lesser-known companion piece in mathematics: in 2012, a paper apparently authored by Marcie Rathke but in fact created by using Mathgen, an online random generator of mathematical papers, was accepted in the journal *Advances in Pure Mathematics* (Eldredge 2012). While mathematicians were quick to point out that this is not a top-tier journal, this "landmark event in the history of academic publishing" shows that determining intellectual rigor in mathematics may be more difficult than scientific realists in the Science Wars implied and the broader public may expect (Taylor 2012).

The twenty-first century sees further explorations of the historical, social, and cultural conditions of mathematics that question traditional assumptions of its objectivity, transcendence, and unchanging truth. *Where Mathematics Comes From: How the Embodied Mind Brings Mathematics into Being* (2000) by George Lakoff and Rafael Núñez presents mathematics as, like everyday language, "grounded in sensory-motor experience. Abstract human ideas make use of precisely formulated cognitive mechanisms such as conceptual metaphors that import modes of reasoning from sensory-motor experience" (Lakoff and Núñez 2000, p. xii). Using cognitive science to investigate mathematical thinking, Lakoff and Núñez conclude that what they call the "romance" of mathematics (p. xv)—namely, its image as a disembodied, transcendent, true language of nature—is wrong and that mathematics originates in embodied experience and "a great many of the most fundamental mathematical ideas are inherently metaphorical in nature" (p. xvi). Where this cognitive approach to mathematics advances a project also pursued by Brian Rotman in Ad Infinitum; *The Ghost in Turing's Machine; Taking God Out of Mathematics and Putting the Body Back In* (1993) and stresses mathematics' connections to language, and metaphor in particular, a 2009 special issue of the journal *Configurations* is dedicated to the imagination as a realm in which mathematics and the arts meet. The contributions to the special issue *Mathematics and the Imagination* bring together scholarship from different disciplines to examine "how/what mathematicians imagine when they do math, and how mathematics is imagined by mathematicians and nonmathematicians alike" (Saiber and Turner 2009, pp. 12–13). Similarly, the collection of essays *Circles Disturbed: The Interplay of Mathematics and Narrative* (2012), edited by Apostolos Doxiadis and Barry Mazur, is consciously designed as "a two-way interaction between mathematics and narrative"

(Doxiadis and Mazur 2012, p. xvi) and adds a focus on narrative to the earlier work on metaphor and imagination.

Less focused on theoretical examination, *Mathematics in Popular Culture: Essays on Appearances in Film, Fiction, Games, Television and Other Media* (2012), edited by Jessica K. Sklar and Elizabeth S. Sklar, takes account of the growing presence of mathematics in popular culture—a trend that is still ongoing with recent cinema films including *The Imitation Game* (2014), *The Man Who Knew Infinity* (2015), and *Hidden Figures* (2016). Recent developments in research on interactions between mathematics, language, literature, and art have seen Lynn Gamwell's monumental *Mathematics+Art; A Cultural History* (2016) that discusses examples from the Stone Age to the present day, and an emerging focus area on mathematics and modernism (Hickman 2005; Cliver 2008; Tubbs 2014; Brits 2017; Engelhardt 2018). The digital humanities constitute another area of interest in twenty-first-century literature and mathematics studies. Computing technologies, based on mathematical processes, allow for quantitative analyzes and considering big data sets and are used to increase the reach and relevance of research in the humanities. At the same time, this initiates renewed discussion of the relations between literature and mathematics and of the value of mathematics and mathematization for the humanities. Matthew Handelman's (2019) *The Mathematical Imagination: On the Origins and Promise of Critical Theory* presents an alternative strand to the well-known position in critical theory that originated with Theodor Adorno and Max Horkheimer who "steadfastly opposed the mathematization and quantification of thought. For them, the equation of mathematics with thinking … provided the epistemological conditions leading reason back into the barbarism and violence that culminated in World War II and the Holocaust" (Handelman 2019, p. 2). Fears that too great a reliance on mathematics could compromise the humanities, for example, by prioritizing quantitative over qualitative interpretations, have reappeared in the twenty-first century and make worthwhile Handelman's project of exploring the more positive role of mathematics in other twentieth-century thinkers' cultural and aesthetic theories (2019, p. 19). At the same time, as this introductory chapter has begun to argue and the following chapters illustrate in much greater detail, mathematics is not a monolithic system of thought and, though lending itself to repressive and reductive thinking, holds surprising potential for paradox, imagination, creativity, and freedom.

NOTE

1. The mathematical symbol ∩ is like the plus sign, +; it is a binary operation that acts on sets—given two sets A and B, the A ∩ B is the set containing the elements that are contained in both A and B. It is called the intersection of A and B.

REFERENCES

Beer, Gillian. [1983] 2000. *Darwin's Plots; Evolutionary Narrative in Darwin, George Eliot and Nineteenth-Century Fiction.* Cambridge: Cambridge University Press.

———. 1990. "Translation or Transformation? The Relations of Literature and Science." *Notes and Records of the Royal Society of London* 44, no. 1: 81–99.

Bernstein, Charles. 1996. "Erosion Control Area 2." In *Experimental – Visual – Concrete: Avant-Garde Poetry Since the 1960s*, edited and translated by David Jackson, Eric Vos, and Johanna Drucker, 17–20. Amsterdam and Atlanta: Editions Rodopi B. V.

Blacklock, Mark. 2018. *The Emergence of the Fourth Dimension; Higher Spatial Thinking in the Fin de Siècle.* Oxford: Oxford University Press.

Bloor, David. 1973. "Wittgenstein and Mannheim on the Sociology of Mathematics." *Studies in History and Philosophy of Science* 4, no. 2: 173–91.

———. [1976] 1991. *Knowledge and Social Imagery.* Chicago and London: The University of Chicago Press.

Brits, Baylee. 2017. *Literary Infinities, Number and Narrative in Modern Fiction.* London: Bloomsbury.

Cliver, Gwyneth E. 2008. *Musil, Broch, and the Mathematics of Modernism.* St. Louis: ProQuest Dissertations Publishing.

Coleridge, Samuel Taylor. 1840. *The Poetical Works of S. T. Coleridge.* Vol. 1. London: William Pickering.

Dawson, Gowan. 2007. *Darwin, Literature and Victorian Respectability.* Cambridge: Cambridge University Press.

De Morgan, Augustus. 1849. *Trigonometry and Double Algebra.* London: Taylor, Walton, and Maberly.

Doxiadis, Apostolos, and Barry Mazur, eds. 2012. *Circles Disturbed: The Interplay of Mathematics and Narrative.* Princeton, NJ: Princeton University Press.

Eldredge, Nate. 2012. "Mathgen Paper Accepted!" *That's Mathematics!* Accessed May 2, 2020. https://thatsmathematics.com/blog/archives/102.

Engelhardt, Nina. 2018. *Modernism, Fiction and Mathematics.* Edinburgh: Edinburgh University Press.

Epple, Moritz. 2010. "'Links and Their Traces: Cultural Strategies, Resources, and Conjunctures of Experimental and Mathematical Practices." In *Science as Cultural Practice*, vol. 1, edited by Moritz Epple and Claus Zittel, 217–40. Berlin: Akademie Verlag.

Galilei, Galileo. 1960. The Assayer. In *The Controversy on the Comets of 1618*. Edited and translated by Stillman Drake and C. D. O'Malley, 151–336. Philadelphia, PA: University of Pennsylvania Press.

Gamwell, Lynn. 2016. *Mathematics + Art; A Cultural History.* Princeton and Oxford: Princeton University Press.

Gray, Jeremy J. 2004. "Anxiety and Abstraction in Nineteenth-Century Mathematics." *Science in Context* 17: 23–47.

———. 2008. *Plato's Ghost; The Modernist Transformation of Mathematics.* Princeton and Oxford: Princeton University Press.

Handelman, Matthew. 2019. *The Mathematical Imagination: On the Origins and Promise of Critical Theory.* New York: Fordham University Press.

Henderson, Andrea K. 2018. *Algebraic Art: Mathematical Formalism and Victorian Culture.* Oxford: Oxford University Press.

1 INTRODUCTION: RELATIONSHIPS AND CONNECTIONS ... 19

Henderson, Linda Dalrymple. 1983. *The Fourth Dimension and Non-Euclidean Geometry in Modern Art.* Princeton: Princeton University Press.

Hickman, Miranda B. 2005. *The Geometry of Modernism; The Vorticist Idiom in Lewis, Pound, H.D., and Yeats.* Austin: University of Texas Press.

Jenkins, Alice. 2017. "Mathematics." In *The Routledge Research Companion to Nineteenth-Century British Literature and Science*, edited by John Holmes and Sharon Ruston, 217–34. New York: Routledge.

Jouet, Jacques. 2000. *Poèmes de métro.* Paris: P.O.L.

Kerkhove, Bart Van, and Jean Paul Van Bendegem, eds. 2007. *Perspectives on Mathematical Practice: Bringing Together Philosophy of Philosophy of Mathematics, Sociology of Mathematics, and Mathematics Education.* Dordrecht: Springer.

Khlebnikov, Velimir. 1989. *Collected Work of Velimir Khlebnikov, vol. II: Prose, Plays, and Supersagas.* Edited by Ronald Vroon. Translated by Paul Schmidt. Cambridge, MA and London: Harvard University Press.

Knorr-Cetina, Karin. 1981. *The Manufacture of Knowledge: An Essay on the Constructivist and Contextual Nature of Science.* Oxford: Pergamon Press.

Kuhn, Thomas. [1962] 1970. *The Structure of Scientific Revolutions.* Chicago: University of Chicago Press.

Küppers, Bernd-Olaf. 2018. *The Computability of the World; How Far Can Science Take Us?* Translated by Paul Woolley. Cham: Springer.

Lakatos, Imre. [1976] 2015. *Proofs and Refutations; The Logic of Mathematical Discovery.* Edited by John Worrall and Elie Zahar. Cambridge: Cambridge University Press.

Lakoff, George, and Rafael Núñez. 2000. *Where Mathematics Comes From: How the Embodied Mind Brings Mathematics into Being.* New York: Basic Books.

Latour, Bruno, and Steve Woolgar. [1979] 1986. *Laboratory Life; The Construction of Scientific Facts.* Beverly Hills and London: Sage.

Levine, George. [1988] 1991. *Darwin and the Novelists: Patterns of Science in Victorian Fiction.* Chicago and London: The University of Chicago Press.

Mandelbrot, Benoît. [1977] 1982. *The Fractal Geometry of Nature.* San Francisco: W. H. Freeman.

Mehrtens, Herbert. 1990. *Moderne Sprache Mathematik; Eine Geschichte des Streits um die Grundlagen der Disziplin und des Subjekts formaler Systeme.* Frankfurt a.M.: Suhrkamp.

Mühlhölzer, Felix, Ulrich Stuhler, and Yuri Tschinkel. 2008. "Mathematics Between the Sciences and the Humanities." Accessed May 25, 2020. https://www.uni-goettingen.de/en/mathematics+between+the+sciences+and+the+humanities/98259.html.

Nicolson, Marjorie Hope. [1950] 1960. *The Breaking of the Circle; Studies in the Effect of the "New Science" Upon Seventeenth-Century Poetry.* New York: Columbia University Press.

O'Connor, Ralph. 2007. *The Earth on Show: Fossils and the Poetics of Popular Science, 1802–1856.* Chicago, IL: University of Chicago Press.

Pickering, Andrew. 1995. *The Mangle of Practice: Time, Agency, and Science.* Chicago: University of Chicago Press.

Pickering, Andrew, and Adam Stephanides. 1992. "Constructing Quaternions: On the Analysis of Conceptual Practice." In *Science as Practice and Culture*, edited by Andrew Pickering, 139–67. Chicago: University of Chicago Press.

Poincaré, Henri. [1902] 2015. *The Foundations of Science: Science and Hypothesis, The Value of Science, Science and Method.* Translated by George Bruce Halsted. Cambridge: Cambridge University Press.

Queneau, Raymond. 1986. "Potential Literature." In *Oulipo: A Primer of Potential Literature*, edited and translated by Warren Motte, 51–64. Lincoln: Nebraska University Press.

Rogers, Janine. 2014. *Unified Fields: Science and Literary Form*. Montreal: McGill-Queen's University Press.

Rotman, Brian. 1993. *Ad Infinitum; The Ghost in Turing's Machine; Taking God Out of Mathematics and Putting the Body Back In*. Stanford, CA: Stanford University Press.

Saiber, Arielle, and Henry S. Turner. 2009. "Mathematics and the Imagination: A Brief Introduction." *Configurations: Special Issue: Mathematics and Imagination* 17, nos. 1–2: 1–18.

Schlegel, Friedrich. [1797] 2003. "From 'Critical Fragments' (1797)." In *Classic and Romantic German Aesthetics*, edited by J. M. Bernstein, 239–45. Cambridge: Cambridge University Press.

Shuttleworth, Sally. 1984. *George Eliot and Nineteenth-Century Science*. Cambridge: Cambridge University Press.

Sklar, Jessica K., and Elizabeth S. Sklar, eds. 2012. *Mathematics in Popular Culture: Essays on Appearances in Film, Fiction, Games, Television and Other Media*. Jefferson, NC: McFarland.

Smith, Barbara Herrnstein, and Arkady Plotnitsky. 1997. *Mathematics, Science, and Postclassical Theory*. Durham, NC: Duke University Press.

Sullivan, J. W. N. 1933. *The Limitations of Science*. New York: Viking.

Taylor, Paul. 2012. "Stochastically Orthogonal." *London Review of Books*, 17 October. https://www.lrb.co.uk/blog/2012/october/stochastically-orthogonal.

Tubbs, Robert. 2014. *Mathematics in Twentieth-Century Literature and Art: Content, Form, Meaning*. Baltimore: Johns Hopkins University Press.

Turner, Henry S. 2006. *The English Renaissance Stage: Geometry, Poetics, and the Practical Spatial Arts 1580–1630*. Oxford: Oxford University Press.

———. 2010. "Lessons from Literature for the Historian of Science (and Vice Versa): Reflections on 'Form.'" *Isis* 101, no. 3: 578–89.

Weizsäcker, Carl Friedrich von. [1971] 1980. *The Unity of Nature*. Translated by Francis J. Zucker. New York: Farrar, Straus and Giroux.

Willis, Martin. 2015. *Literature and Science*. London and New York: Palgrave.

Wittgenstein, Ludwig. 1983. *Remarks on the Foundations of Mathematics*. Translated by G. E. M. Anscombe. Edited by G. E. M. Anscombe, G. H. von Wright, and Rush Rhees. MIT Press.

Open Access This chapter is licensed under the terms of the Creative Commons Attribution 4.0 International License (http://creativecommons.org/licenses/by/4.0/), which permits use, sharing, adaptation, distribution and reproduction in any medium or format, as long as you give appropriate credit to the original author(s) and the source, provide a link to the Creative Commons license and indicate if changes were made.

The images or other third party material in this chapter are included in the chapter's Creative Commons license, unless indicated otherwise in a credit line to the material. If material is not included in the chapter's Creative Commons license and your intended use is not permitted by statutory regulation or exceeds the permitted use, you will need to obtain permission directly from the copyright holder.

PART I

Mathematics in Literature

CHAPTER 2

Numbered Possibilities: Chaucer and the Evolution of Late-Medieval Mathematics

David Baker

Much more than a poet, Geoffrey Chaucer (1343–1400) was an amateur astronomer and philosopher, author of an instruction manual on the astrolabe that bears marks in places of an impressive degree of mathematical interest and competence for a man who, as far as we can tell, had no university education, and translator of one of the most influential philosophical texts in the later Middle Ages.[1] Chaucer also dedicated his finest completed work, *Troilus and Criseyde* (c. 1380s), to a certain "philosophical," or as two manuscripts have it, "sophistical" (that is, logical), Strode (Chaucer 1987, p. 585 [Book v, line 1857]): in all likelihood, this is Ralph Strode, the most influential Oxford logician of his day.[2] Yet in the late Middle Ages, the disciplinary boundaries between arithmetic, geometry and the language-oriented study of logic had begun to break down. This convergence was remarkably fruitful, allowing highly imaginative problems to be posed and sometimes solved, and Chaucer's literary work has a great deal of fun with the coalescence of medieval arithmetic, geometry and logic into a single discipline more recognizable today as mathematics, while delighting in the opportunities and limitations of mathematics to provide analogies for complex human experiences.

D. Baker (✉)
Staffordshire, UK

© The Author(s) 2021
R. Tubbs et al. (eds.), *The Palgrave Handbook of Literature and Mathematics*, https://doi.org/10.1007/978-3-030-55478-1_2

The Mathematical Melting Pot

What for most modern readers is vaguely conceived of as the single academic category of mathematics was in the medieval scholastic system a thread that weaved in and out of distinct disciplines. In the later Middle Ages, a pre-algebraic age, problems that may now be considered essentially mathematical were often treated as questions of logical linguistics because those problems were expressible only in ordinary sentences. For example, problems to do with *maxima* and *minima* were often addressed in terms of finding the point at which certain statements became true or false (Spade 1998, p. 402). The fourteenth-century English logician Roger Rosetus applied what would now be regarded as limit mathematics to the ethical conundrum of at what point a monk's claim to have obeyed an instruction to read the Bible would become true:

> One of the arguments ... is that there is neither a maximum act of studying which would conform to the prelate's precept nor a minimum act of studying which would not so conform, since the intensity of the act of studying may be increased in infinitum. (Wilson 1960, p. 101)

Rosetus was here developing the prior application of a similar approach to theological questions. The thirteenth-century Paris theologian Henry of Ghent had invoked a similar method in his discussion of what would later be formulated as the question of the Immaculate Conception, whether the Virgin Mary was sanctified in the very moment of the infusion of her rational soul, or (however infinitesimally) later. Henry invoked conventional mechanical examples, such as the motion through the air of a rising bean and a falling millstone, and appealed to Aristotle's *Physics* (Aristotle 1984), in his discussion of the possible understandings of instantaneous change and the difficulties of attempting to divide up a continuous process into individual instants of time (Brower-Tolland 2002, pp. 34–37). Logico-mathematical approaches were also applied to what we might think of as human psychology: for example, Nicole Oresme, the fourteenth-century French logician and mathematician, considered which of two human experiences of pain would bring the more suffering by means of a geometrical analogy (as I discuss later). Thus the fourteenth century is in two ways an exciting period for those interested in the history of mathematics as a cultural phenomenon: not only was the discipline of mathematics itself increasingly colonising other fields and beginning to draw into itself the rich range of sub-disciplines that form a more modern conception of the subject, but also by so doing it was interacting with concerns very much human, philosophical and linguistic; and it is in the context of this logico-mathematical-philosophical-theological-linguistic interchange that Chaucer's work must be read in order to grasp both its full intellectual weight and its artistic delicacy.

The Winning Numbers

The mathematical concerns of Chaucer's work have by no means yet been fully explored, but one apparently throwaway remark in "The Pardoner's Tale" (c. 1395) has been noted by historians of mathematics as extremely significant. According to David Bellhouse and James Franklin, in Middle English, "when two players, say A and B, play at dice, the events by which A or B win are called A's chance and B's chance." They give us an example, the Pardoner's imitation of gambling parlance, "Sevene is my chaunce, and thyn is cynk and treye" ("Seven is my chance, and yours is five and three" [Chaucer 1987, p. 198. l. 653]).

> Here Chaucer must be referring to the sum of the faces which show [on two dice], since seven is listed as one of the chances. Now the event of a seven showing has the same probability as the event of five or three showing. (Bellhouse and Franklin 1997, p. 77)

The probability of rolling a total of seven with two dice is $6/36 = 1/6$; the probability of rolling a total of five is $4/36 = 1/9$; and the probability of rolling a total of three is $2/36 = 1/18$. This means that the probability of rolling a 3 or 5 is $1/18 + 1/9$, a sum that equals $1/6$, which is the same as the probability of rolling a seven. Of course, a modern mathematical expression of the gamblers' chances needs to be a little more precise in the choice of conjunction (Bellhouse and Franklin replace Chaucer's "and" with "or" in their analysis); but leaving this detail aside, it nevertheless seems likely that Chaucer was capable of using basic combinatorics to calculate the likelihood of certain totals.

This may not sound impressive, but it is significant: as James Franklin has noted, some twentieth-century historians of mathematics tended to be somewhat reductive in dating the emergence of what we would recognize as probability theory to 1654, the date of an influential correspondence between the French mathematicians Blaise Pascal and Pierre de Fermat (Franklin 2015, p. 362). However, since Daniel Garber and Sandy Zabell (1979) began to question the absoluteness of the alleged 1654 watershed, and especially over the last two decades, there has been increasing recognition of the emergence of basic probabilistic ideas in the thirteenth and fourteenth centuries.[3] These analyzes have focused on two key areas: games of chance and financial investment. The majority of the evidence that late-medieval gamblers knew enough about combinatorics to attempt basic probabilistic assessments has come not primarily from mathematical but from literary sources, such as the thirteenth-century poem *De Vetula*, and the fourteenth-century Italian commentaries on Dante.[4] At the same time, scholastic thinkers, such as the thirteenth-century Franciscan Peter Olivi, were beginning to rethink the idea of future risk in quasi-probabilistic terms for economic and ethical reasons, in order to separate the idea of investment (involving risk and return) from simple usury.[5]

26   D. BAKER

Chaucer's reference to a proto-probabilistic view of rolling dice in the context of "The Pardoner's Tale", however, may well be intended to do more than simply add verisimilitude to his depiction of tavern conversation. In the tale, three riotous drunkards set out to destroy Death, and rudely accost an old man, demanding to know where they can find Death. The old man tells them they will find it beneath a certain tree. The gambling motif returns at the conclusion of the tale, as the men, who have in fact found a pot of treasure beneath the tree, draw lots to determine which of them will return to fetch three bottles of wine with which to celebrate their good fortune. While the one is away, the other two plot to murder him on his return; but he, unknown to them, poisons two of the three bottles, intending to keep the clean one for himself. Upon his return, the two together murder the one; but afterwards, they share a bottle of wine and "par cas" ("by chance"), as Chaucer notes, they choose one of the two poisoned bottles (Chaucer 1987, p. 201, l. 885). Chaucer thus seems to emphasize the ideas of chance and risk at the conclusion of the tale, and particularly to repeat the motif of two-against-one, just as in his earlier allusion to dice he posited a scenario in which one gambler had two possible winning combinations and his opponent just one, although their chance of success was the same: here, it is two-to-one against being chosen by lot; it is two-to-one (literally) in the fight; it is two-to-one against choosing the unpoisoned bottle. The typical Chaucerian twist is that despite various possible calculations of the risk, all three men end up as losers: the only reliable assessment of their chances of success turns out to be that given by the old man, based on probabilities of human psychology and behavior, rather than any more mathematical basis—they all do find death beneath the tree.

A TRICKY DIVISION

The problem of how to divide up a pot of winnings also underlies Chaucer's "Summoner's Tale" (late 1390s), in which a greedy friar disturbs a bedridden man with his begging. The man, Thomas, promises the friar a certain contribution that he has hidden under the bedcover, but only on condition that the friar will distribute what he receives evenly among the twelve members of his convent. The friar agrees, but when he feels under the cover, Thomas farts in his hand. Disgusted, the friar complains to the lord of the manor, who, instead of reprimanding Thomas, is fascinated by the puzzle of how one could divide a fart up equally. In one of Chaucer's most famous scatological jokes, the lord remarks that the problem is so difficult that "In ars-metrike [arithmetic] shal ther no man fynde, / Biforn this day, of swich a question" ("no-one will find such an arithmetical question before now" [Chaucer 1987, p. 135, ll. 2222–23]). Thomas's conditional donation raises a range of logico-mathematical issues.

First, does Chaucer have anything specific in mind when he refers to earlier "ars-metrike"? The friar's original begging patter gives a clue:

A, yif that convent half a quarter otes!
A, yif that convent foure and twenty grotes!
A, yif that frere a peny, and lat hym go!
Nay, nay, Thomas, it may no thyng be so!
What is a ferthyng worth parted in twelve?

(Ah, give that convent half a quarter of oats!
Ah, give that convent twenty-four groats!
Ah, give that friar a penny, and let him go!
No, no, Thomas, it may not be so!
What is a farthing worth divided into twelve?)

(Chaucer 1987, p. 132, ll. 1962–1967)

The first two divisions here seem relatively straightforward and are typical of earlier arithmetical collections, such as the *Propositiones ad acuendos iuvenes*, a highly influential eighth- or ninth-century text attributed to the English polymath and adviser to Charlemagne Alcuin of York.[6] Given the circulation of that text, it is possible that Chaucer specifically knew some of the problems contained in the *Propositiones*, whether directly or indirectly, and it is even possible that he knew them as the "arismetrica" of Alcuin, as one fourteenth-century English library catalogue seems to have called it.[7]

The *Propositiones* contain a number of problems that bear striking similarity to the sort of division that the friar was expecting to perform. *Propositio* 53, for instance, concerns an abbot dividing eggs equally between the twelve monks of his monastery (Alcuin (ascr.) 2003, p. 75). *Propositio* 47 concerns a bishop dividing loaves between his twelve clergy (Alcuin (ascr.) 2003, p. 72). In fact, division into twelve is a staple of the *Propositiones* and the instruction "divide in XII partes" occurs repeatedly (for example, *Propositiones* 22–25: Alcuin (ascr.) 2003, pp. 57–59). The friar's expectations are also highly conventional in terms of *what* is to be divided. *Propositiones* 32–34 concern the division of grain, just as the friar asks for "otes" (Alcuin (ascr.) 2003, pp. 63–65). *Propositio* 35, in which a dying man leaves money in his will on condition of it being shared out in various combinations of twelfths, is just one of a number of division exercises in the collection which involve money, paralleled in the friar's hope that the sick Thomas will give him "foure and twenty grotes" to divide (Alcuin (ascr.) 2003, p. 65). There is also a possible verbal allusion to the formulations of the *Propositiones* at the end of "The Summoner's Tale". The lord asks who can solve Thomas's problem of dividing the fart, and his steward suggests an intricate solution involving a cartwheel. The steward's form of words in replying to the lord's "Tel me how" with "I koude telle" ("I could tell") perhaps reflect the standard challenges at the conclusion of problems in the *Propositiones*: "dicat qui potest" or "dicat qui valet" ("let him tell who can"). Thus Chaucer probably meant the friar's begging to exemplify the older, simpler arithmetical problems that seem basic in comparison to Thomas's conundrum of dividing the fart (Chaucer 1987, p. 136, ll. 2253–86, 2230, 2247).

28   D. BAKER

Yet scholars have noticed that "The Summoner's Tale" is also littered with the terminology of late-medieval scholastic logic, terms such as "probleme," "question," "ymaginacioun" ("imagination"), and "demonstration" (O'Brien 1990, pp. 14–17). The lord of the manor actually refers to Thomas's problem as an "inpossible" (Chaucer 1987, p. 135, l. 2231), which, as Roy Pearcy first noted, is an Anglicisation of the Latin term "*impossibilium*," a specific genre of the *sophismata*, the logical puzzle-sentences that were the building blocks of late-medieval treatises on logic (Pearcy 1967). How, then, is Thomas's problem related to late-medieval scholastic logic, and why does the lord of the manor consider it to be more difficult than any arithmetical question "biforn this day"? Even in the friar's initial begging speech, Chaucer drops two hints of what is to come.

First, the friar finishes his list of examples with that of a "ferthyng ... parted in twelve." A "ferthyng," worth a quarter of a "peny," was the smallest whole denomination of currency. As with all medieval coins, they were at times cut in four; but as for a farthing cut in twelve, as Glending Olson points out, "no such coin existed for exchange purposes" (Olson 2009, p. 422). The friar's point is that things cannot be endlessly divided and subdivided: and Thomas precisely promises to give the friar a unitary "thyng" (Chaucer 1987, p. 134, ll. 2142–43). Olson explores Thomas's problem in terms of the late-medieval controversy over the divisibility of continua: in other words, whether a continuum, such as a mathematical line, can be divided infinitely, or there remain basic, indivisible units, or points, which together make up the line. The question seemed to have been decisively settled by means of a geometric proof (which I discuss later) by Thomas Bradwardine, one of the so-called "Oxford Calculators," the famous mid-fourteenth-century group of logicians, mathematicians and natural scientists, and a thinker some of whose work was undoubtedly known to Chaucer, as he cites him by name elsewhere ("The Nun's Priest's Tale" [c. 1380s–1390s]).[8] However, the issue briefly became live again later in the century, largely through the work of the Oxford logician and heterodox theologian John Wyclif, who maintained an indivisibilist position (that is, he argued that an apparent continuum must be constituted of indivisible points), in part for the theological reason that an omniscient God must know each individual point of his creation, which seemed impossible to him if further points could be added ad infinitum through continuous division (Kretzmann 1986, p. 45). So one of the problems Thomas's condition raises is how, if the fart, like the farthing, is a single "thyng," it can be subdivided "in twelve."

A second question concerns what kind of "thyng" a fart is in the first place. That the friar expects a concrete "yifte" ("gift") he has already made clear: he wants something more than a "ferthyng," which cannot easily be "parted in twelve" (Chaucer 1987, p. 134, l. 2146; p. 132, l. 1967). Chaucer's pun (farthing/farting) is not exactly subtle, and critics have tended to comment on it, as on other examples of Chaucer's toilet humor, with something approaching a sigh.[9] However, many of the most frequently debated of the logical *sophismata* turned, essentially, on what we might call "puns," which draw their power from the problematic ambivalence of language. Furthermore, Chaucer's point is not merely that the word, "ferthyng" is similar to the word "ferting," but that the word "ferthyng" itself can be "parted" into a compound word: "fart-thing." Allowing for the consonantal interchange that often took place between "f" and "p" in the evolution of the English language, Chaucer's line carries another humorous pun in "parted."[10] Thus Chaucer has the friar finish his greedy list with the question, "What is a fart-thing farted between twelve?" Thomas simply addresses the friar's hypothetical question in more practical terms.

Yet "the soun and savour of a fart" ("the sound and smell of a fart") is not the sort of thing that can easily be measured or divided up, as the lord of the manor muses:

> The rumblynge of a fart, and every soun,
> Nis but of eir reverberacioun
> And evere it wasteth litel and litel awey[.]

> (The rumbling of a fart, and every sound,
> Is only a reverberation of the air,
> And, little by little, it continuously dies away.)
>
> (Chaucer 1987, p. 135. ll. 2226, 2233–35)

In other words, a fart is not an object or quantity, but either a continuous event in time or a temporary quality of something else, the air. Perhaps the most important way in which medieval thinkers evolved the physical problems of Aristotle was their determination to quantify or find the limits of processes of continuous change. John Murdoch described their approach as "the near frenzy to measure everything imaginable" (Murdoch 1975, p. 287). As Edith Sylla explains:

> According to Aristotelian theory, quantities and qualities belong to separate categories. One might suppose, therefore, that Aristotelian theorists would not attempt to quantify qualities. During the later Middle Ages, however, theorists who were basically Aristotelian in their approaches did attempt to quantify qualities. (Sylla 1971, p. 9)

William Heytesbury, another of the Oxford Calculators, lists just a few of the qualities that thinkers of his time wrangled with:

> Whether, if Socrates will be bigger than Plato, or better or whiter or hotter, or colder, or healthier, or faster, and now he is not bigger than Plato, or faster and so on, therefore he [Socrates] begins, or at some point will begin, to be bigger than Plato, or better and so on[.] (Murdoch 1979, pp. 143–44)

Bigger, better, whiter, hotter, colder, healthier, or faster: size, goodness, color, temperature, health, and speed all varying over time—and that is just a small sample. The solutions to these problems were often essentially mathematical. For example, the Mean Speed Theorem (the result that a body A subject to uniform acceleration will travel the same distance over a period of time as a body B traveling constantly at the mean speed of body A), one of the most important mathematical breakthroughs of the fourteenth century, was formulated by the Oxford Calculators precisely in response to problems concerning processes of continuous change like those listed above (Longeway 2007).

A fart, according to the lord in "The Summoner's Tale," is merely the speed and smell of the air. The problem of how to measure it out is thus structurally very similar to those that fourteenth-century logicians, physicists, and mathematicians were actually addressing. Chaucer was not the only late-medieval writer to make this connection. When the early sixteenth-century French author François Rabelais constructed a parody of a scholastic question by stringing together the typical terms of the late-medieval *sophismata*, it concerns a chimera farting in a vacuum.[11] Parody aside, the fourteenth-century French logician Jean Buridan discussed belching in his *Sophismata* (Buridan 1966, pp. 67, 161–62), although the context concerns problems of intended meaning; and it is possible that Chaucer uses a belch as an allusion to those problems elsewhere in "The Summoner's Tale" (Baker 2013b, pp. 92–99). Thus fart-division as a parody of fourteenth-century logic seems not so much derisive hyperbole as plausible, albeit humorous, imitation.

Yet when the lord's steward puts forward his solution to the problem of how to divide the fart, it is hailed as worthy of "Euclide" or "Ptholomee" ("Euclid" or "Ptolemy": the first a geometer and the second an astronomer) (Chaucer 1987, p. 136, l. 2289). Chaucer's own work reveals that he is aware of one of the astronomical applications of the controversy over the divisibility of the continuum, since, as Olson has noted, "Chaucer says when describing the markings of the astrolabe, the degrees of the zodiac are divided into 'mynutes,' the minutes into 'secundes, and so furth into smale fraccions infinite'" ("seconds, and so forth into small fractions, infinitely") (Olson 2009, p. 424, n. 9; Chaucer 1987, p. 664 [*Treatise on the Astrolabe*, I., para. 8, ll. 11–13]). The lord's reference to Euclid is even more significant, since it was by means of perhaps the most famous of Euclid's geometric propositions, the forty-seventh proposition of Book I of the *Elements*, better known to us The Pythagorean Theorem, that Bradwardine had convincingly proven the incoherency of the indivisibilist position.[12] Chaucer's treatment of the fart as "ars-metrike" thus reveals a concern with the *evolution* of logico-mathematical puzzles in the later Middle Ages, and a recognition that the arithmetic and geometry of his time was steadily insinuating itself into a broader and more complex umbrella of disciplines, most notably scholastic logic.

Getting Physical

Perhaps the Chaucerian character who is used most tragically to demonstrate the human danger of simplistically chopping up and enumerating the fluidity and endless possibilities of life is poor Troilus. In *Troilus and Criseyde*, Chaucer presents a male protagonist who attempts to treat love itself in logico-mathematical terms, and through his incompetence as a natural philosopher, fails to understand women or even love itself. When writing a love letter, Troilus is warned by his friend Pandarus precisely against using "termes of phisik / In loves termes" (Chaucer 1987, II., p. 503, ll. 1038–39); that is, conflating the language and categories of the natural sciences (and specifically the mathematical physics of the *sophismata physicalia*) with the language and categories of human love.[13] Yet throughout the poem, Chaucer makes many allusions to the scholastic logic, mathematics and physics of his century, and, as noted above, dedicates the poem to an Oxford logician, whom Chaucer asks to "correcte" it as needed (Chaucer 1987, V., p. 585, l. 1858). There is no evidence that Strode did so: rather the appeal would seem to be Chaucer's self-deprecatory means of indicating that there is material in the poem that would be of interest to a logician, and incidentally an avowed admirer of Bradwardine, as Strode makes clear in one of his logical treatises (Spade 1988, p. 116 footnote). Chaucer's choice to self-consciously flout Pandarus's rule and mingle fourteenth-century mathematics and physics into his account of the pain of unrequited love reflects the fact that the logicians of his time were applying mathematical methods to internal qualities and emotions, including pain and love.

Nicole Oresme, who began to develop a geometric approach which would eventually flower in the analytic geometry of the seventeenth century, wrote:

> let A and B be two pains, with A being twice as intensive as B and half as extensive. Then they will be equal simply[.] ... But these two equal and uniform pains when mutually compared are differently figured, ... so that if pain A is assimilated to a square, then pain B will be assimilated to a rectangle whose longer side will denote extension, and the rectangle and square will be equal.[14]

A modern reader can (admittedly anachronistically) comprehend Oresme's analogy by visualizing a Cartesian plane where the x-axis represents time and the y-axis intensity of pain. On that plane, pain A is represented by the line $y=2$ between $x=0$ and $x=1$; and pain B is represented by the line $y=1$ between $x=0$ and $x=2$. The areas beneath the two lines, which represents the total pain experienced, are equal. Oresme also applied his geometric approach to similar problems involving "joy" (Grant 2001, p. 172). Olson notes that fourteenth-century logicians applied the "analytical terminology used to discuss such subjects as proportion, infinity and continuity, and local motion not only to problems in logic and natural science, but also to philosophical and theological questions," and applied "the language of intension

32   D. BAKER

and remission of forms (acceleration/deceleration, or increase/decrease in qualities such as heat)" to questions of love and the movement of the will (Olson 2009, pp. 414–15). The problems of intension and extension concern the ability to quantify phenomena of varying intensity in terms of a "mean," the most famous example being the Mean Speed Theorem, mentioned above. A similar method was used to quantify, for example, the overall temperature of a body whose heat varies across its area (Murdoch 1975, p. 318, n. 45). In another anticipation of later analytic geometry, and specifically asymptotic analysis, yet another of the Oxford Calculators Richard Swineshead imagined

> a given subject ... hot in degree 1 over its first half, in degree 2 over its next quarter, in degree 3 over its next eighth, in degree 4 over its next sixteenth, and so on in infinitum. As a whole, the subject is hot in degree 2. That is, it is finitely hot as a whole even though the heat throughout it increases infinitely.[15]

Chaucer frequently uses heat as a metaphor for love in *Troilus and Criseyde*. This is not in itself surprising, since such a metaphor was highly conventional, but some of Chaucer's formulations of the metaphor seem decidedly mathematical. Take, for example, his treatment of one of the conventional paradoxes of love in his description of Criseyde as, "Now hoot, now cold; but thus, bitwixen tweye" ("Now hot, now cold; but then between the two" [Chaucer 1987, II., p. 500, ll. 811–12]). The convention with which Chaucer is playing here is that of feeling "hot and cold all over," a mysterious phenomenon apparently accompanying the experience of love. In Criseyde's case, however, the paradox is mathematically *resolved*, by figuring her contrasting emotions as contraries on a continuous line of intension and remission (that is, of greater and lesser heat) and assigning her a mean temperature "bitwixen tweye." Similarly, Chaucer hints at the application of a more rigorous quantification of the heat of love in Troilus's exclamation to Pandarus, "I hadde it nevere half so hote as now" ("I never had it half as hot as now" [Chaucer 1987, III., p. 535, l. 1650]), a markedly numerical translation of the Italian of Chaucer's source, Giovanni Boccaccio's *Il Filostrato* (c. 1338), "Io ardo più che mai" ("I burn more than ever" [Boccaccio 1929, pp. 266–67]).

Another set of problems about the continuum involved beginnings and endings, and such questions seem to have been of especial interest to the authors of the *sophismata physicalia*. Heytesbury dedicated a whole chapter of his *Rules for Solving Sophismata* to such problems, which stem from Aristotle's *Physics* and concerned the ability to identify a *precise instant* at which a continuous process, such as heating up or turning white, "begins" or "ends" (Wilson 1960, p. 6).[16] There are several occasions throughout the poem when Chaucer describes an emotional change in terms of a physical or physiological change, both in terms of heat (for example, Chaucer 1987, III., p. 524, l. 800) and color (for example, Chaucer 1987, III., p. 515, l. 82): and it is worth noticing that these changes are frequently qualified with the adverb "sodeynly" or "sodeynliche" ("suddenly"). The image of sudden (that is instantaneous), physical and, by implication, emotional, *mutation* (to use the Aristotelian term) seems to be recurrent in *Troilus and Criseyde*.[17] In fact, Chaucer explicitly addresses the issue of instantaneous change, employing logico-physical metaphors to do so.

While Troilus is first wooing Criseyde, Pandarus counsels him to hope for a sudden and irrevocable change in her feelings toward him:

> Thenk here-ayeins: whan that the stordy ook,
> On which men hakketh ofte, for the nones,
> Receyved hath the happy fallyng strook,
> The greete sweigh doth it come al at ones,
> As don thise rokkes or thise milnestones;
> For swifter cours comth thyng that is of wighte,
> Whan it descendeth, than don thynges lighte.

> (Think on the contrary: when the sturdy oak,
> At which men are repeatedly hacking, for a while,
> Has received what chances to be the felling stroke,
> The great momentum comes all at once,
> As it does with these rocks or these millstones;
> For a heavy thing has a swifter motion,
> When it descends, than do light things.)
>
> (Chaucer 1987, II., p. 508, ll. 1380–86)

The oak and the falling rock or millstone are both images found in fourteenth-century discussions of problems of intension and remission or beginning and ending.[18] Pandarus's argument is that there must be a significant point of "beginning" in a physical change and that such changes are discrete and permanent, rather than continuous and capable of remission. In addition to an apparent sympathy with the indivisibilist understanding of continua, Pandarus's grasp of physics is deliberately simplistic. Aristotle believed that heavier objects fell faster than lighter objects, but this and other aspects of Aristotle's mechanics were being critically rethought in the fourteenth century. Bradwardine, for example, began to challenge this notion by speculating about how fast objects would fall in a vacuum (Grant 1974, pp. 305, 348–49). Pandarus argues that Criseyde might fall instantaneously and permanently in love. He seems to view her as a discrete variable like Troilus himself, whose heart "with a look ... wax a-fere" ("became aflame" [Chaucer 1987, I., p. 476, l. 229]), who "wax sodeynly moost subgit unto love" ("became suddenly most subject unto love" [Chaucer 1987, I., p. 476, l. 231]) and who "sodeynly ... wax ther-with astoned" ("suddenly became therewith amazed" [Chaucer 1987, I., p. 477, l. 274]). Troilus's application of the scholastic maxim that "diversitas requirit distinctionem" ("difference demands distinction") in a later discussion of ethics and human motivation suggests that, like Pandarus, he prefers to divide human behaviors and

34   D. BAKER

motivations discretely or qualitatively, rather than seeing them on a continuum, or as it may be put in modern parlance, as part of a spectrum (Chaucer 1987, III., p. 519, ll. 404–6).[19]

However, if Troilus is a discrete variable, his tragedy is that Criseyde is not. In fact, Chaucer takes pains to dismiss in advance Pandarus's expectation of an instantaneous change of heart when describing how Criseyde falls in love with Troilus. An intrusive narrative parenthesis specifically addresses the problem of sudden emotional change in a passage that again has no original in Chaucer's source. Boccaccio's Criseida "subitamente presa fue" ("suddenly was she captivated" [Boccaccio 1929, II., pp. 204–5, l. 83]). Chaucer's narrator, on the other hand, anticipates an objection from his audience to such an instantaneous emotional change, or "sodeyn love" (Chaucer 1987, II., p. 498, l. 667). Replying that "every thing a gynnyng hath it nede" ("everything needs a beginning" [Chaucer 1987, II., p. 498, l. 671]), the narrator answers the charge of Criseyde's change from not loving to loving Troilus being discrete and instantaneous with the explanation that "she gan enclyne / To like hym first" ("she inclined to like him first" [Chaucer 1987, II., p. 498, ll. 674–75]) and came to fully love him "by proces" and "in no sodeyn wyse" (["by course of events" or "over time," and "not in a sudden way" [Chaucer 1987, II., p. 498, l. 678–79]). Later, even at the very moment of Criseyde's decision to take the one undeniably discrete, objective and physical step in that whole process, the sexual act, Chaucer further emphasizes the continuum of Criseyde's "proces" of falling in love with Troilus with a striking mathematical reference. Expressing her uncertainty what is the best thing to do, Criseyde laments that she is "at dulcarnoun, right at my wittes ende" (Chaucer 1987, III., p. 526, l. 931). "Dulcarnoun," as *The Riverside Chaucer* notes, was the late-medieval nickname for none other than the forty-seventh proposition of Book I of Euclid's *Elements*, the Pythagorean Theorem, and the very same geometric proposition that provided Bradwardine with his clinching argument against the indivisibilists. "Dulcarnoun," in other words, was the lynch-pin of Bradwardine's geometric proof of the continuist view of processes of change, which Chaucer has associated throughout with Criseyde, as opposed to the more simplistic, indivisibilist approach, associated with Troilus. Wyclif, the man largely responsible for the fact that the controversy was still a live issue in Chaucer's time, was aware of Bradwardine's argument, of course, but argues against it with vigor; ultimately, however, he appeals to the fact that "Deus ordinat istos propter melius ordinis universi" ("God orders these things for the better order of the universe") (Wyclif 1893–1899, III., pp. 53–54, 56).

Wyclif's overriding faith in the ordinance of God also lies behind his reputation during his lifetime and immediately thereafter as an extreme necessitarian when it comes to future events. There was one particular proposition already condemned by certain ecclesiastical and scholastic authorities that Wyclif repeatedly asserted in his writings, and which Chaucer probably knew through a correspondence on the subject between Wyclif and Ralph Strode, the dedicatee of *Troilus and Criseyde*, if he did not already know it from elsewhere: that is, that "*omnia que evenient de necessitate evenient*"; or as Chaucer quotes it in a parodically simplistic soliloquy on the subject of future contingency and necessity given by a despairing Troilus, "al that comth, comth by necessitee" ("everything that happens, happens by necessity").[20] Chaucer continues Troilus's meditation with a deliberately bungling rehash of Boethius's famous and much more nuanced treatment of the problem, designed to justify Wyclif's heterodox proposition, and illustrating Troilus's incompetence as a philosopher.[21] Thus Troilus seems satirically to embody both Wyclif's necessitarian view of the future and his indivisibilist view of continua—but why?

The characters of Troilus and Criseyde are vitally and philosophically opposed. Chaucer presents a dichotomy between Troilus's simple or "discrete" character and experience of love, and Criseyde's more "continuous" or fluid personality and emotions. Chaucer's description of Troilus emphasizes his steadfastness of character as "trewe as stiel" ("faithful as steel"). Criseyde, on the other hand, is famously described as "slydynge of corage" ("variable of determination"): when she is forced to leave Troy, she slowly falls in love with the Greek Diomede, leaving Troilus first to suspect and then to lament her betrayal. The adjective "slydynge" implies not merely Criseyde's changeableness, but fluidity of change; and with it the register of natural philosophy and the processes of continuous change (Chaucer 1987, V., p. 571, ll. 831, 825).[22]

Troilus's moment of final realization comes while he is making an almost Euclidean study of the "cote" ("coat") of Diomede, Criseyde's new lover, "avysyng of the lengthe and of the brede" ("considering the length and breadth" [Chaucer 1987, V., p. 582, l. 1657]). On the "cote" he sees the brooch he gave Criseyde, and "ful *sodeynly* his herte gan to colde" ("very *suddenly* his heart became cold" [Chaucer 1987, V., p. 582, l. 1659]: my emphasis). Again, the temperature image is Chaucer's own. The evidence finally convinces him of Criseyde's unfaithfulness, and prompts him to complain to Pandarus of "hire hertes variaunce" ("her heart's variance" [Chaucer 1987, V., p. 582. l. 1670]. In contrast, his own love remains true to his indivisibilist mathematics, as he vows that, despite her unfaithfulness, he could never "unloven" Criseyde (Chaucer 1987, V., p. 583, l. 1698): for Troilus, love functions in either the on or off mode, but nothing in between. Thus to Troilus, whose philosophy is not sophisticated enough to appreciate that her emotions have always been the result of "proces," Criseyde's feelings seem to have instantaneously and inexplicably changed. There have been a range of explanations why exactly Troilus's suffering is described by Chaucer in the opening line of the poem as his "double sorwe" ("double sorrow" [Chaucer 1987, p. 473]). One possibility, however, is that the tragedy of Troilus is the tragedy of a man who, through his unnuanced philosophy, cannot even make human sense of his own plight.

36   D. BAKER

Ultimately, this may perhaps explain Chaucer's own "double" attitude toward his century's astounding mathematical advances and the incredibly fertile broadening out of mathematical interests at the time. Undoubtedly he took a keen amateur interest in the development of the natural sciences, and it seems, in their logico-mathematical elements in particular. No single writer has, I would contend, more sheer fun with the logic and mathematics of his day than Chaucer does. If there is a degree of wariness in his treatment of them, it is generally a distrust of simplistic reasoning, rather than of sophistication, and of the clumsy application of hypothetical scholastic rules to the true fluidity and surprising contingency of human experience. Describing a large wood in *The Book of the Duchess*, one of his dream-vision poems, Chaucer makes a characteristically mathematical joke: even Algus, the mathematician known as the inventor of Arabic numerals, could not count the "wondres" that wood contained (Chaucer 1987, pp. 335–36, ll. 434–42). Even in such a passing reference, it is the limit of mathematics to number all the wondrous possibilities of life that fascinates him, and in spite of the depth of his intellectual curiosity, he is equally aware of the danger of missing the wood for the trees.

NOTES

1. *A Treatise on the Astrolabe* (1391), and Boethius' sixth-century treatise *The Consolation of Philosophy* (c. 1380). Unless otherwise specified, all works by Chaucer are to be found in Chaucer (1987) and all line references and approximate datings of Chaucer's works likewise generally follow those of this edition. Some scholars have recently revisited the possibility of Chaucer's attendance at university. See, for example, Lynch (2000, pp. 18–20).
2. The two manuscripts in question are BL MS Harley 3949 and Oxford, Bodleian Library Rawl. poet 163. See Lynch (2000, p. 9) and Reichl (1989, p. 134); and on the use of *sophistae* to mean logical disputants, see Sylla (1982, p. 545). On Strode, see Delasanta (1991) and Baker (2013b, pp. 57–59).
3. For a more detailed summary of these developments, see Baker (2013b, pp. 215–29). Franklin himself distinguishes between logical (or epistemic) probability and factual probability, and cautions against discounting the ideas of early writers as merely "confused 'anticipations' of later mathematical discoveries" (2015, pp. xviii–xix). Even so, as he states, "a numerical concept of factual probability did develop before the seventeenth century," although the mathematics employed were often faulty (pp. 330–31).
4. On the combinatoric digression in *De vetula*, see Robathan (1957, 1968) and Bellhouse (2000); on the frequent appearance of combinatorics in Italian commentaries on Dante, see Kendall (1956, pp. 5–6) and Ineichen (1988), who corrects Kendall's dating; and Girotto and Gonzalez (2006). See also Franklin (2015, pp. 393–94).
5. See Meusnier and Piron (2007), Piron (2007), Ceccarelli (2007); for more on Olivi's work, see Kaye (1998). See also Franklin (2015, pp. 262–69).

6. For an edition of the *Propositiones* (including a discussion of their authorship and influence), see Alcuin (ascr.) (2003).
7. Folkerts (2003, pp. 4–5). See also the relevant entry in James ([1903] 2011, p. 27).
8. Chaucer (1987, p. 258, l. 3242). Also see Baker (2013a).
9. See, for example, Donald R. Howard (1976, p. 257). For different treatments of Chaucer's pun, see Travis (2010, p. 228) and Allen (2007, pp. 141–46).
10. On the etymology of "fart" and the medieval slippage between "f" and "p", see Allen (2007, pp. 127–31) (although she does not seem to notice Chaucer's pun in parted/farted).
11. Rabelais (1991, p. 152 [Book II, chapter 7]). Edith Sylla provides an interesting history of similar pseudo-sophisms (2005, pp. 249–71). Peter W. Travis also briefly discusses this passage in relation to Chaucer (Travis 2010, pp. 303–4).
12. It is essentially a proof by contradiction: if a horizontal line were drawn from each indivisible "point" on one side of a square to the other side, these lines would each pass through the diagonal at a single "point", and so the diagonal would be have as many "points" (and therefore the same length) as the sides, which contradicts Pythagoras's Theorem. See Olson (2009, p. 417).
13. Although "phisike" has often been interpreted as "medicine" here (Chaucer 1984, p. 205), it is more probable that it is intended as a reference to what we would call "physics". See Burnley (1983, pp. 157–66) and Baker (2013b, pp. 133–36).
14. Grant (2001, p. 172), citing Chapter 39 of Oresme's "Tractatus de configurationibus." See also Clagett (1959, pp. 331–46).
15. Murdoch (1982, pp. 588–89). As Murdoch explains, although anachronistic, the easiest way for a modern reader to grasp why the subject is hot in degree 2 is in terms of the infinite series "$1/2 + 2/4 + 3/8 + 4/16 + \ldots + n/2^n + \ldots$", which converges to 2. As Murdoch notes, Swineshead explicitly refers to the problem as a sophism, although his treatment of it is "distinctively mathematical" (Murdoch 1982, pp. 588–90).
16. Murdoch (1979, pp. 118–119). For further analysis of such problems, especially in the case of Richard Kilvington, one of the early Oxford Calculators, see Kretzmann (1977). See also Aristotle (1984, p. 397 [Book VI, col. 235b]; and p. 440 [Book VIII, col. 263b–264a]).
17. And in Chaucer's work more generally: for a reading of lines 3187–91 of "The Nun's Priest's Tale" in the light of such problems, see Travis (2010, pp. 287–97).
18. The oak is used, for instance, in Heytesbury's discussion of intension of size (i.e. growth) in his *Rules for Solving Sophismata*; see Livesey (1979, p. 13). The falling millstone had been a conventional example for centuries: Grant (2001, pp. 170–71).
19. *The Riverside Chaucer* cites Aquinas ("Summa theologica 1-1.31.2": Chaucer 1987, p. 1039) for an example of the maxim; Bert Dillon cites Duns Scotus ("Expositio in Metaph. Arist. 10.2.1.30": Dillon 1974, p. 85).
20. Chaucer (1987, IV., p. 550, l. 958). See also Wyclif (1913), although Wyclif uses the phrase repeatedly throughout his writings. The first to notice the connection was Bennett (1974, p. 64). For my own extended analysis of Troilus's

38 D. BAKER

soliloquy in relation to the Wyclif-Strode correspondence, see Baker (2013b, pp. 150–74).

21. See Utz (1990), Huber (1965, pp. 120–25), and Baker (2013b, pp. 155–56).
22. Take, for instance, one of the Middle English Dictionary's examples, from an early fifteenth-century manuscript, Glasgow, University Library MS Hunterian 95: "synewes of her owne naturel complexioun ben sliden to naturel colde fro attemperaunce" (McSparran 2013, "sliden," v. 3d and 3f).

REFERENCES

Alcuin (ascr.). 2003. "'Propositiones ad acuendos iuvenes' ('Die älteste mathematische Aufgabensammlung in lateinischer Sprache')." In *Essays on Early Medieval Mathematics: The Latin Tradition*, edited by Menso Folkerts. Aldershot: Ashgate.

Allen, Valerie. 2007. *On Farting: Language and Laughter in the Middle Ages*. Basingstoke: Palgrave Macmillan.

Aristotle. 1984. "Physics." Translated by R. P. Hardie and R. K. Gaye. In *The Complete Works of Aristotle: The Revised Oxford Translation*, edited by Jonathan Barnes, i, 315–446. Princeton: Princeton University Press.

Baker, D. P. 2013a. "A Bradwardinian Benediction: The Ending of the Nun's Priest's Tale." *Medium Aevum* 82, no. 2: 246–53.

———. 2013b. *Literature, Logic and Mathematics in the Fourteenth Century*. http://etheses.dur.ac.uk/7716/.

Bellhouse, D. R. 2000. "De Vetula: A Medieval Manuscript Containing Probability Calculations." *International Statistical Review / Revue Internationale de Statistique* 6: 123–36.

Bellhouse, D. R., and J. Franklin. 1997. "The Language of Chance." *International Statistical Review / Revue Internationale de Statistique* 65: 73–85.

Bennett, J. A. W. 1974. *Chaucer at Oxford and at Cambridge*. London: Oxford University Press.

Boccaccio. 1929. *The Filostrato of Giovanni Boccaccio*. Translated with parallel text by Nathaniel Edward Griffin and Arthur Beckwith Myrick. Philadelphia: University of Pennsylvania Press.

Brower-Tolland, Susan. 2002. "Instantaneous Change and the Physics of Sanctification: 'Quasi-Aristotelianism' in Henry of Ghent's Quodlibet XV q. 13." *Journal of the History of Philosophy* 40 no. 1: 19–46.

Buridan, John. 1966. *John Buridan: Sophisms on Meaning and Truth*. Translated by Theodore Kermit Scott. New York: Appleton-Century-Crofts.

Burnley, J. D. 1983. *A Guide to Chaucer's Language*. London: Macmillan.

Ceccarelli, Giovanni. 2007. "The Price for Risk-Taking: Marine Insurance and Probability Calculus in the Late Middle Ages." *Electronic Journal for History of Probability and Statistics* 3. http://www.jehps.net/Juin2007/Ceccarelli_Risk.pdf. Accessed 13 September 2010.

Chaucer, Geoffrey. 1984. *Troilus and Criseyde: A New Edition of 'The Book of Troilus'*. Edited by B. A. Windeatt. Harlow: Longman.

———. 1987. *The Riverside Chaucer*. Edited by Larry D. Benson. Boston, MA.: Houghton Mifflin.

Clagett, Marshall. 1959. *The Science of Mechanics in the Middle Ages*. London: Oxford University Press.

Delasanta, Rodney. 1991. "Chaucer and Strode." *Chaucer Review* 26: 205–18.

Dillon, Bert, ed. 1974. *A Chaucer Dictionary*. Boston: Hall.

Folkerts, Menso. 2003. "The *Propositiones ad acuendos ivuenes* Ascribed to Alcuin." In *Essays on Early Medieval Mathematics: The Latin Tradition*, edited by Menso Folkerts. Aldershot: Ashgate.

Franklin, James. 2015. *The Science of Conjecture*. Baltimore: The Johns Hopkins University Press.

Garber, Daniel, and Sandy Zabell. 1979. "On the Emergence of Probability." *Archive for the History of Exact Sciences* 21: 33–53.

Girotto, Vittorio, and Michel Gonzalez. 2006. "Norms and Intuitions in the Assessment of Chance." In *Norms in Human Development*, edited by L. Smith and J. Vonèche, 220–36. Cambridge: Cambridge University Press.

Grant, Edward, ed. 1974. *A Source Book in Medieval Science*. Cambridge, MA: Harvard University Press.

———. 2001. *God and Reason in the Middle Ages*. Cambridge: Cambridge University Press.

Howard, Donald R. 1976. *The Idea of the Canterbury Tales*. Berkeley and Los Angeles: University of California Press.

Huber, John. 1965. "'Troilus' Predestination Soliloquy: Chaucer's Changes from Boethius." *Neuphilologische Mitteilungen* 66: 120–25.

Ineichen, Robert. 1988. "Dante-Kommentare und die Vorgeschichte der Stochastik." *Historia Mathematica* 5: 264–69.

James, Montague Rhodes. [1903] 2011. *The Ancient Libraries of Canterbury and Dover: The Catalogues of the Libraries of Christ Church Priory and St. Augustine's Abbey at Canterbury and of St Martin's Priory at Dover*. Cambridge: Cambridge University Press.

Kaye, Joel. 1998. *Economy and Nature in the Fourteenth-Century: Money, Market Exchange and the Emergence of Scientific Thought*. Cambridge: Cambridge University Press.

Kendall, M. G. 1956. "Studies in the History of Probability and Statistics: II. The Beginnings of a Probability Calculus." *Biometrika* 43: 1–14.

Kretzmann, Norman. 1977. "Socrates Is Whiter Than Plato Begins to Be White." *Noûs* 11: 3–15.

———. 1986. "Continua, Indivisibles, and Change in Wyclif's Logic of Scripture." In *Wyclif in His Times*, edited by Anthony Kenny, 31–65. Oxford: Clarendon.

Livesey, Steven J. 1979. "Mathematics *Iuxta Communem Modum Loquendi*: Formation and Use of Definitions in Heytesbury's *De Motu Locali*." *Comitatus* 10: 9–20.

Longeway, John. 2007. "William Heytesbury." In *Stanford Encyclopaedia of Philosophy*. http://plato.stanford.edu/archives/win2010/entries/heytesbury/. Accessed 10 November 2012.

Lynch, Kathryn. 2000. *Chaucer's Philosophical Visions*. Cambridge: Brewer.

McSparran, ed. 2013. *Middle English Dictionary*. http://quod.lib.umich.edu/m/med.

Meusnier, Norbert, and Sylvain Piron. 2007. "Medieval Probabilities: A Reappraisal." *Electronic Journal for History of Probability and Statistics* 3. http://www.jehps.net/Juin2007/MeusnierPiron_intro.pdf. Accessed 13 September 2010.

40 D. BAKER

Murdoch, John E. 1975. "From Social into Intellectual Factors: An Aspect of the Unitary Character of Late Medieval Learning." In *The Cultural Context of Medieval Learning*, edited by John Emery Murdoch and Edith Dudley Sylla, 271–348. Dordrecht: Reidel.

———. 1979. "Propositional Analysis in Fourteenth-Century Natural Philosophy: A Case Study." *Synthese* 40: 117–146.

———. 1982. "Infinity and Continuity." In *The Cambridge History of Later Medieval Philosophy*, edited by Norman Kretzmann and others, 564–91. Cambridge: Cambridge University Press.

O'Brien, Timothy D. 1990. "Ars-Metrik: Science, Satire and Chaucer's Summoner." *Mosaic* 23: 1–22.

Olson, Glending. 2009. "Measuring the Immeasurable: Farting, Geometry, and Theology in the Summoner's Tale." *Chaucer Review* 43: 414–27.

Pearcy, Roy J. 1967. "Chaucer's 'An Impossible' ('Summoner's Tale' III, 2231)." *Notes and Queries* 14: 322–25.

Piron, Sylvain. 2007. "Le traitement de l'incertitude commerciale dans la scolastique médiévale." *Electronic Journal for History of Probability and Statistics* 3. http://www.jehps.net/Juin2007/Piron_incertitude.pdf. Accessed 13 September 2010.

Rabelais, François. 1991. *Gargantua and Pantagruel*. Translated by Burton Raffel. New York: Norton.

Reichl, Karl. 1989. "Chaucer's Troilus: Philosophy and Language." In *The European Tragedy of Troilus*, edited by Piero Boitani, 133–52. Oxford: Oxford University Press.

Robathan, Dorothy M., ed. 1957. "Introduction to the pseudo-Ovidian De Vetula." *Transactions and Proceedings of the American Philological Association* 88: 197–207.

———. 1968. *The pseudo-Ovidian De Vetula*. Amsterdam: Hakkert.

Spade, Paul Vincent. 1988. "Insolubilia and Bradwardine's Theory of Signfication." In *Lies, Language and Logic*, edited by P. V. Spade, Ch. 4. London: Variorum.

———. 1998. "Late Medieval Logic." In *Medieval Philosophy*, edited by John Marenbon, 402–25. London: Routledge.

Sylla, Edith. 1971. "Medieval Quantification of Qualities: The 'Merton School'." *Archive for History of Exact Sciences* 8: 9–39.

———. 1982. "The Oxford Calculators." In *The Cambridge History of Later Medieval Philosophy*, edited by Norman Kretzmann, Anthony Kenny, and Jan Pinborg, 540–63. Cambridge: Cambridge University Press.

———. 2005. "Swester Katrei and Gregory of Rimini: Angels, God, and Mathematics in the Fourteenth Century." In *Mathematics and the Divine*, edited by T. Koetsier and L. Bergmans, 249–71. Amsterdam: Elsevier.

Travis, Peter W. 2010. *Disseminal Chaucer: Rereading the Nun's Priest's Tale*. Notre Dame, IN.: University of Notre Dame Press.

Utz, Richard. 1990. *Literarischer Nominalismus im Spätmittelalter: Eine Untersuchung zur Sprache, Charakterzeichnung und Struktur in Geoffrey Chaucers Troilus and Criseyde*. Frankfurt am Main: Lang.

Wilson, Curtis. 1960. *William Heytesbury: Medieval Logic and the Rise of Mathematical Physics*. Madison: University of Wisconsin Press.

Wyclif, John. 1893–1899. *Johannis Wyclif, Tractatus de Logica*. Edited by Michael Henry Dziewicki. London: Trübner.

———. 1913. "Responsiones ad argumenta Radulfi Strode." In *Johannis Wyclif, Opera Minora*, edited by Johann Loserth, 175–200. London: Paul.

CHAPTER 3

Mercantile Arithmetic, Financial Profit, and Ben Jonson's *The Devil Is an Ass*

Joe Jarrett

"Remember that Money is of a prolific generating Nature. Money can beget Money, and its Offspring can beget more, and so on. Five Shillings turn'd, is Six: Turn'd again, 'tis Seven and Three Pence; and so on 'til it becomes an Hundred Pound. The more there is of it, the more it produces every Turning, so that the Profits rise quicker and quicker" (Franklin 1961, p. 305). By 1905, the German sociologist, philosopher, and political economist Max Weber felt that Benjamin Franklin's words cited here captured better than any others the very "spirit" of capitalism (Weber 2002, pp. 13–38). There is no doubt as to how accurately they represent the mentality which maintains our own economic lives today. The human profit motive, in some form, must be as old as humanity itself, but in the rapidly developing mathematical economics of the sixteenth and seventeenth centuries an expression of it emerges which is substantially more familiar to the ears of modernity than anything before it. In the history of profit, these centuries signposted a monumental shift in the nature of how money could be made, and who was entitled to make and calculate it. The potential for profit-making expanded beyond the wealthy landowners of medieval feudalism and into the hands of a rapidly expanding class of numerate merchants and tradesmen. Capitalism was, in essence, born, and terms such as "profit," "business," "enterprise," "project" and "monopoly" became part of a semantic web increasingly shot through with economic colorings. It is no coincidence that this period also represented the burgeoning of what we might call a specifically social, or mercantile, mathematics. In a new world of economic exchange, knowing your

---

J. Jarrett (✉)
Cambridge University, Cambridge, UK

© The Author(s) 2021
R. Tubbs et al. (eds.), *The Palgrave Handbook of Literature and Mathematics*, https://doi.org/10.1007/978-3-030-55478-1_3

numbers might prove vital for your professional success. From the second half of the sixteenth century onwards in England, mathematical knowledge became increasingly accessible from a range of sources and institutions which I will consider in more detail later in this chapter, and terms such as "reckoning," "summing," "accounting," and "computation" became the arithmetical catchwords of everyday business, developing technical senses within the environs of increasingly mathematicised early modern markets.

The dramatic stage, that playground of terminological exploration, innovation, and abuse, was directly affected by both this increased numerical awareness and the stirring interest in the profit motive. The infamous "Diary" belonging to the theatrical impresario Philip Henslowe (ca. 1550–1616), that recorded performances at the Rose theatre, should perhaps more accurately be known as Henslowe's account-book, and Andrew Gurr has made it quite clear that acting companies "were independent commercial organizations, not doing what pleasure-bent lords or royalty commanded but going where and doing what brought most money and best audiences" (Gurr 2009, p. 39). The theatres in Shoreditch, and later in the Blackfriars, were only a short walk from the financial center of London, so that any participant in one of these arenas must have been at the very least subliminally aware of the presence of the other. In recent years, Shakespeare critics in particular have taken notice of the influence early modern mathematics might have exerted on the language and dramaturgy of his plays. Shankar Raman, for instance, has considered both how the numbers of actuarial science might be associated with death and time in *The Winter's Tale* (Raman 2008), and how the relationship between algebra and legal personhood manifests in *The Merchant of Venice* (Raman 2011). Edward Wilson-Lee has argued that mathematical paradoxes provided Shakespeare with metaphors with which to create an aesthetic of ontological crisis in *Troilus and Cressida* (Wilson-Lee 2013). My own recently published book, *Mathematics and Late Elizabethan Drama*, includes a chapter which explores how the infinitely large and the infinitely small permeate the aesthetic of *Hamlet* (Jarrett 2019, pp. 149–190). In this *Handbook*, Travis D. Williams's chapter "King Lear, Without the Mathematics: From Reading Mathematics to Reading Mathematically" explores explicit and implicit mathematical references and contexts in Shakespeare's *King Lear*. No doubt the volume which opened the door to all of this work was a collection entitled *Money and the Age of Shakespeare: Essays in New Economic Criticism* (Woodbridge 2003). The essays in this volume, although not always pertaining directly to mathematical topics, renewed interest in early modern economics and accountancy, and inspired other scholars to reorient economics toward numbers more precisely.

The function of this chapter is to emphasize the notion that economics became inseparable from mathematics in the early modern period, and to highlight the fact that plays other than Shakespeare's were not only responding to, but directly participating in, a new mathematical-economic order. In

Ben Jonson's city comedy of monetary cozenage, *The Devil is an Ass* (1616), profit-targeted mathematics is used somewhat uniquely as a backbone of pertinent stage business. The play's characters are obsessed with various forms of quantification, and its "projector," Merecraft, conjures a number of entrepreneurial pursuits, or "projects," whose persuasive potential is based as much upon the computational as it is upon the rhetorical. An obsession with the justifiable parameters of profit-making, I will argue, becomes a source of earnest ethical reflection in Jonson's play. What follows in this chapter is divided into four sections, followed by a short coda. The first two sections, in turn, consider how profit was increasingly justified in sixteenth- and seventeenth-century England, and how such justifications acted as an important catalyst in the development of mass numeracy. This provides an important contextual framework with which to consider the specific details of Jonson's play, which are the subject of the chapter's second two sections.

JUSTIFYING PROFIT

"For what is a man profited, if he shall gain the whole world, and lose his own soul? Or what shall a man give in exchange for his soul?" (Matthew 16:26).[1] There was and is a deep-seated Christian suspicion of gain: materialism is contrary to spiritualism, and salvation cannot be bought. Despite Weber's claim that the emergence of capitalism was linked to the Protestant work ethic, the Protestant reformer Martin Luther underplayed the importance of vocational industriousness and wealth-accumulation in favor of a more contemplative life, just as Thomas Aquinas had done three centuries before (see Worden 2010). The poet and cleric John Donne formulated an apposite English expression of these concerns in a sermon of 1615, in which he delineated a "right inventory" of human value: "First, His Soul, then His life; after his fame and good name: And lastly, his goods and estate" (Donne 1953, p. 157). His exegesis focused on Isaiah's "Ye have sold your selves for nought, and ye shall be redeemed without money" (52:3), and Donne continued the economic language he found there into his own paraphrase of Matthew 16:26: "this man [the sinner] should thus abandon this God, and exchange his soul for any thing in this world, when as it can profit nothing, to gain the whole world and loose your own soul, and not exchange it, but give it away, thrust it off, and be a devil to the devil" (Donne 1953, p. 157). The thrust of Donne's argument was that both prodigality and profiteering (especially where the soul was concerned) were two sides of the same sinful coin, and that Christian salvation, as it was articulated in the Bible, was a premise based on the symmetrical equilibrium of fair "exchange," not on its illegitimate distortion.

Such a conflation of justice with the formal proportionality of mathematical equilibrium was not entirely singular to the Bible, and Renaissance fascinations with it most likely had their roots in ancient ethical philosophy

44   J. JARRETT

and utopian politics as much as they did in Biblical hermeneutics. The most important classical account of a specifically quantitative justice appears in Aristotle's *Nicomachean Ethics*. For Aristotle, justice was not simply a case of the basic, *lex talionis* reciprocity he associated with the Pythagorean account, but rather a more complex "reciprocity based on proportion" (Aristotle 2004, p. 124). Justice could be either broadly political, or strictly legal, and each required a different kind of mathematical execution. Distributive justice employed a "geometrical proportion" which "involves at least four terms," where "the ratio is the same between the first two and the last two" (Aristotle 2004, p. 124). This kind of justice could insure fairness in voluntary exchanges such as the trading of commodities and the distribution of common property. If the proportions, however, were subjected to unlawful intervention (by fraud, rape, or murder, for example), rectificatory justice would be required to actively reinstate equality. It would do so by calculating how much had been lost by one party at the hands of a second, and restore that exact amount to the loser by taking it back from the illegitimate gainer. Aristotle called this an "arithmetical proportion," in which the "equal is a mean between the greater and the less" (Aristotle 2004, p. 121). Aristotle's arithmetical and geometrical proportions are not arithmetical and geometrical proportions in the modern sense (they are merely proportions), but the inclination to maintain and/or restore equilibrium by mathematical means is what is most pertinent. For Aristotle, a truly just society was one based on absolute, quantifiable fairness.

This, at least, was the most important message Renaissance readers seem to have taken from the *Ethics*. John Dee (1527–1608/9) for example, in his lengthy "Mathematicall Preface" to Henry Billingsley's 1570 edition of Euclid's *Elements of Geometrie*, emphasized the utility of the mathematical arts to the "Scholemasters of Justice," for whom "Iustice and equity might be greatly preferred, and skilfully executed, through due skill of Arithmetike, and proportions appertainyng" (Billingsley 1570, a1^r). He was following Aristotle carefully here, and soon invokes him directly: "*Aristotle* in his *Ethikes* (to fatch the sede of Iustice, and light of direction, to vse and execute the same) was fayne to fly to the perfection, and powers of Numbers: for proportions Arithmeticall and Geometricall" (Billingsley 1570, a1^v). Less academic readers than Dee could also access Aristotle's text after a drastically simplified English version of the *Ethics* was printed in 1547. This edition spared its audience from the original's complex mathematical terminology but retained its emphasis on a measurable equality:

> the Lord of the lawes laboreth to bring eueri thing equal. Wherefore he killeth one and scourgeth another, & other he lendeth to prison, vntil the partie haue satisfied, & so laboreth to bring to muche and to litel into a meane. Therefore he taketh from one and geueth to another, til they be egal, and therefore the forme therof is to be known, to the intent that his subiectes may liue stedfastly in the middest. (Aristotle 1547, F1^v)

This description of the philosophical underpinnings of the penal system was concerned primarily with "forme," and it was the handing down of formal cognitive models which was perhaps Aristotle's most significant early modern legacy. The form at work here, in some ways exemplified by the passage's tripartite syntactical structures, is one that finds the "meane" betwixt two extremes: a life "in the middest" is the life of honest, ethical justice, and the political and legal structures which regulate the form of any ideal state should do their best to retain that temperate middle ground.

It should follow, then, that early modern economic theory would also resemble this essential model of political and legal equilibrium: the parameters of trade, exchange, buying, and selling should surely not permit any form of profiteering, or any other kind of action which might upset the scales. For Aristotle (and perhaps also for the Bible), economics was indeed inseparable from justice, especially in its political, distributive context: how could a system based on careful quantification survive in the absence of some concrete scale with which to quantify it? This is where money comes in, and in the *Ethics* money acts as little more than a tangible manifestation of the abstract concept of number. In the wording of the English 1547 edition, money is described (perhaps punningly) as "a *mean* by the which a man may bryng euery vnegal into egal"; it is "an instrument, wherby iudges may do iustice" (Aristotle 1547, F2^r). In the Aristotelian vision and its early modern incarnations, money is a facilitator of political and legal equality and equilibrium, a "liuelless lawe" (Aristotle 1547, F2^r) brought to life by living men. It is not a rule unto itself, or a mechanism of personal profit.

As the sixteenth century continued into the seventeenth, however, economic theory developed toward a recalibration of the mathematical parameters of a "just" equality, so that a fair price could begin to include a profit margin, and so that the ethics underpinning exchange began to contradict traditional Biblical and classical morals. By 1588, when the dramatist Thomas Kyd translated Torquato Tasso's *Padre di Famiglia* into English as *The Housholders Philosophie*, there was still some reservation regarding mercantile pursuits of profit. "The Housekeeper ought to handle these things like a Husbandman," Kyd's text reads, "and not like a Merchaunt ... [T]he Merchant preposeth for his principall intent, the encrease and multiplying of his stock, which is doone by traffique and exchaunge, by means wherof, he many times forgets his house, his Children and his Wife" (Kyd 1588, F3^v). These complaints, however, were less ethical in character than they were practical, and *The Housholders Philosophie* was in fact very careful not to entirely condemn the concept of business profit. Within its pages, the reader could even find a sensitive conceptual separation of the dangerous and unnatural sin of usury from the legitimacy of the profit margin:

> Yet betwixt Exchange & Vsury there is some difference. Exchange may be retained, not only for the custome it hath taken and obtained in many famous Citties, but for the force of reason that it seemes to beare. For exchange is vsed

in steede of our transporting and conueighing Corne from place to place, which beeing hardlie to be doone without great discomoditie, and perill, it is reason that the party that exchaungeth may haue some sufficient gaine allowed. (Kyd 1588, G1^v)

This rationalization of "sufficient gaine" may be a long way from a manifesto for the profit motive, but it repositions the structure of exchange to include a labor cost, and attempts to legitimize the concept of charging more for a good than was initially paid for it. This idea tied in neatly with the overarching "philosophie" the book advocates: that the good householder divides his efforts between "*Conseruation*, and *Encrease*" (Kyd 1588, B4^v). "Encrease" is gentler than "profit," but the notion still relied on an arithmetic which facilitated the constant accumulation of wealth. The book's advice to the householder is to "know the quallitie, and quantity of his reuenues and expences," and after to insure that "his expence may prooue the least, making that proportion with his comings in, as *foure* to *eight*, or *sixe* at least" (Kyd 1588, E2^v). With this in mind, a new motivation to understand numbers, and how to manipulate and maneuver them, revealed itself: how else could an early modern citizen make sure to keep his personal wealth above zero? *The Housholders Philosophie* was symbolic of a cultural movement toward the acknowledgment of the social benefit of wealth-accumulation as a structuring mechanism of economics, and the correlative importance of a societal understanding of certain mathematical principles.

LEARNING MATHEMATICS

Accounting for the profit motive was an important catalytic agent in the explosion of mathematical learning that occurred from the second half of the sixteenth century onwards; this period truly can be deemed the crucial first stage in the development of an English mass numeracy. Mordechai Feingold has systematically debunked the scholarly commonplace that only the most insignificant amounts of mathematical pedagogy took place in the Universities at Oxford and Cambridge before the establishment of the Savilian Professorships in geometry and astronomy in Oxford in 1619. "[T]he stereotyped view of Oxford and Cambridge as institutions devoid of mathematical instruction and inimical to new scientific modes of thought," he rightly argues, "is unfounded" (Feingold 1984, p. 21). But individuals living in sixteenth- and early seventeenth-century London could have received a cutting-edge mathematical education without ever setting foot in the elite universities. Indeed, the merchant class had for some time been practicing their own brand of vernacular mathematics to assist them in their daily professional lives. Dee's preface to the *Elements* had neatly outlined the arithmetical "frute" received by "all kynde of Marchants":

> How could they forbeare the vse and helpe of the Rule, called the Golden Rule Simple and Compounde both forward and backward? How might they misse *Arithmeticall* helpe in the Rules of Felowshyp either without tyme, or with tyme and betwene the Marchant & his Factor, The Rule of Bartering in wares onely or part in wares, and part in money, would they gladly want?' Our Marchant ventureres, and Trauaylers ouer Sea, how could they order their doynges iustly and without losse, vnleast certaine and generall Rules for Exchaunge of money, and Rechaunge, were, for their vse, deuised? (Billingsley 1570, *2^v–a1^r)

Dee highlighted, then, as early as 1570, the practical utility of specific mathematical rules for mercantile pursuits, but he did not specify how and where such rules were to be learned. An important and progressive precursor to the official instatement of the mathematical Professorships at Oxford was the establishment of Gresham College in Bishopsgate Street, London, in 1597 (the very street Shakespeare was known to be living in at the time). The College was set up at the bequest of Sir Thomas Gresham, the founder of the Royal Exchange, and appointed professors in mathematics (often poached from Oxford and Cambridge) to give public lectures in English in the afternoons, with the particular intention of teaching merchants and mariners skills useful to their business activities. The writer and educator John Tapp claimed that these lectures were poorly attended: "what good doth these publique readings which hath now beene a reasonable time continued in this Cittie, with great charge, to good purpose, but little profit as may be guessed, by the little Audience which doe commonly frequent them" (Tapp 1613, A2^v). But Tapp may have had an ulterior motive for undermining the College, for his words (which nevertheless do still prove that the lectures had at least *some* audience) appeared in the introductory material to his own mathematical textbook, *The Path-Way to Knowledge Concerning the Whole Art of Arithmeticke* (1613), which he no doubt considered as in competition with the courses available at Gresham.

Whether Tapp's words were entirely true or not, Tapp was almost certainly correct to assume that the printed book was the most popular and influential place for ordinary citizens to learn mathematics. Instructional mathematics books had been on the stalls of London booksellers since the 1530s, and greatly increased in availability as the decades rolled on. Before 1550, two arithmetical books in English came off London presses: the anonymous *An Introduction for to Lerne to Rekyn With the Pen and With Counters* (1537), and Robert Recorde's *The Ground of Artes* (1543). By 1629, *An Introduction* had gone through eight editions; by 1631, the *Ground* had gone through twenty. After 1550, these early pioneers of the genre were joined by a multitude of similar, competing books: Humfrey Baker's *The Welspring of Sciences* (1564); Thomas Masterson's three *Booke[s] of Arithmeticke* (1592–95); Thomas Hood's *The Elements of Arithmeticke* (1596); Thomas Hylles' *The Arte of Vulgar Arithmeticke* (1600); Tapp's already mentioned *Path-Way to*

48   J. JARRETT

Knowledge (1613). All of these books contained explanations and demonstrations of the specific rules Dee mentions in his "Preface" as being useful to merchants—the Golden Rule, the Rules of Fellowship, the Rule of Bartering, the Rules for Exchange—but that is not to say that all of the books were aimed at a readership of merchants. Indeed, some of them clearly wished to teach any reader, regardless of their profession or motivation for learning arithmetic, precisely how to think and calculate *like* a merchant, even when they were not one.

Hylles' was the first mathematical text to denote mercantile arithmetic as its own sort of genre of mathematics, and to devote a cohesive section of the book specifically to mercantile arithmetic, which it entitled "*Musa Mercatorum.*" Continuing the teacher-student dialogue that the text utilizes throughout, it begins with Philomathes declaring to Eumathes that the material they are about to embark upon is

> an example of the common practise of Merchants, who vsing the art generally to no other purpose, but to cast, or foreknow, their gaynes or losse, haue by their dayly practise deuised many kindes of rules and formes of application of numbers, which the learned Mathematician, neyther occupieth nor scarcely knoweth. (Hylles 1600, 225^r)

The divide created here between learned, scholarly mathematics and formulaic, mercantile arithmetic puts neither at a precedent. Rather, Eumathes resolves to "chuse whether [he] will vse the scholers order or the merchants fashions," for he has heard that "the merchants rules are merueilous speedie" (Hylles 1600, 225^v). Speed was of the essence in what was a purely practical mathematics, and intellectual foundations were far less important than the real world "application" of calculating previous or potential "gaynes or losse." The "*Musa Mercatorum*" section at large offered a vast compendium of problem questions and their solutions to its reader, the majority of which were geared toward computing a profit margin. Here is one example: "if a cloath cost me 6.s.8.d. the yarde, at what price may I retayle it, to gaine after 27 $^1/_2$ P. in the 100. pounds disbursed?" (Hylles 1600, 239^r). If for some reason a calculative error was made, the reader could rectify the situation by recalling one of Hylles' memory-assisting poetical aids:

> But if the first sale bring losse, mine aduise,
> Is that from the 100 that losse you abate,
> Dividing the rest by that very price,
> Which bred the losse by selling at that rate,
> Whose quotient then, the latter price infolding,
> Shewes the thing sought for to all mens beholding.
> (Hylles 1600, 243^r–243^v)

Hylles book, then, and the many others like it, was, in a certain way, a handbook of the methods of profit calculation, and there was little room there for concerns about ethics. Indeed, the lack of ethical reflectivity in these popular books regarding profit must have contributed to the very normalization of profit. By 1635, William Scott was entirely secure in justifying not only profit-making but a life dedicated to the pursuit of wealth, all within a moral framework: "Let him labour and consider it for wealth," he wrote in his *Essay of Drapery*, for "the monyed man is the mighty man: Honour, Liberty, and Royalty attend on Riches; Logick faileth, Rhetorick fainteth, when Gold pleads the cause" (Scott 1635, F3^v–F4^r). The profit margin had by this time become an accepted feature of the new economic order, and the old formal proportionality of mathematical equilibrium had shifted into a new form which permitted imbalance and growth. The age of Shakespeare represented the most important decades in this shift, when the profit motive and its associated mathematics exploded well beyond the domain of an elite and distant stratum of society and became an integral part of mass culture. With this context in mind, we can now turn to a consideration of what role Jonson's drama played in this shift.

Staging Summation

Early modern drama was far more than a reactive phenomenon within society: it was a serious force for change. Despite constant censorship, the large audiences that acting companies could command, and their consistent linkage with royal and parliamentary authority, gave them an unusual amount of power. Information could be transferred quickly, and in a variety of forms, so that the average play-goer (or reader) of Elizabethan and Jacobean London must have considered a trip to the theatre (or the purchase of a play-text) an opportunity for pedagogy as much as an opportunity for entertainment. Indeed, when in Jonson's *Devil* Merecraft mentions to Fitzdottrel that he seems "cunning i'the chronicle," Fitzdottrel answers: "No, I confess I ha't from the playbooks, /And think they're more authentic" (Jonson 1994, 2.4.12–14).[2] Fitzdottrel is an avid enthusiast of the theatre, and in what is perhaps the most striking metatheatrical maneuver in any of Jonson's plays, he spends much of the plot preparing to attend a play within the play, at the Blackfriars (where *Devil* was first performed), called (strangely enough) "The Devil is an Ass" (1.4.21). Fitzdottrel, like the real audience who watched him (the distinction is blurred by the play), must have learnt much more than historical narrative at the theatre, and especially at the Blackfriars whose repertory was in fact entirely devoid of history plays. Blackfriars plays specialized rather in biting, often risky satire, and Gurr has claimed that those plays written for the boy companies there between 1601 and 1607 "were the most radical ever staged in London between 1574 and 1642" (Gurr 2009, p. 71). It was

Jonson who had initiated this era of plays with *Cynthia's Revels* (1600), and he continued to concentrate on this kind of playwriting for a number of years.

But critics have had trouble locating *Devil*, written for the King's Men well after the dissolution of the boy companies, and performed only a few months after the publication of Jonson's collected *Works* in the summer of 1616. David Riggs considered Jonson's decision to write the play "perplexing," believing that the *Works* should have provided an obvious opportunity for Jonson to retire from the stage and to narrow his literary efforts toward the printed page (Riggs 1989, p. 240). Anne Barton was similarly surprised that Jonson bothered to write anything at all after the masterpiece of *Bartholomew Fair*: "the effect [of that play] is dazzling, even a little overpowering, but it also signals the end. After this total illumination of space, what next?" (Barton 1984, p. 219). In terms of its plot, *Devil* does not sound too atypically Jonsonian. Fitzdottrel, a foppish Norfolk squire, comes to the city to present himself at fashionable events and to try his hand at some financial speculation. Instead of enhancing or enriching himself, his self-deceptive ignorance and misplaced avarice lead to the nearly successful courtship of his wife by another man, and his exploitation at the hands of a "projector" whose blustering plethora of business ideas are quite clearly too good to be true. All of this is framed by an underworld plot in which a lower devil named Pug convinces his master to let him take human shape for a day in the hope of causing chaos on Earth and raising the Devil's profile. The framing device aligned the play with other recent devil plays such as William Haughton's *Grim the Collier of Croyden; or The Devil and his Dame* (1600?) and Thomas Dekker's *If This Be Not a Good Play the Devil is in It* (1611–12), but Jonson was careful to distinguish himself from a vogue he considered outdated. In the first scene of *Devil*, when Satan tries to warn Pug that a visit to Earth may be a more perilous venture than he has imagined, Jonson has Satan actively locate the play's modernity:

> ... Remember,
> What number it is. Six hundred and sixteen.
> Had it but been five hundred, though some sixty
> Above - that's fifty years agone, and six,
> When every great man had his Vice stand by him,
> In his long coat, shaking his wooden dagger—
> I could consent ...
> But Pug,
> As the times are, who is it will receive you? (1.1.80–86, 88–89)

The Tudor interlude, which brought personified "Vice[s]" onto the stage— perhaps in "long coat[s]" with "wooden dagger[s]"—was still popular by 1560, and its legacy could still be felt in the seventeenth-century devil plays. Jonson, however, implies that both modes of drama, and their resonances with the "old Iniquities" (1.1.118) of the medieval vice plays, were no longer

3 MERCANTILE ARITHMETIC, FINANCIAL PROFIT ... 51

relevant to the new world of sin audiences inhabited in 1616. Despite affiliations with plays like Haughton's and Dekker's, *Devil* is really a kind of anti-devil play, in which the supernatural elements of the drama are constantly suppressed for comic effect by the social realism of the Fitzdottrel plot.

Devil also marks a noticeable departure from the kind of comedy Jonson had hitherto practiced. Although it utilized many of the formal techniques of his earlier plays, the political expediency of *Devil* lay not in the bitter social cynicism audiences encountered in *Volpone* (1606) or *The Alchemist* (1610). The pension of 100 marks a year King James granted to Jonson in February 1616 (in essence conferring upon him the status of poet laureate) must have given him an increased sense of entitlement to explore dramatic characters whose motives for monetary gain were not simply defined as criminal forms of cozenage (as is the case with Volpone, or Subtle and Face in *The Alchemist*), but were in fact legitimized by state authority (as is the case with Merecraft).[3] As a result, *Devil* represented a more active and ambivalent engagement with the actual political, cultural, and ethical ramifications of a changing economic order. Rather than simply critique the profit motive, *Devil* attempts to impart to his audience a knowledge of and terminology for it, including, of course, a representation of its underpinning mathematics. It can now be noted, then, that Satan's chronological reminders cited above act (beyond their metatheatrical function) to initiate the play's arithmetical engagement. In the folio text of *Devil* printed in 1631, over which Jonson had a pervasive influence, each of the numbers involved in the passage were italicized, as if to make their mathematical function prominent in what was an unnecessarily wordy sum. Perhaps the intention was to associate numbers with Hellish sensibilities early on in the play and, in this respect, it is significant that the Devil's chronology is used as opposed to Anno Domini: years are computed from a base of 1000 rather than 0, in a method not unfamiliar to the mercantile convention of working out "Gaynes and losse in the 100. pound of money" (Hylles 1600, 238^v–244^r). But Jonson is forcing his audience to perform mathematical procedures in a foreshadowing of the many sums they will be presented with later in the play.

Numbers and their arithmetical manipulations abound in *Devil*, and their functions are primarily ones of accounting or exchange, not chronology. When the gallant Wittipol first enters the stage, for instance, he encourages Engine to show himself "a mathematical broker" (1.4.4). Wittipol is punning on the mechanical aspect of Engine's name, but he is also prompting Engine to use a specifically arithmetical argument in order to win some gain from Fitzdottrel. In exchange for Wittipol being granted fifteen minutes to woo Fitzdottrel's wife, Engine offers Fitzdottrel a cloak: "It was never made, sir /For threescore pound," he tells him, but "'twill yield thirty. /The plush, sir, cost three pound ten shillings a yard!" (1.4.38–40). The numbers are enough for Fitzdottrel to agree to the exchange, who, when the time comes to explain it to his wife, reiterates the profit margin: "Here is a cloak cost fifty

52   J. JARRETT

pound, wife, /Which I can sell for thirty" (1.6.28–29). The slippage from "threescore" to "fifty" is testament to how quantities are often imperfect or imprecise in the play. Indeed, in the hands of Merecraft and his "Champion" Everill, numbers are often effaced and replaced with vague "proportions." Boasting of his profit-making capabilities, Merecraft exclaims "You will wonder /At my proportions, what I will put up /In seven years!" (2.1.89–91). Everill, outlining a scheme in which he offers his services as a kind of commercial arbitrator (or "Master of the Dependances" [3.3.662]), explains: "If we do find, /By our proportions it is like to prove /A sullen, and black business … then /We file it a dependance" (3.3.126–30). In both uses, "proportions" suggests calculation, but also estimation.

Merecraft is well placed to blur the distinction between the carefully calculated and the frivolously estimated, because he successfully fashions an identity whose singular skill is arithmetic. Only moments after Everill explains his dependency scheme, Fitzdottrel looks to Merecraft to authorize his own attempt at calculation: "That and the ring, and a hundred pieces, /Will all but make two hundred?" (3.3.151–52). Merecraft replies: "No, no more, sir. /What ready arithmetic you have!" (3.3.152–53). The line must be delivered glibly, perhaps even followed by a knowing eye-roll to the audience, because the sum Fitzdottrel has effected is a comically simple one compared to those Merecraft seems to be capable of. From the moment Merecraft enters the stage, the numbers he uses grow larger and larger within only seconds of stage time. Upon what is Fitzdottrel's first encounter with Merecraft, the former witnesses the latter instruct one of his minions: "Commend me to your mistress, /Say, let the thousand pound but be had ready /And it is done" (2.1.10–12). The minion exits the stage, but Merecraft goes on: "I would but see the creature /Of flesh and blood, the man, the prince, indeed, / That could employ so many millions /As I would help him to" (2.1.12–15). Fitzdottrel is flabbergasted by the rising scale of these figures: "How talks he? Millions?" (2.1.15).

When Fitzdottrel and Merecraft first converse directly, the numbers only get even larger. Here is Merecraft explaining to Fitzdottrel a scheme to drain fenland:

MERECRAFT	The thing is for recovery of drowned land,
	Whereof the Crown's to have his moiety
	If it be owner; else, the Crown and owners
	To share that moiety, and the recoverers
	T'enjoy the tother moiety for their charge.
ENGINE	Throughout England?
MERECRAFT	Yes, which will arise
	To eighteen millions, seven the first year:
	I have computed all and made my survey
	Unto an acre. (2.1.45–53)

By 1603, the total coinage available was less than 3.5 million pounds, and by 1640, this figure had risen to somewhere in the region of 10 million pounds (see Challis 1978, pp. 246–47; Kerridge 1988, p. 94). The "eighteen millions" Merecraft has "computed," then, would have seemed an astronomical sum to an audience in 1616. The obvious comic function of this is to show simultaneously Merecraft's tenacity and Fitzdottrel's gullibility, but there was perhaps a serious side too, for the rapidity of inflation in these years must have blunted the joke's preposterousness, even if only a little. The world Merecraft inhabits is one of prolific, ever-enlarging numbers, and the play shows how their strange abstraction is oddly responsible for the success or demise of those they enrich or engulf. *Devil* is careful to represent a society where numbers are as essential as they are bewildering or absurd, and in which an individual's fortunes are bizarrely at the mercy of their own (in)abilities to manipulate numbers persuasively.

PROJECTING, PROFITING, PROFITEERING

Devil's lexicon of numbers and sums, and representations of calculations and miscalculations by characters looking to get rich quick, is the very driving force of the play's comic economy. Such a comic economy was of course heavily invested in the real Jacobean economy of which it was a part, but it was also, to adapt Bradley Ryner's phrase, performing its *own* form of economic thought (Ryner 2014). *Devil* does not just represent an economy, it explores and questions that economy's components, procedures, and consequences. In this respect, Merecraft was undoubtedly the play's most important dramatic legacy, for he introduced the "projector" to the English stage. Later plays such as Philip Massinger's *The Emperor of the East* (1631) and Richard Brome's *The Court Beggar* (1640?) were heavily indebted to *Devil*, featuring projector characters whose business efforts were similarly directed toward the production of profit for both themselves and the parties they attached themselves to. But while Ryner is right to say that projectors were "common targets for derision on the Stuart stage" (Ryner 2014, p. 55), Merecraft represents a more ambivalent and nuanced figure than anything we find in those plays he came to influence. The majority of those in the first audiences of *Devil* would not have ever come across the term "projector" before, so Jonson had to devote a lengthy section of stage time to its explanation. In the moments immediately preceding Merecraft's first appearance, Fitzdottrel asks Engine, "What's th'affair? More cheats?" (1.7.4). Then, the following dialogue ensues:

ENGINE	No sir, the wit, the brain, the great projector,
	I told you of, is newly come to town.
FITZDOTTREL	Where, Engine?
ENGINE	I ha' brought him - he's without -
	Ere he pulled off his boots, sir, but so followed

	For businesses.
FITZDOTTREL	But what is a projector?
	I would conceive.
ENGINE	Why one, sir, that projects

Ways to enrich men, or to make 'em great,
By suits, by marriages, by undertakings
According as he sees they humour it. (1.7.5–13)

As well as his dramatic legacy, Merecraft would have an exciting lexical legacy, for the OED cites Jonson's usage of "projector" here as the first to mean "a promoter of bogus or unsound business ventures" (OED, "projector," definition 1b). The OED's somewhat invidious definition, however, does not entirely coincide with Engine's, whose rhetorical rule of three ("the wit, the brain, the great projector"; "By suits, by marriages, by undertakings") acts to persuasively aggrandize Merecraft's status, careful not to reduce him to a common swindler. Indeed, much of the interest of *Devil* lies precisely in trying to ascertain the social, intellectual, and moral statuses of Merecraft.

Jonson never permits these to become entirely clear-cut, but they are linked in the play to the status of another, associated new term: "business." Inclusive of the pluralized example cited above, the word and its variants appear a total of thirty-four times in the play. "Business" as pertaining specifically to "commercial activity" (now the most common sense) was still developing in the early seventeenth century, so that when Richard Percyvall and John Minsheu defined the Spanish verb "negociar" as "to deale in busines, to follow a trade" in their *Dictionary in Spanish and English* (1599) they were being forward-thinking (Percyvall and Minsheu 1599, p. 175). The predominant sense of "business" at this point was still simply "something with which a person is busy or occupied," (OED, "business," definition 2) but Jonson was clearly exploring the nascent, more cutting-edge sense, for Merecraft's busy-ness is always and only aimed at monetary gain. This proliferation of new terminology would undoubtedly have made it harder for contemporary audiences to judge Merecraft. On the one hand, Engine links Merecraft's drive for profit with the supposedly positive attributes many still assume are prerequisites of the most successful business careers: wit and intelligence. By the time of Massinger's *Emperor*, the quick "brain" of the projector becomes a point of obvious irony and laughter, but the cleverness of Merecraft is undeniable, and the semantic web associated with him is one of swift ingenuity ("He does't by engine and devices" [2.3.46]), creativity ("invention" [2.1.92]), and practical effectiveness ("I have effected /A business" [2.1.146–7]). On the other hand, wit and intelligence are not in and of themselves laudable. How were audiences to make sense of the kind of business projects Merecraft describes and promises Fitzdottrel he will initiate? Are they a case of legitimate profiting, or dubious profiteering?

In many ways, the answers to these questions are mathematical ones. How much profit is too much? Where does one draw the arithmetical boundary between legitimate and excessive gain? If the seventeenth-century represented the period in which profit became increasingly morally and mathematically justified, it also represented the period in which there was disagreement over whether and where an individual's profit-making potential should be capped. In Jacobean England, projector figures became increasingly associated with monopolists and monopolies. Introduced under Elizabeth I, monopolies were considered a morally ambiguous method of regulating exchange, by which monarchical authority would imbue total rights upon an individual (or body of individuals) for the selling of a given product or service. The financial rewards for patentees could be enormous, and the projector's role was to secure them for both his clients and himself. But there were heated disputes in the period between monarchical and parliamentary authorities regarding the political and ethical legitimacy of those profits and their distribution. The first landmark event in this respect came in 1602–3, when the case of *Darcy v Allein* proceeded in the Queen's Bench court. Judges ruled that a patent granted to Sir Edward Darcy by Elizabeth I for the production, importation, and vending of playing cards was void, on the grounds that "it is a monopoly, and against the common law," and that "the incident to a monopoly is, that after the monopoly granted, the commodity is not so good and merchantable as it was before: for the patentee having the sole trade, regards only his private benefit, and not the common wealth."[4]

Darcy v Allein initiated what would become modern competition law, its judgment based on the assumption that excessive private profits would inspire attitudes disadvantageous to the successful commonwealth. If the Aristotelian notion of distributive justice permitted in the distribution of wealth a kind of inequality justified by meritocracy (those who are worth more, get more), the case's judgment asked where, in a system of financial capital, the boundaries of inequality should equitably lie. For the anxiety surrounding monopolies was essentially an anxiety regarding how to keep mathematical ratios in the correct balance: if one gains too much, the judges presumed, the many will suffer a loss. But the subjectivity regarding the precise calibration of such mathematical ratios made the case's outcome anything but simple to arrive at, the decision taking well over two terms to come to a resolution. The infamous English lawyer Sir Edward Coke had acted on behalf of the plaintiff, and had given a number of arguments for the legal existence of monopolies, such as that they regulate and suppress merchandise that is potentially harmful (like gambling ephemera), and that the Queen, not parliament, should have the legal prerogative over decisions regarding matters of recreation and pleasure. If not so strong as to win the case, it seemed Coke's arguments were convincing enough to exert dominance outside of the courtroom, for despite the outcome of *Darcy v Allein*, the early years of James's reign saw a massive spike in the number of patents granted. The notion that monopolies were against the common law went largely ignored until a 1624 statute finally rendered all such patents void (see Davies 1959, pp. 24–27).

56   J. JARRETT

Throughout all of James's reign, then, monopolies were a politically sensitive and divisive subject. It may have been at royal request that *King Lear*'s "monopoly of fools" was removed before the publication of the folio text (see Taylor and Warren 1983, pp. 75–120), and it could have been the reason why, according to William Drummond of Hawthornden, "[t]he King desired [Jonson] to conceal [*Devil*]" (Drummond 1966, p. 17). When Jonson's play was being performed for the first time in 1616, there was no easy answer as to the ethical, economic, or mathematical legitimacy of monopolies and monopolists, and business practice often diverged from judicial authority. Jonson's representation of his projector character's monopolistic intentions, then, was able to remain entirely ambivalent. This is best embodied by Merecraft when he suggests an ambiguous and bipartite function for his partnership with another projector, Lady Tailbush:

> MERECRAFT   … She and I now
>   Are on a project for the fact, and venting
>   Of a new kind of fucus (paint, for ladies)
>   To serve the kingdom: wherein she herself
>   Hath travailed specially by way of service
>   Unto her sex, and hopes to get the monopoly
>   As the reward, of her invention.
> WITTIPOL   What is her end in this?
> MERECRAFT   Merely ambition,
>   Sir, to grow great, and court it with the secret,
>   Though she pretend some other. For she's dealing
>   Already upon caution for the shares. (3.4.48–58)

Either side of Wittipol's question, Merecraft produces two entirely different motivating factors for the pair's project for the singular rights to manufacture ("fact") and sell ("venting") women's cosmetics. On the one hand, the project's "end" is "To serve the kingdom," whereby Lady Tailbush's efforts are characterized by hard work ("travail"), and her potential compensation framed as deserved "reward." On the other hand, Lady Tailbush's motives are described as "merely ambition," her project a selfish scheme to "grow great" and enter the world of the financial elite ("court it with the secret"). The play shows ingeniously how these two seemingly antithetical forces driving the characters' business activities are not necessarily mutually exclusive ones. For instance, Merecraft's most promising project in the eyes of Sir Paul Eitherside, the play's lawyer and Justice, appears to have both a profit-potential and a social conscience:

> My plot for reformation of these follows:
> To have all toothpicks brought unto an office,
> There sealed; and such as counterfeit 'em, mulcted.
> And last, for venting 'em, to have a book
> Printed to teach their use, which every child
> Shall have throughout the kingdom that can read
> And learn to pick his teeth by. (4.2.45–51)

As always, Merecraft's grammar is precise, utilizing punchy, paratactic syntactical structures to delineate his intentions with business-like brevity. That Jonson chose to write *Devil* in verse is curious, especially after *Bartholomew Fair* had been composed entirely in prose, but it seems likely that the decision is linked to the play's obsession with numbers and numbering. For although Merecraft is rarely poetic, as such, the government of his speech by such careful metrical constraints gives it a dignity which is never derisive, or bombastic. Merecraft's delineation of his toothpick project seems entirely feasible and reasonable, as do many of the other projects he suggests in the course of the play. In and of themselves, draining fenland, reducing the cost of cork for bottle-ale, and making wine from raisins instead of grapes are not necessarily villainous business ideas; in fact, they are all examples of actual seventeenth-century pursuits.

Some critics have been too quick to include Merecraft within the corpus of Jonson's more obvious villains. Brian Gibbons, recalling Robert Greene's underworld pamphlets, called him the leader of the play's "Cony-Catchers" (Gibbons 1980, p. 153). There is, of course, much validity to such an assignation: Merecraft is continually disingenuous, and he ruthlessly exploits Fitzdottrel, whose money he has no genuine intention of investing in the business projects he claims to be pursuing. But there is a designed moral complexity to Merecraft that is lacking in some of Jonson's more famous anti-heroes, in no small part because the sheer proliferation of money-making schemes he purports to be involved in— and the varying and ambiguous sense as to whether these schemes are legitimate or illegitimate, ethical or unethical—renders even what should be his most obviously morally despicable acts questionable in their moral status. The poetic justice in place in *Devil* enables Merecraft to leave the stage in relative silence. When the Epilogue announces that "the projector here is overthrown," we are made aware immediately of its disingenuousness, the very next line in the couplet referring to Jonson's intention to reveal "a project of [his] own" (Epilogue, pp. 1–2). There is obvious sympathy between the poet-dramatist and the projector, in their parallel endeavors for royal patents and aristocratic patronage: Shakespeare was surely being provocative when he began his sonnets with the line "From fairest creatures we desire increase" (Shakespeare 1609, B1^r), and Jonson, as explained earlier, had been granted financial security for life at the generosity of King James only months before finishing *Devil*. Merecraft's moral status is left ambiguous in a way Volpone's or Subtle's are not, primarily because the potential political validity of his "projects" and calculations render the ethical legitimacy of the totality of his actions unclear. He was, indeed, an emblem for an emerging economic order where profit-making was increasingly cleansed from old associations with the criminal underworld, and being provided a new, ethically, and arithmetically justifiable framework.

Coda

It would not be too crude or hyperbolical to argue that *Devil* attempts to draw a broad picture of a shift from an old economic order to a new one. These lines of Merecraft, in which he explains to Fitzdottrel how he may become a duke, serve as a useful metonym for such a shift, and how the play represents it:

> ... Keep you the land, sir,
> The greatness of th'estate shall throw't upon you.
> If you like better turning it to money,
> What may not you, sir, purchase with that wealth?
> Say you should part with two o' your millions
> To be the thing you would, who would not do't? (2.1.117–22)

Merecraft's words here exemplify a movement from a medieval, feudal, land-based economy to a new economic environment of pure mathematical capital. His audiences, who really were embroiled in the complete disintegration of the former model in favor of the latter, would have recognized the force and validity of Merecraft's argument that a great quantity of land might imbue its owner with a dukedom, but that the successful monetization of that land could buy any title, and indeed anything. *Devil*, quite uniquely among the canon of early modern drama, took it upon itself to represent the workings and vocabulary of this new economic order, putting in full view its projects, businesses, monopolies, and, of course, the arithmetical procedures which underpinned them. Much like Benjamin Franklin, whose words began this chapter, Merecraft does not wish to interrogate the ethical dimension of a system based upon the mathematics of exponential profit, but in portraying him in this way, Jonson, with subtlety and ambivalence, asks his audience to consider the corollaries of a world in which the power of abstract numbers transcends the importance of land, of rhetoric, and, perhaps, even, of morality.[5]

Notes

1. All Biblical citations are from the King James version.
2. All citations from Jonson's play are from Jonson 1994. Act, scene and line numbers appear parenthetically within the text.
3. On Jonson's royal pension, see Riggs 1989, p. 220.
4. See *English Reports*, Vol. 77, pp. 1260–66; 11 Co. Rep. 84b.
5. The research leading to these findings has received funding from the European Research Council under the European Union's Seventh Framework Programme (FP7/2007–2013)/ERC grant agreement no. 617849.

References

Aristotle. 1547. *The Ethiques of Aristotle, ... Newly Translated into English*. London: Longmans, Green & Co.

———. 2004. *The Nicomachean Ethics*. Translated by J. A. K Thomson. London: Penguin.

Barton, Anne. 1984. *Ben Jonson: Dramatist*. Cambridge: Cambridge University Press.

Billingsley, Henry, ed. and trans. 1570. *The Elements of Geometrie of the Most Auncient Philosopher Euclide of Megara*. London.

Challis, C.E. 1978. *The Tudor Coinage*. Manchester: Manchester University Press.

Davies, Godfrey. 1959. *The Early Stuarts 1600–1603*. Oxford: Oxford University Press.

Donne, John. 1953. *The Sermons of John Donne*. Vol. 1. Edited by George R. Potter and Evelyn M. Simpson. Berkeley, CA: University of California Press.

Drummond, William. 1966. "Notes of Conversation with Ben Jonson Made by William Drummond of Hawthornden." In *Ben Jonson: Discoveries 1641/ Conversations with William Drummond of Hawthornden 1619*, edited by G. B. Harrison. Edinburgh: Edinburgh University Press.

Feingold, Mordechai. 1984. *The Mathematicians' Apprenticeship: Science, Universities and Society in England 1560–1640*. Cambridge: Cambridge University Press.

Franklin, Benjamin. 1961. "Advice to a Young Tradesman [21 July 1748]." In *The Papers of Benjamin Franklin*, edited by Leonard W. Labaree, vol. 3, 304–8. New Haven: Yale University Press.

Gibbons, Brian. 1980. *Jacobean City Comedy*. London and New York: Methuen.

Gurr, Andrew. 2009. *The Shakespearean Stage 1574–1642*. Cambridge: Cambridge University Press.

Hylles, Thomas. 1600. *The Arte of Vulgar Arithmeticke*. London.

Jarrett, Joseph. 2019. *Mathematics and Late Elizabethan Drama*. London and New York: Palgrave Macmillan.

Jonson, Ben. 1994. *The Devil Is an Ass*. Edited by Peter Happé. Manchester and New York: Manchester University Press.

———. 1995. *The Alchemist and Other Plays*. Edited by Gordon Campbell. Oxford: Oxford University Press.

Kerridge, Eric. 1988. *Trade and Banking in Early Modern England*. Manchester: Manchester University Press.

K[yd], T[homas], trans. 1588. *The Housholders Philosophie*. London.

Massinger, Philip. 1632. *The Emperour of the East*. London.

Percyvall, Richard, and John Minsheu. 1599. *A Dictionary in Spanish and English*. London.

Raman, Shankar. 2008. "Death by Numbers: Counting and Accounting in *The Winter's Tale*." In *Alternative Shakespeares 3*, edited by Diana E. Henderson, 158–80. London: Routledge.

———. 2011. "Specifying Unknown Things: The Algebra of *The Merchant of Venice*." In *Making Publics in Early Modern Europe: People, Things, Forms of Knowledge*, edited by Bronwen Wilson and Paul Yachnin, 212–31. London: Routledge.

Riggs, David. 1989. *Ben Jonson: A Life*. Cambridge, MA: Harvard University Press.

Ryner, Bradley D. 2014. *Performing Economic Thought: English Drama and Mercantile Writing 1600–1642*. Edinburgh: Edinburgh University Press.

Scott, William. 1635. *An Essay of Drapery, or the Compleate Citizen*. London.

Shakespeare, William. 1609. *Shake-speares Sonnets*. London.

Tapp, John. 1613. *The Path-Way to Knowledge Containing the Whole Art of Arithmeticke*. London.

Taylor, Gary. 1983. "Monopolies, Show Trials, Disaster, and Invasion: *King Lear* and Censorship." In *The Division of the Kingdoms: Shakespeare's Two Versions of King Lear*, edited by Gary Taylor and Michael Warren, 75–117. Oxford: Clarendon Press.

Weber, Max. 2002. *The Protestant Ethic and the Spirit of Capitalism*. Translated by Stephen Kalberg. Los Angeles: Roxbury.

Wilson-Lee, Edward. 2013. "Shakespeare by Numbers: Mathematical Crisis in *Troilus and Cressida*." *Shakespeare Quarterly* 64: 449–72.

Woodbridge, Linda (ed.). 2003. *Money and the Age of Shakespeare: Essays in New Economic Criticism*. New York and Basingstoke: Palgrave Macmillan.

Worden, Skip. 2010. *Godliness and Greed: Shifting Christian Thought on Profit and Wealth*. Lanham, MD: Lexington Books.

CHAPTER 4

Mathematics and Poetry in the Nineteenth Century

Daniel Brown

In his 1865 "On the Origin of Beauty: A Platonic Dialogue," Gerard Manley Hopkins had one of the personae, an Oxford student called Hanbury, recite some poetry by William Wordsworth and ask his professor: "is there not something mystical there, or is it all in plain broad daylight?" The professor, who is elaborating a formalist theory of aesthetics, responded to this challenge by addressing the assumptions the student brings to his question: "A mathematical thing, measured by compasses, that is what you think I should make of it, do you not?" Hanbury agrees, "Well yes ..." (Hopkins 1959, p. 95). Echoing the criticism that John Keats made in his poem *Lamia* (1820) of the Newtonian "cold philosophy" that works to "unweave the rainbow," Hanbury assumed that poetry needs to be protected from the incursions of a rationalist and reductive mathematics (Keats 1978, pp. 472, 473). His romantic ideology regards mathematics as antithetical to poetry, while the professor's theoretical aesthetics conversely appreciates it as integral to verse. The dialectical form of the "Platonic Dialogue" makes it representative for the tensions, the restless relations, of poetry and mathematics across the nineteenth century.

As the extracts from Hopkins's "Platonic Dialogue" indicate, the stances that the romantics took to mathematics early in the century shaped subsequent discussions. The following chapter begins by introducing contrastive attitudes to mathematics among first- and second-generation romantic poets, and the uses that some made of the mathematics they knew from their

---

D. Brown (✉)
University of Southampton, Southampton, UK

© The Author(s) 2021
R. Tubbs et al. (eds.), *The Palgrave Handbook of Literature and Mathematics*, https://doi.org/10.1007/978-3-030-55478-1_4

61

62   D. BROWN

classical educations, principally Euclid but also the Pythagoreans and Newton. As the century progressed, however, the cultural centrality of Euclid was displaced by new forms of mathematics and the rise of a professional science that encompassed them. Lewis Carroll, James Clerk Maxwell, and other scientists recorded and responded to these changes in verse. The mathematization of the sciences from around mid-century not only marks their new professionalism but also (and, at this time, almost synonymously) an increasingly masculinist identity. The chapter goes on to consider the place of women in relation to mathematics through discussions of poems by William Whewell, W. J. M. Rankine, and Constance Naden. It concludes by considering another aspect of mathematics' cultural prominence from mid-century, its application to poetry in new theories of prosody.

Romantic Responses to Euclid and Newton

Keats's criticism of the "cold philosophy" can be traced back to the dinner that the painter Benjamin Haydon held in his studio on December 28, 1817 for the poet to get to know Wordsworth. As well as describing the evening in his *Autobiography*, Haydon also depicted the two poets in his vast painting, *Christ's Entry into Jerusalem*. The recently completed picture dominated the gathering in the studio, "towering up behind us as a background" (Haydon 1950, pp. 316–17). Along with Keats and Wordsworth, the parallel party of the painting also includes Isaac Newton among the onlookers. Charles Lamb, another member of the dinner party, took affable exception to this:

> He then, in a strain of humour beyond description, abused me for putting Newton's head into my picture, "a fellow," said he, "who believed nothing unless it was as clear as the three sides of a triangle." And then he and Keats agreed he had destroyed all the poetry of the rainbow by reducing it to prismatic colours. It was impossible to resist him, and we all drank "Newton's health, and confusion to mathematics." It was delightful to see the good humour of Wordsworth in giving into all our frolics without affectation and laughing as heartily as the best of us. (Haydon 1950, p. 317)

"Newton's head" emerges from the passage as an emblem of rationalist abstraction, of a faith founded not in the loving Christ of Haydon's painting, but in geometrical proof, the pagan gospels of Euclid's *Elements*. It accords with the determination of Keats's "cold philosophy" to "[c]onquer all mysteries by rule and line" (Keats 1978, p. 472). The principal, usually the only, scientific text on classical school and university curricula at this time, the *Elements* is often seen, as it is here, by those with such an education as synonymous with science. Alfred Tennyson, for example, characterises science in "Locksley Hall" by the irreducible Euclidean element: "Science moves, but slowly slowly, creeping on from point to point" (Tennyson 1987, p. 127).

The clarity of "the three sides of a triangle" Lamb refers to represent a mathematics that is antithetical to Keats's principle of "Negative Capability," which he described in a letter that his editor confusingly dates to both the week before and the day after the dinner at Haydon's. Keats explained Negative Capability in this letter to his brothers George and Tom as a capacity for "being in uncertainties, Mysteries, doubts, without any irritable reaching after fact & reason" (Keats 1958, I, pp. 193–4). Hopkins's Hanbury similarly advocated "something mystical" against "plain broad daylight." Drunken Lamb and tipsy Keats drank to dry sober mathematics, wishing that it succumb to uncertainties, as they toasted "Newton's health, and confusion to mathematics." Such wished-for anarchy suggestively anticipated non-Euclidean geometry and other apparently preposterous developments in mathematics later in the century and the topsy-turvydom of Carroll's poems and Alice books.

In a postscript to this episode, another guest, a comptroller for stamps, blundered in after dinner with a question: "Don't you think Newton a great genius?" The comptroller's advocacy associates Newton with a mercantile mathematics, its instrumental uses by finance, commerce, and industry. It galvanised the bohemian bonds of Lamb and Keats against Newton and mathematics, but put Wordsworth in an awkward position. The comptroller, who was so eager to meet Wordsworth that Haydon invited him to join the party, is revealed to be the boss and correspondent of the poet. Supplementing his income, and supervised by the comptroller, Wordsworth had been the distributor of stamps for Westmoreland since 1813. With this farcical dénouement the older poet is suddenly aligned with the financial officer and his admiration for Newton.

While Wordsworth did not answer the comptroller's question, he did acknowledge Newton's genius on other occasions. Furthermore, like his friend Samuel Taylor Coleridge, the sole example Keats gave of a person who is incapable of Negative Capability, Wordsworth was also partial to "the three sides of a triangle" and its Euclidean entourage. They each knew Euclid from their schooling, Wordsworth at Hawkshead Grammar, which was known for its specialist focus on mathematics, and Coleridge at Christ's Hospital. Each went on to study at Cambridge, where all undergraduates were examined on the *Elements*.

While still at school Coleridge proved himself to be literally well versed in Euclid. In March 1791, he took the first of the propositions from Euclid's *Elements*, which describes how to construct an equilateral triangle from a finite straight line, and cast it as a poem. He then sent this untitled poem, first published in 1834 as "A Mathematical Problem," to his brother George as a "Prospectus and Specimen of a Translation of Euclid in a series of Pindaric Odes." He writes to his brother that he is "surprised that Mathematics, the quintessence of Truth, should have found admirers so few," and sets out to remedy this lack with poetry: "To assist Reason by the stimulus of the Imagination" (Coleridge 1912, I, p. 21).

Coleridge's willingness to render Euclid into verse follows from the logic of the classical education that he, along with other privileged boys and young men, future poets and scientists alike, received in the late-eighteenth through to the mid-nineteenth century. While the *Elements* were studied in English translation, by being taught alongside other classical texts they were presented as integral to literary and philosophical culture. Conversely, this education gave such future scientist-poets as Wordsworth's and Coleridge's young friend, the mathematician and astronomer William Rowan Hamilton, and after him the physicist Maxwell, practice writing verse, usually in Latin, and indeed greater exposure overall to poetry than mathematics.

Hamilton wrote that the clearest formulation of his greatest scientific achievement, quaternion operations, occurs in a poem he wrote in 1846, "The Tetractys":

> I have never been able to give a clearer statement of their nature and their aim than I have done in two lines of a sonnet addressed to Sir John Herschel:
> "And how the one of Time, of Space the Three,
> Might in the Chain of Symbols girdled be."

Hamilton wished to extend the reach that geometry had in manipulating complex numbers (numbers composed of both real and imaginary numbers) from two to three dimensions. His efforts to use numbers with three components ("triplets") having proved futile, his work was at a standstill, until the moment when, crossing a bridge on Dublin's Royal Canal on October 16, 1843, he was struck with the solution to his problem. This was to work not with three- but four-dimensional numbers, which he named "quaternions." These operators accordingly provide an algebraic means of expressing and multiplying geometrical relations of points in three-dimensional space. "The quaternion," Hamilton explains in an 1855 letter to Herschel, was "born, as a curious offspring, of a quaternion of parents, say of geometry, algebra, metaphysics, and poetry" (Hankins 1977, p. 176). The physicist Iggy McGovern used this quotation as the epigraph for his 2013 sonnet sequence on the life of Hamilton, *A Mystic Dream of 4*, and the four parental categories it outlines as section headings for the collection.

While the quadruple coordinates of quaternions furnish an algebra of four-dimensional space, in his poem Hamilton nominates the fourth coordinate as time. Hamilton's friends Wordsworth and Coleridge had an abiding interest in the mathematical relations of space and time from their studies of Newtonian dynamics and fluxions at Cambridge (where Wordsworth was enrolled from 1787 to 1791, Coleridge from 1791 to 1795). Fluxions is the method that Newton devises for measuring subtle rates of change or gradient, which accordingly introduces the element of time to considerations of space—a moving point. It is a version of calculus that he discovers simultaneously with the differential and integral calculus proposed by Gottfried Wilhelm Leibniz.

The literary critic Dometa Wiegand Brothers reads Coleridge's "Kubla Khan" (1816, comp. 1797) not only through his appreciation of Euclidean geometry, but also Newtonian fluxions. She argues that the poem presents "two worlds," one "marked by Euclidean geometry with its walls, domes and towers, the linear spatial conceptions of static art or science superimposed on [the] surface of [the other world, which is represented in] ... a dynamic, productive, curving, infinite natural landscape" that is enabled by the concept of infinitesimals (Wiegand Brothers 2013, p. 183). The curves and continuities that concern fluxions also characterise nature in "Kubla Khan," which is presided over by Alph "the sacred river," ".... meandering with mazy motion" down to "caverns measureless to man." The ruler of the territory, the measure of the place, described by Coleridge's poem, the Khan, provides a proxy figure for the poet as he brings this world into being simply with words, by "decree." The pure form of space is the a priori medium in which the Khan articulates Euclidean mathematical forms, which here repudiate the empirical dictates of gravity and heat transfer: "I would build that dome in air, /That sunny dome! those caves of ice!" Nor did Coleridge allocate feeling and mathematics to separate realms in "Kubla Khan," but with the oddness and assurance of a Salvador Dali composition summons an image of a "woman wailing for her demon-lover" in the midst of its Euclidean space and Newtonian flows, its crystalline and arabesque forms (Coleridge 1912, I, pp. 297–98).

RINGS AND ROUNDS

Coleridge explained in his 1791 letter to his brother George that Euclidean geometry is unsatisfactory when served *au naturel*, for "whilst Reason is luxuriating in its proper Paradise, Imagination is wearily travelling on a dreary desert" (Coleridge 1912, I, p. 121). For Wordsworth, however, neither had a monopoly on such bleakness. Both geometry and poetry traverse a desert in the Arab's Dream he describes in the *Prelude*, a passage that has attracted more critical commentary than any other nineteenth-century poem that engages with mathematics (see, for example, Miller 1972, Kelley 1982). Having put aside the copy of Cervantes's *Don Quixote* he was reading, the poet turns his "eyes toward the wide sea" and thinks "On poetry and geometric truth, /And their high privilege of lasting life," before "Sleep seized me, and I passed into a dream." The poet finds himself in a desolate existential scene, "Of sandy wilderness, all black and void," with "Close at my side" a Bedouin "Upon a dromedary, mounted high":

> A lance he bore, and underneath one arm
> A stone, and in the opposite hand a shell
> Of a surpassing brightness ...
> ... the Arab told me that the stone
> (To give it in
> the language of the dream)

66   D. BROWN

> Was "Euclid's Elements," and "This," said he,
> "Is something of more worth;" and at the word
> Stretched forth the shell, so beautiful in shape,
> In colour so resplendent, with command
> That I should hold it to my ear. (Wordsworth 1979, p. 157)

That Euclid and poetry are transposed into stone and shell "in the language of the dream" sanctions efforts to interpret them as symbols. Imaged as a stone, Euclid's *Elements* offers solidity, foundational substance. Like its counterpart, the stone in the dream may also emerge from the residue of the poet's recent waking life, as a seaside substance, such as limestone or marble, composed of ancient shells (cf. Bernhardt-Kabisch 1984, p. 485). These types of stone are also materials indigenous to Greece, which are engineered and shaped in its classical architecture and sculpture on principles found in Euclid's *Elements*. The principle of the golden section, traditionally attributed to the Pythagoreans, is defined by Euclid in Book VI, of the *Elements*: "A straight line is said to be cut in extreme and mean ratio, when the whole [line] is to the greater segment as the greater segment is to the less" (Todhunter 1869, p. 173). A happy convergence of aesthetics and mathematical science, this proportion is not only instanced in limestone temples and marble sculptures, but also in the successive spirals of certain shells that may compose them. The shell in the dream is evidently open and spiral as it yields echoic sounds, the murmuring song of our circulating blood that can be heard when it is held against the ear, another aspect that makes it apt as a vital symbol of poetry and prophecy, the Ode that is referred to later in the section. The foundational Euclidean substance made light and airy, the shell represents the poetic possibilities of geometry, "something of more worth"; Euclidean substance actualised aesthetically by poetry as Pythagorean proportion.

Described in the passage on the Arab's Dream as "The one that held acquaintance with the stars," Euclid's *Elements* is attributed with a cosmological significance. Building upon the Pythagorean principle of the golden section, which, as was noted earlier, this synoptic text includes, the line can be read as referring to the doctrine of the harmony of the spheres. Probably taking its cue from observable relationships that exist between number, such as the length of a taut string, and musical pitch, the note that this string makes when plucked or bowed, the Pythagorean doctrine identifies the regular courses of heavenly bodies with musical intervals, so that together their ratios yield harmonies. Like the golden section, the harmony of the spheres was thought to demonstrate that mathematical form is not only ontologically significant but also aesthetically powerful. This gives it an affinity with the patterns of rhyme, line lengths, and metres that compose lyric poetry and are similarly mobilised to musical effect. Principles of mathematical ratio structure the macrocosm of the universe and the microcosm of the lyric poem alike. Translating this into Wordsworth's symbolic iconography, the microcosm of the shell's spiralling form can be seen to gesture out- and upwards to the harmony of the spheres, an impassioned Pythagorean expression in the countenance of Euclidean science.

Romantic poets were particularly drawn to the doctrine of the harmony of the spheres. Basically a phenomenon of geometry and mechanics, suggesting a composite orrery and music box, the Pythagorean idea is nevertheless assimilated into romantic poetry and ideology by Percy Bysshe Shelley, who in *Queen Mab* figured it by analogy with organic life, as untamed and sublime: "The circling systems formed /a wilderness of harmony" (Shelley 1943, p. 767). Coleridge similarly elaborated a loose pantheistic principle of Pythagoreanism in his poem "The Eolian Harp." Placed in an open window, "that simplest Lute" of the title is played by the wind, "by the desultory breeze caress'd" (Coleridge 1912, I, p. 100). The first draft dates from 1795, while the second, from 1797, gives the first version of a question that emerges definitively in the following lines:

> And what if all of animated nature
> Be but organic Harps diversely fram'd
> That tremble into thought, as o'er them sweeps
> Plastic and vast, one intellectual breeze,
> At once the Soul of each, and God of all? (Coleridge 1912, I, p. 102)

This principle of being is modified in the following lines that Coleridge added to his poem in 1817:

> O! the one Life within us and abroad,
> Which meets all motion and becomes its soul,
> A light in sound, a sound-like power in light,
> Rhythm in all thought, and joyance every where–
> Methinks, it should have been impossible
> Not to love all things in a world so fill'd;
> Where the breeze warbles, and the mute still air
> Is Music slumbering on her instrument. (Coleridge 1912, I, p. 101)

While the Eolian Harp is a stringed instrument embodying ratios that can produce correlative sounds, this principle is expanded cosmologically as "the one Life" along the lines of a Pythagorean idealism, as a "Rhythm in all thought." By hypothesising that the mathematical ratios of the orbiting planets are musical values, the doctrine of the harmony of the spheres makes the visual and the auditory transposable. While it was prompted by contemporary ideas in natural philosophy (Levere 1981, pp. 156–67), the poem's description of "the one Life" as "A light in sound, a sound-like power in light," is also Pythagorean in its quality of synaesthesia. Later, in 1853, Maxwell described this Pythagorean relation succinctly in a pun on "rounds," as both planetary orbits and musical canons, in his poem "A Student's Evening

68   D. BROWN

Hymn": "And the stars most musically /Move in endless rounds of praise" (Campbell and Garnett 1882, p. 594).

Like Maxwell, Hopkins co-opted the pagan doctrine of the harmony of the spheres for Christian faith, and like Shelley and Coleridge extended its reach to organic nature, in his poem "Let me be to Thee." The poem dates from 1865, the same year as "On the Origin of Beauty," and like the dialectic of this Platonic dialogue reconciles mysticism with mathematics, as the microcosms of nature echo the grand Pythagorean cosmology:

> Let me be to Thee as the circling bird,
> Or bat with tender and air-crisping wings
> That shapes in half-light his departing rings,
> From both of whom a changeless note is heard.

Capable of being crisped, the air is a substantive medium that is shaped both by the flights of the "circling bird" and bat and by their calls, "a changeless note," the correspondent sound-waves that radiate outwards, "departing." A Pythagorean pun to parallel Maxwell's "rounds," Hopkins's "rings" describes the circling flight of the creatures and the notes they emit, sound-waves rippling outwards in the air. Parallel to its pantheist deployment in Coleridge's "Eolian Harp," the Pythagorean principle becomes incarnational in Hopkins's poem, which concludes with the lines, "I have found the dominant of my range and state– /Love, O my God, to call Thee Love and Love" (Hopkins 1990, p. 84). Love is the dominant, which has its modes in the act of its enunciation and the acts of Love itself. Similarly, in her 1854 poem "The Dream of Pythagoras" Emma Tatham, clothing Victorian piety in classical garb, has the Presocratic philosopher declare "that music-without end / ... Is the celestial language, and the voice / Of love; and now my soul began to speak / The speech of immortality" (Tatham 1854, p. 12).

PROFESSIONALISATION

The pact that classical education made between mathematics and poetry, the curricular parity it gave to Euclid's *Elements* and such texts as Horace's *Odes*, was overtaken by the rise of professional science. Begun in 1831, in protest against the aristocratic Royal Society, the British Association for the Advancement of Science (BAAS) placed the twin Newtonian sciences of mathematics and physics as its Section A, the paradigm for the new British professional science that the Association was engineering. Meanwhile, the Cambridge Mathematics Tripos was transformed by a series of formal reforms that were initiated by William Whewell and Charles Babbage in 1827 (Rouse Ball 1889, p. 211). The course that emerged from this process over the following decades was so demanding, subjecting its students to such theoretical rigour and competitive pressure, that, in summer 1853, it pushed even such a surpassingly able student as Maxwell into nervous collapse (Campbell and

Garnett 1882, p. 173). Maxwell and other survivors of the renovated Tripos established modern physics as an unflinchingly mathematical discipline. His friends William Thomson and P. G. Tait, who were like Maxwell both wranglers (that is, top place holders in the Tripos examinations), provided the first systematic mathematical exposition of thermodynamics in their 1867 textbook, *A Treatise on Natural Philosophy*.

The use of Euclid as a foundational text for mathematics came under attack from Tait and others during the 1860s, and gathered momentum with the rise of projective and non-Euclidean geometries in the decades that followed. In his opening address to the mathematical and physical sciences section of the BAAS in 1869, the mathematician James Joseph Sylvester urged that "Euclid [be] honourably shelved or buried 'deeper than did ever plummet sound' out of the schoolboy's reach" (Sylvester 1870, p. 119). Offering a surprising renovation of the old schools' curriculum, Sylvester argued in 1871 that Horace's *Odes* would make a better mathematical textbook than Euclid's *Elements* (Sylvester 1870, pp. 43, 49). An "anti-Euclid Association" was also established in this year, the Association for the Improvement of Geometrical Teaching. Its first report listed modern alternative teaching texts to the *Elements*, including eight English, ten French, and six German titles (Moktefi 2011, p. 329). Meanwhile Charles Dodgson, writing from "the great cause which I have at heart – the vindication of Euclid's masterpiece" critiques many of these textbooks in his dramatic dialogue *Euclid and his Modern Rivals* (Dodgson 1879, p. x).

While Dodgson championed the study of classical geometry he was also understanding, if sceptical, of the contemporary mathematical discoveries and theories that, as Gillian Beer superbly demonstrates in her *Alice in Space*, his alter ego Lewis Carroll explores wittily in the Alice books. Carroll's 1876 poem *The Hunting of the Snark* presents a similarly disordered world. Its protagonists, a sympathetic Ship of Fools adrift on fluctuant foundations, often appeal to number as a principle of certainty, a correlate point that might give them a bearing (none being forthcoming from the Bellman's accessible but blank "large map representing the sea" [Carroll 2012, p. 241]). The Bellman's rule of three is the clearest and best known of these invocations: "What I tell you three times is true" (Carroll 2012, p. 238). Announcing the principle in the second stanza, he also deploys it later in the poem:

> 'Tis the note of the Jubjub! Keep count, I entreat;
> You will find I have told it you twice.
> 'Tis the song of the Jubjub! The proof is complete,
> If only I've stated it thrice. (Carroll 2012, p. 249)

One of the principal arguments used to justify teaching Euclid was, as Whewell put it, that "Geometry really consists entirely of manifest examples of perfect reasoning" (qtd. in Moktefi 2011, p. 323). The Bellman's principle burlesques such reasoning with an arbitrary and insistent appeal to number.

70 D. BROWN

Its tripartite sequence suggests a satire on the syllogism, as a tautology that repeats the same proposition from slightly different aspects and then awards itself the palm of Truth on the occasion of its third statement (for example, "Socrates is a Man"). But rather than simple repetition, "'Tis the note of the Jubjub!" gives way to the more encompassing "'Tis the song ....", nominating a relation of sounds that itself suggests rhyme, and so corresponds to the progressive sequence from "twice" to "thrice" that fulfils its truth claim. "The proof is complete," but transposed into the formal relation of rhyme. As befits the Bellman's advocacy of it, this principle reduces itself to a form of chiming. More clangorous than enlightening, its assurance of truth is not logical but numerological, like the Christian equation of Trinity with Truth, a fetish for threes.

Carroll's carnivalesque proof in *The Hunting of the Snark* is consistent with his good-natured play with Euclid's modern rivals a few years later. He allows the rivals a defendant in the figure of the German Professor Niemand, while also making fun of them through his proxy, Minos: "Did you ever see one of those conjurers bring a globe of live fish out of a pocket-handkerchief? That's the kind of thing we have in Modern Geometry" (Dodgson 1879, p. 4). Dodgson's equitable disposition to contemporary mathematics is also clear from his 1867 *Elementary Treatise on Determinants*, an exposition of the algebraic and the linear geometry that Sylvester and his close friend and collaborator Arthur Cayley saw to have superseded the plane geometry of *The Elements*. "Determinants" is a term that Cayley coined in the 1840s to describe a number calculated from "a group of quantities arranged in square order," a matrix (Sylvester 1904–1912, I, p. 284). Maxwell mobilises them at the opening of his poem "To the Committee of the Cayley Portrait Fund": "First, ye Determinants! in ordered row / And massive column ranged, before him go, / To form a phalanx for his safe protection" (Campbell and Garnett 1882, p. 636). Both Carroll and Maxwell played with the contrastive mathematical ontologies, principally audacious new conceptions of geometrical space that proliferated during the mid-Victorian period, when the professional mathematical sciences became fully established.

While early in the century Coleridge's "A Mathematical Problem" furnishes an emblem for the equable treatment of poetry and Euclidean mathematics current at the time, Maxwell's 1854 poem, "A Problem in Dynamics" records the arrival of mathematical specialisation, sundered from classics and poetry. This verse version of an answer to a representative Tripos problem calculates the forces that apply to a heavy chain on a horizontal plane. The fiftieth of its sixty lines initiates a brief respite from its stream of mathematical technicalities, which rely upon a diagram and an appendix of eight "*Equations referred to*" that are included as infrastructure for the poem. Coming at this late stage in the poem, the comparison that these lines make of the plight of its subject chain to Hector's body dragged behind Achilles's chariot in Book XXII of Homer's *Iliad* is all the more bathetic, indeed grotesque:

The chain undergoes a distorting convulsion,
Produced first at A by the force of impulsion.
In magnitude R, in direction tangential,
(7) Equating this R to the form exponential,
Obtained for the tension when a is zero,
It will measure the tug, such a tug as the "hero
Plume-waving" experienced, tied to the chariot.
But when dragged by the heels his grim head could not carry aught,
(8) So give a its due at the end of the chain,
And the tension ought there to be zero again. (Campbell and Garnett 1882, p. 627)

Maxwell mischievously poses a Tripos problem in another undergraduate poem, a parody of Robert Burns's "Comin' Thru the Rye," "In Memory of Edward Wilson, Who repented of what was in his mind to write after section." While less technical than "A Problem in Dynamics," it similarly draws an analogy between Tripos phenomena and the human body. It provocatively parallels the bodies studied by mechanics with the volatile human bodies documented by Burns, thoughts of which evidently distracted the young physicist Edward Wilson during a mathematics and physics section paper at the BAAS around 1870, and no doubt Maxwell's Tripos peers during the early 1850s too, when he first drafted the poem. Its subtitle "*Rigid Body (sings)*" usurps the premise of this Tripos problem, in which a generic Rigid Body *swings*. The lyricism of this very individual object, its poetic prerogative, precludes mechanical causation, as it sings and swings freely. Both coquettish and pendulous, it defies the students' reductive efforts to take the measure of it, fix its movements mathematically and so solve the assigned problem:

> Gin a body meet a body
> Flyin' through the air,
> Gin a body hit a body,
> Will it fly? and where?
> Ilka impact has its measure,
> Ne'er a ane hae I,
> Yet a' the lads they measure me,
> Or, at least, they try. (Campbell and Garnett 1882, p. 630)

In Maxwell's undergraduate poem, "A Vision Of a Wrangler, of a University, of Pedantry, and of Philosophy" (1852) a cast of actual and allegorical figures emerge from the pages of the protagonist's Tripos problems. Prominent among them, the bad Alma Mater of Pedantry warns the student away from poetry in favour of mathematics, which she recommends to him for being well rewarded financially, presumably through its expanding actuarial, technological, and mercantile applications at this time:

72   D. BROWN

> As for Poetry, inter it
> With the myths of other days.
> "Cut the thing entirely, lest yon
> College Don should put the question,
> Why not stick to what you're best on?
> Mathematics always pays." (Campbell and Garnett 1882, p. 616)

Poetry was a strong and central value for Maxwell, which he associated with play, free will, and scientific discovery against the pedantry, determinism, and reductionism he identified with the mathematics he met in the Cambridge Tripos.

WOMEN IN MATHEMATICS AND MATHEMATICIANS IN LOVE

Maxwell's parody of Burns, "In Memory of Edward Wilson, who repented of what was in his mind to write after section," is written from and for the exclusively masculine cultures of the Mathematical Tripos and Section A of the BAAS during the mid-Victorian period. Women were not permitted to graduate from the Mathematical Tripos, which from the 1850s, when the original version of the poem was written, became a prerequisite for a professional career in mathematics or physics. The well-known examples of Ada Lovelace and Mary Somerville demonstrate that there had been more opportunities for women in mathematics earlier in the century. In the 1840s, Lovelace helped Babbage with the mathematics for his Analytical Engine, and was described by him with quaint gallantry as the "Enchantress of Numbers" (Forbes-Macphail 2013, p. 148). Her friend and tutor, the polymath Somerville, was the more celebrated and respected of the two, with the coinage "scientist" being first published and explained by Whewell in a review of her book *On the Connexions of the Physical Sciences* in 1834. The review closes with two sonnets to Somerville, one of them based on John Dryden's "Lines on Milton" (Neeley 2001, p. 4). It presents Somerville as living in an enlightened age that is accordingly able to recognise her genius, and places her in a canon of female mathematicians, with the classical Greek Hypatia and the eighteenth-century Italian Maria Agnesi. "Rare as poetic minds" these women are respectively paralleled to the figures of Homer, Virgil, and Milton in Dryden's poem:

> Three women, in three different ages born,
> Greece, Italy, and England did adorn;
> Rare as poetic minds of master flights,
> Three only rose to science' loftiest heights.
> The first a brutal crowd in pieces tore,
> Envious of fame, bewildered at her lore;
> The next through tints of darkening shadow passed,
> Lost in azure sisterhood at last;
> Equal to these, the third, and happier far,

4 MATHEMATICS AND POETRY IN THE NINETEENTH CENTURY 73

Cheerful though wise, though learned popular,
Liked by the many, valued by the few,
Instructs the world, yet dubbed by none a Blue.(Whewell 1834, p. 68)

By the time that the Scottish physicist and engineer John Macquorn
Rankine wrote his poem "The Mathematician in Love" (c. 1860s), the simple juxtaposition of a woman and a mathematician was evidently seen as a
recipe for mirth. The title of the poem presents itself as announcing a preposterous predicament. Like Maxwell's Burns parody, "The Mathematician
in Love" applies the apodeictic premises and methods of the mathematical
sciences to the complex and unpredictable field of human sexual and emotional behaviour:

A Mathematician fell madly in love
    With a lady, young, handsome, and charming:
By angles and ratios harmonic he strove
Her curves and proportions all faultless to prove,
    As he scrawled hieroglyphics alarming. (Rankine 1874, p. 3)

"A Mathematician," who is sufficiently important (or self-important)
to demand a capitalised title, is nevertheless attributed with a lack of measure, the clichéd and precipitous intemperance of falling madly in love. The
hyper-rationalist form that this insanity takes is demonstrated by his use of the
mathematical proof to verify both the lady's attractive appearance, a process
that renders it unsightly in "hieroglyphics alarming," and that she will agree
to marry him:

"Let x denote beauty,—y, manners well-bred,—
"z, Fortune,—(this last is essential),—
"Let L stand for love"—our philosopher said,—
"Then L is a function of x, y, and z,
"Of the kind which is known as potential."

"Now integrate L with respect to $d\,t$,
"(t Standing for time and persuasion);
"Then, between proper limits, 'tis easy to see,
"The definite integral *Marriage* must be :—
"(A very concise demonstration)." (Rankine 1874, p. 5)

The mathematician's criteria for a wife, of beauty, manners, and "Fortune,"
are conventional and reductive, and aptly represented in the desiccated form
of mathematical symbols. His a priori proof is, however, contradicted by
empirical fact. The intended fails to comply with the a priori narrative furnished by the mathematician's equations, but instead follows one of her own,
reminiscent of another formulaic genre, the popular romance: "– But the

74   D. BROWN

lady ran off with a dashing dragoon, / And left him amazed and afflicted" (Rankin 1874, p. 5).

While "every science lends a trope" to the campaign that the male persona of Constance Naden's poem "Scientific Wooing" (1887) makes to win the hand of his beloved Mary Trevelyan, he turns to mathematics for his *coup de grace* at the close of the poem:

> Or Mathematically true
> With rigorous Logic will I woo,
>   And not a word I'll say at random;
> Till urged by Syllogistic stress,
> She falter forth a tearful "Yes,"
>   A sweet „*Quod erat demonstrandum!*" (Naden 1894, p. 310)

By dispensing with poeticism and paring his expression and strategy down to the "Mathematically true," the persona plans to punningly propose to Mary Trevelyan. His proposal of marriage took the rigorous form of a proof for a geometrical proposition. This, he believes, must necessarily precipitate a tearful acquiescence from her, "A sweet '*Quod erat demonstrandum!*'" that, parallel to the use of the phrase by Euclid and others, declares the successful completion of a proof, that "which needed to be demonstrated." Soft femininity will, according to this final fantasy, be vanquished by masculine hard science, Q.E.D.

The respective protagonists of Rankine's and Naden's poems entertain a degraded mathematical interest in the quantitative principle of "Fortune." Unlike "beauty" and "manners," the other variables in his marriage equations, the word "Fortune" is, like his own name, capitalised by the "Mathematician" in Rankine's poem. This reveals his true match and intended equation, naming what he goes on to describe as the only "essential" variable, a reductive principle of quantity, a mercantile and mercenary mathematics. While the persona of Naden's poem declares "I covet not her golden dower," this nevertheless leads him to reflect that his sentiment disregards a law parallel to that of gravity: "Yet surely Love's attractive power /Directly as the mass [of the dowry] must vary" (Naden 1894, p. 309). Naden's poem marks the advent of the New Woman in the closing decades of the century, with a protagonist who not only understands the current male gendered prerogative of mathematics, but also the correlative trivialising of the female gendered realm of emotion and love romances, a dialectic that also impels Rankine's satire.

"A Mathematical Thing": Prosody

Nineteenth-century British poetry is marked by its metrical experimentation and variety, and correspondingly sophisticated and diverse theories of prosody. As Hanbury observes in Hopkins's "Platonic Dialogue," such emphases upon form makes poetry mathematical. As the "Dialogue" develops it places the elements of prosody within its larger principle of poetic parallelism, a figure drawn from the "metaphor of parallel lines" (Hopkins 1959, p. 113) and applied to such constituent elements as metaphor, rhyme, assonance, alliteration, and overarching verse forms. The nineteenth-century renaissance in metrical experimentation and prosody aligns with other contemporary forms of quantification, across the domains of commerce and science and into philosophical materialism, positivism, utilitarian ethics, formalist aesthetics, political economy, and statistical sociology.

Yopie Prins observes that "[t]he measurement of utterance by division and quantification turns voice into an abstract pattern: a series of intervals for enumeration rather than enunciation" (Prins 2000, p. 90). This is most clearly apparent from the revival in quantitative classical metres and their domestication in Victorian Britain, the "hexameter craze," which Prins also writes about. The English hexameter movement extends the classical synthesis of mathematics and poetry, not through the conjunction of Euclid with such poets as Pindar and Horace, nor Pythagoreanism, but by transferring and translating the quantitative metres of classical Greece into English verse. Occurring during a time of nationalism and national pride under Queen Victoria, this movement can be compared to the first "hexameter craze," which shaped the poetry of Edmund Spenser, Thomas Campion, and their peers in the time of Elizabeth, when Britain could similarly see itself as heir to classical civilisation, and to cohere a national identity around this claim.

Prins takes note of "an influential 1847 anthology, *English Hexameter Translations*, prefaced by elegiac couplets, that asked the 'lover of Song' to read these hexameters as if they appealed naturally to the listening ear" (Prins 2005, p. 236):

Art thou a lover of Song? Would'st fain have an utterance found it
True to the ancient flow, true to the tones of the heart,
Free from the fashions of speech which tinsel the lines of our rhymesters?
Lend us thy listening ear: lend us thy favouring voice.

These original verses were printed opposite a translation of the lines they are modelled on, "From Schiller," which begin: "Muse, from Teutonic lyres who hast drawn forth the cadences of Hellas" (Whewell 1847, pp. viii, 1). "True to the ancient flow," the analogy that such poetry makes of classical Greece to Victorian Britain is for this poet mathematical although not quantitative. Prins does not identify the anonymous author of the short poem and a majority of the translations. This is Whewell. He also originated and edited the collection.

Recalling that "[t]he subject of English Hexameters engaged so much of Dr Whewell's attention" the Cambridge mathematician Isaac Todhunter devoted a chapter of his 1876 intellectual biography of the polymath to this topic (Todhunter 1876, I, p. 283). In it Todhunter systematically identified

76   D. BROWN

all of Whewell's elegiacs, mostly translations from Greek and German, along with his writings on Hexameters. These include three *Letters* published in *Blackwood's Magazine* in 1846, in the third of which he expressed his aim to have "proved the hexameter to be a good genuine English verse" (Todhunter 1876, I, p. 284). He believed that he has achieved this primarily by making the hexameter accentual in English, arguing that such verse has been "hitherto prevented from having fair play among our readers of poetry, mainly by the classical affections of our hexameter writers – by their trying to make a distinction of long and short syllables, according to Latin rules of quantity; and by their hankering after spondees, which the common ear rejects as inconsistent with our native versification" (Todhunter 1876, I, p. 285).

Whether quantitative or accentual, traditional prosodies effectively reduce the sound patterns of verse to simple arithmetical arrangements of binary values, syllables classed as long or short, stressed or unstressed. Frustrated with these metrics, such poets as Edgar Allen Poe, Sylvester, and Hopkins all looked to mathematics to find more supple means of extending and appreciating the sound patterns of poetry.

Poe wrote in "The Rationale of Verse" that "[Verse] is exceedingly simple; one tenth of it, possibly, may be called ethical; nine tenths, however, appertain to mathematics" (Poe 1927, p. 205). The reformed prosody that Poe outlined in his essay allows a graduated spectrum of quantitative measures based, like classical prosody, upon the unit of the long syllable (which he designates 1), but which can be successively halved (2) and then quartered (4), and so on. Poe's scheme parallels modern musical scansion, while exceeding its range. Musicography provides the benchmark for judging prosody not only for Poe, but also for Hopkins (who composed music) and Sylvester and his colleague at Johns Hopkins University, Sydney Lanier, who applied musical notation directly to the explanation of verse.

Akin to Poe, Hopkins expressed the ratios of metrical feet as fractions in his lecture notes on Rhetoric (Hopkins 1959, p. 272). He embarks upon an original study of "the Dorian and Aeolian Measures or Rhythms" in October 1886, but writing to his friend Baillie on December 23 finds that "I cannot do what I would for want of mathematics" (Hopkins 1956, pp. 228–29, 275). Working methodically to address this want, in May 1887 Hopkins wrote to another friend, Coventry Patmore, the poet and author of the influential "Essay on English Metrical Law" (1856), that, having "a good deal written of a sizable book" on metre, "[f]or the purpose of grounding the matter thoroughly I am subjecting the terms of geometry, line, surface, solid and so on, many others to a searching examination" (Hopkins 1956, p. 379). This extract suggests that in this unfinished and now lost manuscript book Hopkins built upon his earlier mechanical knowledge and the analogies of stress, solids, and points of gravity he uses in his 1860s writings on metaphysics and prosody.

Sylvester knew "The Rationale of Verse" and agreed with Poe's attempt to represent the continuous musical flows of poetry, although not the limited arithmetical means it deploys. The title of Sylvester's 1870 book *The Laws of Verse* suggests a rejoinder to Poe's essay. Rejecting all such arithmetical models, Sylvester's book elaborates a prosody based on calculus. "My chief business is with Synectic," he declares, adopting a term from the French mathematician Augustin Cauchy: "Metric is concern[ed] with the discontinuous, Synectic with the continuous, aspect of Art" (Sylvester 1870, pp. 11, 10).

Continuity prevailed across the sciences during the second half of the nineteenth century, its most well-known and accessible instances being evolutionary biology and the energy principle. In mathematics it focuses upon new forms and articulations of space, such as projective geometry, in which shapes are seen to be preserved through various processes of spatial deformation, and non-Euclidean geometry, which posits spatial continuities beyond those of the Euclidean point, line, plane, and three-dimensional figure. Sylvester followed Carl Friedrich Gauss, who, working in the 1820s, considered space not to be universal, as Euclid presupposes, but local and differential, so that we can have, for examples, idiosyncratic spaces of sustained curvature. Following upon Gauss, Bernhard Riemann theorised space as a manifold of n-dimensions, any number of dimensions. Sylvester outlined the mathematical principle of Continuity in his Presidential Address to Section A at the 1869 BAAS meeting in Exeter. Once again demonstrating the integral nature of his poetry and mathematics, two years later he included a specially annotated version of the Exeter Address in *The Laws of Verse*.

Whether classical quantitative or modern accentual, traditional metrical prosody divides lines of verse arithmetically into segments. It accordingly does not lend itself to recognising principles of continuity in poetry: "Metric is concern[ed] with the discontinuous." Sylvester contested the separatist focus upon vowel sounds in traditional prosodies with his Synectic principle of Phonetic Syzygy, which brings forward connective consonants that facilitate the musical flow of lines of poetry. "Such a succession of consonant-colors," Lanier explains approvingly in his 1880 book *The Science of English Verse*, "has been called Phonetic Syzygy (syzygy, from *sunsugia*, yoking together) by Professor Sylvester, in his *Laws of Verse*" (Lanier 1880, p. 307). Lanier adopts the term, and Carroll developed the principle in his word-puzzle game of Syzygies (Carroll 1899, pp. 289–303). Sylvester specifies his mathematical application for the term in an 1850 paper on conics: "The members of any group of functions, more than two in number, whose nullity is implied in the relation of double contact ... must be in syzygy. Thus PQ, PQR, QR, must form a syzygy" (Sylvester 1904–1912, I, p. 132).

Sylvester's 1876 *Fliegende Blätter: Supplement to the Laws of Verse* includes his longest poem, "To Rosalind." Taking his cue from Shakespeare's *As You Like It* Sylvester applies Orlando's ill-advised conceit to the 268 lines of his poem, making each of them rhyme with "Rosalind" (see Shakespeare act 3, scene 2, ll, 86–92; 94–95). Despite its length "To Rosalind" is, as Sylvester said, "one single sentence," within which all of its rhymes are internal, for, as he writes in *The Laws of Verse*, "the office of rhyme [consists] in marking off a line as a sort of compound foot (like a bar in music)" (Sylvester 1876, p. 6, 1870, p. 44). The rhymes on "Rosalind" form a rhythm around which the subtle harmonies of syzygy are arranged.

The achievement of "To Rosalind" was for Sylvester its "[a]naconda-like sinuosities," its "one single sentence" crowded with curves (Sylvester 1876, p. 6). His Synectic prosody appreciates rhythm and syzygy as subtly sustained in the manner of curves, tracing them in the manner of differential calculus as the continuous motion of a point (Brown 2013, pp. 207-33). By the time that Sylvester came to Cambridge in the early 1830s, the reforms led by Babbage and Whewell had edged out the Newtonian fluxions that Coleridge and Wordsworth had studied in the late-eighteenth century in favour of Leibnizian calculus, "Continental analysis." Sylvester's Exeter Address habitually described the Continuity that calculus engenders and can plot with tropes drawn from hydrodynamics: "Space is the *Grand Continuum* from which, as from an inexhaustible reservoir, all the fertilizing ideas of modern analysis are derived" (Sylvester 1870, p. 125). *The Laws of Verse* is also full of such figures. Indeed he described the continental fluxions of his own river Alph, its sinuous articulation of differential space demonstrating the Continuity of poetry and mathematics: "Synectic leads down from the Alps of Cauchy and Riemann to the flowery plains of Milton and Byron" (Sylvester 1870, p. 64).

Neither poetry nor mathematics lends itself naturally to narrative, and even in offering a selective account of the two that is grounded and bounded historically, as this chapter does, their relations are best understood by spatial metaphors, as constellations of shared concerns that variously cohere and dissipate across time. Poetry and mathematics converge across the century through their engagements with abstract space and time, as poetry renders it in prosody, and the objective correlatives of mental terrains, such as those of Coleridge's "Kubla Khan" and Wordsworth's Arab dream. While earlier in the century mathematicians and poets shared cultural common ground in the accessible texts of Euclid and Pythagoras, the increasing prevalence and professionalisation of complex mathematics created conceptual and social divides between the two, which were exploited in comic and satirical poetry by such figures as Maxwell, Rankine, and Naden.

BIBLIOGRAPHY

Beer, Gillian. 2016. *Alice in Space*. Chicago: University of Chicago Press.

Brown, Daniel. 2013. *The Poetry of Victorian Scientists*. Cambridge: Cambridge University Press.

Bernhardt-Kabisch, ''Ernest. 1984. "The Stone and the Shell: Wordsworth, Cataclysm, and the Myth of Glaucus." *Studies in Romanticism* 23: 455–90.

Campbell, Lewis, and William Garnett. 1882. *The Life of James Clerk Maxwell.* London: Macmillan.

Carroll, Lewis. 2012. *Jabberwocky and Other Nonsense: Collected Poems.* Edited by Gillian Beer. London: Penguin.

———. 1899. *The Lewis Carroll Picture Book.* Edited by Stuart Dodgson Collingwood. London: T. Fisher Unwin.

Coleridge, Samuel Taylor. 1912. *Complete Poetical Works of Samuel Taylor Coleridge.* 2 vols. Edited by E. H. Coleridge. Oxford: Clarendon Press.

Dodgson, Charles L. 1879. *Euclid and his Modern Rivals.* London: Macmillan.

Forbes-Macphail, Imogen. 2013. "The Enchantress of Numbers and the Magic Noose of Poetry: Literature, Mathematics and Mysticism in the Nineteenth Century." *Journal of Language, Literature and Culture* 60: 138–56.

Hankins, Thomas L. 1977. "Triplets and Triads: Sir William Rowan Hamilton on the Metaphysics of Mathematics." *Isis* 68: 175–93.

Haydon, Benjamin Robert. 1950. *The Autobiography and Journals of Benjamin Robert Haydon.* Edited by Malcolm Elwin. London: Macdonald.

Hopkins, Gerard Manley. 1956. *The Further Letters of Gerard Manley Hopkins.* 2nd ed. Edited by Claude Colleer Abbott. London: Oxford University Press.

———. 1959. *Journals and Papers of Gerard Manley Hopkins.* 2nd ed. Edited by Humphry House and Graham Storey. London: Oxford University Press.

———. 1990. *The Poetical Works of Gerard Manley Hopkins.* Edited by Norman Mackenzie. Oxford: Clarendon Press.

———. 2017. "Mathematics." In *The Routledge Research Companion to Nineteenth-Century British Literature and Science,* edited by John Holmes and Sharon Ruston, 217–34. Abingdon, Oxford: Routledge.

Keats, John 1958. *The Letters of John Keats.* 2 vols. Edited by H. E. Rollins. Cambridge: Cambridge University Press.

———. 1978. Th*e Poems of John Keats.* Edited by Jack Stillinger. London: Heinemann.

Kelley, Theresa M. 1982. "Spirit and Geometric Form: The Stone and the Shell in Wordsworth's Arab Dream." *Studies in English Literature, 1500–1900* 22 (4): 563—82.

Lanier, Sydney. 1880. *The Science of English Verse.* New York: Charles Scribner's Sons.

Levere, Trevor H. 1981. *Poetry Realized in Nature Samuel Taylor Coleridge and Early Nineteenth-Century Science.* Cambridge: Cambridge University Press.

Moktefi, Amirouche. 2011. "Geometry: The Euclid debate." In *Mathematics in Victorian Britain,* edited by Raymond Flood, Adrian Rice, and Robin Wilson, 321–38. Oxford: Oxford University Press.

McGovern, Iggy. 2013. *A Mystic Dream of 4.* Dublin: Quaternia Press.

Miller, J. Hillis. 1972. "The Stone and the Shell: The Problem of Poetic Form in Wordsworth's Dream of the Arab." In *Mouvements premiers: Études critiques offertes á Georges Poulet,* 125–47. Paris: José Corti.

Naden, Constance. 1894. *The Complete Poetical Works of Constance Naden.* London: Bickers.

Neeley, Kathryn A. 2001. *Mary Somerville: Science, Illumination, and the Female Mind.* Cambridge: Cambridge University Press.

Patmore, Coventry. 1961. *Coventry Patmore's "Essay on English Metrical Law."* Edited by Mary Augustine Roth. Washington: Catholic University of America Press.

Prins, Yopie. 2000. "Victorian Meters." In *Cambridge Companion to Victorian Poetry,* edited by Joseph Bristow, 89–113. Cambridge: Cambridge University Press.

80   D. BROWN

———. 2005. "Metrical Translation: Nineteenth-Century Homers and the Hexameter Mania." In *Nation, Language, and the Ethics of Translation,* edited by Sandra Bermann and Michael Wood, 229–56. Princeton: University of Princeton Press.

Poe, Edgar Allan. 1927. *Works of Edgar Allan Poe.* London: Oxford University Press.

Rankine, W.J.M. 1874. *Songs and Fables.* Glasgow: James Maclehose.

Rouse Ball, W.W.W. 1889. *A History of the Study of Mathematics at Cambridge.* Cambridge: Deighton Bell.

Shakespeare, William. 1975. *As You Like It.* Edited by Agnes Latham. London: Methuen.

Shelley, P. B. 1943. *The Complete Poetical Works of Percy Bysshe Shelley.* Edited by Thomas Hutchinson. London: Oxford University Press.

Sylvester, James Joseph. 1870. *The Laws of Verse or Principles of Versification Exemplified in Metrical Translations Laws.* London: Longmans.

———. 1876. *Fliegende Blätter: Supplement to The Laws of Verse.* London: Grant & Co.

———. 1904–1912. *The Collected Mathematical Papers of James Joseph Sylvester.* 4 vols. Edited by Henry F. Baker. Cambridge: Cambridge University Press.

Tatham, Emma. 1854. *The Dream of Pythagoras and Other Poems,* 2nd ed. London: Longman.

Tennyson, Alfred. 1987. *The Poems of Tennyson.* 3 vols. Edited by Christopher Ricks. London: Longman.

Todhunter, Isaac. 1869. *The Elements of Euclid.* London: Macmillan.

———. 1876. *William Whewell, D. D.: An Account of his Writings.* 2 vols. London: Macmillan.

Whewell, William. 1834. Review of on the Connexion of the Physical Sciences. *Quarterly Review* 51: 54–68.

———. 1847. *English Hexameter Translations from Schiller, Homer, Callinus, and Meleager.* London: Murray.

Wiegand Brothers, Dometa. 2013. "The Mathematics of Dreams: The Psychological Infinity of the East and Geometric Structures in Coleridge's 'Kubla Khan'." In *Coleridge, Romanticism and the Orient: Cultural Negotiations,* edited by David Vallins, Kaz Oishi, and Seamus Perry, 177–190. London: Bloomsbury Academic.

Wordsworth, William. 1979. *The Prelude: 1799, 1805, 1850.* Edited by Jonathan Wordsworth, M. H. Abrams, and Stephen Gill. New York: Norton.

Wordsworth, William, and S. T. Coleridge. 2007. *Lyrical Ballads.* 2nd ed. Edited by Michael Mason. London: Routledge.

CHAPTER 5

Non-Normative Euclideans: Victorian Literature and the Untaught Geometer

Alice Jenkins

In Thomas Love Peacock's *Nightmare Abbey* (1818), a highly unconventional novel which mocks fashionable ideas as much as the individuals in Peacock's circle, Mr. Flosky is a ludicrous philosopher and a caricature of Samuel Taylor Coleridge. A self-described "transcendental metaphysician, who has pure anticipated cognitions of every thing," one of Flosky's pompous and ridiculous boasts is that he "carries the whole science of geometry in his head without ever having looked into Euclid" (Peacock [1818] 1986, p. 85). Flosky makes this boast as bluster to compensate for the fact that he is being forced to admit his ignorance—albeit on such a mundane matter as the emotions of one of the other characters. In the context of the plot, his claim to comprehensive knowledge of geometry but total ignorance of Euclid is inflated and incredible.[1]

In reality, Coleridge did not claim to be unacquainted with Euclid: on the contrary, he had a better mathematical education than many of his contemporary writers. As a schoolboy he had learned celestial navigation with the distinguished astronomer William Wales, and later mathematics, including geometry, as a student at Cambridge, where it was still compulsory even for those whose main interest was in classics (Owen 2019, p. 26). Coleridge was profoundly interested in philosophical aspects of mathematics, particularly geometry, and, through his reading of Immanuel Kant, came to see geometry as depending on an intuition which is innate in the human mind. Flosky's boast thus does not fundamentally misrepresent Coleridge's philosophical position on geometry, though it misrepresents his Euclidean education.

A. Jenkins (✉)
University of Glasgow, Glasgow, UK

© The Author(s) 2021
R. Tubbs et al. (eds.), *The Palgrave Handbook of Literature and Mathematics*, https://doi.org/10.1007/978-3-030-55478-1_5

Flosky is an example of an untaught geometer, someone who claims or appears to have either innate, intuitive, or instinctual geometrical knowledge but who has never "looked into Euclid." This essay examines some late eighteenth- and nineteenth-century textual representations of characters, both fictional and historical, and human and animal, who are untaught geometers, but whose claims to this status, unlike Flosky's, are supported by the texts in which they appear. The repeated figuration of the untaught geometer in nineteenth-century writing highlights the extraordinary place of Euclid, and his *Elements of Geometry*, in nineteenth-century British culture. No other branch of mathematics was so wholly identified with one author, one book, as geometry; and no other branch was imbued with such broad and deep cultural and moral importance. To know geometry other than through the relatively narrow range of normative Euclidean experiences was indeed to be unorthodox.

In many ways, Euclidean geometry and Victorian high-mindedness were perfectly complementary. Geometry offered a model of all that was most earnest, self-denying, diligent, methodical, and, for these reasons, beautiful. It was also available to everyone, at least in principle; so it could be understood as promoting the cultural unity which, though largely imaginary, was nonetheless crucially important in relieving contemporary fears about the effects of education on non-elite groups. Victorians argued that Euclidean geometry conferred benefits in several areas of life as well as the purely intellectual. As Joan L. Richards writes in her seminal study *Mathematical Visions*, "[i]n England, the essential connection of mathematics to all other aspects of human development was a constant theme" (Richards 1988, p. 7). Geometry was argued to inculcate moral virtue, good character, and estimable habits such as a determined insistence on truth.

Geometry and Victorian earnestness seemed complementary because Victorian writers remade geometry for their own times, taking up a long legacy of admiration and adaptation of Euclid and broadening and diversifying it to fit the cultural imperatives of the age of mass literacy, utilitarianism, and imperialism. In the broad generic spectrum of writing which contributed to this process, directly mathematical texts are only a relatively small segment. The diffusion of geometry through Victorian culture was immensely wide. Allusions to and invocations of geometry are spread through religious, political, and sentimental writings, as well as many other kinds of text. Geometry is thus a key locus of Victorian literature's interaction with mathematics.

In Britain in the nineteenth century, Euclidean geometry did not belong to the few hundreds of publishing mathematicians; it was below their notice as researchers, though as educators, some of them—most famously, Charles Dodgson—were profoundly interested and active in it. Equally, it did not enter into the schooling of the least privileged, for whom the four rules of arithmetic were likely to be the limit of mathematics education. But for a surprisingly large section of the population in between, Euclidean geometry was an important, respected, and all but essential part of education. As a result both of the respect it was accorded, and the large number of people who were exposed to it, geometry spilled out into non-mathematical culture in a way and to an extent that other branches, as for example algebra, did not.

In this imbrication of Euclidean geometry and nineteenth-century British culture, each was used to reinforce the other. The high prestige of Euclid was often considered particularly British, a national characteristic and evidence of British superiority.[2] The "British Isles," wrote the Scottish mathematician John Leslie in 1811, are the "favoured retreat" of Euclidean geometry, but "the followers of the modern system" are endeavoring to drive it out (Leslie 1811, p. 186). The *Elements* continued to be the mainstay of British mathematical education long after it had been abandoned in the rest of Europe; and geometry remained dominant in British curricula in the face of more advanced European analytical methods. Nearly sixty years after Leslie's comments, and despite the production of textbooks determinedly designed to be radical revisions to the *Elements*, or even downright replacements for it, the mathematician J. J. Sylvester famously told the British Association for the Advancement of Science that there were still some among them "who rank Euclid as second in sacredness to the Bible alone, and as one of the advanced outposts of the British Constitution" (Sylvester 1870, p. 126).

This tendency to identify Euclideanism with Britishness reflected the central place which geometry held in the education of British boys and young men of the well-to-do classes. In some ways, this centrality remained surprisingly stable for at least another hundred years, despite immense changes in the reach, philosophy, and content of education. Far from being an exclusive form of knowledge, doomed to dwindle away as access to education widened, geometry's prestige remained very much alive, piggybacking on the most up-to-date methods in technology and pedagogy to spread rapidly to an ever-larger segment of society, as the production of cheap geometry textbooks aimed at autodidacts and children of the middle and lower classes indicates (Jenkins 2007a).

The result was a landscape of very uneven access, in which geometrical knowledge retained immense cultural prestige and was subject to democratizing efforts via publishing and educational provision, at the same time as it continued to serve as a gatekeeper for entrance to elite institutions and professions. Partly to explain and navigate through this uneven landscape, a number of micronarratives about geometry developed which were repeated many times in different versions for different audiences. Some of them seek to make the experience of learning geometry appear timeless and ahistorical by connecting present-day learners with their predecessors, producing highly moralized accounts of access: one, for example, is the story of Ptolemy, the king of Egypt, asking Euclid "if there was no easier mode of becoming a geometrician than by studying his Elements. "There is no *royal road*," was his memorable answer" (Holyoake 1848, p. 58). This particular version of the "royal road" micronarrative, by the great Victorian freethinker George Holyoake, casts the problem of access to geometry in the image of his democratizing political project: "Science owns no idle votaries. The only condition on which it yields its secrets is that of attention" (Holyoake 1848, p. 58). Although stories such as this were retold many times and became somewhat formulaic, as this example from Holyoake illustrates, they were still capable of carrying distinctive and nuanced meaning. Explorations of Victorian mathematics' literary production often focus on writing by major individual figures such as Charles Dodgson and James Clerk Maxwell, but in terms of volume and spread, these micronarratives are perhaps Victorian mathematics' most important engagements with literature.

Many nineteenth-century accounts of learning geometry emphasize the long time it takes, the effort to memorize, sometimes without understanding, and the discipline required to follow rigorously the order of the Euclidean text, the "condition" of "attention" which Holyoake insists on. These accounts see the experience of reading Euclid properly as a corrective to the practices of scattered reading, reading for entertainment, and reading uncritically which many authors associated with reading literature, particularly poetry and fiction. The advice that the poet Henry Kirke White gave his friend Benjamin Maddock in 1805 is typical of many of these prescriptions of Euclidean reading: "you ought to strengthen your mind a little with logic, and for this purpose I would advise you to go through Euclid with sedulous and serious attention .... You are too desultory a reader, and regard *amusement* too much" (Kirke White 1834, pp. 221–222). High-mindedly urging Euclid on those who preferred fiction was so widely endorsed that Dickens derided its ungenerosity in his description of the tired working-class readers of Coketown, who, looking to literature to provide self-understanding and affirmation, "took De Foe to their bosoms, instead of Euclid" (Dickens [1854] 1995, p. 53). If reading Euclid properly was understood as providing necessary discipline for the mind, untaught geometers deviate from this norm; they appear to know geometry without engaging in this disciplining process, and thus they could represent a challenge to established educational assumptions, and perhaps even to social order.

Narratives of the "untaught geometer" use this figure to foreground some of the problems of access and prestige while also blurring or even concealing some of the politics of education which affected the acquisition of geometrical knowledge. In the argument that follows, I examine some of the ways in which Victorian writers considered and dealt with the ideological threats with which they associated non-normative Euclideanism.

Belief in untaught geometers was at least partly a construction of inherited and repeated narratives. This literary figure has a lineage reaching as far as the uneducated slave boy in Plato's *Meno*, who is drawn by Socrates into displaying apparently innate geometrical knowledge. The boy had never been taught any geometry, but under close questioning, he shows a sophisticated understanding which he must have acquired, according to Socrates, before he was born. Some Victorian commentators noted that Socrates's argument is unconvincing, but their reasons for being unconvinced differed, and were part of a much larger argument about what geometry tells us about the human mind. The question is whether human minds contain inherent geometrical knowledge, a priori ideas quite independent of sense experience, let alone of formal education (we could sum this up as the rational approach); or whether we acquire our first knowledge of space, size, and other geometrical concepts through our physical encounter with objects in the world (the empirical approach). The most pointed and well-documented example of a clash between these two schools of thought in Victorian philosophy is that between William Whewell, the Master of Trinity College, Cambridge, surely the period's most polemical traditionalist Euclidean, and John Stuart Mill. As Laura J. Snyder points out in her history of the Whewell-Mill debate, their argument about geometry was just one wing of their long conflict about the fundamentals of scientific method in the service of social reform (Snyder 2006, p. 4).

The patrician nineteenth-century view of geometry for which Whewell was an untiring advocate argued that geometry depends on and draws out ideas, particularly the idea of space, which do not derive from experience. "Since [the mind] can know what it has not learned from the senses, it must have some other source of knowledge," Whewell argued, summing up Plato's view (Whewell 1855, p. 12). This other source was the creator, who had equipped human minds with a priori knowledge, ideas of space, form, and magnitude which exist in the mind before experience furnishes them through sensory perception. Without these pre-existing ideas, indeed, sensory experience would be unintelligible. Granted the existence of these ideas, geometry begins to appear to be independent of experience (at the level of individuals) and of history (at the level of societies). The trope of a man locked in a windowless room—or sometimes a blind man—who would nonetheless be able to reason out mathematics and geometry from innate first principles, was a kind of companion-piece in nineteenth-century educational writing to the figure of the untaught geometer, and it emphasized the view that mathematical thought was based not on sense data but on inherent ideas.[3] For this thought-experiment, it would not matter at all whether the man in the room were a contemporary or an ancient Greek, or of any period between; Victorian traditionalists typically urged that plane geometry transcended history, having reached such a degree of perfection in Euclid's work that it had not since been, and could not be, improved upon. Geometry, they claimed, is derived only from facts which were equally indisputable in Euclid's time as in their own, and is produced only by operations of the human reason which are quite unaffected by changes in the outside world. According to this school, the untaught slave boy in Plato's experiment is able to answer questions about geometry because they prompt him to recover knowledge he always already had. "Whatever we may think of the conclusiveness of this particular experiment," wrote the Victorian classicist (and Whewell's successor as Master of Trinity) W. H. Thompson, evidently rather skeptical about the plausibility of the slave boy demonstration, "the general doctrine it was intended to illustrate is still maintained to be sound doctrine by many modern metaphysicians" (Thompson 1868, p. 168).

In opposition to this rationalist approach to geometry, the empiricist approach advocated by John Stuart Mill argued that not only *could* geometry be drawn from sense data about the physical world—"Every theorem in geometry is a law of external nature, and might have been ascertained by generalizing from observation and experiment" (Mill 1843, 2, p. 173)—but it actually *is* so: "The points, lines, circles, and squares, which any one has in his mind," he argues, "are (I apprehend) simply copies of the points, lines, circles, and squares, which he has known in his experience" (Mill 1843, 1, p. 298). According to this view, the slave boy's knowledge of geometry does not have to come from either formal lessons or an inbuilt idea of space; it comes as a result of having lived in the physical world long enough to have understood principles of size and scale. "If Plato had taken pains to study the early life of the untaught slave," noted the eminent Victorian classicist George Grote, "with its stock of facts, judgments, comparisons, and inferences suggested by analogy, & c., he might easily have found enough to explain the competence of the slave to answer the questions" (Grote 1865, 2, p. 22).

Plato's slave boy is an important example of the untaught geometers available to, and remade by, Victorian writers. But much more recent instances were also discussed. One was that of the Scottish mathematician Colin Maclaurin (1698–1746). Accounts of Maclaurin often highlighted an early accidental encounter with a copy of the *Elements*, when, aged twelve, Maclaurin made himself the "master, in a few days, of the first six books, without any assistance" (*Biographical Dictionary* 1856, p. 530). The story of Maclaurin's Euclidean experience is one of recognition: Maclaurin sees the value of the *Elements*, and engages precociously and powerfully with it. It is easy to tell this story in a way which, though it reverences Maclaurin's innate gift, continues to venerate Euclid. But a number of famous stories circulated throughout the nineteenth century of individuals who, unaware of the existence of the *Elements*, or refusing to study it, had achieved the extraordinary feat of producing a system of geometry in some way comparable with Euclid's, or—in less extreme cases, had simply found Euclid too obvious to be worth bothering with. Some accounts of Newton saw him as an example of the latter type. One of the major sources for this representation of Newton was the much-cited remark by the French mathematician, historian, and biographer, Bernard le Bovier de Fontenelle, that Newton did not study Euclid because "il le savait presque avant que de l'avoir lu" (he knew him almost before he read him) (Fontenelle 1818, p. 387). This remark is repeated, often skeptically, in many Victorian British sources. A typical example comes in an 1858 article on genius, which remarks that "Euclid was to [Newton], not an object of study, but of simple and constant perception" ([Anon] 1858, p. 237).

But making the claim that Newton was an untaught geometer was not without ideological risk. Rebekah Higgitt suggests that the story of Newton's youthful refusal to spend time on Euclid was suppressed by one Victorian biography of Newton, since it did not encourage the working men who were the book's target audience to engage with Euclid and the "steady and dedicated learning" he epitomized for many Victorian educators (Higgitt 2007, p. 38). Another concern was that driving a wedge between Newton and Euclid risked undermining some of the foundations of the British mathematical tradition, which revered Newton and his geometrical methods. Some nineteenth-century writers found it important to stress that despite his juvenile dismissal of the *Elements*, the mature Newton greatly respected Euclid. By re-aligning Newton with Euclid, they supported British mathematics in the face of the powerful challenge from its continental European rival, with its more forward-looking algebraic methods. The highly influential mathematical educator, historian, and first professor of mathematics at University College London, Augustus De Morgan, provides an example of this re-alignment of Newton with Euclid in his review of David Brewster's landmark *Memoirs of the Life, Writings and Discoveries of Sir Isaac Newton* (1855). De Morgan emphasizes that even though Newton as a youngster "threw away his Euclid at first, as very evident," nonetheless at Cambridge he came "to know better." De Morgan insists that Newton's *Principia Mathematica* "is the work of an inordinate Eucleidian [sic]" and that Newton, in fact, "regretted that he had not paid *more* attention to Euclid" (De Morgan 1855, pp. 312, 313). As in this example, stories of Newton's non-engagement with Euclid were often cited only to be quarrelled with or corrected.

Blaise Pascal, on the other hand, was a much safer instance of the invention of Euclidean geometry by a mind that had never come into contact with Euclid. The story of Pascal's childhood invention of geometry from first principles was a favorite of early Victorian exhortatory works. Because his father refused to let him study geometry, hoping, as one rendition gave it, that "balls and hoops" would chase "circles and triangles from his brain," the eleven-year-old Pascal set to deriving his own geometrical system ([Anon] 1846, p. 107). One of the scientific biographies that Mary Shelley wrote tells the story of Pascal's father discovering that his son had "established a chain of [geometrical] propositions deduced from axioms and definitions of his own adoption," which had taken him as far as the knowledge presented in the 32nd proposition of the first book of the *Elements of Geometry*. "The father was struck almost with fear at this exhibition of inborn genius," Shelley comments ([1839] 2002, p. 74), and, relenting gave him a copy of the *Elements* to study.

Most retellings of this story stressed that Pascal's achievement was the result of this individual quality of "genius," though occasionally it was noted that the "natural connection" which links geometrical truths together makes them relatively easier to deduce from first principles ([Anon] 1858, p. 6). Similarly, Victorian stories of untaught geometers such as Maclaurin and Newton generally represent untaught geometrical ability as a particular and individual gift, not as a universal quality as it is in *Meno*, where the point of Socrates's demonstration is that *even* a slave boy knows basic geometry, not that the particular slave boy in question is a natural geometrical genius.

Between these two poles—natural geometrical genius and anonymous example of universal geometrical ability—Victorian fiction gives a number of examples of characters with no philosophical knowledge or extraordinary genius at all who nonetheless display a natural affinity with or ability to perform practical feats which depend on instinctual geometrical knowledge. These characters are typically children or even animals. "Consider how wonderful is the phenomenon of a boy's throwing, successfully, at a mark," exhorted the sometime President of Harvard, Thomas Hill, in his short book on geometry and religious faith:

> The epicycloidal theories of Hipparchus, the Newtonian theory of gravitation, the resolution of centripetal and centrifugal forces, the conic sections of Apollonius, the modifications of those curves by the resistance of the air, – all these are involved in the problem, and must be practically solved, with considerable accuracy, before the school-boy can give his fellow a good ball, or catch one on the fly. (Hill 1874, p. 35)

The boy's throw invokes immensely complex calculations, far beyond the cognitive capacities of the boy himself.

Tom in George Eliot's *Mill on the Floss* (1860) is a sophisticated instance of the figure of the instinctual geometer: despite his utter inability to follow his tutor's Euclidean lessons, he can "throw a stone right into the centre of a given ripple," and "draw almost perfect squares on his slate without any measurement" (Eliot [1860] 1979, p. 147). Unhappily for Tom, his tutor does not recognize these evidences of geometrical ability, precisely because they depend on instinct and not on theory: "Mr. Stelling took no note of these things: he only observed that [Tom] was in a state bordering on idiocy with regard to the demonstration that two given triangles must be equal – though he could discern with great promptitude and certainty the fact that they *were* equal" (Eliot [1860] 1979, p. 147). Unlike Mr. Stelling and the Euclidean traditionalists he represents, Eliot sees Tom's physical skill as legitimate geometrical knowledge. Eliot's Euclideanism, in contrast to that of William Whewell, for example, was distinctly progressive, influenced by her geometry teacher, Francis Newman, whose unusual views on teaching geometry I have discussed elsewhere (Jenkins 2008). A key aspect of Newman's and Eliot's unorthodox Euclideanism was their support for women's access to it. *The Mill on the Floss* provides a contrast to Tom's instinctive geometry in his sister, Maggie, who has a hungry intellectual curiosity and who sees Euclid as a step toward "masculine wisdom – ... that knowledge which made men contented and even glad to live" (Eliot [1860] 1979, p. 299). Even after the bitter humiliation of being told by Mr Stelling that girls "could pick up a little of everything ... but they couldn't go far into anything," (Eliot [1860] 1979, p. 158), Maggie tries again with Euclid, but with no one to help or encourage her, she struggles and ultimately gives up. Neither Tom's instinctive geometry, nor Maggie's inability to submit to the rigors of normative Euclideanism, provides anything like the intellectual and moral strength which was so central to the defense of Victorian Euclidean traditionalism.

The miserable contrast in *The Mill on the Floss* between the damaging imposition of normative Euclideanism on boys and the total absence of support for girls' access to it is rewritten in a very different key in a novel of 1878 by the husband and wife team Mortimer and Frances Collins. *You Play Me False* mixes an elaborate marriage plot revolving around secret identities with warnings against the conversionary work of Roman Catholic clergy. Much of the novel focuses on little Clara and her brother Charlie who are left by their self-indulgent aristocratic parents to bring themselves up surrounded by the natural beauty of their estate. Clara combines Tom Tulliver's instinctual geometry with Maggie's childish impatience with Euclidean method. Sitting by the river, Clara throws a stone at a water-rat and misses, but is enchanted by the circular ripples her stone creates. "Oh, but look at those beautiful rings in the water, all even, dying lower and lower! How wonderful it is!" she cries, and is overheard by a Roman Catholic priest (Collins 1878, I, p. 107). Advised by his charismatic but untrustworthy superior, he sets out to use her affinity with geometry to win her over to Catholicism through the study of Euclid (Collins 1878, I, p. 109). It quickly becomes clear that Clara has natural talent: "She saw through propositions in Euclid which he expected her to find difficult." Clara's Euclideanism is undisciplined, though, and driven by curiosity and delight in sensory experience. "[S]he was in a tremendous hurry to get outside the A, B, C, of the old Greek geometrician, and to find out the application of all his problems and theorems to all the manifold wonders of earth and heaven" (Collins 1878, I, pp. 112, 113). Though naturally able as a practical geometer, she turns out to be immune not only to Euclid, but also to all attempts at religious conversion.

Tom and Clara are examples of the instinctive Victorian untaught geometer: unacquainted or unacquaintable with Euclid, they nonetheless have geometrical ability. Even more instinctive than children, animals were sometimes represented as untaught geometers, and discussion of their geometrical behaviors played an important role in Victorian attempts to determine whether geometrical thought is an essential or exclusive part of the human mind. It was crucial for the cultural prestige of geometry, its status as the highest and purest kind of truth (barring divine revelation), that though other creatures might perform geometrical tasks by instinct, geometrical reasoning should be a distinctively human capacity, but one that reflected in a small way the nature of the divine mind. Geometrical forms occurring in nature were relatively easy to explain within this system, because they could be understood as simple adaptations to universal physical laws, which could themselves be referred explicitly or tacitly to divine provision. Rather harder to explain, however, were geometric objects constructed by non-human animals.

Discussions of geometrical work done by other creatures centered on accounts of bees and, to a much lesser extent, social wasps, and the combs and hives they build. Since classical times, the geometry of bee architecture has been a subject of recurrent scientific inquiry. Pliny the Elder, for example, thought that hive cells are hexagonal because bees build one wall with each of their six legs. The perfectly tessellating hexagonal cells of a honeycomb, and the way in which they are joined so as to maximize space while minimizing on material, came to be seen as a triumph of mathematical planning.

The problem of accounting for the habit of bees to build combs in regular geometrical patterns, and above all in hexagonal cells, recurred frequently through the nineteenth century. The hexagonal shape produced a structure thought to combine the greatest strength and the least waste in a way that seemed remarkable and admirable, so much so that explaining the principles and advantages of this structure was sometimes set as a question in Victorian mathematical examinations (for example Frost 1849, p. 61). Writings for children sometimes played with the idea that this perfection is due to the fact that bees are indeed conscious geometers. One example, which was frequently reprinted in Christian periodicals around the middle of the nineteenth century, is a short dialogue comparing animals and insects to various human trades: "[t]he bee is a professor of geometry, for he constructs his cells so scientifically, that the least possible amount of material is formed into the largest spaces with the least waste of room. Not all the mathematicians of Cambridge could improve the construction of his cells" ([Anon] 1852, p. 247).

However, it was generally agreed that though bee geometry illustrated excellence in design, the bees could not themselves be consciously doing the designing. Thus, for example, the great eighteenth-century Scottish geometer Colin Maclaurin—who had been an untaught geometer himself, as I have discussed—devoted a paper to investigating the supposed excellence of hive design, demonstrating over several pages of dense geometrical working that "the *Bees* do truly construct their Cells of the best Figure" (Maclaurin 1743, p. 571). Their geometrical working "could not have been more Perfect," he concluded, had they benefitted from "the greatest Knowledge in *Geometry*" (Maclaurin 1743, p. 571). For him, though, it was obvious that bees did not actually understand geometry: rather, they had a "wonderful Instinct" which dictated their geometrical behavior. He refused to speculate about the origin of bees' geometrical instinct; he considered this a theological rather than a scientific topic, "a Question of a higher Nature" than his mathematical investigation could support (Maclaurin 1743, p. 571). He was prepared to go no further than to refer bee geometry to the important principle that "what is most beautiful and regular, is also found to be most useful and excellent," an example of what Claire Preston, in her study of bees in culture, calls "beautiful utility – that is, not merely good practical and pleasing design, but rather the indivisible moral union of the aesthetic and the pragmatic" (Preston 2006, p. 100).

By the late eighteenth century, bee geometry had been so much discussed as an example of successful animal instinct that it was an obvious and all but unavoidable example in natural theological texts. Maclaurin's paper was discussed by the Scottish enlightenment philosopher Thomas Reid in a key passage of his highly influential Common Sense treatise *The Active Powers of Man* (1788), a passage which became perhaps the most widely disseminated explanation of animal geometry in the nineteenth century. Reid argued that "the work of every animal is indeed like the works of nature, perfect in its kind," and capable of bearing "the most critical examination of the mechanic or the mathematician" (Reid 1788, p. 106). Citing Maclaurin's demonstrations, Reid rehearsed the familiar claim that hexagons are the ideal shape for a hive cell. He played with the possibility that bees are conscious and deliberate in their adoption of this design; they build in hexagons "as if they knew" their properties, and "as if acquainted with these principles of solid geometry" (Reid 1788, pp. 106, 107). "If a honey-comb were a work of human art," Reid wrote, "every man of common sense would conclude, without hesitation, that he who invented the construction must have understood the principles on which it is constructed" (Reid 1788, p. 107). But in fact, he was quite sure that bees "know none of these things" (Reid 1788, p. 108). We are not obliged to extend the same argument by design beyond the realm of human manufacture to the productions of animals. Rather, bees "work most geometrically, without any knowledge of geometry": "when a bee makes its combs so geometrically, the geometry is not in the bee, but in that great Geometrician who made the bee, and made all things in number, weight, and measure" (Reid 1788, p. 108).

Reid's pious reattribution of geometrical intelligence from the bees to their Creator became influential in confirming the natural theological belief that because geometry reflects the perfect order of the divine mind, thinking about geometry connects us with God ([Duncan] 1840, p. 108; Bevan 1827, p. 340). Since animals cannot be connected with God's mind in the same way, they must not be able to think geometrically; but their geometrical instincts remind us of the place of geometry in the divine creation. Reid's comments on bee geometry were widely repeated in eighteenth- and nineteenth-century natural history texts, and especially in practical manuals of bee-keeping. This was a sub-genre of real practical importance to many Victorian readers, and Reid's views may well have reached more readers through these manuals than through natural theological texts. In 1842, for example, the heavyweight *Quarterly Review* published a fifty-page review of nine different bee-keeping books, opening with an acknowledgement of the significance of bee science ([James] 1842). Of the nine books reviewed in this article, five discussed bee geometry, and three of them quoted Reid's dictum

92   A. JENKINS

that "the geometry is not in the bee, but in that great Geometrician who made the bee."

The popularity of this reading of animals' instinctive geometry was challenged when the problem of accounting for bee behavior re-emerged in evolutionary theory. In *Origin of Species*, Charles Darwin devoted a section of the chapter on "Instincts" to cell-making. The geometrical architecture of honey-bees, having been so long and widely accepted as, in Darwin's words, "the most wonderful of all known instincts," provided him with an excellent test-case for evolutionary explanation (Darwin [1859] 2008, p. 162).[4] Darwin praised the honey-bee's comb as "absolutely perfect in economising labour and wax," and went on to show that far from being a direct reflection of divine design, the development of this perfection was entirely explicable through a lengthy process of natural selection, "the bees, of course, no more knowing that they swept their spheres at one particular distance from each other, than that they know what are the several angles of the hexagonal prisms and of the basal rhombic plates" (Darwin [1859] 2008, p. 175). Darwin's account of instinctual behavior was met with considerable hostility and ridicule, part of which took the form of a reinstatement of the argument about bees as untaught geometers. Bees "do not understand geometry, in this we should all agree," wrote Robert Mackenzie Beverley, but they *do* behave intentionally: "they know what they have to do before they begin" ("A Graduate of Cambridge" 1867, p. 91). Untaught geometers, even in insect form, challenge structures of order and their non-normative Euclideanism has to be accounted for within those structures.

Oren Harman and Michael R. Dietrich, the editors of a key collection in the historiography of science's relationships with disciplinarity and identity, distinguish between "outsider scientists," by which they mean people trained in one "intellectual community" who migrate into another scientific field, and "scientific outsiders," those outside or excluded from the professional scientific world altogether (Harman and Dietrich 2013, p. 4). This second group would include the "fringe physicists" interviewed in Margaret Wertheim's *Physics on the Fringe* (2011). Can we identify analogous outsiders in the world of nineteenth-century mathematics? Daniel J. Cohen discusses a type of mathematical enthusiasts he calls "circle squarers," people fascinated by the problem of constructing a square that has the same area as a given circle, using only compasses and a straight edge (Cohen 2007, pp. 138–46). Though "circle squaring" long predated the emergence of structures of professional mathematics in the later nineteenth century, Cohen argues that it became "something of a cottage industry" thanks to widening access to mathematical knowledge in this period (p. 140). "Victorian circle squarers tended to be upper-middle-class men with some engineering, medical, or technical expertise involving mathematics, but little knowledge of modern mathematical methods," Cohen notes (p. 141). These men have some of the characteristics of Harman and Dietrich's "outsider scientists," in that they bring a specialist training in another field of knowledge to their mathematics, but also of Wertheim's "fringe physicists," since they remained outside the mathematical establishment, taught their place by attacks from professional mathematicians, such as the attack Cohen describes Augustus De Morgan making on one of the best known Victorian mathematical outsiders, James Smith. De Morgan invented the term "*pseudomath*" for Smith and his ilk: "a person who handles mathematics as the monkey handled the razor," that is, in a way that unintentionally shames his ignorance of correct procedure (Cohen 2007, p. 145).

Among the untaught geometers I have discussed in this essay, Newton, Maclaurin, and Pascal, and in their different way, Clara and Tom from *You Play Me False* and *The Mill on the Floss*, can all be thought of as outsider mathematicians. Their lack of familiarity with Euclid is a signal marker of outsider status, since it was all but impossible to study mathematics to any level even approaching the professional in the nineteenth century without such familiarity. But they do not fit either of Harman and Dietrich's categories of outsider. All were very young when they demonstrated untaught geometrical knowledge; they thus had not received a training in another discipline before they encountered Euclid. Having had this encounter, they did not remain on the fringes of mathematics: Newton, Maclaurin, and Pascal moved inside the professional group (very different as the definitions of professional mathematicians were in these various periods). Clara and Tom abandon geometry as soon as the adult world allows them to. My third group of untaught geometers, bees, is as radically "outside" the mathematical establishment as it is possible to be. In reality bees could not, of course, be called mathematical outsiders, because they do not maintain any kind of relationship with mathematics; but *in the texts* I have discussed, in which they are imagined to have a relationship with mathematics, albeit a contested one, they can be considered as belonging to this group.

Victorian writers told and retold the stories of these human and animal untaught geometers in order to highlight their outsiderhood, dramatizing the clash between the process of reading Euclid in the careful and systematic way conventionally deemed normative, and the unexpected ability to solve problems and generate proofs without engaging with Euclid at all. But they are not allowed to remain in this extraordinary position of non-normative Euclideanism. Instead they are disciplined, to use a Whewellian term in a Foucauldian sense, by Euclid, some of them adapting themselves to him, others glancing off the *Elements* for the first and last time. The human untaught geometers I have discussed in this essay are not allowed to remain outsiders; even those who leave geometry behind altogether relinquish their outsiderhood at that point, since they cease to have a relationship with geometry. Outsiderhood is precarious, requiring very substantial investment from the moment at which it is claimed or revealed if it is not to resolve itself into either insiderhood or exile.

Figures of the non-normative Euclidean are used in Victorian literature to investigate relationships between rational and empirical knowledge, as well as between genius and discipline, and ultimately, in many cases, to enforce cultural submission to the imaginary ideal of Euclidean geometry as a perfect, complete, and unarguable system of truth, and to the models of moral and social stability which it underpinned. But the stability of this ideal Euclideanism, and the alternative Euclideanisms which are narrated in contrast and in support of it, did not outlast the Victorian period. The spread of knowledge of non-Euclidean geometry in Britain, from the 1860s onwards, disturbs this ideal, and challenges these stories by offering an engagement with Euclid which is both non-normative and mathematically valid and accepted.

NOTES

1. As the critic Réné Wellek noticed, Coleridge is not the only target of Peacock's satire: the passage is also attacking Kant. Flosky's claim to untaught mastery of Euclid is an unacknowledged quotation from an expostulation against Kantian philosophy by the Scottish diplomat, politician, and philosopher William Drummond, who mocks 'philosophers who know metaphysics a priori, who possess an intuitive faculty, who see visions of pure reason, and who carried the whole science of geometry in their heads before they ever looked into Euclid' (Drummond 1805, I, p. 358; Wellek 1931, p. 182).
2. 'Britishness', the term very often used by Victorian writers in this context, elides the differences between the English and Scottish educational systems and the geometrical traditions associated with each. George Elder Davie's account of mathematics in Scottish Universities during the nineteenth century has been very influential in the history of education, though less so in the history of mathematics; Davie sees the story as one of decline produced by an abandonment of the traditional Scottish emphasis on philosophy in mathematics in favour of English conventions in teaching the subject (Davie 1961, pp. 105–124). But one of the effects of Davie's influence has been, as Jane McDermid points out, 'to obscure the variations and the inferiority in girls' experience of schooling, notably in terms of provision and curriculum, in nineteenth-century Scotland relative to boys' (McDermid 2005, p. 140).
3. For discussion of examples of this trope in texts by Henry Brougham and John Herschel, see Jenkins 2007b, pp. 94–95.
4. Frederick R. Prete argues that Darwin was sensitive to 'the important position these insects held in the intellectual milieu of the time'; the importance of bee products to the British economy, as well as the re-emergence of bees as objects of major public curiosity following the discovery of parthogenesis, and the effects of royal jelly on worker bees, meant that 'if [Darwin] was to convince his intended audience of the plausibility of his theory of evolution, that theory would have to include the honey bees' (Prete 1990, p. 273).

Works Cited

"A Graduate of Cambridge." 1867. *The Darwinian Theory of the Transmutation of Species.* London: Nisbet.

[Anon]. 1846. "The Young Philosopher." *Chambers' Edinburgh Journal* 5: 106–109.

[Anon]. 1852. "Trades Carried on by Birds, Beasts and Insects." In *The Child's Companion, and Juvenile Instructor.* London: Religious Tract Society. 245–48.

Anon]. 1858. "What is Genius?" *The Institute: A Monthly Literary Journal,* 3: 236–39.

Bevan, Edward. 1827. *The Honey Bee, its Natural History, Physiology, and Management.* London: Baldwin, Cradock and Joy.

Biographical Dictionary of Eminent Scotsmen. 1856. Vol. 3. Edited by Robert Chambers; revised edition by Thomas Thomson. Glasgow, Edinburgh, and London: Blackie.

Cohen, Daniel J. 2007. *Equations from God: Pure Mathematics and Victorian Faith.* Baltimore: Johns Hopkins University Press.

Collins, Mortimer and Frances. 1878. *You Play Me False: A Novel.* 3 vols. London: Bentley.

Darwin, Charles. [1859] 2008. *On the Origin of Species.* Edited by Gillian Beer. Oxford: Oxford University Press.

Davie, George Elder. 1961. *The Democratic Intellect: Scotland and her Universities in the Nineteenth Century.* Edinburgh: Edinburgh University Press.

de Fontenelle, Bernard le Bovier. 1818. "Eloge de Newton," in *Oeuvres,* I.1. Paris: Belin. 387–403.

[De Morgan, Augustus]. 1855. "Sir David Brewster's Life of Newton." *North British Review* 23: 307–338.

Dickens, Charles. [1854] 1995, rev. 2003. *Hard Times.* Edited by Kate Flint. Harmondsworth: Penguin.

Drummond, William. 1805. *Academical Questions.* London: Cadell and Davies.

[Duncan, James]. 1840. *The Natural History of Bees.* Edinburgh: Lizars.

Eliot, George. [1860] 1979. *The Mill on the Floss.* Edited by A. S. Byatt. Harmondsworth: Penguin.

Frost, A.H. 1849. *Mathematical Questions of the Senate-House Examination Papers, 1838–1849.* Cambridge: Hall.

Grote, George. 1865. *Plato, and the Other Companions of Sokrates.* 3 vols. London: John Murray.

Harman, Oren, and Michael R. Dietrich. 2013. "Introduction: Outsiders as Innovators in the Life Sciences." In *Outsider Scientists: Routes to Innovation in Biology,* edited by Harman and Dietrich. Chicago: University of Chicago Press. 1–24.

Higgitt, Rebekah. 2007. *Recreating Newton: Newtonian Biography and the Making of Nineteenth-Century History of Science.* London: Pickering and Chatto (Reprint in 2016, London: Routledge).

Hill, Thomas. *Geometry and Faith.* 1874. *A Fragmentary Supplement to the Ninth Bridgewater Treatise.* Rev. and enlarged ed. New York: Putnam.

Holyoake, G. J. 1848. *Mathematics no Mystery; or, the Beauties and Uses of Euclid,* 2nd ed. London: Watson.

[James, Thomas]. 1842. "The Honey-bee and Bee-books." *Quarterly Review* 71: 1–54.

Jenkins, Alice. 2007a. "Geometry," in Leslie Howsam, et al. "What the Victorians Learned: Perspectives on Nineteenth-century Schoolbooks." *Journal of Victorian Culture* 12: 262–85.

Jenkins, Alice. 2007b. *Space and the elsewhere'March of Mind': Literature and the Physical Sciences in Britain, 1815–1850.* Oxford: Oxford University Press.

Jenkins, Alice. 2008. George Eliot, Geometry and Gender. *Literature and Science* 61: 72–90.

Kirke White, Henry. 1834. *The Life and Remains of Henry Kirke White.* London: Scott, Webster, and Geary.

[Leslie, John]. 1811. Review of J. B. J. Delambre, *De l'arithmétique des grècs. Edinburgh Review* 18: 185–213.

Maclaurin, Colin. 1743. "Of the Bases of the Cells wherein the Bees deposite their Honey." *Philosophical Transactions* 42: 565–71.

McDermid, Jane. 2005. *The Schooling of Working-Class Girls in Victorian Scotland: Gender, Education and Identity.* Oxford: Routledge.

Mill, John Stuart. 1843. *A System of Logic.* 2 vols. London: Parker.

Owen, Thomas. 2019. *Wordsworth, Coleridge, and the 'Language of the Heavens'.* Oxford: Oxford University Press.

Peacock, Thomas Love. *Nightmare Abbey* (1818). In *Nightmare Abbey and Crotchet Castle.* Edited by Raymond Wright. Harmondsworth: Penguin, 1969. 37–124 (Reprint in 1986).

Preston, Claire. 2006. *Bee.* London: Reaktion.

Prete, Frederick R. 1990. "The Conundrum of the Honey Bees: One Impediment to the Publication of Darwin's Theory." *Journal of the History of Biology* 23: 271–290.

Reid, Thomas. 1788. *Essays on the Active Powers of Man.* Edinburgh and London: Bell and Robinson.

Richards, Joan L. 1988. *Mathematical Visions: The Pursuit of Geometry in Victorian England.* San Diego: Academic Press.

Shelley, Mary. [1839] 2002, repr. 2016. "Pascal." In *Lives of the Most Eminent Literary and Scientific Men of France.* In *Mary Shelley's Literary Lives and Other Writings,* edited. by Clarissa Campbell Orr. Vol 3, *French Lives.* London: Pickering and Chatto; repr. Abingdon: Routledge. 71–92.

Snyder, Laura J. 2006. *Reforming Philosophy: A Victorian Debate on Science and Society.* Chicago: University of Chicago Press.

Sylvester, J. J. 1870. "Inaugural Presidential Address to the Mathematical and Physical Section of the British Association at Exeter, August, 1869." In *The Laws of Verse, or, Principles of Versification Exemplified in Metrical Translations.* London: Longmans, Green. 99–129.

Thompson, W.H. 1868. *The Phaedrus of Plato.* London: Whittaker.

Wellek, Réné. 1931. *Immanuel Kant in England, 1793–1838.* Princeton: Princeton University Press.

Whewell, William. 1855. On the Influence of the History of Science upon Intellectual Education. *Lectures on Education Delivered at the Royal Institution of Great Britain,* 3–36. London:Parker.

CHAPTER 6

Mathematical Contrariness in George Eliot's Novels

Derek Ball

In an introduction to George Eliot's penultimate novel *Middlemarch* (1871–1872), Felicia Bonaparte notes that Eliot "was an excellent mathematician" (Bonaparte 1996, p. ix). *Middlemarch*, set in the early 1830s, is about the interactions between various groups of people who live in and around the provincial Midland town of Middlemarch and how these interactions change over time. While Bonaparte does not go on to discuss the implications that Eliot's mathematical ability may have for this novel, other critics have described some of the mathematics in Eliot's novels, and the purposes it might serve. I want to take a broader view and discuss the way Eliot's mathematics and her tendency for logical thinking shape the way she worked as a novelist. I shall argue that mathematics and logic influence Eliot's writing on several levels. Firstly, there is explicit inclusion of mathematics within the novels, particularly the early novels; Eliot even challenges readers to solve mathematical problems. Secondly, there is her use of mathematically based images and ideas which occur through all the novels, and that are frequently used to encourage readers to see things differently. And, finally, Eliot makes use of the logic inherent in mathematics when structuring her novels, particularly the later novels, so as to challenge accepted ideas.

The middle of the nineteenth century was a time of turmoil in British mathematics and philosophy that centred mainly on Euclidean geometry. Different mathematicians suggested diverse ways of explaining why Euclid's axioms had to be true (Richards 1988, pp. 3, 24–25), and the use of the certainty of Euclid's axioms as an argument for moral certainties was being

---

D. Ball (✉)
Leicester, UK

© The Author(s) 2021
R. Tubbs et al. (eds.), *The Palgrave Handbook of Literature
and Mathematics*, https://doi.org/10.1007/978-3-030-55478-1_6

97

scrutinised (Mill 1852, pp. 350–51). At the same time, Euclid was no longer universally seen as the best textbook for teaching Euclidean geometry (Price 1994, p. 19), and mathematics came to be seen as more of a formal tool and less as exemplifying absolute truth. A similar development characterised the field of logic: Augustus De Morgan introduced algebra into logic, deciding to detach from his logic any discussion of the truth of premises and to focus specifically on the *means* by which correct conclusions could be inferred from true premises (Rice 2011, p. 7). Moreover, in the 1860s non-Euclidean geometry arrived in England, and Euclid's axioms, at least for some mathematicians and philosophers, were no longer seen as obviously universally true. This was a revolution in thinking that profoundly shook former certainties in a way that Joan L. Richards compares to the Darwinian revolution (Richards 1988, p. 1).

All of Eliot's novels document her interest in philosophical and scientific upheavals and the need for social change, and she taps into the period's "revolutionary spirit" in mathematics and logic, using it not to reinforce widely accepted beliefs but to challenge them, not to reduce uncertainty but to increase it. This essay will demonstrate how Eliot used her understanding of the place of mathematics in mid-Victorian thinking to raise issues about social attitudes and beliefs, not least by making a clear distinction between what is mathematically or logically certain and what is a matter for dispute.

ELIOT'S MATHEMATICS

What we know of Eliot's knowledge of mathematics is fragmentary but compelling. She spoke highly of her father's ability with practical mathematics; he could "calculate with almost absolute precision the quantity of available timber in a standing tree," and he "had large knowledge of building, of mines, of plantations, of various branches of valuation and measurement" (Cross 1885, 1, p. 11; Haight 1954–1978, 3, pp. 168–69). In view of the close relationship the young Marian Evans (she later adopted the pseudonym George Eliot) had with her father, he may have been responsible for initiating her life-long interest in mathematics. In her early letters to her former teacher Maria Lewis, Marian Evans manifested an interest in mathematics and the mathematical sciences, both through her interest in astronomy and through the metaphorical use of the language of mathematics. Evans lapsed into mathematics and mathematical physics when writing about unrelated matters in her letters, in much the same way that she lapsed into French.

There is sporadic but repeated evidence that Eliot continued her study of mathematics throughout her life. In 1848–1849 Marian Evans spent several months in Geneva, during which time she attended "a course of lectures on Experimental Physics by M. le professeur de la Rive"; a course consisting of twenty-six lectures (Haight 1954–1978, 1, p. 325). And, in a letter to her friends the Brays, she described her life in Geneva: "I take walks, play on the piano, read Voltaire, talk to my friends, and just take a dose of mathematics every day to prevent my brain from becoming quite soft" (Haight 1954–1978, 1, p. 321). She must have learned an impressive amount of mathematics during her five months in Geneva, but Evans needed more: shortly after moving to London in 1851 she enrolled in a geometry course at the Ladies' College, a course that Francis Newman had just taken over from Augustus De Morgan (Jenkins 2008, p. 78). She told the Brays: "I am attending Professor Newman's course of lectures on Geometry at the Ladies' College every Monday and Thursday. You will say that I can't afford this, which is 'dreadful true' ... wherefore I must stint myself in some direction – clearly in white gloves and probably in clean collars" (Haight 1954–1978, 1, p. 343).

In spite of attending lectures in experimental physics while in Geneva, Eliot's principal mathematical interests seem to have been in pure mathematics. She appears to have had an enduring interest in algebra, proposing studying it with her step-son in 1860, studying three algebra books with George Lewes, Eliot's lifelong partner, in 1871 and, late in her life in 1879, recording studying algebra on her own (Haight 1954–1978, 3, p. 216; Pratt and Neufeldt 1979, p. 70; Haight 1954–1978, 7, p. 209). In an article about George Eliot's life, the Victorian historian Lord Acton remarked on Eliot's knowledge of the calculus, and Edith Simcox, a friend of Eliot, records that Eliot said in 1880 that she was intensively studying conic sections, because "she didn't want to lose the power of learning" (Acton 1885, pp. 471–72; Haight 1954–1978, 9, p. 293). And according to John Cross, Eliot esteemed her own ability in geometry highly, saying that "she might have attained to some excellence in [geometry] if she had been able to pursue it" (Cross 1885, 3, p. 423).

Eliot's belief in the value of mathematics for understanding the world is demonstrated in her novels, for example in the repeated approving references to the importance of mathematics by characters in the first two novels, *Adam Bede* (1859) and *The Mill on the Floss* (1860). An example occurs in the first novel when Bartle Massey, Adam's night-school teacher, hears of Adam's promotion at work and compliments him on having "an eye at measuring, and a head-piece for figures" ([1859] 1998, p. 259). Moreover, the narrator quotes Adam describing the new rector: "as for math'matics and the natur o' things, he was as ignorant as a woman" ([1859] 1998, p. 180). In spite of Adam's apparently throw-away remark about women, Eliot surely shared Adam's belief about the significance of mathematics for understanding the scientific and social world, and she used mathematics to challenge widely accepted ideas.

Mathematics and Social Certainties

Eliot's interest in mathematics chimed with mathematical and philosophical developments in the middle of the nineteenth century. When she undertook Francis Newman's geometry course, she was about to take on the editorship of the radical *Westminster Review*, which brought her into contact with the ideas of reformist thinkers such as J. S. Mill and Herbert Spencer, who were challenging traditional moralists like William Whewell. Working with the ideas behind some of their articles would have required an understanding of Euclid, that prestigious subject customarily studied mainly by men. For example, Mill's 1852 article in the *Westminster Review*, criticising William Whewell's moral philosophy, described Whewell, a leading polemicist for the a priori nature of Euclid's axioms, as a writer who manufactures even moral "propositions which… may be known independently of proof" because Whewell regards them as "self-evident truths" (Mill 1852, pp. 350–51). Under Eliot's editorship, the Westminster's readers encountered allusions and references to mathematical writers and topics such as these. What Newman taught on this course is not known, but the geometry text book he wrote adopted an axiomatic approach, while questioning some of the conventional presuppositions behind the established geometrical education (Newman 1841, especially pp. 1–2). Such theoretical geometry was relevant for Eliot's role as editor of essays which were, in one way or another, questioning the basis of Euclid's postulates or questioning the philosophical and moral uses to which certainty about these was put. And it would have made her more able to challenge some of the apparently scientifically based claims of certainty in social science, which thinkers like Whewell espoused.

Direct Reference to Mathematics in the Novels

The most straightforward way in which the influence of mathematics is apparent in Eliot's novels is the inclusion of direct references to mathematics and mathematical problems, particularly in her first two novels. In *Adam Bede*, Eliot makes her eponymous hero a mathematically minded carpenter and land manager, who, like Eliot's father, can ascertain, "with only looking at it, the value of the chestnut tree that was blown down" ([1859] 1998, p. 97). The narrator tells us that, when Adam is looking at beech trees, his "perceptions" enable him to calculate "the height and contents of a trunk to a nicety, as he stood looking at it" ([1859] 1998, p. 295). In the second chapter, Eliot poses a mathematical problem while describing the publican, Mr. Casson:

> Mr. Casson's person was by no means of that common type which can be allowed to pass without description. On a front view it appeared to consist principally of two spheres, bearing the same relation to each other as the earth and the moon: that is to say, the lower sphere might be said, at a rough guess, to be thirteen times larger than the upper…. But here the resemblance ceased, for Mr. Casson's head was not at all a melancholy-looking satellite, nor was it a "spotty globe," as Milton has irreverently called the moon. ([1859] 1998, p. 14)

The reference to "thirteen times" sets the reader a mathematical puzzle: in what way is the earth thirteen times larger than the moon? The puzzle is only solved when the reader, having looked up data concerning the sizes of the earth and the moon, and finding that the ratio of their diameters is approximately 3.7 to 1, realises that the comparison offered between the earth and the moon must be made on the basis neither of diameters nor of volumes, since this ratio would be $(3.7)^3$ to 1 (or about 50 to 1), but of surface areas. Eliot might be seen to offer a clue to this when she refers to the moon as a "spotty globe," thus drawing attention to its surface.

Eliot learned much of her mathematics through her own unguided efforts and this too is reflected in her first novel. When Adam meets Arthur Donnithorne, the squire's grandson, Arthur tells him: "I think your life has been a better school to you than college has been to me." In reply, Adam tells Arthur something Bartle Massey, the night-school master, has told him: "college mostly makes people like bladders – just good for nothing but t' hold the stuff as is poured into'em" ([1859] 1998, pp. 167–68). This, like some other of Eliot's pronouncements about the learning of mathematics, echoes points made by Augustus De Morgan in an essay written in the 1830s:

> the pupil having worked unmeaning and useless questions by slateful for some four or five years, comes out master of a few methods, provided he knows what rule a question falls under, which is not always sure to be the case, for in all probability, the first application which it is necessary he should make will be a combination of more rules than one, and, therefore, not exactly to be solved by the rule in his book. (De Morgan 1831, p. 268)

Bartle Massey expresses a similar point of view when lambasting the two youths who are learning the mathematics of accounting: "you'll come and pay Bartle Massey sixpence a-week, and he'll make you clever at figures without your taking any trouble. But knowledge isn't to be got with paying sixpence, let me tell you: if you're to know figures, you must turn'em over in your own heads…. There's nothing you can't turn into a sum." Bartle challenges the youths to invent questions such as this: "if my fool's head weighed four pound, and Jack's three pound three ounces and three-quarters, how many penny-weights heavier would my head be than Jack's?" ([1859] 1998, p. 236). This is not only Bartle challenging the youths: it is also Eliot challenging the reader. Here is another mathematics problem for the reader to solve, and, of course, the reader is to solve the problem "in his head." This is an exhortation to readers to understand the mathematics, and other things, for themselves.

Understanding things for oneself, particularly in the context of mathematics, is an issue that Eliot continues to pursue in her second novel, *The Mill on the Floss* (1860). Tom Tulliver has an intuitive understanding of mathematical ideas, similar to the understanding possessed by Adam Bede:

> Tom could predict with accuracy what number of horses were cantering behind him, he could throw a stone right into the centre of a given ripple, he could

102 D. BALL

guess to a fraction how many lengths of his stick it would take to reach across the playground, and could draw almost perfect squares on his slate without any measurement. ([1859] 1998, p. 139; Jenkins 2008, p. 81)

Tom Tulliver's tutor, Mr. Stelling, completely ignores Tom's skills when teaching him Euclid. When Tom's sister Maggie visits Tom at his tutor's house, she picks up his copy of Euclid and suggests she would be able to understand it "if I'd learned what goes before, as you have." Tom puts Maggie right: "it's all the harder when you know what goes before: for then you've got to say what definition 3 is and what axiom V. is" ([1860] 1998, p. 147). Once again, Eliot displays her awareness of issues concerning mathematics education, at least in the 1820s when the book is set, and once again she echoes De Morgan:

> The propositions are also said by rote, for the convenience of those who find their memory in a better state than their reason.... But the prime feature of the system, though now somewhat obliterated, was the necessity of recollecting the numbers of all the propositions; for it could clearly be of no advantage to know, that three angles of a triangle are equal to two right angles, unless it was also known that this is the thirty-second of the first book. (De Morgan 1831, p. 269)

Eliot would doubtless agree with De Morgan that mathematics is a subject particularly suited to self-reliant learning, because "the data or assumptions… are few, understandable and known to the student from the beginning" (De Morgan 1831, p. 265).

Discussion of Euclid during Maggie's visit to Tom introduces another issue, repeatedly raised by Eliot in her novels: the issue of the education of girls. Maggie asks Tom's tutor whether she could learn Euclid if he taught her instead of Tom and receives the reply that girls "can pick up a little of everything …. They've a great deal of superficial cleverness …. They're quick and shallow" ([1860] 1998, p. 150). By this point in the book, Mr. Stelling has already been presented as a tutor who takes no account of his pupils but only of his own outdated theory of education. Once again Eliot is questioning social beliefs based on insufficient evidence or on no evidence at all.

Eliot's Use of Images and Ideas Based on Mathematics

Specific references to mathematics also occur in later works, but more prominent throughout all of Eliot's novels is the use of images deriving from mathematical and logical thinking. One example of Eliot's concern with apparent facts and the misleading sense of objectivity to which they might give rise, is her fascination with light and darkness and with the use of phenomena from optics. This fascination is apparent in *Adam Bede*. When we are introduced by the narrator to the Hall Farm, we are invited to become, in our imagination, licensed trespassers and to enter the front garden. The narrator then enjoins us: "Put your face to one of the glass panes in the right-hand window. What do you see?" ([1859] 1998, p. 72) Readers are instructed to press their noses against the window, because otherwise they would not see what is inside the room. Ordinary window glass is transparent but also reflects, and so, on a bright day, the viewer from outside will only see in by being so close to the window that the light from outside is prevented from being reflected into the viewer's eyes. Once readers are logically able to see into the room, we are enjoined by the narrator not to misinterpret what we see. The view is of an abandoned "parlour" and the true function of the building can only be ascertained by changing the point of view and focusing on the place from which life now radiates, which is no longer the parlour but "the kitchen and the farmyard" ([1859] 1998, p. 72).

The image based on window glass transmitting but also reflecting light occurs repeatedly in another three of Eliot's novels: *Romola* (1863), *Felix Holt* (1867), and *Middlemarch* (1871–1872). In *Romola* it is used to explore the connection between characters in different conceptual spaces. Tito has betrayed his stepfather Baldassarre by disowning him and effectively stealing his fortune. Plotting revenge, Baldassarre stalks Tito to a glamorous party to celebrate Plato's birthday.

> He paused among the trees, and looked in at the windows, which made brilliant pictures against the gloom. He could hear the laughter; he could see Tito.... But the men seated among the branching tapers and the flashing cups could know nothing of the pale fierce face that watched them from without. The light can be a curtain as well as the darkness. ([1863] 1998, p. 325)

Both literally and metaphorically, Baldassarre can see Tito, but Tito is unable to see Baldasarre, being dazzled by the occasion and by his growing fame. Tito's reflected status obscures Baldassarre. A similar instance of light acting as a curtain occurs in *Middlemarch*, when Sir James Chettam is informed by Mrs. Cadwallader that he can no longer propose to Dorothea Brooke, because she is engaged to the elderly Mr. Casaubon. Mrs. Cadwallader tells him: "Come, come, cheer up! you are well rid of Miss Brooke, a girl who would have been requiring you to see the stars by daylight" (1871–1872, p. 54). Stars still emit light during the day, but we cannot see them since sunlight is so much stronger. For Mrs. Cadwallader, Dorothea is a rather foolish dreamy idealist, yet, the reader, aware of Eliot's interest in astronomy, which is particularly apparent in *Middlemarch* as well as in her last novel *Daniel Deronda* (1876), might conclude that, while we cannot actually "see" the stars by daylight, we can know where they are. While Dorothea's idealism and reflection on the world might be absurd, the author hints at a wider vision than the narrow view displayed by Mrs. Cadwallader.

It is not only sun that can dazzle us. We can be wilfully blinded by big ideas. In Eliot's novel *Felix Holt* (1866), Felix rather unkindly criticises the book Esther is reading:

104   D. BALL

Look here! "Est-ce ma faut, si je trouve partout les bornes, si ce qui est fini n'a pour moi aucune valeur." [Is it my fault, if I find limits everywhere, if that which is finite has no value for me?] Yes it is your fault because you're an ass. Your dunce who can't do sums always has a taste for the infinite. Sir, do you know what a rhomboid is? Oh no, I don't value those things with limits. ([1866] 1998, p. 108)

Giving easy assent to universal ideas is no substitute for working on the detail of what a rhomboid is. While Eliot might not approve of Felix's lack of charity, she would broadly agree with his scorn for people who, as Felix tells Esther, have "taste" rather than "a sensibility to facts and ideas. If I understand a geometrical problem, it is because I have a sensibility to the way lines and figures relate to each other" ([1866] 1998, p. 107). Some time later Felix complains about people who have "no particular talent for the finite, but a belief that the infinite is the right thing" ([1866] 1998, p. 220). Eliot repeatedly uses characters in her novels to cut pretentious people down to size, and she satirises those who have grandiose ideas that distract their attention from grappling with real problems. The narrator of *Middlemarch* criticises Dorothea's clear belief that Mr Casaubon will make a wise and instructive husband: "Signs are small measurable things, but interpretations are illimitable" ([1871–1872] 1998, p. 23). These examples suggest that we should distrust ideas not tied to facts and quantities, and Eliot's novels are scathing about the drawing of extravagant conclusions from limited or no evidence. As Felix suggests, we must learn to do the "sums" instead.

By the time Eliot was writing *Daniel Deronda* in the 1870s, she was well acquainted with non-Euclidean geometry and the furore it was generating among mathematicians and philosophers in Britain. She had come to know William Kingdon Clifford, a polemical advocate for taking non-Euclidean geometry seriously, and presumably supported her partner George Lewes in his writing of *The Foundations of a Creed* (1875) with its appendix on non-Euclidean geometry (Haight 1954–1978, 5, p. 344). Eliot uses the doubts sown concerning the universality of the truth of Euclid's axioms in *Daniel Deronda* when explaining how Daniel cannot be sure whether Mordecai is "worth listening to":

Suppose [Mordecai] had introduced himself as one of the strictest reasoners: do they form a body of men free from false conclusions and illusory speculations? The driest argument has its hallucinations, too hastily concluding that its net will now at last be large enough to hold the universe. Men may dream in demonstrations, and cut out an illusory world in the shape of axioms, definitions, and propositions, with a final exclusion of fact signed Q.E.D. No formulas for thinking will save us mortals from mistake in our imperfect apprehension of the matter to be thought about. And since the unemotional intellect may carry us into a mathematical dreamland where nothing is but what is not, perhaps an emotional intellect may have absorbed into its passionate vision of possibilities some truth of what will be. ([1876] 1998, p. 438)

Eliot warns readers against assuming that methodical reasoning produces truth. Apparently objective facts and even mathematical reasoning can produce illusions, and she suggests that the apparently illogical mixture of emotion and reason might produce something more valuable.

The Influence of Mathematical Logic on the Structuring of the Novels

Perhaps the most significant but least obvious effect of mathematics on Eliot's novels is not the specific mentioning of mathematics, nor the use of mathematical metaphors and imagery, but the results of the logical and methodical mind of the novelist. These are most apparent in *Middlemarch* where Eliot employs logical devices to draw attention to similarities between apparently different events and apparently different characters, and in this way comments on human behaviour and human psychology. One such device, which Eliot uses for the first time in *Middlemarch*, is chapter epigraphs, with which she signals her intention to structure the novel in view of scientific concepts. For example, the epigraph for Chapter 31 is about resonance, the one for chapter 46 is based on the idea of a theorem and its converse, and two epigraphs are about the second law of thermodynamics and reflections in double mirrors. Epigraphs may be seen to function as a background structuring device, suggesting that the novel is structured, in part, through the use of concepts from science, mathematics, and logic.

Another device Eliot employs to draw attention to structural aspects of the novel is to offer the reader of *Middlemarch* two "parables." Parables are short stories with wide application, and the narrator of *Middlemarch* emphasises this transferability:

> [S]ince there never was a true story which could not be told in parables where you might put a monkey for a margrave, and vice versa – whatever has been or is to be narrated by me about low people, may be ennobled by being considered a parable; so that if any bad habits and ugly consequences are brought to view, the reader may have the relief of regarding them as not more than figuratively ungenteel, and may feel himself virtually in company with persons of some style. Thus while I tell the truth about loobies, my reader's imagination need not be entirely excluded from an occupation with lords; and the petty sums which any bankrupt of high standing would be sorry to retire upon, may be lifted to the level of high commercial transactions by the inexpensive addition of proportional ciphers. ([1871–1872] 1998, p. 320)

This parable is presented with Eliot's characteristic irony. For example, it is clear that when similarities are seen between an apparently respectable and an apparently disreputable person, the effect is as likely to bring down the one, as to raise up the other. It is a parable about how people can be grouped together in certain ways or for certain purposes, a parable that proposes, and

106   D. BALL

simultaneously challenges, the notion of classification. It can be interpreted in connection with an earlier epigraph introducing Chapter 13 of *Middlemarch*:

1st Gent.	How class your man?—as better than the most,
	Or, seeming better, worse beneath that cloak?
	As saint or knave, pilgrim or hypocrite?
2nd Gent.	Nay tell me how you class your wealth of books,
	The drifted relics of all time. As well
	Sort them at once by size and livery:
	Vellum, tall copies and the common calf
	Will hardly cover more diversity
	Than all your labels cunningly devised
	To class your unread authors. ([1871–1872] 1998, p. 115)

This epigraph both proposes and challenges the notion of allocating people to classes, in order to make judgements about them. A comparable idea emerges in nineteenth-century logic, as Andrea Henderson describes in an article about Lewis Carroll, author of *Alice in Wonderland*, and his use of logic (Henderson 2014). Henderson sets out how how, in the first half of the nineteenth century, Augustus De Morgan and George Boole worked on reconceiving logic, detaching it from the content of what is being discussed and focusing on rules of deduction (Henderson 2014, p. 83). In 1847, Boole published a book in which he developed an algebra of logic based on the notion of classes of objects; classes which might be combined and which might intersect (Grattan-Guinness 2011, pp. 361–62). Henderson emphasises and perhaps celebrates De Morgan and Boole's detaching form from meaning. Quoting Boole, she explains that the interpretation of elements of language "is purely conventional: we are permitted to employ them in whatever way we please" (Boole 1854, p. 4). This is best illustrated in what we are told to regard as classes in Boole's mathematical logic:

> Let us employ the symbol 1, or unity, to represent the Universe, and let us understand it as comprehending every conceivable class of objects whether actually existing or not, it being premised that the same individual may be found in more than one class, inasmuch as it may possess more than one quality in common with other individuals. Let us employ the letters X, Y, Z, to represent the individual members of classes, X applying to every member of one class, as members of that particular class, and Y to every member of another class as members of such class, and so on, according to the received language of treatises on Logic. (Boole 1847, p. 14)

Henderson summarises: "The goal of classification is not the discovery of the essence of a thing but its distinction from other similar things" (2014, p. 84). Eliot's novels, and particularly *Middlemarch*, also demonstrate this logic: the reader is not to make assumptions about the essence of a character on the basis of social class, age, occupation, or especially gender; rather, characters are to be classified by the differences in the ways they behave in relation to other characters. In this way, Eliot's novels exhibit shifts in perspective also visible in the work of De Morgan and Boole.

In a foretaste of her parable about the monkey and the margrave, Eliot gives us the following episode in *Felix Holt*, describing the ridiculing of opponents voting in the Great Reform Bill election:

> [T]he bodily blemishes of an opponent were a legitimate ground for ridicule; but if the voter frustrated wit by being handsome, he was groaned at and satirised according to a formula, in which the adjective was Tory, Whig or Radical, as the case might be, and the substantive a blank to be filled up after the taste of the speaker. ([1866] 1998, pp. 253–54)

In order to demonstrate that such insults are equally fatuous, irrespective of the party allegiance of the ridiculers, Eliot offers the reader an algebraic formulation. The insult the narrator suggests can be written as, "You X Y!," where Y can stand for any offensive word—"hypocrite," for example—and X can be "Tory," "Whig," or "Radical," according to requirement. The parable of the monkey and the margrave invites readers to regard different characters as the same for some purposes, since they have some characteristics in common. Just as it makes no difference which party a voter or his detractors support if the detractors' behaviour is equally unpleasant, neither do we need to distinguish between a lord and a looby when making an assessment of them, if they behave in the same way.

The implications of this parable pervade *Middlemarch*. The most immediate application of the parable is to the characters Raffles and Mr Bulstrode, who can be considered members of the same class of "men to be despised," since they may have different social positions yet are both morally bankrupt as a result of depriving Will Ladislaw's grandmother of her fortune. Some applications are more frivolous. Doctors in Middlemarch may all be said to belong to the same Boolean class by each being in a class of their own: "everybody's family doctor was remarkably clever... [t]he evidence of his cleverness... lying in his lady-patients'" immovable conviction, and was unassailable by any objection except that their intuitions were opposed by others equally strong" ([1871–1872] 1998, p. 133). Eliot clearly enjoys the logical absurdity here. The narrator of *Middlemarch* establishes an association between class and arithmetic when the mourners at Peter Featherstone's funeral are compared with "the animals enter[ing] the Ark in pairs"; all, having "used their arithmetic," deploring their multitude as there is so little "fodder" to go round ([1871–1872] 1998, pp. 310, 311). The narrator's insinuation is that in this Boolean class, as in any other, any one member might be replaced by any other, which proves true, since none of the mourners will inherit but the estate goes to the "frog-faced" stranger, Joshua Rigg ([1871–1872] 1998, p. 311). The mourners are observed by the squirearchy, who themselves form a class of people that is ridiculed by the narrator for their sense of superiority, precisely because they describe the funeral attendees as classless: "monsters – farmers without landlords – one can't tell how to class them," Mrs. Cadwallader characterises the mourners ([1871–1872] 1998, p. 306). This ironic reference to social class points back to the epigraph quoted above concerning men being no easier to class than books and shows that the novel calls into question the relevance of social class in assessing the value of different characters, whatever Mrs. Cadwallader may wish were the case.

In the previous examples Eliot signals the commonality of members of a class more or less explicitly. She does so less obviously yet makes a more serious moral point when relating the events involving the Methodist preacher, Flavell, and the son of the poor and disreputable tenant farmer, Dagley. After Dagley's boy is caught stealing a hare, and Mr. Brooke, who is a magistrate, has had him locked in a barn, Mr. Brooke goes to see Dagley with the clear idea that Dagley should beat his son. On the way Mr. Brooke tells Dorothea about Flavell, who *also* stole a hare and came up before Mr. Brooke when he was on the bench, Mr. Brooke letting him off because he gave a witty speech in his defence. Mr. Brooke does not notice the connection between the two poaching incidents, but, having finished recounting this story about Flavell, says apparently without any irony: "But here we are at Dagley's" ([1871–1872] 1998, p. 369). The reader is invited to draw the conclusion that Mr Brooke, influenced by the relative status of the people involved, seems unable to see, namely that Flavell and Dagley's son belong to the same class of miscreants who have committed an identical crime and should therefore receive the same punishment.

In *Middlemarch*, Eliot repeatedly warns the reader against making hasty judgements about characters, by placing the reader in the same class as those the reader might be inclined to stigmatise. When describing the mean and vindictive Peter Featherstone's lack of insight when he leaves his property Stone Court to the unattractive Joshua Rigg and fondly imagines that Rigg would be delighted to live there, the narrator identifies the reader with Featherstone's ignorance: "how little we know what would make a paradise for our neighbours!" ([1871–1872] 1998, p. 488). And describing the hypocrisy underlying Bulstrode's moral bankruptcy that eventually comes to light, the narrator opines: "If this be hypocrisy, it is a process which shows itself occasionally in us all" ([1871–1872] 1998, p. 581). The narrator here invites readers not to assume that their status makes them more acceptable than characters whose foibles they share.

The identification of the reader with characters in the novel is developed more starkly in connection with the second parable. A further instance of drawing on optics, the parable has a similar theme as the first one discussed above, but it takes things further when it addresses not only similarity but also difference:

> An eminent philosopher among my friends, who can dignify even your ugly furniture by lifting it into the serene light of science, has shown me this pregnant little fact. Your pier-glass or extensive surface of polished steel made to

be rubbed by a housemaid, will be minutely and multitudinously scratched in all directions; but place now against it a lighted candle as a centre of illumination, and lo! the scratches will seem to arrange themselves in a fine series of concentric circles round that little sun. It is demonstrable that the scratches are going everywhere impartially, and it is only your candle which produces the flattering illusion of a concentric arrangement, its light falling with an exclusive optical selection. These things are a parable. The scratches are events and the candle is the egoism of any person now absent – of Miss Vincy, for example. ([1871–1872] 1998, p. 248)

The parable is based on an optical illusion. If a point light source, a candle, is held up to a pier glass, the strength of illumination of scratches on the glass depends on the direction of the scratches; those perpendicular to the rays emanating from the candle are most strongly illuminated, which produces the illusion of concentric circles. This parable occurs in the novel when the narrator is about to discuss the selfishness of Rosamond Vincy, who is resolved to marry Dr. Lydgate. Rosamond sees the world from her own point of view and arranges events in her mind to produce a "flattering illusion" of her own virtue. As her situation shows, the circles seen on the pier glass are both real and unreal: the scratches, or events, that make up the circles are real enough, but the selection process of Rosamond's perception distorts these into a particular, one-sided view.

However, the pier-glass metaphor is not only about Rosamond's egoism, but it is clear that everybody holding a candle up to the pier glass will see the circles. The message that all of us have a distorted view of the world as a result of our own egocentricity, reverberates throughout the novel (for discussions of the pier-glass metaphor in secondary literature see, for example, Miller 1975, p. 138; Brody 1987, p. 45; Payton 1991, p. 174). So we are all the same, but we are also all different, each of us seeing the world in our own way. The pier-glass parable is a warning against stigmatising others, and it is also a warning against thinking we understand others—and more generally, thinking we understand situations, when all we are doing is seeing them from our own point of view. As the first parable then, the second parable draws on a mathematical or scientific concept to show the dangers of making assumptions about the essence of characters or reality.

Conclusion

In a recent essay, Daniel Wright describes what he calls "George Eliot's vagueness" and explores and celebrates her use of "blurred concepts." He suggests that "we read Eliot not just as a novelist of ethical clarity or "sharpness" but also as an artist who registers the sheer difficulty of the kind of self-understanding required to give our.... lives a meaningful shape" (Wright 2014, p. 640). Wright contrasts the vagueness of ordinary language with the "artificial language of logic" which is "a tool customised for mathematical work," reporting that George Boole and John Venn "insisted that logic should be a branch of mathematics" (Wright 2014, pp. 632–33). Wright thus suggests that the mathematical mind is inimical to the necessary ethical vagueness that he sees expressed in Eliot's novels. Here, mathematics is cast as the villain because of its precision. And yet it is the very precision of mathematics that encourages acceptance of vagueness. Mathematicians understand the arbitrary status of axioms and definitions—"if we assume this then that follows," but we might just as well assume something else. As mathematicians are trained to identify fallacious logical arguments, they are more likely to challenge certainty when there are insufficient grounds for it. Exhibiting similar caution, Eliot repeatedly challenges certainties that lead to moral prescriptions or apparent truths, especially when these are based on scientific reasoning. During the time Eliot was writing, there was increasing realisation that mathematical and philosophical postulates were agreed starting points, rather than self-evident truths. According to Joan Richards, during the second half of the nineteenth century the "silence which initially greeted Lobachevskii's and Bólyai's works was replaced by a cacophony of diverse voices," engaging in "philosophical, psychological and educational discussions" about the implications of non-Euclidean geometry for understanding the world (Richards 1988, p. 6). In Eliot's view, as her novels show, no matter how strong the consensus in their favour, apparently self-evident starting points could be challenged, and she uses logic and mathematics to subvert social dogmas. Not unlike the way in which Galileo used mathematics to advocate a new structure for the solar system, Eliot employed mathematics to promote her own structure for the social and moral universe, one in which certainty is replaced by questioning.

References

Acton, Lord [Sir John Emerich Edward Dalberg]. 1885. "George Eliot's "Life" 1". *Nineteenth Century* 17: 464–85.

Bonaparte, Felicia. 1996. "Introduction" to *Middlemarch*. Oxford: Oxford University Press, vii–xxxviii.

Boole, George. 1847. *The Mathematical Analysis of Logic*. London: Bell.

———. 1854. *An Investigation of the Laws of Thought*. London: Macmillan.

Brody, Selma. 1987. "Physics in *Middlemarch*: Gas Molecules and Ethereal Atoms." *Modern Philology* 85 (1): 42–53.

Cross, J. W. 1885. *George Eliot's Life as Related in her Letters and Journals*. 3 vols. Edinburgh and London: Blackwood.

Eliot, George. (1859) 1998. *Adam Bede*. Oxford: Oxford University Press.

———. (1860) 1998. *The Mill on the Floss*. Oxford: Oxford University Press.

———. (1863) 1998. *Romola*. Oxford: Oxford University Press.

———. (1866) 1998. *Felix Holt, the Radical*. Oxford: Oxford University Press.

———. (1871–1872) 1998. *Middlemarch*. Oxford: Oxford University Press.

———. (1876) 1998. *Daniel Deronda*. Oxford: Oxford University Press.

[De Morgan, Augustus]. 1831. "On Mathematical Instruction". *Quarterly Journal of Education* 1: 264–69.

Grattan-Guinness, Ivor. 2011. "Victorian Logic: From Whately to Russell." In *Mathematics in Victorian Britain*, edited by Raymond Flood, Adrian Rice and Robin Wilson, 359–76. Oxford: Oxford University Press.

Haight, Gordon S., ed. 1954–1978. *The George Eliot Letters*. 9 vols. London: Oxford University Press.

Henderson, Andrea. 2014. "Symbolic Logic and the Logic of Symbolism". *Critical Inquiry* 41 (1): 78–101.

Jenkins. 2008. "George Eliot, Geometry and Gender." In *Literature and Science*, edited by Sharon Ruston, 72–90. Cambridge: Brewer; The English Association.

[Mill, J. S.]. 1852. "Whewell's Moral Philosophy". *Westminster Review*, 2 n.s. 349–85.

Miller, J. Hillis. 1975. "Optic and Semiotic in Middlemarch." In *The Worlds of Victorian Fiction*, edited by Jerome H. Buckley, 125–45. Cambridge MA and London: Harvard University Press.

Newman, Francis William. 1841. *The Difficulties of Elementary Geometry: Especially Those which Concern the Straight Line, the Plane, and the Theory of Parallels.* London: Ball.

Paxton, Nancy. 1991. *George Eliot and Herbert Spencer: Feminism, Evolutionism and the Reconstruction of Gender*. Princeton: Princeton University Press.

Pratt, John Clark and Victor A. Neufeldt. 1979. *George Eliot's Middlemarch Notebooks.* Berkeley and London: University of California Press.

Price, Michael H. 1994. *Mathematics for the Multitude?*. Leicester: The Mathematical Association.

Rice, Adrian 2011. "Introduction" to *Mathematics in Victorian Britain*, edited by Raymond Flood, Adrian Rice and Robin Wilson, 1–15. Oxford: Oxford University Press.

Richards, Joan L. 1988. *Mathematical Visions: The Pursuit of Geometry in Victorian England*. Boston: Academic Press.

Wright, Daniel. 2014. "George Eliot's Vagueness". *Victorian Studies* 56 (4): 625–648.

CHAPTER 7

Mathematics in Russian Avant-Garde Literature

Anke Niederbudde

This chapter discusses the importance of mathematics in the historical Russian avant-garde movement using examples taken from different authors. I will limit myself to three writers: Velimir Khlebnikov (1885–1922), Evgenij Zamyatin (1884–1937), and Daniil Kharms (1905–1942). Each author differs in their specific approach to incorporating mathematical themes into their literary work, as I will show with a focus on numbers and geometry. To situate the analyses, I will start by introducing the general character of the authors and their works.

Velimir Khlebnikov was one of the avant-garde artists whose innovative poetic language caused an uproar in prerevolutionary Russia. In 1913, he was involved in the publication of manifestos stating poets' rights to autonomous creation. One of the manifestos' main demands was to free the most basic unit of language—the word, or rather the phoneme/letter—from its practical use in everyday speech, creating a new, transrational language (*zaum'*) (see the manifestos "The Word as Such" and "The Letter as Such" [Lawton 1988, pp. 57–64]). Khlebnikov's ambitions did not concern poetry alone; he was also interested in science, making it particularly interesting to consider his work in relation to mathematical ideas. Before becoming one of the main voices of the Russian avant-garde, he had studied mathematics at Kazan University (where Nicholai Ivanovich Lobachevsky had been active in the nineteenth century). Although Khlebnikov did not complete his studies, in his social circle he was known as a poet who had a particular interest in numbers and geometry. One of the peculiarities of his body of work is its inclusion, in addition to literary texts in the stricter sense (poetry, drama, prose, supersagas), of comprehensive studies of time and history and of

---

A. Niederbudde (✉)
LMU Munich, Munich, Germany

© The Author(s) 2021
R. Tubbs et al. (eds.), *The Palgrave Handbook of Literature and Mathematics*, https://doi.org/10.1007/978-3-030-55478-1_7

113

114   A. NIEDERBUDDE

quasi-academic works exploring linguistic projects (creating a universal language, what he called the "language of the stars"; see Vroon 1983, pp. 168–81; Gretchko 1999). Not only did these research projects shape the poet's perception of his surroundings, they also left enduring traces in his poetry. Eventually, in *Zangezi* (1922), one of his major works, Khlebnikov created a prophet character to give voice to the results of his own research projects.

The second author addressed here, Evgenij Zamyatin, was a prose writers of the Russian avant-garde. During the 1920s—the second heyday of the Russian avant-garde following the October Revolution—he joined a group of writers in Petrograd (known as the "Serapion Brothers") who sought to use innovative techniques to create narrative forms. His best-known work, and the one most interesting to consider in terms of mathematics, is the dystopian novel *We* (*My*), written in 1920. In *We* Zamyatin envisions a fictional futuristic state where life is organized according to exacting scientific criteria. Not only does the novel anticipate the emerging totalitarian power structures in the Soviet Union, it also functions as a satire and parody of modern culture's faith in science.

The late 1920s and 1930s are considered the later phase of the Russian avant-garde, a period when its activities were increasingly marginalized by the official doctrine of art and literature known as Socialist realism. During this time, a group of poets coalesced that became known as OBERIU; today, its members are characterized as "authors of the absurd" (see Roberts 1997). The work of Daniil Kharms, the best-known representative of this group, has also been incorporated into this study since there are two aspects in which mathematical themes and motifs appear in his writings. Firstly, Kharms wrote a number of philosophical, or rather parascientific, texts in which he both presents and parodies an academic discourse. Secondly, the poetry, dramatic works, and prose of his absurdistic period are interspersed with geometric and numerical motifs as elements of his poetic world.

NUMBERS AND MATHEMATICAL DISCOURSE IN THE LITERATURE OF THE RUSSIAN AVANT-GARDE

In a letter to Khlebnikov in 1914, Roman Jakobson described numbers as a "double-edged sword," "extremely concrete and extremely abstract," but also "arbitrary and fatally precise, logical and absurd, limited and infinite" (Jakobson and Pomorska 1983, p. 172). These are the poles between which numbers move as a basic semiotic structure of culture and that make them interesting for writers of the avant-garde. It is no coincidence that Jakobson expressed these thoughts in a letter to Khlebnikov, for he knew of his interests in mathematics in general and in numbers in particular. But what are the reasons behind the use of numbers and calculations, or reference to mathematical theorems, in the time of the avant-garde and beyond? For example, what distinguishes the numerical references in the art creation of the avant-garde from a Pythagoreanism also found with symbolist artists? While symbolist circles took special interest in numbers as signs of a divine meta-physical world, the avant-garde was interested in numbers as the cipher of modernity, with mathematical aspects playing a special role. I will show that in the cases of the three authors discussed here, the specific relationship between numbers and mathematics is particularly important.

THE PROPHECY OF NUMBERS AND THE POETICS OF THE CALCULATION EQUATION (VELIMIR KHLEBNIKOV)

In the poem "Numbers" (1912), Khlebnikov addresses numbers directly: "I see right through you, Numbers .... You offer us a gift: unity between the snaky movement of the backbone of the universe and Libra dancing over-head" (1997, p. 39; see Lönnqvist 1979, pp. 34–39; Ivanov 1986, p. 395). For the inquisitive writer, numbers are not only a means of universal knowl-edge, which allow for connections between different manifestations of the universe, they also have the potential to facilitate the communication and exchange of ideas. The author's notes from 1915 include a suggestion to abolish words and replace them with a language of numbers: "All the ideas of Planet Earth ... should be designated by ... numbers" (1987, p. 358).

While numbers compete with verbal language in utopian projects, in his poetic work Khlebnikov undertakes a poetization of numbers and math-ematical forms of thought. The starting point for this is the collection of historical data (especially of wartime events) which Khlebnikov worked on throughout his life and on the basis of which he wanted to use mathemat-ics to establish the laws of history. In his early research, Khlebnikov starts from a world formula 365 ± 48: everything in time, therefore, runs in peri-ods of 365 (the number of days of the year), with $365 + 48$ (or equations like $Z = [365 + 48y]x$) marking the period of creation of states and 365 - 48 the period of decline (see Khlebnikov in "Teacher and Student" 1987, p. 281). In later works Khlebnikov considers the numbers 2 and 3 as con-stants of the world and establishes a connection with the "natural" constant 365 by performing computations: for example, he praises the beauty of the series $3^2 \, 2^3 \, (3+2) + 3 + 2 = 365$ and $3^5 + 3^4 + 3^3 + 3^2 + 3^1 + 3^0 + 1 = 365$ (2005, p. 90; 2008a, p. 93). A fundamental idea repeatedly found throughout both Khlebnikov's theoretical texts and his poetic works is that every major process that exists in time may be viewed as an oscillatory phenomenon. This idea was usually applied to acoustic and visual phenomena (for example sounds, radiation), but in Khlebnikov's thinking it could equally well be applied to historical events. Khlebnikov thus claimed that history may be mastered by studying the oscillatory structure of time. The manner in which he applied the methods and procedures of the natural sciences to history expose the utopian inclinations of his worldview.

116   A. NIEDERBUDDE

By 1920, Khlebnikov believed that the duality of 2 and 3 formed the foundation of the world of numbers as it expressed the binary opposition of even and odd numbers and carried semantic charge as, respectively, positive (2) and negative pole (3) (see Khlebnikov 2008b, p. 308). In addition, 2 stands for a two-dimensional surface while 3 is connected to three-dimensional space. However, Khlebnikov was not primarily interested in space but time. To him, time is not the fourth dimension of space (as it was for Albert Einstein), but is n-dimensional and an upside-down space: n^2 and n^3 (equations of space) become 3^n and 2^n (equations of time). Although Khlebnikov emphasized the applied aspects of his theory, he also allowed himself to become completely lost in his universe of power calculations. He mustered an incredible amount of enthusiasm for simple laws of numbers and arithmetic, believing that mathematics is guided by the principle of reduction, the law of "most frugal outlay on ink" (Khlebnikov 2008a, p. 127). Naturally, he was enchanted by any method that could be used to transform the numbers of an equation and reveal new interpretations and insights about the inner relations between them. In his texts, the central calculation method is exponentiation. Exponentiation is based on the multiplication of the same factors ($2 \times 2 \times 2 = 2^3$) and in this way can be seen as related to the reflexive of literature. Indeed, exponentiation had long been used as a poetological metaphor for poetry, not least by the poets of German romanticism (Niederbudde 2006, p. 81).

It is not easy to judge the significance of the time calculations within Khlebnikov's wider oeuvre. In particular, it remains controversial whether they should be viewed as part of his artistic or extra-artistic oeuvre (see Gricanchuk et al. 2008). Despite being artistic texts, they are essentially montages of different elements: factographic/historiographic data, as well as computations and the imagery that Khlebnikov developed from them (Hacker 2008). In the "Tables of Fate" (*Doski Sud'by*, 1920–1922), the centerpiece of Khlebnikov's theory of time, the poet presents his computational discoveries with prophetic pathos. Tables and spreadsheets span large sections of the text, displaying his passion for collection and research. The sheer scope of his unfinished project can also be recognized in the fact that the poet not only collected historical dates but also studied other parameters, such as oscillations and cosmological events. But, above all else, the text is filled with a multitude of artistic metaphors, inspired by the visual appearance of the computations, describing the specific interactions between variables and constants in Khlebnikov's calculus of time. The equations are compared to cascading waterfalls ("like a mountain waterfall, falling from a height" [2008b, p. 297]), to camels that carry sea salt in the humps of their exponents ("$3^9 \pm 3^6$ is the very … camel of two hills of threes that carried the cargo [of a sea festival] (salt) of naval glory" [2008b, p. 297]), and to animal or human figures whose skeletons are formed by the constants and whose muscles constructed from the variables (2008a, p. 97). Khlebnikov also frequently uses mythical imagery to describe the act of calculation, such as the metaphor of a heavenly ladder with numbers for rungs (2008c, p. 407), or the river of destiny whose banks represent the fundamental constant numbers of the equation (3 and 2) while the variable exponents stand for the freely flowing water (2008a, p. 123). His enthusiasm for the visual beauty of the equation is evident:

> This [exponentiation] equation is very pretty if it is written as chain of descending powers of three. Accordingly, the departing exponents nod their heads like feather-grass, like the tops of grasses and ripple like rye fields of numbers, like a kind of rye of threes. I will write down our law in chains of threes, so as to obtain a spectator's joy at the sight of these endless chains of stout numerical ears. (Khlebnikov 2008c, p. 381)

This way of integrating the world of numbers and calculations into his poetic world dominates Khlebnikov's late work. At this point, he had apparently dropped earlier utopian ideas, such as his plan to replace the word language with a language of numbers (1987, p. 358).

Numbers and Other Mathematical Symbols in *We*

Zamyatin's novel *We* uses mathematics to build the narrative of a seemingly perfect futuristic world. Each figure in this novel is identified by a letter-number combination, for example D-503, I-330, O-90, R-13, S-4711. The absence of verbal names corresponds to the ideal of collectivity propagated in the unitary state, but also highlights the exaggerated idea of order and control prevailing there that numerous other numerical images further emphasize. Thus, the characters' daily routine is dictated by statutory schedules, which, with the exception of two "personal hours," regulate their work routine in detail. Numbers also serve as a moral argument for the execution of deviants, since individual cases are negligible in the statistics of the collective (Zamyatin 1993, p. 79; see Cooke 1988, p. 154; Hersh 1993). The number-letter combinations not only fulfill the function of names in the novel, they also indicate a (ornamental) subtext: in the letter components of the names, the visual form of the graphemes pictographically reproduces the outer appearance of the persons (O is round, S is curved). However, it is also possible to read the letters as abbreviation marks of mathematical signs in the context of integral calculus. For example, D-503 associates the emergence of S-4711 (a state security officer) with the (double curved) integral sign (Zamyatin 1993, p. 35). On this basis it is possible to interpret D as differential, and the female figures O and I as 0 and 1 (numbers of the segment to be integrated). While the numerical component of characters' names thus cannot be deciphered unambiguously (for an attempt to decrypt the number code, see Lahusen et al. 1994, pp. 19–27), it can be considered as part of the topic of integral calculation, which the novel addresses at various levels. In the geometric imagery of the novel, the aim of the collectivistic state is to

118 A. NIEDERBUDDE

straighten the wild curves (Zamyatin 1993, p. 4). The scientific recording of curved lines can be accomplished with integral calculus, which in the novel enables the construction of the spaceship "Integral." The ultimate purpose of constructing this spaceship is to control (or, integrate in the figurative sense) the entire universe. On the plot level, but also according to the above interpretation of characters' names, the main problem in reaching this goal is the character I-330, the revolutionary leader of the anti-utopian counterworld. After meeting her, D-503 moves further and further away from the form of narration and life that he strove for at the beginning of the novel. As I detail in the next section, I-330 is a complex figure: the "I" in her name is not only the symbol of 1, but also denoting the imaginary number i—a central mathematical motif in the novel.

THE IMAGINARY NUMBER—DOUBLING THE I AND AN ADDITIONAL DIMENSION OF THE WORLD OF NUMBERS

Mathematician D-503 has an irrational fear of imaginary numbers, mathematically expressed as i or $\sqrt{-1}$. In the novel, the imaginary number is an image of breaking out of the limited world controlled by the totalitarian state. More broadly, in the artistic circles of the Russian avant-garde, the imaginary number works as a numerical emblem of freedom: of science, thought, or literature and art. For Khlebnikov, the imaginary number was a code that bore witness to the power of mathematics and poetry to craft symbolic worlds that are completely independent of the reference world. In his work, $\sqrt{-1}$ is not only the number that represents "freedom from all things," but also the numerical root of the tree of knowledge (2005, p. 24). The mathematical concept of the root of minus one—$\sqrt{-1}$—is likened to the organic roots of trees and the morphological roots of words. The arboreal imagery connects mythological concepts (for example of the tree of life, the tree of paradise) with the notion of a new type of knowledge, interweaving mythological and mathematical mindscapes into a single image.

The use of the imaginary number in Khlebnikov's literary work is diverse: in one of his prose texts $\sqrt{-1}$ represents a futuristic machine (that is, an object) aboard which one can climb; in another text the imaginary number is employed as the title of an exhibition that deals with questions about the origin of humanity (1989, pp. 82, 83). In these short pieces from the years 1915–1916, Khlebnikov uses imaginary numbers to sketch images of human curiosity or time and space travel. Other prose texts establish a connection of $\sqrt{-1}$ to the mythopoetic world. "The Scythian Headdress: A Mysterium" (1916) presents the narrator's travels and his encounter with the "Number-God," who appears as a reflection on the surface of a lake and declares that humans are reflections of numbers (1989, p. 94). The surface of the lake of numbers is a mythological realization of the complex plane, which combines real numbers and imaginary numbers, and the theme of material and immaterial aspects doubling each other reoccurs in the text when Ka, the Egyptian personification of the soul, is the narrator's doppelgänger (Ivanov 1986, pp. 404–5).

While for Khlebnikov, humans who extract the root of minus one from themselves can discover a mythological universe (1997, p. 180; Koll-Stobbe 1986), Zamyatin links the imaginary number with the unconscious of the I or the soul of the narrator D-503 (1993, p. 98). Key motifs in *We* are the mirror through which the I becomes double (1993, p. 59) but also self-reflection, which likewise causes a doubling of the I (Niederbudde 2011, p. 443). D-503's individual I emerges from his reflections, which increasingly stand in opposition to the collective (anti-individual) world of "*We.*" In the eyes of the totalitarian state, the soul ($=$I) is a disease that threatens the "we" of the collective. The imaginary number as a disturbing symbol in the narrator's mind emerges at the same time as I-330, showing that that I (character and imaginary number) is used as a double symbol in the novel.

NUMBER SEQUENCES AND ZERO IN THE WORK OF DANIIL KHARMS

Numbers can be found in Kharms's work in various, sometimes surprising places. Why do the characters in his dramas suddenly and for no apparent reason start counting or calculating (Kharms 1997a, pp. 204, 266)? Why does the crow in Kharms's short story have four legs, and the human in a poem fifteen arms? In these cases, the traditional assignment of a number (as a quantity) to an object is consciously undermined. The wrong number reveals that the number is independent from the object (Niederbudde 2006, pp. 405–11). Most notable, however, is the use of numbers for enumerating and counting objects, persons, or numbers—sometimes apparently based on verses or counting-out rhymes for children where numbers serve as rhyme or rhythm specification (Kharms 1997b, pp. 31, 60). In prose for adults, Kharms uses the enumeration of events as a method for creating a text that not only loses any kind of probability as events are repeated again and again, but actually consists entirely of enumerating ordinary and outrageous events. For example, in "Tumbling Old Women" (1939), one woman after the other falls out of a window (2009, p. 47), and the repetition calls into doubt the uniqueness of the outrageous event, which, it is hinted, could go on to infinity. Other stories operate in similar ways when a hero is repeatedly hit on the head with a brick or a character falls so often that he loses the last vestige of his appearance and his identity ("There once lived a man, and his name was Kuznetsov ..." (1935) 2017, p. 100; "The Carpenter Kushakov" (1935) 2017, p. 98). Kharms chooses an event that could emerge as an unusual, unique element in a realistic prose text, takes it out of context, and presents it as a sequence. The story thus becomes a serial repetition—an enumeration. The texts are also abruptly interrupted and break off without offering any resolution or conclusion (Hansen-Löve 2001, pp. 153–59). The question of the beginning and end of the sequence of numbers arises in this context as a question about the beginning and end of a narrative. That narrative rather than number sequence determines order is also the idea in "Numbers Cannot

120   A. NIEDERBUDDE

be Defined by their Sequence ..." (1932), where Kharms emphasizes the autonomy of numbers from both objects and arithmetic order (2017, p. 47), and in the prose text "Sonnett" (1939), in which the narrator forgets the correct order of the numbers (2009, p. 48).

A further topic with references to mathematics is the "absurdist" logic in Kharms's work. The opening address of his philosophical text "The Infinite: that is the Answer to All Questions" (1932) already announces one of his major tenets on the subject of infinity, namely the (totalitarian) claim of completeness ascribed to it. Infinity claims to contain everything or to be everything—but answering a question with "infinity" automatically leads us to question the nature of this infinity. Accordingly, Kharms's text only asks a single question: "What is infinity?" (2017, p. 48). The tautological relationship between the answer and the question (of infinity) does not discourage Kharms from expounding his thoughts—on the infiniteness of geometric shapes, the infiniteness of physical reality, and the infiniteness of the natural numbers. In Kharms's opinion, the sequence of natural numbers has to be extended to compensate for the one-sided infinity that only runs in the positive direction, and he therefore argues for the need to extend the number system with negative numbers. According to his view, zero was introduced into the world of numbers as a symbol of nothingness that links positive and negative numbers and brings the sequence into an artificial equilibrium (2017, p. 50). It becomes clear that, as far as Kharms is concerned, 0 represents a brilliant mistake in the number system: zero denies and at the same time establishes symmetry in the realm of integers.

Zero is also central to Kharms's writings on what he calls *cisfinity*—a concept that resides at the heart of the intersection of mathematics and literature in his poetic world (see Jakovljevic 2009, pp. 131–43). For Kharms, cisfinity is not just a realm "on this side of" finiteness, but is a realm that commands its own logic (*cisfinite logic*) and as such is by its nature the realm of art, both entangled with and isolated from mathematics. It seems likely that the concept of cisfinity was inspired by Georg Cantor's transfinite numbers. Kharms was fascinated by the creation of new numbers—the extension of the number system—and by the new and unusual rules in the realm of transfinite numbers. In "Cisfinitum: The Falling of a Stem," Kharms devotes six paragraphs to constantly shifting back and forth between abstract pseudo-scientific discussion and concrete illustrative systems, that is, between reflections on infinite sequences and their substitutions (such as the substitution of the sequence E, E', E"... by P_1, P_2, P_3..., which Kharms simply refers to as "postulates") and a stem which grows and then falls. The stem can be viewed either as an image representing a falling mathematical sequence or as the concrete stem of a tree. Finally, the artist-mathematician Kharms proposes his theory of the cisfinite realm as a new approach or "stem," founded on the concept of zero (1997a, p. 331).

Geometric Discourse in the Russian Avant-Garde

In addition to numbers, all three authors discussed here employ geometric themes in their works. In the early twentieth century, geometric figures and motifs were closely tied to questions of dimensionality and (physiological) perceptions of space, topics which were of particular interest to visual artists. The pronounced early-twentieth-century interest in geometry is also deeply bound up with the rise of non-Euclidean geometry and new conceptions of physical space associated with Einstein's theory of relativity. While Cubist painters were drawn to the new geometry as an inspiration for justification for deforming objects, Kazimir Malevich—one of the most important proponents of the artistic avant-garde in Russia—presented non-objective images entitled "in the fourth dimension" in 1913 and 1914 (see Henderson 2013, p. 422). The direct interaction between Euclidean shapes and newer geometries is specific to the avant-garde. This applies not only to visual artists like Malevich but also to poets, writers, and particularly the authors who are the focus of this chapter.

Poetic Linguistic Experiments and Geometry (Velimir Khlebnikov)

In Khlebnikov's poetry, Lobachevsky represents a new way of thinking about and experimenting with language and space. Khlebnikov developed his most fundamental connection between language and space in the consonants of his universal "language of the stars," which he describes in his super-saga *Zangezi* as an "algebra of words" (1989, p. 344). Building on the assumption that phonemes contain elementary units of meaning, he crafted his universal language as a kind of spatial semantics in which every consonant corresponds to an abstract image of movement in space. In this way, Khlebnikov relocates images of space from points and lines to the semantics of words. The battle of letters in *Zangezi* can be interpreted as movement in space: the struggle between R and G appears as the vertical and horizontal movement injected into the semantics of the phonemes (see Vroon 1983, pp. 171–73), as G signifies the movement of a point at right angles and R represents a point intersecting a perpendicular surface. In this sense, "Alphabet is the echo of space" (1989, p. 338). One of the most interesting phonemes in Khlebnikov's interpretation of space is L, evoking Lobachevsky.

In the text "Battle for Priority," the poet claims the ultimate goal of Lobachevsky's geometry to be the desire to create an "other, non-existent real world" (2005, p. 292). Similarly, in the utopian and visionary world of the poem "Lightland," Lobachevskian space is seen as a hope for the future: "Let Lobachevskian space stream from the flagpoles of night-loving Petrograd. ... Let Lobachevsky's curves descend as ornaments over all the city" (1997, p. 169). Equally important was the geometer as a figure

122   A. NIEDERBUDDE

of identification for the poet's persona as Khlebnikov associated Russian avant-garde poets' creation of a new language with the Russian mathematician: while other nations could merely emulate Euclidean geometry, Russians were capable of creating a "new" language comparable to Lobachevsky's geometry (2005, p. 25). The poet saw himself as an alter ego of the innovative mathematician, rising up against the poetic tradition by creating a new linguistic code.

Spaces and Geometric Images in Zamyatin's We

The characters in Zamyatin's novel *We* live in a world of geometric forms. All aspects of urban space are structured geometrically, in accordance with the tradition of utopian city-states: streets in straight lines, round public squares, and the meeting place is called "Cube Square," on which the face of the benefactor appears as a square (Zamyatin 1993, p. 46). *We* repeatedly emphasize the ideal of the quadratic. The "quadratic" harmony of the futuristic world that the narrator D-503 observes at the beginning of the novel refers not only to the geometric shape but also to a structural equilibrium associated with the number four, or rather the idea of quaternity. Contemporary points of reference for the novel include the blueprints for geometric urban architecture in post-revolutionary Russia. A preoccupation with geometric shapes in space was an integral component of *Vkhutemas*, the Russian "Higher Art and Technical Studios" founded in 1920. The workshops, guided by a Constructivist agenda, sought to create rational, functional spatial design (Vöhringer 2007, pp. 35–75). Theatrical sets influenced by geometry and designed by contemporary Constructivists also encompassed the human body, for example in costume design and sketches. Contemporary critics of *We* justly considered the geometric ideal of the totalitarian, centralized state in the novel to be a satirical exaggeration of how Constructivists and their predecessors understood and produced art (Klosty Beaujour 1988, pp. 51–53). And it is not only urban space that is designed geometrically in *We*. The narrator measures the world around him in terms of the mathematical world familiar to him, with geometric figures playing a significant role. This influences both the characters and their interpersonal relationships as well as their external appearances: body parts—which are perceived as abstracted or "fragmented"—are reminiscent of circles, spheres, triangles, or squares. The narrator describes relationships between the characters as triangular (see Barker 1977) and laments them to be unstable. Indeed, in *We*, the world of Euclidean geometric shapes does not form a harmonious unit; in particular, the triangle (which, as a leitmotif, is associated with the female rebel I-330) and the circle (associated with O-90) reveal themselves to be threats to the quadratically ordered world. In this sense, flocks of birds flying in "Vs" (as triangles) portend the calamity about to befall the totalitarian state (Zamyatin 1993, p. 115), and the spiral shape that depicts human history as progression likewise indicates the possibility of imminent revolution.

Yet, the new geometries are also part of the novel and Zamyatin's vision of revolution and literature. The apartment of the poet R-13 is a place that the narrator explicitly associates with non-Euclidean geometry (Zamyatin 1993, p. 41), and in his essay "On Literature, Revolution and Entropy" (1923), Zamyatin describes Lobachevsky's development of non-Euclidean geometry as a revolution par excellence (1970, p. 107; on dimensionality in Zamyatin see Clarke 2001, pp. 202–7). In other pieces, he writes of "four-dimensional" spaces that arise in the literature (1970, p. 244) and afford it revolutionary potential. *We* contrasts the totalitarian state with the alternative space of the Ancient House in which an alternate world (an additional dimension) is revealed to the narrator. The central mathematical motif behind this additional dimension is not non-Euclidean geometry however, but imaginary numbers whose very name suggests a relation to the imaginary possibilities of fictional literature.

GEOMETRIC SYMBOLS OF INFINITY IN THE WORK OF DANIIL KHARMS

In the 1920s, Kharms was in close contact with visual artists such as Malevich who attempted to transpose the new geometry into their work (Jaccard 1991, p. 81). A diary entry expresses Kharms's wish to be in life what Lobachevsky was to the field of geometry (2013, p. 496). This observation does not necessarily mean that Kharms had a deep engagement with geometry; the comment more immediately points to Khlebnikov, the "Lobachevsky of the word" (Tynjanov 2000, p. 219), who was an important role model for him (Jaccard 1991, pp. 32–39).

Kharms presented reflections on geometric shapes in pseudo-scientific texts, for example exploring the point and the line as symbols of infinity and particularly focusing on the circle to which he dedicated an entire text (1931). In eight paragraphs, he considers the circle as an infinity problem, while stating from the outset that he has no interest in the problem of the radius or what he calls "traditional" considerations regarding the circle. Instead, he focuses on the question of the (im)possibility of a perfect, complete figure. Kharms approaches the creation of the circle from a line that becomes "broken" at ever more points and thus forms into a curve that becomes increasingly smoother (1989, p. 85). A limited line thus becomes the infinity of the circle. It is precisely this inability to find an end to this process of becoming a circle that, for Kharms, is the essence of the complete or the perfect—to him, completion or perfection itself would be a flaw, in the sense that it is not possible to add anything to something that is complete. While in Kharms's pseudo-scientific texts, the circle is the dominant geometric figure and associated with the subjects of infinity and zero in a variety of ways (1997a, pp. 311–12), his poetic and narrative works feature the sphere—the circle's three-dimensional counterpart—as the preeminent

124   A. NIEDERBUDDE

geometric motif. In several texts relating to spheres, Kharms takes up ideas of mystics and turns to perception in his use of geometric figures, leaving the reality of everyday life behind in experiments with perception and deception (hallucination) (Niederbudde 2006, pp. 293–97; Jampol'skij 1998, pp. 196–223). Although there is no theoretical exploration of the sphere in his oeuvre, it is still an infinite figure as the sphere wanders from text to text like a being from another world.

The only instance in Kharms's work when a mathematician appears as a quasi-representative of the profession (the profession of geometric figures) is the well-known drama "The Mathematician and Andrei Semyonovich" in the collection *Cases* (2009, pp. 59-61). The mathematician removes a sphere from his head and describes the action, repeating the description four times. The then following dialog between the mathematician and Andrei Semyonovich both displays infantile characteristics and recalls a machine. The mathematician's half of the conversation is focused on negation:

ANDREI SEMYONOVICH: Put it [the sphere] back. Put it back. Put it back. Put it back. MATHEMATICAN: No, I won't put it back! No, I won't put it back! No, I won't put it back! No, I won't put it back! (2009, p. 59)

In the second half of the dialog, structure and tempo change, as Andrei Semyonovich is freed from the forced repetition of statements after he declares his surrender. The mathematician persists with a structure of negation and repetition, showing his inability to communicate:

ANDREI SEMYONOVICH: I'm sick and tired of squabbling with you. MATHEMATICAN: No, you're not sick and tired! No, you're not sick and tired! No, you're not sick and tired! (2009, p. 61)

The dialog does not close with a "result," but rather with a frustrated hand gesture and the departure of Andrei Semyonovich. By this point, the sphere that the mathematician removed from his head and that began the whole thing has been thoroughly forgotten. Indeed, the drama gives no information about the sphere: the reader might view it in light of other passages by Kharms that introduce the sphere as outer space, a ball, an object without gravity, and a symbol of infinity. In any case, the sphere as an abstract figure of infinity is presented in an entirely different light in this drama than in Kharms's philosophical texts. While, in the latter, the author conducts a long-winded parascientific discourse on the comprehensibility and incomprehensibility of ideal, infinite figures, in the drama, the sphere serves as a starting point to repetition from which the mathematician—its creator—has no reasonable means of escaping. Finally, I would like to draw attention to the connection between the geometric discourse, the numbers, and other mathematically related topics in the works of the three authors discussed here. For Khlebnikov, numbers serve as a universal pattern of order that is not only useful for the study of the laws of time, but also shape his writing and thinking. The poet himself acts as a prophet and researcher who believes in the possibilities of mathematical abstraction and in unification with the help of numbers and geometric images. A universal semantics serves as the basis of his poetry in which spatial structures emerge through the smallest units of language. Zamyatin's mathematical world in *We* ironically refers to utopian projects in the early Soviet Union, based on the great optimism for science. The numbers, geometrical and other mathematical motifs in the text serve as the ornamental basis of the novel. But on another level, *We* functions as a prose text centered around the life and thoughts of a mathematician, who uses mathematical images even as he fights against intrusions into the utopian world from an anti-utopian counterworld (feelings, fantasies, souls, etc.). Obviously, Zamyatin's use of mathematical figures creates not only a world oriented toward purely functional criteria, but also shows a mathematician questioning just this world.

Kharms combines geometric figures and number considerations in his paraphilosophical texts, while the play "The Mathematician and Andrei Semyonovich" illustrates in a different way how number pattern and geometric motifs connect in Kharms's poetic world. The repetitive structure of the dialog, the tautological sentences, the "empty" communication between the characters, and the mysterious geometric sphere are all ways in which the text points to a dimension beyond the everyday world. While mathematical objects and repetitive, self-similar patterns might indicate a mysterious dimension beyond for the reader, the mathematician in this absurdist text is trapped in an endless loop that resists any and all explanation.

REFERENCES

Barker, Murl G. 1977. "Onomastics and Zamiatin's *We.*" *Canadian-American Slavic Studies* XI, no. 4: 551–60.

Clarke, Bruce. 2001. *Energy Forms. Allegory and Science in the Era of Classical Thermodynamics.* Ann Arbor: University of Michigan Press.

Cooke, Leighton Brett. 1988. "Ancient and Modern Mathematics in Zamyatin's *We.*" In *Zamyatin's We. A Collection of Critical Essay,* edited by Gary Kern, 149–67. Ann Arbor: Ardis.

Gretchko, Valerij. 1999. *Die Zaum'-Sprache der Russischen Futuristen.* Bochum: Projekt.

Gricanchuk, Nikolaj, Nikita Sirotkin, and Vladimir Feshchenko, eds. 2008. *"Doski sud'by." Velimira Khlebnikova. Tekst i konteksty. Stat'i i materialy.* Moskva: Tri Kvadrata.

Hacker, Andrea. 2008. "Introduction to Velimir Chlebnikov's *Doski Sud'by* (I)." *Russian Literature* 63, no. 1: 5–55.

Hansen-Löve, Aage A. 2001. "Zur Poetik des Minimalismus in der russischen Dichtung des Absurden." In *Minimalismus: Zwischen Leere und Exzeß: Tagungsbeiträge des Internationalen Wissenschaftlichen Symposiums am Institut*

126   A. NIEDERBUDDE

für Slawistik der Humboldt-Universität zu Berlin vom 11. bis 13. November 1999, edited by Mirjam Goller, 133–86. Wien: Otto Sagner.

Henderson, Linda Dalrymple. 2013. *The Fourth Dimension and Non-Euclidean Geometry in Modern Art*. Revised edition. Cambridge, MA: MIT Press.

Hersh, Marion. 1993. "Zamyatin's 'We': A Mathematical Model Society." *Rusistika* 7: 19–26.

Ivanov, Vjacheslav V. 1986. "Khlebnikov i nauka." *Puti v neznaemoe: pisateli rasskazyvajut o nauke* 20: 382–440.

Jaccard, Jean-Philippe. 1991. *Daniil Harms et la fin de l'avant-garde russe*. Bern: Lang.

Jakobson, Roman, and Kystyna Pomorska. 1983. *Dialogues*. Cambridge, MA: MIT Press.

Jakovljevic, Branislav. 2009. *Daniil Kharms: Writing and the Event*. Evanston: Northwestern University Press.

Jampol'skij, Mikhail. 1998. *Bespamjatstvo kak istok (Chitaja Kharmsa)*. Moskva: Novoe literaturnoe obozrenie.

Kharms, Daniil. 1989. *The Plummeting Old Women*. Dublin: Lilliput Press.

———. 1997a. *Polnoe sobranie sochinenij. Tom 2. Proza i scenki. Dramaticheskie proizvedenija*. Sankt-Peterburg: Akademicheskij proekt.

———. 1997b. *Polnoe sobranie sochinenij. Tom 3. Proizvedenija dlja detei*. Sankt-Peterburg: Akademicheskij proekt.

———. 2009. *Today I Wrote Nothing. The Selected Writings of Daniil Kharms*. New York: Ardis.

———. 2013. *"I Am a Phenomenon Quite Out of the Ordinary." The Notebooks, Diaries, and Letters of Daniil Kharms*. Boston: Academic Studies Press.

———. 2017. *Russian Absurd. Selected Writings*. Evanston: Northwestern University Press.

Khlebnikov, Velimir. 1987. *Collected Works of Velimir Khlebnikov. Vol. 1: Letters and Theoretical Writings*. Cambridge, MA: Harvard University Press.

———. 1989. *Collected Works of Velimir Khlebnikov. Vol. 2: Prose, Plays, and Supersagas*. Cambridge, MA: Harvard University Press.

———. 1997. *Collected Works of Velimir Khlebnikov. Vol. 3: Selected Poems*. Cambridge, MA: Harvard University Press.

———. 2005. *Sobranie sochinenij v shesti tomach. Vol. 6,1: Stat'i (nabroski), uchenye trudy, vozzvanija, otkrytye pis'ma, vystuplenija: 1904 - 1922*. Moskva: IMLI RAN [et al.].

———. 2008a. "Doski Sud'by. Otryvok I. Zarej venchannyj / The Tables of Fate. Fragment I. Crowned by Down." *Russian Literature* 63, no. 1: 71–166.

———. 2008b. "Doski Sud'by. Otryvok II. Vzor na 1923-ij god / The Tables of Fate. Fragment II. A Glance at the year 1923." *Russian Literature* 64, nos. 2–3: 269–372.

———. 2008c. "Doski Sud'by. Otryvok III. Azbuka neba / The Tables of Fate. Fragment III. The Alphabet of the Sky." *Russian Literature* 64, nos. 2–3: 374–441.

Klosty Beaujour, Elisabeth. 1988. "Zamiatin's *We* and Architecture." *The Russian Review* 47: 49–60.

Koll-Stobbe, Amei. 1986. "Cognition and Construction: Chlebnikov's $\sqrt{-1}$ as a Metaphoric Process." In *Chlebnikov 1885–1985*, edited by Johanna Renata Döring-Smirnov, Walter Koschmal, and Peter Stobbe, 107–15. München: Otto Sagner.

Lahusen, Thomas, Elena Maksimova, and Edna Andrews. 1994. *O sintetizme, matematike i prochem.... Roman "My" E. I. Zamyatina. – On Sintetizm, Mathematics and Other Things... E.I. Zamyatin's Novel* We. Sankt Peterburg: Astra-Ljuks [et al.].

Lawton, Anna, ed. 1988. *Russian Futurism through its Manifestoes. 1912–1928*. Ithaca and London: Cornell University Press.

Lönnqvist, Barbara. 1979. *Xlebnikov and Carnival. An Analysis of the Poem Poèt*. Stockholm: Almqvist & Wiksell.

Niederbudde, Anke. 2006. *Mathematische Konzeptionen in der Russischen Moderne. Florenskij – Chlebnikov – Charms*. München: Otto Sagner.

———. 2011. "Mathematisches Erzählen im antiutopischen Text. Evgenij Zamyatins Roman Wir (*My*)." In *Zahlen, Zeichen und Figuren. Mathematische Inspirationen in Kunst und Literatur*, edited by Andrea Albrecht, Gesa von Essen, and Werner Frick, 437–52. Berlin: De Gruyter.

Roberts, Graham. 1997. *The Last Soviet Avant-garde: OBERIU—Fact, Fiction, Metafiction*. Cambridge: Cambridge University Press.

Tynjanov, Jurij N. 2000. "O Khlebnikove." In *Mir Velimira Khlebnikova. Stat'i, issledovanija 1911–1998*, edited by Vjacheslav V. Ivanov, Zinovij S. Papernyj, and Aleksandr E. Parnis, 214–23. Moskva: Jazyki russkoj kul'tury.

Vöhringer, Margarete. 2007. *Avantgarde und Psychotechnik. Wissenschaft, Kunst und Technik der Wahrnehmungsexperimente in der frühen Sowjetunion*. Göttingen: Wallstein.

Vroon, Ronald. 1983. *Velimir Xlebnikov's Shorter Poems: A Key to the Coinages*. Ann Arbor: University of Michigan.

Zamyatin, Yevgeny. (1970) 1993. *A Soviet Heretic: Essays by Yevgeny Zamyatin*. Chicago: University of Chicago Press.

———. 1993. *We*. Translated by Clarence Brown. New York: Penguin Books.

CHAPTER 8

Uses of Chaos Theory and Fractal Geometry in Fiction

Alex Kasman

There are many reasons why authors of fiction take the unusual step of including references to mathematics in their writing. For example, authors can utilize stereotypes of mathematicians to quickly communicate certain character traits to a reader (like genius or insanity) by indicating that the character is a professional mathematician. In many of those cases, the specific area of mathematics in which the character works is unimportant and may even go without mention. In contrast, the effectiveness of some literary references to mathematics relies on properties unique to specific topics within mathematics. This chapter surveys references to the areas of mathematics known as chaos theory and fractal geometry which depend on aspects of those particular sub-disciplines.

Loosely put, chaos theory studies the dynamics of mathematical objects that change with the passage of time in a way that makes accurate long-term predictions impossible because small variations at one time grow into much larger differences later. This property of chaotic systems is sometimes referred to as "the butterfly effect," in an allusion to the idea that the flap of a butterfly wing in one location could later result in a hurricane at another because of the chaotic nature of weather phenomena. However, the more technical name of this defining property of mathematical chaos is "sensitive dependence on initial conditions." Unlike the familiar objects from high school geometry (one-dimensional lines, two-dimensional surfaces, and three-dimensional objects like a cube), when measured in one of the standard ways, the numbers of dimensions occupied by the objects in fractal geometry

---

A. Kasman (✉)
Department of Mathematics, College of Charleston, Charleston, SC, USA

© The Author(s) 2021
R. Tubbs et al. (eds.), *The Palgrave Handbook of Literature and Mathematics*, https://doi.org/10.1007/978-3-030-55478-1_8

129

are not whole numbers. Etymologically, "fractal" can be thought of as being "fractionally dimensional," but the beautiful images of fractals that are produced by computers can be appreciated by people who do not understand that abstract definition.

Based on the preceding paragraph, it is not clear that the two subjects of chaos theory and fractal geometry are related, but they are nearly inseparable because chaotic dynamical systems are often used in the creation of fractals and because studying a chaotic system often involves analyzing associated fractal objects. The purpose of this survey is to develop an understanding of how these two mathematical ideas have functioned in literature.[1] This essay divides the key literary texts it discusses into four categories: (I) literary references to chaos theory and fractal geometry that take advantage of some non-mathematical properties of those areas of math; (II) instances in which the actual mathematical content of these theories justifies or motivates a plot development; (III) cases in which the events in the fiction are metaphorically related to (but not directly impacted by) the mathematical properties of chaos and fractals; and finally; (IV) works of fiction in which the abstract ideas of chaos theory and fractal geometry are manifested either in a fictional physical form or in the structure of the work itself.

Category I

The names that happen to have been chosen for mathematical objects and results are generally not considered to be important in themselves. The discipline known as "chaos theory" could have been called "sensitively dependent dynamical systems" after the property described above. Similarly, the mathematical objects known as "strange attractors," which are "attractors" in the sense that points in the phase space of a chaotic dynamical system are pulled into them and are also "strange" in the sense that the region of attraction happens to take the form of a fractal, could instead have been eponymously called "Ruelle-Takens domains" after the researchers who first studied them. Such name changes would not alter the mathematical properties of chaos theory or strange attractors, but if they had those other names then that *would* actually have impacted their use in some works of fiction. This is because in addition to their mathematical definitions, each of these phrases has an alternative, non-mathematical meaning that can be utilized by authors. Leonard Rosen's thriller *All Cry Chaos* (2011) begins with an explosion which (seemingly) kills a mathematician as he is about to speak at a meeting of the World Trade Organization. His mathematical research does indeed involve fractal geometry and chaotic dynamical systems, but the fact that "chaos" can also be used to describe the violent consequences of the murder and its disruption of society is not a coincidence. Surely, if this field of mathematics were known as "sensitively dependent dynamical systems," that phrase would not have been included in the title of this novel. Readers might learn a little about the mathematics behind the word "chaos" by reading this book, but the title and other references to chaos will mean something to them even if they do not.

Some authors make use of the term "chaos" in describing the work of fictional mathematicians without mentioning what the term really means in that context. In Bill Napier's *The Lure* (2002), mathematician Tom Petrie is said to work on "chaos theory," but as the book gets more specific about his work it becomes clear that Napier is using the term as if it were synonymous with pattern recognition. So, why does the author emphasize that he works at the "Chaos Institute"? The author is capitalizing on the readers' familiarity with the common English meaning of the word "chaos," which is separate from its technical mathematical definition. This may also explain why Michael Crichton's novel *Jurassic Park* (1990) features a mathematician who studies chaos theory. The park is presented as being safe and beautiful at the beginning, but having a concerned expert warning about chaos creates the tension expected in a thriller without relying on any knowledge of mathematics.

At the date of writing this essay, there are four works of mathematical fiction with the phrase "strange attractor" in their titles. Its use in titles would be mysterious if it was not apparent that authors simply like the sound of it. The graphic novel *Strange Attractors* (Soule and Scott 2013) is indeed about chaos theory in that the main character utilizes the sensitive dependence in everyday situations to the benefit of humanity. However, neither fractal geometry nor the concept of a strange attractor is directly relevant to the plot at any point. Similarly, in the young adult novel *Strange Attractors* (Sleator 1991), "strange attractors" are people whose charisma (i.e., attractiveness) causes bad things to happen. The reference to chaos theory in this novel is essentially a metaphor based on the common meanings of the words "chaos" and "attractor" that largely ignores the actual mathematical meanings.

The detective in the British novel *The Strange Attractor* by Desmond Cory (1991) explains his use of mathematics to solve the mystery in a brief paragraph utilizing terminology from logic, set theory, mathematical physics, and chaos theory, but the resulting combination is nonsense. The book really has nothing to do with chaos theory, and so the words "strange attractors" appearing in that explanation and the title are presumably only there because the phrase sounds so interesting. When the novel was released in the United States the title was changed to *The Catalyst*, which also does not seem to have anything to do with the content.

The mathematician in the collection of linked literary stories *Strange Attractors* by Rebecca Goldstein (1993) works on the geometry of bubbles. The words "strange" and "attractor" certainly resonate with the fiction in that many of the characters are both odd and romantically appealing. However, there is no obvious connection between bubble geometry (better known as the study of minimal surfaces) and chaos theory. So, it is something of a *non sequitur* when a mathematician with whom she is discussing her research mentions "strange attractors." That one instance is the only mention

132   A. KASMAN

of this phrase in the work and yet "Strange Attractors" was chosen as the title for both the story and for the entire collection. If the mathematical objects had been named "Ruelle-Takens domains," Goldstein would probably not have mentioned them at all. In that sense, the story's reference to an object from chaos has nothing to do with the mathematical content of the theories, only to its nomenclature.

A Recognizable Modern Area of Mathematics Research

In the late 1980s and early 1990s, chaos theory and fractal geometry were popular, receiving an unusually large amount of attention from the general public. From best-selling non-fictional books like James Gleick's *Chaos* (1987) and Benoit Mandelbrot's *The Fractal Geometry of Nature* (1982) to the images of fractals adorning calendars, mugs, and t-shirts, for a while in the 1990s they were a part of popular culture. Consequently, at least during the decades surrounding the turn of the twenty-first century, chaos and fractals could fulfill a unique role in works of fiction as reference to mathematics that was recognizably current and even fashionable. Given nearly any other description of an area of mathematics research, a general reader near the turn of the millennium would probably not be able to tell whether that area was new, whether it was first studied in the eighteenth century, or if it was something that even Archimedes would have known about. But reading that a fictional mathematician was working on chaos theory or fractal geometry would be a clear indication to the reader that this person was working at the cutting edge of mathematical knowledge. Consider the Euclidean geometer Professor Giacopo Tigor in Peter Stephan Jungk's novel *Tigor* (2003). Since Euclidean geometry is clearly an old subject, and one that is taught in school, merely knowing that this is his area of research already sets up our expectation that he will be stodgy and boring. In contrast, the character of Ian Malcolm in the film adaptation of *Jurassic Park* (Dir. Spielberg 1993) is portrayed by Jeff Goldblum as a leather pants-wearing chaos theory expert who confidently flirts with the paleobotanist Ellie Sattler while demonstrating sensitively dependent dynamics via the way water droplets roll off her hand.

Tom Stoppard's *Arcadia* (1993) makes great use of chaos theory's unique recognizability. This play takes place in a British mansion during two different time periods. In the modern period, Valentine Coverly is a mathematician who studies the chaotic dynamics of population sizes using the hunting logs that his family has kept for centuries. These portions of the play are interwoven with scenes of a young girl named Thomasina Coverly who is learning mathematics and Latin from a rakish tutor in the year 1809. As the modern mathematician conducts his research, he begins to find clues that someone who lived in the house long ago knew about both chaos theory and fractals. He eventually realizes that it was Thomasina herself, a teenage girl, who had single-handedly discovered those branches of mathematics. However, for the audience to share his amazement at her brilliance, they need to have some knowledge of the chronology. If she had rediscovered a known result like the Fundamental Theorem of Calculus on her own that would have been impressive, but not nearly as amazing. To appreciate this story fully, it is necessary for the audience to recognize that chaos theory and fractal geometry were unknown in Thomasina's time. In that way, Thomasina appears not simply clever, but at least a century ahead of her time. Stoppard capitalizes on the common knowledge that chaos theory and fractal geometry are twentieth-century inventions, and this aspect of the play would be lost it was rewritten to be about some other area of mathematics.

To appreciate the significance of Category I, it is necessary to be able to differentiate the actual mathematical content of a subject and other superficial features. For example, consider the result called the Pythagorean Theorem which was discovered by ancient Greek geometers. Its actual content (a relationship satisfied by the lengths of the sides of a right triangle) is true regardless of that choice of name or the era in human history in which it was first proved. From a purely mathematical point of view, it is the mathematical content and not the terminology or history that is important. By contrast, this section has illustrated ways in which authors of fiction make use of precisely those non-mathematical features of chaos theory and fractal geometry.

Category II

In the references to chaos theory and fractals to be addressed in this section, the actual mathematical content of these disciplines, i.e., the dynamical properties of chaotic systems or the geometric properties of fractals have a direct impact on the course of events in a work of fiction. In a sense, this is like applied mathematics. However, here we are not concerned with how realistic the applications are. In fact, many of them are entirely fantastical, but they are applications of these mathematical concepts nonetheless, even if only within the reality of the story.

Justification for Startling Predictions

In some forms of fiction, no justification is required for a character's ability to do amazing things. In a work of fantasy, a character could simply be said to have the ability to predict the future and nothing more needs to be said about it. However, in more realistic forms of fiction, even in science fiction, the author is usually expected to provide some sort of explanation to accompany an occurrence that would be normally be considered impossible. This subsection will consider works of fiction in which amazing predictions about the future are justified by references to chaos theory.

There is no doubt that mathematics can be useful in making some accurate predictions such as the date of an eclipse or the percentage of heads obtained in a large number of coin flips. But chaos theory does not allow us to make accurate long-term predictions about chaotic systems like the weather or the stock market. The theory certainly demonstrates that even very complicated-looking systems can be entirely deterministic. Moreover, knowing the limitations and dangers that chaos imposes on our computational predictions allows us to quantify their reliability. It is through these sorts of applications of chaos theory that we can have relatively accurate ten-day weather forecasts and make use of low energy transfer orbits to place unmanned space probes exactly where we want them. It is also possible to make broad and qualitative predictions about chaotic systems (not entirely unlike Ian Malcolm's warnings about the dangers of resurrecting dinosaurs in *Jurassic Park*). However, our inability to make measurements with absolute precision, combined with the fact that small variations lead to eventual big difference in chaotic dynamical systems, means that detailed, accurate, and long-term predictions in chaotic systems are essentially impossible.

Nevertheless, several otherwise realistic thrillers are based on the idea that mathematicians with expertise in chaos theory would be able to make spectacularly accurate long-term predictions about the world of finance. In the novels *All Cry Chaos* (Rosen 2011), *The Fractal Murders* (Cohen 2002), and *Kapitoil* (Wayne 2010), and the film *The Bank* (Dir. Connolly 2001), predictions are so good they are literally worth killing for. Each of these works of fiction presents some general information about chaos theory and fractal geometry and involves a character or group of characters with the ability to apply these concepts to achieve financial gains. The role of the mathematics in the story, then, is as a justification. The plot depends upon the idea that a small number of people have knowledge that would allow them to predict financial markets much more successfully than other people, and a superficial familiarity with these mathematical ideas is supposed to convince the reader that this makes sense.[2]

However, some innovation beyond the existing results in these fields would be necessary to achieve the sort of success presented in the above-mentioned fictional thrillers. Most of them choose not to address the question of precisely what the characters do that is mathematically new and different. Mark Cohen's *The Fractal Murders* stands out in that it attempts to explain the original idea that allows the mathematician in this story to succeed where others have failed. Her innovation is to compress time when little happens and expand it when there is a lot of activity. We are told that applying the fractal techniques with this one additional twist of scaling time appropriately allows one to predict exactly what will happen to the stock market. This is not exactly believable, but the author's attempt to provide a mathematical justification is sure to be appreciated by those readers for whom the mathematical ideas themselves are part of the appeal of mathematical fiction.

Even more outlandish chaos-based predictions of the future are made by Doctor Spencer Brownfield in the graphic novel *Strange Attractors* (Soule and Scott 2013). Brownfield uses his "powers" to save New Yorkers from disasters they will never know about, disasters that are avoided only because of Brownfield's intricately planned but seemingly bizarre behaviors. Mathematics student Heller Wilson contacts Brownfield, who once worked for the mathematics department where Heller is pursuing a Ph.D., in the hopes that Brownfield can help him finish his thesis. Instead, the older man recruits Heller to help pour paint at the entrance to Grand Central Station and to pile garbage up near Grant's Tomb. Brownfield claims he has determined that these things are mathematically necessary to keep the city functioning. As for how these miraculous predictions are made, all we are told is this:

> *Heller*: All that random crap you've been making me do... you think you're actually –
> *Brownfield*: I don't *think* anything! The results speak for themselves. New York City is a classic complex system, and in a complex system a small change in one part can cause a dramatic effect in another. If you know what to change you can make anything happen and I know, Mister Wilson. I speak the city's language...
> *Heller*: But the best people in the field have only been able to do things like that in controlled labs and computer simulation, and with much smaller systems. This is a city – it's an entirely different level.
> *Brownfield*: Did it ever occur to you, Mister Wilson, that I'm just much better at this than anyone else? (Soule and Scott 2013, pp. 47–48)

Readers are likely to share Heller's skepticism. As with the financial predictions mentioned earlier, one character's expertise in the mathematics of chaos theory is being called upon to justify the reader's suspension of disbelief.

Justification/Motivation from Fractal Geometry and Self-Similarity

Fractals are visually appealing. It is therefore no surprise that they appear in the artwork of graphic novels like Soule and Scott's *Strange Attractors*, on the cover of novels like Kat Wilhelm's *Death Qualified: A Mystery of Chaos* (1992), and very effectively in the opening credits of the Australian film *The Bank* (Dir. Connolly 2001). Geometric features of fractals and their self-similarity are also utilized in non-graphical forms of fiction.

One of the most famous fractals is the Mandelbrot set. Although the mathematics necessary to define it is elementary (involving little more than squaring and adding numbers), its appearance is anything but simple. As illustrated in the accompanying Fig. 8.1, magnifying the image at points along its boundary results in an image just like the whole Mandelbrot set, and magnifying a tiny piece of the boundary of that smaller image produces another smaller copy, and so on ad infinitum. This property, which is a hallmark of fractal geometry, is known as "self-similarity."

Just as the works of fiction described in the previous section exaggerate the role of chaos theory in allowing us to predict the future, some novels take

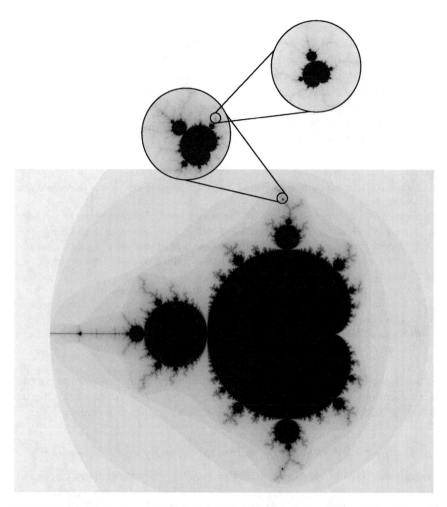

Fig. 8.1 Repeatedly "zooming in" on the Mandelbrot Set confirms that arbitrarily small pieces of it are similar in shape and complexity to the entire set. This "self-similarity" is a hallmark of fractal geometry (*Source* Produced by author)

the undeniable appeal of fractal images to a new level by showing characters whose psychology is changed by them. In *An Angel of Obedience* (Giessmann 2010), a thirteen-year-old music prodigy becomes so obsessed with fractal images that he stares at them for hours, leaves his music career, and enrolls as a mathematics major at Harvard. In Arthur C. Clarke's *Ghost from the Grand Banks* (1990), a mathematician literally loses her mind by staring deeply into images of the Mandelbrot set. And, in *Death Qualified: A Mystery of Chaos* (Wilhelm 1992), the traditional murder investigation that occupies most of the novel takes a turn into science fiction at the end when the motive is revealed to be the stunningly expanded mental abilities of experimental subjects who were shown fractal images. In each of these instances, the role of fractal geometry is again justification. The concept of fractals is introduced into these books to provide an explanation, even if not an entirely believable one, of a dramatic change in a character's mental state.

Not all of the direct uses of fractal geometry in fiction are fantastical; their role in the plot of *Arcadia* (Stoppard 1993) is entirely realistic. It has to do with the relationship between fractal geometry and the geometry of the natural world, which Mandelbrot himself described (Mandelbrot 1982). If we look back at the image of the Mandelbrot set we may notice that it looks more like a biological creature—perhaps an insect with a snout sticking out to the left and legs coming out around its thorax—than many of the geometric objects that are familiar from mathematics. Indeed, one of the uses of fractal geometry is in creating computer graphics that look believably natural. To create landscapes, clouds, and plants for computer animated films, graphic artists often make use of fractal geometry. And, conversely, fractal geometry can be useful in understanding natural objects such as the human circulatory system, rivers and streams, or plants, which also exhibit several levels of self-similarity. Note the branching pattern in the leaf detail shown in Fig. 8.2. The large vein branches into smaller ones that come off it at a characteristic angle and then the smaller veins branch off similarly forming at least three visible levels of self-similarity. By manipulating a mathematical model of a leaf, we can see that many of the different leaf shapes on plants in nature can be obtained by changing the frequency and angle of the branching of the veins. This provides us with a tool for understanding the geometry of leaves.

In Stoppard's play, the young Thomasina notices that the geometry she is learning about from her tutor, Septimus, does not appear useful for understanding natural objects like plants. She vows to do something about this:

Thomasina: Each week I plot your equations dot for dot, xs against ys in all manner of algebraical relation, and every week they draw themselves as commonplace geometry, as if the world of forms were nothing but arcs and angles. God's truth, Septimus, if there is an equation for a curve like a bell then there must be an equation for one like a bluebell, and if a bluebell, why not a rose? Do we believe nature is written in numbers?

Septimus: We do.

Thomasina: Then why do your equations only describe the shapes of manufacture?

Septimus: I do not know.

Thomasina: Armed thus, God could only make a cabinet.

Septimus: He has mastery of equations which lead into infinities where we cannot follow.

Thomasina: What a faint heart! We must work outward from the middle of the maze. We will start with something simple. I will plot this leaf and deduce its equation. (Stoppard 1993, p. 37)

Fig. 8.2 Plant leaves exhibit self-similarity in that the branching patterns of the smaller veins look like little copies of the branching of the larger veins (*Source* Produced by author)

We later learn that she followed through on her promise, producing what she called "The New Geometry of Irregular Forms Discovered by Thomasina Coverly," but which we would know as fractal geometry (Stoppard 1993, p. 43). And so, an actual property of fractal geometry—the fact that it is better-suited to biological objects like leaves than the classical geometry of Euclid—has a direct impact on the plot of *Arcadia*. It is the motivation for Thomasina to begin her own mathematical investigations and leads to her brilliant work that is rediscovered by other characters in the twentieth-century scenes.

Category III

In some cases, the role of a reference to chaos theory or fractal geometry is not directly to contribute to the plot of the work, but rather to shed a different light on it. In these most sophisticated uses of mathematics for literary purposes, the reader's reaction to some elements of the plot can be influenced by a comparison or contrast with some information about chaotic dynamics or fractals that is presented alongside it.

Sensitive Dependence on a Human Scale

It has only been since the nineteenth century, with the work of Henri Poincaré, that sensitive dependence on initial conditions has been recognized and studied in mathematics. However, the concept that one small change can lead to a big difference is not necessarily a mathematical one. A work of fiction can address this general idea, the butterfly effect, without really referencing mathematics at all. On the other hand, if a work of fiction focuses on dramatic consequences of small actions and *also* specifically discusses the mathematics of chaos theory, then it may be reasonable to assume that this is not a coincidence. In such cases, the author intends the reader to make a connection between the plot and mathematics.

One good example of this is Robert Hellenga's novel *The Fall of a Sparrow* (1998) which looks at the ways that different family members deal with the death of an American law school student in a terrorist attack in Italy. As the family learns about the last few days in her life, the reader becomes aware of all of the things that led up to her being at the site of the explosion at that precise moment. The death seems more tragic because we are aware that very small changes in the events would have prevented it. Most of her family members are shown to have been damaged by the tragedy, but not a surviving sister who just happens to be curating an exhibit on chaos theory at Chicago's Museum of Science and Industry. A reader could be forgiven for thinking that it is an unrelated tangent, but it is also possible to see this focus on the chaotic nature of reality as providing the philosophical underpinning of the character's successful method of coping with her sister's pointless death. Rather than seeking meaning in tragedy or obsessing over how it could have been avoided, someone who has developed an expertise in chaos theory and taken its lessons to heart might be able to simply accept the tragedy and move on.

Phillip Persinger's *Do the Math: A Novel of the Inevitable* (2008) uses references to mathematics to highlight the way that life's great joys can also be a manifestation of the butterfly effect. Persinger, who explains in the "About the Author" biography that his own life contains fortuitous coincidences, explores this concept in a novel about a romance novelist married to a mathematician. The story of the two characters and their marital problems itself would not necessarily be very different if the mathematician character studied

140   A. KASMAN

some other field, such as number theory, or even if he were not a mathematician. However, the mathematician's area of expertise is precisely this "sensitive dependence" phenomenon through which small variations lead to large differences in chaotic systems. Indeed, in the novel he is famous for a theorem that demonstrates its role in every person's life. Thus, the reader is led to view the characters' relationship differently by thinking of it in terms of dynamical systems and sensitive dependence on initial conditions, even though the book never uses those technical terms.

Arcadia (Stoppard 1993) also features demonstrations of sensitive dependence on a human scale. As the play jumps between the two time periods, the audience has the task of making sense of the storyline. Some major plot points end up being determined by seemingly tiny details. A poorly worded letter accidentally left in a book in the nineteenth century, a scribbled picture of a hermit by Thomasina on a landscape plan of her parents' home, a monkey bite, and a single candle left burning each lead to much more confusion, tragedy, and death than one might expect.

A skeptic could certainly argue that important events in people's lives really are influenced by little things that at first seem insignificant, and that such occurrences in literature are therefore not necessarily references to chaos theory. Still, the fact that these authors also chose to write about chaos theory in the same work of fiction suggests that they were thinking of the way that the plots themselves illustrate the notion of "sensitive dependence on initial conditions."

CATEGORY IV

Up to this point in this essay, I have discussed texts in which the mathematics of chaos theory and fractal geometry are a means to an end. Whether it is to explain a character's ability to predict the future, to impress us with a character's brilliance, or to better help us understand the grief associated with a tragic and avoidable death, the mathematical references are arguably in support of some greater literary goal and not a goal in themselves. In this last section, I focus on texts which, rather than using mathematics to achieve a literary effect, use literature to say something about the mathematics of chaos and fractals. By giving a fictional reality to some of the abstract concepts from these branches of mathematics, we are able to enjoy them in a new way, and perhaps are even left with a better understanding of them.

Chaos Theory Manifested in Other Worlds

The famous "three-body problem" is a canonical example of a chaotic dynamical system. In general, three massive objects in space moving according to the laws of Newtonian physics will exhibit complicated dynamics, the motions of which are practically impossible to predict due to sensitive dependence on initial conditions. In Cixim Liu's novel *The Three-Body*

Problem (2014), this makes life very difficult on a planet that has three suns. Unlike the reliably predictable seasons and days on Earth, the temperature and light levels on this fictional world do not follow any recognizable pattern as they are a physical manifestation of this famous chaotic system. The question of how life that evolves on such a world would differ from life here on Earth is a major focus of the book.

A person who is especially charismatic or well spoken is generally believed to be able to influence the opinions of others. Like an ordinary attractor in dynamical systems theory, that effect is predictable in that others will invariably be led to agree with that person. In Greg Egan's short story "Unstable Orbits in the Space of Lies" (1992), the narrator is someone who also influences the beliefs of those around him, but unpredictably. This manifests the abstract ideas of chaos theory in a non-mathematical context that may be easier for readers to imagine. They may be able to identify with this human strange attractor who knows that interaction with him will cause people's views to change randomly.

The novel *Mathematicians in Love* (2006) by Rudy Rucker is different because rather than illustrating an example of chaotic dynamics, it presents a fictional world without chaos. Rucker has a Ph.D. in mathematics and worked for a time as a mathematics professor. His novel clearly takes place in a universe that is similar to and yet noticeably different from our own. For example, the city of Berkeley, California is called "Humelocke," presumably after George Berkeley's fellow empiricists John Locke and David Hume. Relatively early in the novel, two mathematicians prove a theorem that allows them to predict the future completely by decomposing any dynamical system into a combination of five "morphons" that are entertainingly named after the items balanced by Dr. Seuss' Cat in the Hat:

> Each morphon had a characteristic activity – the fish swam; the dish shattered; the teapot poured; the rake dragged; the cake blazed with candles. And implicit in these behaviors were five fundamental processes: rhythm, fracture, flow aggregation, transformation. I hadn't been quite sure that these basic modes were sufficient to model everything; but now, in just a few lines of symbols, Paul confirmed my intuition. (Rucker 2006, p. 41)

As with the examples discussed earlier in this essay, it is not realistic to think that mathematics could be used to produce such accurate predictions of the future. However, in Rucker's novel the reader is not supposed to believe that advanced mathematical knowledge can allow a character to make accurate predictions despite chaos. Instead, the universe in which this novel takes place is described as being one that is not governed by chaotic laws. The characters eventually meet mathematicians from other worlds more like our own in which such predictions are not possible. Rather than using the mathematics of chaos to tell a story, this novel uses fiction to create a situation in which the reader can compare worlds that are governed by chaotic dynamical systems with others that are not.

Fractals Manifested in a Fictional World

The Koch Snowflake is a fractal geometric object that was studied before the term "fractal" was coined and before the connections to chaos theory were discovered. It is made by repeatedly placing smaller and smaller equilateral triangles on the edges of an initial triangle until it displays complete self-similarity at all levels (see Fig. 8.3). When William Grey Walter wrote *The Curve of the Snowflake* (1956), the Koch Snowflake was not thought of as a "fractal" but rather was viewed as an intriguing counter-example from the field of topology demonstrating that a simple closed curve could be drawn inside a circle and have infinite length. In the book, a "flying saucer" is discovered whose surface is a two-dimensional analogue of the Koch Snowflake and therefore has an infinite surface area.

Rucker's short story "As Above, So Below" (2000) also features a fractal ship, this one being a Mandelbrot set that has literally come to life, allowing the reader to vicariously experience physical interaction with this abstract mathematical object:

Fig. 8.3 Start with an equilateral triangle and then repeatedly add another equilateral triangle to the center third of each edge of the resulting polygon. The object that is the limit as the process is iterated infinitely many times is the fractal known as the Koch Snowflake (*Source* Produced by author)

Since this whole ship was a fractal, each of the firefly globs was a three-sectioned thing like the main body: each of them was a dimpled round ass part with a little antennaed head-sphere stuck onto it ....

I took one of the baby Mandelbrot sets in my hands and peered at it. It was warm and jittery as a pet mouse. Even though the little globster was vague at the edges, it was solid in the middle. Better than a graphic. I cradled it and touched it to my face. (Rucker 2000, p. 40)

Tourists in the land of Fractalia in Clifford Pickover's *Sushi Never Sleeps* (2002) can take a train with vibration dampeners shaped like Koch snowflakes to see a performance of the "fractal dance" in which smaller fractalians dance around larger ones. And Alex Rose's short story "Goliijo" (2007) describes an island whose fractal geometry is a source of confusion to tourists who get lost on its infinitely dividing streets because each intersection looks like every other. Ironically, the same self-similarity is precisely why locals never get lost:

There is no need for maps because the city is itself a map. Every detail of Goliijo is an inscription of its own layout. (Rose 2007, p. XXX)

Piers Anthony's fantasy novel *Fractal Mode* (1992) takes place in a universe with the structure of a Mandelbrot set. The main character does not recognize this at first because:

[P]eople had plowed out most of the knobs, near the village, so as to use the land for crops. Man always did mess up the scenery. Here the mountains were shaped like boulders with smaller boulders perched on them, and smaller ones on the smaller ones, and so on without end. It was weird — but also true to the Mandelbrot set as she remembered it. True to the entire science of fractals. (Anthony 1992, p. XXX)

None of these manifestations of fractal geometry is necessary to the plots of their respective works of fiction. Instead of having the mathematics introduced to advance the story in any way, the reverse seems true in these cases; these aspects of the story exist for the purpose of being able to present something interesting about fractal geometry.

Self-Similarity in Fictional Structure

There is a *fictional* work of fiction that was supposedly written in such a way that its very structure is a reference to fractal geometry. Mathematician John Allen Paulos wrote a piece that takes the form of a book review of a novel called *Rucker—A Life Fractal* (Paulos 1991). The novel, which does not actually exist, is implied to have been written by the also non-existent Fields medalist Eli Halberstam. The review notes:

144   A. KASMAN

> To those of us in mathematics, Halberstam seems to be saying that human consciousness – like endlessly jagged coastlines, or creased and varicose mountain surfaces, or the whorls and eddies of turbulent water, or a host of other "fractured" phenomena – can best be modeled using the geometrical notion of a fractal. The definition isn't important here, but unlimited branching and complexity are characteristic of the notion as is a peculiar property of self-similarity, whereby a fractal entity (in this case, the book) has the same look or feel no matter on what scale one views it (just the main events or finer details as well). (Paulos 1991, p. XXX)

However, that work of fiction said to have a fractal literary structure is merely described in another work of fiction. An actual literary text that arguably references fractal self-similarity in its very structure is Stoppard's *Arcadia*.

There are many coincidences and similarities in the two plotlines of *Arcadia* that are taking place in different centuries. Others have already pointed out that "[t]hese striking similarities of the treatment of chaos theory in the scenes of the past and those of the present are comparable to the self-similar structures of fractal images" (Vees-Gulani 1999, p. 416). Indeed, the parallelisms in the dialogues involving Thomasina Coverly in the nineteenth century and her relative Valentine Coverly in the twentieth century are like the very similar looking shapes found around the boundary of the Mandelbrot set. However, it is not really similar shapes of the same size that characterizes fractal geometry but rather similarity at *different scales*. This would seem to argue against the notion that *Arcadia* incorporates fractal geometry into its own structure.

It is important to note that there is another coincidence in *Arcadia* that is manifested at different scales. At about the same time that Thomasina is undertaking her mathematical researches, a new trend in landscaping is becoming popular in England. The previously popular topiary gardens, in which plants were cut into the boxy and cylindrical shapes of classical geometry, are being replaced by a more natural aesthetic that looks like wilderness. Those transitions from classical to natural are occurring at very different scales: in Thomasina's notebook in the case of the mathematics and across the gardens of her family's estate and across all of England in the case of the landscaping. Since fractal geometry is a major theme of the play, it is reasonable to assume that this is not merely a coincidence but a rare instance in which multi-scale self-similarity has been successfully incorporated into the structure of a work of literature, and what is more, one in which one side of the similarity is the very concept of fractal geometry itself.

CONCLUSION

It is possible that an author who mentions chaos theory or fractal geometry in a work of fiction really only needed to refer to *some* area of mathematics and selected one of these almost at random. However, in most cases the selection seems to serve a literary purpose that would not be achieved equally well if a different area of mathematics were used instead. As I have argued in this essay, these intentional uses of chaos and fractals in fiction fall into four broad categories.

First, there are reasons that have nothing to do with the actual mathematical content of these theories but are nevertheless specific to chaos theory and fractal geometry. For example, I believe that the emotionally potent terminology of chaos theory is often selected by authors because of the words' ordinary English meanings, distinct from their technical mathematical definitions. This would help account for the popularity of the title "Strange Attractors" even in works of fiction that otherwise seem to have little or nothing to do with attraction basins of fractional dimension. Moreover, these topics are recognizably modern, having only really been active areas of research since the late twentieth century and receiving more public attention than other active areas of modern mathematical research. This fact can also be utilized by authors who wish the reader/audience to think of mathematician characters or their research as avant-garde, when the general impression of mathematics is precisely the opposite.

Second, there are works of fiction that make direct use of the mathematical content of these theories or their applications to justify and motivate the plot. Most of these justifications fit into one of two categories: either that knowing about chaos theory allows one to make highly accurate predictions about the future, or that there are dramatic psychological consequences of looking at images of fractals.

Next, these subjects can be used in fiction because of what they say about how numbers, functions, and other mathematical objects resonate in a less-than-rigorous way with some of our feelings and concerns about the human condition. The unpredictability, sensitive dependence, and even self-similarity of the mathematical abstractions seem familiar to people as they are not entirely unlike the world we live in. And so, regardless of whether those similarities exist because reality is truly described by the nonlinear equations of a chaotic dynamical system or whether it is just a useful analogy, talking about these mathematical topics in a work of fiction can enhance its ability to address those aspects of our daily existence.

The last category is somewhat different, since rather than using the concepts of chaos theory and fractal geometry to achieve a separate literary goal, they use literature to say something about chaos and fractals by giving some of the abstract ideas from those fields a fictional existence, much as a sculpture or painting can represent a mythological creature. The literary purpose here may be purely aesthetic, but the goal may also be didactic in that such fictional representations can sometimes leave the reader with a better understanding of the mathematical ideas themselves.

In this essay, literary uses of chaos theory or fractal geometry have been fit into these four categories. I acknowledge that the categorization scheme is not ideal. The distinctions between the categories are not always clear. For instance, a case could be made that the chaos theory discussions in Liu's *The*

Three-Body Problem and Egan's "Unstable Orbits in the Space of Lies" can be interpreted as political metaphors—in the former case, a metaphor for the difficulty of living in an unstable society and in the latter for the formation of echo chambers of like-minded people in democratic societies—in which case they should have been listed in Category III instead of IV. Moreover, although a quality that might be desired in a categorization scheme is that each item should appear in exactly one category, these categories are not distinct. In fact, according to my classification, Stoppard's brilliant play *Arcadia* uses mathematical ideas in all four of these ways. However, I can offer some defense of the categorization scheme, imperfect as it may be. Even if there is room for disagreement about the ways that individual instances are categorized, it still provides a framework for discussion. There may also be value in the fact that the categories are not entirely disjoint. Perhaps the fact that *Arcadia* appears in all four categories is itself a measure of that play's intricacy as a work of art. In any case, it is my hope that this survey and its use of this categorization scheme will leave the reader with an impression of the entire landscape of uses that references to chaos theory and fractal geometry have served for authors of fiction.

NOTES

1. A link in the browse section of the website (Kasman 2017) allows a visitor to see a list of works tagged with the phrase "Chaos/Fractals."
2. Some real attempts have been made to apply these branches of mathematics to investment, such as "A Multifractal Walk Down Wall Street" (Mandelbrot 1999) and *The Misbehavior of Markets: A Fractal View of Financial Turbulence* (Mandelbrot and Hudson 2004).

REFERENCES

Anthony, Piers. 1992. *Fractal Mode*. New York: Ace Books.
Clarke, Arthur C. 1990. *The Ghost from the Grand Banks*. New York: Bantam Books.
Cohen, Mark. 2002. *The Fractal Murders*. Boulder: Muddy Gap Press.
Connolly, Robert. 2001. *The Bank*. Film. Directed by Robert Connolly. Melbourne: Arenafilm.
Cory, Desmond. 1991. *The Strange Attractor*. Basingstoke: Macmillan.
Crichton, Michael. 1990. *Jurassic Park*. New York: Alfred A. Knopf.
Crichton, Michael, and David Koepp. 1993. *Jurassic Park*. Film. Directed by Steven Spielberg. New York: Universal.
Egan, Greg. 1992. "Unstable Orbits in the Space of Lies." *Interzone* 61: 53–60.
Giessmann, John. 2010. *An Angel of Obedience*. Bloomington: Author House.
Gleick, James. 1987. *Chaos: Making a New Science*. New York: Viking Press.
Goldstein, Rebecca. 1993. *Strange Attractors*. New York: Viking Press.
Hellenga, Charles. 1998. *The Fall of a Sparrow*. New York: Charles Scribner's Sons.
Jungk, Peter Stephan. 2003. *Tigor*. New York: Other Press.
Kasman, Alex. 2017. *Mathematical Fiction Homepage*. http://kasmana.people.cofc.edu. Accessed on July 12, 2017.

Liu, Cixin. 2014. *The Three-Body Problem*. New York: Tor Books.

Mandelbrot, Benoit. 1982. *The Fractal Geometry of Nature*. London: W.H. Freeman.

———. 1999. "A Multifractal Walk Down Wall Street." *Scientific American* 280 (2): 70–73.

Mandelbrot, Benoit, and Richard L. Hudson. 2004. *The Misbehavior of Markets: A Fractal View of Financial Turbulence*. New York: Basic Books.

Napier, Bill. 2002. *The Lure*. London: Headline.

Paulos, John Allen. 1991. "*Rucker—A Life Fractal* by Eli Halberstam." In *Beyond Numeracy*, 108–11. New York: Random House.

Persinger, Phillip. 2008. *Do the Math: A Novel of the Inevitable*. Bloomington: iUniverse.

Pickover, Clifford. 2002. *Sushi Never Sleeps*. Mobile: Lighthouse Press.

Rose, Alex. 2007. "Goliijo." In *The Musical Illusionist and Other Tales*. Brooklyn: Hotel St. George Press.

Rosen, Leonard. 2011. *All Cry Chaos*. Sag Harbor: Permanent Press.

Rucker, Rudy. 2000. "As Above So Below." In *Gnarl*, 397–411. New York: Four Walls Eight Windows.

———. 2006. *Mathematicians in Love*. New York: Tor Books.

Sleator, William. 1991. *Strange Attractors*. London: Puffin Books.

Soule, Charles, and Greg Scott. 2013. *Strange Attractors*. Los Angeles: Boom Entertainment.

Stoppard, Tom. 1993. *Arcadia*. London: Faber & Faber.

Vees-Gulani, Susanne. 1999. "Hidden Order in the Stoppard Set: Chaos Theory in the Content and Structure of Tom Stoppard's *Arcadia*." *Modern Drama* 42 (3): 411–426.

Wayne, Teddy. 2010. *Kapitoil*. New York: Harper Perennial.

Wilhelm, Kate. 1992. *Death Qualified: A Mystery of Chaos*. New York: Fawcett Crest.

CHAPTER 9

Mathematical Clinamen in the Encyclopedic Novel: Pynchon, DeLillo, Wallace

Stuart Taylor

Twenty-five years after C. P. Snow observed the division of intellectual discourse into "Two Cultures"—the "literary" and the "scientific"—Thomas Pynchon believed the split to be an outdated simplification:

> Since 1959, we have come to live among flows of data more vast than anything the world has seen …. Anybody with the time, literacy and access fee these days can get together with just about any piece of specialized knowledge s/he may need. So, to that extent, the two-cultures quarrel can no longer be sustained. (Pynchon 1984, p. 1)

Reading Pynchon's fiction today is an experience akin to being swept away on tides of information. Indeed, his texts—which display a seemingly limitless scope through prose that draws from both literary and scientific "cultures"— are often described as literary encyclopedias. According to Hilary Clark, such an encyclopedic novel betrays its encyclopedic impulse when it "gathers and hoards bits of information and pieces of wisdom following the logic of their conventional (metonymic) associations in the writer's culture" (Clark 1990, pp. 4–5). The reader of *Gravity's Rainbow*, *Moby-Dick*, *Ulysses*, and other such fictions encounters these varied, interdisciplinary associations in a manner that Tom LeClair argues is analogous to "systems theory"—a "mathematical and speculative" "approach to living systems" in which models, including literary models, relate to scientific discourses through "isomorphisms or

---

S. Taylor (✉)
University of Glasgow, Glasgow, Scotland, UK

© The Author(s) 2021

R. Tubbs et al. (eds.), *The Palgrave Handbook of Literature and Mathematics*, https://doi.org/10.1007/978-3-030-55478-1_9

149

homologies" (LeClair 1987, p. 3). Yet such texts, while evoking the encyclopedia's hermetically sealed vessel of knowledge, resist and challenge ideas of epistemological totalisation. Pynchon's use of mathematics, in particular, promotes opportunities of freedom within his literary encyclopedic systems. In *Gravity's Rainbow* Pynchon appeals to a freedom analogous to that provided by the "incompleteness theorems" of twentieth-century logician Kurt Gödel: "Gödel's Theorem – when everything has been taken care of, when nothing can go wrong, or even surprise us ... something will" (Pynchon 2000, p. 275). Such surprises, unexpected swerves in the flow of data, become the loci for individual free will within vast, all-encompassing information systems. In this chapter I will consider how mathematics provides such a swerve, or clinamen, in recent encyclopedic novels by Pynchon, Don DeLillo, and David Foster Wallace.

Mathematics and encyclopedias can be compared through analogies of abstract, universalising geometric shapes, specifically the circle and the line. Etymologically, the encyclopedia derives from the circle of learning (Kuusisto 2009, p. 147) while the closed curve was fundamental to the growth of European mathematical geometry, from the astronomical circular motions of the "closed universe" observed by Classical mathematicians Eudoxus of Cnidus and Claudius Ptolemy, to the seventeenth-century developments of Johannes Kepler's "elliptical orbits" and the mathematical calculus of Isaac Newton and Gottfried Wilhelm Leibniz (Koestler 1959, pp. 22, 63–69, 77, 504–5).[1] Since the Enlightenment, both the encyclopedia and mathematics as an accretive science *par excellence* have also contributed to ideas of the rectilinear progress of rationalism, a teleological narrative of unerring determinism. Between these two imperial images—circle and line—operates the encyclopedic novel. Through cyclical narrative lines, these novels engage with the "circle of learning"—that storehouse of immutable facts by which, through its memorisation, knowledge propagates from one generation to the next. Thus, Pynchon's *Gravity's Rainbow* opens with the launch and closes with the fall of a V-2 missile (Pynchon 2000, pp. 5, 760); the protagonist of DeLillo's *Ratner's Star* begins the novel with a bandaged thumb, inexplicably corresponding to a wound suffered in the final pages (DeLillo 1991, pp. 8, 423), while the opening chapter of Wallace's *Infinite Jest* is chronologically last in the novel's timeline (Wallace 1997, pp. 3–9, 223).

Rather than being looping, hermetically sealed texts, however, these circling narratives encourage subsequent generative readings, emphasising their resistance to encircled totalisation and rectilinear determinism. This motion of resistance evokes the Epicurean clinamen—the unexpected swerve of atoms—as recounted by Lucretius (Lucretius 1977, pp. 216–24). For Lucretius, this swerve was a result of the inherent indeterminacy of the physical universe. This indeterminacy was crucial to Lucretius's argument for individual free will. The clinamen has since been regarded as "the locus and the guarantor of free will" within systems where "unvarying linearity" is expected (Motte 1986, p. 263). I argue in this chapter that, by activating this clinamen using mathematics, Pynchon, DeLillo, and Wallace encourage liberating interdisciplinary reading practices. Examining their respective mathematical imagery of precession, reflection, and inversion will reveal how the mathematical clinamen can operate as a liberating movement within an apparently closed literary system.[2]

Precession in *Gravity's Rainbow*

Pynchon's encyclopedic novels are characterised by "extraordinary incoherence" (Levine 1978, p. 181). They produce a reading experience which recalibrates the balance between the conception of scientific knowledge as the pinnacle of epistemology and the recognition of the violence and trauma inflicted and rationalised in the name of inevitable technological progress—whereas scientific tropes in some less critical generic science fiction might be used to legitimate inexorable technological progress. Pynchon's literary appropriation of science, particularly through mathematical equations, denies such determinacy. Instead it contributes to the "self-figuring turn" of his "encyclopaedic discourse" (Clark 1990, p. 11). Pynchon uses mathematics as a scientific "metadiscourse": a "self-representation [which] generates an excess, opening outward onto the freedom of infinite *speculation* on knowledge," thereby creating new knowledge (Clark 1990, p. 98, original emphasis). In the 1920s formalist mathematician David Hilbert proposed his own program of "metadiscourse" in order to secure the logical foundations of mathematics. His "metamathematics" considers descriptive logical statements made about and outside of a specified logical system to determine the consistency and completeness of that system. Analogous to Hilbert's "metamathematics," Pynchon's "metadiscourse" determines the transmission of internal textual content to an external reader. But, unlike Hilbert's aim of using metamathematics to prove the consistency of a system, Pynchon's discourse in *Gravity's Rainbow* seeks "resistance to totalizing structures" through narrative "line[s] of flight" (Drake 2010, pp. 224–25). Unmoored by multiple metaphorical resonances and levels of self-reflexivity, these lines of discourse—liberated from the confines of a single definitive interpretation—challenge consistent, totalising orders.

The legitimacy of Pynchon's deployment of mathematical scientific concepts—including physical entropy, projectile aeronautics, quantum mechanics, calculus and geometry—tends to be grounded by critics in his early career at the Boeing Company (Friedman and Puetz 1974, pp. 345–46). Initially enrolled in Cornell University's Engineering Physics programme, Pynchon switched to study English literature. Both scientific and literary interests and training gained him subsequent employment at Boeing as an "engineering aide" writing collaborative "technical documents" (Winston 1975, pp. 284–85). Pynchon's anonymous contributions to the *Minuteman Field Service* and *Bomarc Service News* letters have been partially identified. Betrayed by the singular voice of their author while blending technical and poetic registers, these

152   S. TAYLOR

pieces can be seen as embryonic of Pynchon's encyclopedic prose (Wisnicki 2000–2001; Muth 2015). Pynchon's background, then, is used by Wisnicki, Muth, Friedman and Puetz, and other critics to support the seriousness with which to regard his literary appropriations of mathematical sciences.[3]

The most famous, and critically exhausted, of such appropriations is that of entropy. Since Anne Mangel's early essay, Pynchon's literary appropriation of entropy has been regularly celebrated as his signature "ingenious use of scientific-technological concepts as the basis for his fiction" (Mangel 1971, p. 194). From Peter Abernethy (1972) to David Cowart (1981) and David Seed (1981) to more recent studies such as Steve Vine (2011), critical consensus views Pynchon as an authority on the science(s) behind "entropy," despite the author himself noting his own "shallowness of ... understanding" (Pynchon 1985, pp. 12–14). The idea and consequences of Pynchon's allegedly scientifically inadequate understanding of entropy have recently been explored by David Letzler (2015).

Other critics have focused on Pynchon's deployment of mathematical equations, specifically those in *Gravity's Rainbow*. Pioneering this approach, Joseph Slade's 1974 monograph considered Pynchon's fertile mathematical analogies: the double integral sign $\int\int$; the notion of measurable changes in time Δt; the continuity of the parabola. "[J]ust as humans divide up a parabola into fragments for convenience and precision in measurement," Slade writes, "so they divide up their lives into artificial frames to comprehend them. In each case they sacrifice the whole for its parts, but the continuity remains, whether they see it or not" (Slade 1974, p. 220). Lance Ozier developed Slade's approach to consider Pynchon's interest in statistics and the "calculus of transformation" evident from *Gravity's Rainbow*'s mathematical imagery (Ozier 1974, 1975). Following Slade and Ozier, Edward Mendelson was among the first to connect this "expert[ise] in ... some very advanced mathematics" to encyclopedic narrative (Mendelson 1976, p. 1270). Lance Schachterle and P. K. Aravind proceeded to examine the equations in *Gravity's Rainbow*, arguing that Pynchon "turns mathematical expressions into rhetorical tropes which complicate the text by playing upon the authority the equations convey" while providing suggestive scientific contexts for each of the three equations in the novel (Schachterle and Aravind 2000, p. 157). They note that, mathematically literate or not, readers must be frustrated in their "efforts to eke out all the meaning compressed" in the formulae (Schachterle and Aravind 2000, p. 161). Stefan Mattessich has also argued that Pynchon's mathematical analogues specify "the impossibility of any ultimate answers or ends" in a novel governed by "truth as 'unconcealment'" (Mattessich 2002, p. 201). Most recently, Nina Engelhardt has explored the utopian resonances in Pynchon's deployment of mathematics in *Gravity's Rainbow* and *Against the Day*. According to Engelhardt, nineteenth-century physicist William Hamilton's development of quaternions—extending imaginary and complex numbers—features in *Against the Day* to "allow for the freedom to move on imaginary axes[: a] mathematical concept [which] counteracts the reduction of choices under earthly conditions" (Engelhardt 2013b, p. 146). She also considers the opposing explanations of gravity from the fathers of the mathematical calculus, Newton and Leibniz, highlighting that "the historical context of competing concepts of gravity shows that it is not a force that is unproblematically in control" (Engelhardt 2014, p. 8).

In fact, Pynchon presents gravity as a force analogous to imperial conquest. The mathematics with which this is effected, however, resists subjugation, turning or swerving against and between encircled knowledge and deterministic progress. The encyclopedic and mathematical images of circle and line describe the relationship between content and form in Pynchon's major novels—a relationship of deviation and resistance: by bending straight narrative lines into parabolic curves and by denying simple, linearized relationships between signifier and signified to promote a returning, cyclical reading of his text, Pynchon's "metafictional or metacritical language" introduces an unexpected swerve into the closed hermetic system of history (Drake 2010, p. 224).[4] In *Gravity's Rainbow* this clinamen appears as the precession associated with the V-2 rocket—the dominant object of the text and "the embodiment of a totally contained system" (Drake 2010, p. 237). By bending the line and breaking the circle of a rectilinear narrative inside a closed textual system of meaning, Pynchon's clinamen promotes "infinite *speculation*" while undermining encyclopedic authority (Clark 1990, p. 98). Foregrounding incomplete and undetermined mathematical objects and equations within his texts, he emphasises alternate configurations of epistemology and the moral limits of scientific rationale. The result provokes a reconsideration of the casualties of technological progress as inevitable collateral damage.

The parabola of the V-2—as a "Rainbow"-arc under gravity—is a curved image which illustrates the novel's clinamen. *Gravity's Rainbow* repeatedly shows how inextricably bound the natural phenomenon of gravity is to technologically facilitated conquest—of forces and people—by abstract mathematical control. One character, the wife of a Foreign Office agent, Nora Dodson-Truck, believes she is the "Force of Gravity" incarnate: "*I am Gravity, I am That against which the Rocket must struggle, to which the prehistoric wastes submit and are transmuted to the very substance of History. …*" (Pynchon 2000, p. 639, original emphasis). In the novel, engineer Franz Pökler, along with other scientists, was tasked by the Nazis with "invading Gravity itself" (Pynchon 2000, p. 404). Pynchon's architect of the Mittelwerke V-2 factory, Ernst Ölsch, observes these themes in the mathematical symbol of the double integral, $\int \int$. One of the major operations of calculus, integration, facilitates the enumeration of functions—for example, velocity. The double integral represents a similar operation, though more complex in that it deals with functions of two variables. Through double-integration, a dynamic process is consolidated in a static point or value. For Ölsch, this symbol represents

154 S. TAYLOR

the method of finding hidden centers, inertias unknown, as if monoliths had been left for him in the twilight, left behind by some corrupted idea of "Civilization," ... in which imaginary centers far down inside the solid fatality of stone are thought of [as] a point in space, a point hung precise as the point where burning must end, never launched, never to fall. (Pynchon 2000, p. 302)

In as much as it contributes to the fantasy of an eternal world-order following conquest, gravity and its associated mathematics tend to serve conquering powers. This association of mathematics with the divine Elect appears to deny calculated resistance (Pynchon 2000, p. 486). The novel describes how German rocket strikes "*are* distributing about London just as Poisson's equation in the textbooks predicts" (Pynchon 2000, p. 54, original emphasis). While the strike-patterns obey the law of small numbers as proposed by nineteenth-century French mathematician Siméon Denis Poisson, and the novel's leading statistician Roger Mexico seems to appear "more and more like a prophet," Poisson's distribution remains an "equation only for angels"—there is no mathematical equation with which to prevent death by predicting where future rockets will explode. Misunderstanding the affective potential of the data's mathematical organisation, Mexico's lover Jessica Swanlake laments: "Couldn't there be an equation for us too, something to help us find a safer place?" (Pynchon 2000, p. 54). On the other hand, the mathematics of those deploying the V-2 rocket enables "bourgeois" exploitation of forces in a campaign of violent domination to maintain the linear, conservative trajectory of the Elect. Engineer Roland Feldspath introduces a particularly intimidating and frustrating equation to describe the mechanics and motion of the V-2:

the Rocket's terrible passage [is] reduced, literally, to bourgeois terms, terms of an equation such as that elegant blend of philosophy and hardware, abstract change and hinged pivots of real metals which describes motion under the aspect of yaw control:

$$\theta \frac{d^2\phi}{dt^2} + \delta^* \frac{d\phi}{dt} + \frac{\partial L}{\partial \alpha}(s_1 - s_2)\alpha = -\frac{\partial R}{\partial \beta}s_3\beta$$

preserving, possessing, steering between Scylla and Charybdis the whole way to Brennschluss. (Pynchon 2000, p. 284)

Pynchon's equation uses the presumed precision of mathematical notation to gesture implicitly towards the novel's clinamen. As its terms are undefined, however, the meaning of the equation is, to quote the critic Scott Drake, "absence that fuels the search for an authority" even while "[t]he object of this search ... is constantly deferred" (Drake 2010, p. 230). The equation's nonlinear form combined with a constantly deferred meaning characterises it as a clinamen, a locus of free will. Because Pynchon's narrator does not explicitly state its relation to swerving in the plot, or to imagery of gravity and parabolic curves, the equation keeps the encyclopedic system incomplete, inviting creative interpretations. Such interpretations of the implicit suggestions it evokes draw further attention to the equation as a narrative clinamen: it gestures towards mathematically described physical processes whereby straight lines are bent or corrupted. This is immediately suggested by noting that it is a particular type of differential equation, one that deals with partial derivatives (denoted by the symbol ∂). Essentially, partial differential equations are differential equations in which "the unknown function depends on several variables" (Pikulin and Pohozaev 2001, p. 1). These particular variables both frustrate a satisfactory solution and yet gesture towards a meaning resonant with the novel's corruption of the straight line. While critics have been unable to identify the source of the equation's verisimilitude, the narrative context suggests the gyroscopic controlling mechanisms of the rocket.[5]

The equation raises the stakes of mathematical knowledge in an increasingly complex and weaponised world. As Schachterle and Aravind put it, the equation "expresses in highly compact symbolic form, how various influences or driving forces combine (or conspire) to cause the state of the primary entity to change usually over time" (Schachterle and Aravind 2000, p. 161). Crucially, however, the reader is not given enough information to make sense of these symbols: terms are undefined and the relationships between them are unclear in this one line of calculation. Unless values are given, this equation can neither be solved nor used to provide the precise technical information its symbolism proposes. Thus, the reader is here exposed to obscure and therefore potentially threatening symbols—the same cognitive challenge that faced the protagonist Tyrone Slothrop's friend Dumpster Villard, "who'd tried suicide last semester [because] the differential equations ... would not weave for him into any elegance" (Pynchon 2000, p. 76).

Nevertheless, the equation is not entirely meaningless. Pynchon uses its very indeterminacy to develop his narrative of resisting simplified views of complex systems, both historical and contemporary. This simplification can be thought of as analogous to linearisation: the process by which nonlinear systems are mathematically approximated (and simplified) by linear algebraic equations. Pynchon's equation is appropriately *non*linear: it does not describe a straight-line change, but one that—like the novel's master image of a parabolic arc—curves throughout the novel. This curve, as a "steering between," also evokes the sense of clinamen resisting the determinism Feldspath wishes from his equation (Pynchon 2000, p. 284). Surprisingly, the mathematics of control here serves as a metaphor for the unexpected within the novel's encyclopedic scope, frustrating claims to an encircling totalisation of knowledge. Pynchon's equation occupies a liminal space in his text between its promised calculated precision and its indeterminate meaning. This liminal space allows a blend of mathematics and poetics in which the metaphorical resonances of precession can function as a clinamen within the text. By first seducing the reader with such mathematical symbolism, Pynchon then frustrates their appetite for precise information, before suggesting resistant narratives that swerve beyond and between totalising systems of coercion and control.

The V-2 rocket's automatic pilot mechanism "comprise[d] two electrically driven displacement gyros, one acting as a roll and yaw datum with its rotor axis at right-angles to the plane of the target and to the fore-and-aft axis" (Richardson 1954, pp. 353–54). This gyro ensures both roll and yaw angles are corrected as necessary during flight to keep the rocket on "the plane of the target"—the path that leads to its intended destination (ibid., see Savet 1961, pp. 77–78). The role of the gyroscope in correcting these angles as it guides the rocket towards the target is therefore suggested as the ostensible subject of the partial differential equation: "describ[ing] motion under the aspect of yaw control" (Pynchon 2000, p. 284). This revelation of the gyroscope's connection to the equation is corroborated during the interrogation of rocket engineer Horst Achtfaden, who reveals that the Rocket 00000 was to be adjusted to accommodate its human payload: Gottfried, the sacrificial lover of Achtfaden's boss, Nazi officer Weissmann. The extra weight, Achtfaden recalls, meant the rocket "was assymetrical about the longitudinal axis. Toward Vane III. That was the vane used for yaw control" (Pynchon 2000, p. 541).

The implications of reading the gyroscope through Pynchon's equation with regard to our reading of the novel as a whole are far reaching. The controlling power, or force, acting on the gyroscope which gives it its ability to control is gravity, the ubiquitous spectre of the novel, a force technically described, in gyroscopic theory, through "precession." This word derives from astronomy, denoting "the gravitational pull of the moon, and to a lesser extent of the sun" that slowly changes the axis about which the Earth spins.[6] This "precession of the equinoxes" demonstrates, albeit on a much larger scale, the precession of a gyroscope, "[t]hus the properties of the gyroscope are exhibited by the earth" (Richardson 1954, p. 37). Through his partial differential equation, Pynchon connects the tiny gyrocompass with the motion of the earth, as part of his critique of those who aspire to global domination through the force of gravity. In doing so he identifies the precession of V-2 rocket technology with free will. Whereas simple primitive projectiles follow the linear parabolic curve effected by gravity, the gyroscope's spinning disk allows navigation and control within the route described by gravity. Similarly, in the novel, "precession" is described as a phenomenon where the spinning disk is "moving invisibly but *felt*, terrifically arousing" (Pynchon 2000, p. 613, original emphasis). This affirmation through the hardware of the gyroscope suggests an emotional sensitivity against the machinations of oppressive deterministic agents. Although the gyroscopic device is installed to keep the rocket on a preordained course, in the novel, paradoxically, the cyclical motion of the gyroscope's spinning disk is equally suggestive of a rotational resistance to linear determinism—"a parallel movement that resists the rocket's dominance." What Drake identifies as Pynchon's "movement away from the determined flight of the rocket that ultimately makes the flight of the rocket as contingent and open to possibility as the moment itself" is a specific resistance against totalising systems that are emblematic of the encyclopedic fictional enterprise in general (Drake 2010, p. 238). As Leni Pökler, wife of a German chemical engineer, says in the novel: "There is the moment, and its possibilities" (Pynchon 2000, p. 159). By using mathematics to announce his clinamen of bent lines Pynchon provides, in *Gravity's Rainbow* and other encyclopedic fictions, "that all important ninety-degree twist," a liberated movement within and between encircled systems of linear deterministic force (Pynchon 2006, p. 134). As Brian McHale observes, this contributes to his construction of "subjunctive space, the space of wish and desire, of the hypothetical and the counterfactual, of speculation and possibility" (McHale 2000, p. 44).

Reflection in *Ratner's Star*

Pynchon's encyclopedic form has proved influential. For Don DeLillo, Pynchon "more than any other writer" of his generation "set the standard" and "raised the stakes" of fiction (DeLillo, qtd. in Harris 1982). Like Pynchon's "subjunctive space[s]," DeLillo's fourth novel, *Ratner's Star* (1976), also explicates a spatial concern through its encyclopedic material. DeLillo situates a simple science fiction plot (uncovering the meaning of an extraterrestrial communication) within an elaborate formal architecture. The narrative recounts the quest of a prodigal adolescent mathematician, Billy Twillig, as he attempts to

> decipher what the residents of this planet [in the orbit of Ratner's Star] are saying, [which] may mark the beginning of an exchange of information that could eventually tell us where we are and what the universe looks like. It's safe to assume the Ratnerians are superior to us. They may help us draw a picture. A seamless figure no less perfect than its referent. (DeLillo [1976] 1991, p. 50)

As the semantic term "referent" suggests, this simple plot supports DeLillo's exploration of linguistic interrelations between mathematics and knowledge. This strategy draws on two major intertexts: the history of mathematics from Ancient Egypt to twentieth-century analysis, and the Victorian-era *Alice* books by mathematician-writer Lewis Carroll. Through these intertexts DeLillo's bifurcated novel, which is split into two parts that "mirror" each other, turns conventional narrative lines into feedback loops and boomerang bends. By situating his space-age plot in a formally complex narrative site, DeLillo engages with fundamental aspects of our topological imagination— how mathematics affects our visualisation of space. Where *Gravity's Rainbow* deploys its clinamen partly through Pynchon's evasive equation, the swerve of *Ratner's Star* is made apparent through traversing its complex topological space. The mathematical mystery which drives the plot obscures the true mathematical clinamen of the novel, the boomerang swerve by which we traverse DeLillo's topological structure. Furthermore, the (mani-)folding of subterranean and extraterrestrial spaces in the novel bends the linear trajectory of the history of mathematics into a complex cycle that frustrates encyclopedic totalisation.

Tom LeClair reads *Ratner's Star* as "the metasystem of the systems novel, a device by which we can see this form's most fundamental, science-influenced relations and recognize its cultural achievements" (LeClair 1987, p. 136).[7] DeLillo's systemic approach can be seen clearly in LeClair's table which correlates each chapter of the first part of *Ratner's Star* with a period, a mathematician and a concept in the history of mathematics, from the Ancient Egyptian mathematical-scribe Ahmes and the earliest concepts of number in chapter 1 to Georg Cantor's late-nineteenth-century development of set theory in chapter 12 (LeClair 1987, p. 125). The second part of the novel (explicitly titled "Reflections") reverses this historical progression, concluding the novel with the mathematical references with which it began. Crucially, this intricate and encompassing systemic structure does not dominate and determine its contents. Like *Gravity's Rainbow*, DeLillo's encyclopedic novel resists the very hermetic and deterministic ideas with which it engages. The "systems methods" which LeClair finds in the novel render it a "looping reciprocal whole, open in its complex diversity, differentiation of structure, and plenitude, equifinal and homeostatic in its confessed limitations"; *Ratner's Star*'s "simultaneous reciprocity and inventive unpredictability" renders it both whole and equifinal—in the words on the novel's eponymous astronomer, a work of "strict mystery" and in its protagonist Billy's, "zorgasmic" (DeLillo 1991, pp. 227, 438; LeClair 1987, p. 123).

The third chapter of *Ratner's Star*, "Shape"—a title which refers to Kepler's planetary ellipse (LeClair 1987, p. 125)—explicates many of the novel's important ideas and motivations, from topological enquiries into space and the objects which inhabit and transform it, to the supposedly extraterrestrial code which drives the entire plot. Glen Scott Allen observes that "DeLillo rewrites the global plots of *Gravity's Rainbow* onto the larger stage of the Universe" (Allen 1994, p. 11). While Pynchon emphasises the role of power structures in constitution of space, DeLillo's spaces are threatening ultimately because they are determined by language: fluid, they are subject to subjective transformations. Here the mathematics of non-Euclidean geometry functions to complicate those transformations, further undermining epistemological stability. In a later chapter "Rearrangement" (referring to Carl Friedrich Gauss's early-nineteenth-century conception of curved-space geometry [LeClair 1987, p. 125]), Billy and some other scientists at the research "complex" (DeLillo 1991, p. 75) are disturbed by the elevators in which both seeming to move, and actually "not moving," result in "no change in sensation. It's absolutely the same whether we move or stand at rest. Something is being violated here. Some rule of motion or logic, no?"

(DeLillo 1991, p. 124). Leaving the elevator to take to his mathematical work, Billy "was distracted ... by the awareness that the sheet of paper on which he was calculating was not perfectly flat, containing many distortions in the form of furrows and grooves, meagre ravines, curvature rampant from point to point" (DeLillo 1991, p. 127). The curvature of space and text, in the textual space of the page, is a reflective feature of DeLillo's mathematical structure which suggests opportunities for swerving within its system. In mathematics, reflection denotes an operation by which elements are "transferred perpendicularly" through a plane to relative points "on the other side." This operation provides DeLillo with a figurative means to illustrate the novel's clinamen as reflection—a "recurvation" of "bending" or "folding back" (*OED Online* 2017 "reflection, n.").

The topological instability and its resultant "generative dread" illustrated in this early chapter determine the resulting incompleteness of Billy's search for meaning in a mathematical code. In "Shape" we are told that Billy must decipher a presumably extraterrestrial "transmission" of "fourteen pulses, a gap; twenty-eight pulses, a gap; fifty-seven pulses" (DeLillo 1991, p. 47). The "problem" facing the researchers is that they "don't know what the transmission means. Space Brain has printed out hundreds of interpretations without coming up with anything we can call definitive" (DeLillo 1991, p. 47). In the same chapter, Billy encounters a woman who explains that "[i]n the science of subjective mind-healing, both cause and effect exist in the full and perfect idea that the mind is Mind. Everything depends on mental typography ... in the way we picture our thoughts" (DeLillo 1991, pp. 56–57). What is hinted at here, and gradually emerges throughout the remainder of the narrative, is a focus beyond the code itself to the ratiocinations of calculating entities. The "adventures" of the first part of the novel excavate an archaeology of such entities from the history of mathematics, while the "reflections" in the second part name these protagonists. What renders this encyclopedic enterprise incomplete is the absence of a satisfactorily significant meaning of the code necessitated in "Shape." Rather than failing to determine this code (and leaving it a meaningless amusement) DeLillo voids it of contextualised meaning—it is merely a clock-time that does not provide the picture of the universe it seemed to promise (DeLillo 1991, p. 385). The journey of this code and the mathematician who pursues it—a traversal doubled in the reflective structure of the novel—instead provides a more satisfactory meaning to the code: as a mathematical clinamen which dynamises DeLillo's elaborate topological structure through reflection operations.

In his investigations, Billy utilises a "mathematical entit[y] named in his honor," the "stellated twilligon" (DeLillo 1991, p. 116), which "represents several of the book's formal and thematic properties" of reflecting dualities of progress/regress, rise/fall (LeClair 1987, p. 127). This entity, resembling a sort of pointy "boomerang," can be pictured by "[i]magin[ing] two triangles sharing the same base. With one abnormality: the base is invisible" (DeLillo 1991, pp. 118, 181). The stellated twilligon, which some characters believe to model the shape of the universe, is reproduced twice in DeLillo's novel as a graphic (DeLillo 1991, pp. 117, 241). Through this doubling (and doubled) design DeLillo pairs dualities, including his characters, in "specific mirror-image or bilateral relationship[s]": thus Billy's left testicle droops, while for Robert Softly it is his right (DeLillo, 1991, pp. 38, 376); Jean Venable has a star-shaped birthmark on her left buttock, and an unnamed female experimental subject on her right; the initials of Walter X Mainwaring are the mirror-image of those of Maurice Xavier Wu (DeLillo 1991, pp. 280, 336–37, 428).[8] These reflections follow the same recurvation of the code within the narrative: from extraterrestrial to subterranean, futuristic to ancient. The reflective clinamen, visualised as the mathematical stellated twilligon, charts a boomerang swerve, where distance is traversed in a productive, though elliptical, return. This is a shape that permits the novel's seemingly paradoxical archeological movement wherein humanity is found to be "more advanced the deeper we dig" (DeLillo 1991, pp. 321, 360).

DeLillo thus counters a projective, linear view of progress—wherein everything pre-Enlightenment was savage darkness and the future promises to continue to advance civilisation—with a spiritual reflection: a meditation and return to sites of ancient cultural memory. The boomerang shape of the stellated twilligon brings together the present and the past while proposing an optimistic, creative, anti-deterministic future. The mathematical image signals the novel's clinamen: a reflection-swerve which moves within a topologically rendered, anachronistic, mirror-space. As the universe depicted by the novel comes to resemble the mathematical shape named in honour of its protagonist, LeClair observes, mathematics (its practice and its history) conforms to the looping feedback of the system, coming to resemble "a boomerang that threatens its 'throwers,' a whirling, circling metaphor for the historical pattern *Ratner's Star* composes" (LeClair 1987, p. 127). Like Pynchon, DeLillo frustrates the circular and the linear with the parabolic image that swerves between both. Where *Gravity's Rainbow* features the V-2 Rocket as the locus for mathematics of control via gyroscopic technologies that bend around gravity, *Ratner's Star* uses a more "primitive missile"—the boomerang— to conflate the linear and the circular images of progress and encyclopedic knowledge (Cowart 1999, p. 603). Crucially, the reflection movement within DeLillo's novel is emphasised by its mathematical imagery—specifically the stellated twilligon—through which he maintains an encyclopedic wealth of information in a dynamic, liberating structure.

Inversion in *Infinite Jest*

The topological influence of *Ratner's Star* is evident in David Foster Wallace's *Infinite Jest* (1996). The architectural structure of Wallace's tennis academy, where most of the plot is developed, is described early in the novel as a "cardioid" designed by "the topology world's closed-curve-mapping-Übermensch A. Y. ('Vector-Field') Rickey" (Wallace

1997, p. 983n3). This cardioid structure is later revealed to be an architectural homage to a "non-Euclidean figure on a planar surface, i.e., a cycloid on a sphere" (Wallace 1997, p. 502) that foregrounds the influence of mathematical imagery from the novel's father figures: both the fictional James Incandenza and the influential literary forefather Don DeLillo, in whose *Ratner's Star* the "cycloid" forms the "dominating shape" of the novel's research facility (DeLillo 1991, p. 15). This overt geometric engagement is just one example of many throughout *Infinite Jest* that illustrate Wallace's technical and literary interest in mathematics. As with Pynchon, Wallace's interest has often been traced back to his undergraduate studies. "For most of my college career I was a hard-core syntax wienie," he told Larry McCaffery, "a philosophy major with a specialization in math and logic" (McCaffery 2012, p. 34). His philosophy dissertation—which employed "semantic modality" to systematically defend free will against the determinism argued by American philosopher Richard Taylor—is an early example of Wallace's interest in the liberating potential of logico-mathematical language systems (Wallace 2011). Wallace's dissertation, then, defended a model of the universe in which free will, like Lucretius's atomistic clinamen, could be found in seemingly deterministic systems. Following *Infinite Jest* Wallace composed a popular science history of Cantor's mathematical innovations, *Everything and More: a Compact History of Infinity*. Developing such mathematical interests, Wallace came to appreciate

> that a mathematical experience was aesthetic in nature, an epiphany in Joyce's original sense. These moments appeared in proof-completions or maybe algorithms. Or like a gorgeously simple solution to a problem you suddenly see after filling half a notebook with gnarly attempted solutions. (McCaffery 2012, p. 34)

Michael Silverblatt was one of the first to observe Wallace's mathematical aesthetic in the "fractal" structure of *Infinite Jest*. In the interview, Wallace confirmed that the novel was initially "structured like something called a Sierpinski Gasket"—the fractal, attractive fixed set named in honour of Polish mathematician Wacław Franciszek Sierpiński—becoming "lopsided" only after "mercy cuts" from editor Michael Pietsch (Silverblatt 1996, 01:53–2:20). This geometric structure emphasises Wallace's aesthetic interest in mathematics as a means of describing his surroundings. The Gasket (or Triangle), Wallace tells Silverblatt

> looks basically like a pyramid on acid with certain interconnections between parts of them that are visually kind of astonishing and then the mathematical explanations are interesting. ... I would expect that somebody who's a mathematician or a logician ... might be interested in some of the fractal structures of [*Infinite Jest*]. For ... [i]t seems to me that so much of premillennial life in America consists of enormous amounts of what seem like discrete bits of information coming and that the real ... intellectual adventure is finding ways to relate them to each other and to find larger patterns and meanings – which of

course is essentially narrative but that structurally it's [sic] a bit different – and since fractals... its chaos is more on the surface, its bones are its beauty, that it would be a more interesting way to structure [the novel]. (Silverblatt 1996, 2:42–2:53)

Other geometric facets of Wallace's narrative structures have been discussed by Stephen J. Burn. In his *Reader's Guide*, Burn reads the novel's circles—and related processes of annulation (Wallace 1997, pp. 570–71), addiction (53) and reckoning (713)—alongside its encyclopedic scope and qualities. Unlike positively developmental "circles of learning" suggested by the encyclopedia, "the many circles in *Infinite Jest* ... bring little real knowledge." The book's final pages are "an invitation to circle back to the beginning of the narrative disk to review the crucial information" that for Burn highlights that "completing this circle of learning from the novel still leaves the reader's knowledge incomplete," suggesting that "part of Wallace's aim seems to be to break the closed circle and direct the reader outside of the book, to find what has escaped the encyclopedia" (Burn 2012, p. 29). After Burn, mathematician and literary critic Roberto Natalini has so far provided the most sustained mathematical reading of Wallace's work. Natalini argues that Wallace envisioned a mathematical "model for the possibility of direct communication" and that his "writing might be seen as a serious attempt to create a sort of mathematics of human thought" (Natalini 2013, p. 44).[9] According to Natalini, Wallace expresses this mathematical ratiocination with the imagery of inversion; I develop Natalini's reading by arguing that inversion functions as the mathematical clinamen of *Infinite Jest*.

In *Infinite Jest* mathematical inversions can be read as analogies for successful communication between author and reader. As a mathematical clinamen, the inversion becomes a locus for free movements against the terribly limiting circles—those of addiction, self-referentiality and solipsism—that abound in Wallace's encyclopedic novel. Natalini argues that the "main goal" of *Infinite Jest* is to animate "an inversion of our point of view." This, he argues, Wallace achieves by performing a mathematical "inversion process" which "creat[es] an exchange between the *inside* and the *outside*, so that the world enters the mind and, at the same time, the mind invades the world" (Natalini 2013, p. 51). James Incandenza's suicide in which he explodes his head with a microwave oven is such an inversion (Wallace 1997, p. 248). In doing so, James escaped the cycle of addiction which prevented him from conversing with his son. In another grotesque scene that takes place outside the narrative frame of the novel, Hal Incandenza (James's son) exhumes his father's skull in order to look inside it for the "Infinite Jest" master-tape—a fatally addictive movie with which James wanted "to bring [Hal] 'out of himself'" in order to converse (Wallace 1997, p. 839). Unfortunately, it is suggested, the container-content of that skull has already been inverted and its contents once again removed (Wallace 1997, p. 934). Ultimately, then, James's plot does not succeed—and is not completed—in this novel, originally titled *A Failed Entertainment* (Lipsky 2012, p. 172). Nevertheless, the inversion intervention prevents narrative closure, the sealing of the circle that maintains the open-possibilities of free will characteristic of the encyclopedic novel.

Wallace's intended perpetual systemic incompleteness in *Infinite Jest* can be read in the narrative of Michael Pemulis, a character whose mathematical abilities make him an avatar of the novel's deceased yet supernaturally present auteur James Incandenza. As recipient of "the coveted James O. Incandenza Geometric Optics Scholarship," Michael Pemulis is the most vocal champion of mathematical potential within the novel (Wallace 1997, p. 154). He uses his "real enduring gift for math and hard science" to bring a terrifying realism to the fictional game of Eschaton which leads to a cataclysm at the centre of the novel, precipitating much of the action to follow (Wallace 1997, pp. 154, 321–42). Pemulis comforts younger student Todd Possalthwaite, who is undergoing an epistemological crisis, assuring him. "Todder, you can trust math .... The axiom. The lemma .... Always and ever. As in puts the *a* in a priori. An honest lamp in the inkiest black, Toddleposter" (Wallace 1997, p. 1071). But, as Wallace told Caleb Crain, Pemulis was intended "to be 'one of the book's Antichrists,' and so readers shouldn't take anything he said unskeptically" (Crain 2008). Michael Pemulis's function within the system of *Infinite Jest* can be described by the anagrams—that is, mathematical inversions of the order of letters—found within his name: *alchemi* and *impulse*. As an anthropomorphisation of *alchemi* (or "alchemy"), he is both capable of "seemingly magical or miraculous power of transmutation" of the kind that produces gold from trash, yet also of "deceptive cleverness" (*OED Online* "alchemy"). Through this alchemical antichrist, Wallace inverts the notion of a mathematically guaranteed ontological stability, exposing its comforts as deceptive.

Furthermore, Pemulis functions very much like the *OED*'s sense 3a of "impulse," that is, the force or influence of a supposedly supernatural entity, whether through "suggestion, incitement, instigation," exerted upon the mind by some external stimulus. In this role, he influences the various inversions of the novel's main character Hal. The hyper-articulate Hal becomes monstrously inarticulate, perhaps as the result of psychotropic drugs procured by Pemulis (Wallace 1997, pp. 211–12). These drugs, street-named Madame Psychosis or M.P., share Michael Pemulis's initials and his ability to dissemble (Wallace 1997, p. 215). Pemulis's dissembling inclination, or "impulse," is further suggestive of that "sudden or involuntary inclination or tendency to act, without premeditation or reflection" (*OED Online* "impulse"). The same sense of impulse also characterises atoms that swerve in Lucretius's clinamen, an "injection of the aleatory into the motivated, upon the insertion of an element of chaos into a determinist symmetry" (Motte 1986, p. 264). And so, Pemulis, the mathematical antichrist of *Infinite Jest*, draws further attention to the novel's inversion clinamen. Beyond the novel, he summons the spirit of a figure whom Wallace identified as "the devil, for math": Kurt Gödel (Crain

164 S. TAYLOR

2012, p. 126). In 1931, Gödel published two mathematical logic theorems that proved the inherent limitations of any formal, arithmetic-inclusive axiomatic system. Henceforth known as his incompleteness theorems, Gödel's results are generally interpreted as fatal to Hilbert's project to find a complete and consistent set of axioms for the entirety of mathematics. In what can be considered an allusion to an intertextual inversion, Gödel's incompleteness theorems, it has been claimed, "turned not only mathematics but also the whole world of science and philosophy on its head" (Casti and DePauli 2001, p. 211).

Conclusion

Gödel's incompleteness theorems bring this study of mathematics as a means of illustrating systemic incompleteness in the contemporary encyclopedic novel full circle. For Wallace

> after Gödel the idea that mathematics was not just the language of God but a language that we could decode to understand the universe and understand everything ... that doesn't work anymore. It's part of the great postmodern uncertainty that we live in. (Crain 2013)

Wallace identifies the incompleteness theorems as one of the "Ur-texts for *Gravity's Rainbow*" (Crain 2013). This identification confirms Wallace's place in a lineage that extends from Pynchon, through DeLillo, one that employs mathematics as a structural clinamen to enable a creative swerve against systemic encyclopedism, even as it evokes such enclosures. The mathematical clinamen functions as a literary analogy to Gödelian systemic-incompletion: "when everything has been taken care of, when nothing can go wrong, or even surprise us ... something will" (Pynchon 2000, p. 275).

Or someone will. The mathematics explored in these encyclopedic novels have emphasised the role of subjective free will in resistance to enforced, totalising systems of knowledge that restrict action to a linear course of inevitable, deterministic progress. By reclaiming mathematics from such systems by figurative figurations—where mathematical symbolism and concepts foreground the novels' dynamic processes of, respectively, precession, reflection, and inversion—*Gravity's Rainbow*, *Ratner's Star*, and *Infinite Jest* each emphasise creative resistance as a fundamental, almost axiomatic, aspect of human knowledge and ontology. These authors not only bridge a perceived gap between the two cultures: by engaging mathematics within literature they empower readers with liberating operations of precession, reflection, and inversion.

NOTES

1. Pekka Kuusisto has also recently considered this famous mis-translation of Quintilian's phrase *"enkyklios paideia"* as being more accurately rendered as a "choric" or harmony-based educational program (Kuusisto 2009, p. 147).
2. The evident lineage these writers form suggests Harold Bloom's conception of the *clinamen* as part of artistic influence. This chapter, however, shall focus on the mathematical clinamen and its associated imagery instead of its broader legacy role. In this sense, it evokes the work of mathematically-inspired OuLiPo, particularly Georg Perec's call for the *clinamen* to resist totalising coherence (Perec and Pawlikowska 1983, pp. 70–71).
3. This also places greater significance on L. E. Sissman's assessment of Pynchon as "almost a mathematician of prose, who calculates the least and greatest stress each word and line, each pun and ambiguity, can bear[...]" (Sissman 1973, pp. 138–40).
4. Pynchon's alternative histories are, according to Linda Hutcheon, emblematic of "Historiographic Metafiction" by displaying a "theoretical self-awareness of history and fiction as human constructs ... [which] is made the grounds for its rethinking and reworking of the forms and contents of the past" (Hutcheon 1988, pp. 5–6).
5. Nina Engelhardt claims that "the equation is not a description of a rocket's flight for which Schachterle and Aravind try to find a matching formula, but ... it *does* constitute a valid equation for determining processes to control the direction of a rocket's flight." Engelhardt emphasises the performative aspects of Pynchon's equation: that "not able or willing to easily verify the equation, most readers will take its correctness on trust and thus unquestioningly bow to the authority of mathematics" (Engelhardt 2013a, p. 135). In a later essay, Nina and Harald Engelhardt trace Pynchon's equation to Otto Müller's 1957 essay "The Control System of the V-2." Yet, while they have managed to locate the source of Pynchon's equation, their reading discounts the ambiguity that remains in its symbolism (Engelhardt and Engelhardt 2018, p. 24).
6. Imagine a spinning-top wobbling as it begins to slow down.
7. Despite later readings by Glen Scott Allen (1994) and David Cowart (1999), LeClair's remains the most substantive account of this, one of DeLillo's most critically neglected novels.
8. DeLillo's notes on the structure of Part II, Ratner's Star Notebook (1975). Container 51.1. Don DeLillo Collection, Harry Ransom Center, University of Texas at Austin.
9. Natalini has since edited a "David Foster Wallace" special edition of *Lettera Matematica Pristem* (December 2015).

REFERENCES

Abernethy, Peter. 1972. "Entropy in Pynchon's The Crying of Lot 49." *Critique: Studies in Modern Fiction* 14 (2): 18–33.

Allen, Glen Scott. 1994. "Raids on the Conscious: Pynchon's Legacy of Paranoia and the Terrorism of Uncertainty in Don DeLillo's Ratner's Star." *Postmodern Culture: An Electronic Journal of Interdisciplinary Criticism* 4 (2): 1–28.

Burn, Stephen J. 2012. *David Foster Wallace's Infinite Jest: A Reader's Guide*. 2nd ed. London: Continuum.

Carus, Titus Lucretius. 1977. *The Nature of Things*. New York: Norton.

Casti, John L., and Wernder DePauli. 2001. *Gödel: A Life of Logic, the Mind, and Mathematics*. Cambridge, MA: Basic Books.

Clark, Hilary. 1990. *The Fictional Encyclopaedia: Joyce, Pound, Sollers*. London: Garland.

Cowart, David. 1981. "Science and Arts in Pynchon's 'Entropy'." *CLA Journal* 24 (1): 108–15.

———. 1999. "'More Advanced the Deeper We Dig': *Ratner's Star*." *Modern Fiction Studies* 45 (3): 600–20.

Crain, Caleb. 2008. "The Great Postmodern Uncertainty That We Live In." *Steamboats Are Ruining Everything*, September 14. steamthing.com/2008/09/the-great-postm.html.

———. 2012. "Approaching Infinity." In *Conversations with David Foster Wallace*, edited by Stephen J. Burn, 121–26. Jackson: University Press of Mississippi.

———. 2013. "Interview with David Foster Wallace, 17 October 2003, side 2." *Steamboats Are Ruining Everything*, July 24. steamthing.com/2013/07/audio-files-of-my-2003-interview-with-david-foster-wallace.html.

DeLillo, Don. 1991. *Ratner's Star*. London: Vintage.

Drake, Scott. 2010. "Resisting Totalizing Structures: An Aesthetic Shift in Thomas Pynchon's *The Crying of Lot 49* and *Gravity's Rainbow*." *Critique: Studies in Contemporary Fiction* 51 (3): 233–40.

Engelhardt, Nina. 2013a. "Formulas of Change, Alternative, and Fiction: Mathematics in Thomas Pynchon's *Gravity's Rainbow* and *Against the Day*." *Variations* 21: 131–44.

———. 2013b. "The Role of Mathematics in Modernist Utopia: Zamyatin's *We* and Pynchon's *Against the Day*." In *Utopianism, Modernism, and Literature in the Twentieth Century*, edited by Alice Reeve-Tucker and Nathan Waddell, 130–47. Basingstoke: Palgrave MacMillan.

———. 2014. "Gravity in Gravity's Rainbow—Force, Fictitious Force, and Frame of Reference; or: The Science and Poetry of Sloth." *Orbit: Writing Around Pynchon* 2: 1–26.

Engelhardt, Nina, and Harald Engelhardt. 2018. "The Momentum of Pynchon's Secret Formula: *Gravity's Rainbow's* Second Equation Between Archival Sources and Fiction." *Orbit: A Journal of American Literature* 6 (1): 1–27.

Friedman, Alan J., and Manfred Puetz. 1974. "Science as Metaphor: Thomas Pynchon and 'Gravity's Rainbow'." *Contemporary Literature* 15 (3): 345–59.

Harris, Robert R. 1982. "A Talk with Don DeLillo." *The New York Times*, October 10. nytimes.com/books/97/03/16/lifetimes/del-v-talk1982.html.

Hutcheon, Linda. 1988. *A Poetics of Postmodernism: History, Theory, Fiction*. London: Routledge.

Koestler, Arthur. 1959. *The Sleepwalkers: A History of Man's Changing Vision of the Universe*. London: Hutchinson.

Kuusisto, Pekka. 2009. "Gregory Benford's Against Infinity and the Literary, Historical and Geometric Formation of the Encyclopedic Circle of Knowledge." In *Science Fiction and the Two Cultures: Essays on Bridging the Gap Between the Sciences and the Humanities*, edited by Gary Westfahl and George Slusser, 140–59. Jefferson, NC: McFarland & Company.

LeClair, Tom. 1987. *In the Loop: Don DeLillo and the Systems Novel.* Urbana: University of Illinois Press.

Letzler, David. 2015. "Crossed-Up Disciplinarity: What Norbert Wiener, Thomas Pynchon, and William Gaddis Got Wrong About Entropy and Literature." *Contemporary Literature* 56 (1): 23–55.

Levine, George. 1978. "V-2." In *Pynchon: A Collection of Critical Essays*, edited by Edward Mendelson, 178–91. Englewood Cliffs, NJ: Prentice-Hall.

Lipsky, David. 2012. "The Lost Years and Last Days of David Foster Wallace." In *Conversations with David Foster Wallace*, edited by Stephen J. Burn, 161–81. Jackson: University Press of Mississippi.

Mangel, Anne. 1971. "Maxwell's Demon, Entropy, Information: The Crying of Lot 49." *TriQuarterly* 20: 194–208.

Mattessich, Stefan. 2002. *Lines of Flight: Discursive Time and Countercultural Desire in the Work of Thomas Pynchon.* London: Duke University Press.

McCaffery, Larry. 2012. "An Expanded Interview with David Foster Wallace." In *Conversations with David Foster Wallace*, edited by Stephen J. Burn, 21–52. Jackson: University Press of Mississippi.

McHale, Brian. 2000. "Mason & Dixon in the Zone, or, A Brief Poetics of Pynchon-Space." In *Pynchon & Mason & Dixon*, edited by Brooke Horvath and Irving Malin, 43–62. Newark: University of Delaware Press.

Mendelson, Edward. 1976. "Encyclopedic Narrative: From Dante to Pynchon." *MLN* 91 (6): 1267–75.

Motte, Warren F., Jr. 1986. "Clinamen Redux." *Comparative Literature Studies* 23 (4): 263–81.

Muth, Katie. 2015. "Archival Report." *American Studies in Britain: The BAAS Newsletter.* baas.ac.uk/muthfoundersaward2015.

Natalini, Roberto. 2013. "David Foster Wallace and the Mathematics of Infinity." In *A Companion to David Foster Wallace Studies*, edited by Marshall Boswell and Stephen J. Burn, 43–57. London: Palgrave Macmillan.

———. 2015. "David Foster Wallace e la Matematica." *Lettera Matematica Pristem* 95: 3–4.

Ozier, Lance W. 1974. "Antipointsman/Antimexico: Some Mathematical Imagery in Gravity's Rainbow." *Critique: Studies in Modern Fiction* 16 (2): 73–90.

———. 1975. "The Calculus of Transformation: More Mathematical Imagery in Gravity's Rainbow." *Twentieth Century Literature* 21 (2): 193–210.

Perec, Georg, and Ewa Pawlikowska. 1983. "Entretien." *Littératures* 7: 69–76.

Pikulin, E.P., and Stanislav I. Pohozaev. 2001. *Equations in Mathematical Physics: A Practical Course.* Basel: Birkhäuser Verlag.

Pynchon, Thomas. 1984. "Is It O.K. to Be a Luddite?" *New York Times*, October 28:1, 40–41.

———. 1985. "Introduction." In *Slow Learner: Early Stories*, 1–23. Boston: Back Bay Books.

———. 2000. *Gravity's Rainbow.* London: Vintage.

———. 2006. *Against the Day.* London: Jonathan Cape.

Richardson, Kenneth Ian Trevor. 1954. *The Gyroscope Applied.* London: Hutchinson.

Savet, Paul H. 1961. *Gyroscopes: Theory and Design with Applications to Instrumentation, Guidance and Control.* London: McGraw-Hill.

Schachterle, Lance, and P. K. Aravind. 2000. "The Three Equations in Gravity's Rainbow." *Pynchon Notes* 46–49: 157–69.

Seed, David. 1981. "Order in Thomas Pynchon's 'Entropy'." *Journal of Narrative Technique* 11 (2): 135–53.

Silverblatt, Michael. 1996. "David Foster Wallace: Infinite Jest." In *Bookworm*. KCRW.

Sissman, L. E. 1973. "Heironymus and Robert Bosch: The Art of Thomas Pynchon." *The New Yorker*, May 19: 138–40.

Slade, Joseph W. 1974. *Thomas Pynchon*. New York: Warner Paperback Library.

Vine, Steve. 2011. "The Entropic Sublime in Pynchon's the Crying of Lot 49." *Interdisciplinary Literary Studies: A Journal of Criticism and Theory* 13 (1–2): 160–77.

Wallace, David Foster. 1997. *Infinite Jest*. London: Abacus.

———. 2011. "Richard Taylor's 'Fatalism' and the Semantics of Physical Modality." In *Fate, Time, and Language: An Essay on Free Will*, edited by Steven M. Cahn and Maureen Eckert, 141–216. New York: Columbia University Press.

Winston, Matthew. 1975. "The Quest for Pynchon." *Twentieth Century Literature* 21 (3): 278–87.

Wisnicki, Adrian. 2000–2001. "A Trove of New Works by Thomas Pynchon? Bomarc Service News Rediscovered." *Pynchon Notes* 46–49: 9–34.

CHAPTER 10

Squaring the Circle: A Literary History

Robert Tubbs

One ancient geometric problem that has continued to engage the imaginations of writers, mystics, and amateur mathematicians is the problem of squaring the circle—the problem of geometrically constructing, using only a straightedge and compass, a square whose area equals that of a given circle. Mathematically this problem, the so-called *quadrature of the circle*, was resolved in the nineteenth century when this construction was shown to be impossible using the prescribed tools. But before that resolution, going back to at least the fifth-century BCE, mathematicians and non-mathematicians alike attempted to discover a construction for transforming a circle into a square. No hopeful circle-squarer ever succeeded in obtaining a square whose area is exactly that of a given circle using only a compass, which allows you to draw a circle once you have its center and its radius, and a straightedge, which allows you to draw a line segment between any two points. These failures led many to believe that such a construction might not be possible; indeed, as we will see, one reason authors invoked the problem of squaring the circle was to allude to impossibility.

In this chapter I will examine how three authors, Dante Alighieri, Margaret Cavendish, and James Joyce, employed this classical problem in their writings against the background of what was known mathematically about solving the problem when they wrote—respectively, in the fourteenth, seventeenth, and twentieth centuries. I will also discuss, as well as can be determined, what each author believed, knew, or could have known about the mathematical status of the problem and what this reveals about their use of the construction in their writings. My assertions about what each of these

---

R. Tubbs (✉)
University of Colorado Boulder, Boulder, CO, USA

© The Author(s) 2021
R. Tubbs et al. (eds.), *The Palgrave Handbook of Literature and Mathematics*, https://doi.org/10.1007/978-3-030-55478-1_10

169

170   R. TUBBS

authors knew or believed about the quadrature of the circle will be grounded in evidence: the evidence in Dante's case is compelling; in Cavendish's case, circumstantial; and in Joyce's case, slightly speculative but plausible.

THE FOURTEENTH CENTURY

One of the most dramatic appeals to squaring the circle appears in the closing lines of *Paradiso*, the third book of Dante's *Divine Comedy*, at the end of the pilgrim's (Dante's) journey through the nine circles of hell in the *Inferno*, the nine levels of sinfulness in the *Purgatory*, and ten heavens of *Paradiso*. This appeal occurs when the pilgrim is finally being united with God in the Empyrean, in Canto XXXIII. I will examine what has been written about Dante's use of this image, but I first want to look at something the pilgrim had learned earlier in *Paradiso* when he was in the fourth heaven—the sphere of the sun.

Upon their arrival in the sphere of the sun, in Canto X, Dante and Beatrice are encircled by twelve ephemeral beings. Among these souls is St. Thomas Aquinas (1225–1274), who speaks to Dante and Beatrice, first telling Dante that his eventual fate is to be united with God and then explaining that the other eleven are the "Fathers of the Church": philosophers and theologians whose ideas were central to the development of the foundations of the Catholic Church. Aquinas starts with the soul on his right, Albert the Great (before 1200–1280), who plays a role in this chapter for the influence of his writings on Dante, and then introduces each of the others in order. When he arrives at the fifth soul along the circle, Aquinas tells the travelers that this soul is "the loveliest here" and goes on to say:

> within it is the mind to which were shone
> > such depths of wisdom that, if the truth be true,
> > > no mortal ever rose to equal this one. (Alighieri 2003, pp. 681, ll. 109, 112–14)

This soul is that of the biblical King Solomon. Then, in Canto XIII while Dante and Beatrice are still in the fourth heaven, Aquinas elaborates on what he had said about Solomon earlier: "he was a king, and asked the Lord for wisdom / in governing his people and his land" (Alighieri 2003, pp. 707, ll. 95–96).

In particular, Aquinas explains that Solomon did not ask for answers to obscure fourteenth-century scholastic questions.[1] Aquinas then contrasts Solomon with others, for example, men who "fish for truth" but do not know "the art" so "return worse off than they were before." He continues:

> Of these, Parmenides and Melissus bear
> > their witness to all men, along with Bryson,
> > > and others who set out without knowing where. (Alighieri 2003, pp. 708, ll. 122–23, 124–26)

It is Aquinas's reference to Bryson that brings us back to our investigation into what Dante might have known or believed about the possibility of squaring the circle when he invoked it at the end of the *Divine Comedy*.

Bryson was a fifth-century BCE Greek philosopher and mathematician, one of several from that century who attempted to square the circle. The information that Bryson worked on this problem comes to us from comments by Aristotle in his *Posterior Analytics* and *On Sophistical Refutations*. Aristotle's remarks do not offer us any insights into how Bryson attempted to square the circle; instead they are critical of Bryson's understanding of the problem (as alluded to by Aquinas in the above lines from *Paradiso*). Since we are trying to determine what Dante might or might not have known about Bryson's attempt, we turn to two later commentators on Aristotle, both of whom Dante had studied, that gave alternate descriptions of Bryson's method.

One description of Bryson's approach was given by the second-to-third century CE commentator on Aristotle, Alexander of Aphrodisias. Alexander's portrayal of Bryson's solution can be summarized as follows: given a circle C, a square inscribed inside C has an area less than the area of C and a square circumscribed around C has an area greater than the area of C. It follows that there is a square between the inscribed and circumscribed squares whose area equals the area of the circle (Heath 1988, p. 223). Aquinas explicitly described the vague, non-geometric principle Bryson appealed to in this squaring of the circle of which Aristotle was critical: "[Bryson] showed that some square is equal to a circle, using common principles in the following way: In any manner in which it is possible to have something greater and something less than something else, one can find something equal to it" (Aquinas 1970, p. 55). The historian of mathematics T. L. Heath wrote that, more basically, Aristotle's criticism of Bryson was that he "made use of assumptions which were not confined to … geometry, but were equally applicable to other subjects" (Heath 1981, p. 48).

The second description of Bryson's construction, one which is not open to the Aristotelian criticism that it brings non-geometric ideas to bear on the problem, was given by another commentator on Aristotle that we have already mentioned, Albert the Great. The method described by Albert is not just important in helping us understand what Dante knew about the problem of squaring the circle, but it is important for our further discussions of mathematical progress toward understanding the problem in later centuries. Albert's version of Bryson's construction can be distilled down to three steps:

Step 1. Divide the circumference of the circle into four equal arcs.
Step 2. Straighten out each of these four arcs to obtain four equal line segments.
Step 3. Use these four equal line segments to form a square.

According to Albert, Bryson claimed that the area of this square equals the area of the original square (it is a simple calculation using the modern formulas

172   R. TUBBS

for the area and circumference to see that the produced square is a bit too small). Albert was critical of this construction, offering a curious rebuttal to its correctness, saying that the ratio of the chord to the arc is not the same as the ratio of the square to the circle (Albert the Great 1951, p. 18).[2]

My point is that Dante more than likely knew both of these versions of Bryson's quadrature—one importing a general principle that is not specific to geometry to claim that there exists a square whose area equals that of a given circle; and another that constructs a square by straightening-out (rectifying) arcs, each of which is one-fourth the circumference of the circle, and using those segments to form a square. We know from the *Divine Comedy*, *Paradiso* lines 125–26, above, that Dante was critical of the first method ascribed to Bryson, because he did not understand that the problem required a geometric construction, but we also know from another of Dante's writings that he would not have accepted the second method.

Before he wrote the *Divine Comedy* (c. 1308–1320) Dante wrote the *Convivio* (between 1304 and 1307) in which he had already concluded that a circle cannot be squared for a reason that undermines the second approach attributed to Bryson. In *Convivio* Dante wrote that "it is impossible to square the circle perfectly because of its arc, and so it cannot be measured exactly" (qtd. in Dasenbrock and Mines 2002, p. 84). By having the circle "measured" Dante could have meant one of two things: that a line segment cannot be constructed whose length exactly equals the circumference of a circle or that the area of the circle cannot be determined exactly because it is not a rectilinear figure. But for reasons that will become clear Dante almost certainly meant the former. This sense of measuring the circle implies, in particular, that a line segment cannot be constructed that exactly equals one-quarter of the circumference; so, Dante believed that Bryson's Step 2, above, was not possible. Dante's conclusion that the arc of a circle cannot "be exactly measured" may very likely have come from what Aristotle had written about the matter in *Categories*:

> the square is no more a circle than the rectangle, for to neither is the definition of a circle appropriate. In short, if the definition of the term proposed is not applicable to both objects, they cannot be compared (Ross 1949, p. 27).

Now that we have an understanding of what Dante believed about the possibility of squaring the circle, I will discuss how and what he might have been trying to accomplish by appealing to the problem at the end of *Paradiso*. This discussion is clearer if we examine *Paradiso*'s last fourteen lines with some care. The first seven of these lines begin to compare a potential circle-squarer, searching for the necessary principle to square the circle, to Dante's inability to understand what he sees when he encounters God:

> Like a geometer wholly dedicated
>   to squaring the circle, but who cannot find,
>   think as he may, the principle indicated—

so did I study the supernatural face.
  I yearned to know just how our image merges
  into that circle, and how it there finds place;
but mine were not the wings for such a flight. (Alighieri 2003, pp. 893–94,
ll. 133–39)

In these lines Dante strives to understand how the image of humanity can be, as John Ciardi wrote in the notes to his translation, "woven into the very substance and coloration of God," but he does not have the tools (Alighieri 2003, p. 895). This is the question of the Incarnation—how an infinite God could be present in the finite human form of Christ—and the overwhelming conclusion of Dante scholarship is that when he invokes the impossibility of squaring the circle he is comparing it to our inability to comprehend the Incarnation.

But the scholarship varies on exactly what Dante sought to convey when he appealed to the work of the geometer trying to find the secret to squaring the circle in his effort to understand the Incarnation. We will look at three different takes on Dante's appeal to a struggling geometer that can be placed on a scale of increasing geometrical germaneness. The first of these is from one of the leading twentieth-century Dante scholars, Charles Singleton. In his commentary on the closing lines of *Paradiso*, Singleton writes that "No poet was ever more daring in his final simile." He also tells us *why* Dante appealed to the quadrature of the circle at this stage of the pilgrim's journey, just as the circle and square cannot be related geometrically because they are incommensurable and so too are the "Word (divinity) and flesh" (Singleton 1975, pp. 584, 585). For Singleton, as scholars Ronald Herzman and Gary Towsley point out, Dante's appeal to the geometric problem was "no more than a comparison between two problems … each very difficult, if not impossible to human reason" (1994, p. 97).

In their own analysis Herzman and Towsley go further than Singleton in stressing the geometric aspects of Dante's use of the quadrature of the circle—specifically examining how Dante, while contemplating how the human figure can reside within the divine, is comparable to the geometer trying to square the circle. On this contemplation, Herzman and Towsley make two points: the first is the reason given by Singleton, that in trying to square the circle the geometer is trying to bring "the infinite circle and the finite square into a commensurable relationship with each other," just as the Incarnation did with "the infinite God and the finite humanity." But a second way in which Dante is like a geometer goes beyond the specific problem of squaring the circle—he is trying to bring reason to bear in trying to understand "a Christian reality that transcends reason." For Herzman and Towsley it is not just that the geometric impossibility of squaring the circle is a metaphor for Dante's inability to reconcile the mysteries of the Incarnation; it is, more generally, that mathematical reasoning, indeed rational thought, will

174   R. TUBBS

never lead to a comprehension of the transcendental nature of Christian ideals (Herzman and Towsley 1994, pp. 97, 112–13, 114).

Before we describe the third and most geometrically germane view of Dante's appeal to a geometric problem at the end of *Paradiso*, we need to see what happens in the lines following Dante's yearning for understanding, in lines 136-37, above. That yearning does not remain unfulfilled for very long. In the two lines following those quoted above we learn:

> Yet, as I wished, the truth I wished for came
> cleaving my mind in a great flash of light.

Following the "great flash of light," Dante's "instinct" balances with his "intellect" and he feels he is "being turned ... / as in a wheel whose motion nothing jars – / by the Love that moves the Sun and the other stars" (Alighieri 2003, p. 894, ll. 140–41, 144–46). With these concluding lines in mind, I now turn to the third analysis of Dante's appeal to squaring the circle. This one, by the poet Richard Levine, goes further than either Singleton or Herzman and Towsley in ascribing a role to geometry in aiding Dante's obtaining the truth he wished for at the very end of *Paradiso*. According to Levine, Dante was aware of Boethius's claim in his commentary of Aristotle's *Categories* that the "march of science" had availed him of a means to square the circle, and further, that Dante also had access to mystical Christian writings that ascribed to the Virgin Mary extraordinary geometric skills. Levine claims that at the end of *Paradiso* Dante "did not hesitate to draw upon both of these solutions" to square the circle "to establish his final harmony among the visionary company of love" (Levine 1985, pp. 281, 284).

Although Levine's conclusion is a mathematically interesting one, the text does not appear to support it. The truth Dante wished for was to comprehend his vision of humanity within God, "how our image merges / into that circle," an understanding he has as *Paradiso* closes without having to have squared the circle. As the Dante scholar John Freccero explained in his introduction to Ciardi's translation of *Paradiso*, Dante's final revelation "is not simply Beatific Vision, but a vision of the principle that renders intelligible the union of humanity and the divinity in the person of Christ" (Alighieri 2003, p. 593). This vision is not based on Dante's intellect or reason; it is a divinely inspired mystical insight—an insight not available to the struggling geometer applying reason to the problem.

Before we move on to a discussion of the problem of squaring the circle in seventeenth-century poetry, we look at an important transformation of the problem that Dante could have known about and that will inform our later discussions. In the third-century BCE the Greek mathematician Archimedes established that *if* the circumference of a circle could be rectified, that is, straightened-out into a line segment, the circle could be squared.[3] This observation follows from an amazingly simple result Archimedes established that, in principle, offered another geometric approach to squaring a circle.

10 SQUARING THE CIRCLE: A LITERARY HISTORY 175

If we begin with a circle whose radius is R and whose circumference is C, Archimedes established that the area of the circle equals the area of a right triangle whose legs have lengths R and C—so demonstrating that the area of the circle is ½ RC. A proposition in Euclid's *Elements* (c. 300 BCE) shows how to construct a square whose area equals that of any triangle. Putting Archimedes and Euclid's ideas together, we can describe the geometric constructions that *could*, in principle, produce a square whose area equals that of a given circle: construct a line segment of length C, and on this segment construct a right triangle whose legs are of lengths C and R. (By Archimedes's result the area of this triangle is equal to the area of the original circle.) Then use Euclid's proposition to construct a square whose area equals the area of the triangle. The area of this last square equals the area of the circle. The missing piece in this construction is, of course, the very first step—one that both Dante and Aristotle thought to be impossible because of the different natures of arcs and lines—to produce a line segment whose length equals the circumference of a given circle.

THE SEVENTEENTH CENTURY

In the seventeenth century several poets appealed to the problem of squaring the circle in their writings. Famously, John Donne (1572–1631) relied on the assumed impossibility of squaring a circle as a metaphor for his inability to express the nature of God in his writing, for example in "Upon the Translation of the Psalms by Sir Phillip Sydney, and the Countess of Pembroke, his Sister" (1621). Andrew Marvell (1621–1678), in "Upon Appleton House" (c. 1650), contrasted those with the vanity to try and square the circle with the humility of Lord Fairfax, the subject of his poem. But neither of these poets have easily discernable connections to mathematical developments in the seventeenth century. One aim of this essay is to view mathematical thinking about the quadrature of the circle through the lenses of authors who could have known something about the status of the problem. So instead of looking more closely at the thoughts of Donne or Marvell, I turn to another seventeenth-century poet, Margaret Cavendish (1661–1717), who appealed to squaring the circle and had demonstrable connections to some of the leading mathematicians and their thoughts about the possibility or impossibility of the construction.

Cavendish was not formally educated at a university, such an education was not available to women in seventeenth-century England, but the status of her father, the Duke of Newcastle, allowed her access to a well-stocked library. And she reportedly had many discussions on intellectual matters with her brother, John, who had studied law, philosophy, and the natural sciences. Her contact with specifically mathematical ideas likely came later, when she was married to William Cavendish who held salon-type gatherings of his Parisian intellectual acquaintances (Hutton 2009, pp. 198–99).

Before we get into any specific mathematical ideas Cavendish may have been exposed to, we note that two things happened between the fourteenth and seventeenth centuries that changed how mathematicians viewed their discipline. One was the development of algebra beyond what Europeans had learned from late first millennium Arabic documents, which, when combined with a coordinate system for geometry (what became the Cartesian plane), allowed mathematicians to use algebraic methods to examine geometric problems. The other was the bold acceptance of infinite process and infinite representations of numbers or other quantities long before these methods or representations were known to be mathematically sound. We begin with an infinite process that influenced René Descartes's view of the possibility of squaring the circle.

In 1628 Descartes (1596–1650) investigated the relationship between the diameter and circumference of a circle. He began with a line segment representing the circumference of a circle with the goal of finding a sequence of geometric steps that would yield the diameter (a reversal of what would be needed to use Archimedes's idea to square the circle). Descartes found a procedure which, when applied to the circumference-length segment, produces a somewhat shorter segment; and when this procedure is then applied to this shorter segment produces an even shorter segment. He established that if this process were to be iterated repeatedly and endlessly, an infinite number of times, the sequence of lengths produced would more and more closely approximate the diameter of the circle. We note that if it were possible to produce the diameter from the circumference in a finite number of steps, rather than Descartes's infinitely many, then this process could be reversed and the circle could be squared. Perhaps influenced by the infinitude of geometric steps that his method required to connect the circumference and diameter of a circle, in March 1638 Descartes wrote to another important seventeenth-century mathematician, Marin Mersenne (1588–1648), that "some problems are impossible like the quadrature of the circle." This indicates that by 1638 Descartes was convinced the circle could not be squared. Mersenne was not as certain as Descartes on this matter; four years earlier he had written that "geometers are split on the problem" (qtd. in Mancuso 1996, pp. 78–79).

We can take these two mathematicians to represent the uncertainty among early seventeenth-century mathematicians on the possibility of squaring the circle; and the reason Descartes and Mersenne's views of the quadrature of the square are important to our discussion of Cavendish's use of squaring the circle in her writings, is because they were both among the intellectuals with connections to the Cavendish family (Hutton 2009, p. 297). Cavendish very well may have heard through either of them that squaring the circle is more-likely-than-not impossible or that it might be possible. She certainly would have known that the matter had not yet been settled.

In 1653 Cavendish published a collection of poems and short prose pieces, *Poems and Fancies*, which reflected her philosophical and scientific ideas.

We want to focus on one section of the book that contains a poem and a prose piece that reveal a great deal about her understanding of the quadrature problem. We first look at the prose piece "A Circle Squared in Prose" (Cavendish 1653, pp. 48–49), especially at its first two sentences:

A Circle is a Line without Ends, and a Square is foure equall Sides, not one longer, or shorter than another. To square the Circle, is to make the Line of the Square Figure to be equall with the Round Figure.

Cavendish's description of the problem of squaring the circle is not the one we have been considering; hers is to produce a square whose "Line," that is the sum of the lengths of its four sides (its perimeter), equals the circumference of the circle. This is precisely the square Bryson's second construction, as reported by Albert the Great, would yield if it were possible to carry it out. If this had been the final status of Cavendish's writing about the quadrature of the circle we would not be able to hold her up as a seventeenth-century writer who was influenced by the developments in mathematics. Fortunately, one scholar, Liza Blake, has studied how Cavendish's writings changed over various editions of the texts (Blake 2017). By looking at Blake's examination of *Poems and Fancies* we learn that Cavendish corrected this initial misrepresentation of the classical problem in its two later editions, in 1664 and 1668. In each of these editions the second sentence reads: "To Square the Circle is to make the Square Figure to be equal with the Round Figure," so that the square has an area equal to that of the circle (Cavendish 1664, p. 50). We note here that this change in wording by Cavendish provides a bit more circumstantial evidence that she had engaged with someone about the problem of squaring the circle.

In the next couple of sentences of "A Circle Squared in Prose," Cavendish equates the circle with "Honesty" and writes that an honest person "is honest for Honesties sake." It follows, using her metaphorical thinking, that it is difficult to square a circle because you cannot have four sides without "Factions"—and where there are factions there cannot be only one truth, as is required by honesty (at least in Cavendish's thinking). Continuing with this metaphor, Cavendish does believe it is possible to square the circle representing honesty: to accomplish this one must replace the factionist sides of a square by sides representing the qualities of "Prudence, Temperance, Fortitude, and Justice," where these are mediated by the corners of the square representing "Obedience, Humility, Respect, and Reverence" (Cavendish 1653, pp. 48–49).

But in a poem from the same collection, "The Circle of the Brain cannot be Squared," Cavendish invokes the impossibility of squaring the circle. Since we just saw that Cavendish's thinking about the problem changed between the 1653 edition and the 1664 and 1668 editions of her collection, we look at the poem's text from these later editions. This twelve-line poem begins:

178 R. TUBBS

> A Circle round divided in four parts
> Hath been a Study 'mongst Men of Arts,
> Since Archimed's or Euclid's time, each Brain
> Hath on a Line been stretched, yet all in vain.(Cavendish 1664, pp. 58–59,
> ll. 1–4)

In the first couplet Cavendish refers to attempts to square the circle perhaps along the same lines as in Bryson's construction: an attempt to square the circle begins with dividing the circumference into four equal arcs. But in the second couplet, when she mentions "Since Archimed's [sic] and Euclid's time," she reveals a possibly deeper understanding of the problem. Cavendish appears to be aware that the circle could be squared if it could "on a line [be] stretched," rectified. She also seems to understand that attempts to square the circle using the Archimedes/Euclid result have been "all in vain."

In the next four lines of her poem, Cavendish describes various ideas that had been brought to bear on the problem of squaring the circle. Cavendish follows this with:

> But none hath yet by Demonstration found
> The way by which to Square a Circle round.
> For, while the Brain is round, no Square will be:
> While Thoughts divide, no Figures will agree.(1664, pp. 58–59, ll. 9–12)

wherein she employs the impossibility of squaring the circle as a metaphor for humankind's attempts to use rational thought, the square, in trying to understand the brain, representing human nature. More specifically, in the words of one scholar, in these last four lines Cavendish rejects as impossible "man's attempt to take control over [square] *irrationalia* [the circle] such as fancy and female nature" (Bertuol 2001, p. 21).

We can see in these two pieces how Cavendish's thinking reflects the ambivalence of her time on whether or not the circle could be squared. Yet it seems that she came down on Descartes's side and believed it was impossible to square the circle. In her prose piece she gives a metaphoric squaring of the circle, but she does not give any indication that the geometric circle is being squared; rather she is philosophizing about how different human values can be forged together to produce honesty. Then, in her poem, Cavendish offers the impossibility of squaring the circle as a metaphor for the inevitable failure of the seventeenth century's increasingly rational examination of nature.

What Cavendish probably did not know, and could not have known because she died in 1673, is that two of the most well-known mathematicians of the second half of the seventeenth century, Isaac Newton (1643–1727) and Gottfried Leibniz (1646–1716), employed infinite processes and representations in attempts to prove the impossibility of squaring the circle. Their demonstrations fell short of rigorously establishing this impossibility,

10 SQUARING THE CIRCLE: A LITERARY HISTORY   179

but they are evidence that a mathematical consensus was emerging that Dante and Descartes's as yet unproven conclusion that the circle cannot be squared was correct.

Before we move on to the twentieth century, we need to look at a less-geometric approach to the problem which lent evidence to mathematicians in the seventeenth and eighteenth centuries that the construction might not be possible, and ultimately led to a proof in the nineteenth century that the construction is indeed impossible. This different way to think about the quadrature problem is a consequence of Archimedes's result combined with yet another proposition from Euclid's *Elements*—to the effect that for any two circles, the ratios (area)/(diameter)^2 are equal. It is possible to deduce from these two ideas that for any two circles, the ratios (circumference)/ (diameter) are equal. Since these ratios are always equal, in the modern conception of the (real) number line as a continuum, there is a number that all of these latter ratios equal. We now denote this number by the symbol π; so for any circle we have the equation (circumference)/(diameter)$= \pi$. In particular, if we have a circle with a diameter of 1, its circumference will be π, and that particular circle can be squared if it is possible to start with a line segment of length 1 and construct a line segment of length π.[4] This is the conclusion that offered mathematicians an alternate path for attempting to understand whether it is possible to square a circle. Instead of working to find some elaborate geometric construction that transforms a circle into a square, one could try to determine if a line segment of length π can be constructed from a segment of length 1. And this question can be approached using methods not available in Euclidean geometry.

To see what this last claim means it is important to know that in the early eighteenth century, it was realized, although it had been suspected earlier, that the nature of a number might determine whether a line segment can be constructed that has a length equal to that number, and earlier than that mathematicians were beginning to believe that certain lengths could not be constructed. A number that is the length of a line segment that can be constructed, using only a straightedge and compass from a segment of length 1, is called a *constructible* number. Every rational number is constructible; so, if π were a rational number, for example, equal to $22/7$, the circle could be squared. But it is more complicated than that because even if π is not a rational number but an *irrational* number, it might be constructible. For example, $\sqrt{2}$ is an irrational number but it can be constructed: it is the diagonal of a one-by-one square so it can be constructed with a straightedge by just drawing a line from one corner of the square to the opposite corner.

At least one seventeenth-century mathematician, John Wallis (1616–1703), understood that the nature of the number π might play a role in settling whether or not a circle could be squared; and took up the challenge to use an infinite geometric process to calculate it. In 1656 Wallis sought to determine the area of a circle of radius 1 by using a number of tall, thin

180   R. TUBBS

rectangles. He then let the number of rectangles become infinite, so their widths became zero! He found that this area can be given by an infinite product of fractions. Specifically,

$$\frac{Area}{4} = \frac{2}{3} \times \frac{4}{3} \times \frac{4}{5} \times \frac{6}{5} \times \frac{6}{7} \times \frac{8}{7} \times \cdots$$

Using slightly more modern notation, and knowing that the area Wallis attempted to calculate equals π we obtain the representation

$$\pi = \frac{4}{1} \times \frac{2}{3} \times \frac{4}{3} \times \frac{4}{5} \times \frac{6}{5} \times \frac{6}{7} \times \frac{8}{7} \times \cdots$$

Wallis concluded, not just from this infinite product representation for the area but from how he obtained it, that this infinite product of fractions "cannot be forced out in numbers according to any method of notation so far accepted" (Wallis 2004, p. 161). While Wallis recognized that, in our notation, π might be a different type of number than mathematicians were encountering in algebra, he did admit that if it were possible to simplify his expression it might be possible to square the circle (Lutzen 2014, pp. 218–19, 221). As we will see in the next section, it was not until the nineteenth century that mathematicians discovered that Wallis's expression, and so π, cannot be simplified down to some sort of algebraic expression.

THE TWENTIETH CENTURY

In 1702 a translation from French of Ignace Gaston Pardies's geometry book was published in London with the title *Short, but yet Plain Elements of Geometry*. The English translation of Pardies's book went through many editions, at least to an eighth in 1748, but the one that interests us most is the 1711 edition. Our interest in this particular edition is not due to how the text might have changed over editions, which was crucial to our discussion of Cavendish's understanding of the quadrature of the circle, rather it is because of who has a copy of this edition on his bookshelf in 1904: Leopold Bloom from James Joyce's novel *Ulysses* (1922). It is not surprising that Bloom might have a geometry book in his possession; after all, we know from the narrator of the *Ithaca* episode of *Ulysses* that Bloom's bookshelf holds a fairly eclectic collection of books (Joyce [1922] 1961, p. 709). We also do not know if Bloom has studied the book, but it is reasonable to assume we are supposed to think he has. This is especially reasonable in light of the revelation (from the narrator of the *Ithaca* episode) that in the summer of 1886, when he was 20, Bloom was "occupied with the problem of the quadrature of the circle" (Joyce 1961, p. 699). Whether or not we are to imagine that Bloom consulted this particular book while working on the problem, we do know that Joyce purposely describes it as standing on Bloom's bookshelf.

10 SQUARING THE CIRCLE: A LITERARY HISTORY   181

This was probably not a randomly chosen geometry book; perhaps it is one Joyce had at least looked at.

Had Bloom looked at Pardies's text he would have seen that it was not a straightforward approach to Euclid's *Elements*. For example, many of the results are left as exercises for students.[5] More germane to our interests is that he would have discovered that Pardies's book includes a proof of Archimedes's result, equating the area of a circle with the area of a right triangle whose legs equal the radius and circumference of the circle; and he went on to show what we claimed above, namely, that Archimedes's result implies that one can square the circle if one can rectify its circumference. He concludes this discussion with: "But such an Equality is not to be found Geometrically," reflecting the mathematical consensus that had begun to emerge in the seventeenth century when Pardies's French text was first published in 1671 (Pardies 1734, pp. 51–52).[6]

That this book on his shelf is all the evidence Joyce gives us for what Bloom might know about the quadrature problem, but we know more from *Ulysses* about *when* Bloom had originally planned to work on the problem and *why*, and this provides a small insight into what Joyce might have known. In the *Circe* episode, on Bloomsday a couple of hours before the *Ithaca* episode, Bloom is visited by, or imagines being visited by, the ghost of his grandfather. The ghost tells Bloom: "You intended to devote an entire year to the study of the religious problem and the summer months in 1882 to square the circle and win the millions.... From the sublime to the ridiculous." Then two hours later in that same day, and two hundred pages later in the text, the narrator of the *Ithaca* episode answers a series of questions, one of which is: "What rapid but insecure means to opulence might facilitate immediate purchase [of a country home for Bloom]?" After discussing a few possibilities, the narrator ends with two others: "A proposed scheme based on the laws of probability to break the bank at Monte Carlo. [Or finding a] solution of the secular problem of the quadrature of the circle, government premium £1,000,000 sterling," Thus, according to the text, Bloom simply attempted to square the circle to win a one million British pound prize in order to acquire a country home (Joyce 1961, pp. 515, 717, 718).

It is interesting that, at least in the early editions of *Ulysses*, as quoted above, Joyce had the ghost say that Bloom had planned to work on squaring the circle in 1882, four years before the date given in the final edition (Kenner 1987, p. 167). So, Joyce originally was going to have Bloom toil away during the same year as the mathematician Ferdinand von Lindemann (1852–1939) published a short note proving that π is not a constructible number, and consequently showing that it is impossible to square the circle.[7] It is unlikely Bloom could have known about this result either in 1882, when it was published in a mathematical research journal in Germany, or four years later when he devotes a summer to the problem in the final version of *Ulysses*. We also do not know if we are supposed to imagine that Bloom knew of

182   R. TUBBS

Lindemann's result on Bloomsday in 1904 (there is no indication either way). However, Joyce himself could easily have known about it in 1914 when he started writing *Ulysses*, and even more easily by 1920 when he was writing the *Ithaca* episode that contains the reference to Bloom's efforts in the summer of 1886. So, the reference to 1882 might have been Joyce's hinting at the irony of Bloom struggling with the problem in the same year it was shown to be impossible to solve.[8]

It would have been convenient if we had found some support for the above claims about Joyce's awareness of Lindemann's proof in a geometry book that belonged to Joyce when he lived in Trieste, *A Text-Book of Euclid's Elements: For the Use of Schools*, by Hall and Stevens (1904) (Ellmann 1977, p. 111). This text, at least the 1904 edition we looked at, turns out to be a straightforward presentation of Books I-VI and Book XL of Euclid's *Elements*, as well as one elementary eighteenth-century mathematical result, with exercises added for students. It is likely that Joyce had carefully worked through some of this text—in a letter dated 28 May 1929 he wrote: "I have had too much to do, being up sometimes till 1:30 fooling over old books of Euclid and algebra" (qtd. in Ellmann 1975, p. 341). So we know that at least a decade after finishing *Ulysses*, Joyce was studying geometry.

But there are a few other paths that could have led Joyce to believe, or know, of the impossibility of squaring the circle. For one, Joyce could have believed that it is impossible to square the circle for the same reasons as Dante; after all, Joyce not only knew the *Divine Comedy* well but he "knew the sources of Dante's mathematical ... thinking from the inside" (Dasenbrock and Mines 2002, pp. 84–85). Beyond this, even if Joyce did not know of Lindemann's result by 1920, he could have known that mathematicians had long thought the construction impossible.

Additionally, by the beginning of the twentieth century Joyce could have learned of Lindemann's theorem from a popularization of mathematics by the German mathematician Hermann Schubert, *Mathematical Essays and Recreations* (first published in English in 1898). This short, 149–page book covers a wide range of topics and contains a concluding chapter, "The Squaring of the Circle," in which Schubert both discusses a bit of the history of the problem and Lindemann's result. I am not suggesting the Joyce read or even knew about Schubert's book; I mention this book to indicate that the popularization of mathematical ideas in the early twentieth century was not restricted to the widely discussed non-Euclidean geometries or the fourth dimension. It is highly unlikely that Joyce would have been unaware of these popularizations.[9]

Whether Joyce knew Lindemann had shown squaring the circle to be impossible or not, why do we believe Joyce presents Bloom as having spent an entire summer trying to solve the problem? The most likely reason is Dante's appeal to the problem at the end of *Paradiso*. Reed Dasenbrock and Ray Mines, a Joyce scholar and a mathematician, respectively, write that "Joyce's ambition for *Ulysses* was to write the [*Divine Comedy*] of the twentieth century" (Dasenbrock and Mines 2002, p. 80). This claim may or may not be an exaggeration: if it is, it is not exaggerating much. As another Joyce scholar, Mary Reynolds, has noted, Joyce "announced" that Homer and Shakespeare were his guides while writing *Ulysses*. Reynolds also argues that the next writer on Joyce's list of guides would certainly have been Dante; elements of the *Divine Comedy* are infused throughout *Ulysses* (Reynolds 1981, p. 3).

The difficulty facing anyone writing about the influence of the *Divine Comedy* on *Ulysses* is how to "describe an influence that is elusive yet pervasive" (Reynolds 1981, p. 4). I will not survey the research that supports these pervasive influences of the *Divine Comedy* on *Ulysses* and Joyce's other writings; hundreds of these influences are annotated in the fascinating appendix to Reynolds's book. It is possible that Joyce's allusion to squaring the circle is just another snippet from the *Divine Comedy* that made its way into *Ulysses*. But, owing to that problem's striking and memorable appearance at the end of *Paradiso*, Joyce most likely employed this problem to make an indelible connection between *Ulysses* and the *Divine Comedy* and thus an irrefutable connection between himself and one of his literary guiding lights, Dante Alighieri.

Conclusion

It is inarguable that each of the authors we have discussed, Dante Alighieri, Margaret Cavendish, and James Joyce, employed the ancient problem of squaring the circle in at least one of their writings. The point of this investigation was to try and answer the question of *why* each of them did so, and one approach to answering that question, the one I have taken here, was to try to determine what they each might have thought about the possibility or impossibility of constructing the appropriate square and letting that determination influence our thinking about the *why* question. As to these authors thinking about the quadrature of the circle: Dante explicitly wrote that he thought it was impossible, for somewhat categorical reasons; Cavendish was in a position to know that many mathematicians were beginning to believe it was impossible, and she used that assumed impossibility as a potent metaphor. Joyce is the only one of these three authors who could have, and I argued should have, known that mathematicians had established in the year of his birth that squaring the circle is impossible. Interestingly, Joyce is the only one of these authors who does not appeal to this problem to allude to the notion of impossibility. Joyce's use of the problem of squaring the circle is in support of a higher cause—to strengthen his connection with Dante to help secure his place in the pantheon of our most significant authors and observers of the human condition.

NOTES

1. For example, knowing how many angels are required to move each of the celestial bodies or "if in the space of a semicircle a non-right triangle/may be drawn with a diameter as its base." Lines 101−2 (p. 707). This last question about the semicircle is asking about the truth of a proposition proved in Euclid's *Elements*. We will take up a discussion of this reference elsewhere.
2. I thank my colleague Professor Robert Pasnau (Department of Philosophy, University of Colorado, Boulder) for help with Albert the Great's description of Bryson's construction and his commentary on it.
3. Dante most certainly knew of Archimedes result. One source for him could have been Leonardo de Pisa's (Fibonacci's) *De Practica Geometry* (1220), wherein he discusses Archimedes' result without attribution (Fibonacci 2008, pp. 152−53).
4. The British mathematician William Jones introduced the symbol π for this ratio in 1706.
5. See Wardhaugh's chapter of this *Handbook*.
6. Later in the text Pardies performs the deduction we did not include above, that for any circle the ratio of the circumference to the diameter yields the same number. Pardies denoted this number by e and derived the formula: Area $= e$(radius)^2; this was before Jones' symbol π was widely used (Pardies 1734, p. 105).
7. Lindemann actually proved the stronger result that π is a *transcendental* number.
8. Alternatively, it might have been a nod to the year, 1882, of Joyce's birth.
9. In his notes for the *Ithaca* episode, Joyce mentions Bolyai and Lobachevsky, who independently discovered non-Euclidean geometry in the early nineteenth century (Dasenbrock and Mines 2002, p. 87).

REFERENCES

Albert the Great. 1951. *Physica*. Edited by B. Geyer et al. Münster: Aschendorff.

Alighieri, Dante. 2003. *The Divine Comedy: The Inferno, the Purgatorio, and the Paradiso*. Translated by John Ciardi. New York: New American Library.

Aquinas, Thomas. 1970. *Commentary on Posterior Analytics of Aristotle*. Translated by F. R. Larcher. Albany, NY: Magi Books.

Bertuol, Roberto. 2001. "The *Square Circle* of Margaret Cavendish: the 17th-century Conceptualization of Mind by Means of Mathematics." *Language and Literature* 10: 21−39.

Blake, Liza. 2017. *Margaret Cavendish's Poems and Fancies, Part I*. http://poemsandfancies.rblake.net/. Accessed on July 12, 2017.

Cavendish, Margaret. 1653. *Poems and Fancies*. London: Printed by T. R for F. Martin and F. Allestrye.

———. 1664. *Poems and Fancies. The Second Impression, much Altered and Corrected*. London: Printed by William Wilson.

Dasenbrock, Reed Way and Mines, Ray. 2002. "'Quella Vista Nova': Dante, Mathematics and the Ending of *Ulysses*." In *Medieval Joyce*, edited by Lucia Boldrini, 79−91. Amsterdam and New York: Rodopi.

Ellmann, Richard (ed.). 1975. *Selected Letters of James Joyce*. New York: The Viking Press.

———.1977. *The Consciousness of Joyce*. Toronto and New York: Oxford University Press.

Fibonacci. 2008. *Fibonacci's De Practica Geometry*. Edited by Barnabas Hughes. New York: Springer.

Hall, H.S., and F.H. Stevens. 1904. *A Text-Book of Euclid's Elements*. London: Macmillan.

Heath, Thomas. 1981. *A History of Greek Mathematics*, vol. I. New York: Dover.

———. 1988. *Mathematics in Aristotle*. Bristol, England: Thoemmes Press.

Herzman, Ronald B. and Gary W. Towsley. 1994. "Squaring the Circle: *Paradiso 33* and the Poetics of Geometry." *Traditio* 49: 95–125.

Hutton, Sarah. 2009. "In Dialogue with Thomas Hobbes: Margaret Cavendish's Natural Philosophy." In *Ashgate Critical Essays on Women Writers in England, 1550–1700: Volume 7 Margaret Cavendish*, edited by Sara H. Mendelson, 192–208. Surrey, England: Ashgate Publishing Limited.

Joyce, James. 1961. *Ulysses*. New York: Vintage Books.

Kenner, Hugh. 1987. *Ulysses: Revised Edition*. Baltimore: Johns Hopkins University Press.

Levine, Robert. 1985. "Squaring the Circle: Dante's Solution." *Neuphilologische Mitteilungen* 86: 280–84.

Lutzen, Jesper. 2014. "17th Century Arguments for the Impossibility of the Indefinite and Definite Circle Quadrature." *Revue d'Histoire des Mathématiques* 20: 211–51.

Mancosu, Paolo. 1996. *Philosophy of Mathematics and Mathematical Practice in the Seventeenth Century*. New York and Oxford: Oxford University Press.

Pardies, F. Ignat. Gaston. 1734. *Short, but yet Plain Elements of Geometry*, 7th ed. Translated by John Harris. London: D. Midwinter.

Reynolds, Mary T. 1981. *Joyce and Dante: The Shaping Imagination*. Princeton, NJ: Princeton University Press.

Ross, W.D. 1949. *Aristotle's Prior and Posterior Analytics*. Oxford: Oxford University Press.

Schubert, Hermann. 1898. *Mathematical Essays and Recreations*. Translated by Thomas J. McCormack. London: Kegan, Trench, Truebner, & Co.

Singleton, Charles. 1975. *The Divine Comedy: Paradiso, Volume 2, Commentary*. Princeton, NJ: Princeton University Press.

Wallis, John. 2004. *The Arithmetic of Infinitesimals*. Translated by Jacqueline A. Stedall. New York: Springer.

PART II

Mathematics and Literary Forms

CHAPTER 11

Mathematics and Poetic Meter

Jason David Hall

This chapter explores meter in poetry and its relationship to mathematics, specifically to theories of ratio, harmony, and abstraction. We find, from the earliest attempts to define rhythm in poetry to more recent accounts of verse's measure and proportion, poets and metrical theorists borrowing from both mathematical vocabulary and paradigms in order to explain the way verse lines are segmented and how their units can be counted and measured. Indeed, etymologically, meter and mathematics are closely related. Among the various aspects of poetic form signalled by the cognate yet distinct terms *meter*, *prosody*, and *versification*, it is the first—derived from the Greek μέτρον, meaning variously meter in poetry, measure, length, or rule—that signals most explicitly poetry's connections with mathematics. Perhaps because of this connection between meter and mathematics, the well-known twentieth-century prosodists Karl Shapiro and Robert Beum, in their influential *Prosody Handbook* (1965), preferred to discuss the structural elements of verse under the aegis of *prosody*, emphasizing poetry's connection with song and music, rather than *metrics*. For them the term *prosody* is appropriately poetic: it suggests "interiority" and something "primordial." *Meter*, by contrast, "has a slight but still undesirable overtone of the mathematical or the clinical": like the science-sounding *versification*, it suggests something "contrived," calculated, inertly available for scientific analysis (Beum and Shapiro 1965 [2006], p. x). Yet it is precisely the measuring imperative of meter, with its more or less explicit mathematical leanings, that has proved to be among the most enduring and contested aspects of poetic practice and criticism. Indeed, it persists in mid-century "practical" accounts of meter, including Beum and Shapiro's, in spite of putative disavowals. In the tradition

J. D. Hall (✉)
University of Exeter, Exeter, UK

© The Author(s) 2021
R. Tubbs et al. (eds.), *The Palgrave Handbook of Literature and Mathematics*, https://doi.org/10.1007/978-3-030-55478-1_11

of English verse, as well in its classical antecedents, the question of how to measure has been central to poetic epistemology: how we know what a poem is and what it does largely comes down to how we measure it or describe its metrical constitution. Cutting across theories that prioritise the vocal and musical and ones that assert abstract ratios of verse, metrics is a study grounded, right from the tradition of Pythagoras, in mathematical laws, formulas, and computational exercises. Thus, meter, in a manner of speaking, *is* verse's mathematics. And we can learn much more about the history of meter if we take time to consider how the mathematics of meter has been expressed from classical to modern times. By enhancing our metrical numeracy, we will perceive how complex the question of how to measure and define the verse line has been and what answers have been given—answers that have shaped how poets composed and how readers have sought to explain how poems work.

The equation *meter=mathematics* has a mythology as well as a history. For the sixteenth-century Italian humanist Polidoro Virgili (*c.* 1470–1555), the original metrician was none other than God, whose world is a testament to measurement, harmony, and "formal concord": "The beginner of Meter [was] the true God, which proportioned the world; with all the contents of the same, with a certain order as if it were a Meter: for there is none (as *Pythagoras* taught) that doubteth, but that there is in things Heavenly and Earthly, a kind of harmony; and unless it were governed with a formal concord and described-number, how could it long continue?" All the particular metrical formations devised by human poets—the *Iambus*, the *Hexameter*, the *Pentameter*—were merely the manifestations of a "spiritual influence" of this divine proportioning (Virgili [1663] 1868, pp. 25–26). Pythagoras himself—whose "doctrine of *emanations*" posited "God is the soul of the universe, pervading all things incorporeal"—placed numbers and their proportional relations at the cornerstone of his philosophy, and these ideas underpinned his conception of a universe "maintained by the laws of harmony" (Mills 1854, p. 423). The conventions of classical quantitative meter as established by the Greeks and perpetuated through Roman versification were founded upon these principles of harmony and proportion. Thus, we often find references to poetic "numbers," another word "for metrical feet, meter, and hence verse." As Terry Brogan, one of the great historians of meter, observes, "the notion of [numbers] involved the idea of mathematical proportion in meter (chief among which was 1 long=2 shorts) and was linked to the proportions of musical harmonies" (Brogan 1993, p. 845). Not only were these features foundational for classical metrics; they would also become key ingredients for the modern English metrical tradition that began to define itself during the Renaissance. Citing Richard Willes's *De Re Poetica* (1573), the literary critic and historian of meter Derek Attridge demonstrates how the "numerological thought" of Pythagoras, as embodied by classical quantitative theories of meter, "encouraged poets like [Edmund] Spenser and [Philip] Sidney to organise poems on strict numerological principles."

11 MATHEMATICS AND POETIC METER   191

An appreciation of "the harmonious patterning of syllables," often arrived at via a process of counting and "mechanical calculation" quite distinct from an ordinary reading of verses, was central to this early modern numerological metrics. Attending to the abstract mathematical patterns of verse was, for the early architects of English metrics, a part of the classical inheritance that meshed with "the Elizabethan desire for 'artificiality'" (Attridge 1974, pp. 115, 116, 114).[1]

This numerological tradition, modelled on classical quantitative verse, prevailed in English poetry, marginalizing the older tradition of accentual verse as exhibited by Old and Middle English texts such as the fourteenth-century romance *Sir Gawain and the Green Knight*. From the sixteenth to the eighteenth centuries, meter was more or less dominated by a prosody of proportion, harmony, and "numbers." John Milton's "*English* Heroic Verse" in *Paradise Lost* (1667), for example, asserted itself in "full harmonic number[s]" (Book 4, line 687), placing itself in the tradition "of *Homer* in *Greek*, and of *Virgil* in *Latin*." It was a feature that the poet felt it necessary to gloss in his note on "The Verse" as "apt Numbers, [a] fit quantity of Syllables," unmarred by the "defect" of "Rime" (Milton 1971, pp. 249, 250, 334). Several later commentators seized on Milton's choice of blank-verse numbers—that is, his use of unrhymed iambic pentameter—as a hallmark of the poem's greatness. Writing in his *A Critique and Notes upon the Paradise Lost* (1764), Joseph Addison (1672–1719) identifies Milton's ability to "var[y] his numbers in such a manner" as a beneficial means of mitigating the monotony that otherwise would have arisen from "the same uniform measure" recurring line after line. By abandoning end rhyme and attending to "the length of his periods," argues Addison, Milton makes his poem more engaging for the reader (Addison 1764, p. 23). Indeed, the true merit of a poet, for Addison and many of his contemporaries, lay in his command of numbers. Perhaps the most deliberate and exemplary statement of this Augustan axiom came in 1711, with the publication of Alexander Pope's (1688–1744) prescriptive *An Essay on Criticism*, where the poet elevates a mastery of numbers above all other aspects of the artform:

> But most by *Numbers* judge a Poet's Song,
> And *smooth* or *rough*, with such, is *right* or *wrong*,
> In the bright *Muse* tho' thousand *Charms* conspire,
> Her *Voice* is all these tuneful Fools admire;
> Who haunt *Parnassus* but to please their Ear,
> Not mend their Minds; as some to *Church* repair,
> Not for the *Doctrine*, but the *Music* there.
> These *Equal Syllables* alone require,
> Tho' oft' the Ear the *open Vowels* tire[.] (Pope 1711, p. 21, ll. 339–347)

The emphasis here—literally in Pope's own italics—is on numerical proportion ("*Equal Syllables*"), which may (or may not) "please [the] Ear." Pope even encourages his readers to count syllables in his performance of ponderous metrical movement, where we find what the respected twentieth-century literary critic Northrop Frye once called the "*reductio ad absurdum* of heavily stressed monosyllables" (Frye 2006, p. 243): "And ten low Words oft' creep in one dull Line" (Pope 1804, l. 347).[2] As we can see from these examples, the two senses of *meter* that Shapiro and Beum had wished to disentangle—its mathematics and its music—are very much imbricated, with tunefulness, harmony, and numeration all forming part of Pope's encomium to "right" versification.

If one feature of eighteenth-century mathematical thought was the "language of proportion," largely associated with the geometry of Euclid's *Elements*, then another was an increasing use of scientific instruments. It was a century in which instruments and processes of mechanization began to assert a new calculating imperative. As the seventeenth century drew to a close, Gottfried Wilhelm Leibniz (1646–1716) was busy imagining a brave new world of mechanical mathematics: "the entire arithmetic could be subjected to a … kind of machinery so that not only counting, but also addition and subtraction, multiplication and division could be accomplished by a suitably arranged machine easily, promptly, and with sure results." Even "the major Pythagorean tables" (e.g., ones setting out squares and cubes) could benefit from mechanical assistance (Leibniz 1959, pp. 173, 181). Throughout the eighteenth century, "artisans, engineers, and natural philosophers" attempted to perfect variations on this theme (Jones 2016, p. 129). It is, therefore, not surprising that we should find at the same point in history examples of a rather more mechanical attention to poetic numbers, erring in the direction of the overly contrived approach to meter that Shapiro and Beum would later attempt to hold at arm's length. In his 1737 book *Stichology: Or, a Recovery of the Latin, Greek, and Hebrew Numbers*, the prosodist Edward Manwaring (fl. 1737–1744) presents an enquiry into "the Numbers of Poetry," their harmonies, and proportions, asking how we ought to understand the metrical "systems" of the Ancients. In doing this, Manwaring looks back directly to the mathematical underpinnings of metrics, pointing to the "metrical Species of Geometry" advanced by Aulus Gellius (*c.* 125–180 A.D.), where the "geometrical Ratio" of verses involved, for hexameters, "the Proportion of Squares"—the six-foot lines of the ancients being conceived of as having a cognate pattern in the self-multiplying structures of square numbers (Manwaring 1737, pp. 1, 2). Manwaring is looking for nothing less than a correct equation that resolves the inconsistencies he perceives in extant theories of meter through a more deliberate numerology—hence his book's title *Stichology*, a "science or theory of poetic metres" (*OED*). What begins, however, as a means of establishing a ratio of quantity proportions for the classical verse line becomes a more narrowly focused exercise in counting. The ratios of feet (e.g., pyrrhic, spondee, dactyl) are presented in a table of proportions that can be doubled (or followed through other multiples) according to "the harmonic Rule, invented by *Pythagoras*" (Manwaring 1737, p. 17). Even

11 MATHEMATICS AND POETIC METER   193

more mechanically enumerative is the "Artificial Versifying" table included in Manwaring's text (Fig. 11.1). Here we have poetic numbers reduced to actual integers. Whether or not a reader has a grasp of the "Latin Tongue" or its meters is irrelevant: an ability to count and make selections from the table based on this counting will enable him or her to "make both Hexameter & Pentameter Latin Verses" (Manwaring 1737, n.p.).

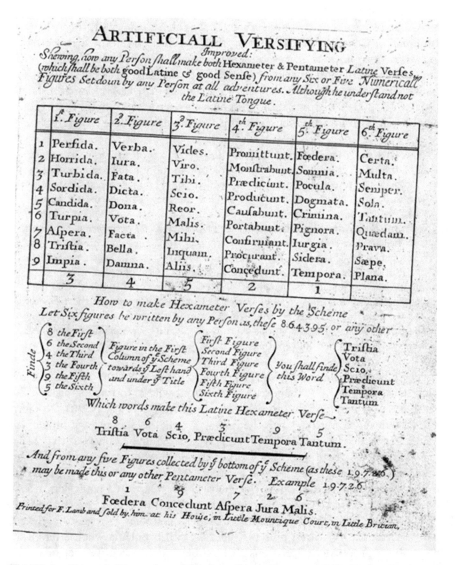

Fig. 11.1 Edward Manwaring, *Stichology: Or, a Recovery of the Latin, Greek, and Hebrew Numbers* (London: n. pub., 1737), n. p.

194   J. D. HALL

While the term "numbers" began to fall out of regular use among poets and prosodists in the nineteenth century, the mathematical impulse that governed eighteenth-century metrics did not. If anything it became more entrenched. Questions of proportion and harmony remained at the heart of English metrical discourse, and the musical and measurement aspects of *prosody* and *meter* continued to interact. Maintaining the association of meter and music, Edgar Allan Poe (1808–1849), in his 1843 "Notes upon English Verse," offers to complicate the prioritizing of harmony in English metrics by asserting a mathematics of his own, one that accepts a proportional arrangement but places more emphasis on melody. Taking issue with the American grammarian Goold Brown's assertion that "versification is the art of arranging words into lines of correspondent length, so as to produce harmony by the regular alternation of syllables differing in quantity," Poe "dispute[s] the essentiality of any alternation, regular or irregular, of syllables long and short" (Poe 1843, p. 103).[3] Instead, he proposes an attention to meter that recombines music with measurement, in terms that align "numbers" with rhythm:

> Now rhythm, from the Greek ἀριθμος, *number*, is a term which, in its present application, very nearly conveys its own idea. No more proper word could be employed to present the conception intended; for rhythm, in prosody, is, in its last analysis, identical with *time* in music. For this reason I have used, throughout this article, as synonymous with rhythm, the word metre from μετρον, *measure*. Either the one or the other may be defined as *the arrangement of words into two or more consecutive, equal, pulsations of time.* These pulsations are *feet*. Two feet, at least, are requisite to constitute a *rhythm* .... The syllables of which the foot consists, when the foot is not a syllable in itself, are subdivisions of the pulsations. No equality is demanded in these subdivisions. It is only required that, so far as regards two consecutive feet at least, the sum of the times of the syllables in one, shall be equal to the sum of the times of the syllables in the other. Beyond two pulsations there is no *necessity* for equality of time. All beyond is arbitrary or conventional. (Poe 1843, p. 105)

In Poe's metrical sums, rhythm in poetry is equated "with *time* in music"; there is a glimmer of Pythagorean proportion, where "numbers" carry "the attributes and ratios of the musical scales" (Aristotle 1987, p. 96). Notably, Poe also leans toward, rather than away from, the measuring impulse of meter. Yet he departs from this thinking as well. While he concedes that a certain nominal proportioning is at work in English, where consecutive feet are more or less equal in time to one another, he remains unconvinced that the principle of harmony on which the metrical convention of equivalence rests need be adhered to slavishly. Harmony and proportion play a role in the metrical structure of a verse line, but there are also melodies in verse to consider, whose "irregularity" often presents metrists with problems when attempting to establish proportional relations that go beyond consecutive feet or across several lines of verse (Poe 1843, p. 106). The deliberate counting involved in scanning verses—a counting given hyperbolic expression in Manwaring's tables—can compound this problem by encouraging readers to divide lines into too-regular periods—that is, scanning can encourage readers to exaggerate the "equal" spacing between accents in feet or fall into sing-song delivery, where every unit of verse monotonously echoes those before and after it. To simply count periods and add them up, as though they were merely integers, is to neglect the music of verse.

However much Poe may have lamented the prevailing insistence on counting and enforced regularity, it nonetheless remained central to nineteenth-century metrical discourse, as did the language of proportion and harmony. In what is among the more influential prosody treatises of the mid-Victorian period, the 1857 "Essay on English Metrical Law," Coventry Patmore (1823–1896), remembered primarily for authoring the poem *The Angel in the House* (1854–1862), established his own method of counting or measuring the verse line. This essay inaugurated what the later prosodist T. S. Omond would term the New Prosody, which was characterized by its designation of meter as an abstract "modulus" or measurement index against which the delivery of verse (for instance when it is read aloud or recited from memory) can be measured. For Patmore, as for some of his contemporaries and successors, English metrics was fundamentally temporal, rather than accentual, in character. Here again was the familiar notion—a variation on the proportional element we observed in Renaissance metrics—that musical harmonies underpinned English metrics. Yet unlike those other theories, which attempted to transpose the features of classical quantitative verse onto English, Patmore's metrical "law" all but disregarded the duration of actual syllables in favor of a mental "beat." According to Patmore,

> English metre depends, in contradistinction to the syllabic metre of the ancients, … [on] the marking, *by whatever means*, certain isochronous intervals. Metre implies something measured … . The thing measured is the time occupied in the delivery of a series of words. But time measured implies something that measures, *and is therefore itself unmeasured*; an argument before which those who hold that English accent and long quantity are identical must bow. There are two indispensable conditions of metre, – first, that the sequence of vocal utterance, represented by written verse, shall be divided into equal or proportionate spaces; secondly, *that the fact of that division shall be made manifest* by an "ictus" or "beat," actual or mental, which, like a post in a chain railing, shall mark the end of one space, and the commencement of another …. Yet, all-important as this time-beater is, I think it demonstrable that, for the most part, *it has no material and external existence at all*, but has its place in the mind, which craves measure in everything, and, wherever the idea of measure is uncontradicted, delights in marking it with an imaginary "beat". (Patmore 1961, p. 15)

Patmore's metrics of abstract proportion is, to an extent, in keeping with Poe's musical metrics; however, here the emphasis is very much upon lines divided into not only regular but equally proportionate periods, as marked by the mental beating of the ictus. For Patmore, this perfectly proportioned meter is indebted to the idealist theories of the philosopher G. W. F. Hegel (1770–1831), themselves inspired by the Pythagorean tradition of harmonic proportions.[4] Where Poe discussed two units forming one rhythmical *number*, Patmore, who is more interested in harmonies than melodies, establishes "the isochronous *bar* for the metrical integer" (Patmore 1961, p. 33). Here, interwoven, are *meter*—the numerical equivalence of measurement—and *prosody*—the inherently musical constitution of the verse line.

Via the New Prosody the idea of proportion and ideal ratios played a significant role in Victorian metrical theory and practice, as historical prosodist Yopie Prins and others have demonstrated.[5] As we will see below, a strain of this metrics persisted into the twentieth century, where its abstract proportions were inherited and modified by the architects of Practical Criticism (in Britain) and the New Criticism (in America). But in the intervening period, both metrics and mathematics would also take a turn away from the ideal and immaterial toward the decidedly mechanistic and material, away from abstract measurement and toward an ostensibly more decimalized, instrumental form of numeration. To an extent, this story might be understood as a fulfilment of Leibniz's fantasy about a machine that might revolutionize mathematics. Just such a machine began to take physical form in the first half of the nineteenth century when the British mathematician and engineer Charles Babbage (1791–1871) began to assemble his Difference Engine. Before he took up the Lucasian Chair of Mathematics at the University of Cambridge in 1828, Babbage had been at work on a device which would facilitate an automatic calculation process. This machine was envisaged as ameliorating the burdensome and error-prone work of human "computers," whose job was to compile extensive tables of mathematical calculations. In addition to Leibniz, whose ideas of mechanical calculation directly informed this project as well as his contribution to mathematics at Cambridge, Babbage was also influenced by the eighteenth-century French mathematician Gaspard de Prony, who had brought an assembly line process of human application to the production of mathematical tables.[6] Blending Leibniz's mechanical imagination with de Prony's large-scale project of tabular compilation, Babbage embarked on what he himself described, in his 1832 book *On the Economy of Machinery and Manufactures*, as a "science of *calculation*" (Babbage 1842, p. 316). Babbage never completed his Difference Engine, abandoning it into work on an even more ambitious project: the Analytical Engine, which promised not only a "simplification of the arithmetical processes to repeated addition and subtraction" (Swade 2003, p. 71) but more complex calculations where the machine itself "decided what formula to use" (Abbas 2006, p. 52).

Around the same time, a meter machine—conceived on similar principles—was being built by a Somerset inventor named John Clark. This device, known as the Eureka, was, like Babbage's Difference Engine, also partly a means of mechanizing tables, specifically "versifying" tables such as those included in Manwaring's eighteenth-century treatise on prosody. The process of counting numbers to form a line of verse found curious instantiation in Clark's machine, which expressed the tabular data as a series of lettered staves and revolving drums (Fig. 11.2). When set in motion, the Eureka could produce around twenty-six million lines of verse—all of them in Latin and all of them in the hexameter measure of the ancient epics. Yet while words in the right metrical order were the end product of Clark's mechanical versifier, the function of the device, again like Babbage's calculating machine,

Fig. 11.2 View of gears from John Clark's Eureka Latin Hexameter Machine (*Source* Photograph by the author, 2015)

was fundamentally mathematical. The Eureka's "process of composition," as explained by one of Clark's contemporaries in the *Illustrated London News*,

> is not by words already formed, but from *separate letters*.... The machine contains *letters* in alphabetical arrangement. Out of these, through the medium of *numbers*, rendered *tangible* by being expressed by indentures on wheel-work; the instrument selects such as are requisite to form the verse conceived; the *components* of words suited to form hexameters being alone previously calculated, the harmonious combination of which will be found to be practically interminable. ([Anon.] 1845, p. 37)[7]

Alleviating the labor-intensive activities of another class of Victorian human "computers"—boys in classical schools whose day-to-day task it was to "manufacture" Latin hexameters—the Eureka's true medium was not the metrical foot but its mathematical proxy: the numerical integer. Whole lines of verse could be reduced to a numerical code, with the machine capable of combining a vast number of combinations, each one different from the last. Thus, Clark's mathematical meter machine not only facilitates the "composition" of Latin verses; it also belongs to the history of applied mathematics, specifically to the development of code by the process of random-number generation. Because the Eureka combines numbers (letters) at random (only the movement of the apparatus's moving parts determined which letter came next, not some pre-determined or sequential pattern), one cannot predict which will come next in a given sequence. Such devices would play an important role in the applied mathematics of computer science and cryptography, particularly in relation to the work of the twentieth-century British mathematician Alan Turing and his successors. Long before the code-breaking work undertaken by Turing and his colleagues at Bletchley Park, the Eureka exhibited a mechanized mathematics predicated on randomness that hinted at the artificial intelligence machines Turing would propose in his 1950 work "Computer Machinery and Intelligence." With the aid of a mathematical "program," it might be possible to make a machine capable of some limited verse "thought."

"'Mathematics'," as Vladimir I. Arnold has pointed out, "means 'precise knowledge'," and machines played an increasing role in allowing nineteenth- and early twentieth-century mathematics to realize the potential of its name (Arnold 2005, p. 225). Measurement devices began to feature more and more centrally in a number of sciences, where precise numbers, rather than merely notionally equivalent proportions, were part of an accelerated agenda of exactness and empirical proof. This shift has been read by Friedrich Kittler as a transition from "logic" to "physics," where the idealized mathematics of Pythagoras gave way to the language of "frequencies," which new apparatuses enabled researchers to measure with greater accuracy (Kittler 1999, p. 24). No longer were ratios and fractions thought to be precise enough when instruments could measure a physical response—such as pressure or vibration—in numbers expressed to three decimal digits. This empirical, mechanized form of numeration was especially crucial to the development of sciences such as physics, physiology, and psychology, where measurement devices and laboratory experiment produced ground-breaking work from about the 1840s. A leading light in this field of enquiry was Hermann von Helmholtz (1821–1894), whose work on matters of acoustics would directly inform methods and machines later used to provide an ostensibly more accurate measure of the verse line. For Helmholtz, like many of his experimental contemporaries and successors, the language of harmony, as inherited from classical idealist philosophy, would no longer suffice. "What have the ratios of small whole numbers to do with harmony?" queried Helmholtz in "The Physiological Causes of Harmony in Music" (1857). "This is an old riddle, propounded by Pythagoras and hitherto unsolved. Let us see whether the means at the command of modern science will furnish an answer" (von Helmholtz [1857] 1971, p. 76). To better understand musical harmonies, Helmholtz used this "modern science" and its various instruments—some of which he invented himself—to produce precise measurements of sound vibrations, underwritten by graphic records that devices could inscribe. Instead of the comparably rough proportions of Pythagorean ratios, Helmholtz offered a quantifiable language, intervening in existing debates about matters such as isochronism, which Patmore was at the same mid-century moment theorizing in relation to English verse.

Influenced by this and related laboratory work, the next generation of empirical researchers, most of them operating in the fields of experimental phonetics and psychology, would attempt to find a more precise way to measure verse lines. By the 1880s, one of the pioneers of modern psychology, Wilhelm Wundt, was working in his laboratory in Leipzig "to determine the accuracy with which the ear can distinguish musical intervals" (Cattell 1888, p. 43). These experiments would feed directly into instrument-assisted calculations of the "rhythms of the poetic text,"[8] enabling other verse scientists, many of them also working in psychological laboratories to replace assertions about meter's ratios and proportions with exact numbers—specifically in relation to such features of versification as accent, temporality, and the function of pause. Edward Wheeler Scripture, Warner Brown, and Ada Snell, to name just three, would turn their attention to questions of proportion, isochrony, and related matters, favoring the extreme accuracy of measurement their devices provided: the wavelengths of voice vibrations could be measured and graphed (using instruments such as kymographs) with "results in thousandths of a second" ([Anon.] 1893, n.p.). Often suspicious of "the ordinary routine scansions" of verse, which too often lapsed into the vague terms of proportion and harmony, these experimental prosodists gave versification a new computational medium: decimal notation. Instead of assuming the ratios based on rational numbers might provide the keys to meter's periods, they speculated that, in some instances, irrational numbers (those that cannot be expressed as a ratio of two integers) might hold a clue to the

200   J. D. HALL

"real" movement of the verse line—when it is voiced, that is, as opposed to the "imaginary" metrics that had been central to Patmore's verse theory. Their so-called specimen records—data and graphs of a subject reading in a given meter (e.g., iambic, trochaic, or dactylic)—allowed them to both see and accurately measure "the length of time between successive movements" of verse (Scripture 1902, p. 523). From this the instrumental metrists were able to offer evidence supporting a challenge to the longstanding theory of meter as a matter of ideal ratios. Discussing rhythm in speech, Scripture contests the assumption that a

> series of vocal sounds is divided into relatively large portions of equal lengths, or "measures"; these measures are divided either into two proportions, thesis and arsis, bearing simple relations of length, or into a number of small proportions termed "morae" .... The relations of length between thesis and arsis will be as 2 : 1, 1 : 1, etc. Such a mathematical relation was called by the Greeks ... "rational." A rhythm with such simple relations of length may be called a "rational rhythm", or, perhaps preferably, an abstract rhythm. In vocal music we would expect to find the nearest approach to the relations of abstract rhythm, although even here measurements will show that the actual relations are not exact or constant. (Scripture 1902, pp. 551–52)

Flying in the face of the abstract theories of Patmore and other nineteenth-century proponents of the New Prosody, Scripture here rejects "rational," evenly proportionate verse in favor of "the actual concrete rhythm of a particular piece of verse," which is "irrational" because it is "a compromise between the natural lengths"—which he describes as "voiced syllables inflected by 'usage ... accent, emotion, etc.',—and those required by abstract rhythm" (Scripture 1902, p. 552).

The laboratory work of Scripture and his contemporaries promised to reveal the verifiable, mathematically supported "truths" of meter, but it did no such thing. Some researchers used the numbers to support Scripture's assertion that there is no actual proportional consistency between feet of the same measure, while others, such as Snell, would use machines to hold on to the ratios inherited from classical prosody—in this case the proportionate syllabic units of the verse foot. "The foot is a fact," Snell would write in her 1918 study *Pause*. As laboratory metrics began to wane in the 1920s, another, ostensibly less scientific, method was on the rise. In 1929 the influential English literary critic I. A. Richards, who expressed reservations about laboratory assessments of meter, published his landmark book *Practical Criticism*. Though Richards denied that a kymograph could resolve the contest between the abstract pattern of meter and what he called the "actual sounds in verse," he in fact demonstrated agreement with the axiom of Scripture that rhythm is an "irrational" and complex comprise between "natural" speaking and abstract metrical patterning (Richards [1929] 2004, p. 219). So while he holds such experimental methods and the mathematics and sciences associated with them at bay, exhibiting dissatisfaction with the fantasy of accurate measurement and the promise of machines to disclose the numerical "truths" of meter, he nonetheless brokers a compromise of his own: between the more rigorously abstract theories of Patmore, for example, and the messier "actual" rhythms that verse scientists had found it necessary to look beyond simple ratios to express. This position is very much in keeping with Richards's general inclination toward a "practical" or pragmatic and less "scientific" approach to prosody, and also of the emergence of a new era of metrical abstraction that would attempt to disavow its affiliations with the overtly scientific approach to poetry that dominated the turn of the century. Signally, Richards was instrumental in articulating an "art of responding to the form of poetry," laying the groundwork for the central metrical claims of the American New Critics and providing a template for classroom practice (Richards [1929] 2004, p. 214). Poetry, for Richards, is first and foremost about a dialogue between poet and reader, and the dialectic between meter and rhythm is merely a subordinate means of establishing clear lines of communication. Positions regarding meter were inseparable from larger questions regarding textual stability and the determinability of "meaning." The fundamentals of metrical patterning, in such a context, and in particular the idea that meter provides "support" or "orientation" for poets and readers alike, are extensions of Richards's belief that clear, unequivocal communication is not only desirable but possible. In *Practical Criticism* Richards gives coherent expression to the concept of metrical "orientation," made possible by a poet's adoption and a reader's recognition of abstract metrical patterning. For Richards, meter constitutes a "pattern" (perhaps "only a convenience, though an invaluable one") that is at once "inherent in" the sounds and rhythms of verses themselves and "*ascribed* to verses" by readers. Meter, as Richards imagines it, is "the skeleton upon which the reader casts flesh and clothing": "it gives both poet and reader a firm support, a fixed point of orientation in the indefinitely vast world of possible rhythms" (Richards [1929] 2004, pp. 219, 218, 219).

This concept of metrical "orientation," which explains a poem's rhythm in terms of a given line's vacillation between "departures from and returns to the [poem's prevailing metrical] pattern," is integral not only to Richards's verse theory but also to many mainstream twentieth-century prosodies.[9] Sometimes articulated in terms of "expectation-based tension" or "expressive variation," it forms the bedrock of conventional post-1900 Anglo-American metrics, though its roots reach back much further, and it enjoyed "pedagogic perpetuation" (Holder 1995, p. 23) throughout the twentieth century, becoming, in some prosodists' estimation, a hegemonic metrical doctrine.[10] As Alan Holder observes, influential textbooks—for example, Cleanth Brooks and Robert Penn Warren's *Understanding Poetry* (first published in 1938 and reissued in numerous editions) and Karl Shapiro and Robert Beum's *A Prosody Handbook* (mentioned above), among others—followed Richards's

202   J. D. HALL

lead in promoting a "compromise" between metrical pattern and rhythm, conceived of in terms of "rhetorical variation." Shapiro and Beum went so far as to set out a "right way to read English verse," which stressed the give and take between "a 'natural' reading in which one delivers the verse as if it were prose or ordinary speech, simply observing conventional phrasing and logical emphases; and a metrical reading," which pays attention to "the general metrical pattern" (Beum and Shapiro [1965] 2006, p. 30).

Probably the most enduring expression of the extent to which New Critics at the mid-century espoused theories of metrical abstraction is an essay titled "The Concept of Meter: An Exercise in Abstraction," published in *PMLA*, the Modern Language Association of America's field-leading journal, in 1959 by W. K. Wimsatt and Monroe C. Beardsley. Wimsatt and Beardsley shared Richards's insistence on limiting, on the one hand, arcane scientific critiques of meter—New Critics were quick to stress that literary criticism was not properly a branch of linguistics, much less experimental rhythm science as practiced by psychologists—and, on the other hand, overly subjective interpretations. Linguistics-based criticism they deemed too "needlessly contrived," and mechanically aided studies of meter—they cite an "analysis of eight tape-recorded readings of a short poem by Robert Frost"—are rejected for focusing too much on "the stress-pitch-juncture elements in spoken English" at the expense of "the normative fact of the poem's meter" (Wimsatt and Beardsley 1959, p. 585). Where researchers such as Scripture had looked to mathematically demonstrable facts (the data of their kymographs), and where the more linguistics-focused critics, such as Trager and Smith, had read linguistic analysis as the mathematics of criticism,[11] the New Critics invested the abstraction of meter with its own matter-of-fact factual status. Theirs was not a mathematical but a "good sense" metrics, and a variety of Richards's "orientating" metrics was, for Wimsatt and Beardsley as well, a more productive way of thinking about meter. A good sense reading does not need a scientist's numbers to show that there is a slippage between the ideal proportions of meter in the abstract and rhythmical voicings of it because the abstract pattern of meter was there to limit the extent of licence possible. The abstraction of meter is the "truth" of the poem and authorizes (or not) possible performances of it. As Wimsatt and Beardsley write: "Each performance of the poem is an actualization of it, and no doubt in the end everything we say about the poem ought to be translatable into a statement about an actual or possible performance of it. But not everything which is true of some particular performance will be necessarily true of the poem" (1959, p. 587). Only the unchanging, abstract pattern of meter can underwrite the many possible performances, and form a basis for rhythmical interpretations. Meter, in other words, is the structure that enables the poem to function. In a nod toward Richards's metrics of "orientation," they uphold the metrical poem as "a public linguistic object, something that can be examined by various persons, studied, disputed – univocally" (Wimsatt and Beardsley 1959, p. 588).

Even though there are similarities between these theories of metrical abstraction and Patmore's idea of metrical modulus (that meter is a regular series of time-keeping, imaginary beats), just as there were points of connection between the idea of metrical "orientation" and Scripture's "irrational" metrics, it is clear that the "newer" and more avowedly "practical" methods that dominated the Anglo-American study of poetry for much of the twentieth century wanted to distance themselves from both the sciences and mathematics on which the connections rest. To understand poetic meter, one does not need exact numbers, theories of isochronous temporality, or received quantitative models. Like Shapiro and Beum, Wimsatt and Beardsley treat with suspicion the measuring capacity of meter. Admitting that "[m]eter involves measurement, no doubt" (Wimsatt and Beardsley 1959, p. 590), they prefer a method that emphasizes "the recognition of the few homely and sound, traditional and objective, principles of prosody." That their essay should uphold this particular term, *prosody*, aligning it with a "good sense" approach to verse form demonstrates the extent to which they, like Shapiro and Beum, feel at odds with the mathematical heritage of metrics, in spite of the continuities that their theory has with the several strains of metrical measurement that this essay has discussed. In this reluctance to compute meter, to put numbers against its units and values, and in the tendency to rejoice in its "tension" and "interplay," we can discern a rhetorical strategy where authenticity triumphs over artificiality, where the mathematical aspect of meter is devalued in favor of a more common-sense appreciation (Wimsatt and Beardsley 1959, pp. 596, 597). Yet in identifying the crucial role that the abstract metrical pattern plays in a poet's or reader's experience of a poem, the New Critical prosodists nevertheless leave a glimmer of mathematical code, even if not acknowledged as such, in the background as a reminder of the tradition of counting and measurement that has figured so centrally to the history of meter.

NOTES

1. Attridge points out how "the comparison of the artist's creation of harmony with God's creation of the universe," such as we saw in the quotation from Polidoro, "was a commonplace ... and had been used by Augustine with specific reference to quantitative metre" (Attridge 1974, p. 115).
2. Diaeresis is metrical phenomenon that occurs when syntactical and metrical units coincide, here to exaggerate the plodding movement of the line.
3. Poe's short essay is considered to be a trial run of the ideas he would develop in more detail in his later and longer essay, "The Rationale of Verse" (1848).
4. See, for example, Schnädelbach (2010, p. 77).
5. See Taylor (1989), Prins (2000), and Rudy (2009).
6. On Babbage and Leibniz, see Higham (2015, p. 65).
7. For a more in-depth account of the Eureka and its relationship with machinery and mathematics, see Hall (2017, pp. 11–163).
8. See Golston (2008), p. 71.

204   J. D. HALL

9. See also Cureton (1992, pp. 8–9).
10. See Finch (1993, pp. 6–10).
11. Wimsatt and Beardsley quote disapprovingly Harold Whitehall's overview of the linguistic metrics of Trager and Smith. As Whitehall notes, "as no science can go beyond mathematics, no criticism can go beyond linguistics" (qtd. in Wimsatt and Beardsley 1959, p. 585).

REFERENCES

Abbas, Niran. 2006. *Thinking Machines.* Berlin: Lit. Verlag.

Addison, Joseph. 1764. *A Critique and Notes upon the Paradise Lost. From the Spectator.* London: n. pub.

[Anon.]. 1845. "The Eureka." *Illustrated London News,* 19 July: 37.

[Anon.]. 1893. "Measurement of Thought: Yale's Psychological Laboratory and Its Work" *New York Times,* December 10: n.p.

Arnold, Vladimir I. 2005. "Mathematics and Physics." In *The Role of Mathematics in Physical Sciences: Interdisciplinary and Philosophical Aspects,* edited by Giovanni Boniolo et al., 25–233. Dordrecht: Springer.

Aristotle. 1987. "The Pythagoreans." In *The History of Mathematics: A Reader,* edited by John Fauvel and Jeremy Gray, 1–2. Basingstoke: Palgrave Macmillan.

Attridge, Derek. 1974. *Well-Weighed Syllables: Elizabethan Verse in Classical Metres.* Cambridge: Cambridge University Press.

Babbage, Charles. 1842. *On the Economy of Machinery and Manufactures.* London: Knight.

Beum, Robert, and Karl Shapiro. [1965] 2006. *The Prosody Handbook: A Guide to Poetic Form.* Mineola: Dover.

Brogan, Terry. 1993. "Number(s)." In *The New Princeton Encyclopedia of Poetry and Poetics,* edited by Alex Preminger and T. V. F. Brogan, 845. Princeton: Princeton University Press.

Cattell, James McKeen. 1888. "The Psychological Laboratory at Leipsic." *Mind* 13 (49): 37–51.

Cureton, Richard. 1992. *Rhythmic Phrasing in English Verse.* London: Longman.

Finch, Annie. 1993. *The Ghost of Meter: Culture and Prosody in American Free Verse.* Ann Arbor: University of Michigan Press.

Frye, Northrop. [1956] 2006. "Introduction: Lexis and Melos." In *Collected Works of Northrop Frye,* vol. 21, edited by Germaine Warkentin. Toronto: University of Toronto Press.

Golston, Michael. 2008. *Rhythm and Race in Modernist Poetry and Science.* New York: Columbia University Press.

Hall, Jason David. 2017. *Nineteenth-Century Verse and Technology: The Machines of Meter.* London: Palgrave Macmillan.

Higham, Nicholas J., ed. 2015. *The Princeton Companion to Applied Mathematics.* Princeton: Princeton University Press.

Holder, Alan. 1995. *Rethinking Meter: A New Approach to the Verse Line.* Cranbury: Associated University Presses.

Jones, Matthew L. 2016. *Reckoning with Matter: Calculating Machines, Innovation, and Thinking About Thinking from Pascal to Babbage.* Chicago: University of Chicago Press.

Kittler, Friedrich. 1999. *Gramophone, Film, Typewriter.* Translated by Geoffrey Winthrop-Young and Michael Wutz. Stanford: Stanford University Press.

Leibniz, G. W. 1959. "On His Calculating Machine," translated by Mark Kormes. In *A Source Book in Mathematics,* edited by David Eugene Smith, 173–81. New York: Dover.

Manwaring, Edward. 1737. *Stichology: Or, a Recovery of the Latin, Greek, and Hebrew Numbers.* London: n. pub.

Mills, Abraham. 1854. *The Poets and the Poetry of the Ancient Greeks; with an Historical Introduction, and a Brief View of Grecian Philosophers, Orators, and Historians.* Boston: Phillips, Sampson.

Milton, John. [1667] 1971. *Paradise Lost.* In *The Complete Poetry of John Milton.* Rev. ed., edited by John T. Shawcross. New York: Anchor Books-Doubleday.

Patmore, Coventry. 1961. *Coventry Patmore's "Essay on English Metrical Law": A Critical Edition with a Commentary.* Edited by Mary Augustine Roth. Washington, DC: Catholic University of America Press.

Poe, Edgar A.. 1843. "Notes upon English Verse." *The Pioneer,* March.

Pope, Alexander. [1711] 1804. *An Essay on Criticism.* In *The Works of the Poets of Great Britain and Ireland with Prefaces, Biographical and Critical,* vol. 6, edited by Samuel Johnson. Dublin: Pat Wogan.

Prins, Yopie. 2000. "Victorian Meters." In *The Cambridge Companion to Victorian Poetry,* edited by Joseph Bristow, 89–113. Cambridge: Cambridge University Press.

Richards, I. A. [1929] 2004. *Practical Criticism: A Study of Literary Judgment.* New Brunswick: Transaction.

Rudy, Jason R. 2009. *Electric Meters: Victorian Physiological Poetics.* Athens, OH: Ohio University Press.

Schnädelbach, Herbert. 2010. "Hegel." In *Music in German Philosophy: An Introduction,* edited by Stefan Lorenz Sorgner and Oliver Fürbeth, 69–94. Chicago: University of Press.

Scripture, Edward Wheeler. 1902. *The Elements of Experimental Phonetics.* New York: Charles Scribner's Sons.

Swade, Doron David. 2003. "Calculation and Tabulation in the Nineteenth-Century." PhD diss., University of London.

Taylor, Dennis. 1989. *Hardy's Metres and Victorian Prosody.* Oxford: Clarendon Press.

Virgili, Polidoro. [1663] 1868. *Polydori Virgilii de Rerum Inventoribus; Translated into English by John Langley; with an Account of the Author and His Works, by William A. Hammond, M.D.* New York: Agathynian Club.

von Helmholtz, Hermann. [1857] 1971. "The Physiological Causes of Harmony in Music." In *Selected Writings of Hermann von Helmholtz,* edited by Russell Kahl, 75–108. Middletown: Wesleyan University Press.

Wimsatt, W. K., and Monroe C. Beardsley. 1959. "The Concept of Meter: An Exercise in Abstraction." *PMLA* 74 (5): 585–98.

CHAPTER 12

Randomizing Form: Stochastics and Combinatorics in Postwar Literature

Alison James

si c'était / le Nombre / ce serait / le Hasard
[if it were / the Number / it would be / Chance]
—Mallarmé, *Un coup de dés jamais n'abolira le hasard*

The intersection of mathematics and literary form often involves the harnessing of chance. What is sometimes called "procedural" writing aims in general to produce unforeseen results. It seeks the novelty of unexpected combinations by suspending or displacing authorial intention. The genealogy of such procedures includes the combinatorial poems of the French *rhétoriqueurs*, such as the fifteenth-century poet Jean Meschinot (1420–1491), whose *Litanies of the Virgin* (1461–1464) allow more than 36,000 combinations. The seventeenth-century German philosopher Gottfried Wilhelm Leibniz found inspiration for his *Dissertatio de Arte Combinatoria* (1666), which proposes a universal logical system based on combinatorics, in the "Protean" permutational poetry of Latin and German baroque poets (Bernardus Bauhusius (1576–1619), Thomas Lansius (1577–1657), Georg Philipp Harsdörffer (1607–1658), and Julius Caesar Scaliger (1484–1558); Quirinus Kuhlman (1651–1689) might be added to the list) (Berge 1986, p. 118). Within a fixed poetic form, the permutation of linguistic elements leads to a set of results that are finite, but extremely large, in relation to the time of reading. There is a mystical as well as a playful dimension to such endless litanies. These poems echo the philosophical method proposed by the Catalan

A. James (✉)
University of Chicago, Chicago, IL, USA

© The Author(s) 2021   207
R. Tubbs et al. (eds.), *The Palgrave Handbook of Literature and Mathematics*, https://doi.org/10.1007/978-3-030-55478-1_12

logician Ramon Llull (circa 1232–1315) in his *Ars magna* (1305), in which a system of diagrams and rotating circles simultaneously gestures toward the infinite nature of God and claims to exhaustively explore the possibilities of finite human knowledge (Gardner 1958, pp. 1–27). Medieval and early modern philosophy and literature, in producing generative forms that configure hierarchized yet mobile elements, thus offer some early glimpses of the structural possibilities that Umberto Eco associates with twentieth-century "open works," which "are brought to their conclusion by the performer at the same time as he experiences them on an aesthetic plane" (Eco [1962] 1989, p. 3).

The experimental works (of music, literature, or visual art) that interest Eco nevertheless respond to a specifically twentieth-century preoccupation with contingency and randomness as central existential and artistic problems. They turn to combinatorics among other procedures, including but not limited to mathematical ones, in order to produce ambiguous works that generate, formalize, and contain indeterminacy. Within the general "dialectics of oscillation" between openness and closure, order and chaos, identified by Eco (Eco [1962] 1989, p. 65), there is room for the differentiation of specific methods and divergent aesthetic aims. In order to elucidate what is at stake in the turn to mathematical models of randomness in postwar experimental literature, I will consider three main examples: first, the "stochastic literature" developed by the Stuttgart school in the 1950s and 1960s; second, the "chance procedures" favored by Fluxus writers in the 1960s and 1970s, and finally the permutational algorithms favored by the Paris-based Oulipo group from 1960 onward. These cases bring to light the possibilities and limitations of literature's scientific aspirations—that is, art's attempt to imitate nature "in her manner of operation," in constant dialogue with scientific discoveries (Cage 1961, p. 194)—as well as its varied responses to new technological possibilities, in particular the advent of digital computers.

If I limit myself in this chapter to the literary realm, while acknowledging that the methods under consideration operate in a vaster intermedia landscape (particularly in the case of Fluxus), it is in order to bring into focus the particular problems associated with literary experimentation. The semantic dimension of literature produces a specific mode of resistance to both mathematical formalization and chance operations. Aleatory writing, insofar as this is a meaningful term—it is, as we shall see, an ambiguous one—seems both to thwart and to revive the "literary absolute" of the Romantics which, according to Philippe Lacoue-Labarthe and Jean-Luc Nancy, understands literature as endlessly self-generative, "producing itself as it produces its own theory" (Lacoue-Labarthe and Nancy 1988, pp. 11–12). This romantic absolutization of literature "aggravates and radicalizes the thinking of totality and the Subject" (Lacoue-Labarthe and Nancy 1988, p. 15), elevating the artist to the status of absolute mediator of the divine (p. 70), beyond the bounds of mere subjectivity (p. 104). In the twentieth century, the turn to the aleatory projects the text's meaning beyond the finitude of authorial intention, while also transforming the work of art into an imitation or formalization of chance itself.

ROLLING THE DICE: FORM, NUMBER, CHANCE

Setting aside the unfinished project of the Livre (the Book), in which Stéphane Mallarmé envisaged a mobile form subject to multiple permutations (Mallarmé 1957), we may consider Mallarmé's last published poem *Un coup de dés jamais n'abolira le hasard* (*A Throw of the Dice Never Will Abolish Chance*) as the inaugural work in an experimental tradition that posits and interrogates the interrelatedness of literary form, number, and chance. First published in 1897, Mallarmé's most experimental poem dissolves traditional verse form to produce a visual "score" (Mallarmé (1897) 1914, preface), but still addresses the problem of number as a metrical principle, following the "crisis in verse"—the disintegration of the traditional 12-syllable alexandrine—that Mallarmé had diagnosed earlier (Mallarmé [1886] 2003). The scenario dramatized in *A Throw of the Dice* involves a storm at sea and a captain (the Master) attempting to avoid shipwreck. The outcome hinges on a possible dice-throw that might or might not produce the "unique number"—the ideal number that could not be any other, symbolizing absolute necessity. The Master holds the dice but hesitates to throw them, observing sardonically that "*IF / IT WERE / THE NUMBER, / IT WOULD BE / CHANCE*" ("*SI / C'ÉTAIT / LE NOMBRE / CE SERAIT / LE HASARD*"). According to most readings, this hypothetical statement represents the Master's recognition of his inevitable failure to produce a necessary event: even if the dice-throw could produce the "unique number," this number would still be a result of chance, and the aleatory can therefore never be overcome (Murat 2005, p. 175). The Master disappears into the waves; the poem tells us that nothing has taken place "except / perhaps / a constellation" ("*excepté / peut-être / une constellation*"). In Quentin Meillassoux's reading, however, the apparent alternative between the Number and Chance mutates into a conception of the Number as a figure of infinite chance— that is, a number "that would possess the inalterable eternity of contingency itself" (Meillassoux 2011, p. 43). The poem then finds its secret justification in an encrypted number—according to Meillassoux, the number 707—that structures the poem formally and thematically (even determining the number of words), and is figured by the exception of the final constellation, the Septentrion.

Whether the outcome of the Master's action represents the failure or triumph of poetic modernity—a mere effect of chance or the codification of absolute chance—it places literary form under the sign of contingency, and glorifies the gesture of the dice-throw that determines but does not control the distribution of outcomes. At the same time, Mallarmé's commitment to

210   A. JAMES

formal rigor places him on one side of an aesthetic debate that would play out in various forms in the course of the century. It finds its informal, "anti-art" counterpart in Dadaist practices of cut-up and collage, which turn against the Romantic glorification of the artist while seeking in chance a "primeval magic power" (Richter 1965, p. 59). Following suit, the surrealists aim through automatic writing and chance encounters to uncover the operations of the unconscious. Later, in the post-Second World War moment, the legacy of these avant-garde movements was reshaped by new developments in technology, cybernetics, and information theory.

STOCHASTICS IN STUTTGART

The group that became known as the "Stuttgart School" in the 1960s was based at the Technische Hochschule of Stuttgart and had at its central figure Max Bense (1910–1990), a professor of the philosophy of scientific theory and mathematical logic.[1] Bense's fields of interest included mathematics, physics, aesthetics, semiotics, information theory, and cybernetics. He insisted on the link between mathematical thinking and the aesthetic consciousness of form, and developed (by analogy with generative linguistics) a project of "generative aesthetics" (Bense 1965), for which the Hochschule became a testing ground. There is also a political context to the Stuttgart group, which celebrates its commitment to progress and its break with the national past, and attempts to restore the connection with an avant-garde legacy that had been interrupted by Nazism (Döhl 1987). The group's celebration of experimental art is comprehensive and internationally oriented. Their journals and other publications feature collage and montage writing (by Reinhard Döhl, Helmut Heissenbüttel, and others), and highlight approaches such as the concrete poetry of the Brazilian Noigandres group (around Haroldo and Augusto de Campos), the French spatialist poetry of Pierre and Ilse Garnier, Hansjörg Meier's experiments with typography, Rolf Garnich's work in industrial design, and the computer graphics of Georg Nees.

Literature, and specifically poetry, is at the center of the Stuttgart group's concerns, despite the latter's intermedial and interdisciplinary orientations. In their manifesto "Zur Lage" ("On the State of Affairs"), Max Bense and Reinhard Döhl herald a literary revolution that leaves behind the illusions of national poetry in favor of a rational and methodical art, perfected in both its mechanical and human realization. They argue for a poetry of mixed forms that dissolves barriers between visual arts, music, and poetry, and they treat art as a material entity to be organized through rational techniques. Bense and Döhl identify six main tendencies in the contemporary field: letter pictures (*Buchstaben-Bilder*); pictures composed of graphical signs (*Schrift-Bilder*); serial and permutational production (*serielle und permutationelle Realisation*); phonetic poetry (*phonetische Poesie*); stochastic and topological poetry (*stochastische und topologische Poesie*); and cybernetic and material poetry (*kybernetische und materiale Poesie*) (Bense and Döhl 1964).

The general criteria involved are "experiment and theory, demonstration, models, pattern, play, reduction, permutation, iteration, randomness (disruption and deviation), series and structure" (Bense and Döhl 1964, n.p.).

At first glance, this list of criteria might seem to contrast the ordering principle of models, series, and structure with the opposing principle of "disruption and deviation" brought by randomness. The relationship between structure and randomness, however, is more complex. One of the most important directions explored by the group was stochastic literature, which involves the use of random variables to generate permutations of words. This approach linked the avant-garde's fascination with chance to a scientific, cybernetic impulse (Bense had invited Norbert Wiener to lecture at the Technische Hochschule in 1955 [Walther 1999]). It was also influenced by the aleatory works of contemporary composers, such as John Cage (Döhl n.d.), and brought new technological means to bear on the exploration of chance. The Hochschule had a computing center from 1959, during a period when the development of programming languages such as FORTRAN (1956), ALGOL (1958), and COBOL (1959) was making it possible to process letters and images. In particular, computer-based random number generators—or rather, pseudo-random number generators—allowed the basic mechanism of the dice-throw to be replaced with longer series. But stochastic processes also offer models for generating, analyzing, and managing specific types of randomness. Theo Lutz, one of Bense's students, developed a number of experiments in generating "stochastic texts," based on probability distributions. Bense initially proposed to Lutz the idea of generating basic sentences (subject, verb, object) from a repertory of 100 words from Franz Kafka's *The Castle* (1926) and linking these sentences through logical conjunctions like negation, conjunction, or disjunction. Beginning initially with a smaller sample of 16 subjects and 16 predicates, Lutz programmed a generator of pseudo-random numbers which produced sentences from the repertories using a Zuse 22 computer. Here are four lines from the results, published in Bense's review *augenblick* in 1959:

NICHT JEDER BLICK IST NAH. KEIN DORF IST SPAET.
EIN SCHLOSS IM FREI UND JEDER BAUER IST FERN.
JEDER FREMDE IST FERN. EIN TAG IST SPAET.
JEDES HAUS IST DUNKEL. EIN AUGE IST TIEF. (Lutz 1959, pp. 8–9).

(NOT EVERY GLANCE IS CLOSE. NO VILLAGE IS LATE.
A CASTLE IN THE OPEN AND EVERY FARMER IS FAR.
EVERY STRANGER IS FAR. A DAY IS LATE.
EVERY HOUSE IS DARK. AN EYE IS DEEP.)

This set of "elementary sentences" that are syntactically correct (if sometimes semantically odd), result from a straightforward permutational operation that recalls earlier combinatorial forms, as well as bearing a certain resemblance to automatic writing, and perhaps especially to the "exquisite corpse" game of sequential assemblage favored by the surrealists. (In a different context, the Oulipian Jean Queval will later point out the similarities between surrealist "psychic automatism" and "mechanical automatism" that seems to produce equally random results [qtd. in Bens 1980, p. 136]). In this basic model from 1959, Lutz stochastically inserts words from a given repertory into a fixed syntactic template, according to certain relative frequency rules, in order to produce meaningful language. Words are selected and combined randomly from a certain number of subjects, predicates, logical operators, conjunctions, and the word "*ist*" ("is") (Lutz 1959, pp. 3–5).

Presented as the first step in exploring the possibilities of "electronic calculating machines" (Lutz 1959, p. 3) such texts illustrate the possibility of creating meaning through a probability matrix that distributes a field of words. Lutz would later develop Markov processes for dealing with larger quantities of text, training the computer to build its probability distributions from an analysis of the text, rather than fitting words into a pre-established grammatical template (Walther and Harig 1970, pp. 105–110). This generative process brings to the combinatorial impulse a cybernetic frame, involving a feedback loop that modifies the initial system. In its probabilistic modeling of natural language that draws its repertoire from a canonical literary work, Lutz's approach lies at the intersection of literature and computational linguistics.

"Chance Operations": Fluxus

Lutz, in his text "Zum 'Problem des Cicero'" ("On 'Cicero's Problem'") revives a question from Cicero's *De natura deorum* (*On the Nature of the Gods*): could a meaningful text result from an arbitrary collection of letters (Walther and Harig 1970, p. 105)? Lutz brings a computational answer to the question without addressing Cicero's broader anti-Epicurean analogy (how could the world be the result of a fortuitous collision of atoms?): Lutz's point is that the computer can in fact approximate natural language. The artists associated with the Fluxus group, on the other hand, are interested not in computational approximations of human language, but rather in the centrality of chance as a principle in nature, beyond human control. For instance, the composer and artist George Brecht uses the term "chance-imagery" to refer to "our formation of images resulting from chance, wherever these occur in nature" (Brecht [1966] 2004, p. 11).

Taking shape in the early 1960s, Fluxus draws dual inspiration from the Dada anti-art legacy and from the compositional methods of Cage. Its strategies for generating randomness include rolling dice, flipping coins, or turning to the hexagrams of the ancient Chinese *I Ching*. But "chance operations" is a very broadly deployed term, as is the alternative concept of "alea" invoked by Pierre Boulez (Boulez [1957] 1991). Yayoi Uno helpfully distinguishes between "chance as a generating mechanism for 'aleatoric processes,' indeterminacy as uncertainty in outcome resulting from them, and randomness as a type and a measure of uncertainty" (Uno n.d., par. 1). Aesthetic divergences tend to concern the ways and places in which aleatory processes intervene, either in composition or in performance, and the extent to which the author-composer manages the resulting indeterminacy. Boulez accepts a degree of indeterminacy determined by the choices of the performer, for instance, but is dismayed by Cage's incorporation of chance operations such as coin-flipping into the process of composition, which he sees as an abandonment of aesthetic criteria (Boulez and Cage 1993, p. 112). Cage's approach can also be distinguished from the stochastic methods developed by the composer and music theorist Iannis Xenakis, who turns to probability calculations as a means of organization and control of his material, partly as a reaction against the *effect* of randomness produced by serial music (Xenakis 1994, pp. 39–43). Cage does sometimes turn to the computer for random number generation, for instance in creating the work *HPSCHD* (1969), in collaboration with the composer LeJaren Hiller, which used recordings of sound generated by the ILLIAC II supercomputer. But for Cage the computer serves largely as a tool that emulates and extends the operations of the *I Ching* hexagrams, generating rather than limiting indeterminacy (see Joseph 2012, p. 165).

The Dada poet Tristan Tzara had in 1922 described Dada as "a return to a quasi-buddhist religion of indifference" (Tzara 1992, p. 108). With Cage's music and writings the reference to Zen Buddhism and to East Asian thought in general becomes even more central. Cage's paradoxical pursuit of "a purposeful purposelessness or a purposeful play" (Cage 1961, p. 12) explores the terrain of nonintention without entirely excluding choice in the selection of words and phrases. The term "chance operations" both ascribes agency to chance and suggests that the artist is an impersonal "operator" (a term that Mallarmé had already used to designate the reader-choreographer of his projected Book). The work becomes a machine to be put in motion. The texts in La Monte Young's proto-Fluxus compilation *An Anthology of Chance Operations* (1963), for instance, emphasize the place of interdeterminacy in performance, where the texts function as notations or event scores. Chance operations here include the shuffling of instruction cards, in George Brecht's 1960 "Motor Vehicle Sundown (Event)," Earle Brown's introduction of variable elements into musical notation in "25 Pages for Piano(s)," or Young's instruction that "everyone may do whatever he wishes for the duration of the composition" (Young 1963, n.p.). Staged as an event or a situation subject to variation, the artistic work is thus called into question, but it is not formless; it incorporates chance elements into a process or defined procedure—what the poet Jackson Mac Low calls "chance regulation" (qtd. in Young 1963, n.p.)—or simply into a delimited space and duration.

An Anthology of Chance Operations includes in its full title "concept art, anti-art, indeterminacy, plans of action, diagrams, music, dance constructions, improvisation, meaningless work, natural disasters, compositions, mathematics, essays, poetry." These categories often overlap. The texts designated as

214   A. JAMES

"poems" vary in their emphasis on the textual, the visual, or the performative, but they have in common a disruption of the naturalness of speech or the legibility of writing. For instance, Joseph Byrd's "Homage to Jackson Mac Low" (1961) asks the reader to choose five words and read them according to a set of intructions, breaking them down into sequences of vowels and consonants; Yoko Ono's "To George, Poem no. 18" (1961) is a kind of palimpsest in which illegible fragments of words and a few Japanese characters appear on white scraps of paper under a black surface (in Young 1963, n.p.). As for mathematics, it appears as one field or realm of action among others, associated with the conceptual and the formal but also with the informal and the indeterminate. Notably, Young's anthology contains Henry Flynt's essay "On Concept Art" (1961), which describes the origins of concept art as "structure art" on the one hand (formalist music and art) and mathematics on the other: "since the value of pure mathematics is now regarded as aesthetic rather than cognitive, why not try to make up aesthetic theorems, without considering whether they are true" (qtd. in Young 1963, n.p.). That is, mathematics is considered as formal invention rather than discovery, and concept art "straightens out" traditional mathematics by removing its Platonic vestiges. The reference to mathematics thus justifies aesthetic form even as it denies this form any ontological grounding in reality or truth.

Text-based chance operations often involve the manipulation of a source text, which they disrupt and reorganize to produce a new form. In Cage's "mesostic" writings, such as his 1988–1989 Charles Eliot Norton Lectures (Cage 1997), he uses computer-aided *I Ching* chance operations (Jim Rosenberg's computer program Mesolist) to help generate vertical "writings through" a range of source materials, including newspaper reports, philosophical texts, and Cage's own writings. In the following lines, for instance, the source material comes from Marshall McLuhan, Henry David Thoreau, and the *Christian Science Monitor*, among other sources, and the mesostic word read vertically is "structure":

> he iS
>   To
> muskRats
>   bUt he stopped short
> who paniC now
>   if we were To observe
>   Using
>   vasteR in
>   may bE totally involved (Cage 1997, p. 12)

This excerpt by itself does not give a sense of the complex patterns of the whole, where words and phrases recur and are reorchestrated; it is also important to consider the effects of performance (see Perloff 1991, p. 215).

Among other Fluxus-associated writers who turn to mathematical and computational ways of generating randomness, we might cite Alison Knowles, whose 1967 "poem in progress" *The House of Dust*, created in collaboration with the composer and programmer James Tenney, is a computer-generated work programmed in FORTRAN from a set of four word lists (Higgins 2012, p. 195). The result is a set of quatrains that "build" the house by selecting "a material, a site or situation, a light source, and a category of inhabitants" (p. 195):

A house of wood
   in a metropolis
   using electricity
   inhabited by friends and enemies. (qtd. in Higgins 2012, p. 196)

The computer does the combinatorial work of selecting and sequencing, but based on a human-generated repertoire. As with other combinatorial works, Knowles's poem gestures toward the possibility of exhaustive enumeration. As Hannah Higgins notes: "The built house is a manifestation of just one set of permutations whose realization implies the possibility of building, or attempting to build, them all" (pp. 198–199). While there is some resemblance here to Theo Lutz's Kafka permutations, discussed earlier, Knowles's aim is not to use statistical probability to model human language processing, but rather to explore the tension between a minimalist structure and the seemingly endless possibilities of recombination. This dynamic also lies behind the multimedia reworkings of the poem, in the form of performances, public installations, and collaborative responses.

Mac Low also uses chance methods as a compositional principle. Influenced by Cage and inspired by Zen Buddhism, Mac Low's techniques include an "acrostic reading-through" method, which consists in searching a source text for words and other verbal units which have as their first letter the successive letters of a "seed text" (for example *Stanzas for Iris Lezak* [1960], *Asymmetries* [1960–1961]). His "diastic reading-through" method selects words containing the letters of the seed text in the same positions that they occupy in the seed text (for example *The Pronouns* [1964], *Words nd Ends from Ez* [sic; 1981–1983]). As in the work of Cage, computer programs are later brought into the process to automate and extend already existing manual methods. Mac Low's *42 Merzgedichte in Memoriam Kurt Schwitters* uses the program DIASTEXT, developed by Charles O. Hartman, as well as Hugh Kenner and Joseph O'Rourke's "pseudo-text-generating program" TRAVESTY, to select and process linguistic units of an existing poem composed for the Dada artist Kurt Schwitters (Mac Low 2008, p. 307; Bootz and Funkhouser 2014, p. 84). These "Merz" poems (taking up the nonsense word that Schwitters used to describe his method) are examples of "collage art" with both a visual and a verbal component; they are also "musical compositions in that words, phrases, sentences, and other linguistic elements are treated like the tones or intervals of scales or of tone rows, melodic themes or motifs, or rhythmic figures,

216   A. JAMES

recurring again and again … in various combinations and concatenations" (Mac Low 2008, pp. 307–308). The results are very varied in visual format and syntactic structure. Here is the beginning of the first *Merzgedicht*, a prose poem with typographical variations that suggest different voices:

> First word for Kurt Schwitters: *Merz*. **If I ever move from Hannover, where I love and hate everything, I will lose the feeling that makes my *world* point of view.** 1887. Now I call myself MERZ. A *degenerate artist*. (Mac Low 2008, p. 308)

As a verbal assemblage, this work is an apt homage to Schwitters's *Merzbilder*, which were collage pictures that incorporated found objects. Yet Mac Low's methods are more mechanical, and arguably less open to chance than those of Schwitters. Mac Low would in fact come to distinguish true chance operations from deterministic text-selection methods such as diastics (Tardos 2008, p. xix). However, both chance-based and deterministic methods have for Mac Low the same essential function of reducing dependence on the author's ego, since the author cannot predict the output in either case.

These examples reveal the difficulty of a strict distinction between aleatory and deterministic methods, especially when what is at stake is less the definition of chance in some pure form, than the role of intention or nonintention in the creative process. The same difficulty lies at the heart of the Oulipo group's ambivalence toward chance, which it strongly rejects in the name of "voluntary literature" (Queneau, qtd. in Bens 1980, p. 35)—even as it adopts methods that deliberately displace or disrupt the author's agency.

Permutation and Potentiality: Oulipo

When the French writer and Oulipo founder Raymond Queneau declares in a 1964 presentation that the Oulipo "is not aleatory literature," he aims explicitly to distinguish himself from the work of Bense and the Stuttgart school (Queneau 1965, p. 322). This is a somewhat surprising move. Queneau was familiar with Bense's work, had been translated by Ludwig Harig, a member of the Stuttgart group, and had in several early Oulipo meetings mentioned Bense's research as part of the international work of *Lipo* (potential literature) (Bens 1980, pp. 103, 130). In distancing himself from Bense in his 1964 talk, he insists on the specificity of the Oulipo and hints at a gap between literary value and scientific rationality. Speaking on this particular occasion to the participants of a seminar in quantitative linguistics, Queneau positions the Oulipo somewhat at the margins of science, claiming that the Oulipo's aim is to propose mathematical structures and mechanical procedures to aid literary creation (Queneau 1965, p. 321), yet also emphasizing the modest, artisanal, and playful nature of the group's activities.

Queneau had founded the Oulipo in November 1960, along with the chemical engineer and polymath François Le Lionnais. Initially named the

Seminar on Experimental Literature (*Séminaire de littérature expérimentale*, shorted to *Sélitex*), the group was rebaptised the following month as *Ouvroir de littérature potentielle*, or Oulipo ("Workshop for Potential Literature"). The name change indicates the group's ambivalent relationship toward science: the archaic term *ouvroir* suggests an artisanal workshop—in fact, a sewing circle—rather than a scientific laboratory; and "potential" bears fewer scientific connotations than "experimental." Despite what is sometimes asserted, the relationship between literature and mathematics was not initially at the center of the group's discussions. The stated goal was more general: to bring a methodical approach to literary creation. Eventually, the group came to be associated with the rediscovery or invention of "formal constraints," and the illustration of these constraints in writing. These constraints are not necessarily mathematical, but encompass a large variety of fixed forms, formulas, rules, and restrictions, including syntactic or alphabetic ones. Most famously, Georges Perec's 300-page novel *La Disparition* (*A Void*, 1969), is a lipogram that does not contain the letter "e." More recent Oulipian works often have recourse to temporal constraints that determine the parameters of writing, such as Jacques Jouet's "metro poems," in which lines of poetry are mentally composed during the interval that separates two metro stops, and transcribed when the train is stationary (Jouet 2000).

According to the Oulipo poet and mathematician Jacques Roubaud, the Oulipo is composed of four types of people: (1) composers of literature who are not mathematicians; (2) mathematicians who are not composers of literature; (3) composers of literature *and* mathematics; (4) composers of mathematics *and* literature (Roubaud 1995, pp. 202–203). Roubaud places himself in the last group, while Georges Perec belongs to the first group. The second group includes Claude Berge, a mathematician whose research dealt with hypergraph theory, topology, and combinatorics (Mathews and Brotchie 1998, p. 80). Among the more recent recruits to the group, Valérie Beaudouin ("co-opted" by the Oulipo in 2003) has applied text-mining methods to literary analysis, while the mathematician and writer Michèle Audin, who joined the group in 2009, often uses numerical and geometrical constraints in her literary works.

The group's relationship to mathematics is connected in complex ways with its avowed aversion to chance. When Berge asserts in the group's April 1962 meeting that "we are essentially anti-chance" ("*nous sommes essentiellement anti-hasard*") (Bens 1980, p. 136), it is in order to distinguish Oulipian algorithms from surrealist automatism. At issue, in particular, is the $N+7$ ($S+7$) method, which consists in replacing each noun in a given text with the seventh one that follows it in a given dictionary. When Queneau distances himself from "aleatory literature," as we have seen, he has in mind the stochastic computation practiced by the Stuttgart group, which attempts to model systems of random variables (even as the computer, strictly speaking, can only produce pseudo-random sequences). It seems that for the Oulipo, mathematics comes to represent at once an absolute signifier of anti-chance (as in the mechanical determinism encapsulated by the $N+7$ algorithm) and, on the contrary, to entail a dangerous form of fascination with chance, in the form of "aleatory" or "experimental" literature.

Queneau's own prior involvement with surrealism is a factor here, as is his lifelong fascination with number theory. In a significant episode, the narrator of Queneau's autobiographical novel *Odile* (1937) expresses a belief in the superior reality of numbers (Queneau [1937a] 1997, p. 33), and on the basis of his mathematical interests is invited to join an avant-garde group. The association is based on an intellectual misunderstanding, as is revealed when the group leader Anglarès (a stand-in for André Breton, the founder of the surrealists) asks Travy (Queneau) for examples of chance phenomena (*faits de hasard*) in mathematics (Queneau [1937a] 1997, p. 49). At a loss, Travy offers the fact that every natural number is the sum of at most nine cubes, as well as the still unproven (in 1937) conjecture that 23 and 239 are the only integers requiring nine cubes.[2] For Travy, these theorems point to the existence of a hidden order that is not wholly irrational, but both intrigues and exceeds finite human reason. When Anglarès/Breton responds to these examples by asking, "then there is in fact something like a mathematical unconscious?" ("*il y a donc bien comme un inconscient mathématique?*") (Queneau [1937a] 1997, p. 50), he conflates the strange properties of numbers with the notions of chance, irrationality, and the unconscious. Anglarès humanizes mathematics, but also psychologizes and diminishes it.

Queneau's fascination with number theory inflects the Oulipo's initial explorations of mathematical forms, including his own exploitation of mathematical exponents in the inaugural Oulipian work *One Hundred Thousand Billion Poems* (*Cent mille milliards de poèmes*, 1961). This book, which Claude Berge calls the "first complete literary work of frankly combinatory nature" (Berge [1973] 1986, p. 116), presents ten initial sonnets whose lines can be combined to produce the number of poems stated in the title (10^{14}). A deterministic composition in that all possible outcomes are (in theory) exhaustively enumerable, contained in virtual form in the initial structure, it can also be described as an aleatory work in that it incorporates indeterminacy into the reading process. Any given poem can be selected at random by the reader/user. Queneau's book, in its very unreadability (many lifetimes would be necessary to read all of the poems), exemplifies the Oulipian conception of potential literature—a literature that exceeds the intentions and the timespan of any individual human life.

Other mathematically inspired Oulipian forms include the "quenina" (a generalization of the medieval sestina form, in which rhyme schemes are permuted according to a rule), Paul Braffort's Fibonacci poems (whose syllable count is based on the Fibonacci number series), Jacques Bens's "irrational" sonnets (with a verse structure based on the number π), and the "Boolean poetry," proposed by François Le Lionnais, which invokes the set-theory notions of union and intersection (if each poem is considered as a set of words, an "intersection" of two sonnets uses the words that these have in common to produce a new poem). At a more ambitious level, Georges

Perec's combinatorial novel *Life A User's Manual* (*La Vie mode d'emploi*, 1978) takes as its point of departure a table of themes, descriptive elements, and intertextual allusions. With the help of Claude Berge, Perec used a Latin bi-square of order 10 to distribute a different combination of initial elements in each chapter (Berge [1973] 1986; Perec 1993). Harry Mathews's eponymous algorithm is another permutational system: it requires at least two sets of heterogeneous elements that are then superimposed to form a table of sets (each element in a set having a function, whether grammatical or semantic, that is equivalent to the corresponding element[s] of the other set[s]). A two-dimensional shift operation within the table then allows the production of new sets—that is, new permutations of the elements of the original sets (Mathews 2003a, pp. 303–304). This is a simple procedure that does not exhaust all possible combinations, but operates on the basis of slight shifts. The method can be applied at different levels of the text: letters, words, themes, or narrative episodes. Mathews uses this or similar algorithms for various purposes, for instance to compose a new poem from lines of Shakespeare's sonnets (Mathews 2003a, pp. 306–307), to invent the vocabulary of a tribal dialect (p. 306), or, more opaquely, to determine the narrative structure of his novel *Cigarettes* (1987) by pairing characters with other elements in a "permutation of situations" (Mathews and Brotchie [1998] 2005, p. 126). As Mathews notes, such methods introduce into literature the disquieting sense that no piece of writing can be taken for granted: "Beyond the words being read, others lie in wait to subvert and perhaps surpass them" (Mathews 2003a, p. 301).

As Mathews puts it, the Oulipian approach "guarantees that the unforeseen will happen and keep happening" (Mathews 2003b, p. 81)—even as it tries to exclude chance by bringing the possibilities under the control of the author who selects the initial algorithm. The emphasis on combinatorial text generation, as opposed to stochastic processes, is significant in this regard. Few Oulipians go as far as Italo Calvino in his 1969 essay, "Cybernetics and Ghosts," which attributes agency to the machine. Indeed, Calvino suggests that literature may already be thought of as a combinatorial game—a machine that multiplies stories until a shock connection occurs between the story and the reader's unconscious (Calvino [1969] 1986, p. 22). In a later essay on combinatorics (written after he joined the Oulipo in 1973), Calvino concludes, rather differently, that the computer liberates the artist from the need for a combinatory search "allowing him also the best chance of concentrating on this 'clinamen' which alone can make of the text a true work of art" (Calvino [1981] 1986, p. 126). In associating the Lucretian *clinamen* with the creative act, Calvino is perhaps less asserting the Lucretian chance principle against rigid structuralism, as has been claimed (Duncan 2012, p. 96), than he is abandoning his earlier cybernetic view of combinatorial systems in favor of a more limited approach to contemporary computational possibilities.

The computer-assisted writing pursued by the Oulipo from the 1970s through the 1990s, with Braffort serving as the main programmer of the group, tends to be limited to graph structures and other combinatorial ordering techniques, without recourse to the Markov models favored by the Stuttgart group. A computer version of Queneau's *Hundred Thousand Billion Poems* was implemented for the 1975 Europalia festival in Brussels, and a 1977 exhibit at the Pompidou Center in Paris offered interactive versions of other Oulipian experiments (Fournel 1986). Queneau's *A Story as You Like It* (*Un conte à votre façon*), a narrative where the reader has the choice of different outcomes at various points in the story, anticipates hypertext fiction in its use of graph structure—narrative nodes linked by directed connections (Queneau 1986). Marcel Bénabou's aphorisms were also adapted for computer: this constraint involves applying an algorithm to a repertory of syntactic formulas drawn from adages or well-known sayings, along with a repertory of meaningful words and common pairs of words (whether associated or opposite). One example transforms Clausewitz's famous statement on war (as the "continuation of policy by other means") to the significant statement: "Art is the continuation of chance by other means" (*"L'art est la continuation du hasard par d'autres moyens"*) (Bénabou [1980] 1987, p. 268). In 1981, Paul Braffort and Jacques Roubaud created the ALAMO (*Atelier de Littérature Assistée par la Mathématique et les Ordinateurs*; "Workshop for Literature Assisted by Mathematics and Computers"), which built on earlier Oulipian ideas such as Le Lionnais's Boolean poetry to produce such experiments as "Rimbaudelaire" poems, which insert vocabulary drawn from Baudelaire into the syntactic and metrical structure of Rimbaud's poems. Thus the first line of Rimbaud's "Dormeur du val" ("The Sleeper in the Valley," 1870), "C'est un trou de verdure où chante une rivière"/"There is a green hollow where a river sings") can become, "C'est un roi de campagne où roule une fleur d'ombre" ("There is a countryside king where a shadow flower rolls") (Wolff 2007, par. 7). Nevertheless, as Mark Wolff notes, the group dislikes using random number generation to produce instantiations of poems (Wolff 2007, par. 8), preferring to privilege the inventor's role in choosing material and/or the reader's engagement in manipulating the input.

Beyond simply applying discrete algorithms, Paul Braffort and Jacques Roubaud have suggested that a mathematical "formalization" of the Oulipian practice might be desirable. Roubaud offers two related propositions: "A constraint is an axiom of a text" and "[w]riting under Oulipian constraint is the literary equivalent of the drafting of a mathematical text, which may be formalized according to the axiomatic method" (Roubaud 1986, p. 89). The Oulipo often cites as one of its influences the Bourbaki group; although as Roubaud observes, it is not just an homage to but also a parody of Bourbaki's "megalomaniac" plan to rewrite mathematics (Roubaud 1998, p. 38). Bourbaki was the pseudonym of a group of French mathematicians who, from the late 1920s onward, attempted to formalize the whole of mathematics by following the axiomatic method and taking set theory as their basis. In this they were inspired by the German mathematician David Hilbert's project, in the *Foundations of Geometry* (1899) of finding a solid logical foundation for geometry based on a complete and consistent system of axioms. While attempting an axiomatic approach to mathematics, Bourbaki brackets the question of truth (seen as a "metaphysical" notion); only the rules of mathematical syntax matter, not their content (Bourbaki 1970, E.I.8). In a sense, mathematics is for Bourbaki a formalist system, treating rigorous rules of deduction as an end in themselves. The *Oulipo Compendium* (Mathews and Brotchie [1998] 2005), a dictionary of Oulipian constraints edited by Mathews with the Atlas Press founder Alastair Brotchie, takes a Bourbakist, non-Platonic view of the axiom, giving the dictionary definition of "self-evident proposition, requiring no formal demonstration," but adding that "in modern mathematical logic axioms are no longer regarded as being self-evident or necessarily true" (Mathews and Brotchie [1998] 2005, p. 54). The analogy with this Bourbakist idea of the axiom provides a means of defending arbitrarily chosen constraints against the idea of a "natural" literary expression, or against the neo-classical acceptance of existing conventions, since conventions are themselves marked as arbitrary (Roubaud 1986, pp. 90–91). Queneau's "Les Fondements de la littérature d'après David Hilbert" ("The Foundations of Literature According to David Hilbert"), a playful transposition of Hilbert's axiomatics into linguistic rules, may be read as an acknowledgment of this point, as well as of the limitations of any attempt to mathematize literature (Queneau 1976). Roubaud acknowledges that his own axioms of poetry are ultimately pseudo-axioms without any "imposing and severe" logical meaning (Roubaud 1995, p. 77). Poetry is not a complete formal system and does not respect the principle of non-contradiction (Roubaud 1995, p. 85).

Yet, as Christelle Reggiani points out, it cannot be denied that the Oulipian dislike of chance motivates a personal investment in the superior reality of numbers (Reggiani 1999, p. 62)—as if numbers stood for a universal necessity beyond human finitude. This investment lies behind Roubaud's efforts in the 1970s to develop a general mathematical theory of rhythm (Roubaud 1979), as well as Queneau's Platonic fantasy, already expressed in the 1930s, that a novel shaped by the "virtues of the Number" would participate in the eternity of the forms ([1937b] 1965, p. 33). That is, the Oulipian mathematical imaginary, like Mallarmé's invocation of the "unique Number," involves a fantasy of a mathematical order that would elevate the contingencies of literature, language—and perhaps even of human life itself—to a higher order of reality (Reggiani, 1999, p. 68). However, as the Master of Mallarmé's *Coup de dés* realizes as he hesitates to throw the dice, the problem remains that there is no way of completely eliminating arbitrariness from creation: if it were the Number, it would be Chance.

Conclusion: The Subject of Chance

We have seen that applications of aleatory processes in the literary text are broadly split between cybernetic approaches, which investigate and exploit a machine's capacities to generate, model, and regulate randomness (notably through stochastic processes), and combinatorial approaches that incorporate an element of unpredictability, but ultimately rely on the writer's or the reader's selection among the possible permutations. While the former may include a combinatorial element (as in Theo Lutz's Kafka-based permutations), they develop models for purposive systems that delegate agency to the machine. Non-cybernetic combinatorial systems offer ways of configuring a given number of objects, but semantic and aesthetic purpose remain on the side of the human creator. Beyond this basic cybernetic/combinatorial distinction, it is possible to offer further refinements or complications. For instance, within the field of combinatorial literature, Fluxus artists' pursuit of nonintention and purposelessness contrasts with the tension in Oulipian practice between the notion of potentiality and the insistence on the voluntary. These aesthetic differences hinge on conflicting understandings of the function of art, and of its attempts to mimic or organize nature, to express a meaning, or to exceed the human intention in which it originated.

In opening this essay with Mallarmé's *Un coup de dés*, my aim was to inscribe the complex relationship between chance, numbers, and literary form within the long trajectory of modernism, without limiting it to the more recent preoccupations with conceptual writing or computational inspiration. The central question staged in Mallarmé's poem—can a number-based order abolish chance, or is it simply another effect of chance?—is recast in the second half of the twentieth century in terms of the paradoxes of deterministic versus chance procedures. To what extent do formal rules and constraints simply produce the equivalent of chance, from the point of view of the creator or audience? Conversely, does apparent chance simply mask a hidden determinism? In twentieth-century literature, the mathematical fantasy of overcoming chance has as its obverse side the elevation of the random number to a figure of absolute chance—and thus of all that exceeds authorial intention. Nevertheless, the manipulation of chance procedures remains a means of re-inscribing subjectivity within the impersonal order of the world.

NOTES

1. Reinhard Döhl, a close associate of Bense, prefers the term "group" to school, emphasizing mutual cooperation rather than pedagogical instruction (Döhl 1987).
2. This is part of Waring's Problem. In 1939, two years after the publication of Queneau's novel, L. E. Dickson proved the conjecture that the only integers requiring nine cubes are 23 and 239 (Weisstein, n.p.).

REFERENCES

Bénabou, Marcel. (1980) 1987. "Un aphorisme peut en cacher un autre." In *La Bibliothèque oulipienne*, vol. 1, 251–269. Paris: Editions Ramsay.

Bens, Jacques. 1980. *OuLiPo 1960–63*. Paris: Christian Bourgois.

12 RANDOMIZING FORM: STOCHASTICS AND COMBINATORICS ... 223

Bense, Max. 1965. "Projekte generativer Ästhetik." *Computer-Grafik.* Edited by Georg Nees and Max Bense. Rot no. 19. Stuttgart: Walther.

Bense, Max, and Reinhard Döhl. 1964. "Zur Lage." *Als Stuttgart Schule machte: Ein Internet-Reader.* Edited by Reinhard Döhl, Johannes Auer, and Friedrich Block. www.stuttgarter-schule.de/zur_lage.htm. Accessed on August 15, 2017.

Berge, Claude. (1973) 1986. "For a Potential Analysis of Combinatory Literature." In Motte 1986, 115–125.

Bootz, Philippe, and Christopher Funkhouser. 2014. "Combinatory and Automatic Text Generation." In *The Johns Hopkins Guide to Digital Media,* edited by Marie-Laure Ryan, Lori Emerson, and Benjamin J. Robertson, 83–89. Baltimore, MD and London, UK: Johns Hopkins University Press.

Boulez, Pierre. (1957) 1991. "Alea." In *Stocktakings from an Apprenticeship.* Edited by Paule Thévenin. Translated by Stephen Walsh. Oxford: Clarendon Press.

Boulez, Pierre, and John Cage. 1993. *The Boulez-Cage Correspondence.* Edited by Jean-Jacques Nattiez. Translated and edited by Robert Samuels. Cambridge and New York: Cambridge University Press.

Bourbaki, Nicolas [pseud.]. 1970. *Éléments de Mathématique. Théorie des Ensembles.* Livre I. Paris: Hermann.

Brecht, George. 1966. *Chance-Imagery.* New York: Something Else Press.

Cage, John. 1961. *Silence: Lectures and Writings.* Middletown, CT: Wesleyan University Press.

———. 1997. *I–VI.* Charles Eliot Norton Lectures, 1988–89. Hanover, NH: University Press of New England.

Calvino, Italo. (1969) 1986. "Cybernetics and Ghosts." In *The Uses of Literature,* 3–27. Translated by Patrick Creagh. New York: Harcourt Brace.

———. (1981) 1986. "Prose and Anticombinatorics." In Motte 1986, 143–52.

Döhl, Reinhard. N.d. "Exkurs über Aleatorik." *Als Stuttgart Schule machte: Ein Internet-Reader.* Edited by Reinhard Döhl, Johannes Auer, and Friedrich Block. http://www.stuttgarter-schule.de/aleatori.htm. Accessed on August 16, 2017.

———. 1987. "Die sechziger Jahre in Stuttgart. Ein Exkurs." *Als Stuttgart Schule machte: Ein Internet-Reader.* Edited by Reinhard Döhl, Johannes Auer, and Friedrich Block. http://www.stuttgarter-schule.de/stutt60.htm. Accessed on August 16, 2017.

Duncan, Dennis. 2012. "Calvino, Llull, Lucretius: Two Models of Literary Combinatorics." *Comparative Literature* 64 (1) (December): 93–109. https://doi.org/10.1215/00104124-539226.

Eco, Umberto. (1962) 1989. *The Open Work.* Translated by Anna Cancogni. Cambridge, MA: Harvard University Press.

Fournel, Paul. 1986. "Computer and Writer: The Centre Pompidou Experiment." In Motte 1986, 140–152.

Gardner, Martin. 1958. *Logic Machines and Diagrams.* New York: McGraw-Hill.

Higgins, Hannah. 2012. "Introduction to The House of Dust." In Higgins and Kahn 2012, 195–199.

Higgins, Hannah, and Douglas Kahn, eds. 2012. *Mainframe Experimentalism: Early Computing and the Foundations of the Digital Arts.* Berkeley and Los Angeles: Univ of California Press.

Joseph, Branden W. 2012. "HPSCHD: Ghost or Monster?" In Higgins and Kahn 2012, 147–69.

224 A. JAMES

Jouet, Jacques. 2000. *Poèmes de métro*. Paris: P.O.L.

Lacoue-Labarthe, Philippe, and Jean-Luc Nancy. 1988. *The Literary Absolute: The Theory of Literature in German Romanticism*. Translated by Philip Barnard and Cheryl Leser. Albany, NY: State University of New York Press.

Lutz, Theo. 1959. "Stochastiche Texte." *Augenblick* 4 (1): 3–9.

Mac Low, Jackson. 2008. *Thing of Beauty: New and Selected Works*. Edited by Anne Tardos. Berkeley: University of California Press.

Mallarmé, Stéphane. (1886) 2003. "Crise de vers." In *Œuvres complètes*, vol. 2, edited by Bertrand Marchal. Bibliothèque de la Pléiade, 239–52. Paris: Gallimard. .

———. (1897) 1914. *Un coup de dés jamais n'abolira le hasard*. Paris: Gallimard.

———. 1957. *Le "Livre" de Mallarmé*. Edited by Jacques Schérer. Paris: Gallimard.

Mathews, Harry. 2003a. "Mathews's Algorithm." In *The Case of the Persevering Maltese: Collected Essays*, 301–20. Normal, IL: Dalkey Archive Press.

———. 2003b. "Translation and the Oulipo: The Case of the Persevering Maltese." In *The Case of the Persevering Maltese: Collected Essays*, 67–82. Normal, IL: Dalkey Archive Press.

Mathews, Harry and Alasdair Brotchie, eds. (1998) 2005. *Oulipo Compendium*. Rev. ed. London: Atlas Press.

Meillassoux, Quentin. 2011. *Le nombre et la sirène: un déchiffrage du coup de dés de Mallarmé*. Collection ouvertures. Paris: Fayard.

Motte, Warren F., ed. and trans. 1986. *Oulipo: A Primer of Potential Literature*. Lincoln: University of Nebraska Press.

Murat, Michel. 2005. *Le Coup de dés de Mallarmé: un recommencement de la poésie*. Paris: Belin.

Perec, Georges. 1969. *La Disparition*. Les lettres nouvelles. Paris: Denoël.

———. 1978. *La Vie mode d'emploi*. Paris: Hachette/P.O.L.

———. 1993. *Cahier des charges de* La Vie mode d'emploi. Edited by Hans Hartje, Bernard Magné, and Jacques Neefs. Paris: CNRS Éditions and Zulma.

Perloff, Marjorie. 1991. *Radical Artifice: Writing Poetry in the Age of Media*. Chicago: University of Chicago Press.

Queneau, Raymond. (1937a) 1997. *Odile*. Collection l'imaginaire 276. Paris: Gallimard.

———. (1937b) 1965. "Technique du roman." In *Bâtons, chiffres et lettres*, 27–33. Paris: Gallimard-Idées.

———. 1961. *Cent mille milliards de poèmes*. Paris: Gallimard.

———. 1965. "La littérature potentielle." *Bâtons, chiffres et lettres*, 317–45. Paris: Gallimard-Idées.

———. 1976. "Les Fondements de la littérature d'après David Hilbert." Paris: La Bibliothèque oulipienne, no. 3.

———. 1986. "A Story as You Like It." In Motte 1986, 156–58.

Reggiani, Christelle. 1999. *Rhétoriques de la contrainte: Georges Perec – l'Oulipo*. Saint-Pierre-du-Mont: Éditions InterUniversitaires.

Richter, Hans. 1965. *Dada: Art and Anti-Art*. London: Thames & Hudson.

Roubaud, Jacques. 1979. "Le silence de la mathématique jusqu'au fond de la langue, poésie." *Po&sie* 10: 110–124.

———. 1986. "Mathematics in the Method of Raymond Queneau." In Motte 1986, 79–96.

———. 1995. *Poésie, Etcetera: Ménage*. Paris: Stock.

————. 1998. "The Oulipo and Combinatorial Art." In *Oulipo Compendium*, edited by Harry Mathews and Alastair Brotchie, 37–44. London: Atlas Press.

Tardos, Anne. 2008. Foreword, *Thing of Beauty: New and Selected Works*, by Jackson Mac Low, xvi–xxvi. Berkeley: University of California Press.

Tzara, Tristan. 1992. *Seven Dada Manifestos and Lampisteries*. Translated by Barbara Wright. Illustrated by Francis Picabia. London and New York: Calder Publications.

Uno, Yayoi. n.d. "Aleatoric Processes." In *Encyclopedia of Aesthetics*, edited by M. Kelly. *Oxford Art Online*. Oxford University Press. http://www.oxfordartonline. com/subscriber/article/opr/t234/e0016. Accessed on August 14, 2017.

Walther, Elisabeth. 1999. "Max Bense Und Die Kybernetik." *Computer Art Faszination* (10): 360. http://www.stuttgarter-schule.de/bensekybernetik.htm. Accessed on August 16, 2017.

Walther, Elisabeth, and Ludwig Harig (eds.). 1970. *Muster Möglicher Welten. Ein Anthologie für Max Bense*. Wiesbaden: Limes Verlag.

Weisstein, Eric W. n.d. "Waring's Problem." From *MathWorld–A Wolfram Web Resource*. http://mathworld.wolfram.com/WaringsProblem.html. Accessed on August 15, 2017.

Wolff, Mark. 2007. "Reading Potential: The Oulipo and the Meaning of Algorithms" 1 (1). http://www.digitalhumanities.org/dhq/vol/1/1/000005/000005.html. Accessed on August 17, 2017.

Xenakis, Iannis. 1994. *Kéleütha: écrits*. Edited by Alain Galliari. Paris: L'Arche.

Young, La Monte (ed.). 1963. *An Anthology of Chance Operations*. New York: La Monte Young and Jackson Mac Low.

CHAPTER 13

Oulipian Mathematics

Warren Motte

In a country with a legacy of mathematical inquiry as strong and venerable as that of France, it is inevitable that mathematics should have influenced other cultural traditions, here and there, and that certain figures in the history of French culture should have mixed their mathematics and their aesthetics very liberally indeed. One thinks in the first instance of René Descartes, who is broadly acknowledged as the founder of analytical geometry. Blaise Pascal wrote an influential treatise on projective geometry when he was merely sixteen. As a young man, Jean le Rond d'Alembert authored groundbreaking work on dynamics, and later assumed editorial responsibility for everything pertaining to mathematics and science in the *Encyclopédie*. Isidore Ducasse, better known as the Comte de Lautréamont, interpolated in his *Chants de Maldoror* an "Ode to Mathematics" which begins, "O stern mathematics, I have not forgotten you since your learned teachings, sweeter than honey, filtered through my heart like a refreshing wave. From the cradle I instinctively aspired to drink from your spring more ancient than the sun, and, most faithful of your initiates, still I continue to tread the sacred court of your grave temple."[1] Paul Valéry remarked of mathematics: "I worship this most beautiful subject of all and I don't care that my love remains unrequited" (Fujiwara 2017). Robert Tubbs notes that "the work of the Parisian surrealists illustrates that some of the mathematical ideas influencing artists and writers were not simple geometric or numerical notions; instead, they were sometimes fairly sophisticated and even highly abstract modern mathematical concepts" (Tubbs 2014, p. 1).

More fundamentally, one may recognize that mathematics—or at least arithmetic—is deeply bound up in certain kinds of poetics. At a basic level,

W. Motte (✉)
University of Colorado, Boulder, CO, USA

© The Author(s) 2021
R. Tubbs et al. (eds.), *The Palgrave Handbook of Literature and Mathematics*, https://doi.org/10.1007/978-3-030-55478-1_13

228   W. MOTTE

poetic meter clearly wagers upon a mathematization of time and speech. Moreover, the organizational principles of many fixed poetic forms rely on mathematics. Forms like the sestina, the triolet, the virelay, the rondel, the villanelle, and the sonnet put into play a transparent or more intricate artic-ulation of numbers and words. Raymond Queneau pithily reflects on the essential relation of mathematics and poetry: "The poet, however refractory toward mathematics he may be, is nonetheless obliged to count up to twelve in order to compose an alexandrine" (Queneau 1986, p. 55).

I would like now to invoke a more recent, and more programmatic, body of work, that of the Ouvroir de Littérature Potentielle, the "Workshop of Potential Literature," or "Oulipo" for short. The group was conceived (if not yet birthed) during a ten-day colloquium devoted to the work of Queneau at Cerisy-la-Salle, in Normandy, in September 1960. It held its first offi-cial meeting in November of that year, in Paris. Originally composed of ten members, it numbers forty today (of whom seventeen are now deceased—but not thereby excused from membership).[2] It is generally understood that Queneau co-founded the group with his friend François Le Lionnais. With just a dash of historical perspective, it is now possible to argue that Queneau was the most important "writer's writer" of the French twentieth century, in view of his influence on two (and perhaps three) generations of French writ-ers working in his wake. Like his friend Le Lionnais, Queneau was a poly-math whose intellectual curiosity embraced a staggeringly broad horizon of cultural achievement. More particularly, granted my present purposes, both men were devoted and accomplished amateur mathematicians—and in fact Le Lionnais was something more than an amateur, in view of his early train-ing as a chemical engineer and his writings on the history of mathematics.[3] From the outset, Queneau and Le Lionnais insisted that the membership of the Oulipo include not only writers, but also mathematicians; and indeed the group has always included practicing, professional mathematicians in its numbers.[4] Claude Berge, for instance, a founding member of the Oulipo, was a pioneer in graph theory and combinatorics; Paul Braffort was trained in mathematics and contributed significantly to the then-emerging field of artificial intelligence; Jacques Roubaud taught mathematics at the University of Paris for many years; Pierre Rosenstiehl is a distinguished figure in graph theory; Olivier Salon, a specialist in number theory, has taught at the École Normale Supérieure de Cachan; and Michèle Audin, who has worked in alge-braic topology, symplectic geometry, integrable systems, and the history of mathematics, has taught at universities in Geneva, Paris, and Strasbourg.

All of this is to say that the Oulipo's interest in mathematics goes well beyond the level of mere flirtation. From its very beginnings, one of the group's foundational principles has been that literature and mathematics entertain profound affinities—and in the best cases such active reciproci-ties that the one cannot be imagined without the other. In 1964, Queneau described the Oulipo's goals in terms of a formalist quest: "What is the objec-tive of our work? To propose new 'structures' to writers, mathematical in nature, or to invent new artificial or mechanical procedures that will contribute to literary activity: props for inspiration as it were, or rather, in a way, aids for creativity" (Queneau 1986, p. 51). What Queneau is talking about is, precisely, structures of formal constraint, structures that the group elaborates, but does not necessarily illustrate in finished works—thus the notion of *potential* in "potential literature." Queneau foresaw the Oulipo proposing such structures to writers who were beached, blocked, or brutalized by the false prophets of genius and inspiration. That is what led Le Lionnais to speak (only half-jokingly) of the Oulipo as a kind of "Institute for Literary Prosthesis" (Le Lionnais 1986a, p. 31).

In order to illustrate the kind of dynamic the Oulipo has in mind, it is useful to call upon what can be thought of as the group's seminal text, Queneau's *Cent mille milliards de poèmes*, or "One Hundred Trillion Poems" (1961). It is a collection of ten sonnets constructed in such a way that any line in any sonnet may be substituted for its opposite number in any of the other sonnets. Given ten possibilities for the first lines,[5] there are one hundred possibilities for the first two lines, a thousand for the first three, and so forth. Since a sonnet has fourteen lines, there are ten to the fourteenth power, 10^{14}, or one hundred trillion possibilities in all. Queneau points out in his preface to the book that if one were to read a sonnet per minute, eight hours a day, two hundred days a year, it would take a bit more than a million centuries to read the entire text. (Actually, he was off by an order of ten; it would in fact take a bit more than ten million centuries. But who's counting?) Le Lionnais remarks in his postface: "Thanks to this technical superiority, the work you are holding in your hands represents, itself alone, a quantity of text far greater than everything people have written since the invention of writing, including popular novels, business letters, diplomatic correspondence, private mail, rough drafts thrown into the wastebasket, and graffiti."[6]

Clearly enough, *Cent mille milliards de poèmes* is a text different from most; and it proclaims its difference boldly, even ostentatiously. Its very strangeness may cause some readers to rebel. Is this *really* a literary text? A legitimate aesthetic artifact? Or is it a mere toy, a curiosity? An example of empty pyrotechnics? A symptom of literary madness? Must we take it *seriously*, or should we dismiss it outright as a joke? It is helpful to imagine that Queneau offers us the text not so much as a collection of sonnets, but rather as a laboratory of poetry, an experiment in literary form intended to shed light upon certain possibilities inherent to literature.

The work is animated by a powerful attention to form, and the nature of that form is rooted in mathematics. It puts into play the notion of literature as a combinatory, permutational system, an idea that enjoyed a great deal of theoretical currency at the time Queneau conceived his project (I am thinking here of the work of Vladimir Propp, Italo Calvino, Umberto Eco, A. J. Greimas, and Tzvetan Todorov, among others). Another principle that subtends the text is that of *difficulté vaincue*, or difficulty overcome. In his conversations with the critic Georges Charbonnier, Queneau remarked that this

230 W. MOTTE

text was undoubtedly the most daunting one he had ever undertaken: "I had written five or six of the *Cent mille milliards de poèmes*, and was a bit loath to continue, actually I didn't have the courage to continue, the more it went along, the more difficult it was to do naturally" (Queneau 1962a, p. 116). Yet, once achieved, that exercise in *difficulté vaincue* (a venerable aesthetic principle infused here with new life) puts difficulty itself into play, prompting it to speak about issues of textual construction and the new possibilities that they offer. In close articulation, we readers are invited to retrace constraint along the grain of those issues—indeed, such a retracing seems unavoidable—in the course of a reading that, *potentially* at least, will never end, a reading that constrains us, in turn, to be as wholly mobile as we possibly can. Indeed, one way of coming to terms with *Cent mille milliards de poèmes* is through constraint—just as Queneau's own reflections on constraint enabled him to write it. It is fair to say, too, that in this text form becomes theme, because, more than anything else, *Cent mille milliards de poèmes* is a meditation upon literary form, one where the reader is inevitably called upon to think about form, and about the ways in which it shapes literature, privileging a vision of literature as something dynamic rather than static, more like a verb than a noun. For whatever else one might say about it, *Cent mille milliards de poèmes* is patently mobile, constantly moving and impossible to grasp. One can take soundings in it here and there, but one can never fully traverse the landscape that it limns. It escapes from us very largely, then, both ineluctably and definitively. Despite that, it has appealed to readers in ways that other combinatory texts, constructed according to principles that are similar ones, have not. There are even websites devoted to it now, where one can instruct the computer to select and print a "personal" sonnet for oneself, among the one hundred trillion potential poems.[7] It *speaks* to us as it whizzes by, issuing invitations that are hard to resist.

One of those invitations involves play. In any but the most cursory considerations of the text, it shortly becomes apparent that there is a clear and most refreshing ludic impulse at work therein. In one of its several dimensions, Queneau proposes his text to the reader as a game—which is not to say it lacks sober import, for to Queneau's way of thinking, one can play very soberly indeed. He intends that the playfulness which animates the process of production should find its double in the process of reception. That ludic reciprocity is related to Queneau's notion of the poet as cruciverbalist or maker of logogriphs. Moreover, for him, it is quite obviously a question of a game solidly based in form:

> My first books were conditioned by the concern for order, more of a arithmomaniacal kind than a mathematical one, I would say, and also by a concern for structure ... there is an aspect of play in that process, a game whose rules one invents and obeys.
>
> Afterward, I think that I freed myself from that arithmomania, while maintaining the concern for structure; and now once again I am deeply interested in

all of these questions, not so much involving the mathematics of language, but rather language as a game with rules, a rational game one might say, or a game of chance with a maximum degree of logic. (Queneau 1962a, p. 56; ellipsis in original)

Queneau's text argues that literature is inherently playful in nature, a ludic dynamic wherein writer and reader find important points of communication—as anyone who has actually played with this text, shuffling verses from one configuration to another, will be obliged to admit. Furthermore, it serves to put the tired binary of playfulness and earnestness on trial. In *Cent mille milliards de poèmes*, those two modes cannot be mutually disentangled, and they collaborate in ways that are highly productive of meaning. In such a light, the text is a test case for what one might call "serious play"—and that, too, is an important part of Queneau's experiment.

If the idea of an experiment can hardly be avoided when discussing *Cent mille milliards de poèmes*, that idea takes on more pungency still in view of the fact that the "object" of experimentation should be a form as noble as that of the sonnet. Furthermore, it puts the notion of the limits of literature radically into question, both insofar as production and reception are concerned. If this text has a lot to say about writing, it comments upon reading as well, beggaring as it does—*exhausting* in fact—the notion that we might read it through to the very end. More particularly still, it serves as a brief for constraint-based writing, for the use of rigorous and systematic artifice in the literary text as an effective way of investing a given piece of writing with form. Here, too, the choice of the sonnet is obviously not an arbitrary one. For of course the sonnet itself is a highly constrained literary form, equipped with its own set of rules. When Queneau grafts additional rules onto that form, he is in a sense exploiting and amplifying possibilities that are latent in the original system, and putting those possibilities into *play*, precisely. Most importantly, *Cent mille milliards de poèmes* serves as a manifesto, a defense and illustration of the Oulipo's foundational principle, that of "potential literature." Beyond question, the vast majority of the textuality that this literary machine generates is fated to remain in the territory of the virtual, never to be made manifest either upon a page or in a human reading. It is consequently not so much about *being* as it is about *becoming*.

In that same spirit, the Oulipo's project takes shape, involving the desire to create structures that put literary possibility on display, that ramify, virtually, into hitherto unsuspected areas of literary *potential*. "One can state with little risk of error," says Jacques Bens of Queneau's poems, "that they constitute the first work of *conscious* potential literature. Or rather: *concerted*" (Bens 1986, p. 66). More recently, Roubaud's appreciation of Queneau's text is more categorical still: "The first properly Oulipian work *par excellence*, claimed as such by the Oulipo, is a work that exhibits *potentiality* in all its force: the *Cent mille milliards de poèmes* by Raymond Queneau. Its constraint is rather elementary, but its *potentiality* is spectacular" (Roubaud 2004, pp.

100–1). Queneau's text can thus be productively viewed as a smoothly functional machine for the production and dissemination of literature, and (not coincidentally) as a sleek advertisement for the nascent Oulipian aesthetic. Moreover, its mechanisms are such that very process calls attention to itself. In other words, it is an *exemplary* text, in the fullest sense of that word.

Among all of the members of the Oulipo, it is perhaps Roubaud who has most consistently put mathematics to work in his texts, arguing throughout his career a compelling brief for systematicity and rigor in literature. Early on, like other Oulipians,[8] he was influenced by Bourbaki, a group of individuals who proposed to rethink and retheorize mathematics through set theory in the mid-twentieth century, in search of more precision and more elegance. Roubaud's conception of Bourbaki as "a sort of mathematical surrealism, but altogether foreign to literature" (Roubaud 1986b, p. 80) has without a doubt informed his approach to writing. For one of the virtues of mathematics, in his view, is that it "repairs the ruin of rules" (Roubaud 1986b, p. 93). One may recall in this perspective Queneau's characterization of language as "a game with rules"; and the importance of the notion of *rule* in Oulipian poetics should also be underscored (see Bénabou 1986). From his very first book forward, Roubaud's work testifies to a sustained, focused attention to rules, an attention that is moreover carefully foregrounded in the text itself. That first book, whose title consists merely of the mathematical symbol for inclusion, $\in$ (1967), contains a first section entitled "Mode d'emploi de ce livre" ("Directions for Use of this Book"). There, Roubaud complaisantly explains that the 361 texts in the volume correspond to the 180 white pieces and the 181 black pieces in the Japanese game of *go*, and sketches out three protocols of reading the book, affiliating mathematics and ludics. That gesture may now be seen as the initial manifestation of what later came to be known as Roubaud's First Principle: "A text written according to a constraint must speak about that constraint" (Roubaud 1981, p. 90).

Throughout his long career, Roubaud has remained faithful to that first principle, and also to a second one that follows immediately upon it: "A text written according to a mathematizable constraint must contain the consequences of the mathematical theory that it illustrates" (1981, p. 90). Thus, for example, he suggests that his *Princesse Hoppy ou le conte du Labrador* (*Princess Hoppy, Or, the Tale of Labrador*, 1990a), based on a group of four elements, takes account of the combinational properties of that group. Other texts are based upon other kinds of mathematical figures, ranging from rather simple ones to far more intricate, developed, and (to many critics) practically impenetrable ones. *Quelque chose noir* (1986a), an elegy for Roubaud's wife, is constructed around the number nine, an integer Roubaud associates with death and mourning; *Trente et un au cube* contains thirty-one numbered texts, and plays both on the notion of the third power and on the geometrical shape that its title invokes; *Parc sauvage* (2008) calls upon graph theory; and so forth.

One of the mathematical figures that has fascinated Roubaud the most enduringly, however, is the one upon which the poetic fixed form known as the sestina is based. Le Lionnais gives a concise definition of that form: "A sestina is a series of stanzas, each of which is composed of six verses ending in six different words. The structure of each stanza is deduced through permutation of that of the preceding stanza, such that the poem can contain only six stanzas, as a seventh stanza would merely repeat the structure of the first" (Le Lionnais 1986b, p. 78).[9] Roubaud has frequently turned toward that structure, notably deploying it as an organizational principle in some of his works in prose, such as his "Hortense" novels and his multivolume project, *le grand incendie de Londres* (2009).[10] Several aspects of that form appeal to Roubaud. First, it is deeply associated with poetic tradition, and notably with the twelfth-century Occitan troubadour Arnaut Daniel, whom Roubaud regards as a precursor figure. As such, the sestina is affiliated with memory, and particularly poetic memory. One might argue that the repetition that the form prescribes can also be associated with the notion of memory, either because it serves as a mnemonic device or because it functions literally to *recall* certain words in the broader context of the poem. The explicitly permutational structure of the sestina puts on display the combinatory character of literature in ways that are inevitable and convincing. Granted that latter point, the form offers a significant ludic potential based on systematic articulation. And finally, the sestina by its very nature argues that mathematics and literature are—or *can be*, when put to good use—mutually enriching discourses, rather than mutually refractory ones.

Perec was certainly less of a mathematician than Roubaud, but his thirst for rigorous literary shape was no less pronounced. Often, he sought to slake that thirst by turning toward literal constraints, that is, constraints based on the alphabetical letter. One of his most notorious accomplishments in that area was a 300-page novel written without using the letter *E*; another was a record-breaking palindrome of 5566 letters.[11] Yet from time to time he would call upon mathematical principles, either ones that he imagined on his own or ones suggested to him by his colleagues in the Oulipo. And sometimes Perec would imagine savant amalgams of both literal and mathematical principles. Such is the case of his poem "Ulcérations" (1980c). Composed in what looks very much like free verse, its first quatrain is eminently readable, and gives no clue about the extremely constraining compositional principles that subtend the text:

> Coeur à l'instinct saoul,
> reclus à trône inutile,
> Corsaire coulant secouant l'isolé,
> crains-tu la course intruse?
> [Heart of drunken instinct,
> recluse on a useless throne,
> Sinking pirate rescuing the lonely,
> Do you fear the intrusive errand?]

234 W. MOTTE

A close inspection of the poem reveals, however, that it consists of 399 anagrams of the title, which is in turn composed of the eleven alphabetical letters that the French language uses most frequently. It is an example of what Perec called "heterogrammatic poetry," a form using anagrams that he practiced frequently after he joined the Oulipo in 1967.[12] Certain features of this poem are especially deserving of mention. In the first instance, a poetics of the letter and a mathematical organization are deeply affiliated here. For one way of scanning "Ulcérations" is as a grid of 399 by 11 integers; and the geometries created by that structure are a crucial, eloquently performative aspect of the poem. The constraint that rules the text is extremely rigorous; yet that constraint is so deeply camouflaged that nothing short of a very attentive—indeed obsessional—reading will make it apparent. The way that Perec wagers upon necessity, as opposed to chance, should also be noted: each heterogrammatic unit in the poem *must* use the same eleven letters, none others, and no more or less. One may recall that, from its earliest days, the Oulipo took a stand against the aleatory,[13] but I think something additional is afoot in Perec's work. For as a war orphan (his father was killed at the front in 1940, his mother was deported to Auschwitz in 1943), chance had not been kind to Perec. It is legitimate to read in "Ulcérations" a discourse of anti-chance, and the construction of a space so tightly controlled as to exclude anything contingent, and to enable only the most deliberate, intentional gestures. The poem speaks, then, in a variety of registers, ranging from the most objectively formalist to the most intensely personal.

In Perec's *Alphabets* (1976), that heterogrammatic strategy assumes a shape still more uncompromising, unrelenting, and radical. Briefly stated, *Alphabets* is a collection of 176 poems, each of which is composed of eleven lines of eleven letters each; its shape is thus that of a square. Every poem puts into play the ten most frequently used letters in the alphabet, plus one variable letter among the remaining sixteen letters. Each of those sixteen letters is used in precisely eleven poems: $16 \times 11 = 176$. The degree of formal constraint that *Alphabets* displays in its process of production is thus very daunting indeed; and the level of linguistic and poetic virtuosity that Perec puts on display as he finds solutions to the problems he has set for himself can leave a reader slack-jawed. Yet beyond those admittedly astonishing pyrotechnics, I would like to suggest that we are dealing with another sort of *difficulté vaincue*. As he cycles through the various permutations of the form he has chosen, exploiting the combinatory potential of the alphabet, and wagering upon the principles of completeness, symmetry, and exhaustion, Perec erects a very impressive sort of literary geometry. That structure is clearly both literal and mathematical in character—that is, it is based both on letters and on numbers. Perec occupies that structure and claims it as his own unabashedly, for every level of *Alphabets* calls our attention to process issues and the question of purpose. One might indeed argue that, more than anything else, *Alphabets* is *about* process and purpose. Those notions, taking material shape in the text, furnish a dwelling for Perec, one that he inhabits in a posture of defense. For whereas things outside the structure may be more than passingly chaotic, disorganized, and unforeseeable, on the *inside* they are to the contrary sleekly coordinated, methodical, and predictable. What more could one ask of a literary world?

The last of Georges Perec's experiments that I will discuss is his novel *La Vie mode d'emploi* (1978), translated into English as *Life A User's Manual: Fictions* by David Bellos in 1987. In that text, the action revolves around a Parisian apartment building. Perec describes that edifice as possessing ten floors vertically and ten units horizontally. Looking at the building's facade, then, one can conceive it as a ten-by-ten grid of squares—much like an expanded chessboard. And indeed Perec will call upon the game of chess in an effort to organize the sequence of chapters in his novel, adapting the classic chess problem of Knight's Tour (which consists of making a knight visit every square of the chessboard once, and only once) for a grid of ten-by-ten, rather than eight-by-eight. Each "square" that the knight visits corresponds to a space in Perec's imaginary apartment building, and to a chapter dealing with the people inhabiting that space. Mathematically, the problem that the Knight's Tour poses can be most usefully addressed through graph theory (though the problem itself is far older than that branch of mathematics).

Another constraint that Perec deploys in *La Vie mode d'emploi* is based upon a more recent mathematical figure that Berge brought to his attention in 1967, in a letter addressed to Perec and Roubaud.[14] Oulipian tradition dubs that figure an "orthogonal Latin bi-square, order 10," but it can also be called a "10×10 Graeco-Latin square." Briefly described, it is a combinatory structure, adapted for a grid of ten squares by ten, each square containing an ordered pair composed, say, of capital letters from A to J and small-case letters from a to j, such that each row in the grid and each column contain one element of the former set and one of the latter, and that no squares contain the same pair. Having imagined for his novel a catalogue of constitutive elements (objects, animals, actions, quotations, style, age and gender of characters, and so forth), and having divided those elements into sets of lists, the mathematical figure provides for the systematic distribution of those elements in the novel. Yet that very systematicity was a bit too rigid for Perec, and in order to render it a bit more supple, he called upon yet another figure, called a *quenine*, or "a generalization by Queneau of the sestina, invented by the troubadour Arnaut Daniel" (Oulipo 1981b, p. 243). Adapting that figure to fit his purposes and calling it a "pseudo-quenine, order 10," it furnishes a bit more creative freedom for Perec, while still preserving the programmatic quality of his project, allowing him "to produce in non-aleatory fashion several different Latin bi-squares by permuting either the columns or the rows of the original model" (Perec 1993, p. 25). In that perspective, it is legitimate to argue that mathematics serves a double purpose in *La Vie mode d'emploi*: that of constructing a coherent, predictable, and orderly space, and also that of rendering such space a bit more comfortable, a bit more livable, in short a bit more human.

236   W. MOTTE

The final Oulipian experiments that I would like to mention are Jacques Jouet's "metro poems." Like many of his colleagues, Jouet has long been fascinated with fixed poetic forms, not only those historically practiced in his own literary culture, but also those arising in other cultures, such as the pantoum. True to the Oulipo's double mission of "analysis" and "synthesis"—that is, on the one hand, the appropriation and rehabilitation of old (and sometimes ancient) literary forms, and, on the other, the elaboration of new ones—Jouet set himself the task in the late 1990s of conceiving a new fixed poetic form. He called it the "metro poem," and the best definition of it is contained in one of his first attempts in that new form, a text entitled "What is a Metro Poem?":

> From time to time, I write metro poems. This poem is one of them.
> Do you want to know what a metro poem is? Let's say that your answer is yes. Here then is what a metro poem is.
> A metro poem is a poem composed in the metro, during the duration of a trip.
> A metro poem has as many verses as your trip has stations, minus one.
> The first verse is composed in your head between the two first stations of your trip (counting the station from which you departed).
> It is transcribed onto paper when the train stops at the second station.
> The second verse is composed in your head between the second and third stations of your trip.
> It is transcribed onto paper when the train stops at the third station. And so forth.
> One must not transcribe when the train is in motion.
> One must not compose when the train is stopped.
> The last verse of the poem is transcribed on the platform of your last station.
> If your trip involves one or more changes of subway lines, the poem will have two or more stanzas.
> If through bad luck the train stops between two stations, that's always a ticklish moment in the writing of a metro poem. (Jouet 2000a, p. 7)

At first glance, the mathematical dimension of the metro poem is not immediately apparent; but as one considers the experiment in some detail, that dimension ineluctably emerges. First and most obviously, it should be noted that Jouet has chosen to connect his form to a privileged site of the quotidian. For most Parisians like himself, the metro is deeply entwined in the fabric of daily life, a site of gestures and experiences so deeply habitual that they tend to be dismissed as unimportant. Jouet's choice is a daring one, then, for it calls into question much of what we have always thought poetry to be: can one imagine a space less apparently suited to poetic contemplation? Yet he is painfully aware that, just as our culture tends to dismiss the significance of sites like the metro, shunting them to the margins of things, so too does it tend to dismiss poetry. In a public lecture he delivered in 1998, Jouet characterized his new form as "a poem struggling against its own marginality: presence in the world, integration of the poem into the repetitive quotidian; use of a place reputed to be non-poetic; a measured poem that measures itself in its non-reading" (Jouet 2001, p. 47). Jouet suffers from no illusions of grandeur, obviously enough. He invites his eventual reader to share the pungency of what is clearly a very quixotic enterprise—and to savor its ludic potential too, because these are texts conceived in a spirit of play, and they are tendered to us in that very same spirit.

Jouet's metro poems moreover put on offer the notion of writing as a daily practice, as an activity deeply imbricated in the quotidian, rather than somehow apart from it. That is an idea to which Jouet himself is profoundly attached. For many years, he made it a practice to write a poem a day, regardless of whatever other writerly projects he may have been engaged in. He mentions, too, that often he would compose those poems while riding in the metro, and that it was that habit which ultimately led him to conceive his new poetic form (Jouet 2001, pp. 44–45). To the extent that we are willing to reconsider literature—both writing and reading—as everyday activities, Jouet argues, so in that same measure can we begin to accept the notion that literature can eventually help us to understand our ways of being in the everyday world. Jouet also wagers on poetry's potential to transform those ways of being. "Going into the metro for a metro poem," he muses, "and going into the metro for something else is not the same thing" (Jouet 2000a, p. 37). It is a question of rethinking those ways of being, of reconfiguring our interpretive habits, and of being willing to look for meaning in those spaces of the ordinary where meaning seems to be most elusive.

Furthermore, Jouet's project is extremely systematic in nature—and this is where the mathematical character of the metro poem may begin to become apparent. Patently, the metro itself is a "system." It is important to recognize, however, that characteristic of the metro provides Jouet's poems in turn with a crucial systematicity. The phenomenon becomes immediately clear when one considers their metrics. The metro has its own particular rhythms, to be sure, rhythms upon which its users rely. "Subway riders," remarks Marc Augé, "basically handle nothing more than time and space, and are skilled in using the one to measure the other" (Augé 2002, p. 8). Jouet, for his part, calls the metro "an excellent metronome" (Jouet 2001, p. 47), recognizing the poetic potential of its rhythms, and appropriating them for his own purposes. Jouet's poems are likewise intersective, combinatory, and permutational. In that light, these poems play out quite explicitly aspects of literature which Jouet sees as absolutely fundamental ones; they consequently, like other texts I have mentioned, *exemplary* works.

The apotheosis of the metro poem is another Jouet text called "Poème du métro parisien" (1998).[15] Having practiced his new form for a while, Jouet conceived the idea of putting it to a test which would both strain its limits and reveal its potential. He imagined a trip on the metro where he would pass through every station once, with a minimum of reduplication. He asked

238   W. MOTTE

his fellow-Oulipian Pierre Rosenstiehl, a mathematician specializing in theories of labyrinths and graphs, to design an itinerary for him—and that mathematical design is what makes the poem possible, in the first instance. Jouet took his trip on April 18, 1996, beginning at half past five in the morning and finishing at nine o'clock in the evening. What resulted was a poem of 490 verses, distributed in 48 stanzas. One of the most striking aspects of the text is the will toward totalization that animates it. Jouet seeks to address the Parisian subway system as a whole, and to exhaust its possibilities—exhausting thereby too, not coincidentally, the possibilities of the poetic form he had postulated. One can also imagine how exhausting that very long day of writing must have been for the poet himself; yet he is quick to point out in his poem that time takes on a very different aspect in an exercise such as this one: "I have no time to dream of anything other than the poem" (Jouet 2000b, p. 70). Poetry itself is thus clearly at stake here: if Jouet wishes to imagine a total voyage, he also wishes to imagine a total poem. A poem that takes as its substance the real world in its crude, unexamined materiality; that finds meaning and order in the symbolic play of that world; and that casts that meaning, very concretely indeed, in a language which boldly proclaims its own materiality, its own quiddity, in a demonstrably systematic, *geometrical* manner. Riding round and round in the metro, Jouet confesses the desire that drives him: "I simply want to work round" (Jouet 2000b, p. 78). It is legitimate to read that statement on several levels. First, geometrically: the circular itinerary that Jouet follows, ending up just where he began, is an image of a certain kind of perfection, when viewed stripped of its trappings. The iterative patterns of his text play out circles within circles, commenting wryly and incisively thus upon the fundamental iterativeness of poetry itself. If we can accept the idea that highly codified structures of repetition in poetry such as rhyme and rhythm can convey meaning, Jouet intimates, why is it that we habitually dismiss the repetitive patterns of our daily existence as meaningless? Finally, his comment can also be read as a very lapidary formulation of a certain theory of literature, one that embraces both "Poème du métro parisien" and Jouet's *oeuvre* as a whole, for both are circular works, elaborated in a deliberate, canny, and exceptionally adroit dynamic of *working round*.

In all of these experiments, the notion of shape is absolutely fundamental, whether it be the barely imaginable shape of a hundred trillion poems, or the fugal intricacies of a sestina, or the sleek symmetries of heterogrammatic verse, or the configuration of the Parisian metro. More than anything else, to the Oulipian mind, mathematics offers a way of thinking about shape—and most particularly literary shape—in productive, programmatic ways. For shape signifies eloquently in their work. It is not a mere vehicle for meaning, it participates in literary meaning—and often enough, it enables meaning. It is true that much literature understates the importance of form, leaves it aside, insisting rather upon message, theme, mood, and so forth. Yet even in those cases, form is latent, one might argue—and it is impossible to imagine a truly shapeless work. So, too, our daily existence may seem largely shapeless to us, until we agree to reflect closely upon the shapes it takes, and upon the way its specific rhythms signify. In such a perspective, the Oulipo's meditations on literary shape may be seen to assume somewhat broader implications, devolving upon the ways we deal with experience and the manners in which we imagine ourselves in the world. For if these experiments are undeniably formalist in character, I contend that they are also—and perhaps chiefly—profoundly humanist.

Notes

1. The passage comes from *Maldoror* II, 10; the English translation appears in https://thanatologist.wordpress.com/ode-to-mathematics.
2. Among the better-known Oulipians are Raymond Queneau, Italo Calvino, Marcel Duchamp, Georges Perec, Harry Mathews, and Jacques Roubaud. A full list of the Oulipo's current membership can be found on the group's website, oulipo.net. The group's key collective publications include *La Littérature potentielle: Créations, recréations, récréations, Atlas de littérature potentielle, Anthologie de l'Oulipo*, and the Bibliothèque Oulipienne, a collection of brief texts published in editions limited to 150 copies, and which includes to date some 225 volumes.
3. Those writings include the edited volume *Les Grands Courants de la pensée mathématique, Dictionnaire des mathématiques*, with Alain Bouvier and Michel George, and *Les Nombres remarquables*. Speaking about Queneau, Le Lionnais remarks, "But this expert amateur was not merely a consumer of mathematics; he was also an authentic mathematician. Not a great one (the pomposity of this adjective would have displeased him) but a real one (this appraisal would have gratified him), whose work was considered by the best to be both valid and interesting" (Le Lionnais 1986b, p. 75).
4. For accounts of the Oulipo's early years, see Jean Lescure, "Brief History of the Oulipo," and Warren Motte, "Raymond Queneau and the Early Oulipo."
5. The last words of those ten first lines are, respectively, *chemise, frise, prise, marquise, éprise, valise, sympathise, prosaïse, friandise*, and *agonise*.
6. My translation, as elsewhere, unless otherwise noted.
7. See for example http://x42.com/active/queneau.html; http://www.bevrowe.info/Queneau/QueneauHome_v2.html; or https://www.pedagogie.ac-aix-marseille.fr/jcms/c_286730/fr/cent-mille-milliards-de-poemes.
8. See for instance Queneau's "Bourbaki et les mathématiques de demain."
9. For a more detailed discussion of the sestina and its posterity, see Audin (2009).
10. The "Hortense" novels include *La Belle Hortense, L'Enlèvement d'Hortense*, and *L'Exil d'Hortense*.
11. See Georges Perec, *La Disparition* and "Palindrome."
12. Other examples of that form can be found in *La Clôture et autres poèmes*.
13. See Jacques Bens (1986, p. 67): "For the members of the Oulipo have never hidden their abhorrence of the aleatory, of bogus fortunetellers and penny-ante lotteries. 'The Oulipo is anti-chance,' the Oulipian Claude Berge affirmed one day with a straight face, which leaves no doubt about our aversion to the dice shaker."

14. See Perec (1993, p. 21). The editors' preface in this volume provides a very clear and detailed account of the novel's formal organization.
15. The text was originally published in a small, limited edition of its own, entitled *Frise du métro parisien*, which appeared as volume 97 of the Bibliothèque Oulipienne (1998).

REFERENCES

Audin, Michèle. 2009. "Poésie, spirales et battements de cartes: d'Arnaut Daniel à Jacques Roubaud en passant par Gaspard Monge et quelques autres." http://images.math.cnrs.fr/Poesie-spirales-et-battements-de.html.

Augé, Marc. 2002. *In the Metro*. Translated by Tom Conley. Minneapolis: Minnesota University Press.

Bénabou, Marcel. 1986. "Rule and Constraint." In Motte 1986: 40–47.

Bens, Jacques. 1986. "Queneau Oulipian." In Motte 1986: 65–73.

Fujiwara, Masahiko. 2017. "Literature and Mathematics." www.asymptotejournal.com/nonfiction/masahiko-fujiwara-literature-and-mathematics.

Jouet, Jacques. 2000a. *Poèmes de métro*. Paris: P.O.L.

———. 2000b. "Poème du métro parisien." In Jouet 2000a: 63–86.

———. 2001. "Avec les contraintes (et aussi sans)." In *Un Art simple et tout d'exécution: Cinq leçons de l'Oulipo, cinq leçons sur l'Oulipo*, by Marcel Bénabou, Jacques Jouet, Harry Mathews, and Jacques Roubaud, 33–67. Belfort: Circé.

Le Lionnais, François. 1948. *Les Grands Courants de la pensée mathématique*. Paris: Cahiers du Sud.

———. 1983. *Les Nombres remarquables*. Paris: Hermann.

———. 1986a. "Second Manifesto." In Motte 1986: 29–31.

———. 1986b. "Raymond Queneau and the Amalgam of Mathematics and Literature." In Motte 1986: 74–78.

Le Lionnais, François with Alain Bouvier and Michel George. 1979. *Dictionnaire des mathématiques*. Paris: Presses Universitaires de France.

Lescure, Jean. 1986. "Brief History of the Oulipo." In Motte 1986: 32–39.

Motte, Warren, ed. and trans. 1986. *Oulipo: A Primer of Potential Literature*. Lincoln: Nebraska University Press.

———. 2006. "Raymond Queneau and the Early Oulipo." *French Forum* 31 (1): 41–54.

Oulipo. 1973. *La Littérature potentielle: Créations, recréations, récréations*. Paris: Gallimard.

———. 1981a. *Atlas de littérature potentielle*. Paris: Gallimard.

———. 1981b. "La Quenine." In Oulipo 1981a: 243–48.

———. 2009. *Anthologie de l'Oulipo*. Paris: Gallimard.

Perec, Georges. 1969. *La Disparition*. Paris: Denoël.

———. 1976. *Alphabets*. Paris: Galilée.

———. 1978. *La Vie mode d'emploi*. Paris: Hachette.

———. 1980a. *La Clôture et autres poèmes*. Paris: Hachette.

———. 1980b. "Palindrome." In Perec 1980a: 43–47.

———. 1980c. "Ulcérations." In Perec 1980a: 55–67.

———. 1993. *Cahier de charges de La Vie mode d'emploi*. Edited by Hans Hartje, Bernard Magné, and Jacques Neefs. Paris and Cadeilhan: CNRS and Zulma.

Queneau, Raymond. 1961. *Cent mille milliards de poèmes*. Paris: Gallimard.

———. 1962a. *Entretiens avec Georges Charbonnier*. Paris: Gallimard.

———. 1962b. "Bourbaki et les mathématiques de demain." *Critique* 176: 3–18.

———. 1986. "Potential Literature." In Motte 1986: 51–64.

Roubaud, Jacques. 1967. ∈ . Paris: Gallimard.

———. 1973. *Trente et un au cube*. Paris: Gallimard.

———. 1981. Jacques Roubaud, "Deux Principes parfois respectés par les travaux oulipiens." In Oulipo 1981a: 90.

———. 1985. *La Belle Hortense*. Paris: Ramsay.

———. 1986a. *Quelque chose noir*. Paris: Gallimard.

———. 1986b. "Mathematics in the Method of Raymond Queneau." In Motte 1986: 79–96.

———. 1987. *L'Enlèvement d'Hortense*. Paris: Ramsay.

———. 1990a. *Princesse Hoppy ou le conte du Labrador*. Paris: Hatier.

———. 1990b. *L'Exil d'Hortense*. Paris: Seghers.

———. 2004. "Perecquian OULIPO." Translated by Jean-Jacques Poucel. *Yale French Studies* 105:100–1.

———. 2008. *Parc sauvage*. Paris: Le Seuil.

———. 2009. *'le grand incendie de Londres'*. Paris: Le Seuil.

Tubbs, Robert. 2014. *Mathematics in 20th-Century Literature & Art: Content, Form, Meaning*. Baltimore: Johns Hopkins University Press.

CHAPTER 14

Mathematics and Dramaturgy in the Twentieth and Twenty-First Centuries

Liliane Campos

> It should be obvious by now that I am only interested in mathematics as a creative art. In that sense it has nothing to do with physical reality.
> G. H. Hardy, quoted in Complicite's *A Disappearing Number* (2008, p. 81)

In his essay *A Mathematician's Apology* (1940), G. H. Hardy defends the idea that mathematics is a creative art, comparable to poetry or painting, which is concerned with "another reality" distinct from "the material world, the world of day and night, earthquakes and eclipses" (Hardy 2006, p. 122). Quoted at length in Complicite's play *A Disappearing Number* (2007), Hardy's remarks condense the ambivalent appeal of mathematics for the stage: on the one hand its aesthetic potential as a creative mode of thought, on the other the challenge of connecting its abstract constructions with the physical bodies of a theatrical performance. If mathematical thought entails, as the American mathematician James Pierpont once suggested, a "total separation from the world of our senses," then introducing it into the theatre is a paradoxical gesture, rich with potential tensions (Pierpont 1899, p. 406).

Mathematicians have increasingly appeared as dramatic characters on screen and stage since the 1990s, a trend which has gradually modified popular stereotypes and public awareness of the profession (Osserman 2005). In recent years they have populated the growing body of "science plays" that were first examined as a genre by theatre scholar Kirsten Shepherd-Barr in her study *Science on Stage,* and whose popularity remains partly indebted to David Auburn's depiction of mathematical genius in *Proof* (2001), winner of

L. Campos (✉)
Sorbonne Nouvelle, Paris, France

Institut Universitaire de France, Paris, France

© The Author(s) 2021
R. Tubbs et al. (eds.), *The Palgrave Handbook of Literature and Mathematics,* https://doi.org/10.1007/978-3-030-55478-1_14

243

the Pulitzer Prize and later adapted for the screen (Shepherd-Barr 2006, p. 129). Mathematics, however, started playing a part in European dramaturgy well before mathematicians became fashionable characters in Western drama. Since the 1950s, both theoreticians and dramatists have found formal inspiration in the dizzying possibilities of combinatorics, the equations of chaotic systems or the patterns of number theory. Whether they are used as model or as metaphor, these mathematical tools have been used to question fundamental theatrical components, such as time and memory, live presence and embodiment.

The analysis presented here is supported by a range of critical studies of individual artists and works, but few researchers have so far examined the relation between theatre and mathematics in a broad comparative approach, with the exception of Shepherd-Barr's analysis. This chapter begins by examining ways in which mathematics has been used to structure dramatic models and theatrical pieces, before turning to thematic uses of mathematical ideas in contemporary performance. Although theoretical attempts to modelize drama have generalized from existing plays, they have much in common with artistic attempts to give performances a mathematical structure. From pre-structuralist analysis to theatrology, and from the Oulipo group's creative constraints to Samuel Beckett's geometrical *Quad* (1982) or Annie Dorsen's *A Piece of Work* (2013), combinatory models have dominated these formal approaches. These theoretical and practical experiments are examined in parallel here so as to highlight their common ambivalence: a recurring tension between reduction and proliferation, between the opening up of possibilities and the hollowing out of theatrical categories.

The final section discusses ways in which contemporary dramatists and directors have embodied mathematical thought on stage. Works such as Tom Stoppard's *Arcadia* (1993) draw on the metaphorical potential of the science that the mathematician and philosopher Henri Poincaré once called "the art of giving the same name to different things" (Poincaré 1947, p. 29). Yet mathematics does not only provide metaphors for human situations, it is also, conversely, embodied by the performers on stage. The productive contradictions arising from the enactment of abstract thought are examined here through the comparison of two recent productions, *A Disappearing Number* (2007), devised by British director Simon McBurney and his company Complicite, and *Le Cas de Sophie K.* (2005), devised by French director Jean-François Peyret and his company Tf2, two plays which set out to visualize on stage the workings of a mathematical mind.

Modelizing Dramatic Form: Combinations and Theatre Trees in Twentieth-Century Theory and Practice

In her defense of "drametrics" and of the potential benefit of mathematical analysis for dramaturgy, dramaturg and theatre theorist Magda Romanska argues that mathematics already appeared in the theory of drama in 1863, when playwright Gustav Freytag described the shape of classical Aristotelian tragedy as a triangle formed by the rising and falling of tragic action, in his treatise *Die Technik des Dramas* (Romanska 2015, p. 439). It is true that Freytag's pyramid-shaped model evokes the temptation of geometrical perfection, but it is only in twentieth-century theory that the appeal of mathematical form became explicit, and tools such as combinatory analysis, Markov chains and Boolean logic were applied to the study of drama. Whereas Freytag's model simplified the shape of the tragic plot to an extreme, twentieth-century modelizations seem to have worked in the opposite direction, depicting dramatic structure as a proliferating set of potential situations through the analysis of possibilities and probabilities. This focus on multiplying possibilities characterizes both theoretical attempts to modelize drama, which reached their peak in the 1970s and 1980s, and creative attempts to give performance a mathematical structure, from the logical constraints explored by the Oulipo group in the 1970s to more recent experiments with texts generated by algorithms.

Combinatorial analysis and combinatory poetics are remarkably prevalent in both theoretical and practical attempts to mathematize dramatic work. In 1895, the French writer Georges Polti set out to identify the thirty-six situations that would encompass all dramatic creation. Polti was a writer and translator of theatre, but also a critic and theoretician, and his attempt was inspired by Goethe's reference to this number. His *36 Situations dramatiques* anticipated structuralist analysis, since he considered characters as acting in a game of forces to create dramatic conflict, and identified the kind of recurrent actions that were later analyzed as functions or actants (De Bary 2004). His work has mostly been influential through that of Etienne Souriau, whose analysis of theatrical "functions" and the "operations" through which they can be combined in his *Deux cent mille situations dramatiques* (1950) paved the way for structuralist and semiotic approaches to dramatic texts. Neither Polti nor Souriau, however, seemed to have been content with analyzing existing texts: their approaches were also driven by the excitement of embracing the vast, as yet unexplored, possibilities of dramatic composition, which led Polti to dream of a totalizing piece of theatre, "*spectacle-algèbre*," which would include all possible situations and emotions (Polti 1900). Both Polti and Souriau thus anticipated creative uses of combinatorics, and particularly the work on "literary generativity and potentiality" carried out by the *Oulipo* group since 1960 (De Bary 2004, p. 187).

Founded by writer Raymond Queneau and mathematician François Le Lionnais, the *Ouvroir de littérature potentielle* group invents formal constraints that will help literature to renew itself.[1] According to Queneau, mathematics are an answer to the exhaustion of traditional literary forms (Roubaud 1981, p. 66). Relatively few theatrical works have emerged from the *Oulipo*'s creative constraints, but they have proposed a number of models, for instance "Boolean theatre," which takes two existing texts and plays with different logical operations, such as *reunion* (both text A and text B are performed on stage simultaneously) or *intersection* (a third text, using lines from both A and B, is performed simultaneously), in order to create a new emergent play (Le Lionnais 1973). While "Boolean" theatre uses existing plays to generate new modes of performance, other *Oulipien* constraints are designed to generate new texts: in 1973, Paul Fournel and Jean-Pierre Enard proposed a combinatory model, *l'arbre à théâtre*, in which the audience would make choices after each scene, thereby directing the actors among the sixteen different possible versions of the play. In the plot imagined by Fournel and Enard, after the first scene in which "The king is sad," the spectators choose between two possible reasons: either his daughter the princess is sad or she has been kidnapped (Fournel and Enard 2005, p. 200). Each choice then determines the following scene among a limited number of paths, leading to only two possible final scenes. Fournel and Enard highlight the challenges of such "combinatorial comedy" for the actors, and indeed the most famous text to embrace this kind of technical challenge, an arborescent play imagined by Alan Ayckbourn, did not give the audience the kind of influence they imagined. In Ayckbourn's *Intimate Exchanges* (1983), the characters' choices lead them along sixteen different versions of the play, in which romance may or may not blossom, chances may or may not be taken, and safety may or may not be preferred over adventure. Although all versions of the play are meant to be performed on different nights, the company chooses the path taken by each performance to ensure that the text only requires two performers, and the audience is encouraged to return to see where different paths will lead within the branching structure (Ayckbourn 1985).

For both practitioners and theoreticians, the main difficulty that arises from these attempts to apply mathematical tools to dramatic texts is the need to simplify in order to be able to systematize. When combinatorial analysis was carried out by Romanian "theatrologists" in the 1970s, mathematical complexification went hand in hand with further simplification of dramatic situation. Solomon Marcus's mathematical-linguistic model used a matrix, whose columns correspond to characters and rows to scenes, to study "configurations" of characters (Marcus 1970, 1971; Carlson 1993, pp. 493–94). This reduction of the scope of analysis to the binary question of absence and presence and to combinations of characters present on stage was equally adopted by analysts following in Marcus's footsteps in the 1970s and 1980s. When Brainerd Barron and Victoria Neufeld developed and discussed his model, adding additional data such as the number of words spoken by characters, they acknowledged that "some information must be sacrificed" but argued that Marcus's model brings out "nuances of plot structure," for instance a contrapuntal structure in Shakespeare's *The Comedy of Errors*, "involving the entrances and exits of characters who are not to meet until the end of the play" (Barron and Neufeld 1974, pp. 41, 73). The separation of dramatic form from thematic, verbal content was a conscious decision for researchers such as Marcus, who suggested that verbal analysis is over-represented in theatre studies (Marcus 1975, p. 89), or Mihai Dinu, who defends the importance of character configuration as a "non-verbal component" (Dinu 1984, p. 67). Dinu has used group theory, topology, Markov models, and Boolean logic to analyze configurations and to calculate the probability of a character's presence (Dinu 1972) or the resulting "weight of stage relations" (Dinu 1977). When he sets out, for instance, to study the stage presence of Edward Albee's characters in *Who's Afraid of Virginia Woolf* (1962) as a "logical function," his model reveals that all scenes in the play have one of the three following properties: either George is present, or Nick and Martha are present, or Martha interacts with George and or Nick in the absence of Honey. These three categories of scenes are then analyzed as an expression of character "strategies," for instance Martha's seduction of Nick in the second type (Dinu 1974).

In the special issue of *Poetics* that he edited dedicated to the formal study of drama, Dinu acknowledges that mathematical investigations will define "general laws" rather than "great" writing, in "a kind of "negative" investigation of the work of art consisting in the separation of the unique kernel which cannot be studied by mathematical means from the features connected with craftsmanship, routine, therefore the ones which can be expressed by algorithms" (Dinu 1977, p. 210). The allure of mathematics is ambivalent: it opens up infinite potential through analysis and combinatorics, yet reduces art to models whose simplification is haunted, in Dinu's terms, by the absent "kernel," the core that cannot be modelized. While the conclusions drawn from such theoretical attempts may disappoint, they reveal a fundamental ambivalence which also characterizes contemporary theatre artists' use of mathematical tools: a tension between generative potential and negative reduction, between the proliferating energy of combinatory poetics and the absences these poetics emphasize.

COMBINATORICS AND ALGORITHMS AS DRAMATURGY: SAMUEL BECKETT, STAN'S CAFE, ANNIE DORSEN

Beckett's fascination with mathematical thought is the object of a growing body of critical work dedicated to the geometrical patterning, calculations, and permutations that can be found in his work.[2] Series and combinations in particular play a key structuring role not only in his use of language, but also in the actions performed by his characters and in his construction of theatrical space. His most mathematical piece, the teleplay *Quad* (1982), almost escapes the category of theatre, since it is designed to be performed on screen. Yet its four "players," who silently move about a square space to the sound of rhythmical percussions, remain haunted by the codes of theatre: entrances and exits, a constrained space of performance, and the possibility, never quite realized, of an encounter. Each player is given a number and anonymized by a cowl hiding their face. They follow a precise "course" which takes them along straight lines from one corner of the square (A, B, C, D) to another: these courses are indicated by letters ("AC, CB, BA...") and the different courses of the four players are combined into series, in which one player sometimes follows their course alone, but two, three, or four players may also be simultaneously following their courses (Beckett 1986, p. 451).

Part of the appeal of mathematics for Beckett lies in the possibility of purifying form almost to the point of abstraction (Casanova 1997, p. 163). Indeed Beckett's reader, when confronted with the abstract diagrams that he provides with the stage directions for *Quad*, may be reminded of Wassily Kandinsky's reflections on the square "basic plane" and its "dramatization" in his influential study of the basic elements of non-objective painting, *Point and Line to Plane* (1947).[3] Kandinsky emphasizes the tensions or "dramatization" that can be created by displacing and decentering lines within the square (Kandinsky 1947, p. 129). Beckett also works with the tensions created by decentered movement, since he indicates that the performers avoid meeting in the center by deviating slightly round the "danger zone" of the central point E. By opposing the "silent lyric" and rigidity of vertical, horizontal, or diagonal lines meeting in the center of a square to the "dramatization" of their displacement, Kandinsky associates dramatization with both avoidance of the edges and acentric structures, noting that an "acentric structure here serves the purpose of increasing the dramatic sound" (Kandinsky 1947, p. 138). If we describe the play in Kandinsky's terms, Beckett's *Quad* thus draws on the silent lyric of the square every time the players reach a corner, yet also dramatizes the space by oscillating around the center.

Quad's combinatory poetics produce both a sense of completeness within the system and an exhaustion of theatrical form. Beckett emphasizes the comprehensiveness of the approach in his directions: "Four possible solos all given," "All possible light combinations given," "All possible percussion combinations given" (Beckett 1986, pp. 451–52). Completing the possibilities of the system is, however, also a form of closure, negating further potential. According to Gilles Deleuze, combinatorics are the "art or science of exhausting the possible," and *Quad* exhausts not only the possibilities it generates, but the key potential of the performance square, the potential for encounter (Deleuze 1992, p. 82). This hindering of theatrical potential—the avoidance of encounter, the exhaustion of presence without ever quite accepting absence—characterizes much of Beckett's work, but *Quad* makes it very clear that he approaches it as a mathematical "problem," or a series of problems. In a detailed analysis of the geometry and combinatorics of *Quad*, mathematician Brett Stevens has translated the shape of the piece into several mathematical problems, including the different sets of its combinatory progression and the oscillations of the center of mass of the space depending on the number of players moving around. He concludes not only that Beckett "has maximized the constraint" but also that mathematics is paradoxically both a structuring force and an obstacle: "[t]he mathematical axioms of the object defined in *Quad* are its own obstacles to its own existence" (Stevens 2010, pp. 176, 179).

Although Beckett's poetics carry the exhaustion of theatrical form to an extreme, the fundamental ambiguity of his mathematical gesture, opening up theatrical possibilities while hollowing out theatrical categories, also characterizes more recent uses of mathematical tools by artists such as Stan's Cafe or Annie Dorsen. Founded in 1991 by Graeme Rose and James Yarker, the British theatre company Stan's Cafe devises very diverse theatrical forms that often turn a simple constraint or set of rules into the inspiration for a performance. Their best-known, much-toured piece, *Of All the People in All the World* (2003–2017), is a theatrical visualization of human statistics, in which a grain of rice represents a human being, and performers measure out piles of rice to represent figures in a striking visual installation. The statistics vary from show to show, to represent a variety of numbers such as victims of the Holocaust, epidemiology figures, military personnel in Iraq, or climate change statistics. Stan's Cafe thus theatricalizes statistics in a way that can be compared to the German company Rimini Protokoll's very successful *100% City* shows, in which 100 inhabitants of a city are invited to embody the statistics of this city on stage (gender, political preferences, means of transport, etc.). *Of All the People in All the World*, however, proposes a much more abstract vision of historical and political issues, since humanity is figured by grains of rice manipulated by scientific-looking performers in white coats. It is an ironic acknowledgment of a world "turned into numbers" (Rey 2016), which performs this transformation yet dissociates numbers from discursive contexts, leaving the audience to weave the juxtaposed figures together into any number of stories.

This preference for juxtaposition is also clear in the combinatory games that have inspired several of their pieces, most evidently in their poem for the radio *Bleak Heart Driver*, and the theatre performance *Simple Maths* (1997). While *Bleak Heart Driver* plays with combinations of words, *Simple Maths* contains no words, but is composed purely of permutations between five performers who are given six chairs and six possible moods (angry, happy, sad, tired, worried, or neutral). The script of this performance contains 286 lines, each of which indicates the initials of the performers (A, S, N, C, J), the chairs they are sitting on, and whether they change moods (in which case the initial of the new mood, a, h, t etc., is indicated). Each new line is therefore a new combination of positions and moods, but moods do not change with every line, and a performer may not always change seats (in which case a square root appears beside his or her initial). Below is a sample of seven lines from the typescript:

250 L. CAMPOS

Ch.1 Ch.2 Ch.3 Ch.4 Ch.5 Ch.6

	Ch.1	Ch.2	Ch.3	Ch.4	Ch.5	Ch.6	
80	N	A		S√	C	J	S tissue. A asks S. A wakes N & chats, N glasses on
81		A	N(ø)	S	C	J	
82		A	N	S	C	J√	C hits J who leaps up. N & A discussion pointing out.
83	C(h) A	N		S		J	J chats to S over C who is tying his laces
84	C	A√(ø)N		S		J	J & S kiss. A & C on the beach pointing out
85	C	A	N		S(t)	J	S head on J. N walks out front has vision.
86	C	A		N	S	J	S headswop J to N. A's head is on J's shoulder

(Stan's Cafe 1997)

While spectators may imagine plot-like elements in the interaction between the performers, these are constantly prevented from developing by the permutations, as for instance when heads shift from shoulder to shoulder in the excerpt above. Mathematics, then, is not just a playful, *Oulipo*-style constraint in *Simple Maths*, it is a tool through which the company resists the narrative bent of theatre. It allows the company to disable overarching narratives and to replace the progression of drama along the lines of a plot with the playful *parataxis*, or non-hierarchical structure, that performance theorist Hans-Thies Lehmann associates with the deferral of meaning typical of post-dramatic theatre (Lehmann 2006, pp. 86–87).

The proliferating energy of combinatorics can thus free performance from the need for verbal text. Yet contemporary practitioners also increasingly apply it to pre-existing texts, using algorithms to transform canonical works of literature into new, computer-generated pieces. These experiments often have non-theatrical texts as their starting-point: French director Jean-François Peyret has used computer-generated dialogue drawn from Thoreau's *Walden* and spoken by chatbots in the virtual environment of *Second Life* in *Re: Walden* (2013), while the American company Elevator Repair Service have performed text produced by algorithms combining sentences from *The Great Gatsby*, *The Sound and the Fury*, and *The Sun Also Rises* in *Shuffle* (2011). Digital technology allows these artists to process classic texts as data, and to draw theatrical performances from the encounter between performers and unpredictable scripts generated by the set of rules defining the algorithm. As a result, live presence is foregrounded by the contrast between human and digital performers, and by the actors' often improvised reactions to the system's output. Although the algorithm itself is not displayed, the projection of its output on screens tend to set it up as a performer in its own right. This was made particularly clear in the American director Annie Dorsen's "algorithmic" show *A Piece of Work* (2013), which used Shakespeare's *Hamlet* as its data, and in which live performers all but disappeared from the stage, leaving the computer-generated text and voices to fill the void in four out of five acts. Here too, the application of combinatorics to theatre generated new possibilities while emphasizing a key absence, in this case the absence of the founding text, and its displacement from the stage to the memory of actors and spectators.

A Piece of Work uses various algorithms to rearrange fragments from Shakespeare's text, which are both read by computer voices and projected onto a screen. The generated text also triggers changes in lighting, sound, and voice effects, thanks to the tagging of words by emotional scores that were used as data for the computer to modulate sound and light. One result of this tagging strategy is an averaging out of emotional value: the scores were attributed by a crowdsourcing exercise, in which anonymous workers were given words via Mechanical Turk, an Amazon Web Service that out-sources digital tasks, and asked to give them emotional scores (from one to five) for anger, joy, sadness, and fear. The results were then averaged out by the creative team, and each word in *Hamlet* thus receives scores that are not correlated to its context in Shakespeare's text. In turn these scores determine light or sound effects that also rely on average perceptions of emotional value: for instance, the Emotional-Based Music Synthesizer designed by Greg Beller used "a generative algorithm that takes an emotional tag as an input and then provides highly archetypal 'emotional' music" (Dorsen 2017, p. xii). This input is visualized for the spectator in the first act, which runs through the first four percent of each scene in the play, voices the text electronically and projects the corresponding instructions onto the screen, including the four emotional scores of each word:

SEND WHO OPHELIA
[...]
As: [-1,-1,-1,-1], if:[0.7,0.5,0.8,0.7], he: [0.2,0.5,0.9,0.5], had:[1.3,0.6,1.1,1.7], been: [0.2,0.7,0.5,0.5], loosed: [2.2,2.4,1.2,2.2], out:[1.0,1.1,1.0,1.8], of:[1.2,0.9,0.3,0.6], hell:[4.4,4.7,0.0,4.2]
SEND LINE 'As if he had been loosed out of hell'
To: [-1,-1,-1,-1], speak: [0.8,0.9,2.1,0.7], of: [1.2,0.9,0.3,0.6], horrors,: [-1,-1,-1,-1], he: [0.2,0.5,0.9,0.5], comes: [1.2,0.5,2.3,0.6], before: [0.6,0.8,0.9,1.3], me.: [-1,-1,-1,-1]
SEND LINE 'To speak of horrors, he comes before me.' (Dorsen 2017, p. 13)

For the spectator of *A Piece of Work*, the decontextualization of disembodied textual fragments is strengthened by the averaged emotional weighting of words, light, and sound effects. The four following acts are produced by different "scrambling/sorting" processes (Dorsen 2017, p. viii): in Act 2, only lines beginning with "what," "I," or "O," two-word lines, and three-word lines are used; in Act 3, soliloquies are rearranged by isolating nouns, noun phrases and common grammatical constructions, or by permuting parts of

252   L. CAMPOS

speech (nouns, verbs…); in Act 4, a Markov model, including a preference for words that score high on one particular emotion, is used to produce new scenes from the texts of groups of original scenes; and in Act 5, excerpts from Act 5 Scene 2 are re-sequenced by Markov chains on the level of individual letters. The published text, which corresponds to the script generated for performance on December 21, 2013, contains many incidences of surprising poetry, comic disturbance, and revealing ironies. Many of Claudius's lines, for instance, reattributed to Hamlet by the scrambling performed in Act 4, not only still make sense but highlight symmetries between the two murderers:

> HAMLET
> Now, mother, what's the matter now?
> O, what form of prayer
> Can serve my turn? Forgive me my foul murder?
> O, what form of prayer
> Can serve my turn? Forgive me my foul murder?
> Yet what can it not?
> That cannot be; since I am slain!
> *(Falls and dies.)*
> *(Enter HAMLET.)* (Dorsen 2017, p. 109)

As the algorithmic selection replaces the progression of plot, the text gets trapped in loops or makes unexpected leaps between characters and situations. The result can be viewed as an assertion of the "fungibility of the text, the extent to which it is material for other uses, the selective and combinatory operation of Dorsen's algorithmic plot" (Worthen 2017, p. 327), but equally as a resistant material, which carries too many associations and cultural resonances for it to be completely fungible for the audience. W. B. Worthen refers to Jerome McGann's concept of "deformance" to argue that *A Piece of Work* deforms the text of *Hamlet*, "visualizing and so opening to critique the operational coding of the text" (Worthen 2017, p. 328). Indeed, encoding the text reveals how encoded a text like *Hamlet* already is, on levels that we do not usually notice, and the psychological and narrative assumptions that this implies. But the algorithmic games played by Dorsen reveal how difficult it is to decode a canonical text, since every element awakens associations in the spectator's mind.

The role of memory in theatrical experience is both questioned and redistributed by these experiments. While Peyret has suggested that his work confronts the memory of actors with the memory of machines, Dorsen refers to her algorithms as "memoryless," since she uses Markov chains, processes which define each new state of the system only according to the one immediately preceding it (Dorsen 2017, p. iii). Yet most spectators of *A Piece of Work*, not to mention its human performer in act 3, will be haunted by the memory of *Hamlet*, rendering many lines resonant with their cultural afterlife, and reintroducing plot and character even where they have been erased.

Although categories such as author, text, plot, and character are questioned and visibly hollowed out, the algorithmic performance still relies on their absence to define it.

STAGING MATHEMATICAL THOUGHT: METAPHOR AND EMBODIMENT IN THE WORK OF TOM STOPPARD, JEAN-FRANÇOIS PEYRET, AND COMPLICITE

While digital technologies are turning mathematical models into a readily available dramaturgical tool, they are also increasing our awareness of algorithms as ubiquitous in contemporary life. This awareness may be one reason for which mathematics is providing thematic inspiration for a growing body of theatrical work, in which abstract thought is enacted on stage.[4] Since Tom Stoppard's very successful *Arcadia* was first staged in 1993, "science plays" exploring mathematics have been dominated by several recurring ideas, often drawn from chaos theory, fractal geometry, and number theory, which have been used to address key dramatic themes. Kirsten Shepherd-Barr has demonstrated that three remarkable theatrical pieces from the 1990s and 2000s, Stoppard's *Arcadia*, Complicite's *Mnemonic* (1999), and Barrow and Ronconi's *Infinities* (2002, 2003) all address questions of time and memory (Shepherd-Barr 2006, pp. 132–54). More generally, contemporary dramatists and directors tend to emphasize unpredictability, paradox and proliferating possibilities in the mathematical ideas they explore—ideas in which are then enacted as metaphors for human experience.

Chaos is a recurring motif in the metaphorical thrust of these plays, which tend to emphasize the ordering, patterning, and visionary power of mathematics in the face of turbulence and disorder. Tom Stoppard stands out among contemporary dramatists as a writer who has consistently drawn metaphors from mathematics, not least in his perception of the theatre as "an equation which is continuously changing" in which "most of the variables are specific to the performance" (Stoppard 1988). *Arcadia*'s witty dialogue keeps his audience alert to metaphorical potential, inviting us to enjoy the ambiguities that even scientific language may not escape: when sixteen-year-old Thomasina is asked by her mother what she is working on, and answers that it is a study of "the action of bodies in heat," the expression refers both to the science of thermodynamics she is studying and to the sentimental and sexual imbroglio of *Arcadia*'s plot (Stoppard 2009, p. 114). As the plot alternates between nineteenth- and twentieth-century scenes, *Arcadia*'s contemporary characters gradually understand the mathematical genius of the nineteenth-century girl who foresaw the principles of chaos theory and the potential of fractal mathematics. Stoppard invites us to see the birth of these ideas in parallel with the birth of Romanticism, and builds many bridges between artistic and scientific representations of disorder. Yet he also chooses mathematical ideas that have inherent affinities with the process of *emplotment*, both in the sense of playwriting and of the historian's emplotment of past events (White 1973). Thomasina discovers the principles of fractal geometry by *plotting* the forms of nature, in the mathematical sense of translating written symbols into visual representation and conversely finding the algorithm that a set of points may represent. She also formulates the unpredictable determinism of chaotic systems, a mathematical idea whose Romantic potential is emphasized by the play's twentieth-century mathematician:

> Valentine: ... The ordinary-sized stuff which is our lives, the things people write poetry about – clouds – daffodils – waterfalls – and what happens in a cup of coffee when the cream goes in – these things are full of mystery, as mysterious to us as the heavens were to the Greeks .... (Stoppard 2009, pp. 64–65)

Stoppard gives his mathematical concept metaphorical resonance by choosing Wordsworthian examples—clouds, daffodils, waterfalls—but also by insisting on its relevance to the human scale, "the ordinary-sized stuff which is our lives." Effectively this sensitivity to initial conditions is enacted in *Arcadia*'s turbulent plot by the catastrophic consequences of small details—the misattribution of a few words leading to a completely wrong version of history, or the forgetting of a candle flame leading to a tragic fire. Stoppard himself has suggested that the play imitates the bifurcations of graphs representing the evolution of chaotic systems (Fleming 2008, p. 52). Unpredictable determinism thus plays not only a metaphorical, but a metadramatic role, as does the process of *iteration* through which Thomasina constructs her graphs. Just as Thomasina's equations progress by feeding the results of an algorithm back into itself, Stoppard's *Arcadia* portrays the creative energy of both art and science as an endless process of repetition and reinvention.[5]

Stoppard's evocation of mathematics is a purely verbal feat, leaving the audience to imagine the graphs and equations that only the characters actually see. Similarly verbal metaphorical uses of mathematics can be found in a range of contemporary plays, such as Ira Hauptman's *Partition* (2003), Kefi Chadwick's *Mathematics of the Heart* (2012), or Itamar Moses's *Completeness* (2013), whose titles announce the figurative role of their key concepts. But some of the most inventive productions in recent years have been created by practitioners who sought to embody mathematical thought, using bodies on stage to give life to abstract ideas that do not obviously apply to them.[6] Two shows in particular stand out through the inventive ways in which they have created and questioned this embodiment of abstract thought: *A Disappearing Number* (2007), devised by the theatre company Complicite and directed by Simon McBurney, and *Le Cas de Sophie K.* (2006), devised by Jean-François Peyret and his company Tf2.

Both McBurney and Peyret choose to examine mathematicians who came from the margins of the academic circles of their time: Sofia Kovalevskaya, the first woman to become a professor in mathematics at Stockholm University, in *Le Cas de Sophie K.*, and Srinivasa Ramanujan, a self-educated Indian mathematician, in *A Disappearing Number*. Yet the two directors take very different approaches to the history of mathematics. *A Disappearing Number* explicitly sets out to explore the collaboration and friendship between Ramanujan and English professor G. H. Hardy, who worked together in Cambridge between 1913 and 1919. This collaboration is examined via a double time frame, as a contemporary character Alex, who is mourning the loss of his wife Ruth, also a mathematician, becomes fascinated by their story while remaining an "epistemological outsider" (Wisniewski 2016, p. 190) whose grasp of mathematics is very basic. The scenes alternate between the First World War period and the contemporary, and between England and India. Peyret on the other hand rejects the biographical form and what he calls the "authority of the biographer" (Campos and Shepherd-Barr 2006, p. 249). Although many of his shows have been built around great figures of the history of science, from Galileo to Alan Turing, he favors tangential approaches, refusing to give them the artificial "intelligibility" of a biography (Peyret 2005). Like most of his pieces, the text of *Le Cas de Sophie K.* is almost entirely composed of quotations and contains neither story nor characters. Although much of the text is drawn from Kovalevskaya's correspondence, they are spoken by three different actresses and one actor.

Despite these radically different dramaturgical approaches, the ways in which Peyret and McBurney approach mathematics have striking similarities. Both pieces were created in collaboration with scientific researchers, and devised through workshops with mathematicians and improvisations based on non-dramatic texts. Mathematicians who have watched these shows or been involved in their creation have emphasized the performative enactment of mathematical thought processes: Steven Abbott points out that Complicite's workshops with Oxford professor Marcus du Sautoy led them to "experience a corporeal manifestation of a mathematical concept at the same time that they [were] trying to understand it" (Abbott 2014, p. 231) and Michèle Audin, who collaborated with Peyret and his company Tf2 on *Le Cas de Sophie K.*, has compared his experimental approach to rehearsals to her own work as a mathematician (Audin 2009, p. 51). Moreover Peyret and McBurney focus their exploration of mathematical thought on figures who refute the idea of a gap separating the scientific and the artistic imagination. Kovalevskaya was also a poet and novelist, who defended the necessity of imagination in mathematical thought (Tf2 2006, p. 10), and Ramanujan's unorthodox methods have been described by Hardy as "a process of mingled argument, intuition and induction, of which he was entirely unable to give any account" (Complicite 2008, p. 68). The text of *Le Cas de Sophie K.* presents Kovalevskaya's "amphibian" brain (Peyret 2005) through a thoroughly hybrid text, drawn from Kovalevskaya's correspondence, but also from mathematical, philosophical, and poetic texts, from Poincaré's reflections on the role of the unconscious in mathematical thought to Lautréamont's nihilistic *Chants de Maldoror*. A similar emphasis on the creativity of mathematical thought is present throughout *A Disappearing Number*, in which quoted passages from G. H. Hardy's book, *A Mathematician's Apology*, suggest an aesthetic vision of science, where "The mathematician's patterns, like the painter's or the poet's, must be *beautiful*; the ideas, like the colours or the words, must fit together in a harmonious way" (Hardy 2006, pp. 84–85).

Mathematics is thus presented as a source of imagination for the theatre, providing poetic schemata, structuring images that may be both visual and metaphorical, for the stage.[7] In *A Disappearing Number* images of flowing equations and mathematical symbols are frequently projected onto the screens and actors. Three key concepts provide structuring images for the human relationships that we observe. Partition, or the number of ways in which an integer can be represented as a sum of integers, was one of Ramanujan and Hardy's main areas of research. The scenography conveys the importance of this image, as mobile screens frequently divide the stage and emphasize the cultural, historical, and psychological divisions separating the characters from one another. Both partition and decomposition—the reduction of a number to the multiplicative products of its prime—are embodied at key moments of the show by actors displacing shoes or books around the stage, as for instance when Ruth discovers she is pregnant while Al is travelling abroad:

> *The sound of a toilet flushing.* Ruth *comes back through the door. She holds something in her hand. She stands to one side of* Aninda. Al *stands on the other side. He is lost in memory.*
>
> Aninda: Three may be written as three or as two plus one or as one plus one plus one. So the number of partitions of three is three. (Ruth *sits on the downstage bed. She slips off her shoes and arranges them on the floor. As* Aninda *speaks, she makes the partitions with her feet and shoes.*) Four may be written as four, as three plus one, as two plus two, as two plus one plus one or as one plus one plus one plus one. So the number of partitions of four is five.
>
> *Suddenly,* Ruth *gasps at the object in her hand. She tries to make a phone call, but the line is engaged. Throughout the scene she continues to try to get through.* Aninda *walks back upstage to where* Hardy *and* Ramanujan *work.* (Complicite 2008, p. 61)

Here the mathematical concept is given human form by the arrangement of shoes and transformed into a metaphor, for both the imagined parenthood and the separation at work in the scene, between two lovers divided by time and space. The concepts of partition and decomposition become structuring images for the isolation, separation, and losses suffered by the characters. The dramaturgical and thematic fragmentation emphasized by these concepts is however counteracted by a third mathematical concept: in a lecture which serves as a prologue to *A Disappearing Number*, Ruth introduces the idea of the series, which may be either divergent or convergent. This image is embodied in the scenography at key points in the show, when the actors form lines linking the present to the past on stage. Complicite's aesthetics of fragmentation are thus compensated by the search for continuities.

Mathematics, which a character specializing in string theory describes as "the structures which bind everything together" (Complicite 2008, p. 87), provides a structuring metaphor for the company's search for an organicist coherence underlying the "shards of stories" that make up their work (McBurney 1999).

In Complicite's work, the bodies on stage become metaphors for the mathematics and the mathematics a metaphor for the plot. This figurative symmetry can be contrasted with Peyret's dramaturgy, in which embodiment is problematized as a locus of potential conflict between body and thought. Mathematical ideas are given visual presence in *Le Cas de Sophie K.*, and tested for figurative potential without being turned into explicit metaphors. Kovalevskaya's work on the rotation of solid bodies around a fixed point, for which she was awarded the Prix Bordin in 1886, provides the show with the central image of the spinning top. Filmed on stage and projected onto screens, real spinning tops are ghosted by their virtual selves, their movement triggered physically by human hands or electronically by the notes played on a piano. Their movements, both smooth and seemingly erratic, balanced and inevitably leading to imbalance, seem to echo the emotional turmoil of Sophie's life, as her bourgeois existence jars with her political ideas, and her abstract thought is contrasted with her nihilistic emotions. In the fourth act they are further materialized in a huge spinning chandelier, which the female actors set in motion and whose bulbs are positioned at asymmetrical points in its frame. Onstage projections invite us to study its movement in parallel with the face of one of the performers', projected alongside it. The metaphorical potential of these images is left open: the spinning chandelier, whose movement is accompanied by quotations from Poincaré's reflections on sensitivity to initial conditions, may be seen as an image of diffraction and unpredictability which further weakens the notion of biographical coherence. In Peyret's avowedly "post-Brechtian" theatre, this scene is also a tongue-in-cheek acknowledgment of Bertolt Brecht's call for a scientific gaze in the theatre. It echoes the famous passage in the "Short Organum for the Theatre" in which Brecht describes Galileo's reaction when confronted with an oscillating chandelier, and suggests that Galileo's surprise and curiosity is the kind of de-familiarizing gaze the theatre needs to provoke (Brecht 1964, p. 192). Peyret's chandelier, however, remains a body on stage juxtaposed and contrasted with other bodies, emphasizing the limits of mathematical prediction as much as its fascinating powers.

In *Le Cas de Sophie K.*, mathematical discourse is an opportunity for friction between the speaking body and abstract thought, so that the theatre, as Peyret has written elsewhere, can become "the objection of the body to thought" (Peyret and Vincent 2000, p. 187). Ironic tensions arise between speaker and discourse, as for example in the prologue, in which the performers discuss the abstraction of mathematics and their "total separation from the world of the senses" according to Pierpont: throughout this exchange, they are filmed off-stage, in close-up, as they prepare in the dressing-rooms, reminding us that we are, in the theatre, very much in "the world of the senses" (Tf2 2006, p. 5). The redistribution of discourse among the performers is in itself ironical, for although most of the scientists quoted in *Le Cas de Sophie K.* are men, their words are mostly voiced by three women. Body and text diverge, as Peyret creates noise within the discourse he explores, and questions the idea, set out in the prologue, that mathematics is a "transparent science" (Tf2 2006, p. 5).

In the range of approaches examined in this chapter, the dramaturgical appeal of mathematical models and metaphors is somewhat paradoxical. By introducing conceptual and formal abstraction into an art form based on presence and immediacy, both theoreticians and artists have questioned inherited theatrical categories, such as characters and plot, as well as fundamental theatrical components, such as live performance and embodiment, which are either absent or challenged by these combinatory poetics and algorithmic performances. Moreover, unpredictability and open-endedness stand out as key features of the mathematical models and metaphors preferred by contemporary practitioners. When mathematical thought is explored as a theme, it seems to be valued both for its visionary power and for the limits of its predictive potential, hence the popularity of models of turbulence and unpredictable determinism. When algorithms are used as a tool, they provide open-ended processes rather than complete models for performance. As digital technology becomes an increasingly central component of twenty-first-century theatrical creation, the search for creativity in algorithms seems likely to generate new forms of improvisation, immediacy, and unpredictable theatricality.

Notes

1. Raymond Queneau gives the following definition on the group's website: "We call potential literature the search for new forms and structures that can be used by writers as they please" ("*Nous appelons littérature potentielle la recherche de formes, de structures nouvelles et qui pourront être utilisées par les écrivains de la façon qui leur plaira*") (Bénabou 2017, my translation).
2. For their relevance to Beckett's theatre in particular, see Ackerley (2010), Culik (1993), Degani-Raz (2008), Klaver (2005), and Stevens (2010). For a concise summary of previous interpretations of *Quad*'s relation to mathematics, see Stevens (2010).
3. See Sutil (2014) for a detailed analysis of the continuity between Kandinsky's conceptualization of visual space and theatrical space. Beckett's *Quad* has been compared to experimental performance created by Bauhaus artists: see Schmid (1988) for a study of *Quad* as a realization of the "abstract synthesis of the stage" imagined by Kandinsky.
4. A list of plays focusing on mathematics as a theme can be found in Shepherd-Barr (2006, p. 224).
5. For further discussions of Stoppard's theatrical use of mathematics, see Angel-Perez (2011), Demastes (2014), Edwards (2001), Jernigan (2003), Melbourne (1998), Shepherd-Barr (2006), and Vees-Gulani (1999).

6. The performance of mathematics is not only due to a newfound artistic interest in the sciences: it is also fueled by the artistic curiosity and creativity of mathematicians such as Marcus du Sautoy and Victoria Gould, who, in 2013, created and performed *X&Y* in the Science Museum in London—a hybrid performance which Kirsten Shepherd-Barr describes as "a mixture of science education, thought experiments, mathematical explanation by speech act and gesture, and acting" (Shepherd-Barr 2014, p. 297).
7. For an analysis of ways in which the poetic schemata of scientific discourse become structuring motifs in contemporary performance, see Campos (2012, 2013).

References

Abbott, Stephen. 2014. "Simon McBurney's Ambitious Pursuit of the Pure Maths Play." *Interdisciplinary Science Reviews* 39 (3): 224–37.

Ackerley, C. J. 2010. "Beckett and Science." In *A Companion to Samuel Beckett*, edited by S. E. Gontarski, 143–63. Chichester: Wiley-Blackwell.

Angel-Perez, Elisabeth. 2011. "Newton/Notnew : comédie et chaomédie dans *Arcadia* de Tom Stoppard." *Etudes Anglaises* 64 (3): 326–38.

Auburn, David. 2001. *Proof.* New York: Faber and Faber.

Audin, Michèle. 2009. "Souvenirs sur *Le Cas de Sophie K.*" *Alternatives théâtrales* 102–3: 51–53.

Ayckbourn, Alan. 1985. *Intimate Exchanges, A Related Series of Plays.* London: Samuel French.

Beckett, Samuel. 1986. *The Complete Dramatic Works.* London: Faber and Faber.

Bénabou, Marcel. 2017. "Historique de l'Oulipo."Source: http://oulipo.net/fr/historique-de-loulipo. Accessed on 20 February 2017.

Brainerd, Barron, and Victoria Neufeld. 1974. "On Marcus's Method for the Analysis of the Strategy of a Play." *Poetics* 3: 31–74.

Brecht, Bertolt. 1964. *Brecht on Theatre: The Development of an Aesthetic.* Edited and translated by John Willett. London: Eyre Methuen.

Campos, Liliane. 2007. "Searching for Resonance: Scientific Patterns in Complicite's *Mnemonic* and *A Disappearing Number*." *Interdisciplinary Science Reviews* 32 (4): 326–34.

———. 2012. *Le Discours scientifique dans le théâtre britannique contemporain.* Rennes: Presses universitaires de Rennes.

———. 2013. "Science in Contemporary British Theatre: A Conceptual Approach." *Interdisciplinary Science Reviews* 38 (4): 295–305.

Campos, Liliane, and Kirsten Shepherd-Barr. 2006. "Science and Theatre in Open Dialogue: *Biblioetica, Le Cas de Sophie K.* and the Postdramatic Science Play." *Interdisciplinary Science Reviews* 31 (3): 245–53.

Carlson, Marvin A. 1993. *Theories of the Theatre: A Historical and Critical Survey, from the Greeks to the Present.* Ithaca: Cornell University Press.

Casanova, Pascale. 1997. *Beckett l'abstracteur, Anatomie d'une révolution littéraire.* Paris: Seuil.

Complicite. 2008. *A Disappearing Number.* London: Oberon.

Culik, Hugh. 1993. "Mathematics as Metaphor: Samuel Beckett and the Esthetics of Incompleteness." *Papers on Language & Literature* 29 (2): 131.

De Bary, Cécile. 2004. "Georges Polti, ou l'anticipation du théâtre potential." *Poétique* 138 (2): 183–92.

Deleuze, Gilles. 1992. "L'épuisé." In Samuel Beckett, *Quad et autres pieces pour la television*, translated by Edith Fournier, 55–106. Paris: éditions de Minuit.

Degani-Raz, Irit. 2008. "Diagrams, Formalism and Structural Homology in Beckett's *Come and Go*." *Journal of Dramatic Theory and Criticism* 22 (2): 133–46.

Demastes, William W. 2014. "Tom Stoppard's Big Picture: The Chaos That Is Our World." *Interdisciplinary Science Reviews* 39 (3): 204–12.

Dinu, Mihai. 1972. "L'interdépendance syntagmatique des scènes dans une pièce de théâtre." *Cahiers de linguistique théorique et appliquée* 9: 55–69.

———. 1974. "La stratégie des personnages dramatiques à la lumière du calcul propositionnel bivalent." *Poetics* 3 (2): 147–59.

———. 1977. "How to Estimate the Weight of Stage Relations." *Poetics* 6 (3–4): 209–27.

———. 1984. "The Algebra of Scenic Situations." In *Semiotics of Drama and Theatre, New Perspectives in the Theory of Drama and Theatre*, edited by Herta Schmid and Aloysius Van Kesteren, 67–92. Amsterdam and Philadelphia: John Benjamins.

Dorsen, Annie. 2017. *A Piece of Work*. New York: Ugly Duckling Presse.

Edwards, Paul. 2001. "Science in *Hapgood* and *Arcadia*." In *The Cambridge Companion to Tom Stoppard*, edited by Katherine Kelly, 171–84. Cambridge: Cambridge University Press.

Fleming, John. 2008. *Tom Stoppard's Arcadia*. London: Continuum.

Fournel, Paul, and Jean-Pierre Enard. 2005. "The Theatre Tree." In *Oulipo Compendium*, edited by Harry Mathews and Alastair Brotchie, 200–1. Revised and updated. London: Atlas Press.

Hardy, G. H. 2006. *A Mathematician's Apology*. Cambridge: Cambridge University Press.

Jernigan, Daniel. 2003. "Tom Stoppard and 'Postmodern Science': Normalising Radical Epistemologies in *Hapgood* and *Arcadia*." *Comparative Drama* 37 (1): 3–35.

Kandinsky, Wassily. 1947. *Point and Line to Plane*. Translated by Howard Dearstyne and Hilla Rebay. Bloomfield Hills, MI: Cranbrook Press.

Klaver, Elizabeth. 2005. "'*Proof*, π, and *Happy Days*': The Performance of Mathematics." *The Journal of the Midwest Modern Language Association* 38 (1): 5–22.

Lehmann, Hans-Thies. 2006. *Postdramatic Theatre*. Translated by Karen Jürs-Munby. Abingdon and New York: Routledge.

Le Lionnais, François. 1973. "Théâtre booléen." In *La Littérature potentielle (Créations Re-créations Récréations)*, edited by the OuLiPo, 263–64. Paris: Gallimard.

Marcus, Solomon. 1970. *Poetica mathematica*. Bucarest: Editura Academiei Republicii Socialiste Romania.

———. 1971. "Ein mathematisch-linguistisches Dramenmodell." Translated by Wolfgang Klein. *Zeitschrift für Literaturwissenschaft und Linguistik* 1 (1–2): 139–52.

———. 1975. "Stratégie des personnages dramatiques." In *Sémiologie de la représentation, théâtre, télévision, bande dessinée*, edited by André Helbo, 73–95. Bruxelles: éditions Complexe.

McBurney, Simon. 1999. "Collisions." In Complicite, *Mnemonic*. London: Methuen.

Melbourne, Lucy. 1998. "'Plotting the Apple of Knowledge': Tom Stoppard's *Arcadia* as Iterated Theatrical Algorithm." *Modern Drama* 41 (4): 557–72.

Moses, Itamar. 2013. *Completeness*. London: Samuel French.

Osserman, Robert. 2005. "Mathematics Takes Center Stage." In *Mathematics and Culture II, Visual Perfections: Mathematics and Creativity*, edited by Michele Emmer, 187–94. Berlin: Springer.

Peyret, Jean-François. 2005. "Note d'intention." Source: http://theatrefeuilleton2.net/spectacles/le-cas-de-sophie-k/. Accessed on 9 February 2018.

Peyret, Jean-François, and Jean-Didier Vincent. 2000. *Faust, Une histoire naturelle*. Paris: Odile Jacob.

Pierpont, J. 1899. "On the Arithmetization of Mathematics." *Bulletin of the American Mathematical Society.* 394–406.

Poincaré, Henri. 1947. *Science et Méthode*. Paris: Flammarion.

Polti, Georges. 1895. *Les Trente-Six Situations dramatiques*. Paris: Mercure de France.

———. 1900. *Timidité de Shakespeare*. Paris: Éditions de l'Humanité nouvelle.

Rey, Olivier. 2016. *Quand le monde s'est fait nombre*. Paris: Stock.

Romanska, Magda. 2015. "Drametrics: What Dramaturgs Should Learn from Mathematicians." In *The Routledge Companion to Dramaturgy*, edited by Magda Romanska, 438–47. Abingdon and New York: Routledge.

Roubaud, Jacques. 1981. "La Mathématique dans la méthode de Raymond Queneau." In *La Littérature potentielle (Créations Re-créations Récréations)*, edited by the OuLiPo, 42–72. Paris: Gallimard.

Schmid, Herta. 1988. "Samuel Beckett's Play, *Quad*: An Abstract Synthesis of the Theater." *Canadian-American Slavic Studies* 22: 263–87.

Shepherd-Barr, Kirsten. 2006. *Science on Stage: From Doctor Faustus to Copenhagen*. Princeton: Princeton University Press.

———. 2014. "Review: *X&Y*." *Interdisciplinary Science Reviews* 39 (3): 297–98.

Stan's Cafe. 1997. *Simple Maths*. Typescript. Stan's Cafe Archive.

Stevens, Brett. 2010. "A Purgatorial Calculus: Beckett's Mathematics in *Quad*." In *A Companion to Samuel Beckett*, edited by S. E. Gontarski, 164–81. Chichester: Wiley-Blackwell.

Stoppard, Tom. 1988. "The Text's the Thing." *Daily Telegraph*, April 23.

———. 2009. *Arcadia*. London: Faber and Faber.

Sutil, Nicolas Salazar. 2014. "Mathematics in Motion: A Comparative Analysis of the Stage Works of Schlemmer and Kandinsky at the Bauhaus." *Dance Research* 32 (1): 23–42.

Tf2. 2006. *Partition 6.1. Le Cas de Sophie K*. Source: http://theatrefeuilleton2.net/spectacles/le-cas-de-sophie-k/. Accessed on 9 February 2018.

———. 2013. *Re: Walden. Partition du spectacle présenté à Avignon, en 2013*. Source: http://theatrefeuilleton2.net/spectacles/re-walden/. Accessed on 9 February 2018.

Vees-Gulani, Susanne. 1999. "Hidden Order in the 'Stoppard Set': Chaos Theory in the Content and Structure of Tom Stoppard's *Arcadia*." *Modern Drama* 42 (3): 411–26.

White, Hayden V. 1973. *Metahistory: The Historical Imagination in Nineteenth-Century Europe*. Baltimore and London: Johns Hopkins University Press.

Wisniewski, Tomasz. 2016. *Complicite, Theatre and Aesthetics: From Scraps of Leather*. London: Palgrave Macmillan.

Worthen, W. B. 2017. "Shakespearean Technicity." In *The Oxford Handbook of Shakespeare and Performance*, edited by James C. Bulman, 321–40. Oxford: Oxford University Press.

CHAPTER 15

Nonlinearity, Writing, and Creative Process

Ira Livingston

In this chapter, I will consider the relationships among nonlinearity, writing and language, and creative process. These relationships are themselves nonlinear, which means that each shapes and is shaped by the others and that intervening in one will have repercussive effects in the others. How can we use engagement with nonlinearity to enhance and even transform what we do as literary critics and theorists, what we do as teachers of literature and writing, how we understand what we do in these cases as resonant with a range of other nonlinear phenomena, and above all, how we think? While some of this range may be surprising in a volume devoted to literature, understanding this resonance is only meaningfully possible through *practice*, so this chapter necessarily explores the nonlinear logics of everyday language together with literary criticism and pedagogy.

We have to start by grappling with the fact that "nonlinearity" is a word—an artifact of language and human thought. It may well be that the linearity of language (that is, the mandate that we put one word after another like beads on a string) is precisely what requires that we grapple and come to terms with its nonlinearity as such. This does not, however, reduce our inquiry to a navel-gazing exercise whereby we can legitimately reflect on language and thought but are forbidden from making any (ontological) claims about any world beyond this—not if our language and thought are themselves aspects of the universe no less than gravity and quarks (each of these having evolved since the first moment of the Big Bang, apparently), not if a universe possessed of language and thought is a different universe than one without them (just as a universe that evolves with gravity and quarks is different than one with other forces and particles), and not if our language and

---

I. Livingston (✉)
Pratt Institute, Brooklyn, NY, USA

© The Author(s) 2021
R. Tubbs et al. (eds.), *The Palgrave Handbook of Literature and Mathematics*, https://doi.org/10.1007/978-3-030-55478-1_15

263

thought are constantly being woven back through the fabric of things, changing their substrates in the process. Even so, my objective here is to explore a much more humble but still important nonlinearity, whereby thinking and tinkering in language with the concept of nonlinearity can have positive repercussive effects on thought, creativity, writing, and teaching.

Linearity, Nonlinearity, and the Sentence

Linearity, in the realm of causes and effects, simply means that A influences B, which in turn influences C (the usual example is one billiard ball hitting another, which hits another in turn). If A changes, one can recalculate B and C accordingly and be done with it. This is also the case in the slightly more complex situation of a zero-sum game, where A, B, and C are inversely proportional to each other: if A gets larger, then B and C must get smaller, as in a pie divided into three pieces. Here again, if A increases, one can simply recalculate the remainder available for B and C.

Nonlinearity derives from *interdependence*, the state of multiple elements or values being shaped in relation to each other, where their relationship is not simply proportional. If A, B, and C are in nonlinear relationships, then changing A changes B and C, but the new B and C will affect A in turn, so we cannot simply recalculate once and be done: for example, think of the relationships among participants in an ecosystem; say, me and my fellow creatures, the creatures we prey upon, and those who prey on us. We will likely have to track how successive waves of repercussions intersect, amplify and/or cancel each other out, and perhaps how they transform the system and its elements altogether (A, B, and C may end up as X, Y, and Z) before equilibrium is restored, if it was ever present in the first place. Repercussions, as opposed to linear percussions, do not simply expand outward as from a pebble tossed into a perfectly still and infinite pond but, like a fish in a pond (and like all ecosystems), interact in a more ongoing and volatile way.

In one of the simplest versions, a variant of what is sometimes called circular causality, A triggers B, which triggers A. For example, a sudden rise in my heart rate triggers the thought that I'm having a heart attack; that thought raises my heart rate. This is the self-escalating logic of a panic attack. Fortunately, causal loops can also be self-limiting: the rising panic triggers the realization that I'm having a panic attack, enabling me to intervene, breathe, talk myself down. Note that this example describes a rudimentary system that begins when that which the trigger triggers triggers the trigger. Not all nonlinearities are knitted into constellations self-sustaining enough to be called systems (the simple panic loop of anxious thinking and rising heart rate might have been a one-off event), though all systems are fundamentally nonlinear (as in, say, the interlocking sets of cycles that comprise our metabolisms). As systems are my primary object of study rather than nonlinearity per se, they tend to be my focus here—though there is also a positive reason for this focus: the entity-likeness of systems makes them easier to approach intuitively. Accordingly, I would like to think that there could be alternate ways of leading children into mathematics through a focus on nonlinear play and systems that would build on this intuitively and help make not just mathematics but conscious cognition in general more *nonlinear-centric*, changing our sense of what is simple and what is complex, and along with that, opening up our approaches to a host of otherwise intractable real-world problems.

A mathematical version of basic nonlinearity is the recursive equation, such as the one that generates the Mandelbrot Set, whereby an output is repeatedly fed back as an input into an equation to generate a new output and so on, and the values of each iteration are mapped according to whether they converge on a specific value, diverge, or vacillate. As the mapping progresses, it begins to generate an infinitely intricate object with ever finer patterns at progressive magnification: zoom in on what looks like a seahorse tail and it turns out to be studded with feathery barbules; zoom in on one of these and it turns out to be serrated with more seahorse tails at an even smaller scale, and so on.

Another way of recursively embedding simple elements to create potentially endless and infinitely complex series is through *language*—or more specifically the set of rules known as *grammar*—as when a clause beginning with "as when" is added to this very sentence, looping it back on itself, or a new sentence beginning with "however" works to take back or radically modify what had just been asserted.

However, as mentioned above, *linearity* is also the leading parameter of language: words must be strung together one after the other. A basic sentence such as the classic example "the cat sat on the mat" is linear, furthermore, in that the action moves exclusively forward from a subject to a verb to an object (although the rhyme keeps our attention looping back to the repeated sound). Notice, though, that embedding even a simple clause introduces a grammatical loop ("the cat, which was grey, sat on the mat"—where the inserted phrase loops back to refer to the subject preceding it) or can modify the assertion retroactively ("the cat sat on the mat, Jane dreamed"). These loops and modifiers can be added indefinitely.

Linear sentences emerge from a largely nonlinear thought process, such as the process that selected the cat as the subject in the first place. People who have trouble grasping nonlinearity may protest that everything—including all our thinking and selecting—takes place in a linear flow of time. For a start, one might point out that, to the contrary, to conceive of time *as space* (as in the assertion that everything "takes place" in time) or worse, as a one-dimensional line or a flow, is to apply a reductive, lower-dimensional metaphor to a higher-dimensional phenomenon. But even if you like to think of a comforting linearity providing a through-line, in telling the story, you will still have to reckon with a continuous range of choices from all sides, starting simply with subjects (the cat? the mat? Jane? the dream? the greyness of the cat—or at night, of all cats?) and the thicket of verbs and objects branching off from and weaving back into them in all directions.

It is easy to complicate the agency of a grammatical subject ("the cat, rendered drowsy by the tranquilizer put into her food by Jane, sat on the mat"), and such complications might even be understood as mandated by the over-attribution of sovereign and singular free will implied (at least provisionally) by grammatical subjectivity. Even Jane, who perhaps "only years later was able to recover a repressed memory of having been traumatized in her childhood by a cat," may still have to cede back some of her agency to cats—or, alternately, her memory could turn out to have been implanted by a too-suggestive therapist. By typically requiring subjects and objects, grammar poses questions of agency and causality but cannot ultimately answer them, ensuring that philosophers—and fiction writers—won't run out of work.

The linearity of language (again, in the sense that words must typically be strung together one after the other) can be said both to enhance and to obscure the nonlinearity of meaning. This linearity works to generate nonlinearity much in the same way that the simple distinction between an inside and an outside is, for all of us living creatures, a way of ensuring, amplifying, and organizing the complex traffic between them. (Nature generates simplicities, by accident if nothing else, and then cannot resist complicating them; the drop quivering on the verge of becoming a sphere splashes into a baroque crown.) Each word in a sentence sets up a forward-facing framework in which the words that follow will have meaning, but each word also re-frames the ones that came before, as one can illustrate in a juvenile but effective way by adding "not" to the end of a declarative sentence. I have been focusing here on the sentence (and will return to a more developed example at the end of this chapter), but nonlinearity also happens at many other levels of language, as in the push-and-pull relationship of any given utterance or text with and against the genre-specific horizon of expectations (and even more broadly, the discursive landscape) that shapes it and is shaped and reshaped by it.

Nonlinearities in meaning cannot always be easily captured by diagrams. These include various forms of self-reference: "in this essay, 'the cat sat on the mat' is not primarily a story about a pet but a grammatical illustration." In the process of using the sentence as an example, I recognized that my use of it to make a philosophical point is at odds with its simple, referential function, and I turned this act of metacognition into an act of metacommentary. Likewise, if I write that "your reading this sentence, right now, fulfills my infantile wish to have people pay attention to me," I have looked backward at my own motivations, and forward to your reading of the sentence, and mapped both of these back onto my writing it.

People write books about self-reference—and this is itself a rather heavy-handed self-reference (Livingston 2006)—but for the moment (I will return to this topic toward the end of this chapter), I only want to call attention to the fact that self-referential sentences complicate the transparent realism of "the cat sat on the mat"—realism in which language offers itself as a more-or-less objective window onto the world to which it refers, its one-way reference constituting the apparent linearity of its meaning.

In the closed systems described by structuralist theory, words have meaning not through reference to the world "outside" but in a cross-referential network of relationships with other words (for example, a noun is a noun not because things in the world are noun-like—which, arguably, they are not—but because it is not a verb, adjective, or other part of speech). In poststructuralist and other open systems, this "internal" network of relationships is itself in a network of "external" relationships, and the two interact so thoroughly and continuously as to embed their binary opposition in endless Mandelbrottian arabesques. Changing the meaning of one word or phrase in relation to others (say, as with the recent rise of the phrases "alternative facts" and "fake news") does not simply alter the meanings of the words involved and have repercussions for a range of related words and concepts; it participates in a range of consequential real-world dynamics, not merely as an index of them but as a full participant. The more that words are regarded not simply as referring to things or to each other but as participating in their worldly contexts, shaping and being shaped by them, the more we can come to grips with their nonlinearity, such as by considering their *performativity*: what words do in the world as speech acts. Speech acts are generated in the contexts of intertextualities (in their ongoing conversations with other texts) and broader discursive ecologies: they have meaning only as they interact with their contexts. For example, the relatively recent insistence on the performativity of language is occasioned by a long history of excessive philosophical focus on representation and referentiality, without which it might well have been unnecessary.

The nonlinearity that is, in any case, always in operation in the writing process, can also be deliberately enhanced.

NONLINEARITY AND CREATIVE PROCESS

Think of how nonlinearity works when you are writing rhymed poetry. If you are a novice, you might think first of what you want to say, more or less, and then start trying to squeeze it into the constraints (such as meter and rhyme) that you have selected for the poem. This mostly one-way process is likely (likely, but not guaranteed) to lead to a stilted performance. On the other hand, if you're inspired or an experienced poet or both, the constraints and what you want to say—the how and the what—can dance improvisationally and nonlinearly with each other.

This suggests more basic questions about language usage. When you speak, do you get some pre-verbal sense of what you want to say and then start selecting words that fit, or, even if you start off like this, don't you proceed to rush headlong forward at some point, finding fully what you want to say only in the process of speaking or writing? Speaking ad hoc in public can be nerve-wracking for this reason: you are never certain exactly what words will come out of your mouth. The fear of public speaking may thus be a cousin of nonlinearphobia, contributing to the counter-intuitivity of the larger realization that, while neither is reducible to the other, language and thought have been co-evolving in a nonlinear fashion for 100,000 years. Language thinks us, too.

If you write or even read enough in an iambic meter, you start to be able to think in iambs. If you write enough in rhyme, your syntax, and word choice, as you approach a rhyme, will adapt in advance, as a baseball player adjusts speed and direction continuously while running to catch a fly ball—arriving, like magic, at just the right time and place. This example is not simply a metaphor because it does not elaborate an "unmotivated" similarity between two otherwise unrelated realms (rhyming and catching) but marks what seems to be a family of nonlinear phenomena, showing (in both cases) how legs, eyes, and brains can intuitively solve nonlinear equations if they are posed in the right way. And this ability is not limited to humans. Even ant colonies, for example, as they allocate some foragers to one food source and some to another, "are solving multivariable maths equations" (Hess 2016, p. 31). All nonlinear systems can be said not just to engage in such computations but *to be* such computations (think of a storm system, for example, as a sustained mathematical orchestration of updrafts, downdrafts, inflows, and outflows).

In the case of rhymed poetry, both the possible path and the destination (that is, both the syntax and rhyme) can change and adjust continuously in relation to each other. We can say that rhymed poetry is a way of making nonlinearity conspicuous: we are enabled to *feel* the knitting-together of a nonlinear constellation, and this may well be what is *most delicious* about it; a poem can be ambiguously a kind of fellow nonlinear entity (it cannot quite be called alive, yet it may seem to have an internal organization intricate and organic enough to arouse our sense of entity recognition). The unresolvable extent to which it is almost a system and almost an entity holds our attention.

There is also considerable nonlinearity in writing fiction, and again, this need not be something subjective or theoretical or even very difficult. I am simply referring to how, as a story evolves, the writer can go back and rewrite what came earlier in the narrative, knitting the story together. When William Godwin wrote the novel *Caleb Williams, or Things as They Are* (1794), he imagined first the situation of the latter part of the novel[1]: a man is being pursued by a relentless, oppressive power. For a radical political philosopher in Britain's repressive 1790s—as no doubt for us—this situation was all too familiar. To spin a novel around the situation, Godwin proceeded to work backwards, going on to posit how such a situation could have come about. He made the narrator a servant who must flee after discovering his apparently humane and rational master's dark secret: that he had killed a rival. The secret of rational and humane modern power is that it too is based on violence it is relentless in covering up. Godwin then spun the narrative forward again, imagining two diametrically opposed but diabolically unhappy endings. In one, the narrator is discredited and imprisoned, and in the other, he succeeds in discrediting his master but understands immediately that he has only played into the system's seamless damage control: either way, the dark secret—the violence at the heart of modern power—could not simply be revealed to transform the system. Godwin needed a nonlinear writing/thinking process to grapple with the nonlinear complexity of modern power and ideology; it is not much of a stretch to imagine the novel being rewritten today as the story of a woman's persecution by and crusade against a man with a secret history as a sexual harasser.

In the 1984 film *Terminator*, a man—Kyle Reese—is sent back in time from a dystopian future where humans are locked in what seems to be a losing battle with insubordinate and ruthless military machines that have taken on a life of their own. Reese carries a Polaroid photo of Sarah Connor, whose son will grow up to be the leader of the human forces. Reese has been sent to protect her from a Terminator robot that the machines have sent back to kill her. The final scene of the film reveals the origin of the photo: Sarah is pregnant and, having been informed of the historical role she and her son are to play, is on the run in Mexico. When she stops for gas—and, as she sits at the wheel lost in thought about the complex crosscurrents of present and future—a boy snaps her photo and proceeds to wheedle her into buying it. I do not propose to analyze here the nonlinear, time-travel causality loop at the heart of the film (Kyle is the father of Sarah's child—who will grow up to be the man who sends Kyle back in time) but to make a practical observation about creative process. The most obvious surmise is that the filmmakers filmed the final gas-station scene before the rest of the film in order to set-up the photo that was to be used as a prop in the rest of the film—but it is also possible that they re-edited the film to insert the photo into earlier scenes: it is, typically, difficult or impossible to resolve exactly how a nonlinear system emerges, but either way, the parts (the plot, props, shots, and scenes) and the whole (the film) emerge in tandem. Or to revisit my initial example: did a sudden rise in my heart rate trigger the thought that I am having a heart attack, or was it the thought that triggered the rise? Once the system has gained a foothold, its own circular causality can render the either/or question irresolvable or even irrelevant.

Was it "cheating" for the novelist or the filmmaker to work backwards (justifying a narrative retroactively much as a criminal might plant evidence after the fact to support an alibi), making it seem as if their plots unfolded with a kind of fateful yet unpredictable inevitability, giving them the feel of real worlds? It does seem to fly in the face of novelistic *naturalism* as developed in the nineteenth century by Emile Zola and others, in which the author is supposed to place various kinds of characters in various situations, then simply run them all forward (see Zola's 1893 "The Experimental Novel"). As the algorithm unfolds, we (along with the author) are supposed to discover—as in a scientific experiment—how each of them fare: who evolves and triumphs, who fails, who persists unchanged. Of course the novelist already had in mind characters differentiated by particular sets of features, and a world that operates by particular logics, rewarding particular sets of features differently—so it is a set-up after all. The cheating was not in nonlinear composition, which has the better claim of evolving as systems really do, by nonlinear knittings-together. Note that such a process does not *represent* the nonlinear logic of emergent systems but genuinely *participates* in it. If there was any sleight-of-hand, it was in pretending the process could be linear in the first place; in other words, that a social-scientific experiment could unfold toward objective results while ignoring the self-fulfilling-prophecy effect of its own initial operating assumptions.

Mathematicians Milan Rajković and Milos Milovanović have developed a computational analysis they claim can distinguish—by scanning and analyzing visual patterns—the copy of a painting from an original, even when the original artist made the copy. This is because the "causal architecture" of an original work involves self-organization, a version of the nonlinear looking-and-adjusting backwards-and-forwards we have just seen in poetry, prose, filmmaking (and baseball). Self-organization

> denotes a spontaneous emergence of structures and organized behavior without any external influence in systems consisting of a large number of interconnected elements. Due to the feedback relations between constitutive components, the dynamics of self-organizing systems is non-linear. Self-organization indicates a spontaneous increase in structural entanglement (complexity) of a system over time. (Milovanović and Rajković 2015, p. 3)

For systems such as the ant colony, self-organization enables complex and interconnected divisions of labor to emerge. For a second-order system (a system with an additional layer that allows self-observation), such as artist and viewer, engaging a work of art involves a nonlinear process of toggling back and forth between predictions and observations in which the "interplay of predictive patterns and unpredictable interruptions and the proportion of their occurrence determines to a large degree the aesthetic experience and gratification" (Milovanović and Rajković 2015, p. 10). Seeing, for viewer and artist alike, is not simply a matter of adjusting to what is actually there, but in the process, creating a new interpretive frame for it, a new relationship between artist, viewer, and work that affects all parties to the interaction.

As Octavia Butler's science-fiction heroine Lauren Olamina wrote in *Parable of the Sower*, "all that you touch / You Change. / All that you Change / Changes you" (Butler 1993, p. 3). We are so thoroughly nonlinear in our capacities to learn that even simple, one-way copying can have educational value: if you hand-copied out or even read aloud all of the epic poem *Paradise Lost*, it is likely that the rhythms of your own writing and speech would be altered by it. The emulative capacities of our brains mean we cannot be perfectly neutral copy machines; even stubbornly to resist being changed is to be shaped in a nonlinear way by the process. Even (or especially) the forger, who must strive continually to put even the expressive flourishes of his or her hand to the service of better forgery, is transformed by the work: forgers, of course, are not born but made by their careers. In most cases, though, mutations introduced in copying creatively bedevil communication, both broadly and deeply: as the history and contemporary landscape of human languages amply demonstrates, when speakers of different languages interact, they do not simply engage in one-way translations back and forth from one language to another, leaving each language intact. Languages change and evolve, often radically, in the process, and cultural DNA is exchanged as promiscuously as viruses exchange actual genetic information.

Nonlinear Pedagogy

Nonlinearity, then, is part of the process by which language and literature are generated—which is, in turn, part of a family of nonlinear phenomena common to systems generally, but understanding this would mean little unless we put this understanding to use. Writers and teachers of writing can work to enhance generative nonlinearities in the writing process, by design. The most familiar way nonlinear process is deployed in the classroom is by "workshopping" and collaborative exercises, in which each participant is engaged in rewriting and editing their own and others' work, a reiterative process in which (ideally, anyway) emulation and learning flow upstream from critique and rewriting to writing, and from group to individual. Participants integrate operating principles they have gained from participating—and then bring their evolved principles to bear on their subsequent participation. This feedback loop is the whole point of the workshopping process.

Likewise, when writing prose, you may begin by writing a provisional premise and an argument, then go back and rewrite the premise in light of what you have learned in making the argument—and then, most likely, you will have to rewrite the argument and conclusion as well (recalling our first examples in which B and C cannot simply be recalculated when A changes, but waves of repercussions will pass through the whole ensemble). Note that I have linearized this process as a sequence of events but some form of it must be happening continuously in the moment of writing as well (which may be what we call *thinking*). Enhancing this process can start as simply as mandating—after a first draft has been submitted—that students highlight the conclusion of the draft and place it at the beginning of a second draft to be written anew—understanding that genuine rewriting may involve scrapping all but the conclusion and effectively starting over, finding out in the process which points from the original draft need to be imported into the new version and which can be left behind: this process allows for real evolution of the essay. It is not by any means a radically new insight or process, but it is radical in the sense that it is based on the nonlinear roots of the learning and creative process.

In beginning writing classes, I use variants of this exercise. In one, students are first asked to condense two pages of writing to a single sentence and then (but only when the condensation has been completed) to build two new pages around the sentence. In another, we study aphorisms (for example, Octavia Butler's succinct encapsulation of nonlinearity, mentioned earlier) and students are asked to try to articulate the principle behind one of the aphorisms they find especially delightful or profound, via the question of how the writer came to generate it. It is important that they not simply paraphrase the aphorism (a linear process) but try to identify the principle at work in it (a process that moves backward to the various reversals, transvaluations, juxtapositions and other logics or conceptual algorithms that generated it), then write new aphorisms that embody the principle. I call this assignment "Wild Seeds" because it is impossible to predict how the next generation will resemble (and differ from) the parents. For example, in one class, we discussed William Blake's injunction, "Listen to the Fool's reproach, it is a kingly title!" (Blake 1988, p. 37). Our discussion focused on the aphorism as a conceptual algorithm for deconstructing conventional wisdom: Blake transvalues criticism into praise, moving from a false ideal of making art as the utterance of universal truths to understanding it as a way of (performatively) making friends and enemies; a dialectical and political process in which meeting with opposition is validating. I have found that these discussions often fall short or fail altogether to nail the principles at work, but this failure also seems to be productive, at least insofar as it motivates the generation of more aphorisms that embody the principles that themselves have not quite been articulated. After the discussion, one student translated the principle behind Blake's aphorism into "you must be doing something right when someone warns you that you're committing professional suicide," while another more economically proposed "thanks, haters!"

A more developed scenario for a collaborative writing and/or design project might involve the following steps (I have kept this general to apply to various kinds of classes): (1) The group is asked to come up with collaborative project ideas; (2) they pitch and discuss the ideas, which begin to evolve, and in the process, (3) participants form small groups around the particular ideas that most interest them. (4) Further discussion, now within each particular group, refines the ideas further and leads to (5) the first iteration. In the process of creating the iteration, (6) the idea changes, and in the process of (7) workshopping the iteration (thinking and talking about what went wrong, how it can be improved, and so on), more changes are incorporated, whether these turn out to be tweaks or remakes or even scrapping the idea and starting over, and so the process continues to the next iterations.

This process can be diagrammed as a series of discrete steps in loops that keep coming back around to the same elements: to the ideas, to the collaborative groups and their dynamics, and to the projects themselves, each of which evolve into new iterations in tandem as they go. Just as in any relationship, the groups do not stand outside but evolve in the process along with the ideas and projects: initial divisions of labor and roles may become too constrictive and have to be revised, new members with new skillsets recruited, and so on. You might find it useful to add specific components such as research, discussion, documentation, and archiving—or in the bureaucratic mode, *goal-setting* and *self-assessment*—feedback loops by which the system can evolve via self-reflection, part of the process whereby its outputs become inputs for the next generation of the process.

At its best—at its most alive—the creative processes described here—whether in writing, filmmaking, collaborative design projects, or other forms of cultural production—are forms of *play*, the ongoing interplay of constraints and freedoms. The fact that we have nouns and noun phrases—such as *design project* or *Mandelbrot Set*—should not obscure our understanding that all these are nonetheless ongoing spatial and temporal *events*. The creative process succeeds not by concluding and spitting out an output but by generating a product that is only successful if it can become an input in turn, as for readers who are transformed or inspired to do more than simply have aesthetic experiences and go on their ways.

Insofar as writers and texts can be said to share and channel agency, then as much as the text wants to encapsulate and become a thing unto itself, it also wants to reproduce itself, to disseminate, to be replanted in new soil, not to reproduce itself precisely but to spark wild variants among whom there will be complex kinships even where there is no mutual recognition.

The process is as elegant in principle (and easy to diagram as a series of loops) as it is productively messy in practice. And it is effective: you can even start with a *bad* idea. If the process is nonlinear enough—if the interaction of its components with each other and with their environments are robust enough—and if it is followed doggedly enough, it will evolve quickly into something more sustainable—necessarily, at least, into something that must sustain the attention of its participants and their interlocutors if it is to survive at all.

Nonlinear Thought and Shapely Sentences

Virginia Woolf famously claimed that Jane Austen rejected the patriarchal sentence: Austen "looked at it and laughed at it and devised a perfectly natural, *shapely* sentence proper for her own use and never departed from it" (Woolf 1929, p. 77; my italics). Woolf describes the typical masculinist sentence in terms of its linearity: "so direct, so straightforward," but subject to "the dominance of the letter 'I' and the aridity, which, like the giant beech tree, it casts within its shade. Nothing will grow there" (Woolf 1929, pp. 99–100). The linearity is to be found in the kind of sovereign subjectivity (the "I") from which authority flows in one direction (a process now generally known as *mansplaining*).

274   I. LIVINGSTON

Jane Austen's most famous shapely sentence begins *Pride and Prejudice*: "It is a truth universally acknowledged that a single man in possession of a good fortune must be in want of a wife" (Austen 1972, p. 51). The chapter that follows—the arc of the whole novel that follows—unpacks the nonlinearities of this initial sentence, a fractal quality that seems to be a signature of systems built and evolved by recursive processes.[2] Here, *fractal* means that the same structure (let's call it *shapeliness*) is played out on at least three scales: sentence, chapter, novel.

Notice first that the passive voice makes it difficult to say who is speaking (who is the narrator) and about whom (the ones doing the *acknowledging*, who would have to be identified if the active voice were being used). This situation is complicated further because it is not a transparent statement (a realistic, one-way reference to the world) but a *statement about statements*; that is, the narrator is *saying something about what people say*. So we begin simply by trying to work out the implied narrator and object of the sentence, a more explicit version of what is likely to be going on automatically in our brains as we read (who is speaking to me? about what?), whether we are pursuing this inquiry at some unconscious level or systematically setting it aside.

The so-called *truth* is *acknowledged*, presumably (and this will be borne out in the chapter that follows), by *marriageable women and their parents*, but in order to reference that acknowledgment at all—and to get some ironic distance on it—the narrator (like Elizabeth and her father in the novel) must stand both inside that social circle and outside it, which is why its "universality" can, paradoxically, become apparent. We catch the ironic wink because the sentence allows us to see that the "want of a wife" does not proceed simply from the young men but must be cultivated in them by marriageable women and their parents—though of course the motivations of the women and their parents must not be original to them, either, but installed in them by the patriarchal logic of the "marriage market" to which all are subject. This, then, is how the sentence rejects the unproblematic subjectivity of "I" via the strategic use of the passive voice: all subjects construct each other and their desires and are constructed by such ventriloquisms in a nonlinear web of relationships (that one could go on to study under such headings as *society, ideology, capitalism, patriarchy, discourse*).

For those in the web—for all those constrained as social/economic actors in a situation in which it is so difficult to discover and/or invent something that might be more like *one's own authentic desire*—is something like *love* even possible? This question drives the novel, which goes on to generate a spectrum of couplings based to different degrees on "external factors" such as economic and social necessity and on "internal" factors such as sexual attraction. The relationship that can be said to transcend—or at least *to displace*—the external and internal factors in operation is the one in which the partners will be constructed *by their relationship*: they will become a nonlinear system unto themselves, learning, and evolving through their interaction—and, in the process, overcoming their initially definitive character traits, as named in the book's title. This becoming-a-system is, of course, the arc of the novel, the often convulsive process whereby the main characters evolve by "getting over themselves." This is not at all a new reading of the novel but an attempt to put any reading back in touch with nonlinear principles in operation in the novel's sentence structures no less than in its overall arc of plot and character development.

Might Austen have come up with the ironic aphorism that is the novel's first sentence and then proceeded to "unpack" it forward into the novel? Or, on the other hand, once she had written enough of the novel to understand the dynamics she was exploring, did she go back and "compress" them into a virtuosic opening sentence? Each of these speculative accounts of her process seem equally reductive. An understanding of nonlinear process allows the realization that even if we had a series of carefully dated drafts to document what she wrote and when, we would likely find that she was engaged in this process at something like every moment of writing, that the whole and its parts emerged and evolved together.

The *bad faith* revealed by Austen's irony is the same as that of transparent (linear) realism: the implied subjects of the sentence purport merely to *acknowledge* something rather than avowing their own performative agency in both being shaped by and shaping their objects. To displace the bad faith requires that which is explicitly—and *generatively*—prohibited by grammar: that we occupy the positions of subjects, objects, and verbs at once.

The novel is a privileged place for nonlinearity for several largely historical reasons: (1) because the novel evolved as a reflexive and self-referential form (as against the thoroughgoing referentiality required of nonfiction [Livingston 2011, pp. 48–49]), (2) because the form of the novel makes it an ideal milieu for exercising "theory of mind" (our own evolved, strategic ability to imagine what others are thinking [Zunshine 2006]), and, accordingly, (3) because the polyvocality of the novel and its typical use of *free indirect discourse* (which moves around seamlessly from the thoughts of one character, to another character, to a broader perspective not available to any of the characters, and so on) makes it an ideal milieu for exorcizing "the dominance of the letter 'I'."

How is it that Jane Austen came to write such a sentence? Again, there are important historicist answers to this question: for a start, being a woman writer in the early nineteenth century requires being both inside and outside the logic of the marriage market. My interest here (as it has been throughout this essay) is not historical knowledge as such but nonlinear know-how: in light of this history, then, how can we better write such sentences ourselves, beyond engaging in nonlinear writing exercises? Who is now historically positioned—and how are each of us positioned—to understand their identities and their agency as taking shape in a web of a complex and nonlinear web of forces? Where can you find this kind of especially nonlinear *intelligence*?

It seems that you might productively begin by understanding your own intersectionality and the ways in which you are at once an insider and outsider, or to elaborate that a little, to be someone whom it behooves to understand how your disempowerment and empowerment intersect and how to speak from that intersection. A good way of being so positioned would be to be a woman, a minority person in a majority culture, a cross-cultural and/or cross-class person, and while you are at it, to live at the cusp of a historical transition. Finally, it would help to be someone so poorly served by the social and ideological forces and frameworks in which you must operate that you are forced to theorize, to play the ruling game as far as you must *and another* at the same time, never forgetting that survival means ruthlessly seeing things as they are no less than continuing to imagine that another future is possible and weaving it back into your present actions.

A compelling contemporary account of such nonlinear intelligence can be found in the 2016 film *Arrival*, which elaborates a 1998 novella, "The Story of Your Life," by Ted Chiang. A linguist, Louise Banks, is called on to establish communication with squid-like aliens who have landed their ships at multiple locations around the planet. The squids live in giant tanks with transparent walls and communicate by a spoken language of sounds (Heptapod A) and by writing feathery, circular ink strokes (Heptapod B)—shapely sentences indeed—in the clear fluid that surrounds them. Banks discovers that Heptapod B is nonlinear and that the kind of thinking that goes with this writing enables one *to see the future*, that nonlinear thinking is essential to the survival of the planet, that the squids have come to earth to teach us this because they know that, for their own survival, they will need to call upon us in the future and therefore need to ensure *our* survival, that what we thought were reveries of her lost daughter are visions of the future loss, to cancer, of the daughter she *will have*. Coming to know all this, Banks is able to get a kind of Rosetta Stone from the aliens before they are forced to leave, which enables her to accomplish their mission by writing and publishing a book that teaches people Heptapod B.

So, can enhanced nonlinear language and thinking save the world? The answer to that baldly phrased question would have to be *no*, since language does not simply shape the world (a linear and idealist account of causality often caricatured as "constructionism"). But because language is, instead, a player with partial agency (like the rest of us) in a web of relationships, it might be more to the point to ask whether the world can be saved *without* enhanced nonlinear thinking and language, and if not, what will it take to foster these and to be fostered by them.

CONCLUSION

As I hope this chapter has demonstrated, the ability to see and interact with the future in a nonlinear loop with one's present actions is not just a science-fictional premise; it's part of the ordinary way we think and speak, write sentences and novels, teach and learn—and one can build on it. We do not need alien squids to show us a future in which pumping vast amounts of carbon into the atmosphere upsets the delicate ecological balance that keeps us alive, or in which various ongoing tendencies of capitalism lead its destruction or, as currently seems more likely, our own. As Louise Banks puts it: "What if the experience of knowing the future changed a person? What if it evoked a sense of urgency, a sense of obligation to act" not opposed to detached fatalism but somehow sustainable only together with it?

Acknowledgements Big thanks to Devoney Looser for suggestions about Austen, and to my co-teachers Youmna Chlala, Duncan Hamilton, and Jennifer Miller (and our students!) for leading me along the collaborative process/pedagogy path.

NOTES

1. As Godwin related in his preface to an 1832 Standard Novels edition: "I invented first the third volume of my tale, then the second, and last of all the first" (Godwin 1988, p. 349).
2. I do not claim originality for my broad-strokes reading of the novel that follows, only that I have more plainly articulated how the novel plays out nonlinear logic at several levels. For a related but more specialized and fine-grained account, see Rinaldi, Della Rossa, and Landi's "A Mathematical Model of Pride and Prejudice" (2014). *For other examples of how novelists have operationalized knowledge that scientists later codify,* Michael Suk-Young Chwe's 2014 *Jane Austen, Game Theorist* follows on the popular success of Jonah Lehrer's 2007 *Proust Was a Neuroscientist.* Hoping to capitalize on this trend, I briefly considered several book projects including "Franz Kafka, Coleopterologist," and "Edmond Rostand was an Otorhinolaryngologist."

REFERENCES

Austen, Jane. [1813] 1972. *Pride and Prejudice.* New York: Viking/Penguin.

Blake, William. 1988. *Complete Poetry and Prose of William Blake.* Edited by David Erdman. Rev. ed. New York: Anchor/Doubleday.

Butler, Octavia. 1993. *Parable of the Sower.* New York: Four Walls Eight Windows Press.

Chiang, Ted. 1991. "The Story of Your Life." In *Starlight 2.* Tor Books. New York. Republished 2002 in *Stories of Your Life and Others.* New York: Macmillan/Picador.

Chwe, Michael Suk-Young. 2014. *Jane Austen, Game Theorist.* Princeton, NJ: Princeton University Press.

Godwin, William. [1794] 1988. *Caleb Williams, or Things As They Are.* London: Penguin.

Hess, Peter. 2016. "Collective Genius." *New Scientist* 9 (16): 29–31.

Lehrer, Jonah. 2007. *Proust Was a Neuroscientist.* Boston and New York: Houghton Mifflin.

Livingston, Ira. 2006. *Between Literature and Science: An Introduction to Autopoesis.* Champaign, IL: University of Illinois Press.

———. 2011. "Chaos and Complexity Theory." In *The Routledge Companion to Literature and Science*, edited by Bruce Clarke and Manuela Rossini, 41–50. New York: Routledge.

Milanovic, Milos, and Milan Rajkovic. 2015. "The Artists Who Forged Themselves: Detecting Creativity in Art." *arXiv:* 1506.04356v1 [csCV], June 14, 2015: 1–26.

Rinaldi, Sergio, Fabio Della Rossa, and Pietro Landi. 2014. "A Mathematical Model of Pride and Prejudice." *Nonlinear Dynamics, Psychology, and Life Sciences* 18 (2): 199–211.

Villeneuve, Denis (dir.). 2016. *Arrival.* Paramount Pictures.

Woolf, Virginia. [1929] 1981. *A Room of One's Own.* New York: Harcourt Brace Jovanovich.

Zola, Emile. [1893] 2012. "The Experimental Novel." In *The Experimental Novel and Other Essays*, 1–56. London: Forgotten Books.

Zunshine, Lisa. 2006. *Why We Read Fiction.* Columbus, OH: The Ohio University Press.

PART III

Mathematics, Modernism, and Literature

CHAPTER 16

Mathematics and Modernism

Nina Engelhardt

MATHEMATICS, MODERNISM, AND REASON

Mathematics and modernism can appear to stand in diametrical opposition to each other. Understood as a reaction against modernity, modernism challenges the Enlightenment celebration of reason and rationality. In 1917, the sociologist Max Weber complained: "The fate of our times is characterized by rationalization and intellectualization and, above all, by the 'disenchantment of the world'" (Weber 1991, p. 155). Mathematics has a prime place in explaining and thus, in Weber's view, disenchanting the world, at least since the publication of Isaac Newton's *Principia Mathematica* in 1687. The full title *Philosophiae Naturalis Principia Mathematica*—Mathematical Principles of Natural Philosophy—signals the crucial place accorded to mathematics in the physical sciences and European Enlightenment explorations of the natural world. Another *Principia Mathematica*, written by Alfred North Whitehead and Bertrand Russell and published in three volumes from 1910 to 1913, is immensely influential in the early twentieth century and points to this period's concerns with re-examining and re-defining the place of Enlightenment traditions and, more specifically, of the principles of mathematics and its place in gaining knowledge about the world: where Newton's aim is the study of the physical universe that would later develop into the natural sciences; Whitehead and Russell's work examines the foundations of mathematics itself. This turn from attending to mathematics' relation to nature towards investigating the principles of the discipline finds a correspondence in modernist literature: reacting against nineteenth-century realism and its ideal of

N. Engelhardt (✉)
Department of English Literatures and Cultures, University of Stuttgart, Stuttgart, Germany
e-mail: nina.engelhardt@ilw.uni-stuttgart.de

© The Author(s) 2021
R. Tubbs et al. (eds.), *The Palgrave Handbook of Literature and Mathematics*, https://doi.org/10.1007/978-3-030-55478-1_16

representing the world with objectivity and fidelity, modernist literature turns to self-referentially examining and experimenting with literary form and expression. Ezra Pound's famous dictum to "make it new" articulates the modernist aim to reconfigure literature, and writers turned not least to mathematics to introduce innovations in literary subject matter, style, and form.

The study of mathematics and modernism comprises different facets, and this chapter is concerned with ways in which literary writers draw on mathematical vocabulary, concepts, and discussions about mathematics in their attempts to renew literature. I focus on canonical modernist writers in English such as D. H. Lawrence, Virginia Woolf, Ezra Pound, and James Joyce, and discuss texts by several well-known modernist figures from the non-English speaking world: Yevgeny Zamyatin, Velimir Khlebnikov, and Hermann Broch. Examining works by these writers allows exploring topics that animate discussion in both mathematical and literary circles in the first half of the twentieth century: language, form, a domain outside of reason, and a shift from representation to creation. Modernist texts that address these issues in relation to mathematics draw on it in contradictory ways, both as a foil against which to define literature and as a mirror of modernist writing.

In the mathematical sphere, the topics identified above—language, form, the non-rational, and a shift from representation to creation—contribute to the development of three accounts of mathematics: logicism, formalism, and intuitionism. Logicists, such as Gottlob Frege in the nineteenth century and Bertrand Russell in the early twentieth century, attempted to deduce the foundations of mathematics from logical principles, ultimately understanding mathematics as logic. The formalist movement, headed by David Hilbert, treats mathematics as a formal game in which symbols have no meaning in reference to the real world but are manipulated according to rules: in Hilbert's words, doing mathematics resembles "a game played according to certain rules with meaningless marks on paper" (Hilbert 1964, p. 143). While formalism is based on the idea that the foundations of mathematics are exclusively determined by its formal rules, intuitionism, founded by L. E. J. Brouwer, instead proposes that numbers are mental constructions, that mathematics originates in intuition, and that this root in the human being guarantees mathematics' relation to the world and therefore its meaning (see Snapper 1979). None of these approaches managed to formulate a complete and consistent basis of mathematical knowledge, but the schools either had to abandon well-working mathematical procedures as they could not be proved intuitively, or they ran into contradictions. The inability to eliminate problems at the foundation of the field gave rise to scepticism regarding the certainty of knowledge as a whole. Hilbert exclaimed: "If mathematical thinking is defective, where are we to find truth and certitude"? (Hilbert 1964, p. 141), and speaking more generally, the historian of mathematics Jeremy Gray maintains that "the mathematics of the nineteenth century is marked by a growing appreciation of error leading to a note of anxiety, hesitant at first but persistent by 1900" (Gray 2004, p. 23). Gray further argues that mathematics not only experienced a major transformation that made it "modern" but that it can also be said to become "modernist", that is, to share characteristics with art that allow understanding mathematics as part of modernist culture. Not least, with anxiety regarding its certainty and the split into three different schools, the situation in mathematics contributes to broader feelings of fragmentation and uncertainty in the early twentieth century that the modernist arts reflect. This chapter does not specifically explore the notion of a modernist mathematics but focuses on ways in which literature engages with mathematics. Nevertheless, examining the role of mathematics for modernist writing can illuminate how art contributes to renegotiating mathematics' place in modernist culture.

Historian of mathematics Herbert Mehrtens introduced the idea of a modernist mathematics in the 1990s, and in literary studies, the modest field of modernism and mathematics has seen a recent growth, with several monograph projects in the last decade or so: Miranda B. Hickman's *The Geometry of Modernism; The Vorticist Idiom in Lewis, Pound, H.D., and Yeats* (2005), Gwyneth E. Cliver's dissertation *Musil, Broch, and the Mathematics of Modernism* (2008), Baylee Brits's *Literary Infinities* (2017), my own *Modernism, Fiction and Mathematics* (2018), and Jocelyn Rodal's project *Modernism's Mathematics: From Form to Formalism*. Robert Tubbs's *Mathematics in Twentieth-Century Literature and Art: Content, Form, Meaning* (2014) and Lynn Gamwell's *Mathematics and Art: A Cultural History* (2015) do not specifically focus on modernism but include modernist texts in their wide-ranging discussions. Yet, much of the literary discussion of modernism and mathematics is dispersed, part of more general explorations of science and modernist literature, and often focused on specific authors. I here examine works by some of the most frequently discussed writers, while others could be added, of course: Bertold Brecht, Thomas Mann, and Robert Musil in the German context, French writers Stéphane Mallarmé, Paul Valéry, as well as the Oulipo movement, and the Russian writers Andrei Bely and Pavel Florensky, to give just some examples. In this chapter, I give an overview of well-known modernist writers and recurring concerns with mathematics. This cannot, of course, be comprehensive but only tentatively categorise matters that were of interest to both the literary and the mathematical sphere and that inspired modernist literature's engagements with mathematics.

Before turning the spotlights on language, form, and a realm outside of reason, and to the logicist, formalist, and intuitionist approaches associated with these, it is important to stress that texts or writers do not necessarily follow any one of the mathematical philosophies: a text can include different, ambiguous, and even contradictory relationships with mathematical positions. Below, I discuss some texts in more than one mathematical context—reflecting the fact that they do not necessarily engage with specific schools but draw on various aspects of and issues in mathematics to explore modernist ideas. The following brief discussions of Yevgeny Zamyatin's *We* (1924) and D. H.

Lawrence's novels *The Rainbow* (1915) and *Lady Chatterley's Lover* (1928) demonstrate that mathematics can play a multifaceted role even in texts that seem to condemn it (for discussions of *We*, also see the chapter "Mathematics in Russian Avant-garde Literature" by Anke Niederbudde in this volume).

We was written in Russian but first published in English; the first official Russian version appeared only in 1988. The novel was hugely influential on dystopian fiction such as Orwell's *Nineteen Eighty-Four* (1949). It presents a state completely ruled by mathematics: citizens have no names but are given numbers, and all areas of life are controlled by calculation. The citizens solve ethical problems with moral mathematics; happiness is defined as a fraction with "bliss and envy [as] the numerator and denominator"; and mathematics equals truth: "The multiplication table is wiser and more absolute than the ancient God. It never – repeat, never – makes a mistake" (Zamyatin 1993, pp. 22, 66). Upon meeting the female revolutionary I-330, the first-person narrator D-503, whose diary-style "records" constitute the novel, learns about the elements of life that the mathematized society excludes: individuality, irrational feelings, and imagination. As D-503 begins to experience irrational desires, the writing changes from objective recording to subjective, fragmented diction. Moving from an objective realist to a fragmented modernist style, the writing reflects D-503's process of challenging the rule of mathematics and presents its rationality and objectivity as a negative foil for modernist literature. Yet, while *We* expresses fear of modern processes of mathematization, it also points to aspects of mathematics that can be used to advance revolution. D-503 takes irrational numbers and imaginary numbers to indicate the existence of irrational and imaginary elements even in this discipline, and I-330 motivates the necessity of perpetual revolution with the concept of infinity, comparing the unattainability of a final state to the impossibility of reaching a final number when counting. The novel thus presents mathematics as an instrument both of rationalization and of resistance against it, and acts as a warning against expecting modernist literature to simply equate mathematics with rationality and the opposite of imaginative literature.

Mathematics can play a role in "making it new" even in texts that unambiguously associate it with problems of modernity. In D.H. Lawrence's work, mathematics features as the opposite of what is presented as valuable: authenticity, human connection, and life. In *The Rainbow* (1915), the description of a "hideous, abstract" town with a "great, mathematical colliery on the other side" (Lawrence 2011, p. 321) firmly situates mathematics in the context of dehumanising industrialization and intellectualization that inhibit true connection of individuals to each other and to the natural world. Mathematics is also part of Lawrence's criticism of twentieth-century gender roles that, he claims, result in men becoming weak and women being forced to take on male characteristics. In *The Rainbow*, Ursula develops a fascination with the "cold absoluteness" of mathematics when she goes to university, aiming to "make her conquest also of this man's world" (Lawrence 2011, 310). While here gendered masculine in the context of a female character, mathematics is associated with emasculation in Lawrence's later novel *Lady Chatterley's Lover* (1928). Impotent Clifford, sitting in a mechanical wheelchair, embodies the overly rational modern man who is hardly alive or human, and his use of mathematical vocabulary sets the discipline at the heart of intellectualism and in opposition to love and life: "I have certain calculations to make in certain astronomical matters that concern me almost more than life and death. ... Marriage might – and would – stultify my mental processes" (Lawrence 2006, p. 33). While here, and in Lawrence's work more generally, there is no doubt on which side mathematics is set in the contrast between detestable mechanism and positive vitalism, mathematics does not stand in pure opposition to modernist literature. Rather, Lawrence employs the vocabulary of calculation and number in developing a literary style designed to make the novel a medium for life and human connection. As Michael Whitworth argues: "A scientific vocabulary enabled Lawrence to delete a religious and romantic vocabulary, exemplified here by 'soul', and replace it with a vocabulary which was still, at the time, mysterious, and which broke with a conventional humanist philosophy" (Whitworth 2001, p. 152). Although the terminologies of radioactivity and relativity theory to which Whitworth refers here are more mysterious than Lawrence's unspecific references to mathematics, the introduction of its terms forms part of his break with romantic diction. In other words, although Lawrence's works oppose mathematics on the level of content, it is part of a vocabulary that allows him to create a new way of writing and distance himself from a literary tradition perceived as no longer adequate to the times.

Having begun with the thought of modernism as opposed to mathematics' role in Enlightenment reason and rationalisation processes, the brief examination of novels by Zamyatin and Lawrence has shown a more complicated picture even in texts that establish mathematics and literature as antitheses, as they also draw on mathematics as a means to resist rationalisation and to add to a modernist renewal of literature. Moreover, as I discuss in the next section, in the decades around 1900 mathematics was not unambiguously associated with the rational, but mathematical and non-mathematical movements questioned its ties to Enlightenment reason and proposed a closer kinship with that which is outside of reason—intuition, the spiritual, and the soul.

Intuitionism: Mathematics and a Dimension Outside of Reason

Weber, who deplored increasing rationalisation and a loss of enchantment, maintained that the "ultimate and most sublime values have retreated from public life either into the transcendental realm of mystic life or into the brotherliness of direct and personal human relations" (Weber 1991, p. 155). Movements shifting attention to mystical and spiritual perspectives gained

286   N. ENGELHARDT

traction more widely in the early twentieth century. Not least, scientific and technological discoveries seemed to point to the limits of rational analysis and strengthened belief in the existence of realms beyond the senses: X-rays and the waves of wireless telegraphy, and storing devices such as the gramophone and photography, inspired attempts to communicate with the spirit realm and the dead and seemed to offer scientific explanations for the occult (see Armstrong 2005). Mathematics, too, was interpreted as indicating realms hidden from everyday experience: four-dimensional geometry was taken to add to the three dimensions of Euclidean space—height, width, and depth—a fourth one that lies outside of sensory reality. Hugely successful popularizations of four-dimensional geometry by Charles Howard Hinton and Edwin Abbott Abbott fuelled hopes to find in the fourth dimension an explanation for supernatural phenomena (see Throesch 2017).

While mystics and spiritualists such as P. D. Ouspensky, and the physicist Johann Karl Friedrich Zöllner's experiments with the medium Henry Slade, presented the geometry of four dimensions as a way to go beyond the restrictions of the superficial material world, the period's tensions between rational and non-rational perspectives also show in discussions about the foundations of mathematics. Brouwer claimed that "mathematics is independent of the so-called *logical laws* (laws of reasoning or of human thought)" (Brouwer 1975, p. 72) and located its origin in intuition. The intuitionist approach thus does not understand mathematics as a solely rational domain; a view that many mathematicians, for example the formalists around Hilbert, vehemently opposed. Foundational discussions in mathematics thus reflected and contributed to a perceived split between rational and intuitionist-spiritual perspectives in the first decades of the twentieth century. The fragmentation of the world and of approaches towards it is a main concern of modernism, as is the attempt to respond to this epistemological crisis by reconciling seemingly opposed perspectives. While, as we have seen above, modernists such as D. H. Lawrence take mathematics to stand only for the side of rational investigation, others explore the idea that mathematics takes part in intuitive, spiritual, or mystical realms.

Woolf's novel *Mrs. Dalloway* (1925) famously contrasts a focus on the material world with attention to inner and psychological dimensions: for example, the characters' time-perception is subjective, including memories of the past that are outside the objectively calculated passing of time, marked in the narrative by the periodic chiming of Big Ben. The novel also opposes rational and non-rational responses to Septimus Warren Smith who suffers from shell shock: the materialist, analytical views of scientist-doctors Holmes and Bradshaw drive him to suicide, while Mrs Dalloway understands the psychological dimension of his suffering. The critique of the "rationalisation of areas where more natural feelings – even simple sympathy – might be more appropriate" (Stevenson 1992, p. 66) might imply a negative view of mathematics, but in *Mrs. Dalloway*, it is instead introduced as linked to the soul and transcendence: watching an aeroplane, Mr. Bentley understands it as "a symbol … of man's soul; of his determination … to get outside his body … by means of thought, Einstein, speculation, mathematics" (Woolf 2000a, p. 24). Mathematics here stands for a way to break out of the restrictions of rationalisation and material reality. In Woolf's earlier novel *Night and Day* (1919), too, mathematics turns out to connect with mysticism. The novel at first pits literature and mathematics against each other when the main protagonist Katharine Hilbery, born into a heavily literature-oriented family, feels that "mathematics were directly opposed to literature" (Woolf 2000b, p. 37). Katharine's affinity with mathematics corresponds to her leaning towards progressive modern values and, in particular, to the rejection of the tradition of marriage. Her mother connects mathematics and the abandonment of conventions when she first says "A plus B minus C equals $x\,y\,z$" and then, referring both to her daughter's mathematical writings and to her suggestion of living with Ralph Denham without marrying him, complains: "It's so dreadfully ugly, Katharine. That's what I feel – so dreadfully ugly" (Woolf 2000b, p. 464). However, when Ralph reads Katharine's mathematical papers and gives her "his own unfinished dissertation, with its mystical conclusion", the two become "a united couple, an indivisible unit" (Woolf 2000b, p. 474), at least temporarily, and decide to marry. At this moment, the text brings several opposites together: it "marries" algebra and mystical writing, male and female, reason and emotion. Woolf's texts thus do not present mathematics as exclusively part of the rational domain but locate in it ways that go beyond reason and the material.

We will encounter work by Ezra Pound in all sections of this chapter, but focus here on how his Vorticist poetry uses the concept of four-dimensional geometry to combine rational-objective and mythical-spiritual perspectives. As Ian Bell points out, Pound explicitly mentions the fourth dimension only twice in his writing and long after the height of its popularity in the first two decades of the twentieth century (Bell 2012, p. 131). In a letter of 1941, Pound writes: "I conceive of a dimension of stillness which compenetrates the [three] Euclidean dimensions" (see Bell 2012, p. 130), and his "Canto 49", part of the long, incomplete poem *The Cantos*, specifies this dimension as "[t] he fourth; the dimension of stillness" (Pound 1996, p. 245). But Bell argues that Pound's earlier work, most importantly Vorticism, can also be related to the fourth dimension: Pound presents the vortex as outside common experience when it combines the opposites of rushing movement in three-dimensional space and the fourth dimension of stillness at its centre. His poem "Dogmatic Statement on the Game and Play of Chess: Themes for a Series of Pictures", published in the second edition of *Blast* (1915), is an example of the short-lived Vorticist movement, and it illustrates its ideals of precision and dynamism when using mathematical terms such as "angles", "lines", "x", "y" in combination with dynamic verbs emphasising motion and force, for example "striking", "falling", "clashing". With this mathematical and dynamic vocabulary, the poem establishes the highly formalised and at the same time energetic movements of a game of chess:

Red knights, brown bishops, bright queens,
Striking the board, falling in strong "l"s of color,
Reaching and striking in angles,
    holding lines in one color.
This board is alive with light;
    these pieces are living in form,
Their moves break and reform the pattern:
    Luminous green from the rooks,
Clashing with "x"s of queens,
    looped with the knight-leaps.

"Y" pawns, cleaving, embanking!
Whirl! Centripetal! Mate! King down in the vortex,
Clash, leaping of bands, straight stripe of hard color,
Blocked light working in. Escapes. Renewing of contest.

A change in the pattern made up by the chess pieces, that is, a change of the formation on the board, is where the action lies, and it determines whether a piece continues "living in form" (l. 6) or, like the king in the last lines, goes down in the vortex. The mathematically precise movements on the chessboard thus equal the absolute dynamism of life. So on the one hand, Vorticism is "an attempt to revive the sense of form" (Pound 1963, p. 47), and pure form is, for Pound and others, expressed in geometric terms. But the poem also suggests that form and a source of life beyond it come together when the king escapes from the vortex and the contest on the chessboard begins again. In Pound's Vorticist poetry, mathematics thus stands for formalist order but also, in the guise of the fourth dimension, for a realm beyond that can engender rebirth and renewal.

In a period that saw a growing split into rationalised and spiritual approaches to knowledge and the world and that witnessed the formation of an intuitionist school announcing mathematics' independence from the logical laws of reasoning, modernist writers do not only treat mathematics as an example of rationality but also reflect on developments that seem to move it into the realm of mysticism and the spiritual. Building on the examination of how mathematics offers modernist writers a vocabulary to announce their departure from literary traditions, the next section focuses on questions of language that receive pronounced attention both in modernist literature and logicist mathematics.

Mathematics and Language—Logicism

Gottlob Frege developed an ideal language based on the model of mathematics in *Concept Notation: A Formula Language of Pure Thought, modelled upon that of Arithmetic*, published in 1879. With this, he founded the mathematical school of logicism and also laid the groundwork for a modern philosophy of language. Building on Frege's work and sharing the assumption that mathematics could be reduced to logic, Russell and Whitehead's *Principia Mathematica* decisively developed analytic philosophy with its mathematical symbolism, technical style, and aim of creating an unambiguous language to express certain knowledge. With his widespread interests and acquaintances, Russell made accessible foundational research in mathematics and philosophical language to laypersons and to discussions about the precision and reliability of literary language. T. S. Eliot, who heard Russell lecture on symbolic logic at university and later lived in his London home, praised Russell's contribution to making English "a language in which it is possible to think clearly and exactly on any subject. The *Principia Mathematica* are perhaps a greater contribution to our language than they are to mathematics" (Eliot 1927, p. 291). Epitomising the objectivity and impersonality that he aims for in poetry, precise mathematical language promises for Eliot a way to "approach the condition of science" (Eliot 1998, p. 30) in literature and thus arrive at a new, modernist way of writing.

Eliot's dissatisfaction with an unclear and imprecise English language that inhibits thinking illustrates that in modernism, language is not perceived as a transparent medium that uncomplicatedly represents the world but that it becomes questioned as a means of understanding and communication. In "How to Read" (1929), Pound complains:

> Save in the rare and limited instances of invention in the plastic arts, or in mathematics, the individual cannot think and communicate his thought ... without words, and the solidity and validity of these words is in the care of the damned and despised "litterati." When their work goes rotten ... when their very medium, the very essence of their work, the application of word to thing goes rotten, i.e. becomes slushy and inexact, or excessive or bloated, the whole machinery of social and of individual thought and order goes to pot. (Pound 1968, p. 21)

Like Eliot, Pound laments the vagueness of literary language and proposes mathematics as a medium for clear expression and communication (also see Jocelyn Rodal's "The Metaphor as an Equation: Ezra Pound and the Similitudes of Representation" in this volume). Woolf's *Night and Day* develops a similar proposal. Katharine differs from her relatives when preferring "the exactitude, the star-like impersonality, of figures to the confusion, agitation, and vagueness of the finest prose" (Woolf 2000b, p. 37). She experiences again and again the limits of everyday speech: "none of her words seemed to her at all adequate to represent what she felt" (Woolf 2000b, p. 230). As characters become aware that language is not a transparent medium but can fail to transmit meaning, they either fall silent or focus on words themselves: Katharine's friend Joan speaks "rather as if she were sampling the word", her friend and later husband Ralph "examined each word with the care that a scholar displays upon the irregularities of an ancient text", and Katharine herself focuses more and more on her mathematics with its "pages of neatly written mathematical signs" (Woolf 2000b, pp. 24, 235, 204). As Jocelyn Rodal argues, "the foreign marks of mathematics ... call attention to the intrinsic strangeness of all written symbols" (Rodal 2012, p. 203), and Woolf's novel strengthens the connection to non-mathematical language when Katharine hides her pages of symbols and figures in a Greek dictionary. The realisation that language becomes strange when viewed through the lens of mathematics also informs Lawrence's *The Rainbow*, which similarly places mathematical symbols in the vicinity of everyday language: Ursula "enjoyed the vagaries of English Grammar, because it gave her pleasure to detect the live movements of words and sentences; and mathematics, the very sight of the letters in Algebra, had a real lure for her" (Lawrence 2011, pp. 310–11). Relating it to vocabulary and grammar as the building blocks of speech, the novels use mathematics to defamiliarize the medium of modernist literature— language itself.

Outside of mathematics, the examination of the basic building blocks of language was greatly advanced by the linguist Ferdinand de Saussure. His work, published posthumously in 1916, introduces language as a formal system of signs with each sign comprising two aspects, the signifier or sound structure, and the mental concept this sound evokes, called the signified. The meaning of a sign is a function of its relation to other signs in the system. To reprise a well-worn example, the signifier "cat" takes its meaning not from any relation to the animal, but from its difference from other signifiers such as "hat" or "mat". The referent, in this case the animal, is of no concern in Saussure's theory, which presents language as a self-referential system that is independent of the physical world. The idea that words only refer to themselves underlies Ralph's need to examine words with care and Katharine's inability to put into language what she wants to express in Woolf's *Night and Day*, and it also manifests in Pound's feeling that "the application of word to thing goes rotten" (Pound 1968, p. 21).

In mathematics, nineteenth-century discoveries, such as of four-dimensional geometry, similarly put into question the connection between words and things, or, more precisely, between mathematical language and the natural world. In the early 1900s, the mathematician Henri Poincaré looked back on the impact of formulating geometries different from the geometry of three dimensions that had been taken for granted since its formulation by Euclid in the third century bce: "If several geometries are possible, they say, is it certain that our geometry is the one that is true"? (Poincaré 1905, p. 48). He concludes that, since different geometries are equally mathematically valid, any particular geometrical system cannot be said to be true but only to be "more convenient" (Poincaré 1905, p. 50). In algebra, too, attention turned to concepts that jarred with the idea that mathematics expresses truths about the natural world. When William Rowan Hamilton used three imaginary values in his formulation of quaternions in 1843, John Graves wrote to him that he was unsure about "the extent to which we are at liberty arbitrarily to create imaginaries, and to endow them with supernatural properties" (qtd. in Baez 2002, p. 146). The term "imaginary number" was coined by René Descartes in the seventeenth century to describe the unit $\sqrt{-1}$ that seemed to be impossible or "unreal" since it does not have a correspondence in nature: while we might point to three apples to visualise the number three, no object in the material world stands for a realisation of imaginary numbers. Graves here questions Hamilton's creating new numbers, namely quaternions that do not represent what can be found in reality but are "supernatural". But in the nineteenth century not only new numbers but all of mathematics came to be understood as not necessarily tied to the natural world: to use Poincaré's terms, mathematics could no longer be understood as physically true but only as convenient. While this raised anxiety about the knowability of the world, the mathematician Georg Cantor's statement that "the *essence* of *mathematics* is its *freedom*" ("das *Wesen* der *Mathematik* liegt gerade in ihrer *Freiheit*") (Cantor 1883, p. 564) employs a positive phrase to describe modern mathematics' developing according to its own rules, uninhibited by reference to physical reality.

Several modernist texts compare the freedom of mathematics to that of literature and employ imaginary numbers as a metaphor for the literary imagination. Velimir Khlebnikov, one of the founders of Russian Futurism, uses the mathematical unit as an image signalling a flight of the imagination: "We climbed aboard our $\sqrt{-1}$ and took our places at the control panel" (Khlebnikov 1989, p. 82), so the translation of a prose poem from 1916 begins. The speaker then goes on to observe the world at different places and ages simultaneously, suggesting that they record a four-dimensional space-time continuum. The term "imaginary number" here expresses the modernist idea that a changed, ungraspable reality—a four-dimensional world—requires a new form of representation: not realist copying of the surface world but an account based in the imagination.

While not employing the metaphor of imaginary numbers, Woolf's *Night and Day* similarly presents mathematics as an imaginary realm that is free from the restrictions of the real world. Mathematics and astronomy lead Katharine to an otherworldly state where she becomes impersonal but also "[s]omehow simultaneously, though incongruously" experiences intimate feelings: "certainly she loved some magnanimous hero" (Woolf 2000b, pp. 184, 131). As in the novel mathematics does not refer to the world of social conventions and marriage but opens up Katharine's "imaginary world" (Woolf 2000b, p. 131), it functions as a means of liberation. As Ann-Marie Priest argues, the very symbols of mathematics provide an alternative to patriarchal language:

> Katharine's algebraic squiggles represent her precisely by not representing her. They convey nothing whatever about her – and thus they do not constrain, construct, or delimit her in any way. Thus, she can accept them – as symbols which simply create a textual space for her, without giving that space any content. (Priest 2003, p. 71)

For Katharine, the essence of mathematics is its freedom indeed, in the sense of liberating her from the restrictions of a society that controls the means of representation. And Katharine is finally able to reveal her sense of self in her mathematical writing: she first hides it but then lets Ralph read her mathematics, and after this "moment of exposure" (Woolf 2000b, p. 474) they decide to marry. As mathematics moves from being antithetical to facilitating marriage, it also emerges not to be opposed to all literature but to share with progressive modern writing a freedom from conventions, not least freedom from realistically representing the given world.

FORMALISM: MATHEMATICS AND FORM

Katharine Hilbery in Woolf's *Night and Day* almost shares her last name with David Hilbert and also has in common with him a focus not on mathematical meaning but on "marks on paper" (Hilbert 1964, p. 143) that have significance only in the system of mathematics itself and no direct relation to reality. While there is no evidence that Woolf knew of Hilbert's formalist approach, he was the most famous mathematician of the time, and she might have encountered his thought through Russell, who was instrumental to the reception of logicism and formalism in the literary sphere. Reviews of the *Principia Mathematica* in prominent places such as on the front page of the *Times Literary Supplement* on 7 September 1911 ensured that its approach was also accessible to persons unfamiliar with its symbolic writing (see Gamwell 2016, p. 212). *Night and Day* connects the formalist notion of mathematics, evoked by the main protagonist's last name and a focus on mathematical symbols, to painting and modernist art theory. Katharine is modelled on Woolf's sister, the painter Vanessa Bell, and the novel draws parallels between painting and mathematics as media through which Katharine could communicate: her romance "was a desire, an echo, a sound; she could drape it in color, see it in form, hear it in music, but not in words; no, never in words" (Woolf 2000b, p. 274). While she finally expresses herself in mathematics, painting also presents itself as a possibility to communicate meaning, and the above phrase echoes the formalist conceptualization of art by Vanessa's husband Clive Bell: "lines and colours combined in a particular way, certain forms and relations of forms, stir our aesthetic emotions" (Woolf 2000b, p. 8). In *Art* (1914), Bell introduces the idea that the form of an artwork is expressive, independently of the work's relation to reality, and fellow Bloomsbury Group member Roger Fry similarly maintained that the form of the artwork, its use of line, colour, and shape, is significant in itself. While Woolf acknowledges in *Night and Day* that formalist art shares with mathematics a focus on form as the locus of meaning, the form of the novel itself remains fairly conventional. The novel's experimentation with new possibilities rather shows in its unusual communication through mathematics that draws attention to the limits of the English language and of realist writing.

16 MATHEMATICS AND MODERNISM   293

An example of modernism's focus on form as itself meaningful is James Joyce's *Ulysses* (1922). Its penultimate episode "Ithaca" is written as a mathematical catechism, that is, as a manual of questions followed by answers traditionally used in religious instruction but here taken out of this context and focusing on rational facts. "Ithaca" features many references to mathematics, which Joyce collected when working through Russell's *Introduction to Mathematical Philosophy* (see transcriptions in Joyce 1972, e.g. pp. 432–36). Before *Ulysses* closes with Molly Bloom's stream of consciousness that sounds the depths of internal thought and personal subjectivity, "Ithaca" presents a catalogue of hard, precise facts. Lists of conversation topics, Leopold Bloom's books, his wife's lovers, and the contents of the kitchen dresser give the impression of scientific examination while also making for laboured reading. With its catechist style and exaggerating precision, "Ithaca", like other episodes of *Ulysses*, draws attention to the fact that reality is perceived through language and that even seemingly objective representations of the world are coloured by it. Moreover, the episode emphasises that even its apparently objective mathematical facts are based on agreements that can change. For example, after a number-heavy paragraph with calculations of the relation between Bloom's and Stephen's ages, the next question reads: "What events might nullify these calculations"? (Joyce 2000, p. 794). Besides death or the end of the world, "the inauguration of a new era or calendar" invalidates the calculations, and the text similarly points to the cultural bases of measuring systems and computations based upon them when Bloom feels the "cold of interstellar space, thousands of degrees below freezing point or the absolute zero of Fahrenheit, Centigrade or Réaumur" (Joyce 2000, pp. 794, 827).

As the episode draws attention to the fact that measurements and applied calculations are not unquestionably valid or eternally true, it might not be surprising that the calculation of the age relation is itself faulty, as are various other computations in the episode. While early criticism took these to be errors on Joyce's part, they can also be seen as intentional and pointing to the pitfalls of a purely rational, mathematical worldview:

> [T]he style gives us the impression that everything is being described quite precisely, while a variety of errors ... suggest that a sort of artistic uncertainty principle is at work here, and that no matter how many measurements we take and how many angles we view something from there will be some element of indeterminacy in every description. (McCarthy 1984, p. 612)

At the same time as relativizing the "truth" of mathematics, "Ithaca" alerts us to the fact that mathematics does not necessarily represent reality but lends itself to examining hypothetical states. A direction reads: "Reduce Bloom by cross multiplication of reverses of fortune ..., and by elimination of all positive values to a negligible irrational unreal quantity" (Joyce 2000, p. 855). The answer then "reduces" Bloom to a pauper, using mathematics to introduce a possibility that does not have a correspondence in reality.

294   N. ENGELHARDT

Producing possibilities is not only a mathematical feat but also a feature of language, and "Ithaca" connects the two when Bloom explains that in Hebrew, each letter corresponds to a number: here, language and mathematics meet in the arithmetical values of the alphabet, suggesting that spoken languages do not carry any more meaning than numbers. Indeed, Bloom and Stephen's knowledge of Irish and Hebrew is "[t]heoretical, being confined to certain grammatical rules of accidence and syntax and practically excluding vocabulary" (Joyce 2000, p. 806). In other words, to Bloom and Stephen, the languages are rule and form, not meaning and content. After this presentation of language as a formal structure not unlike mathematics, "Ithaca" ends in a series of formal word manipulations: "Sinbad the Sailor and Tinbad the Tailor and Jinbad the Jailer ... and Xinbad the Phtailer" (Joyce 2000, p. 871). Reminiscent of Bloom's contemplating mathematics in the episode "Sirens" where he thinks: "Do anything you like with figures juggling" (Joyce 2000, p. 359), "juggling" the phrase "Sinbad the Sailor" leads to nonsense but also to new content when the permutations arrive at the meaningful words "tailor" and "jailer". The episode then ends by juggling words in a way that achieves something impossible, namely the squaring of the circle, a problem that Bloom ponders earlier in the episode: "Going to dark bed there was a square round Sinbad the Sailor's roc's auk's eggin the night of the bed of all the auks of the rocs of Darkinbad the Brightdayler" (Joyce 2000, p. 871). Formal play that creates new words and, through them, new perspectives on reality, here enables a combination of opposites: of round and square, bright and dark, science and myth.

While literary form is of interest in all periods, modernism is particularly intensely concerned with it and formalist criticism posits form as the most important requirement of art. I. A. Richards, laying the groundwork for a formalist movement in literary theory in the middle of the twentieth century, maintained: "A good deal of poetry and even some great poetry exists ... in which the sense of the words can be *almost* entirely missed or neglected without loss" (Richards 1926, pp. 22–23). The symbols of mathematics, which Hilbert called "meaningless" (Hilbert 1964, p. 143), embody the precedence of form, and literary texts engage with mathematics as a way to comment on formalism in art. "The organization of forms is a much more energetic and creative action than the copying or imitating of light on a haystack" (Pound 1970, p. 92), Pound explains in his essay "Vorticism" (1914), with a gibe directed at Claude Monet's impressionist paintings. Pound explains that Vorticist poetry does not aim to merely copy what exists in reality, but that it is aimed at creation. For him, mathematics offers an example of such creation: with analytical geometry "one is able *actually to create*" (Pound 1970, p. 91), as an equation gives a rule that describes a shape and thus "creates" forms. For example, "we learn that the equation $(x-a)^2 + (y-b)^2 = r^2$ governs the circle. It is the circle" (Pound 1970, p. 91). The equation thus brings into being the circle; it creates form. And Pound contends that Vorticist art creates in analogy to this mathematical process: "Great works of art contain this ... sort of equation. They cause form to come into being" (Pound 1970, p. 92). He concludes that "[t]he difference between [Vorticist] art and analytical geometry is the difference of subject-matter only": as analytical geometry deals with form, so "art handles life" (Pound 1970, p. 91), and both involve creating, bringing into being. In the "Ithaca" episode of Joyce's *Ulysses*, the phrase "let H. F. be L. B" (Joyce 2000, p. 851) points to a similar way in which mathematics can be seen to create: the phrase "let x be an even number" defines that, indeed, x is an even number, and similarly in "Ithaca", H. F. *is* L. B once the definition is spoken or put on paper. The mathematical formulation brings into being the identity of Leopold Bloom and his pseudonym Henry Flowers.

The Austrian author Hermann Broch, who studied mathematics at university and was well-informed about foundational questions in the field (see Riemer 1986), takes the process of creation in mathematics as the basis of his short story "Methodically Constructed" (1917), reprinted as a chapter in his novel *The Guiltless* (*Die Schuldlosen*, 1950). As the title suggests, the narrative presents itself as an overtly constructed work of art, and it uses the same mathematical expression as Joyce to signal that it creates the setup by formulating it: "let the hero be localized in the middle class of a medium-sized provincial town, ... let him be called Zacharias" (Broch 2000, pp. 26–27). As "a character constructed of middling qualities", the mathematics and physics teacher knows only "operational problems ... never those of existence" (Broch 2000, p. 27). And since Zacharias is not concerned with existence, he suggests suicide when, in his relationship with Philippine, the opposition of bodily and spiritual love becomes an "insoluble" problem for him. In a metafictional comment, the narrator points to the literary conventions that determine the narrative: when Philippine instructs Zacharias to buy a gun, "[h]e felt, and so do we in constructing the story, that the die had thus been cast" (Broch 2000, p. 35). Breaking narrative conventions when showcasing the constructedness of the story, the meta-comment points to the questions of existence that lie outside Zacharias's interest: it raises awareness of the fact that he "exists" only inside the methodical textual construction, like a mathematical object that is being defined at the outset and then manipulated in calculations. The end of Zacharias's "existence" indeed follows calculable, romantic conventions when the couple kills each other and merges in perfect union with the universe. Yet, after this ending, the narrator brings into existence a different reality in another metafictional moment: in a second possible outcome the couple returns home, marries, and finds not spiritual union but the banality of a "natural, appropriate, but not necessarily happy end" (Broch 2000, p. 38). The surprising second ending questions the literary tradition of naturalism with the "presumptuous fallacy of the naturalists to suppose that they can fully determine a human being on the basis of environment, atmosphere, psychology and similar components, forgetting that it is never possible to know all his motivations" (Broch 2000, p. 37). Introducing ambiguity and "a multitude of possible solutions" (Broch 2000, p. 38), the short story moves from the precision of mathematical definition in its beginning to a closing acknowledgement of unknowability.

In a letter from 1932, Broch stresses the importance of form for gaining new insights: "Writing means aiming to win knowledge through form, and new knowledge can only be created through new form" ("Dichten heißt, Erkenntnis durch die Form gewinnen wollen, und neue Erkenntnis kann nur durch neue Form geschöpft werden") (Broch 1974–1981, p. 223). Mathematics offers him and other modernist writers an example of how form can create knowledge and, indeed, being. Writers investigate a capacity for poetic creation that mathematics and literature seem to share and employ mathematics as a model for abandoning demands for realist representation and instead developing autonomous literary forms.

Conclusion: Mathematics and Modernism

Modernist literature puts mathematics at the pinnacle of problematic excesses of reason and rationalisation, while at the same time, novelists and poets locate in mathematical precision and certainty a model for their own literary endeavours, encouraging responses to the period that build on mathematics to "make it new". I suggested that modernist writers' uses of mathematics to renew literature can be roughly divided into concerns with language, form, and non-rational dimensions and that these issues are also reflected in the three approaches dominating mathematics in the early twentieth century: logicism, formalism, and intuitionism. Brouwer opposes logicism's understanding mathematics as a language and criticises Hilbert's formalist approach as reducing mathematics to a "meaningless series of relations" (Brouwer 1913, p. 83). Instead, he argues that, although communicated through language, mathematics originates in pre-linguistic intuition, and that this human intuition ties mathematics to the world and ensures its meaning and value. The conflict between the schools thus boils down to questions of language (form) and meaning (content)—an abstract concern that modernist literature similarly addresses and which it explores, not least, with reference to mathematics.

When modernist literature draws on mathematics to explore notions of reason, representation, language, and form, it presents mathematics as sharing central concerns with art and as a model for modernist experimentation with style and form. Attending to the roles mathematics plays in modernist texts shows that these do not one-dimensionally condemn processes of modernization and Enlightenment traditions but also locate in these potential for creatively constructing new possibilities. Not least, when modernist texts engage with mathematical concepts and debates in their searches for new style and form, they also formulate—and thus create—the place of mathematics in modernist culture.

REFERENCES

Armstrong, Tim. 2005. *Modernism; A Cultural History.* Cambridge and Malden, MA: Polity.

Baez, John C. 2002. "The Octonions." *Bulletin of the American Mathematical Society* 39 (2): 145–205.

Bell, Clive. 1914. *Art.* New York: Stokes.

Bell, Ian F. A. 2012. "Ezra Pound and the Materiality of the Fourth Dimension." In *Science in Modern Poetry; New Directions,* edited by John Holmes, 130–50. Liverpool: Liverpool University Press.

Brits, Baylee. 2017. *Literary Infinities, Number and Narrative in Modern Fiction.* London: Bloomsbury.

Broch, Hermann. 1974–1981. *Briefe 1.* Edited by Paul Michael Lützeler. Frankfurt am Main: Suhrkamp.

———. 2000. *The Guiltless.* Translated by Ralph Manheim. Evanston, IL: Marlboro and Northwestern.

Brouwer, Luitzen Egbertus. 1907 [1975]. "On the Foundations of Mathematics." In *L. E. J. Brouwer; Collected Works,* vol. 1. Edited by Arend Heyting, 15–101. Amsterdam and Oxford: North-Holland.

———. 1913. "Intuitionism and Formalism." *Bulletin of the American Mathematical Society* 20 (2): 81–96.

Cantor, Georg. 1883. "Über unendliche, lineare Punktmannigfaltigkeiten V." *Mathematische Annalen* 21 (4): 545–91.

Cliver, Gwyneth E. 2008. *Musil, Broch, and the Mathematics of Modernism.* St. Louis: ProQuest Dissertations Publishing.

Eliot, T. S. 1927. "A Commentary." *The Monthly Criterion* 6 (4): 289–91.

———. 1998. "Tradition and the Individual Talent." In *The Sacred Wood and Major Early Essays,* 27–33. Mineola, NY: Dover.

Engelhardt, Nina. 2018. *Modernism, Fiction and Mathematics.* Edinburgh: Edinburgh University Press.

Gamwell, Lynn. 2016. *Mathematics + Art; A Cultural History.* Princeton and Oxford: Princeton University Press.

Gray, Jeremy J. 2004. "Anxiety and Abstraction in Nineteenth-Century Mathematics." *Science in Context* 17: 23–47.

Hickman, Miranda B. 2005. *The Geometry of Modernism; The Vorticist Idiom in Lewis, Pound, H.D., and Yeats.* Austin: University of Texas Press.

Hilbert, David. 1925 [1964]. "On the Infinite." *Philosophy of Mathematics; Selected Readings.* Edited by Paul Benacerraf and Hilary Putnam, 134–51. Oxford: Blackwell.

Joyce, James. 1972. *Joyce's* Ulysses *Notesheets in the British Museum.* Edited by Phillip F. Herring. Charlottesville: University Press of Virginia.

———. 2000. *Ulysses.* London: Penguin Books.

Khlebnikov, Velimir. 1989. *Collected Work of Velimir Khlebnikov, vol. II: Prose, Plays, and Supersagas.* Edited by Ronald Vroon. Translated by Paul Schmidt. Cambridge, MA and London: Harvard University Press.

Lawrence, D. H. 2006. *Lady Chatterley's Lover.* New York and London: Penguin Books.

———. 2011. *The Rainbow*. London: Vintage.

McCarthy, Patrick A. 1984. "Joyce's Unreliable Catechist: Mathematics and the Narration of 'Ithaca.'" *English Literary History* 51 (3): 605–18.

Mehrtens, Herbert. 1990. *Moderne Sprache Mathematik; Eine Geschichte des Streits um die Grundlagen der Disziplin und des Subjekts formaler Systeme*. Frankfurt am Main: Suhrkamp.

Poincaré, Henri. 1905. *Science and Hypothesis*. Translated by William John Greenstreet. London and Newcastle-on-Tyne: Walter Scott.

Pound, Ezra. 1914 [1970]. "Vorticism." In *Gaudier-Brzeska: A Memoir*, 81–94. New York: New Directions.

———. 1963. "Interview with Ezra Pound." Interview by Donald Hall. *Writers at Work: The "Paris Review,"* Second Series. New York: Viking.

———. 1968. "How to Read." In *Literary Essays of Ezra Pound*. Edited by T. S. Eliot, 15–40. New York: New Directions.

———. 1996. *The Cantos of Ezra Pound*. New York: New Directions.

Priest, Ann-Marie. 2003. "Between Being and Nothingness: The 'Astonishing Precipice' of Virginia Woolf's 'Night and Day.'" *Journal of Modern Literature* 26 (2): 66–80.

Richards, Ivor Armstrong. 1926. *Science and Poetry*. London: Kegan Paul.

Riemer, Willy. 1986. "Mathematik und Physik bei Hermann Broch." In *Hermann Broch*, edited by Paul Michael Lützeler, 260–71. Frankfurt am Main: Suhrkamp.

Rodal, Jocelyn. 2012. "Virginia Woolf on Mathematics: Signifying Opposition." In *Contradictory Woolf; Selected Papers from the Twenty-First Annual International Conference on Virginia Woolf*, edited by Derek Ryan and Stella Bolaki, 202–8. Clemson: Clemson University Digital Press.

Snapper, Ernst. 1979. "The Three Crises in Mathematics: Logicism, Intuitionism and Formalism." *Mathematics Magazine* 52 (4): 207–16.

Stevenson, Randall. 1992. *Modernist Fiction; An Introduction*. Lexington: University Press of Kentucky.

Throesch, Elizabeth. 2017. *Before Einstein; The Fourth Dimension in Fin-de-Siècle Literature and Culture*. London and New York: Anthem Press.

Tubbs, Robert. 2014. *Mathematics in Twentieth-Century Literature and Art: Content, Form, Meaning*. Baltimore: Johns Hopkins University Press.

Weber, Max. 1991. "Science as a Vocation." In *From Max Weber: Essays in Sociology*, edited and translated by H. H. Gerth and C. Wright Mills, 129–56. London: Routledge.

Whitworth, Michael. 2001. *Einstein's Wake; Relativity, Metaphor, and Modernist Literature*. Oxford: Oxford University Press.

Woolf, Virginia. 2000a. *Mrs Dalloway*. Oxford: Oxford University Press.

———. 2000b. *Night and Day*. London: Vintage.

Zamyatin, Yevgeny. 1993. *We*. Translated by Clarence Brown. New York: Penguin Books.

CHAPTER 17

Mathematics in German Literature: Paradoxes of Infinity

Howard Pollack-Milgate

Introduction

In his essay, "The Mathematical Man" (1913, "Der mathematische Mensch"), the Austrian novelist Robert Musil (1880–1942) distinguishes two ways of looking at mathematics. The common one views mathematics as an indispensable collection of technical methods, essential for the everyday affairs of human beings in the modern world and, ultimately, underlying the existence of our entire civilization. True mathematicians, however, are completely indifferent to the instrumental use of elementary mathematics. They see mathematics, rather, in its most advanced and specialized form as a special, esoteric passion. Yet, for them, as for Musil, theoretical mathematics, no matter how recondite, deals with the "serious business of our lives" and is "one of the most amusing and intense adventures of human existence" (Musil 1990, p. 41). In particular, initiates of this second point of view are aware of the recent discovery that the whole system of mathematics is "standing in mid-air," without solid foundation, and thus modern civilization is based on the "error" of assuming certainty where there is none. Musil finds courage in such mathematical thinking, the determination to face the truth no matter how dangerous or contrary to our everyday sense of life, and sees such "fantastic, visionary feelings" (Musil 1990, p. 42) as a model for what should be accomplished in broader cultural realms, especially literature.

This dual view of mathematics is clearly related to the so-called "crisis in foundations" which affected mathematics in the early twentieth century

---

H. Pollack-Milgate (✉)
DePauw University, Greencastle, IN, USA

© The Author(s) 2021
R. Tubbs et al. (eds.), *The Palgrave Handbook of Literature and Mathematics*, https://doi.org/10.1007/978-3-030-55478-1_17

299

when the paradoxes of infinite sets were discovered, but it is not unique to early twentieth-century modernism. In 1784, the Royal Prussian Academy of Sciences in Berlin, noting the pervasive use of the mathematical infinite in modern calculus in contrast to its avoidance in Ancient Greek geometry, stated: "the Academy hopes, therefore, that it can be explained how so many true theorems have been deduced from a contradictory supposition" and set a prize for the goal "that a principle can be delineated which is sure, clear – in a word, truly mathematical – which can appropriately be substituted for 'the infinite'" (Grabiner 2005, pp. 41–42). We find here many of the same assertions and contrasts as in Musil: the condition of the modern world, the utility of results versus the problems of a foundation, the notion of the mathematical as, on the one hand, certain and clear, on the other, perplexing and contradictory. In this essay, we will trace these views of mathematics, which go through all modern times, and show how they inform various thematic and formal aspects of German-language literature.

Much of modern-age mathematics was developed by German authors. Famous German-speaking mathematicians include, to name only some of the most prominent representatives, Gottfried Wilhelm Leibniz (1646–1716, co-inventor of calculus together with Isaac Newton), Leonhard Euler (1707–1783, generalist with accomplishments in many fields), Carl Friedrich Gauss (1777–1855, known as the "prince of mathematicians" for his extraordinary accomplishments), Bernhard Riemann (1826–1866, developed non-Euclidean geometry), Georg Cantor (1845–1918, developed set theory and the modern theory of the infinite), Kurt Gödel (1906–1978, logician).[1] The University of Göttingen, founded in 1737 and Gauss's home institution, became in the nineteenth century a center for German, and indeed, international mathematical research. One characteristic of the German approach to mathematics is its close link to philosophy; especially as mathematics became more abstract and, in its abstraction, was applied to the physical investigation of the world, attempts were made to describe the methods and subject matter of mathematics and to investigate its logical foundations. After Leibniz, who was a key philosopher for the writers of the German Enlightenment (*Aufklärung*), the philosophy of Immanuel Kant (1724–1804) was a central influence on many of the literary figures of Romanticism that I will discuss. Although not himself a mathematician, Kant wrote extensively about the nature and role of mathematical knowledge. Later figures to mention in this regard are Gottlob Frege (1848–1925), who made fundamental contributions to logic and to problems of the foundations of mathematics, influencing generations of modernists, and Martin Heidegger (1889–1976), who in the light of the history of the twentieth century critically examined the role of generalized mathematical thinking in the modern age. The relationship between the natural sciences and the humanities has also been a prominent theme in German thought, most famously in the works of Wilhelm Dilthey (1833–1911), and as a consequence of this philosophical mediation, the "two cultures" have traditionally not been as separate in German intellectual history as in other European cultures.

Given the prominence of mathematics in modern German intellectual history, it is no surprise that the modern German literary tradition contains works which address mathematics and mathematicians or use mathematical techniques in their construction. In this essay, we will concentrate on one of the most significant and representative aspects of mathematics in German literature, focused on the dual nature of modern mathematics described above, and especially the logical difficulties which arise through the use of the concept of the infinite (one Romantic writer fancifully describes the introduction of this concept into mathematics as a borrowing from outside the discipline, presumably from theology, "A science grows through ingestion – through assimilation of other sciences, etc. So mathematics e.g. through the ingested concept of the infinite" [Novalis 1960, 3, p. 268]). There are many facets to this question, such as: the theological and mystical associations of the concept of infinity, which will continue to play a role throughout; the relationship of the infinite to the finite, including the Pythagorean heritage of mathematics as key to the creation and structure of the world; corresponding questions of the limits of human knowledge or human morality in the instrumentalization of the infinite; and the appropriate symbolic uses of mathematical concepts.

Though these questions are some of the most fundamental in the German literary tradition, relatively few authors have the necessary mathematical background (relative to the time period in which they lived) to phrase them explicitly. In this essay, the focus will be on those authors who did study mathematics—Novalis, Robert Musil, Hermann Broch, and Daniel Kehlmann—and who are also representative of the specific mathematical forms used in their epochs. There are a number of studies of the role of mathematics in each of these particular authors, or of particular mathematical themes, but few overviews of mathematics in German literature in general.[2]

From Nicholas Cusanus to Gottfried Leibniz
and the *Aufklärung*

Crucial influences on literary treatments of mathematics in the modern German context are the works of the philosopher and theologian Nicholas Cusanus (also known as Nicholas of Cusa, 1401–1464) and Leibniz, both of whom tried to integrate traditional theological ideas with mathematical treatments of the infinite. Cusanus is a key figure in the transition from the medieval view of the cosmos to the infinite universe of modernity and is often taken to be the first person to argue that the universe is spatially infinite, paving the way for Giordano Bruno and others.[3] Born in Germany, Cusanus studied in Italy, and spent his life working as a reformer within the Church. He became a Cardinal in 1448 and a Bishop in 1450 and attempted to mediate the Church's many factional struggles and to reunify Eastern and Western

Christianity. This work led him to insight into the necessity of reconciling different perspectives which he elaborated in a number of original and sometimes controversial works on topics such as theology, philosophy, and mathematics (including the discovery of a more accurate estimate of the value of pi). These works were based in part on the attempt to link the thought of earlier writers, such as the German mystics, with the new tendencies of humanist and scientific thought he encountered in Italy. The fundamental subject of his writings was the examination of the possibilities and limits of human knowledge in a universe created by a God who, as infinite, is incommensurable with finite human understanding, yet whose infinity is manifested in finite appearances.

In his most famous work, *On Learned Ignorance* (*De Docta Ignorantia*, 1440), Cusanus posits that the nearest we can get to knowledge of transcendent truth, such as the nature of an infinite God, is through our recognition that such truth, as infinite, transcends the rules of normal human knowledge, rules that are in some sense mathematical:

> Every inquiry proceeds by means of a comparative relation, whether an easy or a difficult one. Hence, the infinite, qua infinite, is unknown; for it escapes all comparative relation. But since *comparative relation* indicates an agreement in some one respect and, at the same time, indicates an otherness, it cannot be understood independently of number. (Cusanus in Hopkins 1990, p. 50)

The highest knowledge is knowledge of the inability to know, of ignorance. In order to explain this ignorance, the notion of God is stated in quasi-quantitative terms, since knowledge, as described here with explicit reference to Pythagoras, is inherently numerical. Whereas all quantities in the sensory world can be increased and decreased ad infinitum, God is the "Absolute Maximum," that is, greater than all other possible things. However, Cusanus argues that this "superlative" greatness also implies that God possesses all other superlatives as well, such as being also the absolute minimum; in the divine, the maximum and the minimum coincide, so that God is both everything and anything. Thus, the infinite is characterized by the coexistence of oppositions within it and hence violates the normal rules of reason. God is oneness, but, as Cusanus argues, one is not a number; thus, "it is evident that Philosophy ... must leave behind all things imaginable and rational." Yet, as God is fully present in each individual, and each individual contains in a way a representation of the entire universe, we can somehow witness and understand God in created things, and since mathematics provides us with the clearest, most precise, and certain understanding of ordinary creation, it is the best place to begin to explore that which is more speculative; indeed, "mathematics assists us very greatly in apprehending various divine truths" (Cusanus in Hopkins 1990, pp. 56, 53, 60, 61).

Cusanus informs us that to ascend in this way to an understanding, or rather a non-understanding of the infinite God, we must first take a normal mathematical figure and imagine it as extended infinitely. Cusanus begins with the notion of an "infinite line," and provides a series of arguments that such a line will be at once a straight line, a triangle, a circle, and a sphere (for example: as the diameter of a circle increases, its circumference tends to be straighter; or, a finite line can generate a triangle, a circle, which can, in turn, generate a sphere; hence an infinite line, which by definition must include all potentialities of a finite line, includes these figures as well). In the situation of the infinite, as we see, seeming contradictories become one in a quasi-sensible manner, just as all opposites coincide in the divine. In fact, the doctrine of the trinity is suggested by his exploration of the infinite triangle. The most adequate expression of this symbolism is the influential metaphor from the hermetic tradition which Cusanus adapts: "God is an infinite circle whose center is everywhere and whose circumference is nowhere." We should note here that Cusanus does not take infinity to be a mathematical concept (there is no numerical maximum, [Cusanus in Hopkins 1990, pp. 63, 54]), but through the introduction of this philosophical concept, mathematics can help make the divine more in/comprehensible, because mathematics itself as a construction of reason is the most precise form of knowledge.

Two hundred and fifty years later, Leibniz took up the challenging problems and results of Cusanus, and, trying to eliminate paradox from his thought by means of deeper mathematical investigations, he became one of the fundamental sources of the German Enlightenment. Like Cusanus in his involvement in the political world, Leibniz was a polymath, involved in diplomacy, philosophy, and a variety of scientific fields, often planting the seeds for intellectual developments which would only come to fruition in the next few centuries, as well as producing his own philosophical system of optimistic rationalism.

For Leibniz, calculus as a mathematical technique involving infinitesimals was intimately connected with a metaphysics in which the role of the infinitely small—that is, smaller than any quantity which could be given—featured prominently as the infinitely detailed structure of the universe which could be grasped by reason and calculation. Thus, unlike Cusanus, Leibniz's mathematics involved the use of infinite quantities (here, infinitely small ones), though Leibniz saw such small quantities as useful fictions rather than mathematically real (Katz and Sherry 2013, p. 571). Leibniz's metaphysical theories carried further many ideas from Cusanus, including the notion that each individual existent in the universe reflects in its own way the universe as a whole according to its rational capacity. Where Leibniz differs from Cusanus is in his highly developed mathematical notions specifically related to the continuum. The coincidence of opposites is relaxed into a more general notion of the law of continuity in nature, so that one concept, such as rest, appears as a limiting case of its apparent opposite, motion. However, this mathematics of the infinite still remained symbolic. A thoroughly mathematical, hence geometrical and mechanical approach to nature, as in Descartes's theory of matter which attempted to reduce all physical phenomena to the principles of size, shape, and motion, did not appeal to Leibniz. Rather, he sought to supplement mathematics with the idea of "force," in fact, an infinite number of elementary forces (Iltis 1971, p. 23). In all these infinities—the infinite temporal progression of each infinitesimal being, the infinite combination of infinite creatures into perfect harmony, and the even more ungraspable infinity of all of the possible combinations of all possible creatures among which God must select the most commonly possible, the very best of all possible worlds—there is correspondingly the need for an infinite rationality, transcending in power though not necessarily in principle human reason, to comprehend and scrutinize all different combinations and possibilities.

If, for Cusanus, the mathematics of the infinitely large provided an appropriate model for the ungraspable infinity of the divine, for Leibniz, the infinitesimal calculus provided a model (if not a literal one) of a universe which is put together in a metaphysically understandable manner, that is, assembled from simple elements. As for Cusanus, mathematics remained for Leibniz the paragon model for all of knowledge, and Leibniz tried to expand the method which worked here so well—analysis—to all fields of inquiry, coming up with the idea of a logical calculus which could reduce all theoretical questions to a matter of calculation. In this way, he expanded Descartes's idea of a "universal mathematics" into a generative encyclopedia of all knowledge (see Rutherford 1995; Neubauer 1978).

Leibniz set the tone for eighteenth-century philosophy in Germany through the school of the philosopher Christian Wolff (1679–1754), who systematized Leibniz's scattered publications into a system which in true Enlightenment fashion was focused more on practical worldly wisdom than abstruse speculation. Mathematics as a research discipline was not taught extensively in eighteenth-century Germany, hence the appearance of mathematics in the literature of the Enlightenment was not pervasive, except as a model of order and structure which even poetry should attempt to imitate (Radbruch 1997, p. 40). The notion of the infinite appears in literature not in its mathematical form, but rather in the *topos* of the sublime. The scientist and poet Albrecht von Haller (1708–1777, most famous today for his physiological investigations into the concepts of irritability and sensibility) provides an example. His "Uncompleted Poem on Eternity" (1736) continues some of the despair in the seventeenth-century Baroque era of the contrast between the finite and the infinite, with the human mind still unable to conquer the incomprehensibility of infinite eternity:

> I heap up giant numbers
> Pile millions on millions;
> Eon upon eon and world upon world,

And when I am on that endless march
And dizzy on that terrifying height
I seek you again.
The power of numbers, though multiplied a thousand fold,
Is still not even a fraction of you.(Zweig 2002, p. 309)

Mathematics here appears as an insufficient instrument to help our imagination grasp that eternity which swallows all that is human.

In his survey of infinity in eighteenth-century Germany, John H. Smith argues that the infinite becomes a theme of, especially religiously toned, literature, and that, as the century progresses, the infinite becomes progressively "normalized" (Smith 2019). We can see an illustration of this process of normalization in a central figure in the second half of the eighteenth century, the Göttingen mathematics professor Abraham Kästner. He was celebrated for his dual achievements in both poetry and mathematics, though Gauss, one of his later successors in Göttingen, said of him with some irony that he was "among poets the best mathematician, among mathematicians the best poet" (Cantor and Minor 1882, p. 446, my translation). Though forgotten in literary history, his epigrams reproduce standard eighteenth-century tropes and show little influence of his mathematical teachings. Ironically, it is in his mathematical works where he comes closest to a literary treatment of the mathematical infinite. His *Elements of the Analysis of the Infinite* (*Anfangsgründe der Analyse des Unendlichen*, 1761), a popular calculus textbook of the time, begins with a preface in which he declares his intent to dispel all the mysteries which appear to be connected to the use of the infinite. These mysteries, for example that different quantities are actually the same, or that something infinitely large can be infinitely small in comparison to something else, or that curves can be made of straight lines, arise only when the infinite is taken as a real thing rather than a process of approximation. The use of the infinite, he writes, as masterfully accomplished by the genial Leibniz, is akin to the use of the "bold expressions of the poet" which serve to abbreviate the longer mathematical proofs of the philosopher (Kästner 1761, p. iii). Indeed, he is moved in his preface to produce a Latin epigram, making an analogy between the rule which allows one to generate an arbitrary term of an infinite series from the first few terms and a stern admonition against both the atheist and the overly compassionate soul that one's life on earth will determine one's fate in eternity. Unlike Haller, the infinity of eternity does not overwhelm him; it can be overcome via the mathematical rule of the infinite which aligns with rational divine justice. In this way, the Enlightenment, beginning with Leibniz, loses the tension of the mathematical infinite present in Cusanus and simultaneously deprives it of its metaphysical overtones. The Romantic generation will rediscover the unruliness of infinity.

EARLY GERMAN ROMANTICISM: NOVALIS AND FRIEDRICH SCHLEGEL

A high point for the significance of mathematics for German literature was attained in the early Romantic movement,[4] including the writers Novalis (1772–1801, the pen-name of Friedrich von Hardenberg), Friedrich Schlegel (1772–1829), and others. In the English-speaking world, German Romanticism is often associated with a rebellion against a mathematical-scientific worldview, associated for example with the anti-Newtonian *Theory of Colors* by the central German literary figure of the time, Johann Wolfgang von Goethe (1749–1832). In fact, many of the German Romantics were well informed about natural science. Indeed, the Romantic generation developed a speculative theory called *Naturphilosophie*, which attempted to integrate the new scientific revolutions, especially in the emergent fields of chemistry and biology, into a unified theory of nature. *Naturphilosophie* achieved important results but was later thrown into disrepute by approaches which emphasized quantitative experimentation and were critical of the philosophical methods of *Naturphilosophie*.[5] The involvement of Romantic-generation poets in science was not merely speculative; both Goethe and especially Novalis were involved with the administration of mines.

In its explorations of mathematics, German Romanticism was closely dependent on contemporary philosophy, particularly the critical philosophy of Kant. Kant, in his most important work, *The Critique of Pure Reason* (1781/1789), begins with the aim to produce a science of metaphysics which can be as reliable and applicable as ancient geometry and Newtonian physics. He attempts to accomplish this by means of his self-styled "Copernican Revolution" which directs intense scrutiny on the sources of knowledge and its claims to validity, thereby determining the limits of what can be certainly known. Though not himself a productive mathematician, Kant wrote extensively about the nature of mathematical knowledge as well as the centrality of mathematics in the more general scientific understanding of the world. As he writes in the preface to his *Metaphysical Foundations of Natural Science* (1786): "a doctrine of nature will contain only as much proper science as there is mathematics capable of application there" (Kant 2002, p. 186). Key for Kant's understanding is a particular notion of construction, unique to mathematics: "to construct a concept means to exhibit a priori the intuition corresponding to it" (Kant 1998, p. 630). For Kant, the manner of this construction was still largely geometrical and Euclidean, as in the drawing of geometrical figures in a kind of spatial imagination which could achieve universality and necessity; it was left to his followers to integrate more contemporary versions of mathematics. In Kant's view, mathematics becomes a uniquely human science of self-generated constructions within the particular a priori forms of human intuition (space and time). Of fundamental importance for Kant was the distinction between mathematical and philosophical method, in contradistinction to the Wolffian school, which did not recognize a distinctive place for intuition in mathematical reasoning. Theoretical examinations of reality, as in philosophy, can only be done by means of the senses and can have nothing to say about things in themselves, namely, what things would be like independent of human sensibility. Hence, an entire section of the *Critique* is devoted to refuting one by one all of the historical attempts to prove the existence of God by speculative means. God, Kant says, cannot be constructed a priori and hence cannot be proven to exist theoretically. In other words, a mathematics of God is impossible for Kant.

Several of Kant's students attempted a closer integration of his theory with the methods of calculus, for example the self-educated Salomon Maimon, who tried to equate things-in-themselves with infinitesimals as an echo of Leibnizian metaphysics.[6] On the other hand, Kant's later successors, the German idealists Johann Gottlieb Fichte (1762–1814) and Friedrich Schelling (1775–1854), attempted more than Kant to bring the "constructive" method of mathematics into philosophy, as for example evident in Fichte's construction of the self through intellectual intuition as a foundation for an entire philosophical system (see Breazeale 2013). As with Kant though, their view of mathematics was largely shaped by Euclidean geometry and its methodology (see Wood 2012).

Among the Early German Romantics, the poet Novalis had the best knowledge of mathematics. While there is evidence that he studied mathematics at the University of Leipzig (Uerlings 1991, p. 179), most of his education in scientific and mathematical studies was a result of his occupation as an administrator in the salt mines of his native Thuringia, for which he attended the Mining Academy in Freiberg for a year. Novalis died young, and left behind only several collections of poetry (including *Hymns to the Night* and *Spiritual Songs*) and two unfinished novels: *Heinrich von Ofterdingen*, which became one of the signature works of the Romantic movement, and *Disciples of Sais*, a mystical version of his studies at Freiberg. He is, however, equally known for his fragments and notebooks (during his life, he published two sets of fragments, including *Grains of Pollen* which appeared in the *Athenäum* in 1798). The most important notebook for our purposes is the *Allgemeine Brouillon*, a series of notes written during his time at the Mining Academy as part of a project to produce an encyclopedia.[7] The goal of this encyclopedia was not so much to exhaustively classify and outline, but rather to show the interconnection of all forms of knowledge. Even in their edited form, the fragments are difficult to decipher and put together; many are paraphrases, sometimes heavily edited, of books Novalis read; many are idea experiments which are taken up or dropped later. Mathematics plays a significant part in the 1151 notes that make up the collection.

In this volume, two aspects of mathematics are especially notable for Novalis's suggestive use of them. First, Novalis was interested in broader interpretations of mathematical procedure. As he notes while reading Kant, the constructive procedure used in mathematics could be expanded to other fields of inquiry:

With regard to the essential, individual character of the mathematical method, Kant maintains that the mathematician is not discursive like the philosopher – but proceeds intuitively – he doesn't infer from concepts, but constructs his concepts – presents them in a sensible manner – yet actively sensible – or forms pure intuitions.

(Here too I believe that the mathematician's procedure is not unique. He sculpts the concepts in order to fix them ... Shouldn't the philosopher do likewise – or even every scientific expert?) (Novalis 2007, p. 205)

As emphasized by Hans Niels Jahnke, Novalis's view of mathematics was heavily influenced by the combinatorial school which stressed an algebraic rather than an arithmetic approach, and whose most fundamental result was the binomial theorem (Jahnke 1991, pp. 283–84). Indeed, generalizing about mathematical constructions, Novalis writes:

Especially interesting is the philosophical study of hitherto purely mathematical concepts and operations – with powers, roots, differentials, integrals, series – *curves* – and *direct* – functions. The *binomial theorem* – with regard to polarities etc. – could yield a far higher significance. (Novalis 2007, p. 18)

Here, the construction procedure of mathematics involves not so much the mental drawing of geometric diagrams, but the so-called analytic approach, which involves the use and manipulation of a symbol for the unknown, x. Such a symbol, for Novalis, is already a fictional claim that we know the unknown and can work backwards from this assumed knowledge (just as in his other writings, he suggests that a utopian project might itself be motivated by the fictional description of a utopian state, see Novalis 2007, p. 144). Thus, the primary example of such construction is not through spatial diagrams but through symbolic expression. There are echoes of such thought in Novalis's magical theory of language, which concerns the power of the word to summon that which it represents. The parallels between mathematics and language, specifically poetry as a shaped language, are most apparent in another short text written during the time at the Mining Academy (Mahoney 2001, p. 81), the "Monolog":

If only one could make people understand that language works like mathematical formulas – they form a world for themselves – they play only with themselves, express nothing but their own marvelous nature, and thus they are so expressive – thus they reflect in themselves the peculiar play of relationships of things. Only through their freedom are they part of nature and only in their free movement does the world soul express itself and make them into a delicate measure and ground plan of things. (Novalis 1960, 2, p. 672)

Here, the constructive power of mathematics is mixed with the combinatorial (from the Leibnizian tradition of the *ars combinatoria*) and leads in many

17 MATHEMATICS IN GERMAN LITERATURE: PARADOXES OF INFINITY   309

directions, toward avant-garde poetry, for example, or toward a formalist view of mathematics. Even the very definition of the romantic in one of Novalis's most famous notes (written before his studies in Freiberg) is depicted via mathematical concepts:

> The world must be romanticized. Then we will rediscover its original meaning. Romanticization is just a qualitative exponentiation. In this operation, the lower self is identified with a better self. We ourselves are just such a qualitative exponential series. The operation is still entirely unknown. If I give what is common a higher sense, what is usual a mysterious appearance, what is familiar the dignity of the unknown, what is finite the look of the infinite, I romanticize it. – This operation functions in reverse for what is higher, unknown, mystical, and infinite – it is logarithmized through this connection. – It acquires an everyday expression. Romantic philosophy. Lingua romana. Reciprocal raising and lowering. (Novalis 1960, 2, p. 545; my translation)

The Fichtean construction of the self (where the self, by a series of self-reflections, attains a higher intuition of itself) is here seen as a process of expansion in a power series, based on the basic techniques of mathematical analysis, where a function is broken up into a series of exponential powers.[8]

The second aspect of mathematical thought Novalis takes up, also present in this last note, is the idea of the infinite—whose very definition, as well as its usefulness as an abbreviation in mathematical applications puts it into a poetic space which lies between fictionality and reality. Of greatest interest to Novalis is the plurality of approaches to the foundations of calculus and its echo in other realms of thought:

> The difference in the Leibnizian and Newtonian manner of conceiving the infinitesimal calculus rests on the same foundation as the difference between the atomistic theory and the vibration or etheric theory. ... The affinity of geometry and mechanics with the loftiest problems of the human spirit, shines forth from the atomistic and dynamic sectarian strife. (Novalis 2007, pp. 117–118)

Novalis continues the thought of Cusanus in seeing the infinite as involving a coincidence of opposites. Here, however, the opposition is at the level of the theories which explain the operational usefulness of infinitely small quantities in obtaining the results of calculus, rather than in the infinite quantities themselves. The "loftiest problems of the human spirit" are those which Kant also addressed in his *Critique of Pure Reason*, but for Novalis the progress of mathematics suggests, as for Cusanus, the possibility of insight into that which goes beyond reason:

> The *proofs* of God are perhaps worth something *en masse* – as *method* – Here God is something like ∞ in mathematics – or $0°$. (Degrees of zero) (Philosophy of 0.) (God is now $1 \cdot \infty$ now $1/\infty$ – now 0) (Novalis 2007, p. 166)

Here, Novalis suggests, unlike Kant who went through proofs of God one by one, dismissing each, that it is not so much the validity of each proof which is the point of greatest importance, but rather the very plurality of different types of proof, each of which perhaps could shine a different light on that which cannot be comprehended by finite rationality. Novalis died before he was able to implement his literary plans in the second, incomplete part of *Heinrich von Ofterdingen* to "poeticize all the sciences, even mathematics," which was to be Heinrich's last act on earth (Novalis 1960, 1, p. 367).[9]

Novalis's friend and "symphilosopher," the author and critic Friedrich Schlegel, wrote: "Whoever desires the infinite doesn't know what he desires. But one can't turn this sentence around," implying that a precise understanding of the incomprehensibility of the infinite exists (Schlegel 1971, p. 149). Schlegel, author of many literary and philosophical texts, is best known for his theory of Romantic poetry, which in the famous 116th *Athenäum*-fragment he defines as a "progressive universal poetry. … its real essence [is] that it should forever be becoming and never be perfected. … It alone is infinite" (Schlegel 1971, p. 175).[10] Schlegel's interest in mathematics was motivated by his friendship with Novalis, and in his philosophical writings he attempted to apply the mathematics of the time to help explain the numerous references to the infinite which are contained in Novalis's philosophy. In particular, as John H. Smith argues, the mathematical theory of the infinite, with its contradictions and ambiguities helped him to conceptualize—not eliminate—the dualisms endemic to post-Kantian thought through the notion of infinite approximation.[11] Also significant is Schlegel's use of mathematical symbols in his notes, with frequent use of the fraction $1/0$ to indicate the infinite or absolute, as in his formula for Romantic poetry:

$$\text{The poetic ideal} = \sqrt[1/0]{\frac{FSM^{1/0}}{0}} = \text{God (Schlegel 1958, 16, p. 148).}[12]$$

The early Romantics, then, with their visions of interconnectedness and their promotion of the "magic wand of analogy" (Novalis 1996, p. 72), insisted again on the relevance of even the symbolic or technical sides of mathematics, especially where these, like the use of the infinite, seem themselves to have extra-mathematical significance. On the other hand, their treatment of mathematics had more to do with the theoretical consideration of literature than as a thematic part of literary works themselves.

Mathematics and Modernism: Musil and Broch

In the later Romantic period, and later in the nineteenth century, the German literary interest in the mathematics of the infinite notably declined, perhaps connected to the move toward more empirical and specialized fields of the natural sciences. Mathematics was to reemerge as an important theme in literary modernism at the start of the twentieth century. Indeed, some historians of modernism even claim that "the story of Modernism begins with German mathematicians," from there becoming a broad cultural movement (Everdell 1997, p. 11).[13] Mathematics in early twentieth-century German literature is most prominent in the works of two novelists, each of whom had specialist training in the field: Robert Musil (author of "The Mathematical Man," quoted at the beginning of this chapter) and Hermann Broch (1886–1951). Both of these figures are concerned with the relationship between recent developments within the field of mathematics and larger cultural questions of value.

Musil first studied engineering in Austria and then received a doctorate in philosophy and psychology in Berlin, yet, after serving in World War I, ultimately abandoned academics for a career as a writer and critic (for more biographical information, consult Payne 2007). For Musil, as we have seen above from his essay on "The Mathematical Man," mathematics is useful precisely in the attempt to reconnect what he later calls the "ratoïd" (rational) and the "non-ratoïd" (mystical) sides of life, a theme developed at greatest length in his magisterial, unfinished novel *The Man Without Qualities* (*Der Mann ohne Eigenschaften*; 1930/1933/1943), which describes Austria on the eve of the war and the general cultural crisis in Europe.

Most prominently, mathematics takes on an important supporting role in Musil's first novel, *The Confusions of Young Törless* (*Die Verwirrungen des Zöglings Törless*, 1906). Mathematics, here, as in his essay, symbolizes the groundlessness which lies beneath the ordinary world. Törless, a student at an elite boarding school, is confronted with what he sees as the intersection of mutually exclusive planes of existence: in his family situation (his correct and proper parents also being intimate lovers), the situation of society (the harsh, sadomasochistic underbelly of military discipline, shown in the sexual abuse of a classmate), and in human existence in general (the limited reach of rationality and logical expressibility in confrontation with the abysses of natural and social worlds). In his confusions, he turns to mathematics. Two mathematical concepts particularly intrigue him: imaginary numbers and the use of the infinite in calculus. Törless does not understand how these seemingly absurd quantities (which cannot exist, but which are given names) can be used for practical calculations. It is, for him, as if the mechanisms of the world are based upon quicksand:

"Infinity!" Törless knew the word from maths lessons. It had never meant anything special to him. It kept on cropping up; someone or other had invented it at some time in the past, and since then it had become possible to perform calculations using it that were as reliable as those using anything fixed. … And now it hit him all at once that there was something terribly disturbing about the word. He felt it was a tamed concept with which he had daily performed his little tricks and which had now suddenly been unleashed. Something that went beyond understanding, something wild, destructive, seemed to have been put to sleep in it by the work of some inventors and had now suddenly awoken and become terrible again. (Musil 2014, p. 68)

312   H. POLLACK-MILGATE

When Törless approaches the mathematics teacher, he is unable to give Törless a satisfactory account of the use of the infinite or complex numbers, saying only that Törless will understand it after he has become more advanced in mathematics, an explanation which seems like a mere evasion (or, as a fellow student suggests, of a suggestion of future indoctrination). The result here, as in the later essay "The Mathematical Man," is that the mathematics of the infinite provides a stepping stone for those truly destined to become poets, who later use the sensitivity demonstrated in these logical puzzles for more creative ends, to explore artistically possibilities which go beyond apparent realities.[14]

Another Viennese modernist author, sometimes conversation partner, sometimes rival to Musil, Hermann Broch was born and lived most of his life in Vienna, before leaving Austria in 1938 in the wake of the Nazi annexation to emigrate to the U.S. Broch is best known for his novel trilogy *The Sleepwalkers* (*Die Schlafwandler*, 1931/1932), detailing the "disintegration of values" (Broch 1996, p. 373) in the modern world, and his later novel published in exile, *The Death of Virgil* (*Der Tod des Vergil*, 1945). Broch had initial training in mathematics as part of his business education and later abandoning work in his family's factory, decided to reenroll at the University of Vienna in 1928 to receive a doctorate in philosophy and mathematics a program he never completed (Cliver 2019, p. 111). Broch's interests in mathematics were primarily focused on the problem of foundations which had become acute in the 1920s, especially in the German-speaking world where a bitter quarrel arose between the "formalists" (led by David Hilbert) and the "intuitionists" (led by L. E. J. Brouwer, 1881–1966), who rejected certain uses of the mathematical infinite. For Broch, these mathematical questions had a much wider application: the inability to come up with a certain foundation for mathematics was for him intimately related with a crisis of values in general, especially poignant as he watched the growth of the fascist movements around him.[15] In fact, many of his mathematical works were destroyed by the Nazis after they came to power in Austria, and only partially recreated in his ensuing American exile.

Of all his works, mathematics has the most central role in Broch's novel *The Unknown Quantity* (*Die unbekannte Größe*, 1933). The main character is the mathematics student Richard Hieck, whose familial and academic life are depicted over the course of a year. Richard lives at home with his mother and two of his younger siblings, Otto and Susanne. Each of these characters has reacted to the complicated symbolic ambiguity of their parents: Richard's mother has a joyous sensuality and a love of nature, while his father, who died when Richard was young, was fascinated by the darker sides of life: night, shadows, confusion. Otto, the youngest in the family, wants to be an artist, but for lack of money is forced to work as a graphic designer and thinks of stealing money from his mother. Susanne, Richard's younger sister, has devoted her life to religion and wants to become a nun. Richard's study of mathematics is also a sign of resistance against the darkness of his father's path. For Richard, mathematics and its "most wonderful achievement," the theory of groups, is a "crystalline landscape," "that complicated, infinite, and boundless structure of balanced forces which is built up out of nothing but the relations of things to each other" (Broch 2000, pp. 18–19). It points to moments of illumination among the darknesses which dominate existence.

The ultimate goal of mathematics for Richard is a systematic knowledge of the universe, yet, as Richard realizes when confronted with his own clumsy physicality, mathematics only addresses part of the world, namely, the world of the head and not the world of the body. Though mathematics is certainty as such, when applied to reality through the laws of physics, it becomes merely a realm of probability which always leaves open the space for the "incalculable." Unlike his colleague at the university who is content with merely solving whatever problem is in front of him, for Richard, the "ultimate justification of mathematics lies beyond mathematics and yet in it." Like the senior professor Weitprecht who, facing his own mortality, finds the one-sided commitment to knowledge a sort of "wrong" and "evil," mathematics must itself strive to overcome its "isolation," must attempt to include the incalculable. Two events urge Richard on in this direction. The first is the suicide of his brother Otto. The second is his blossoming love for a fellow student, Ilse Nydhalm. In the course of the book, Richard realizes that his love is not necessarily distinct from his theoretical aspirations, moving from an insistence that the mathematician must not be sullied by the "sin" of sex to thoughts of Marie and Pierre Curie who combined research with love. If mathematics is the knowledge of the "holiness of life"; love also delivers a knowledge of the "holiness of death," the latter "valid without proof" but nonetheless the "final proof of the logical, which only from this point finds its justification" (Broch 2000, pp. 160, 147, 155, 169, 176, 169).[16] In different ways, then, Musil and Broch approach the question of those parts of the world which cannot be described mathematically, linking them in both cases to the tenuous bases of mathematics itself.

Mathematics in Contemporary German Literature: Kehlmann

Recently, a new trend in literature that makes mathematicians central characters in literary works has emerged (Albrecht 2011, p. 547). One good example is *Measuring the World*, a worldwide bestseller by the contemporary Austrian writer Daniel Kehlmann (1975–). *Measuring the World* presents a scurrilous view of the historical world of Germany in the early nineteenth century through the contrasting figures of the mathematician Gauss and the naturalist Alexander von Humboldt (1769–1859) on his explorations in South America and beyond. Each character tries to measure the world in his own way. The ill-tempered but brilliant Gauss arrives at his results through theoretical work which gives him insight into the fundamentals of existence without leaving his writing desk or becoming aware of the tumult of the Napoleonic world around him. He muses that "science is a man alone at his desk" (Kehlmann 2006, p. 212), though he is ironically forced for financial reasons to conduct land surveys in Germany. Humboldt travels the world, obsessively measuring everything in his attempt to bring Kantian morality and German order to the Amazonian wilderness. Though sarcastically exaggerated, the book is based on the true stories of the historical figures, mapping with invented, indirect dialog Humboldt's actual itineraries through the Andes (and later, through Russia), and including Gauss's achievements in number theory and as his most important, but unpublished theory: non-Euclidean geometry, specifically the curvature of space and time.

> For Gauss, deep mathematical knowledge has a tinge of blasphemy:
> Was he digging too deep? At the base of physics were rules, at the base of rules there were laws, at the base of laws there were numbers; if one looked at them intently, one could recognize relationships between them, repulsions or attractions. Some aspects of their construction seemed incomplete, occasionally hastily thought out, and more than once he thought he recognized roughly concealed mistakes – as if God had permitted Himself to be negligent and hoped nobody would notice. (Kehlmann 2006, p. 73)

It is a mathematics which detects the defects of the creation and the imperfection of the world, against the backdrop of the antiquated social and physical conditions of Germany in the first decades of the nineteenth century. Gauss, who hates the physical discomfort of travelling, makes the long journey to Königsberg to see Kant, one of the few contemporaries he thinks can understand his new thoughts about space and time, but Kant is too senile to understand any of his theory. In another episode (modeled on Franz Kafka's novel *The Castle*), Gauss, on one of his surveying trips, meets God in the character of the Count von der Ohe zur Ohe. Gauss is insightful enough to realize that he is a fictional caricature of his historical self and seeks an exit:

> Death would come as a recognition of unreality. Then he would grasp what space and time were, the nature of a line, the essence of a number. Maybe he would also grasp why he always felt himself to be a not-quite-successful invention, the copy of someone much more real, placed by a feeble inventor in a curiously second-class universe. (Kehlmann 2006, p. 242)

Yet, for once, the character Gauss does not grasp the true identity of Count von der Ohe zur Ohe, who admits that even he has learned something from Gauss's book. Most importantly, the book itself shapes its narrative time in congruence with Gauss's insight that all parallel lines meet through the inherent curvature of time and space: the parallel endeavors of Humboldt and Gauss converge, as they contemplate the immeasurability of death as a realm beyond number.[17]

Kehlmann's postmodern novel provides a fitting conclusion to this survey of the rich tradition of mathematics in German literature. If the infinite provides the link between the precision of mathematical reasoning and our in/ ability to comprehend God for Cusanus, for Leibniz it is the calculable rule at the basis of an infinitely comprehensible universe and the endpoint of an eternal optimism of gradual progress and providence. In turn, the Romantics, spurred on by the ultimate groundlessness of the endlessly meaningful play of mathematical symbols, again make the infinite the symbol of the incomprehensible, here in the form of the unfathomable openness of possibility, in the infinity of romantic poetry. For the moderns, the infinite figures the abysses in the depths of the world, visible in the disturbing mysteries of sex and death, against the experience of a World War, and, finally, demonstrating the lack of transcendence in a culture without a center. Kehlmann's Gauss will leave us with the final word: Gauss tells Humboldt that one thinks one determines the course of one's own life, but in the light of statistics and probability theory applied to mortality, it is clear that one is only the plaything of mathematical laws. When Humboldt replies that these laws are themselves shaped by reason, Gauss says:

> The old Kantian nonsense. Gauss shook his head. Reason shaped absolutely nothing and understood very little. Space curved and time was malleable. If one drew a straight line and kept drawing it further and further, eventually one would reencounter its starting point. He pointed to the sun which hung low in the window. Not even the rays of this dying star came down in straight lines. The world could be calculated after a fashion, but that was a very long way from understanding it. (Kehlmann 2006, p. 187)

NOTES

1. We can immediately add August Ferdinand Möbius (1790–1868), Karl Weierstrass (1815–1897), Richard Dedekind (1831–1916), David Hilbert (1862–1943), and Emmy Noether (1882–1935) whose names live on for those with interests in mathematics.
2. Works in German include Radbruch 1997 and the collection edited by Albrecht (see Albrecht 2011). John H. Smith has authored articles on a variety of figures. Particular references will be given as each author is discussed.
3. Nicholas of Cusa is the first figure in Alexandre Koyré's famous work *From the Closed World to the Infinite Universe* (Koyré 1958).
4. Early German Romanticism is most often dated from 1797 to 1801 and associated with the city of Jena in central Germany, and the journal *Athenaeum* (1798–1800), edited by Friedrich Schlegel and his brother August Wilhelm Schlegel. The early romantics are celebrated not so much for their literature but for their theory of literature, although more recently arguments have been made that their philosophical importance was much broader (see, for example, Beiser 2003).

5. For a concise summary of *Naturphilosophie* see Beiser (2002, 506–28).
6. See Smith (2014), which particularly considers the influence of Maimon.
7. They have been published in English under the title: *Novalis: Notes for a Romantic Encyclopedia.*
8. See again Jahnke (1991) and Pollack-Milgate (2015) for further connections to the combinatorial school.
9. In some of his later notes, referred to sarcastically by Dilthey as the "hymns to mathematics," Novalis sets up a mystical ideal of the mathematician as magician; these notes clearly differ from the notes about mathematics as such, with statements such as: "There can be mathematicians of the first rank who cannot do calculations"; "Only the mathematicians are truly happy. The mathematician knows everything. He could do it, if he did not know it." (Novalis III, pp. 593–94). These notes perhaps augur a change from writing about mathematics to those who do mathematics.
10. Schlegel adds, in parallel to Novalis's talk of exponentiation: "And it can also — more than any other form — hover at the midpoint between the portrayed and the portrayer, free of all real and ideal self-interest, on the wings of poetic reflection, and can raise that reflection again and again to a higher power, can multiply it in an endless succession of mirrors" (Schlegel 1971, p. 175).
11. John H. Smith has discussed in particular the use of mathematical paradoxes of the infinite in his article "Friedrich Schlegel's Romantic Calculus" (Smith 2014).
12. Here, $F/0$ is absolute fantasy, $S/0$ is absolute sentimentality, and $M/0$ is absolute mimesis. For an attempt to interpret this formula, see Eichner (1956).
13. Everdell claims in particular: "Separating forever the digital from the continuous, at least in arithmetic, Dedekind became the West's first Modernist in 1872" (Everdell 1997, p. 30).
14. For more on mathematics in Musil, see Genno (1986), McBride (2006).
15. See, for example, the discussion of the "antinomies of the infinite" in the excursus on the "Disintegration of Values" in Broch (1996, p. 481), and Könneker (1999a, p. 344).
16. Articles on Broch and mathematics in *The Unknown Quantity* are Könneker (1999b) and Cliver (2013).
17. See Andersen (2008) for a discussion of the mathematics in *Measuring the World.*

References

Albrecht, Andrea. 2011. "'Spuren menschlicher Herkunft': Mathematik und Mathematikgeschichte in der deutschen Gegenwartsliteratur (Daniel Kehlmann, Michael Köhlmeier, Dietmar Dath)." In *Zahlen, Zeichen und Figuren: Mathematische Inspirationen in Kunst und Literatur*, edited by Andrea Albrecht, Gesa von Essen, and Werner Frick, 543–63. Berlin: De Gruyter.

Andersen, Mark. 2008. "Der vermessende Erzähler: Mathematische Geheimnisse bei Daniel Kehlmann." *TEXT+KRITIK* 177: 58–67.

Beiser, Frederick C. 2002. *German Idealism: The Struggle Against Subjectivism, 1781–1801.* Cambridge: Harvard University Press.

—. 2003. *The Romantic Imperative: The Concept of Early German Romanticism.* Cambridge: Harvard University Press.

Breazeale, Daniel. 2013. *Thinking Through the Wissenschaftslehre: Themes from Fichte's Early Philosophy.* Oxford: Oxford University Press.

Broch, Hermann, Willa Muir, and Edwin Muir. 1996. *The Sleepwalkers: A Trilogy.* New York: Vintage Books.

—. 2000. *The Unknown Quantity.* Translated by Willa and Edwin Muir. Evanston: Marlboro Press and Northwestern.

Cantor, Moritz and Jacob Minor. 1882. "Kästner, Abraham Gotthelf." *Allgemeine Deutsche Biographie* 15: 439–51.

Cliver, Gwyneth. 2013. "Landscapes of the Mind: The Spatial Configuration of Mathematics in Hermann Broch's *Die Unbekannte Größe.*" *Seminar: A Journal of Germanic Studies* 49 (1): 52–67.

—. 2019. "Limits of the Scientific: Broch's *Die Unbekannte Größe*" In *A Companion to the Works of Hermann Broch*, edited by Graham Bartram, Sarah McGaughey, and Galin Tihanov, 108–22. Rochester, New York: Camden House.

Eichner, Hans. 1956. "Friedrich Schlegel's Theory of Romantic Poetry." *PMLA* 71 (5): 1018–41.

Everdell, William R. 1997. *The First Moderns: Profiles in the Origins of Twentieth-Century Thought.* Chicago: University of Chicago Press.

Genno, Charles N. 1986. "The Nexus Between Mathematics and Reality and Phantasy in Musil's Works." *Neophilologus: An International Journal of Modern and Mediaeval Language and Literature* 70 (2): 270–78.

Grabiner, Judith V. 2005. *The Origins of Cauchy's Rigorous Calculus.* Mineola, NY: Dover Publications.

Hopkins, Jasper. 1990. *Nicholas of Cusa On Learned Ignorance: A Translation and an Appraisal of De Docta Ignorantia.* Minneapolis: A.J. Banning Press.

Iltis, Carolyn. 1971. "Leibniz and the Vis Viva Controversy." *Isis* 62: 21–35.

Jahnke, Hans Niels. 1991. "Mathematics and Culture: The Case of Novalis." *Science in Context* 4: 279–95.

Kant, Immanuel. 1998. *Critique of Pure Reason.* The Cambridge Edition of the Works of Immanuel Kant. Translated by Paul Guyer and Allen W. Wood. Cambridge: Cambridge University Press.

—. 2002. *Theoretical Philosophy After 1781.* The Cambridge Edition of the Works of Immanuel Kant in Translation. Translated by Gary C. Hatfield, Michael Freidman, Henry E. Allison, and Peter Heath. Cambridge: Cambridge University Press.

Kästner, Abraham Gotthelf. 1761. *Der mathematischen Anfangsgründe, Dritter Teil, Zweite Abteilung. Anfangsgründe der Analysis des Unendlichen.* Vandenhoeck: Göttingen.

Katz, Mikhail G, and David Sherry. 2013. "Leibniz's Infinitesimals: Their Fictionality, Their Modern Implementations, and Their Foes from Berkeley to Russell and Beyond." *Erkenntnis* 78 (3): 571–625.

Kehlmann, Daniel. 2006. *Measuring the World.* Translated by Carol Brown Janeway. New York: Pantheon Books.

Könneker, Carsten. 1999a. "Hermann Brochs *Unbekannte Größe.*" *Orbis Litterarum* 54: 439–63.

—. 1999b. "Moderne Wissenschaft und moderne Dichtung: Hermann Brochs Beitrag zur Beilegung der "Grundlagenkrise" der Mathematik." *Deutsche Vierteljahrsschrift für Literaturwissenschaft und Geistesgeschichte* 73: 319–51.

Koyré, Alexandre. 1958. *From the Closed World to the Infinite Universe*. New York: Harper.

Mahoney, Dennis F. 2001. *Friedrich von Hardenberg (Novalis)*. Stuttgart: Metzler.

McBride, Patrizia C. 2006. *The Void of Ethics: Robert Musil and the Experience of Modernity*. Evanston: Northwestern University Press.

Musil, Robert. 1990. *Precision and Soul: Essays and Addresses*. Translated by Burton Pike and David S. Luft. Chicago: University of Chicago Press.

———. 2014. *The Confusions of Young Törless*. Translated by Mike Mitchell. Oxford: Oxford University Press.

Neubauer, John. 1978. *Symbolismus und symbolische Logik: Die Idee der Ars Combinatoria in der Entwicklung der modernen Dichtung*. Munich: W. Fink.

Novalis. 1960–. *Novalis Schriften: Die Werke von Friedrich von Hardenberg*. Edited by Richard Samuel, H.-J. Mähl, P. Kluckhorn, and G. Schulz. 6 vols. Stuttgart: W. Kohlhammer.

———. 1996. "Christianity or Europe: A Fragment." In *The Early Political Writings of the German Romantics*. Translated by Frederick C. Beiser, 59–79. Cambridge: Cambridge University Press.

———. 2007. *Notes for a Romantic Encyclopaedia: Das Allgemeine Brouillon*. Translated by David W. Wood. Ithaca: State University of New York Press.

Payne, Philip. 2007. "Introduction: The Symbiosis of Robert Musil's Life and Works." In *A Companion to the Works of Robert Musil*, edited by Philip Payne, Graham Bartram, and Galin Tihanov, 1–49. Rochester, New York: Camden House.

Pollack-Milgate, Howard M. 2015. "'Gott Ist Bald 1 · ∞ – Bald 1/∞ – Bald 0': The Mathematical Infinite and the Absolute in Novalis." *Seminar: A Journal of Germanic Studies* 51 (1): 50–70.

Radbruch, Knut. 1997. *Mathematische Spuren in der Literatur*. Darmstadt: Wissenschaftliche Buchgesellschaft.

Rutherford, Donald. 1995. "Philosophy and Language in Leibniz." In *The Cambridge Companion to Leibniz*, edited by Nicholas Jolley. Cambridge: Cambridge University Press.

Schlegel, Friedrich von. 1958–. *Kritische Friedrich-Schlegel-Ausgabe*. Edited by E. Behler, J. J. Anstett, and H. Eichner. 22 vols. Paderborn: Schöningh.

———. 1971. *Friedrich Schlegel's Lucinde and the Fragments*. Minneapolis: University of Minnesota Press.

Smith, John H. 2014. "Friedrich Schlegel's Romantic Calculus: Reflections on the Mathematical Infinite Around 1800." In *The Relevance of Romanticism: Essays on German Romantic Philosophy*. Edited by Dalia Nasser, pp. 239–57. Oxford: Oxford University Press.

———. 2019 "Religion and German Literature, 1700–1770: The Shock and Normalization of the Infinite." In *Religion and Literature in the German-Speaking World 1200–2015*, edited by John Walker and Ian Cooper, 144-203. Cambridge: Cambridge University Press.

Uerlings, Herbert. 1991. *Friedrich von Hardenberg, genannt Novalis: Werk und Forschung*. Stuttgart: Metzler.

Wood, David W. 2012. *"Mathesis of the mind": A Study of Fichte's Wissenschaftslehre and Geometry*. Amsterdam: Rodopi.

Zweig, Arnulf. 2002. "Albrecht Von Haller: 'Uncompleted Poem on Eternity.'" *The Philosophical Forum* 33 (3): 304–11.

CHAPTER 18

The Ghosts of Departed Quantities: Samuel Beckett and Gottfried Wilhelm Leibniz

Chris Ackerley

THE ARCHIVE

The degree to which the writings of Gottfried Wilhelm Leibniz (1646–1716) informed those of Samuel Beckett was largely unrecognized before 1996, when an "archival turn" (Feldman 2013, p. 303) was signaled by the publication of James Knowlson's authorized biography, *Damned to Fame: The Life of Samuel Beckett*, which revealed how Beckett's intense reading and auto-didacticism in the 1930s had generated a largely unrecognized corpus of early writings and informed an aesthetic that shaped the later, better-known works. Knowlson characterized Beckett as a "note-snatcher": one who drew on his sources to create "intellectually complex patterns of ideas and images" (Knowlson 1996, p. 353). Others had intimated the range of Beckett's reading, notably Hugh Kenner and Ruby Cohn, yet this *Life* changed the common perception of Beckett as an ascetic hermit or impenetrable post-modernist to that of one very much the master of the modernist mode of complex allusion. Deirdre Bair's earlier biography *Samuel Beckett* (1978) had revealed much of Beckett's private life, especially as expressed in the hundreds of letters written to Thomas MacGreevy. Bair was pilloried (often justly, sometimes unfairly) for her manifold inaccuracies and intrusions into Beckett's privacy; yet she was among the first to take a serious interest in Beckett's early years and to see these as formative (though she fails to mention Leibniz, whom Knowlson briefly admits, noting Beckett's close reading

---

C. Ackerley (✉)
University of Otago, Dunedin, New Zealand
e-mail: chris.ackerley@otago.ac.nz

© The Author(s) 2021
R. Tubbs et al. (eds.), *The Palgrave Handbook of Literature and Mathematics*, https://doi.org/10.1007/978-3-030-55478-1_18

320   C. ACKERLEY

of him in late 1933 and his active presence in the "German Diaries" of 1936–1937 [Knowlson 1996, pp. 152, 240]).

In 1992, after unseemly legal controversy, Beckett's *Dream of Fair to middling Women* had appeared, with its portrait of a seedy young man who further revealed himself in letters and notebooks (the "Dream Notebook" was published by John Pilling in 1999), and in the pleroma of archival materials first made available in the decade after Beckett's death (1989) at Trinity College Dublin, the Beckett International Foundation at the University of Reading, and the Harry Ransom Research Center at the University of Texas at Austin. The awakening of critical interest in this archive has been called "the empirical turn," as iterated in Matthew Feldman's *Beckett's Books* (2006) to mark the paradigm shift in Beckett studies characterized by the "gray canon": manuscripts, marginalia, drafts, draff, sources, annotations, letters, biographical impedimenta, bibliographical insignia, and post-partum placentae revalued as textual "quantities" (indeed, in genetic criticism, often indistinguishable from them). Holly Philipps charts this archival "swerve" (to appropriate the Lucretian metaphor) in the first chapter ("Locating the Empirical Turn") of her dissertation, "Samuel Beckett and the Nominalist Ethic" (2015), and in an essay derived from this (2016).

Leibniz was "an underestimated presence" (Tonning 2007, p. 17) in Beckett's thought until Jean-Michel Rabaté in "Beckett's Ghosts and Fluxions" (1996a) conjured the calculus, though in its psychological rather than mathematical intensions. Leibniz moved unobtrusively through the post-Cartesian follies of the seventeenth-century nuptials of Philosophy and Philology that intrigued Beckett: Arnold Geulincx (1624–1669), inseparable from his Occasionalism; Baruch Spinoza (1632–1677), in love with his *Ethics*; George Berkeley (1685–1753), torn between *percipere* and *percipi*; Immanuel Kant (1724–1804), categorical and imperative; and Arthur Schopenhauer (1788–1860) glowering over all. Two essays by Garin Dowd, "Nomadology" (1998) and "Mud as the Plane of Immanence" (1999), unmasked the larval consciousness of the "monade nue" in Beckett's *The Unnamable* (1953) and *How It Is* (1964). Intelligent, perceptive, provocative, but refracted through the lenses of continental theory, they drew the wrath of Feldman and the neo-empiricists, who affirmed the scientific methodology of Karl Popper, grounded in principles of empirical verification and validity, and (for Beckett) as mediated through Wilhelm Windelband's *History of Philosophy* (1901), from which he took extensive notes. The debate between Dowd and Feldman, often acrimonious, was for detached spectators valuable and stimulating, not the least of its legacy being a growing recognition of Leibniz within Beckett's philosophical discourse.

My annotations of *Murphy* (*Demented Particulars*, 1998) and *Watt* (*Obscure Locks*, 2005), and entries on "Leibniz" and "mathematics" in *The Grove Companion to Samuel Beckett* (2004) revealed the mind as "pictured" in *Murphy* in terms of Leibniz's monad: "a large hollow sphere, hermetically closed to the universe without" (*Murphy* 1, p. 67).[1] Feldman in *Beckett's Books* (2006) recorded the many Leibniz references in Beckett's "Philosophical Notes" (written in the early 1930s) and showed how his note-snatching (recorded in the "Whoroscope Notebook" of that period) shaped the genesis of *Murphy*. Feldman's further essay, "Samuel Beckett, Wilhelm Windelband and Nominalist Philosophy," appeared in an excellent volume of philosophical essays that he co-edited in 2012. Anthony Uhlmann's *Samuel Beckett in Context* (2013) includes Feldman's essay on "Philosophy," which summarizes these issues; and Hugh Culik's "Science and Mathematics," which plumbs Beckett's nescience from the Greeks to Gödel, but without invoking Leibniz. Another significant book was Erik Tonning's *Samuel Beckett's Abstract Drama* (2007), which reveals how Beckett's reading of Leibniz and his appreciation of Karl Ballmer's "metaphysical concrete" (an aesthetic that Beckett approved for its clear affinities with the monad) infused the later works, as both invisible scaffolding and tangible theme. My three essays, "Samuel Beckett and Science" (2010c), "Monadology" (2012), and "The Geometry of the Imagination" (2013) owe much to Feldman and Tonning, as does Anthony Cordingley's *Philosophy in Translation* (2018), the final chapter of which reads Leibniz against Blaise Pascal and the Enlightenment. This, to date the only substantial attempt to reconcile in Beckett Leibniz's mathematics with (and within) the monad, suggests that Leibniz's calculus of infinitesimals was a specific source for *How It Is*, and one that Beckett drew on for the philosophical concepts that shaped his reconfiguration of an original poetic order (Cordingley 2018, p. 232).

This essay attempts a similar reconciliation, with a different orientation. I offer this overview of the limited critical response to Beckett and Leibniz, to contextualize what follows, as the thrust of this essay is not so much *The Monadology*, which I have treated extensively in the various entries and essays noted above, but rather an attempt to position Beckett's interest in Leibniz with respect to his calculus. Previously, I had intimated the impact of the theory of infinitesimals upon Leibniz's *petites perceptions* (those below the level of consciousness) by drawing attention to Windelband's insistence that these should be considered as "unconscious representations" within the mind (Windelband 1901, p. 462), an insight which I here re-affirm as essential to any understanding of why Leibniz mattered for Beckett. Cordingley endorses this position:

> For Leibniz *petites perceptions* are impressions of the world in the soul, or monad, of which it is not conscious. Leibniz criticized Descartes for failing to account them within his notion of clear and distinct ideas. As imperceptible quantities they express the integration into Leibniz's metaphysics of the infinitesimal quantities of his calculus that are larger than zero but not finite, or smaller in absolute value than any positive real number without being zero. (Cordingley 2018, p. 236)

Windelband's insight, I further contend, underwrites Beckett's appreciation of Leibniz's sense of monads as units of "simple substance": neither abstract like Platonic ideas nor material like the atoms of Democritus, but rather a sense of "neither" (Ackerley 2013, p. 126) that was crucial to Beckett in that (as a mere monad, endowed with appetite and perception) he was neither physical nor metaphysical. This "neither" assumed increasingly a quantum-like simultaneity of "both" and/or "not" (and hence the co-extension of "I" and the "Not-I," and Beckett's life-long sense of an unborn self, a ghostly another). I revisit this vertiginous sense of the unreal self with reference to Berkeley's "ghosts of departed quantities" (Berkeley 1734, #XXXV), but suffice it to say that "neither" (like that other favored word, "perhaps") accounts tolerably for Beckett's sense of the dis-integrated self as a dubious equation of fugitive memories and vestigial experiences that might or might not add up to a life.

This essay tries to anchor Beckett's philosophical themes more firmly in the mathematical *Grundgesetze* (fundamental laws) of their times; but it does not address in detail the precise contribution made by the theory of infinitesimals to the activity of the *petites* perceptions within the monadic mind. Nor does it examine, with reference to *Watt*, how the monad accommodates itself to the rational world, or, in its failure to do so, ends up in the asylum. To do this well would entail a better understanding than I possess of permutations and the mathematics of infinite series, so I can only restate my intuition that in Beckett these led not to a vindication of the pre-established harmony but to the equal and opposite conclusion of a pre-established arbitrary (a phrase used in *Watt*, I, p. 276). This essay offers a hesitant step toward such matters, with its scrutiny of the Western mathematical tradition from the Greeks to Leibniz, and the recognition of how Beckett used Leibniz to gain perspective on a major concern of his times: "Russell's paradox" that threatened to undermine the sovereignty of mathematics by challenging—as irrational numbers had dismayed the Pythagoreans and Berkeley's theological objections to infinitesimals had cursed the calculus—serious attempts made (notably by Bertrand Russell (1872–1970) and Ludwig Wittgenstein (1889–1951) to create an axiomatic and sufficient system of symbolic logic. The fuller calculation must await another day.

BECKETT AND THE CALCULUS

Before embarking, with Alfred North Whitehead, on the unchartered seas of the *Principia Mathematica*, published in three volumes (1910, 1912, and 1913), with a revised edition (1925–1927), and a shorter paperback (1962, the first to earn royalties), Bertrand Russell amused himself with *A Critical Exposition of the Philosophy of Leibniz* (1900). This was new territory, as the standard view of Leibniz was the account, sensible and informative (and as such pillaged by Beckett), by William Sorley, largely unchanged from the ninth edition of the *Encyclopaedia Britannica* (1882) to the fourteenth

(1929). As noted in his preface, Russell was initially skeptical, assuming that Leibniz's *Monadology* (recently edited in English by Robert Latta 1898, with a bow to Sorley) was "a kind of fantastical fairy tale, coherent, perhaps, but wholly arbitrary" (Russell 1900, p. xvii; Ackerley 2012, p. 140). Once he established how the metaphysics derived inexorably from a small set of coherent premises, Russell's appreciation changed radically. The intention of his *Principia* was similar: it is a defense of logicism, that is, the thesis that mathematics in some significant sense is reducible to logic; and an ambitious attempt to ground it in a set of irrefutable axioms to constitute beyond doubt a body of objective knowledge. Russell later described it as an elaborate attempt to prove that $1+1=2$, only to find that the proof was flawed. Mathematics was, he said, "the subject in which we never know what we are talking about, nor whether what we are saying is true" (qtd. in Kline 1953, p. 462). Despite this small deficiency, vaguely reassuring to one of my limited mathematical means, *Principia Mathematica* remains by common consent one of the greatest books on mathematical logic ever written.

The title Whitehead and Russell selected pays homage to an even more celebrated volume, Isaac Newton's *Philosophiae Naturalis Principia Mathematica* (1687). Yet, as Philip Jourdain noted in an early review in *The Monist* (1916), at the heart of the later discourse lies the recognition that logic was the foundation of both the mathematics and the metaphysics of Leibniz, the *Monadology* resting entirely upon those principles (Jourdain 1916, pp. 481, 483). In that issue, one T. Stearns Eliot, exploring the analogy between Leibniz's monads and the "finite centers" of F. H. Bradley's monism, made a like point, calling Leibniz "disquieting and dangerous" (Eliot 1916, p. 566). The *Principia Mathematica* describes in symbolic logic a set of axioms and inference rules from which all mathematical truths might in principle be derived. This is predicated on Gottlob Frege's *Die Grundgesetze der Arithmetik* [Foundations of Arithmetic] (vol. 1, 1884; vol. 2, 1903), Frege acknowledging as master none but Leibniz. Russell discovered in Frege's theory of sets, as volume 2 was going to press, a contradiction thereafter called "Russell's paradox." concerning R, the set of all sets that are not members of themselves: if R is not a member of itself, then by definition R must be a member of itself; but if it is a member of itself, then it contradicts its definition as the set of all sets that are not members of themselves (a restatement of the familiar "Cretan Liar" paradox). This condition, which Russell had identified in Frege, was to cause him profound difficulties.

Kurt Gödel's "incompleteness theorems" of 1931 demonstrated that the Principia Mathematica or any comparable axiomatic system must "necessarily produce true statements that can be neither proven nor disproven"; and that such systems "would necessarily contain contradictions, that is, be inconsistent" (Culik 2013, pp. 354–55). The consequences were, to mix the metaphor, earth-shattering, confounding any easy definition of mathematics as a pure or ideal science, upon which the edifice of Truth might be raised. One clamorous fall from the broken battlements was Ludwig Wittgenstein, whose

Tractatus Logico-Philosophicus (1922) tries to anchor in logic the sludgy relationship of language and reality, to find "atomic propositions that correspond to atomic facts" (Monk 1990, p. 129; cited Ackerley 2016, p. 112). This gave way to the transactional and pragmatic *Philosophical Investigations* (1953) that abandoned any commonality between language and the world, something that Beckett had absorbed not from Wittgenstein but from Fritz Mauthner's *Beiträge zu einer Kritik der Sprache* (1923), from which he took extensive notes. One might note, as a somewhat frivolous footnote to this transition from *Homo sapiens* to *Homo ludens*, that the final words of the *Tractatus*, "Wovon man nicht sprechen kann, darüber muß man schweigen" ("Whereof one cannot speak, thereof must one be silent") match rhythmically the well-known hymn, "Good King Wenceslas looked out / On the feast of Stephen." Their translation also echoes, to my ear, Beckett's cherished dictum of the cardinal virtue of *humilitas* from Arnold Geulincx: "Ubi nihil vales, ibi nihil velis" (Where you are worth nothing, there you should want nothing).

I have chosen this vignette of twentieth-century mathematical history and (with insincere apologies) trivialized the ending, as indicative of a crucial "gap" in Beckett's understanding of mathematics. Critics of Beckett and mathematics generally assume in their subject a mastery of the discipline akin to their own: Reuben Ellis, matrix algebra, to explain the "Matrix of surds!" in the dark zone of Murphy's mind (1993); Hugh Culik, a knowledge of Gödel adequate to the elucidation of Russell's paradox (2013); Brett Stevens, an understanding of the dark calculus and Gray codes that pave *Quad* (2010). Beckett, whose chosen way was that of ignorance and impotence, took one required paper at Trinity on Euclid and Algebra, in which his mark was only average (*Companion*, p. 347).[2] He was not mathematically illiterate, but his was a competence in the elements only: an intelligent awareness but little more. To illustrate this point: Werner Heisenberg's *Physics and Philosophy* (1957) identifies Whitehead and Russell's intention in *Principia Mathematica* as a mode of Realism (in the scholastic sense), one consequence being the rejection of *substance* (Ackerley 2016, p. 111). Beckett, in the 1930s, could not have been familiar with this opinion, but, sensitive to the *Zeitgeist* and its "quantum of wantum" (*Murphy* I, p. 38), he broadly intuited as much. His was a shabby Nominalist irony that paved the only possible philosophical path through the quag of quotidian particulars, but he understood the challenge that Gödel and Heisenberg proposed to the mathematical discourse of his day. Alert to the paradoxes, but lacking the technical mastery to resolve them, he gained perspective on his age by turning to a more familiar past.

This was not untypical. Beckett was fascinated by geology, but by the fossil record, the testimony of the rocks, and the struggle of larval consciousness denied its metamorphosis in the light and so returned to the mineral from which it somehow had arisen (*Companion*, p. 221); he makes no mention of Alfred Wegener, continental drift, or tectonic plates. While accepting evolution, he avoided discussion of it (other than a few jests at Darwin's expense), and remained unmoved by Mendelian genetics and Watson and Crick's decoding of the helical structure of DNA in the 1950s. The biological in Beckett is better explained by a biblical and post-Aristotelian tradition of generation and corruption. He was aware of the radical changes in physics during his early lifetime: relativity, splitting the atom, uncertainty, and/or the quantum universe; yet his *kosmos* is essentially the Newtonian universe (the cosmic billiard-table), with an emphasis on the paradoxes of rationality, motion, and perception that both reset and snooker it. One explanation of this detachment (by choice, not ignorance) appears in the *Ethica* of Geulincx, a sentence defining the little world of Murphy's mind, where alone he might be free: "Sum igitur nudus speculator hujus machinae" ("I am merely the spectator of this machine") (qtd. in Ackerley 2010a, p. 29; the "machine" being the material world of extension). Just as he chose to view the physics of his age, one of uncertainties and quantum flux, by seeing it through the lens of ancient Atomism, the better to gain perspective, so Beckett responded to the most compelling mathematical disjunction of his day by recourse to an earlier theory of infinitesimals, that of Newton and Leibniz, in which the calculus (like the Pythagorean discovery of the surd millennia earlier, or Russell's paradox two and a half centuries later) was perceived as threatening the stability of a theocentric, harmonic, and rational world, in which mathematics, queen of the sciences, was widely held (and to this day often is, the essential flaws within the paradigms of Pythagoras and Russell notwithstanding) to reveal the timeless and eternal Truths of a Platonic world of Ideas.

The calculus is a mathematical analysis of the process of motion and change that, by invoking two complementary operations, differentiation and integration, computes the rate at which one variable changes with respect to another. Newton invented it in 1665, and for the next forty years used it to explain the spectra of light, the movement of heavenly bodies, and the theory of gravity; but from an obsession with secrecy, or perhaps because infinitesimal methods were widely considered disreputable, he did not publish his method. Leibniz, in 1675, invented the calculus independently and published it (1684) before Newton, who, persuaded by Edmund Halley and stung by what he regarded as plagiarism, finally also did so. Newton called it his "science of fluxions," while Leibniz gave the new discipline its accepted name, the Latin *calculus*, "a pebble," invoking its ancestry from the abacus. A vulgar dispute ensued as to precedence, with the consequence that continental scholars, adopting the superior notation of Leibniz, developed it further, while English scholars, "hampered by the less felicitous notation devised by Newton, foundered in a morass of perplexities" (Bergamini 1963, p. 113). Beckett ignores this dispute, instead placing Leibniz and his theory of monads within a broadly Newtonian universe, the rationalist foundations of which he proceeds to undermine. That attempt to subvert was aided by a quality of the calculus that caused much dissent.

Differentiation computes the rate at which one variable changes with respect to another within a process, or fluid situation (Newton had stressed fluency), by dividing a small change in one variable by a small change in

326   C. ACKERLEY

another, and letting these changes shrink until they approach zero, to find the value or "limit" that the ratio between them approaches as the changes become indefinitely small. Integration works the other way: it takes an equation in terms of the rate of change (or derivative) and converts it into another in terms of the variables (the integrals) that change. Newton's key discovery was that the constant factor in many processes of nature is the rate at which another rate (the fluxion of a fluxion) changes. David Bergamini expresses this incremental process plainly:

> [T]he difference between the state of affairs at one moment and the state of affairs at the next moment is an indication of how the situation is shaping up; and if the ratio of the net changes between the two moments is evaluated as a limit – a limit approached when the interval between the two moments is imagined as diminishing toward zero – then that limit shows how fast developments are taking place. The logic of calculus can be applied to moments of time, points on a curve, temperatures in a gas or any state of affairs which can be related by equations; the same rules of differentiation apply to all of them. (Bergamini 1963, p. 109)

Perfunctory as this definition is, it illuminates a process of uncertain incremental motion, gradual disintegration, and/or diminution that is implicit in many of Beckett's works, usually as metaphor rather than equation. A brief account of Beckett's mathematical interests, as bearing upon the calculus, may be useful, given that while a scaffolding of seventeenth-century rationalism is often deployed, his early texts look back to that of Pythagorean and Platonic philosophy (Cordingley 2018, p. 50).

A first consideration might be the Pythagorean harmony that regulated the octave and the music of the spheres; hence the Pythagorean Order's fury at the betrayal of "the incommensurability of side and diagonal" (*Murphy* I, p. 32) when the apostate Hippasos divulged the surd, or irrational number—one (such as π or $\sqrt{2}$) that is inexpressible as the ratio of two integers. The ensuing scandal threatened faith in the rational macrocosm, eliciting what Beckett called "a kind of Pythagorean terror, as if the irrationality of pi were an offence against the deity, not to mention his creature" (*Three Dialogues* IV, p. 563). Irrational numbers were no threat to the calculus, but their acceptance by Newton and Leibniz led to similar criticism.

As of the Big World, so too the little world. Writing to Mary Manning (22 May 1937), Beckett used the phrase "cardiac calculus," or pain in the heart, likening it to the agony of passing a "stone" (*Companion*, p. 278). The phrase defines Murphy's "irrational heart," one that "no physician could get to the root of" (*Murphy* I, p. 4). The sense of "calculus" here, within microcosmic man, is that of mediation between fluctuations, with a sense of the half-light of unreality thus entailed. Murphy calls this the APMONIA, or "harmony" arising from the attunement (or integration) of the movement between discrete points. The jest arises from John Burnet's *Greek Philosophy*, the running header of a page which spells the word thus, in Greek capitals (Burnet 1950, p. 45; Ackerley 2010a, p. 33).

Molloy comments: "Extraordinary how mathematics help you to know yourself" (*Molloy* II, p. 26), an ironic statement given that he has just calculated his (indiscrete) farts, each time rounding the number off, thus compounding the error, so that instead of one fart every 3.62 minutes (*Companion*, p. 353), he produces a more decorous result: "Damn it, I hardly fart at all." His sucking stones, in this light, might be a primitive form of calculus (or abacus), their movements from pocket to pocket exhaustively described (*Molloy* II, pp. 63–69). Molloy has picked them up on the seashore, calling them pebbles, in a nod to Newton's celebrated sentiment (original source unknown) that he sees himself as a boy playing on the seashore, diverting himself now and then with a smoother pebble or prettier shell than ordinary, while the great ocean of truth lay all undiscovered before him. Molloy hobbles inland.

Zeno's paradoxes are another tribulation for Beckett, given their sophistic intent of placing flies in the ointment of Microcosmos (*Murphy* I, p. 108) by challenging the Pythagorean assumption of time and space as "consisting of points and instances" (Boyer 1989, p. 74) yet as having a property of "continuity" (which Newton termed "fluency"). Achilles can never catch the tortoise, because before he can do so the tortoise has moved on, and on, and so on; and since it is impossible to exhaust an infinite number of contacts in a finite time it follows that motion must be impossible (that is, under the assumption of the infinite sub-divisibility of space and time). This paradox was implicit in philosophical objections to Newton's "fluxions."

Closely related to this is the *sortites*, or "heap": the paradox of the part and the whole that in the history of philosophy is one of the earliest manifestations of infinitesimals. Beckett's *Endgame* (1956), set within the skull and enacting the Cartesian dualism of body and mind, or the drama of Leibniz's *petites* perceptions realizing themselves within the monadic mind, begins with Hamm's echo of Christ's *Consummatum est*: "Finished, it's finished, nearly finished, it must be nearly finished"; followed by the statement: "Grain upon grain, one by one, and one day, suddenly, there's a heap, a little heap, the impossible heap" (*Endgame* II, p. 90). This conundrum, attributed to Eubulides of Miletus (Windelband 1901, p. 89) but in the spirit of Zeno, offers such "little catches" as: "Which kernel of grain by being added makes the heap?" or "Which hair falling out makes the bald head?" The opening of the play sets the tone: that on this day, effectively indistinguishable from any other, and marked as always by counting down to the asymptotic "zero" of the dying of the light, suppose, just suppose, perhaps, maybe, that something is at last going to mean something? This hope may be, in the painful pun, "a little vein"; and despite the unexpected apparition of the child, or Clov's decision to leave, this one day may yet add to the infinite series of imperceptibly fading days that fail to mount up to a life (*Endgame* II, p. 103,

141). The play is Beckett's most analytical work (in the mathematical sense of that phrase), with diminishing series of everything (neighbors, biscuits, pain-killers, memories, chess, rats, the light ...), and a natural world (like Paley's clock) that is running down to that impossible zero in accordance with the inexorable Second Law of Thermodynamics (with its echo in Genesis 3:19: "for dust thou art, and unto dust shalt thou return").[3] For dubious consolation, one must return to an early draft where the patterns of attrition are more obviously accentuated and less witty, save for one jest upon the eternal order of things: "Everything has an end ... Except the sausage ... It has two" (*Companion*, p. 34). But there are no more sausages.

Zero was an ancient problem. The Greeks (especially the later school of Alexandria) had considered curves as the trajectories of moving points and had attempted the analysis of curvature by the technique of division into infinitely fine segments (Boyer 1989, p. 80), but they were constrained by a general reluctance to confront the mathematics of infinity and hampered by an imperfect understanding of zero. Yet from the strife of various opposites emerged the problem of how the Limited (that which cannot be divided) gave form to the Unlimited (that which may be infinitely divisible). This ignited a controversy that would endure for two millennia, for Aristotle had contended in Book VI of his *Physics* that a continuum (such as length, time or movement) cannot be constituted by indivisibles nor be resolved into them. Neary's tractatus, *The Doctrine of the Limit*, might shine some light on this intractable problem, were it to be rescued from Miss Counihan (*Murphy* I, p. 33); but until then let the editors of the Loeb *Physics* define the issue:

> The atomists contended that all quanta were reducible to indivisible particles or "monads," and that these constituted the ultimate units from which all quanta are built up. In their view, therefore, all ratios were commensurable.
>
> Aristotle, on the other hand, held that there was no limit to the potential divisibility of a continuum. In his view, therefore, there would be no "ultimate units" of a continuum, and therefore there must be incommensurable ratios. (Aristotle 1957, lxxxix)

Beckett recorded in a footnote to his poem *Whorescope* that Descartes "proves God by exhaustion" (*Whorescope* IV, p. 7). The "method of exhaustion" relates to the comparison of curved and straight-line configurations (Boyer 1989, p. 90), and consists of subtracting from any magnitude a smaller measurable portion, repeating the process until only an infinitesimal or unmeasurable part remains. Consider a circle with a polygon inscribed: the enclosed figure may be calculated, and the more complex the polygon the more accurate is it as the measure of the circle's area; but (as in squaring the circle) however exhaustive, the method is approximate only, for the infinitesimal "gap" remains, like that between God (as sphere or perfect circle) and reason (*ratio*); or that between the Monad of monads of Plotinus and the imperfection of other monads; or even the piano, the octaves whereof must be tempered to distribute more evenly the deficiency of attunement known as the "Pythagorean comma," an irrational gap built into the acoustic scale (Lees 1983, pp. 174–76; the repeated jest about Mr. Knott's piano tuners, "the Galls father and son" (*Watt* I, p. 228), accordingly lacks the comma).

This paradox is implicit in the very definition of the calculus, from which the cynical might assume that any expectation of mathematics as a symbolic system representing the eternal truths of man's relationship to God is (like Mr. Knott's piano) doomed. As Beckett says at the outset of his study of Proust: "The Proustian equation is never simple" (*Proust* IV, p. 511). He accepts regretfully "the sacred ruler and compass of literary geometry" but these are inadequate to analyze such incremental matters as the "poisonous ingenuity of Time," or the contention that love is a "function" of man's sadness, as friendship is of his cowardice (*Proust* IV, pp. 512, 538). A key image is the "*gouffre interdit à notre sonde*" (gulf forbidden to our sounding) (*Proust* IV, p. 522), a recurrent metaphor in Beckett for the infinitesimal but impassible gap separating the finite self from its infinite object of its desire, subject from object, the particular from the Idea, a pot from the Pot: even in the experience of perception, or consciousness, or involuntary memory, at some neurological level the gap between the object and its representation persists. Murphy reluctantly acknowledges the gulf between himself and the inmates of the asylum: "In short there was nothing but he, the unintelligible gulf and they. That was all. All. ALL" (*Murphy* I, p. 143). Beckett was deeply influenced on this theme by Schopenhauer, but also by Jules de Gaultier, whose *De Kant à Nietzsche* (1900) tried to bridge, in language curiously compatible with Leibniz's *Monology* (when describing the commotion of the *petites perceptions*), the way that even at the level of molecular movement in the brain a screen is thrown up between the subject and object of knowledge, or, "to abuse a nice distinction" (*Murphy* I, p. 147), between *percipere* and *percipi*.

Windelband contended that the great problem of Greek philosophy, in the world of ethics as much as physics, was the relationship of the two categories of reality: *ousia*, or Being; and *genesis*, or Becoming (Windelband 1901, p. 73)—the relationship, that is, of the unchanging order of things and the world of change. Heraclitus had defined cosmic process as continuous Becoming; but Parmenides affirmed the centrality of Being; for one, the illusory nature of the senses, and for the other the permanence of τὸ ἕν (the One, or absolute Monad), with an absolute distinction between the world of Being and that of *genesis* and change. Mr. Knott "abides" but his servants come and go; the error implicit in Watt's "ancient labour" (*Watt* I, p. 278) is his desire to know the essence of his master through the enumeration and calculation of his accidentals, but no matter how he tries to square this particular circle, with paradigms that become more monstrous and desperate as they tend to infinity, the configuration cannot be brought together. Consider, too, the picture of the circle and its center on Erskine's wall (pp. 272–73) and Watt's inability to know whether the point is advancing or receding (Watt

330   C. ACKERLEY

knew nothing about physics), or what the artist had intended to represent (Watt knew nothing about painting), or what relation might exist between circle and center (Watt knew nothing about mathematics).

Descartes did. Three strange dreams on the night of 10 November 1619 confirmed his belief that the method of the newly conceived analytical geometry might be extended to other studies. *La Géométrie*, published in 1637 as one of three appendices to the *Discours de la méthode*, proposed to unite the traditional and separate disciplines of geometry and algebra in such a way as to "correct the defects of the one with the help of the other" (qtd. in Kline 1953, p. 165), and in so doing flush away all the encrustations of scholasticism and replace them with a new organum. The analytic method consists in reducing geometry to a form of arithmetic and translating geometric shapes into algebraic equations; as Boyer comments, Descartes's success was such that "today we read xx as 'x squared' without ever seeing a square in our mind's eye" (Boyer 1989, p. 338); that is, those conversant with the notation may form instinctively an abstract understanding without having to visualize the geometrical shape of a square of unit x dimensions (indeed, with equations of higher dimensions it may be necessary to do so, as the shapes cannot so readily be visualized). The analogy may be distant, but consider Augustine's small epiphany when he discovered that his mentor, Ambrose, was reading without moving his lips; that is, he had internalized the art of silent reading and, unremarkable as this later might seem, it initiated a small paradigm shift in Western consciousness.

Near the beginning of *Dream of Fair to Middling Women*, the protagonist, Belacqua Shuah, gazes up at the night firmament, sees an "abstract density of music," and experiences a moment of "tense passional intelligence" when "arithmetic abates" (Beckett 1992, p. 16). This invokes the ancient distinction of the medieval Quadrivium between arithmetic as that (like music) which unfolds in time, and geometry as that (like astronomy) which unfolds in space. The phrase "arithmetic abates" may be glossed as "when time stops," suggesting the stasis of an aesthetic experience; but when Belacqua looks at the stars he sees (or feels) "a network of loci that shall never be co-ordinate," but rather as "figured in the demented performance of the night colander." That is, in terms of the analytical geometry of Descartes, he experiences the moment not as one for whom the arithmetical and the geometrical proportions are atoned but rather as one for whom they coalesce in an image of "astral incoherence" that is later invoked with reference to Beethoven's music as a "punctuation of dehiscence," with "punctuation" asserting its literal sense of movement from point to point (Beckett 1992, pp. 17, 138).

The invocation of analytical geometry is striking, even if the theme struck is that of a "Proustian equation" that has not assumed its rational graphic form but has instead expanded into what Beckett elsewhere called the "incoherent continuum" of a quantum world, its continuity "bitched to hell" in a "blizzard of electrons" pitted with "dire storms of silence" (pp. 138–39).

By means of the co-ordinate geometry Descartes aspired to render complex equations in clear and distinct form but, as Cordingley notes, Leibniz rejected the Pythagorean notion of space as an aggregate of points and time as the accumulation of instants; his dynamic theory instead rested on a continuum of space and time, which he sought to harmonize with his metaphysics (Cordingley 2018, p. 248). Beckett's incoherent continuum has a similar function.

Beckett was not persuaded by Descartes's insistence on the "clear and distinct," and he did not make much obvious use of the new method (which was indispensable to the invention of the calculus). However, there is a moment in *The Unnamable* when the narrator returns from the jungles of Java and the rafflesia stinking with carrion: he moves in a vast gyre that seems to take the form of a Cartesian graph centering upon a small rotunda (*The Unnamable* II, p. 311), "windowless" as indeed a monad should be; only to find that when his "vaguely circular motion" (p. 316) reaches the center (or origin) his family have all died of "sausage poisoning" (everything has an end). There is here at least the intimation of the analytical geometry, even if the co-ordinates are obscure.

At some point along the mortal continuum from vagitus to death-rattle, Beckett's narrative practice took a turn, one not always easily discerned yet finally clear and distinct, from the arithmetical (narrative unfolding in time) to the geometrical (the "shape of ideas"), with the action spatialized on stage or within the theater of the mind. This may be discerned, for instance, in the movement of the prose from narratives of coming and going, as predicated on the "Laetus exitus tristem" dictum of Thomas à Kempis, that a merry outgoing brings forth a sad homecoming (*Companion*, p. 573), to such works as the *Texts for Nothing*, where the "action" (if any) is interiorized, narrative yields to spatial form, and the defining aesthetic is increasingly one of stasis. This trend is evident in his drama: in *Waiting for Godot* (1952), for instance, where the original vaudeville and slapstick elements gave way, under Beckett's later direction, to more iconic features (the raising of Pozzo in Act II as a visual icon of Christ between the two thieves). This may or may not be the legacy of the analytical geometry, but (as Tonning has made clear) Beckett's later works for stage and screen represent in dramatic form the "ghosts and fluxions" of the Leibniz monadology. Consider, in particular, the late television plays, such as "Quad" and "what where," where spectral figures move in exhaustive patterns in the flickering light that represented the then-current technology.

Exigencies of the time-space continuum preclude further exploration of this exponential theme, but perhaps one last instance of the algebraic and the geometrical coming together in a compelling way: in *How It Is*, the protagonist of Beckett's last novel of motion calculates his path through the mud as a zigzag, "advancing two yards, turning 90 degrees, advancing two yards, turning back 90 degrees" (Cordingley 2018, p. 50). This creates geometrical chevrons, or alternating right-angled triangles of unit one yard extending

332   C. ACKERLEY

from a hypothetical straight line, which in turn forms the hypotenuse of each triangle, which must be of length √2, or 1.414241..., an irrational number that compromises any notion of the ideal Triangle of Plato's *Philebus* as the fundamental plane surface. Curiously, Russell called one attempt to cope with the paradox named for him a "zigzag theory" of types (Carey and Ongley 2009, p. 258). The movement of Beckett's narrator is incremental at best, that of a monad in motion struggling to comprehend the mystery of other monads or that of the sacks and tins. Is he "sole elect" (*How It Is* II, p. 416)? In what pre-established sense is there a matching of souls and sardines? The incessant mathematical patterns and series, fugitive memories, and disjunctive motion all testify to a calculus of attrition diminishing at its inevitable limit to a muted scream: "DIE ... I MAY DIE ... I SHALL DIE" (*How It Is* II, p. 521).

Beckett was not necessarily aware of nor applied the calculus to such incidents, but he appreciated its mathematical importance sufficiently to deploy it as image and metaphor. He was enchanted by the self-evident silliness of the entire post-Cartesian enterprise (this by no means incompatible with a recognition of its genius): the apparently bizarre ontology of Leibniz's *Monadology* and the theory of pre-established harmony that it both required and entailed, and how it was made to accommodate a transcendent Monad of monads; or Newton's application of his powerful explanatory mode of analysis to curious alchemical conundra, or his enlisting it in the attempt to reconcile the new cosmology with the widely accepted dating of the Fall as 4004 bce (Bergamini 1963, p. 109). Beckett's willingness to be seduced at the wayside by such passing fancies was tickled by the curious conjectures he found in Pierre Gustave Brunet's *Curiosités théologiques* (1861), for instance that concerning John Craig (1663–1731), a Scottish mathematician and friend of Newton, who was among the first to use Leibniz's notation (*Companion*, p. 112). Craig wrote a few minor works about the new calculus, which he applied in the *Theologiae Christianiae Principia Mathematica* (1699) to define the limits of the Second Coming (Brunet 1861, p. 144). The probability of an historical event (such as the Crucifixion), he believed, depends on the number of primary witnesses, the train of transmission through secondary witnesses, the elapsed time, and the spatial distance; hence, the probability of the truth of the gospels is diminishing to zero by the year 3144 (some say 3150), which thus limits the Second Coming. In Beckett's *Molloy*, Moran asks: "The algebraic theology of Craig. What is one to think of this?" (*Molloy* II, p. 161). Consider, in these terms, Vladimir's story of "Our Saviour" and the "Two Thieves" in *Waiting for Godot* (pp. 6–7): salvation may seem an even chance, as in Pascal's celebrated bet; but of the four Evangelists only one (Luke 23:43) speaks of a thief being saved. If all four were "there" (or in a chilling Craig-like qualification, "thereabouts"), the odds are logically reduced, to one in eight (or worse, depending on "abouts"). Vladimir wants to know: "Why believe him rather than the others?" Estragon replies: "People are bloody ignorant apes"; but Luke had also asked: "when the

Son of Man cometh, shall he find faith on the earth?" (Luke 18:8). The Darwinian thrust in Estragon's reply (present in the English text only) adds another variable to the algebraic theology, perhaps rendering the English odds of salvation even less likely than the French. This way absurdity lies, but it sets the tone for my equally silly but more serious conclusion.

Newton's calculus met with opposition from an unexpected source: Bishop Berkeley, long after its first publication in the *Principia Mathematica* (1687), launched an attack on its basic principles in a tract called *The Analyst* (1734). Berkeley did not deny the utility of Newton's fluxions (derivatives of a function), nor the validity of results thus obtained, but he objected to the assumption that "increments are given to the variables and then [mathematicians] take the increments away by assuming them to be zero" (Boyer 1989, p. 430). Trying to undercut the basic assumptions of the calculus, he expressed his frustrations with the notion of infinitesimal divisibility:

> Now to conceive a Quantity infinitely small, that is, infinitely less than any sensible or imaginable Quantity, or than any the least finite Magnitude, is, I confess, above my Capacity. But to conceive a Part of such infinitely small Quantity, that shall be still infinitely less than it, and consequently though multiply'd infinitely shall never equal the minutest finite Quantity, is, I suspect, an infinite Difficulty to any Man whatsoever. (Berkeley 1734, #V)

The calculus seemed to Berkeley a compounding of errors, and invalid because "no just Conclusion can be directly drawn from two inconsistent Suppositions" (Berkeley Berkeley 1734, #XV); even if, as was apparently the case, "by virtue of a twofold mistake you arrive, though not at Science, yet at Truth" (#XXII). He condemned "Leibniz and his Followers in their *calculus differentialis* making no manner of scruple, first to suppose, and secondly to reject Quantities infinitely small" (#XVIII); of increasing a Quantity by nothing, or dividing by zero, whereby "all had vanished at once, and you must have got nothing by your supposition" (a conflation of the numerical "zero" with the eschatological "nothing"). Instead of providing a mathematical key to unlock "the secrets of Geometry, and consequently of Nature" (#III), Berkeley insisted, the calculus generates unnecessary fluxions of fluxions of fluxions "of a thing which hath no Magnitude," so that our "Sense" is "strained and puzzled" to frame "clear Ideas of the least particles of Time, or the least Increments generated therein" (#IV).

Berkeley was provoked into publication because he was irritated with the Astronomer Royal, Edmond Halley (1656–1742), who had been instrumental in urging (and underwriting) the publication of Newton's *Principia* (1687); the subtitle of his tract was "a Discourse Addressed to an Infidel Mathematician." Halley had encouraged a sick friend to refuse spiritual consolation because of the untenable nature of Christian doctrine; hence Berkeley's true mission to dispute the principles and inferences of *Analysis* as more distinctly conceived or more evidently deduced than religious mysteries

334   C. ACKERLEY

or points of faith (Boyer 1989, p. 430). Who are infidel mathematicians to declaim such absurdities while rejecting the "Authority" (Berkeley 1734, #XIX) of revealed Truth!

In fairness, Berkeley's doubts about mathematical rigor were shared by many eminent calculists long after Newton's death in 1727. Voltaire, for instance, mocked the calculus as "the art of numbering and measuring exactly a Thing whose Existence cannot be conceived" (Kline 1953, p. 232); he should, perhaps, have contented himself with *Candide*, which intimates as much of God. Yet Joseph-Louis Lagrange and Leonhard Euler, a century after Newton and Leibniz, still believed that "the calculus was unsound but gave correct answers only because errors were offsetting each other" (Kline 1953, p. 232). Indeed, Newton and Leibniz were themselves unable to offer proof of their methodology, and attempts by others to formulate a theory of limits and to ground derivatives in logic proved unsuccessful until 1821, when Augustin-Louis Cauchy defined the essential concepts of the limit correctly (Kline 1953, p. 233).

Beckett, I suggest, was seduced by one particularly attractive passing fancy, Berkeley's proposition #XXXV of *The Analyst*, concerning fluxions, so exquisitely dressed as to overcome any rational impulse toward mathematical fidelity:

> And what are these Fluxions? The Velocities of evanescent Increments And what are these same evanescent Increments? They are neither finite Quantities, nor Quantities infinitely small, nor yet nothing. May we not call them the Ghosts of departed Quantities?

My admiration for this seductive sentiment is shared by Jean-Michel Rabaté, who invoked it in "Beckett's Ghosts and Fluxions" (1996a, p. 24). Rabaté noted that the passage had been used, explicitly, in various novels by Wilfred Bion, Beckett's friend and (psycho)analyst at the Tavistock Clinic, which he attended for almost two years (1934–1935). Rabaté identifies Newton's "ghosts" (impalpable traces) and his "fluxions" (derivatives) in a differential calculus of diminution within the mind. He suggests that "ghosts" for Beckett are encrypted words (and, I would add, fugitive memories) emerging from the deepest layers of the unconscious; and that Berkeley replaces these "ghosts" firmly within "their perceptual or phenomenological context," below the threshold of conscious awareness, from whence they might return as images (1996a, pp. 24, 28). For Rabaté, this constitutes the point of departure for Beckett's sense of *nothing* as "a fundamental object of unconscious desire" (1996a, p. 26), such vestigial images creating the insistent voices that trouble Beckett's various protagonists and, more broadly, instantiate the spectral images that haunt the later works: the "boy" of *Waiting for Godot* and "Ghost Trio"; the frequent sense of a "nothing" out of which comes a "something" diminishing *al niente* (*Companion*, p. 61), as in the *Texts for Nothing* or such late plays as " ... but the clouds ... " or "Nacht und

Träume" (or, as I suggested earlier, the flickering figures on the primitive television sets of the day).

Beckett, like Russell, had approached Leibniz with considerable skepticism, describing him to Tom MacGreey as "a great cod, but full of splendid little pictures" (*Letters* I [4 December 1933], p. 172), only to experience a growing respect and admiration for his writing, despite the absurdity of some of the little pictures. As I have argued elsewhere (Ackerley 2013, pp. 101–2), the theory of the *Monadology* allowed Beckett to move from the failed attempt in *Dream* to write a novel premised on the incoherent continuum, by providing, as in *Murphy*, a picture of the monadic mind that became, in time, that of *The Unnamable*, arguably the successful "incoherent" novel that *Dream* was intended to be, its only "unity" that of the monadic consciousness. The calculus of Leibniz gave Beckett a notation for interpreting the dark zone of the monadic mind, with the theory of infinitesimals finding "immediate application in his revision of Descartes's theories regarding matter and motion" (Cordingley 2018, p. 239). Central to this was the crucial insight from Windelband that the *petites perceptions* of Leibniz, those below the threshold (or limit) of perception, may best be understood in our more skeptical age as "unconscious mental states" (Windelband 1901, p. 73). This in turn led Beckett to the conception of the residual self (that remaining after the processes of dis-integration), as "neither" abstract not material, but destined to live (like the protagonist of the *Texts for Nothing*) in the intervals ("Text 6" IV, p. 313), where "you miss the derivatives" ("Text 2" IV, p. 300). For Beckett, the "occult arithmetic" of Leibniz (*Proust* IV, p. 553) does not lead to the conclusion that all is for the best in the best of all possible worlds, but rather to the uncertainty of the monadic variable resenting its differential existence as but a derivative of the function of Time. Leibniz had offered, in support of his theory of pre-established harmony, the splendid little picture of a choir, in which independent monads sing in perfect choral harmony, "a real harmony formed out of the complementary movements of several self-acting units" (Latta 1898, p. 47). The experience of Watt, however, as he tries to accommodate himself to a world of insistent patterns and infinite series, is finally that of complete disintegration; and the protagonist of *How It Is*, trying to ascertain if he is one among "billions" or irreducibly "sole elect," hears not a mixed choir but harsh megaphones, and as he experiences the inevitable "loss of species" the fugitive memories of the "humanities" he once had now condemn him to at best a liminal existence haunted by the ghosts of those departed quantities.

NOTES

1. References to Beckett's primary works, unless otherwise specified, are to the four-volume Grove Press Centenary edition (2006), cited by short title, volume number and page, thus: (*Endgame* IV, p. 89).

2. For convenience, commentary that is now part of a broad tradition is often taken from C. J. Ackerley and S. E. Gontarski's *The Grove Companion to Samuel Beckett* (2004), cited as "*Companion*"; as well as entries for individual works, see also those for: **absurdity; apperception; atomism; Ballmer; Berkeley; Bion; Craig; Descartes; Geulincx; God; greatcoat; incoherent continuum; irrationality; Leibniz; mathematics; mixed choir; monad; Murphy's mind; music; mysticism; physics; pre-established harmony; pre-Socratic philosophy; Zeno.**

3. Biblical references are to the *King James Authorized Version*.

REFERENCES

Ackerley, Chris. 2010a. *Demented Particulars: The Annotated* Murphy. 1998 & 2004; rpt. Edinburgh: University of Edinburgh Press.

———. 2010b. *Obscure Locks, Simple Keys: The Annotated* Watt. 2005; rpt. Edinburgh: Edinburgh University Press.

———. 2010c. "Samuel Beckett and Science." In *The Blackwell Companion to Samuel Beckett*, edited by S. E. Gontarski, 143–64. Oxford: Blackwell.

———. 2012. "Monadology: Samuel Beckett and Gottfried Wilhelm Leibniz." In *Beckett/Philosophy*, edited by Matthew Feldman and Karim Mamdani, 140–61. Sofia: University Press "St. Klimrnt Ohridski.".

———. 2013. "The Geometry of the Imagination." In *Samuel Beckett: Debts and Legacies: New Critical Essays*, edited by Peter Fifield and David Addyman, 85–107. London: Bloomsbury.

———. 2016. "Samuel Beckett and the Physical Continuum." *Journal of Beckett Studies* 25 (1): 110–31.

Ackerley, Chris J., and S. E. Gontarski. 2004. *The Grove Companion to Samuel Beckett*. New York: Grove Press.

Aristotle. 1957. *Physics*. Translated by P. H. Wicksteed and F. M. Cornford. 2 vols. 1929; rev. & rpt. Cambridge MA: Harvard University Press [Loeb Classical Library].

Beckett, Samuel. [1934–1938?]. "Whoroscope Notebook." Beckett International Foundation, University of Reading. RUL MS 4000/1.

———. 1992. *Dream of Fair to middling Women*. Edited by Eoin O'Brien and Edith Fournier. Dublin: Black Cat Press.

———. 1999. *Beckett's* Dream *Notebook*. Edited by John Pilling. Reading: Beckett International Foundation.

———. 2006. *Samuel Beckett: The Grove Centenary Edition*. 4 vols. Edited by Paul Auster. New York: Grove Press.

———. 2009. *The Letters of Samuel Beckett: Volume I: 1929–1940*. Edited by Martha Dow Fehsenfeld and Lois More Overbeck. Cambridge: Cambridge University Press.

Berkeley, George. 1734. *The Analyst, or a Discourse addressed to an Infidel Mathematician. Wikisource*. Accessed 6 March 2018.

Bergamini, David, ed. 1963. *Mathematics*. New York: Time [Life Science Library].

Boyer, Carl B. 1989. *A History of Mathematics*. 1968; 2nd ed.. Revised by Uta C. Merzbach. New York: Wiley.

Brunet, Pierre Gustave. 1861. *Curiosités théologiques par un bibliophile*. Paris: Adolphe Delahays, Libraire-Éditeur.

Burnet, John. 1950. *Greek Philosophy: Thales to Plato*. 1914; rpt. London: Macmillan.

Carey, Rosalind, and John Ongley. 2009. *Historical Dictionary of Bertrand Russell's Philosophy*. Toronto: Scarecrow Press.

Cordingley, Anthony. 2018. *Samuel Beckett's How It Is: Philosophy in Translation*. Edinburgh: Edinburgh University Press.

Culik, Hugh. 1993. "Mathematics as Metaphor: Samuel Beckett and the Esthetics of Incompleteness." *Papers on Language and Literature* 29: 131–51.

―――. 2013. "Science and Mathematics." In *Samuel Beckett in Context*, edited by Anthony Ullmann, 348–57. Cambridge: Cambridge University Press.

Dowd, Garin. 1998. "Nomadology: Reading the Beckett Baroque." *Journal of Beckett Studies* 8 (1): 15–49.

―――. 1999. "Mud as the Plane of Immanence in *How It Is*." *Journal of Beckett Studies* 8, no. 2: 1–28.

Eliot, T. Stearns. 1916. "Leibniz's Monads and Bradley's Finite Centers." *The Monist* 26 (4): 566–76.

Ellis, Reuben J. 1993. "'Matrix of Surds': Heisenberg's Algebra in Beckett's *Murphy*." In *Papers on Language and Literature* (1989); rpt. in *Critical Essays on Samuel Beckett*, edited by Lance St John Butler, 362–65. Aldershot, Hants: Scolar Press.

Feldman, Matthew. 2006. *Beckett's Books: A Cultural History of Samuel Beckett's "Interwar Notes."* London and New York: Continuum.

―――. 2012. "Samuel Beckett, Wilhelm Windelband and Nominalist Philosophy." In *Beckett/Philosophy*, edited by Matthew Feldman and Karim Mamdani, 110–39. Sofia: University Press "St. Klimrnt Ohridski.".

―――. 2013. "Philosophy." In *Samuel Beckett in Context*, edited by Anthony Ullmann, 301–11. Cambridge: Cambridge University Press.

Gaultier, Jules de. 1900. *De Kant à Nietzsche*. Paris: Mercure de France.

Geulincx, Arnoldus. 1891–1893. *Opera Philosophica*, recognivit J. P. N. Land. 3 vols, Hague Comitum: Martinum Nijhoff.

Heisenberg, Werner. 1957. *Physics and Philosophy: The Revolution in Modern Science*, with an Introduction by F. S. C. Northrop. New York: Harper & Row.

Irvine, Andrew David. "Principia Mathematica." In the *Stanford Encyclopedia of Philosophy*. https://plato.stanford.edu/entries/principia-mathematica/. First published 1996; substantive revision March 10, 2015; accessed 12 April 2018.

Jourdain, Philip E. B. 1916. "Gottfried Wilhelm Leibniz (1646–1716)." *The Monist* 26 (4): 481–85.

Kline, Morris. 1953. *Mathematics in Western Culture*. London, Oxford University Press.

Knowlson, James. 1996. *Damned to Fame: The Life of Samuel Beckett*, London: Bloomsbury.

Lees, Heath. 1983. "*Watt*: Music Tuning and Tonality." *Journal of Beckett Studies* 9: 5–24.

Leibniz, Gottfried. 1898. *The Monadology and Other Philosophical Writings*. Translated and edited by Robert Latta. London: Oxford University Press.

Lucretius [Titus Lucretius Carus]. 2002. *De Rerum Natura*. Translated by W. H. D. Rouse (1924); rev. Martin Ferguson Smith (1975). Cambridge MA: Harvard University Press [Loeb Classical Library].

Mauthner, Fritz. 1923. *Beiträge zu einer Kritik der Sprache*. 3 vols, Leipzig: Verlag von Felix Meiner.

Monk, Ray. 1990. *Ludwig Wittgenstein: The Duty of Genius*. London: Jonathan Cape.

Nixon, Mark. 2011. *Samuel Beckett's German Dairies 1936–1937*. London: Continuum.

338   C. ACKERLEY

Phillips, Holly. 2015. "Samuel Beckett and the Nominalist Ethic." PhD dissertation, University of Otago.

———. 2016. "'What I Want Is the Straws, Flotsam, etc': Beckett and the Nominalist Ethic of Humility." *Journal of Beckett Studies* 25 (1): 56–77.

Rabaté, Jean-Michel. 1996a. "Beckett's Ghosts and Fluxions." *Samuel Beckett Today Aujourd'hui*, vol. 5, *Beckett & La Psychoanalyse & Psychoanalysis*, edited by Sjef Houppermans, 23–40. Amsterdam and New York: Rodopi.

——— 1996b. *The Ghosts of Modernity*. Gainesville: University of Florida Press.

Russell, Bertrand. 1900. *A Critical Exposition of the Philosophy of Leibniz, with an Appendix of Leading Passages*. Cambridge: Cambridge University Press.

Sorley, William Ritchie. 1882 [1929]. "Leibniz." In the *Encyclopaedia Britannica*, vol. 13. 1882; rpt. London and New York: Encyclopaedia Britannica, 886.

Stevens, Brett. 2010. "A Purgatorial Calculus: Beckett's Mathematics in 'Quad.'" In *The Blackwell Companion to Samuel Beckett*, edited by S. E. Gontarski, 164–81. Oxford: Blackwell.

Tonning, Erik. 2007. *Samuel Beckett's Abstract Drama: Works for Stage and Screen, 1962–1985*. Berne: Peter Lang.

Windelband, Wilhelm. 1901. *A History of Philosophy, with Special Reference to the Formation and Development of Its Problems and Conceptions*. Translated by James Tufts. 1893; 2nd ed., rev., London: Macmillan.

Wittgenstein, Ludwig. 1922. *Tractatus Logico-Philosophicus*. Translated and edited by Frank F. Ramsey and C. K. Ogden.

———. 1953. *Philosophical Investigations*. Translated by G. E. M. Anscombe. London: Macmillan.

CHAPTER 19

"Numbers Have Such Pretty Names": Gertrude Stein's Mathematical Poetics

Anne Brubaker

Even enthusiasts of Gertrude Stein's work concede that her experimental writing often approaches the limits of readability. So to introduce the subject of mathematics in relation to her work might not seem the ideal way to draw new readership or even the return of the already initiated. And yet, there is already a scholarly precedent for reaching beyond literary history and criticism to other fields in order to make sense of Stein's methods and motivations. Scholars have examined her work in connection with psychology and the work of William James, abstract art and her friendship with Pablo Picasso, anatomy lessons and her training at the Johns Hopkins School of Medicine, and indeed even the philosophy of mathematics and her connection to Alfred North Whitehead (1861–1947).[1] These contexts are crucial for making sense of her highly abstract writing style, but these approaches tend to solidify the relationship of Stein's writing to already codified and competing literary paradigms.

I share with scholars such as Steven Meyer, Maria Farland, and Jennifer Ashton an interest in exploring Stein's writing as conversant with contemporaneous scientific and mathematical developments; however, my goals differ in two significant ways. The first is that my focus is less on tracking the "circulation of intact ideas" between disciplines and more on how Stein's work reinforces literature and mathematics as "open fields"—a term Gillian Beer uses to describe fields whose parameters are continuously under revision and subject to transformation through their interactivity (Beer 1996, pp. 8, 115). Scholars of modernism have long taken an interest in how writers draw

---

A. Brubaker (✉)
Wellesley College, Wellesley, MA, USA

© The Author(s) 2021
R. Tubbs et al. (eds.), *The Palgrave Handbook of Literature and Mathematics*, https://doi.org/10.1007/978-3-030-55478-1_19

339

340   A. BRUBAKER

on particular scientific concepts and developments, but, as Mark Morrison observes, they have tended to "take science as a stable and completed given" and to grant science "a privileged position of autonomy and purity" (Morrison 2002, p. 675). Perhaps more than any other scientific discipline, mathematics has served as the paradigm of an autonomous and pure form of knowledge. And yet, during the late nineteenth and early twentieth centuries, the fields of mathematics and mathematical philosophy experienced a number of significant upheavals and a turn toward disciplinary introspection. Literary figures of this period, like mathematicians, were also seeking a more robust disciplinary foundation.[2] I show how Stein joins her mathematical contemporaries—namely, Bertrand Russell (1872–1970), Alfred North Whitehead, Georg Cantor (1845–1918), Gottleb Frege (1848–1925), and theoretical physicist Niels Bohr (1885–1967)—in a common search for order, exactitude, and determinacy. But, as important, this search reveals to them the representational limits of language and an awareness of the interdependency of ordinary language and mathematics. They are connected more by their discovery of logical ambiguities and linguistic paradoxes than by any of the particular conclusions of their work.[3]

This perspective might thus help to resolve some of the ongoing critical debates about how to categorize Stein's work. Critics have tended to position her fiction as either committed to "the determinacy of meaning" (Jennifer Ashton) or to a "poetics of indeterminacy" (Marjorie Perloff), and accordingly label her either a modernist or proto-postmodernist. However, placing her work alongside these contemporaneous mathematicians reveals the disciplinary thresholds her work crosses; the comparison offers a way to understand this determinacy/indeterminacy dialectic not as a dividing line between modernism and postmodernism but instead as a defining characteristic of early twentieth-century literary and mathematical inquiry.

Secondly, my analysis diverges from approaches that see Stein's engagement with scientific concepts as merely an aesthetic tool or a formalistic end unto itself. Her engagement with mathematics brings the referential aspects of her writing into greater relief. I explore how she draws on concepts from set theory, logic, and arithmetic to critique patriarchal literary standards, to explore the dynamic between individual perception and external reality, and to consider the ontological status of subjects and objects, individuals and things. For Stein, mathematical thought and expression are crucial to how we perceive and describe ourselves as individuals in relations to others as well as how we make sense of external reality.

I begin by tracing Stein's comments on the relationship between her writing and mathematical thought as well as some of the more recent criticism on her scientific impulses. I use this background to contextualize points of commonality between various fields of mathematics and her early and later writing, including *Q.E.D.* (1903), *Tender Buttons* (1914), "Are There Arithmetics," (1923), "Patriarchal Poetry" (1927), *The Geographical History of America* (1936), and *To Do: A Book of Alphabets and Birthdays* (1940).

I show how the shift in her fiction away from more realistic modes of representation (e.g. *Q.E.D.*) toward greater and greater abstraction (e.g. *Tender Buttons*) coincides with her move away from a realist conception of mathematics and toward a non-Platonist, material-discursive view of the discipline.[4] Paradoxically, her move toward abstractionism leads her to engage more fully with questions of subjectivity, materiality, embodiment, and cognition—experiential phenomena that are for Stein indivisible from mathematical reasoning.

The Search for Exactitude

Stein's critical assessment of her own work, to which she devoted much attention throughout her career, stresses the connection between her writing and mathematics. As she notes in *The Autobiography of Alice B. Toklas* (1933), "In Gertrude Stein, the necessity was intellectual, a pure passion for exactitude. It is because of this that her work has often been compared to that of mathematicians" (Stein 1990a, pp. 198–99). Stein explains further her overarching effort to achieve the kind of semiotic precision attributed to mathematics in "How Writing is Written" (1935):

> While I was writing I didn't want, when I used one word, to make it carry with it too many associations. I wanted as far as possible to make it exact, as exact as mathematics; that is to say, for example, if one and one make two, I wanted to get words to have as much exactness as that … I made a great many discoveries, but the thing that I was always trying to do was this thing. (Stein 1975, p. 157)

She not only suggests an effort to achieve a close correspondence between word and referent—so that each word does not "carry with too many associations"—but also offers a basic formula for sentence construction to emphasize the idea that each unique word (each "1"), when put together in a particular sequence, can add up to different meanings (as in $1 + 1 = 2$). It is not just the individual word that counts, but also that its associations with other words combine to create different meanings and effects.

Those familiar with Stein's writing know well that she trades referential transparency for calculated repetition and syntactical rearrangement. She approaches each word as an individual unit that can be combined with other units to form a sequence, even if that sequence does not add up to a cohesive idea: "I took individual words and thought about them until I got their weight and volume complete and put them next to another word" (Stein 1974, p. 15). This metaphor is significant to understanding the mathematical nature of Stein's methodology: meaning and connotation—what she calls weight and volume—are literally translated into a mathematical procedure.

Perhaps this is why some critics, such as Steven Meyer, have argued that her writing is not just informed by science, but "turns out to be a form of experimental science itself" (Meyer 2000, xxi). For Meyer, Stein's "'fundamental intuition' remained scientific," even as she later uses her training to

342   A. BRUBAKER

challenge traditional scientific conventions (Meyer 2000, p. 4). Jennifer Ashton shifts from a biographically centered analysis to a more historicist and formalist approach that considers Stein's work as contiguous with certain aspects of modern mathematics, namely, specific attempts by Bertrand Russell, Alfred North Whitehead, and Gottlob Frege to establish a logical basis for all of mathematics. Ashton compares Stein's interest in what counts as a whole in *The Making of Americans* (1925) and *Lucy Church Amiably* (1930) to Russell and Whitehead's efforts to define an infinite collection without recourse to enumeration. In a later essay, Ashton connects Stein's theories on proper names in *Four in America* (1931–1934) to Frege's theory of linguistic reference in his 1892 essay "Sense and Reference" (Ashton 2002, pp. 581–604). Together these critical essays provide powerful and fascinating moments of convergence between Stein and her mathematical contemporaries. Widening the lens to explore these ideas within the context of mathematical history more broadly helps us see just how important—albeit problematic—ordinary language is to the production of mathematical knowledge.

The mathematical ideas Ashton considers central to Stein's work originate in the basic premises of set theory, introduced primarily by Georg Cantor in the 1880s.

His early work on infinities led him to the general conclusion, offered in an 1883 lecture: "every set of distinct things can be regarded as a unitary thing in which the things first mentioned are constitutive elements" (quoted in Johnson 1972, p. 36). A set is defined by the *elements* that belong to it. A set could contain anything—real numbers, single people, apples, other sets—but it must be well-defined so that there is no ambiguity about whether a given thing *belongs* to the set. The relation of a part to its whole within set theory is called *containment*; for example, the set of all British novelists is contained in the set of all novelists. Set theory in essence involves making relational statements (e.g., one set is a subset of another) and implementing operations to create new sets out of old sets (e.g., the union of two sets). This basic notion—the idea that mathematical objects can be grouped according to sets—inspired new methods for formalizing all mathematical concepts under a unified theory and establishing elemental principles that would provide validity for the whole of mathematics.[5]

Comparably, Stein's experimental fiction attempts to distill the most basic ingredients of composition. She explores the interplay of words, phrases, syntax, and punctuation in an attempt to understand the inner workings of English grammar and to capture the semantic permutations of particular word sequences. Her apparent interest in sets and their constitutive elements arises in *The Making of Americans* and also in her 1923 poem "Are There Arithmetics," an early exercise in considering the relations of smaller to larger sets of linguistic elements—smaller to larger groups of syllables, words to sentences, sentences to lines or sections—all of which are part of the whole piece.[6] An excerpt of the poem reads:

Are there arithmetics. In part are there
arithmetics. There are in part, there
are arithmetics in parts.
Are there arithmetics.
In part
Another example.
Are there arithmetics. In parts.
As a part.
Under.
As apart.
Under.
This makes.
Irresistible.
Resisted.
This makes irresistible resisted. Resisted. (Stein 2008, pp. 198–201)

Lines 2–5 are all "parts of" the first and longest line. While lines 1–5 become progressively shorter, relying on fewer words in each line, each set of words connects to the previous line by way of its relation to the first. The syntactical reordering allows each word to serve different roles within the structure of the sentence so that no hierarchy is created among the words and no regular or fixed pattern dictates their position as the poem "progresses." In line 6, Stein offers "another example" of the "arithmetics" of linguistic expression. "As a part" and "as apart" use the same letters of the alphabet and contain the same number of elements but yield different meanings, demonstrating the possibility of multiple answers to the same grammatical equation (and perhaps why Stein pluralizes the word "arithmetic"). Judy Grahn points out Stein's demonstration of the "correspondence between small numbers being a part of larger numbers, and small groups of syllables being a part of larger groups of syllables" in lines 13–15, as "the 'resist' is part of 'resisted' is part of 'irresistible' – and then irresistible itself is a part of the larger language structure" (Grahn 1989, p. 255). The poem is a self-conscious illustration of the idea that each part or element is contained within the work as a whole.

Considering her work in relation to basic set theoretic ideas also provides context for understanding some of her lesser known books such as *To Do: A Book of Alphabets and Birthdays* (1940). In *To Do*, a children's book, Stein moves through each letter of the alphabet, under which she associates four proper names—"B is for Bertha and Bertie and Ben and Brave and a birthday for each one." Names and birthdays are common denominators of identity, and this is what unifies all the individuals described in the text: "And so each one had to have one birthday … even their mother Bertha had to have one." Stein moves from the largest set (all those with names and birthdays) to subsets (all those whose names begin with a particular letter from the set {A, B, C, …, X, Y, Z}). Each individual is then further distinguished from another by a set of particular characteristics: "Orlando liked to lick stamps but he did not like to keep them, Olga hated stamps, Only liked stamps but he could not read." *To Do* ruminates on universal and particular forms of identity categorization and (non)belonging. Its underlying structure also bears close resemblance to a fundamental question of set theory: whether a set can be characterized based on the properties of its members (Stein 1957, pp. 5, 36).[7]

For Ashton, the correspondence between Stein's literary project and the mathematical philosophies of her contemporaries situates her writing firmly in the context of modernism, complicating recent criticism that claim her as a postmodernist *avant la lettre* or as a pioneer of linguistic indeterminacy. Ashton argues that poststructuralist readings of Stein disregard her commitment to a poetics of determinacy—one that she claims Stein reinforces through her appeal to mathematics. She understands Stein's expressed effort to make her writing "as exact as mathematics" as what Stein "thinks of as its absolute determinacy" (Ashton 1998, p. 4). I argue, however, that Stein, especially in her later fiction, begins to use numbers as descriptive tools in a way that parallels a contemporaneous shift in mathematical philosophy: away from an object-oriented focus—the study of mathematical things and their properties—toward a more complex linguistic foundation—studying the effects of the ambiguous nature of language on mathematics. Stein reinforces the convergence of mathematics and language made apparent through the logicist program of mathematicians such as Frege, Russell, and Whitehead.[8] Exploring what Stein writes *about* mathematics in addition to how she employs some of its concepts suggests that she comes to understand mathematics less as a paradigm of determinacy and more as an expressive tool integral to how we make sense of the natural and social world. Stein admittedly never uncovers some absolute measure of determinacy; instead, her quest for exactitude leads her again and again to the representational paradoxes and limits that both mathematics and writing encounter. While she might aim for an exactitude of reference analogous to a mathematical equation—so that "one and one make two"—her systematic use of self-referential language is in direct tension to the linguistic exactness she seeks.

Consider, for example, her famous line, "a rose is a rose is a rose," which is more than just a simple repetition of phrases.[9] While Stein considers this sentence as an attempt to link firmly the quality of "rose-ness" to the word "rose,"—or to establish what Pound called "the thing-ness of the thing"—the sentence elicits a certain ambiguity with regard to how we read it: either as "a rose is a rose" *is* a rose, or, a rose *is* "a rose is a rose." Because the phrase circles back on itself, it is unclear to the reader which part of the phrase is being described as having a rose-like quality. These self-referential statements and passages create ambiguities that undermine her efforts toward precision and exactitude.[10]

The logicians of Stein's era also come up against the limits of linguistic determinacy in their attempts to create an unambiguous mathematical language. While the problem of assigning truth values to self-referential statements is classical, it became a fundamental concern for mathematicians such as Russell and Frege, who sought to axiomatize set theory—that is, to make more rigorous the naïve form of set theory introduced by Cantor. Statements referring to themselves, such as "this statement is false," are logically problematic because their truth value is indeterminate; to claim the statement is false is to admit its truth, and to say the statement is true is to admit it is false. Ultimately, these mathematicians realized the inevitability of such ambiguous statements, and while these discoveries did not interrupt the everyday practice of "doing mathematics," they had a profound effect on the philosophy of mathematics, which now required a much more careful treatment of language, particularly in dealing with self-referential statements. Comparably, Stein's literary experiments appear to yield different results from those she intended because her commitment to exactitude paradoxically produces a surplus of linguistic meaning. She is, like these mathematicians and philosophers, attentive to the slipperiness of language—to referential excess—*because of*, but not in contradiction to, her efforts toward precision and exactitude. The dual impulses that critics have observed and debated about in Stein's writing—the tension between linguistic exactitude and ambiguity—turn out to be characteristic tendencies of modern scientific inquiry rather than epistemological or historical dividing lines.

Fitting Stein's writing into either modernism or early postmodernism also tends to undermine the changes *within* Stein's writing—her moves from realist modes to non-mimetic ones—imposing a coherency on her writing that suits the agreed upon tenets of that particular literary movement more than it does her actual body of work. Moreover, Stein's convergence with mathematical philosophers of her time reveals her investment in ideas that transcend literary boundaries. Her fiction and non-fiction works alike reveal far less concern with defining a literary aesthetic than with uncovering universal qualities of writing and the engineering of meaning. Mathematics figures centrally in her exploratory writing, both methodologically and philosophically, ultimately providing her with the rhetorical and conceptual tools for exploring social, material, and subjective experience. For Stein, how we come to know and be—and indeed also to express ourselves—are as central to the fundamental questions of literature as they are to mathematics.

FROM *Q.E.D.* TO TENDER BUTTONS

While Stein is known mostly for her radical experimentalism, her earliest attempts at fiction, such as *Q.E.D* and *Fernhurst*, draw significantly from the traditions of American realism and naturalism. To encounter her early fiction is akin to observing, for example, Picasso's early sketches, which illustrate the representational skill and detail that his abstract paintings deliberately obscure. *Q.E.D.* (1903) is considered Stein's first novel.[11] The original title *Q.E.D.*, later changed to *Things As They Are*, refers to the Latin abbreviation of the phrase, "which was to be demonstrated," used at the end of mathematical proofs to indicate that something has been proved definitively. The title, along with the third person, past tense narration, emphasizes the characters' fates as fixed and the story's conclusion as inevitable. The story is divided into three sections, each following one of three women—Adele, Mabel, and Helen—who develop intimate friendships after traveling together on a steamship to Europe. The main character, Adele, who some argue is a stand-in for Stein, develops intense feelings for Helen and burning resentment toward Mabel, who maintains a closer connection to Helen. But *Q.E.D.* is more than simply an ironic title referring to the irresolvable love triangle at the center of the story; it comes to represent the deterministic worldview—figured as a mathematical proof—that Adele adopts by the novel's end. This perspective emerges early in the story when Helen kisses Adele and Adele considers the kiss a result of outside forces rather than inner desires:

> "Haven't you ever stopped thinking long enough to feel?" Helen questioned gravely. Adele shook her head in slow negation. "Why I suppose if one can't think at the same time I will never accomplish the feat of feeling. I always think. I don't see how one can stop it." "Well" Adele put it tentatively "I suppose it's simply inertia". (Stein 2006, p. 186)

For Adele, fate draws two people together just as a mathematical equation works itself out: "Why ... it's like a bit of mathematics. Suddenly it does itself and you begin to see." Like a mathematical solution, Adele's relationship with Helen appears laid out before her, predetermined. Here, Stein represents mathematics as a complete and self-generative system—something that "does itself" prior to human interference. Mathematics is seen as proof of the existence of a constant and transcendent reality—an external world whose force is greater than individual will. Adele reinforces this notion in the novel's conclusion when she responds to Helen's last letter: "Hasn't she learned that things do happen and she isn't big enough to stave them off ... Can't she see things as they are and not as she would make them if she were strong enough as she plainly isn't." Adele now sees the "dead-lock" between these three women as inevitable, as that "which was to demonstrated." In *Q.E.D*, mathematics functions as a metonym for the underlying order of reality, but it also marks the beginning of Stein's investment in mathematics as a subject crucial to any modeling of reality (Stein 2006, pp. 186, 227).

In *Tender Buttons* (1914), one of her early and most recognized experimental works, Stein moves away from the notion of a fixed, determinate reality "behind" language to the notion of a reality defined through language, including the language of mathematics. As Christopher Knight argues, Stein's collection marks her transition from the classical to the modern episteme.[12] According to Knight, this text is "constructed upon some rather classical premises of perception," reinforced by its emphasis on "color, difference, space, time, measure, meaning, substance, use, necessity, etc.," or those qualities "most identifiable with the classic episteme." He also points to Stein's expressed goal of "'really knowing what a thing was'" and defining that object through the "seemingly more material and spatial qualities of number, measure, weight and difference" as evidence of the text's foundation in classical epistemology (Knight 1991, pp. 36, 37). In "Colored Hats," for example, one can see this attention to quantitative description: "Colored hats are necessary to show that curls are worn by the addition of blank spaces, this makes the difference between single lines and broad stomachs, the least thing is lightening." And yet, Stein shows little interest in using these qualities to conjure an accurate picture or a realistic representation of reality; there is hardly a direct correspondence between the objects and their descriptions. While the descriptions emphasize material, physical, and sensory qualities, they read more like passing observations than a list of the object's determinate properties. Consider, for example, the opening sentence of "Milk": "A white egg and a colored pan and a cabbage showing settlement, a constant increase." As her use of the present tense emphasizes, she is primarily interested in capturing phenomena rather than the inherent attributes of a particular thing (Stein 2006, pp. 473, 487).

Knight likewise concedes that Stein does not "do what she set out to do: to make the word be the thing. That she is, though she is hesitant to admit it, much more engaged in the making of metaphors, of substitutions, than she is in the making of nonlinguistic things" (Knight 1991, p. 43). For Knight, *Tender Buttons* both represents "the culmination of the classical episteme" and marks a crucial shift in her conception of language: "Stein is beginning to conceive of language in a different way, not the way of correspondence within which the plane of language stands parallel to that of things; rather, more in the way of language as an enveloping net faithfully contouring the ever various lumpish matter of reality" (Knight 1991, p. 42). What Knight understates, however, is the degree to which Stein self-consciously challenges, rather than simply draws on, classical notions of number, measure, color, and difference as stable, objective qualities independent of language and observation. She identifies these classical modes of observation for the readers, referring to the *acts* of counting, measuring, and differentiating rather than specific instances of these practices, as she does in the following sentences: "A measure is that which put up so that it shows the length has a steel construction"; "Why is there a difference between one window and another, why is there a difference, because the curtain is shorter"; and "Suspect a single buttered flower, suspect it certainly, suspect it and then glide, does that not alter a counting" (Stein 2006, pp. 502–3, 484). More than simply defining an object according to its unique quantitative qualities, Stein is interested in *how* something is described—that is, the *methods* of measurement and how these function semantically and grammatically to shape our perceptions of the objects observed. As she writes in *Lectures in America* (1935), "I began to discover the names of things, that is not discover the names but discover the things to see the things to look at" (Stein 1998, p. 331). Rather than simply call a thing by its name, Stein focuses on the "things to see the things to look at" or the descriptive words that conjure the things "without naming them."

Indeed, while critics emphasize Stein's focus on the thing-in-itself in *Tender Buttons*, the book as a whole is far less concerned with objecthood than it is with the act of describing. The descriptions refer less to the given object than to their own grammatical and syntactical construction. In "Roastbeef," Stein engages in a metadiscourse that playfully explores the function of different parts of speech: "In the inside there is sleeping, in the outside there is reddening, in the morning there is meaning, in the evening there is feeling." The preposition "in" begins each clause and modifies nouns ("inside," "outside," "morning," and "evening") that reinforce the preposition's basic function: to express spatial and temporal relationships. A similar sentence arises later in this section that uses prepositions, nouns, and verbs that express size, position, space, and movement: "Around the size that is small, inside the stern that is the middle, besides the remains that are praying, inside the between that is turning, all the region is measuring and melting is exaggerating." These examples show Stein's interest in exploring the fundamental aspects of grammatical construction. She experiments with the way words can act as multiple parts of speech (by shifting the same word from adjective to noun or from noun to verb) and considers how different modifiers shape our understanding of a noun: "A transfer, a large transfer, a little transfer, some transfer, clouds and tracks do transfer, a transfer is not neglected." While Stein may have set out to define the characteristics of a particular thing, her metadiscourse demonstrates that the object's identity or "thingness" is inseparable from the language that describes it—that is, objects do not have inherent attributes independent from the language that describes them. This shift in perspective has important implications for her representation of mathematics as a paradigm of determinacy (Stein 2006, pp. 475, 479, 481).

In order to demonstrate how her later fiction contributes to rather than resists available models of indeterminacy, I trace key similarities between Stein's perspective in *Tender Buttons* and the philosophical insights of physicist Bohr. While somewhat overshadowed by his protégé, Werner Heisenberg (1901–1976), Bohr was deeply invested in exploring the larger implications of his discoveries beyond the realm of physics. Karen Barad, a physicist and cultural theorist, finds Bohr's "philosophy-physics" essential to current debates about material–discursive relations. As Barad argues in detail, one of Bohr's most important contributions to physics, his complementarity principle, qualifies Heisenberg's more oft-cited uncertainty principle:

> Heisenberg's uncertainty principle is an epistemic principle: it favors the notion that measurements disturb existing values, thereby placing a limit on our knowledge of the situation. By contrast, Bohr's indeterminacy principle … is an ontic principle: the point is not that measurements disturb preexisting values of inherent properties but that properties are only determinate given the existence of

particular material arrangements that give definition to the corresponding concept in question. (Barad 2007, p. 261)

By claiming that objects do have inherently determinate, observation-independent properties, Bohr challenges the long-standing notion that the physical world abides by the laws of Newtonian physics. According to Bohr, there can be no clear distinction "between a phenomenon and the agency by which it is observed ... an independent reality in the ordinary physical sense can neither be ascribed to the phenomena nor to the agencies of observation" (Bohr 1987, p. 54). As Barad glosses, "Phenomena are constitutive of reality. Reality is composed not of things-in-themselves or things-behind-phenomena but of things-in-phenomena." Bohr's own analyses of the epistemological and ontological implications of his findings in quantum theory have not only made his work more accessible to non-specialists but also underline the importance he places on the interpretive aspects of physics. What makes his work particularly useful for cultural and literary theorists is not that it offers a scientific analogy for linguistic indeterminacy, but rather he shows that physical, mathematical, and linguistic indeterminacy are interrelated problems. Bohr not only finds that "things do not have inherently determinate boundaries or properties," but also that "words do not have inherently determinate meanings" (Barad 2007, pp. 139–40, 138).

While neither Bohr nor Stein is evidently aware of each other's writings, they share some notable similarities both in their perspectives on language, including a semiotic perspective on mathematics and the challenges it poses to classical notions of reality as stable and independent. In his claim against the idea that language is secondary, Bohr famously wrote, "We are suspended in language in such a way that we cannot say which is up and which is down. The word 'reality' is also a word, a word which we must learn to use correctly" (quoted in Peterson 1985, p. 302). Like Bohr, Stein begins to embrace the idea that language comprehends reality rather than lies behind it. Reflecting on her work in *Lectures in America*, Stein offers a comparable statement:

> Language as a real thing is not imitation either of sounds or colors or emotions it is an intellectual recreation ... And so for me the problem of poetry was and it began with *Tender Buttons* to constantly realize the thing anything so that I could recreate that thing. I struggled I struggled desperately with the recreation and the avoidance of nouns. (Stein 1998, p. 331)

According to Stein, language does not serve a mimetic or mediating function, but rather a creative one; meaning does not inhere in words or things, but is created over and again through different linguistic arrangements. This helps to explain why certain objects in *Tender Buttons* receive multiple sets of unique descriptions. For example, the first two of four different sections on "chicken" reads:

CHICKEN
Pheasant and chicken, chicken is a peculiar bird.
CHICKEN
Alas a dirty word, alas a dirty third alas a dirty third, alas a dirty bird. (Stein 1990b, pp. 492–93)

Each description alters the reader's impression of the object; the first sentence recalls the chicken as a kind of fowl, the second refers to connotations of the word "chicken." For Stein, as for Bohr, description does not uncover an object's inherent properties but "recreates" qualities observed under specific conditions. Language is the measuring apparatus that cannot be disentangled from the object under observation.

However distant or neutral the narrator of *Tender Buttons* might seem—given that the words appear detached from a perceiving, embodied subject—the text does not represent a world of static objects with observation-independent qualities, but rather a world of things in flux, phenomena to which an observer must be witness. The narrator's presence manifests in observations such as: "The author of all that is in there behind the door and that is entering the morning. Explaining darkening and expecting relating is all of a piece" (Stein 1990b, pp. 498–99). The author "behind the door" recognizes her inseparability from the phenomena described. Indeed, Stein's interest is in experiential phenomena, in capturing material, tactile, and sensory experiences. What might appear as a purely formal exercise turns out to be a reflection on the fundamental aspects of subjective experience. The concluding paragraph of "Roastbeef" insistently distinguishes the experiential and material from the abstract:

There is coagulation in cold and there is none in prudence. Something is preserved and the evening is long and the colder spring has sudden shadows in a sun. All the stain is tender and lilacs really lilacs are disturbed .... There is no superposition and circumstance, there is hardness and a reason and the rest and remainder. There is no delight and no mathematics. (Stein 1990b, p. 482)

Here, Stein arrives at an anti-Platonist perspective that rejects the notion of the physical world as a shadowy reflection of a transcendent, ideal realm. This passage suggests that there is nothing "behind phenomena" but only "things in phenomena." Her claim that there is "no mathematics" rejects the idea of math as something that exists outside of physical reality or apart from human conceptual frameworks. Bohr's quantum theory holds similar implications for mathematics; his indeterminacy principle overturns the Newtonian view that objects have inherently determinate values that need only be revealed mathematically.

Tender Buttons marks a significant shift in Stein's perception of mathematical objects compared to the view represented in *Q.E.D.*—that is, a shift away from the idea that mathematics simply uncovers hidden truths to the idea that it can be a creative practice. The numerals Stein uses in *Tender Buttons* and in her later plays and operas reinforce the semiotic nature of mathematical representation. For example, in the first section titled "Objects" and within the subsection called "A Box," Stein uses numbers as a descriptive, pronominal substitute for a box:

> The one is on the table. The two are on the table. The three are on the table. The one, one is the same length as is shown by the cover being longer .... The other is different and that makes the corners have the same shade the eight are in singular arrangement to make four necessary. (Stein 1990b, p. 465)

She plays around with the idea that these quantifiers have a definite amount and yet remain somewhat ambiguous pronouns, especially without the context the subtitle "A Box" provides. Rather than simply a rote tool for calculating and measuring, mathematics serves as an expressive means for describing and explaining phenomena.

Another crucial similarity that links Stein and Bohr to their mathematical contemporaries, including especially Frege, Russell, and Gödel, is that they all encounter the limits of representation, whether linguistic or mathematical, despite their efforts to achieve determinate results, that is, to attain an exactitude of reference (Stein), to maintain scientific objectivity (Bohr), to establish a robust foundation for mathematics (Frege and Russell), and to uphold the belief in an objective mathematical reality (Gödel).[13] Though their work offers some of the most prominent models of indeterminacy, these scholars were also invested in resolving the representational problems they encountered. These efforts are neither contradictory nor ironic, but rather the outcome of investigating what can be definitively known. In different ways, they sought as much to preserve a certain determinacy of meaning as to account for impediments to it.

Stein's self-analysis reflects this dialectic: while she aspires to make her writing "as exact as mathematics" and to know definitively "what a thing was," she also questions whether these goals can be realized and admits to arriving at unexpected results. She confesses in *Lectures in America* (1935) to being "bothered about something and it had to do as my bother always has had to do with a thing being contained within itself." She also "began to wonder at about this time just what one saw when one looked at anything really looked at something ... did it make itself by description by a word that meant it or did it make itself by a word in itself." She begins to question whether meaning inheres in the word or is instead created through its relation to other words. What "excited [her] so very much" was that the descriptions that conjured a specific object, or the "words that make what I looked at be itself," were never a precise or transparent referent for that object, often having "nothing whatever to do with what any words would do that described that thing." She recognizes the unwieldiness of language, the "difficulty," to use her own term, of pinning language down to an exact science. No longer bound by a classical, deterministic view of mathematics, Stein increasingly approaches mathematical terms and concepts as creative, constructive resources for critiquing gendered power relations and asking questions about "human nature" and "the human mind" (Stein 1998, pp. 305, 303).

Mathematics as Culture in Stein's Late Writing

Nothing in her critical essays or lectures evinces anything like a fully fledged philosophy of mathematics, but Stein's body of work does reveal the extent to which her ideas about mathematics evolved alongside her critical perspectives on language and meaning making. As she reconciled the indeterminacy of language, and her later writing became more evidently referential and culturally conscious, her sense of mathematics as a culturally embedded, linguistically dependent, and symbolically charged subject emerges as well.

In "Patriarchal Poetry" (1927), mathematics serves as an extended metaphor for her literary and cultural critique. In this poem, she explores mathematics not simply as a symbol for an abstract reality (*Q.E.D.*) nor as a descriptive or operational tool (*Tender Buttons*), but rather as a cultural phenomenon, bound to the system of patriarchal order and authority. Stein presents the traditional literary canon as similar to other male-dominated institutions that value order, method, reason, and tradition: "Patriarchal poetry reasonably/Patriarchal poetry administratedly/Patriarchal poetry with them too/Patriarchal poetry as to mind/Patriarchal poetry reserved/ Patriarchal poetry in regular places." Patriarchal poetry, according to Stein, is a masculinist institution ("Patriarchal poetry *their* origin and *their* history") whose standards are militantly upheld ("Patriarchal poetry left left left right left"), resulting in a homogenous body of literature ("Patriarchal poetry is the same"). She associates "patriarchy" with a parodied view of science, technology, and mathematics, and with the kind of authoritative language typical of scientific or technical writing: "Patriarchal in investigation and renewing of an intermediate rectification of the initial boundary between cows and fishes." She sees the literary canon as a business-like institution, one that develops according to a generalized economic principle in which value is determined by comparison to an ideal standard: "Patriarchal poetry makes no mistake makes no mistake in estimating the value to be placed upon the best and most arranged of considerations." Such rigid standards mean that the canon must "include cautiously," proceeding "one at a time." This systematic, quantitative method of inclusion/exclusion—"putting three together all the time two together all the time ... five together three together all the time"—masks the continual reproduction of traditional literary conventions such that we "never ... think of Patriarchal Poetry at one time." As her frequent use of numerical expressions to define patriarchal poetry suggest, Stein mocks the reliance on mathematics as an ideal standard for masculine epistemology and its attendant qualities: order, method, reason, tradition, and precision (Stein 1980, pp. 123, 110–11, 124).

She satirizes mathematics as a paradigm of rationality and objectivity ("There never was a mistake in addition") and also observes its appropriation to preserve hegemonic structures ("Patriarchal poetry makes no mistake"). She compares patriarchal poetry to the whole number system:

> Patriarchal Poetry shall be as much as if it was counted from one to one hundred.
> From one to one hundred.
> From one to one hundred.
> From one to one hundred.
> Counted from one to one hundred.
> (Stein 1980, pp. 106, 115, 125)

Like any number system—defined as a "set of symbols used to express quantities as the basis for counting, determining order, comparing amounts, performing calculations, and representing value"[14]—patriarchal poetry is a set of literary conventions that serve as the basis for determining hierarchies, comparing elements among literary works, and conferring value on these works. The number system is the most recurring metaphor Stein uses to stress the calculated control asserted over the literary canon by its bearers.

More broadly, she offers a model for a feminist poetics that is not simply imitative or reactionary, but rather *illustrative* of the gender biases that inform both literary and scientific praxis.[15] Beyond showing that language, including the language of mathematics, is deeply imbricated in gendered power relations, Stein's deeper aim, as I see it, is also to stress the endless permutations that language offers and to inspire new constructions with the available tools, so that patriarchal poetry "may be finally very nearly rearranged." (Stein 1990a, p. 117)

Stein's turn toward a more material, discursive conception of mathematics, and her related interest in how mathematical thought shapes our perceptions of the social and natural world, culminates in *The Geographical History of America or the Relation of Human Nature to the Human Mind* (1936). The book as a whole is an attempt to distinguish between human nature, or all that has to do with our "hard-wired" instincts and behaviors, and the human mind, which, according to Stein, does not store data, such as emotions or memories, but acts as a processor of immediate sensory data; "it knows what it knows when it knows it" (Wilder 1936, p. 8). Counting, she speculates, is perhaps what connects human nature to the human mind, joining instinct with reasoning: "What is the relation of human nature to the human mind. Has it anything to do with any number." She later suggests an intuitionist model of mathematics, or a kind of innate mathematical sensibility:

That is why numbers really have something to do with the human mind. That they are pigeons had nothing to do with it but that there was one and then that there were three and that then there are four and that then it may not cease to matter what number follows another but the human mind had to have it matter that any number is a number. (Stein 1936, pp. 78, 90)

Humans, she suggests, have an innate number sense, making them capable of distinguishing one from two from three, or performing simple arithmetic. In this way, she aligns with scholars such as George Lakoff and Rafael Nuñez, who argue for a mind-based model of math: "Mathematics is a product of the neural capacities of our brains, the nature of our bodies, our evolution, our environment, and our long social and cultural history" (Lakoff and Nuñez 2002, p. 9). Similarly, Stein suggests that our number sense develops in response to environmental, material stimuli, so that one might determine, for example, the number of pigeons in the sky. Like Lakoff and Nuñez, she thus challenges the Platonist philosophy of mathematics, which holds, as Brian Rotman argues, "that mathematical objects are mentally apprehensible and yet owe nothing to human culture; they exist, are real, objective, and 'out there,' yet without material, empirical, embodied, or sensory dimension" (Rotman 2000, p. 47). Instead, she suggests that the mental apprehensibility of numbers owes something to human culture and cognition.

For Stein, the connection between mathematics and human culture is well illustrated by the language that brings numbers into being, or that gives them "such pretty names": "The thing about numbers that is important is that any of them have a pretty name. Therefore they are used in gambling in lotteries in plays in playing in scenes and in everything. Numbers have such a pretty name." In context, the adjective "pretty" appears to mean that each number has a distinctive, characteristic name. She also suggests that it is only through semiotic or symbolic representation that numbers can have abstract applications, such as in gambling, lotteries, or scenes in a play. The basis of more complex, abstract mathematics is language. Without a representational system, one could do little more than perceive that "there were three and that then there are four." "Pretty" also implies that there is an aesthetic quality to numbers that involves emotion and sensation, a perspective she reinforces in the following lines:

It can bring tears of pleasure to ones eyes when you think of any number eight or five or one or twenty seven, or sixty three or seventeen sixteen or eighteen or seventy three or anything at all or very long numbers, numbers have such pretty names in any language numbers have such pretty names. Tears of pleasure numbers have such pretty names.

This passage emphasizes the embodied, sensory dimension of mathematics—the notion that one can sense that "any number is a number" and that this can produce emotional and physical effects, bringing "tears of pleasure to ones eyes." Numbers both have "something to do with" our most fundamental reasoning skills as well as our instinctual impulses, since numbers "have something to do with the human mind," but "tears have nothing to do with the human mind" (Stein 1936, pp. 78–79).

If our number sense tells us something about how the mind processes information, Stein also theorizes that numbers have "something to do with" how we make sense of the physical world:

> They have something to do with money and with trees and flat land, not with mountains or lakes, yes with blades of grass, not much a little but not much with flowers, some with birds not much with dogs, quite a bit with oxen and with cows and sheep a little with sheep and so have numbers anything to do with the human mind. (Stein 1936, p. 79)

She resists the common belief—one on which *Q.E.D.* was evidently predicated—that mathematics is inherently inscribed in nature. Instead, she implies an observing subject for whom quantitative reasoning is a means of navigating the social and natural world. Mathematics mediates between the external world and our perceptions of underlying form. As important, she insists on the material basis on which mathematics is founded; numbers function as a representational system through which we order the elements of our external world, such as "trees" and a "bit of oxen." Stein's considerations anticipate cultural theorists of science such as Brian Rotman, Michel Serres, and Bruno Latour, who likewise problematize the rigid opposition between the material and the mathematical and argue for mathematics as a semiotic system.

While *The Geographical History of America* reflects Stein's ongoing interest in formal experimentation and abstraction, it is also arguably her most focused study on the nature of the embodied, perceiving subject. Throughout this text, Stein repeats the phrase "I am I because my little dog knows me," a playful rewriting of Descartes's "I think, therefore I am." This belies a more serious attempt to show that identities are not essences, but instead defined by their relation to and effect on other experiencing subjects. This notion compliments Barad's re-articulation of Bohr's insights: humans are "not independent entities with inherent properties but rather beings in their differential becoming" (Barad 2003, p. 818). Stein's turn to mathematics as a means of exploring subjectivity is ultimately neither an attempt to abstract nor stabilize subjectivity, but rather to emphasize the process of "differential becoming." Stein comes to recognize the central role mathematics plays in debating and destabilizing Cartesian oppositions between object and subject, reason and emotion, mind and body, nature and culture. Her interest in mathematics also has important implications for literary studies; in exploring the interdependency of language and mathematics—the idea that "numbers have such pretty names"—Stein challenges one of the most enduring fictions: that literature and mathematics have little to do with one

356   A. BRUBAKER

another beyond serving as disciplinary opposites. What Stein's mathematician friend, Whitehead, said about artistic creation could well apply to how we might approach her challenging body of work: "Art is the imposing of a pattern on experience, and our aesthetic enjoyment is recognition of the pattern" (Whitehead 1954, p. 225). The recognition of mathematics as an enduring motif throughout her *oeuvre* helps make sense of—and might therefore enhance the enjoyment of—her artistic and intellectual contributions.

Notes

1. See, e.g., Farland (2004), Morgan (2006), Meyer (2002), Cope (2005), and Ashton (2002, 1997).
2. The emergence of mathematical formalism—the belief that mathematical signs function according to a set of formal rules, referring only to themselves—as well as the publication of Russell and Whitehead's *Principia Mathematica* (1910–1913), which sought to establish a logical foundation of mathematics, reflected efforts to reconfirm the certainty and precision of mathematics and demonstrate the kind of disciplinary self-examination also fundamental to the burgeoning movement of literary formalism. Russian formalism, an influential literary movement during the interwar period and predecessor to American New Criticism, urged a "scientific" approach to studying literary works and sought to define a text's intrinsic properties, its "literariness," or "the specific properties of literary material…that distinguish such material from material of any other kind" (Eichenbaum 1998, p. 8). Russian formalists considered the literary object, like all supposed mathematical objects (within Classical thought), as an enclosed, perfect system subject to its own internal standards and conventions. These methodological similarities between literary theory and mathematics are not simply an example of one discipline's influence over another, but more significantly reflect concurrently developing interests in questions of form, representation, objectivity, and referentiality—questions fundamental to how both define their disciplinary objectives.
3. I am not so much interested in tracking the degree of Stein's knowledge about specific mathematical theories (though scholars such as Steven Meyer and Jennifer Ashton have written about Stein's friendship with Alfred North Whitehead and her impressions of Bertrand Russell); instead, I am interested in how certain epistemological and ontological question characterized the intellectual "ether" of the time, not only transcending disciplinary boundaries but also inspiring disciplinary cross-pollination.
4. Platonism holds that mathematical objects are eternal and unchanging and exist apart from human experience. Brian Rotman describes, in contrast, a culturally embedded and semiotic-based perspective on mathematics in his book, *Mathematics as Sign: Writing, Imagining, Counting* (2000).
5. Scholars of mathematics refer to these most basic set theoretic principles as naïve set theory. See, e.g., Halmos (1974). It is worth noting that Cantor's ideas were first met with serious skepticism within the math community and his contemporaries, such as Frege, saw the need for a more rigorous treatment of these concepts.

6. Stein explains the concept of wholeness she aimed to explore in *The Making of Americans*: "You see, I had this new conception: I had this conception of the whole paragraph and in *The Making of Americans* I had this idea of a whole thing. But if you think of contemporary English writers, it doesn't work like that at all. They conceived of it as pieces put together to make a whole, and I conceived it as a whole made up of its parts (Stein 1974, p. 153).

7. This is one of Frege's axioms in which he attempts to axiomatize underlying assumptions of set theory—one that was ultimately shown to be inconsistent by Russell.

8. In their aims to axiomatize set theory and to develop a logical foundation of arithmetic principles, Frege and Russell found themselves having to theorize about language, and more specifically, to navigate the difference between what a term *means* and what it *denotes*. As Frege wrote in his 1918 essay "Thoughts": "[O]ne fights against language, and I am compelled to occupy myself with language although it is not my proper concern here." Their mathematical efforts produced important work in the philosophy of language; see, e.g., Frege, "Sense and Reference" (1892), and Russell's "On Denoting" (1905).

9. This line originally appears in her 1913 poem "Sacred Emily" as "Rose is a rose is a rose is a rose," but there are numerous variations of this sentence in her other works. In *Lectures in America*, Stein explains the intention behind this sentence, "When I said. A rose is a rose is a rose is a rose. And then later made that into a ring I made poetry and what did I do I caressed completely caressed and addressed a noun" (1998, p. 231). In *Four in America*, Stein again explains this sentence: "I think that in that line the rose is red for the first time in English poetry for a hundred years," suggesting that the qualities of a rose are inherent in the word 'rose' (Stein 1947, p. vi). Stein also insists that she "never repeats," which hints at her notion that each word functions like an individual, discrete unit that has its own function within a larger phrase or sentence.

10. Though Ashton acknowledges that Stein's writing generates "indeterminate effects," she maintains that these are counter to Stein's intentions, and thus aligns her goals with those of Russell and Whitehead, who likewise sought a determinate logical grammar: "her commitment to the autonomous text is directly bound up with an account of language that insists that, like symbols in Whitehead and Russell's vision of a logically perfect language in *Principia Mathematica*, words and their meanings stand in a relation of one-to-one correspondence," (Ashton 2005, p. 28).

11. *Q.E.D.* remained unpublished until 1950, four years after Stein's death. According to Marianne Dekoven, this was due to its "overt, realistic, autobiographical lesbian content" (quoted in Stein 2006, xi).

12. Knight defines the classical episteme as that "which privileges analysis and discrimination and in which space is conceived as three-dimensional; time as linear; and language as artificial yet determinately connected with the parallel plane of 'reality,'" while the "new modern sensibility" involves a turning away from "the problems of reference and human narrative, and back toward its own plasticity. Art here is autotelic; it serves its own ends" (Knight 1991, p. 45).

358   A. BRUBAKER

13. Though I do not go into detail here about Gödel's work, his Incompleteness Theorem builds on the problems of self-referentiality that these other mathematical logicists had begun. Gödel's theorems can be summarized as follows: "In any formal system adequate for number theory there exists an undecideable formula—that is, a formula not provable and whose negation is not provable...A corollary to the theorem is that the consistency of a formal system adequate for number theory cannot be proved within the system." Like other well-known scientific theories of this period (e.g. Heisenberg's "uncertainty principle" and Einstein's "relativity"), postmodern theorists have been drawn to Gödel's Incompleteness Theorem as another challenge to absolute determinacy and objectivity, but it's also important to note that Gödel remained a mathematical realist and did not necessarily see his ideas as a challenge to the foundation of mathematics. As Goldstein explains, "Gödel's audacious ambition to arrive at a mathematical conclusion that would simultaneously be a metamathematical result supporting mathematical realism was precisely what yielded his incompleteness theorems" (Goldstein 2005, pp. 23, 47).
14. See Lamb and Johnson (2002).
15. Maria Farland's essay, "Gertrude Stein's Brain Work," offers support for this point in her investigation of Stein's exposure to the "sexist assumptions in the realm of professional science," whereby women were seen as "incapable of abstract thought" and approached "brain work as intrinsically masculine" (Farland 2004, pp. 120, 124).

References

Ashton, Jennifer. 1997. "Gertrude Stein for Anyone." *ELH* 64: 289–331.
———. 1998. "Writing that Counts: Gertrude Stein and the Mathematics of Modernism." Ph.D. diss. Johns Hopkins University.
———. 2002. "Rose Is a Rose: Gertrude Stein and the Critique of Indeterminacy." *Modernism/modernity* 9: 581–604.
———. 2005. *From Modernism to Postmodernism: American Poetry and Theory in the Twentieth Century*. Cambridge: Cambridge University Press.
Barad, Karen. 2003. "Posthumanist Performativity: Toward an Understanding of How Matter Comes to Matter." *Signs: Journal of Women in Culture and Society* 28: 801–31.
———. 2007. *Meeting the Universe Halfway: Quantum Physics and the Entanglement of Matter and Meaning*. Durham: Duke University Press.
Beer, Gillian. 1996. *Open Fields: Science in Cultural Encounter*. Oxford: Clarendon Press.
Bohr, Niels. 1987. *The Philosophical Writings of Niels Bohr*. Vol 1. Woodbridge: Ox Bow Press.
Cope, Karin. 2005. *Passionate Collaborations: Learning to Live with Gertrude Stein*. Victoria: ELS Editions.
Eichenbaum, Boris. 1998. "Introduction to the Formal Method." In *Literary Theory: An Anthology*, edited by Julie Rivkin and Michael Ryan, 7–14. Oxford: Blackwell.
Farland, Maria. 2004. "Gertrude Stein's Brain Work." *American Literature* 76: 117–48.
Frege, Gottlob. 1948. "Sense and Reference." *The Philosophical Review* 57: 209–30.

Goldstein, Rebecca. 2005. *Incompleteness: The Proof and Paradox of Kurt Gödel*. New York: Norton.

Grahn, Judy. 1989. *Really Reading Gertrude Stein: A Selected Anthology with Essays by Judy Grahn*. Freedom: The Crossing Press.

Halmos, Paul R. 1974. *Naïve Set Theory*. New York: Springer.

Johnson, Phillip E. 1972. *A History of Set Theory*. Boston: Prindle, Weber & Schmidt.

Knight, Christopher. 1991. "'Tender Buttons,' and the Premises of Classicalism." *Modern Language Studies* 21: 35–47.

Lakoff, George and Rafael Nuñez. 2002. *Where Mathematics Comes From: How the Embodied Mind Brings Mathematics into Being*. New York: Basic Books.

Lamb, Annette, and Larry Johnson. 2002. "Number Systems." *42 Explore*. https://42explore.com/number.htm.

Meyer, Steven. 2000. *Irresistible Dictation: Gertrude Stein and the Correlations of Writing and Science*. Stanford: Stanford University Press.

Morgan, Lynn Marie. 2006. "Strange Anatomy: Gertrude Stein and the Avant-Garde Embryo" *Hypatia* 21 (1): 15–34.

Morrison, Mark. 2002. "Why Modernist Studies and Science Studies Need Each Other." *Modernism/modernity* 9: 675–82.

Perloff, Marjorie. 1999. *The Poetics of Indeterminacy: Rimbaud to Cage*. Evanston: Northwestern University Press.

Peterson, Aage. 1985. "The Philosophy of Niels Bohr." In *Niels Bohr: A Centenary Volume*, edited by A. P. Kennedy and P. J. French, 299–310. Cambridge: Harvard University Press.

Rotman, Brian. 2000. *Mathematics as Sign: Writing, Imagining, Counting*. Stanford: Stanford University Press.

Russell, Bertrand. 1905. "On Denoting." *Mind* 56: 479–93.

Stein, Gertrude. 1936. *The Geographical History of America or The Relation of Human Nature of the Human Mind by Gertrude Stein*. New York: Random House.

———. 1947. *Four in America*. New Haven: Yale University Press.

———. 1957. *To Do: A Book of Alphabets and Birthdays*. Edited by Carl Van Vechten. New Haven: Yale University Press.

———. 1974. "Transatlantic Interview." In *A Primer for the Gradual Understanding of Gertrude Stein*, edited by Robert Bartlett Haas, 31–32. Los Angeles: Black Sparrow Press.

———. 1975. *How Writing Is Written (Volume II of the Previously Uncollected Writings of Gertrude Stein)*. Edited by Robert Bartlett Haas. Los Angeles: Black Sparrow Press.

———. 1980. "Patriarchal Poetry." In *The Yale Gertrude Stein*, edited by Richard Kostelanetz, 106–46. New Haven: Yale University Press.

———. 1990a. "The Autobiography of Alice B. Toklas." In *Selected Writings of Gertrude Stein*, edited by Carl van Vechten, 1–238. New York: Vintage.

———. 1990b. "Tender Buttons." In *Selected Writings of Gertrude Stein*, edited by Carl Van Vechten, 407–52. New York: Vintage Books.

———. 1998. "Lectures in America." In *Stein: Writings 1932–1946*, edited by. Catherine Stimpson and Harriet Chessman, 191–336. New York: Library of America.

———. 2006. *Three Lives and Q.E.D.* New York: Norton.

———. 2008. "Are There Arithmetics." In *Gertrude Stein: Selections*, edited by Joan Retallack, 198–201. Berkeley: University of California Press.

Whitehead, Alfred North. 1954. *Dialogues of Alfred North Whitehead*, Boston: Little, Brown.

Wilder, Thornton. 1936. "Introduction." In *The Geographical History of America or The Relation of Human Nature of the Human Mind by Gertrude Stein*. New York: Random House.

CHAPTER 20

Modernist Literature and Modernist Mathematics I: Mathematics and Composition, with Mallarmé, Heisenberg, and Derrida

Arkady Plotnitsky

INTRODUCTION: MATHEMATICS AND LITERATURE FROM MODERNITY TO MODERNISM

This chapter addresses the relationships between modernist literature, a long-established denomination, and modernist mathematics, a recent and infrequent denomination. *Historically*, both phenomena are commonly understood as belonging to the same period, roughly, from 1900 to the 1940s, but as extending to or continuing to impact literature and mathematics of our own time. *Conceptually*, the thinking concerning each (or modernism elsewhere) is more diverse and, in each case, only partially reflects the nature of each and the relationships between them.[1] This is an unavoidable limitation, and the conception of modernism to be offered here cannot circumvent it either. This conception is grounded in the double movement, found, I argue, in modernist literature and modernist mathematics alike. The first is the movement from ontology to technology, specifically the technology of composition, which can place the ultimate nature of reality (referring, roughly, to what exits) beyond a representation or even conception, and thus beyond ontology (referring to such a representation or at least conception). While, generally, technology is a means of doing something, of getting "from here to there," as it were, this article extends the concept of "experimental technology" in modern, post-Galilean physics, defined by its jointly experimental and mathematical character, to mathematics and literature, and

---

A. Plotnitsky (✉)
Purdue University, West Lafayette, IN, USA

© The Author(s) 2021
R. Tubbs et al. (eds.), *The Palgrave Handbook of Literature and Mathematics*, https://doi.org/10.1007/978-3-030-55478-1_20

361

to theoretical thinking elsewhere. Thus, this extension, when it comes to abandoning ontological thinking, also defines (modernist) theoretical physics, beginning with quantum mechanics, as a fundamentally mathematical project. The second movement is toward independence and self-determination of either field, which does not preclude the relationships between them.

Emerging from this double movement, *modernist* thinking in literature or mathematics, or other fields, is, in the present view, defined by *a realization of at least the possibility* of, ultimately correlatively, the *irreducible unthinkable* in thought, the *irreducible (uncontainable) multiple*, with *irreducible randomness or chance*, often accompanying them. My emphases indicate, first, that realizing this possibility need not mean that one adopts this irreducibility as one's actual philosophical position, in which case I shall speak of "radical modernism"; and secondly, that the *irreducible* nature of each is essential, because the unthinkable, the multiple, and chance are also considered by what I shall call classical thinking, a form of ontological thinking. There, however, each is seen as ultimately reducible to, respectively, accessibility to thought, unity or containable multiplicity, and causality (a connection, given by a rule, by means of which something, *the cause*, leads to something else, *the effect*, and as such is not required in modernist thinking).

Three cases, I argue, offer remarkable and historically important examples supporting my argument: that of Stéphane Mallarmé (1842–1898), marking the onset of literary modernism, contemporaneous with the *rise* of mathematical modernism; that of Kurt Gödel (1906–1978), at the time, in the 1930s, when mathematical modernism had reached its full-fledged stage, which extends to much of the mathematics of our time; and between them, contemporaneously with the height of literary modernism, that of Werner Heisenberg (1901–1976) in the mid 1920s. I shall discuss Gödel's case, in connection to Mallarmé, in the companion article in this volume, and focus here on Mallarmé and Heisenberg. The reasons for this, perhaps unexpected, connection are as follows. First, Heisenberg's thinking was defined by the same double movement from ontology to technology, that of mathematical composition, and toward greater *mathematical* independence of theoretical physics, now freed from the aim of representing nature (ontology), which defined the preceding history of modern, post-Galilean, mathematical physics. Heisenberg's aim was a theory, now known as quantum mechanics, that was a *mathematical technology* able to predict, in probabilistic terms (no other predictions are possible on experimental grounds), certain experimentally observed events, without mathematically representing those aspects of nature that were responsible for these events (Heisenberg [1925] 1968; [1930] 1949). Heisenberg's use of mathematics may be seen as *compositional*. Quantum mechanics was mathematically *composed* by Heisenberg in the manner of abstract compositions in painting, such as those of Wassily Kandinsky (1866–1944), which equally broke with representation in art, as did modernist literary compositions (abstracted from ordinary language, representational in its functioning). Thus, mathematics became compositional in quantum mechanics, following the same trend in modernist mathematics itself, proposed a concept ofhile literary or artistic compositions became akin to abstract mathematics. That is, then, the first reason why I would like to consider Heisenberg's work in the context of the relationships between literature and mathematics. The second reason is the essential relationship between the mathematical (a broader concept, extending beyond mathematics in its disciplinary sense) and the random or (they are not the same either) probabilistic, shared by Mallarmé's and Heisenberg's projects. While in Heisenberg's case these relationships were unavoidable on experimental grounds, bringing them to literature in this proto-quantum-mechanical way was a remarkable and radical feature of Mallarmé's literary practice.

Before I proceed to my argument, I would like to establish more firmly my key concepts. I begin with modernity and modernism, defined here, first, in contradistinction from each other and, secondly, in connection not only to literature or art, but also to philosophy, mathematics, and science, or to theoretical thinking in general. "Modernity" is customarily seen, and will be seen here, as a cultural category. It refers to the period of Western culture extending from about the sixteenth century to our own time, although during the last fifty years or so, modernity entered a new stage, known as postmodernity, brought about by the rise of digital information technology. Modernity is defined by several interrelated transformations, sometimes known as revolutions, although each took a while. Among them are scientific (defined by the rise of modern mathematical sciences and the new cosmological thinking, beginning with the Copernican heliocentric view of the solar system); industrial or, more broadly, technological (defined by the transition to the primary role of machines in industrial production and beyond); philosophical-psychological (defined by the rise of the concept of the individual human self, from the Cartesian cogito on); economic (defined by the rise of capitalism); and political (defined by the rise of Western democracies).

In contrast to modernity, modernism has been primarily used as an aesthetic category, referring to certain developments in literature and art in the first half of the twentieth-century, from roughly the 1900s on. In literature, it is represented by such authors as Marcel Proust, Paul Valery, James Joyce, Franz Kafka, Virginia Woolf, and Samuel Beckett, and earlier figures, such as Charles Baudelaire and Mallarmé, my main literary subject here. His thinking had closer relationships with mathematics than that of other modernist authors, who were more interested in modernist physics, such as relativity and quantum physics, than modernist mathematics. Physics is also important for Mallarmé, especially because of the role of chance in his thinking. Chance is a more physical concept, although it can be applied to events of thought, as it was by Mallarmé, as stated in the final sentence of *Coup de dés* (1897): "Every Thought Emits a Throw of Dice" (Mallarmé 2012, p. 235).[2]

As I said, the term "modernist" has rarely been used in considering mathematics and physics, as opposed to "modern," used, however, with different periodizations. In mathematics, "modern" tends to refer to the mathematics that had emerged in the nineteenth century, while in physics it refers to all mathematical-experimental physics, from Galileo Galilei (1564–1642) and René Descartes (1596–1650) on. This is fitting because this physics emerged along with and had shaped the rise of modernity, making the latter fundamentally scientific. After the discovery of relativity and quantum theory, the term "classical physics" was adopted for the preceding physics. More recently, however, one does find applications of the term "modernism" to mathematics and science. Thus, in *Plato's Ghost: The Modernist Transformation of Mathematics* (2007), Jeremy Gray (1947) proposed a concept of *modernism* in mathematics. The concept covers certain developments in *modern* mathematics (such as topology, set theory, and abstract algebra) that had reached their modernist stage around 1900, contemporaneously with the rise of literary or artistic modernism. Gray mentions the latter, but does not really address the relationships between it and mathematics, my main subject here.

Modernist thinking fundamentally concerns the nature of reality, the nature of that which exists in matter or mind, and what we can or cannot know and, in the first place, think concerning the character, especially the ultimate character, of this existence. The term "ontology," derived from the ancient Greek *on* (to exist), was made prominent during the last century by Martin Heidegger (1889–1976), in particular, his 1927 *Being and Time* (Heidegger 2010). It refers to one's understanding or, as Alain Badiou (1937–) would have it, to a decision or thought, of what exists and (nearly uniformly) *how* it exists, the *character* of this existence or reality (Badiou 2007, p. 94). By reality, I shall understand, very generally, that which exists or is assumed to exist, without necessarily making a claim concerning the character of this existence. I understand existence as a capacity to have effects (to which one cannot always assign causes) on the world with which we interact, the world that also has such effects upon itself. Our phenomenological (representational) or epistemological (cognitive) technologies may be in either harmonious or discordant relationships with the character of reality: from knowledge or representation of this character, to only forming a conception concerning it, to the impossibility of forming such a conception.

As stated, modernism is defined here by the *awareness* of this range, including the last position, adopting which, *in dealing with the ultimate character of reality considered in a given field* (I shall explain this emphasis presently), I call radical modernism. Modernism is not limited to this awareness alone. It involves the concepts of multiplicity and randomness, or chance, and sees them as equally irreducible, although these concepts cannot, by definition, apply to reality, if the latter is assumed, in accordance with radical modernism, to be beyond thought. Reversing a more common concept of that which can be thought but not named or represented, at stake here is that which can be named, as "reality beyond thought," but cannot be thought of and thus represented. This reality is literally unthinkable, ultimately unthinkable even as unthinkable, for to be unthinkable is still a human concept, as are all other concepts, such as character, constitution, and so forth. At the same time, this unthinkable is seen as responsible for everything we can think of or know in a given domain. A reality beyond thought is also a reality beyond realism. Realism, a term more prominent in the philosophy of science (or, in considering literary realism, in literary studies) is defined, more or less equivalently to ontology, as an assumption of either the possibility of a representation or at least conception of the *character* of reality, rather than merely claiming the existence of something. The ultimate constitution of the reality (material or mental) considered is assumed to possess properties and the relationships among them, or, as in so-called structural realism, just relations or structures, which are either known in one degree or another, or representable, or unknown or even unknowable, but still assumed. It is crucial, however, that placing reality beyond thought only concerns the ultimate character of reality, for ontologies are possible and necessary at other levels. Indeed, the decision to adopt this concept of reality is based in, inferred from, ontological structures at these other levels and must be so inferred in order to be theoretically rigorous, rather than arbitrary.

While there are earlier precursors, for example, certain Romantic authors (Plotnitsky 2015), the *possibility* that the ultimate nature of reality is beyond thought comes into the forefront with modernism. I, again, stress "possibility" because not all thinkers associated with modernism ultimately adopted the conception of reality beyond thought and thus radical modernism as their actual philosophical position, in contrast to a more common emphasis on the independence and self-determination of modernist mathematics or literature. One finds ambivalences in this regard in Mallarmé and Heisenberg (especially in his later works), and an outright rejection of anything like "reality beyond thought" in many major modernist figures in mathematics and science, Gödel and Albert Einstein (1879–1955) among them. Gödel would have seen radical modernist thinking as a failure of finding a proper representation of independent mathematical reality, along the lines of mathematical Platonism (explained below). He would have thought of this reality as yet unthought, rather than as never to be reached by thought. This is how he saw his own findings, concerning undecidability and incompleteness. On the other hand, Jacques Derrida (1930–2004), whose reading of Mallarmé metaphorically borrows Gödel's concept of undecidability, may be argued to represent radical modernism in philosophy, also in connecting the irreducible unthinkable (admittedly, not Derrida's term, who indeed appears to have thought more in terms of the irreducibly unrepresentable) to the irreducibly multiple. Both imply each other, for one thing because any form of containment of the multiple would imply that ultimate reality could be given a conception, even if not a specifiable ontology. By the same token, while each time unthinkable,

366 A. PLOTNITSKY

this reality is each time, with each of its effects, different. Derrida's reading of Mallarmé in "The Double Session," defined by this conjunction of the irreducibly unthinkable or at least unrepresentable, associated with what Derrida calls "*différance*," and the irreducibly multiple, associated with what he calls "dissemination" (Derrida 1981, pp. 173–206).[3]

It follows that, while modernist thinking is defined here by realizing the possibility of, correlatively, the irreducible unthinkable in thought, the irreducible multiple, with irreducible randomness or chance, accompanying them, only the irreducibly unthinkable applies to the ultimate nature of reality considered, which, by being beyond thought, cannot be multiple any more than single, or random any more than causal. Irreducible multiplicity or irreducible randomness or chance appear at the level of effects or events. Besides, it is possible to assume that the ultimate nature of reality is random or chance-like, or mixed, and random ontologies have been around since the pre-Socratics, as in Democritus's and then Epicurus's and Lucretius's atomism, also important as ontologies of the irreducibly multiple, but without the irreducibly unthinkable.[4] It is worth qualifying that the ultimate reality at stake here is not merely something that, while as yet unthought, might eventually be reached by thought. If such a conception is formed, one reverts to ontology or realism, at least a nonrepresentational one. Following Einstein, many have hope that this will eventually happen in fundamental physics and some argue that quantum theory can be understood on realist lines in any events. As things stand now, the debate continues with undiminished intensity, and it is anyone's guess whether this will ever be resolved or not.

Probability theory, which mathematically deals with events whose outcome is undetermined, is rarely considered as part of modernist mathematics (say, in Gray's sense). This is not surprising. First, mathematical propositions, including those of probability theory, have strictly ordered structures, which leave no place for indeterminacy in their organization or operation, an important point to which I shall return. Secondly, while modernist mathematics is characterized by its divorce from physics, probability theory was not only initially developed to deal with physical processes and events, beginning with a throw of dice, with Gerolamo Cardano (1501–1576), Blaise Pascal (1623–1662), and other founding figures, but never lost its connection to them. Randomness or chance is most essentially an effect of material reality, even in the case of events of thought, because it involves temporality, and probability always concerns future events. David Hilbert (1862–1944), stating the sixth problem of his famous 1900 list (which helped to usher in mathematical modernism), the problem of the axiomatization of mathematized physical sciences, expressly mentioned probability theory and mechanics. Probability theory may, however, be given a modernist flavor, akin to that of quantum mechanics (a form of probability theory), because it establishes predictive, rather than representational, relationships to the world. Probability theory only a possible outcome of a throw of dice. It does not describe this throw; classical physics does, even if only in principle and not in practice, which is why we cannot predict the outcome of each throw exactly. In classical cases, however, such as a throw of dice, this recourse is necessitated by the mechanical complexities of the systems considered, the behavior of which, while it cannot be tracked in practice, emerges from the behavior of the individual constituents that is causal and, in the first place, amenable to a realist representation by means of classical mechanics. I define causality as an ontological concept according to which the state of a given system is determined at all future moments of time, once it is determined at a given moment of time; and determinism as an epistemological concept, referring to the possibility of predicting the outcomes of the processes and events exactly, which may not be the case in classical physics (when the system considered is complex) and is *never* the case in quantum physics, interpreted on radically modernist lines. There the individual behavior of the ultimate constituents of matter is not amenable to representation and, as a consequence, is never causal. As a result, in contrast to classical physics, even those events that are defined by the individual behavior of these constituents ("elementary particles") are undetermined.

One must keep in mind the difference between randomness or chance, or (a more general category) indeterminacy, and probability. Both indeterminacy and randomness or chance refer to a manifestation of the unpredictable. The uses of these terms fluctuate. I shall use them as follows. I shall use "indeterminacy" as a more general term, referring to any event that cannot be predicted exactly, but at most probabilistically. I shall use "randomness" or "chance" synonymously and speak primarily of chance, which is more pertinent in Mallarmé's case. (In quantum mechanics, randomness is used more commonly). Randomness or chance will refer here to indeterminate events to whose occurrence no probability can be assigned. Either an indeterminate or chance event may or may not result from some underlying causal processes, whether this process is accessible to us or not. The first eventuality defines classical indeterminacy or chance, purely epistemological in nature because the ultimate reality considered is assumed to be causal. The second eventually, which makes indeterminacy and chance irreducible, is found either in an indeterminate or chance ontology, as the ultimate ontology, or in radical modernism, in the absence of an ultimate ontology. The absence of causality is automatic if one adopts either view, in the second case, again, because an assumption of causality (just as that of ontological randomness or chance) makes it available to thought.

Probability and statistics deal with estimates of the occurrences of certain individual or collective events, which defy deterministic handling, in accordance with mathematical probability theory. Probability, thus, introduces an element of order into situations defined by the role of indeterminacy or chance in them. Commonly, "probabilistic" refers to our estimates of the probabilities of either individual or collective events, and "statistical" refers to our estimates concerning the outcomes of multiple identical or similar experiments. In particular, the so-called Bayesian understanding of probability defines probability as a degree of belief concerning an occurrence of an individual event on the basis of the information we possess.

The occurrence of an individual indeterminate event may, however, not allow for an assignment of probability on Bayesian lines, which makes it a random or chance event. In quantum mechanics, while the indeterminate or chance character of elemental individual quantum events is irreducible, it is a matter of interpretation whether it is a random event that can be assigned probability, on Bayesian lines, as in the so-called Quantum Bayesianism or QBism (for example Fuchs et al. 2014), or is a purely chance event that precludes such an assignment (for example Plotnitsky 2016, pp. 173–86). On the other hand, certain collectivities of quantum events are statistically correlated and thus ordered, which is an experimentally confirmed fact. Indeed of the greatest mysteries of quantum physics is that, in certain circumstances, individual random or chance events form statistically correlated and thus ordered multiplicities (of the type not found in classical physics). This makes quantum theory as much about order, a statistically correlated order, as about indeterminacy or chance. It is equally mysterious and fortunate for us that quantum mechanics correctly predicts this order; a pure chance we cannot predict. In fact, this mysteriousness is unavoidable in any radically modernist theory because no order, individual or collective, can be assumed to be inherited from some underlying order, which would imply a conception of the ultimate reality. This mystery is, however, without mysticism, because one does not presuppose any mystical agency governing these processes, say, in the manner of the mystical or negative theology. "*Le mystère précipité*" of *Coup de dés* may be such a mystery without mysticism (Mallarmé 2012, p. 231).

The preceding outline, again, offers a particular conception of modernist thinking, even though this conception, I argue, characterizes the thinking of several key modernist literary figures mentioned above, as well as of most philosophical figures considered here, sometimes associated with postmodernist theory, many of whom aligned their thought with modernist literature and art, as well as modernist mathematics and science.[5] Not all of this thinking was, again, radically modernist. Historically, modernist literature and art, or mathematics and science, have certain additional traits, found in the thinking of these figures as well. Some of these trends do relate to the modernist traits introduced here. The nature of these relations is complex, and addressing them in detail would extend this chapter beyond its intended scope. Most crucial is, again, the project of *independence* (not excluding the political implications of the term).

In mathematics, this was an independence from physics and, with it, from the representation of natural objects, philosophically, the realism, conceptual or mathematical, of physics. On the other hand, mathematical realism and, especially, the so-called mathematical Platonism (a twentieth-century development, not to be identified with Plato's own thought) has been important for the project of the independence of mathematics. This project has been developing as part of *modern* (post-1800) mathematics, but by1900 it became part of the *modernist* project that reached even further by breaking with general phenomenal intuition in all areas of mathematics, notably, in geometry, making it "profoundly counterintuitive." "This realization," Gray argues, "marks a break with all philosophy of mathematics that present mathematical objects as idealizations from natural ones: it is characteristic of modernism" (Gray 2008, p. 20). Indeed, mathematical modernism divorces mathematical concepts from our general phenomenal intuition as well, as was stressed by Hermann Weyl (1885–1955), himself a major figure in mathematical modernism (Weyl [1918] 1994, p. 108).

The same type of divorce has been at work in physics and reaches the same stage with relativity and especially quantum mechanics. Both used modernist mathematics, respectively Riemannian geometry and infinite-dimensional Hilbert spaces, divorced from our phenomenal general intuition. The first still did this in a realist way, as the divorce from our phenomenal intuition does not entail a divorce from realism or ontology, because the latter could be mathematical, and is in all of modern physics. The second, however, used mathematics in a modernist way, by providing the mathematical technology of probabilistic predictions concerning the outcomes of quantum experiments, as events, without providing a representation or even conception of the processes responsible for these events.

In parallel, modernist literature or art aims to make itself independent of representation and realism. It is this independence that, for Mallarmé, makes poetry akin to mathematics. Literature still borrows its symbols from ordinary language, which complicates this analogy, although mathematics, let alone physics, cannot be entirely freed from ordinary language either. However, just as modernist mathematics is above all mathematics and modernist physics is above all physics made mathematics, modernist literature or art are above all literature or art. All three are parallel, if, while keeping their independence, interactive, projects that concern themselves with the irreducibly unthinkable, the irreducibly multiple, and irreducible randomness or chance, but also with new forms of order arising from them. To cite Niels Bohr (1885–1962): "We are not dealing here with more or less vague analogies, but with an investigation of the conditions for the proper use of our conceptual means shared by different fields" (Bohr 1987, p. 2). It is not merely a matter of traffic, for example, metaphorical or narrative, between fields but of parallel modernist situations in each field.

Mathematics and Composition: Mallarmé and Heisenberg, with Derrida

I begin my discussion of Mallarmé with a disclaimer. This chapter offers only a limited engagement with Mallarmé's works, especially his literary works, a subject of numerous readings, including by several major theorists, such as

370   A. PLOTNITSKY

Maurice Blanchot (1907–2003), Derrida, and Badiou, and numerous commentaries on these readings, although next to none (apart from Badiou's) deal with mathematics.[6] I am primarily concerned with Mallarmé's theoretical ideas, and will discuss (deliberately, given my aims here) some of the best known of Mallarmé's passages. It is true that his theoretical ideas are often enacted in his literary works, and his theory of literature cannot be adequately understood apart from these enactments, some of which will be addressed here. Besides, Mallarmé mixed his theoretical and literary works, sometimes reversing their roles, as in *Divagations* (Mallarmé 2009). On the one hand, then, there is Mallarmé's conception of literature, which his literary works practice, and on the other, some of his works are the enactments or allegories of this conception, including as it functions in these works themselves, while, as I shall explain, still escaping being ultimately representational of this functioning.

Mathematics and the relationships between literature and mathematics are central to Mallarmé, aided by his concerns with the question of translation and Descartes, whose mathematics is defined by translation, namely that between geometry and algebra. These translations may be between languages, French and English in particular, different art forms, such as literature and music, or literature and dance, or between different fields, such as literature and mathematics, or literature and philosophy, sometimes in their interaction as in translating Edgar Alan Poe (1809–1849), the author already preoccupied with translations, that between literature and mathematics included. As Mallarmé wrote:

> In others, the great and long period of Descartes.
> Then, in general, some La Bruyère and some Fénelon with a hint of
> Baudelaire.
> Finally, some me – and some mathematical language. (Mallarmé 2013, p. 851)

More than the language of mathematics itself, "mathematical language" here is the language of literature or poetry, poetic language (a later concept with strong affinities with Mallarmé's view), to which Mallarmé wants to give a rigor akin to that of mathematics, without making it mathematics in its disciplinary sense. Conversely, mathematics, even in its disciplinary practice, acquires a dimension of literature, because of the defining role of composition there. It is not clear how much Mallarmé was familiar with contemporary mathematics, but he must have been to some degree, because its various aspects and controversies, such as those surrounding set theory, were extensively discussed at that time in intellectual circles and popular press. More important, however, is, again, to return to Bohr's formulation, "an investigation of the conditions for the proper use of our conceptual means shared by different fields" (Bohr 1987, p. 2).

Descartes and the project of analytic geometry is an intimation of modernist thinking in mathematics and physics at the heart of modernity, an intimation that has its history, too, as part of the history of algebra, especially the concept of equation that emerged in ancient Greek mathematics, especially in Diophantus (around the third century AD). Analytic geometry, however, by in effect making geometry algebraic, gave mathematics its de facto independence of physics and material nature, thus, along with Descartes's contemporary algebraists, such as Pierre Fermat (1607–1665), initiating mathematical modernism within modernity. Descartes's analytic geometry did so because the equation corresponding to a curve, say, $Y = X^2 - 1$ for the corresponding parabola, could be studied as an algebraic object, independently of its geometrical representation or its connection to physics. A curve becomes, in its composition, defined by its equation, divested from its representational geometrical counterpart. It no longer geometrically idealizes the reality exterior to it. It only represents itself, is its own ontology, as does and is a poetic *line* (a rhythmic curve) divorced from its functioning, especially, a representational one, as in ordinary language—a divorce that, as will be seen, is crucial to Mallarmé. The equation is the poetry of the curve.

With his invention of quantum mechanics, Heisenberg brought this independence to physics by using highly abstract (modernist) mathematics, which does not represent any reality, material or mental, exterior to itself, and, in this respect, made quantum mechanics essentially mathematical. Its connections to physical reality were limited to probabilistic predictions concerning quantum events, without representing these events. Heisenberg moved to the algebra of quantum mechanics against the representational geometry of classical physics. Einstein, in this connection, spoke of "the Heisenberg method" as "purely algebraic" (Einstein 1936, p. 75). One could also describe Mallarmé's poetic method in this way, giving algebra a broader sense, yet still applicable in mathematics—namely that of composition. Nothing can describe Heisenberg's method better either: it was compositional without being representational. In literature, mathematics, and physics (at least modern, mathematical physics), alike, ontology or realism is merely a particular way of using the technology of composition.

For Mallarmé, poetry is a kind of abstract algebra, in absolute strictness of its structure, as against the geometrical (representational) arbitrariness of ordinary language, which may well have been the meaning of "mathematical language" in the passage just above. This is not merely a matter of form or genre, which, while sometimes radically transformed, may be conventional, as a genre sometimes was in Mallarmé's compositions, his sonnets, for example. However, the *composition* of his sonnets was far from ordinary language, even, as I shall explain, when it becomes an allegory of its own creation. This poetic "algebra" has its modern history, too, as modern "realist" literature or art (never entirely realist) has modernist dimensions, even if without, in parallel with Descartes's use of algebra, quite divorcing it from realism in the way modernism does. Literary compositions, especially poetry, have always been more akin to a language of its own, with its own vocabulary and grammar, rather than being statements in ordinary language. Thus, metaphor is a great technology for creating new words, giving a new signified to a signifier, not found in any dictionary, although it may eventually end up in a dictionary. Modernist independence reveals an essential element at the core of all art, as Mallarmé realized. As he wrote: "I say: a flower! And from the oblivion to which my voice relegates all contours, as something other than unmentioned calyces, musically arises the idea itself, and sweet, absent from all bouquets" (Mallarmé 2010, p. 851). The flower, thus announced, enters a poetic composition (even if it only consists of one word) and loses its reference to any actual flower, while retaining its singularity and concreteness, as an idea, rather than being a general concept of flower. It is more like an algebraic symbol, thus, giving a new meaning to symbolism and brining it closer to modernism. Badiou relates this passage to mathematics. He says: "If I say 'a flower,' I separate it from every bouquet. If I say 'let a sphere,' I separate it from every spherical object. At that point, matheme and poem are indiscernible" (Badiou 2007, p. 47). This is true. However, the parallel with separating the algebra of the equation from the geometry of the curve is closer to Mallarmé, who, too, belongs, to return to his locution, to "the great and long period of Descartes," still far from over (Mallarmé 2013, p. 851). Poetry is the algebra of language.

If, then, literature or, ultimately, mathematics is defined by the technology of composition, freed from representation and thus ontology, how does one bring this view in accord with several of Mallarmé's statements, such as, famously, "*Le monde est fait pour aboutir à un beau livre*" [the world is made to come up with a beautiful book] (quoted in Huret 1999, p. 45), or with the unfinished long-term project book of the Book [*Le Livre*], which one would assume to be literature? The first statement, akin to a fragment of Heraclitus, is too open and the project of "the Book" is far too unfinished for any definitive interpretation of either or of the concept of book used by Mallarmé. Thus, Blanchot, arguably the most prolific commentator on Mallarmé and a modernist literary author himself, links this project to "the absence of the book" or "the disappearance of the book," and, in juxtaposition to Hegel, to the disappearance of the author, which, too, brings literary compositions closer to mathematical ones (Blanchot 1992, pp. 427–29). Indeed, as Mallarmé argued: "The pure work of art implies the elocutionary disappearance of the poet who yields the initiative to words," just as a mathematical composition yields to the initiative of the (symbolic) elements of mathematical structures (Mallarmé 2010, p. 849). While there are reasons for this claim, which is a commonplace of commentaries on Mallarmé and modernism (and which was taken up by several trends in literary studies), this is not entirely the case, even in mathematics. The author both disappears in the independence of a literary composition and remains a signature underneath it, and the name of the problem it poses. Thus, Quentin Meillassoux (rightly) sees *Coup de dés* as a unique chance event, "a zero-event … a unique, nonreproducible wager … of which Mallarmé is the 'unique Name'" (Meillassoux 2012, p. 166). As Blanchot asserted: "Through the very force of his experience, Mallarmé immediately pierces the book in order (dangerously) to designate the Work whose center of attraction—a center always off center—would be writing" (Blanchot 1992, p. 429). Blanchot in part follows Derrida's argument for "the end of the book and the beginning of writing," in Derrida's special sense in *Of Grammatology* (Derrida 2016, p. 6), discussed below, and Derrida's reading of Mallarmé, which juxtaposed Mallarmé and Hegel along similar lines, including decentering (a major theme in Derrida). There are still books but they are only cases of composition. They have structures but need not have the totalizing aims of the Platonist or Hegelian (dialectical) project of the book.

Both Plato and Hegel are, nevertheless, great precursors of Mallarmé, Plato not the least as concerns mathematical thinking; by contrast, he no longer is a model for philosophical thinking in Hegel. Plato famously condemned, as in Books III and X of *The Republic*, poetry for its bad mimesis, only possible for philosophy and its mathematical thought, which Hegel subtracts from philosophy (Plato 2005, pp. 1022–52, 1199–211). For Plato, as for Parmenides, his pre-Socratic precursor, the ultimate ontology was ideal, a form of *thought* and, in part correlatively, mathematical. Following Derrida, I shall define Platonism, a certain tradition of thought stemming from Plato, thinking based on the claim that one can *imitate* or represent (the Greek *mimesis* has both meanings), that which truly is, that is, to have an ontology of the ultimate constitution of the real. This type of claim is found in Plato, too, but, in addition to the complexities of his argument, one also finds movements away from it, even if against Plato's own grain, not found in most forms of Platonism. According to Derrida: "What is decided and maintained by [the Platonist] ontology … is precisely the ontological: the presumed possibility of the discourse about what is, the deciding and decidable logos of about the *on* (being present)" (Derrida 1981, p. 191).

In contrasting Mallarmé's "mimesis" to the Platonist one, Derrida is adopting, by "analogy or metaphorically," Gödel's concepts of decidable and undecidable propositions in mathematical logic, which, respectively, allow or disallow one to ascertain either their truth or their falsity (Derrida 1981, p. 219). Derrida reads Mallarmé's "mimetic" propositions concerning the relationships between reality and representation or Being and Appearance as undecidable: neither their truth and nor falsity can be ascertained. This still allows for a possibility of ontology insofar that such propositions might still be true even though they cannot be demonstrated to be. If one allows for this possibility, in dealing with Being and Appearance, one is closer to Kant's epistemology than to radical modernism, which places the ultimate nature of reality beyond thought and thus the possibility of any true claim concerning it. In Kant's (ontological) scheme of noumena or things-in-themselves vis-à-vis phenomena or appearances formed in our minds, noumena are unknowable,

374   A. PLOTNITSKY

yet they are still in principle conceivable, especially by what he calls "Reason" [*Vernunft*], a higher faculty than "Understanding" [*Verstand*], which only concerns phenomena (Kant 1997, p. 115). Derrida's overall reading of Mallarmé appears, however, to be closer to radical modernism. Mallarmé's mimesis is only a mimesis of technology, and not of reality, placed beyond thought and thus all mimesis: a technological mimesis relates to this reality non-mimetically. Mallarmé's "Crisis in Poetry" [*Crise de vers*] at least allows for this type of reading:

> Languages, which are imperfect in so far as they are many, lack the supreme language: because thinking is like writing without instruments, not a whispering but still keeping silent, the immortal word, the diversity of idioms on earth [,] prevents anyone from proffering the words which otherwise would be at their disposal, each uniquely minted and in themselves revealing the material truth. This prohibition flourishes expressly in nature (you stumble upon it with a smile) so that there is no reason to consider yourself God; but as soon as my mind turns to aesthetics, I regret that speech [*parole*] fails to express objects by marks that correspond to them in colors and movement, marks that exists in the instruments of the voice, among languages and sometimes in a single language. … *But*, we should note, *otherwise poetry would not exist*: philosophically, it is poetry that makes up for the failure of language, providing an extra extension. (Mallarmé 2010, p. 847)

It is possible to read this elaboration as a form of idealism of poetic language, "the supreme language," beyond ordinary language, in parallel with how modernist mathematics would be viewed by mathematical Platonism. The point, however, may instead be that the "extra extension," defining the supreme language of poetry, allows one to move beyond a referential structure of ordinary languages, and establish, through a poetic composition, as a technological mimesis, a different relation to reality, revealing the material truth of this new, ultimately nonmimetic, relationships. This is closer to the double movement I consider here. The first aims at separating, along the lines of mathematical modernism, mathematics from its referential connections to physical reality, but without assuming that it represents some ultimate mental reality existing independently of thinking of living mathematicians; and the second aims at using this mathematics to relate to the world, material (as in quantum mechanics) or mental, differently. Mallarmé continues these lines of thought as follows:

> Speech has no connection with the reality of things except in matters commercial; where literature is concerned, speech is content merely to make allusion or to distill the quality contained in some idea.
> On this condition, the song burst forth, as a lighthearted joy.
> This ambition, I call Transposition—Structure is something else. …
> An order of the book of verse springs from it, innate and pervasive, and eliminates chance; such order is essential, to omit the author: well, a subject.

20 MODERNIST LITERATURE AND MODERNIST MATHEMATICS I ...   375

Destined, implies amongst the elements of the whole, as to the appropriate place within a volume. ...

Contrary to the facile numerical and representational functions, as the crowd first treats it, speech, ..., finds again in the Poet, by a necessity that is part of an art consecrated to fictions, its virtuality.

The line of several words which recreates a total word, new, unknown to the language and as if incantatory, achieves that isolation of speech; denying, in a sovereign gesture, the arbitrariness that clings to words despite the artifice of their being alternately plunged in meaning and sound, and causes you that surprise at not having heard before a certain ordinary fragment of speech, at the same time as the memory of the named object bathes in a new atmosphere. (Mallarmé 2010, pp. 847–49, 851)

This, too, should or at least could be read as thinking in terms of poetry or literature as liberating language from its "representational function," expressly invoked here, and, while still technologically imitating ordinary language, giving it independence and an ability to relate to the real non-mimetically. Thus, it would be difficult to speak of Platonism or idealism here, notwithstanding the appeal to (the reality of?) ideas versus the reality of (material) things, which may, at most, reflect Mallarmé's ambivalence toward radical modernism. For one is not dealing here with a Platonist mimesis of some primordial or eternal original forms, or again, at the very least such a non-Platonist reading is possible. This is because, while, as against ordinary language, "distill[ing] the quality contained in some idea" and in this sense a poetic composition as a "transposition" (of language) emerges each time anew: as singular, unique each time, making this emergence an event, always a crisis. This is more in accord with William James's (1842–1910) view: "Truth HAPPENS to an idea. It BECOMES true, is MADE true by events" (James 1978, p. 73). Mallarmé's sense of event applies to that of each composition, rather than, as in Badiou, only in a radical transformation or a creation of new theory or way of thinking. But then, for Mallarmé, every poetic composition is such a transformative event. This is one of the meanings of Mallarmé's title, "*Crise de Vers*," a crisis enacted in each poetic composition, "a throw of dice launched ... from the depth of a shipwreck," launched by a poetic thought, a composition that "eliminates chance" in its structure to always remain a chance, never to be abolished, as *Coup de Dés* famously says: "a throw of dice will never abolish a chance" (Mallarmé 2012, pp. 271, 225).

Indeed, the poem may be read as allegorizing this situation, as its last words, "a throw of dice" ("every thought emits a throw of dice") are the same as its first, title words, thus inviting or even compelling us to read the poem anew, but also always differently, beginning with its title (Mallarmé 2012, pp. 225, 271).[7] Every poem is a throw of dice, falling differently with reading, and each fall is just another throw, or the condition of another possible throw, just as in quantum physics any measurement is a condition of a future prediction, in the continuing throw of quantum dice, thus, especially in poetry, never allowing one to abolish the chance. Any such event is, thus, irreducibly discrete and what "connects" them is a reality the ultimate constitution of which is beyond thought and to which, thus, neither continuity not discreteness can be assigned, any more than any other attribute. However, the internal, mathematical-like rigor of poetic compositions gives their future life an order alongside chance, makes this life chaosmic, in Joyce's famous coinage (itself a micro composition), as against the random arbitrariness of the functioning of ordinary language (Joyce 2012, p. 118). In this particular form of chaosmos (there are others), the ultimate reality behind an event cannot be assigned a chaosmos, any more than randomness or order. The chaosmos of events is shaped by the virtuality of poetic thought here invoked by Mallarmé, actualized by an event of reading, or in the first place, that of its creation. A similar conceptuality of the virtual and the actual is found in Gilles Deleuze (1925–1995), arguably influenced by, among others sources (quantum theory among them), Mallarmé. Deleuze and Guattari link the virtual to the movement of Mallarmé's Mime in *Mimique*, which becomes an allegory of the virtual, as it is also, arguably correlatively, an allegory of writing in Derrida's sense, discussed below (Deleuze and Guattari 1994, pp. 169–70).[8] It is true that any thought, which always emits a throw of dice, can be made poetry, but it must be *made* poetry by being "launched ... from the depth of a shipwreck" (Mallarmé 2012, p. 271) like Mallarmé's "flower," when, from this depth, "musically arises the idea itself, ... absent from all bouquets" (Mallarmé 2010, p. 851).

The same *singularity* defines the birth, or, as Badiou would have it, the *event*, of a mathematical composition in mathematics or physics, as in Heisenberg, and makes them akin to poetry or art. Heisenberg's discovery, too, was "a throw of dice launched ... from the depth of a shipwreck" of quantum theory, then in crisis, while also leading to a new deeper crisis of quantum foundations. Similarly, the idea of the different order of infinities by Georg Cantor (1845–1918) or that of the undecidable propositions by Gödel led to a crisis, a shipwreck of the foundations of mathematics.

The difference between poetry and mathematics remains irreducible. One might speak, with Derrida, of the *différance* of literature and mathematics—a play of difference, proximities, and interaction, here under the condition of the unthinkable nature of the ultimate reality responsible for this interplay. In particular, mathematical compositions, while created singularly, uniquely acquire a great degree of universality, which enables their disciplinary functioning in mathematics or physics, while literary compositions never lose their singularity, although they can change their meaning in events involving them. Literary compositions also retain their connection to language, a limitation from which mathematics is (sufficiently) free, helping quantum mechanics to be mathematics divorced from representing reality (Heisenberg 1930, p. 11). In contrast to mathematics, literature retains a *technological* (mathematical-like) mimesis of ordinary language because a poem is still written in or (in some cases) linked to a language. But this is no longer an (ontological) mimesis of reality, either the ultimate reality or even of the reality of the world of events to which this composition can only relate—unlike ordinary language, which can represent the world. Because of this residual connection to language, a reading can, in a reversal, return a representational functioning to a literary composition, but at the cost of losing the compositional mathematics of poetry and, with it, its independence.[9] There is also some mimesis of previous literary or other compositions, just as in mathematics there is a mimesis of previous mathematical or other compositions, but, again, only as a mimesis of technology.

Heisenberg's quantum mechanics was similarly a technological rather than ontological mimesis. He formally borrowed the equations of classical mechanics—his first mathematical mimesis. But, in order to be able to make correct predictions, he changed the variables used in these equations, variables that came from a totally different area of mathematics, namely matrix algebra. This was the second mathematical mimesis in his mathematical compositions. Famously, Heisenberg did not know about the existence of matrix algebra and reinvented it, but this does not affect my point. This reinvention would be impossible without some mathematical mimesis. Heisenberg's technological mimesis was a mathematical composition. There was only mathematics for predicting the outcomes of quantum experiments, without anything to be said or thought about how these data come about. The way in which Heisenberg's abstract mathematical scheme nonrepresentationally relates to quantum events is parallel to the way in which a poetic composition that is abstracted, in a mathematical fashion, from ordinary language nonrepresentationally relates to human events in Mallarmé.

Quantum mechanics, too, announces the end of the book and the beginning of *writing* (in Derrida's sense) in physics, by virtue of relating to events of nature rather than being a representation of "the book of nature," already written in the language of mathematics as envisioned in Galileo's famous statement, cited by Derrida at the same juncture (Galileo 1960, pp. 183–84; Derrida 2016, p. 16). Quantum mechanics tells us that nature may not allow us to "write" its ultimate constitution in the language of mathematics or in any language, or even to conceive of this constitution. But it allows us to write something by experimenting with it by means of our experimental and mathematical technologies, to predict the outcomes and effects of such experiments. Quantum mechanics is a mathematical literature of physics, in the absence of the book of nature. Because it is no longer governed, as is classical physics, by the mimesis of reality, quantum mechanics needed to be invented in the same way as a literary composition is. This invention had its mathematical and experimental basis, but this basis was not a mimesis of a presumed ultimate constitution of nature, just as, in Mallarmé, a literary composition, which "distill[s] the quality contained in some idea," is not based on any ultimate primordial mental reality.

Some mental ontologies are created by a technological mimesis in any field. In physics, they are configured as experimental phenomena, perceived and configured by our mind. In mathematics, they are mental symbols and their structured compositions, a model for Mallarmé's (surface) ontologies of ideas. This is all the idealism there is in Mallarmé, and it certainly does not imply that there is no other reality than that of ideas. Mallarmé does invoke "Idealism," in claiming than one "adopts, as meeting point, an Idealism which (like fugues or sonatas) refuses the natural material and brutally demands an exact thought to put them in order, so as to keep nothing but mere suggestion" (Mallarmé 2010, p. 849). It is, however, the Idealism of "an exact thought," subject to the scheme just outlined, in which this thought, defining a poetic composition, emerges anew each time. It is not the Idealism, archeologically, of a Platonist ideal reality or, teleologically, of a Hegelian absolute Spirit or Idea. If this is alchemy, which Mallarmé liked to invoke (a "glorious," if "premature and louche precursor of political economy" [Mallarmé 2013, p. 400]), it is alchemy without the Philosopher's stone, an alchemy that creates new gold anew each time. And, if "youth, in poetry, where a dazzling and harmonious plenitude imposes itself, has stuttered the magic concept of the Work," this Work is nothing else either (Mallarmé 2010, p. 850). This dazzling and harmonious—or chaosmic—plenitude, without a whole, is arguably best exemplified by Joyce's *Finnegans Wake* (1939), conceivably inspired by Mallarmé's ideas and his project of the Book.

A technological mimesis is always a mimesis of mimesis, Derrida argues via his reading of Mallarmé's *Mimique*, which he sees as enacting or allegorizing this double and, ultimately, interminably multiple mimesis. It technologically refigures an imitation of imitation that Plato famously condemns, referring to art in *The Republic* and to writing in *Phaedrus*, as the loss of the original presence, the presence of the absolute origin. In the case of art, it is a mimesis of nature, *physis*, and not of the true reality of ideas (for example Plato 2005, pp. 1199–202). In the case of writing, it is, still in affinity with painting, an imitation of speech, already an imitation of thought, which and only which could be in a true mimetic proximity to the ideal placed at the absolute origin of everything (Plato 2005, pp. 551–53). Writing, as is art or science, such as the mathematics of the mathematicians, is merely a *tekhne*, an auxiliary and ultimately flawed technology, even as against speech, let alone thought, which is the best technology because it provides the best ontology.

The problem is that, as Derrida shows, nothing can avoid this double and ultimately interminable distance from reality, neither speech nor thought. Any thought is always a double mimesis within a chain or rather a multiple, disseminating network of thoughts. Everything differs from and defers itself—which is what Derrida calls *différance*—under the assumption of a radical alterity akin to a reality beyond thought. In fact, everything also irreducibly, uncontrollably multiplies without allowing for any oneness, thus adding *dissemination* to *différance.* By the same token, there is no absolutely original object of imitation. There is no end of this virtual tracing (which defines what Derrida calls "trace"), only a provisional termination of any mimesis, either literary or mathematical. The ultimate efficacy of this process is the irreducibly unthinkable, which is not an entity to be imitated and not an entity at all. It follows that nothing could, in principle, be structurally different from writing, from the technology of writing. Platonism may denounce it as an imitation of imitation, in favor of the proper mimesis of that which is absolutely original, but it cannot avoid it, because anything claimed to be absolutely original is created by the technology of writing. This technology, however, enables the rise of new compositions, and no compositions have ever emerged otherwise. This is why Derrida gives the name "writing" to any language or thinking, relating writing to technology throughout his work. Mallarmé's writings (in either sense), such as *Mimique,* stage and enact these technological workings of writing as translations without an absolute original, which can serve as a short-hand definition of writing in Derrida's sense.

Furthermore, in a self-doubling (*dedoublement*), the events of literature often appear in Mallarmé (and elsewhere in modernism) in a literature of events: a literary composition becomes a "mimesis" of an event of a literary composition, including its own, as is the case in *Coupe de dés* or *Mimique,* as Derrida's reading would suggest but, in truth, in most of Mallarmé's literary and even some theoretical works. It might appear that this type of claim would contradict the claim of the nonrepresentational nature of a modernist composition. This is not the case, however, especially if one uses allegory in Paul de Man's sense, according to which, in an allegory—in the algebra of allegory, one might say—"the emphatic clarity of representation does not stand in the service of something that can be represented," or in the present view, even thought of at all (de Man 1996, p. 51).[10] In their technological mimesis, Mallarmé's and other modernist works *enact* the unthinkable, also as it figures in the events of literary compositions, possibly their own. Thus, they do not represent the ultimate nature of their emergence.

As always singular and unique, such events involve randomness or chance, although the emergence of a literary (or mathematical) composition is never a strictly random process. This irreducible presence of randomness or chance is consistent with Mallarmé's claim, cited above, that the *order* of a poetic composition "eliminate[s] chance," because this claim only means that the elements of compositions themselves relate to each other within a strict order, as opposed to the arbitrariness inherent in ordinary language, which is no longer part of mathematical compositions. As noted, even the propositions of probability theory, by means of which we relate to indeterminate events, are strictly ordered within themselves, as mathematical structures. This aspect of mathematics is clearly on Mallarmé's mind, as part of the parallel between poetry and mathematics.

380   A. PLOTNITSKY

On the other hand, randomness or chance plays a major role in the emergence of compositions or their use. One might, accordingly, agree with Badiou, when he says:

> For me, Mallarmé represents the person who launched a poetic challenge to mathematics by assuming that the rigour of poetic language could equal the rigour of mathematics, but moreover also assumed the power of Chance, something mathematics cannot do. It is for this reason that the possibility of Igitur's act brings forth the victorious cry, "You mathematicians, expired." When, later on, I made the event the key to the upsurge of truths, I in fact integrated this Mallarméan doctrine in the following form: mathematics thinks being as such, being qua being, but cannot integrate the aleatory notion of the possible that is conveyed by the event. The effective sign of this point is that the event suspends a fundamental axiom of the mathematical theory: the axiom of foundation. This upsurge of the unfounded obviously finds its poetic symbol in Mallarmé's *Coup de dés*. It is thus here that, as the agent of the consequences of the event, the individual can become a subject. (Boncardo and Gelder 2018, p. 85)

The challenge is, I would argue, mutual, given that mathematics still provides the model of compositional rigor. Mathematicians expire only if they believe in the mathematical ontology of everything, even of the emergence of events. But they and mathematics, which does not exist apart from mathematicians, are alive if mathematics is understood for what it is: a technological mimesis. It is a *creative* deployment of elements or compositions of earlier mathematics in creating new mathematical compositions, in mathematics or elsewhere, for example, in physics, as in Heisenberg's invention of quantum mechanics. It is this compositional nature of mathematics that makes it most akin to art.

The emergence of a new artistic or mathematical composition inevitably involves chance, and hence, as Badiou agrees, cannot be captured by a mathematical ontology. The question is whether (a) there is some underlying causal architecture leading to this event; (b) the underlying reality is that of pure chance; or (c) whether this event is a product of the ultimate reality beyond representation or even thought, which, automatically, excluded either the underlying causal or chance ontology of its emergence. The present view is (c), which, while assuming chance or indeterminacy to be irreducible, leaves space for probability, on Bayesian lines, by either allowing for partially causal dynamics or strictly on quantum lines (in a Bayesian interpretation), without any such dynamics, or some mixture of both. While a composition is a cosmos (Deleuze and Guattari 1994, p. 180), the world, as a manifold of its events, is a chaosmos, here, again, under the assumption that the ultimate reality responsible for this chaosmos is beyond thought and, hence, is neither ordered nor chaotic, not any combination of cosmos and chaos that we can conceive of. It precludes an ontological treatment of this reality by any composition, but allows some compositions to technologically relate to the world of events that are effects of this reality. When a literary work (this generally does not happen in mathematics) becomes an allegory of an event (again, in de Man's sense of allegory), possibly an event of its own creation, it becomes a cosmic, organized, allegory of a chaosmic world, the ultimate constitution of which is beyond representation or even conception and hence cannot be captured by this allegory either.

NOTES

1. Some online sources, such as *Modernism, Wikipedia* or *Modernism-art, britanica.com* offer acceptable overviews.
2. I cite *Coup de dés* (Mallarmé 2012) from Meillassoux 2012 which prints it in French and English in the graphic form designed by Mallarmé.
3. Derrida also connects both and their relationships to chance throughout his works, but the subject and his view of chance (vis-à-vis the present one) will be put aside here.
4. In random or mixed ontology, the dynamics leading to random or chance events, such as Lucretius's famous clinamen, or the random swerve of an atom from a causal trajectory, is not given an explanation (Lucretius 2009, pp. 2, ll. 218–19, 42). On the other hand, such events are not seen as effects of the irreducibly unthinkable, which resolves this problem by an essentially modernist concept.
5. Thus, for Jean-François Lyotard (1924–1999), who introduced the term "postmodern" in this context, postmodern thinking is the same as modernist thinking—in effect radically modernist thinking, as defined here—in literature or art and in mathematics and science (Lyotard 1984).
6. For interviews with three prominent commentators, Jacques Rancière, Jean-Claude Milner, and Badiou and a helpful introduction, see Boncardo and Gelder (2018).
7. This reading of the poem is fundamentally different from Quentin Meillassoux's in Meillassoux (2012), which claims that *Coup de dés* is strictly coded, thus, in principle not leaving any space to chance in reading a poem that is about chance. The present reading suggests that the poem enacts or is an allegory of what Derrida would see as the "supplement" of a code, shaping the code but always remaining beyond this or any code, and in the present view, beyond thought, which implies the irreducible potential of chance in any reading.
8. Deleuze and Guattari associate Mallarmé's Mime with philosophy and the invention of concepts (in their sense), but their argument applies to any composition, and concepts in their sense are compositions.
9. This, I would argue, happens in Jacques Rancière's reading, which reduces, or just about, Mallarmé's texts to (barely) indirect representations of the political (Rancière 2011). That is not to say that Mallarmé's is not concerned with the political, quite the contrary: as I indicated, he is deeply concerned with the alchemy of the political (Mallarmé 2013, p. 400). But Mallarmé's *alchemy* (without the Philosopher's stone) and algebra of the political implies that the political has an allegorical relation to a reality beyond thought and thus beyond the political, which is only a set of effects of this reality. I can, however, only mention the subject in passing here.
10. I have considered the connections to quantum mechanics, implied here, in Plotnitsky (2000).

382 A. PLOTNITSKY

REFERENCES

Badiou, Alain. 2007. *Briefings on Existence*. Translated by Norman Madarasz. Albany, NY: SUNY Press.

Blanchot, Maurice. 1992. *The Infinite Conversation*. Translated by Susan Hanson. Minneapolis, MN: University of Minnesota Press.

Bohr, Niels 1987. *The Philosophical Writings of Niels Bohr*. 3 vols. Woodbridge, CT: Ox Bow Press.

Boncardo, Robert, and Christian P. Gelder. 2018. *Mallarmé: Rancière, Milner, Badiou*. London: Rowan and Littlefield.

Deleuze, Gilles, and Félix Guattari. 1994. *What Is Philosophy?* Translated by Hugh Tomlinson and Graham Burchell. New York: Columbia University Press.

de Man, Paul. 1996. *Aesthetic Ideology*. Minneapolis, MN: University of Minnesota Press.

Derrida, Jacques. 1981. *Dissemination*. Translated by Barbara Johnson. Chicago, IL: University of Chicago Press.

———. 2016. *Of Grammatology*. Translated by Gayatri C. Spivak. Baltimore, MD: Johns Hopkins University Press.

Einstein, Albert. 1936. "Physics and Reality." *Journal of the Franklin Institute* 221: 349–82.

Fuchs, Christopher, N. David Mermin, and Rüdiger Schack. 2014. "An Introduction to QBism with an Application to the Locality of Quantum Mechanics." *American Journal of Physics* 82: 749–54. http://dx.doi.org/10.1119/1.4874855.

Galilei, Galileo. 1960. *The Assayer*. In *The Controversy on the Comets of 1618*, edited by Stillman Drake and C. D. O'Malley, 151–336. Philadelphia, PA: University of Pennsylvania Press.

Gray, Jeremy. 2008. *Plato's Ghost: The Modernist Transformation of Mathematics*. Princeton, NJ: Princeton University Press.

Heidegger, Martin. 2010. *Being and Time*. Translated by Joan Stambaugh. Albany, NY: State University of New York Press.

Heisenberg, Werner. (1925) 1968. "Quantum-theoretical Re-interpretation of Kinematical and Mechanical Relations." In *Sources of Quantum Mechanics*, edited by B. L. Van der Waerden, 261–77. New York: Dover.

———. (1930) 1949. *The Physical Principles of the Quantum Theory*. Translated by Carl Eckhart and F. C. Hoyt. New York: Dover.

Huret, Jules. 1999. *Enquête sur l'évolution littéraire*. Paris: José Corti Editions.

James, William. 1978. *The Meaning of Truth*. Cambridge, MA: Harvard University Press.

Joyce, James. 2012. *Finnegans Wake*. Oxford: Oxford University Press.

Kant, Immanuel. 1997. *Critique of Pure Reason*. Translated by Paul Guyer and A. W. Wood. Cambridge: Cambridge University Press.

Lucretius, Titus Car. 2009. *On the Nature of the Universe*. Translated by Ronald Melville. Oxford: Oxford University Press.

Lyotard, Jean-François. 1984. *The Postmodern Condition: A Report on Knowledge*. Translated by Geoff Bennington and Brian Massumi. Minneapolis, MN: University of Minnesota Press.

Mallarmé, Stéphane. 2009. *Divagations*. Translated by Barbara Johnson. Oxford: Belknap Press.

—. 2010. "Crisis in Poetry." In *The Norton Anthology of Theory and Criticism*, edited by Vincent Leitch et al., 847–51. New York: W. W. Norton.

—. 2012. *Coup de dés* [A Throw of Dice]. In *The Number and the Siren* by Quentin Meillassoux. Translated by Robin Mackay, 225–71. New York: Sequence Press.

—. 2013. *Oeuvres Completes (Bibliotheque de la Pleiade)*. Paris: French and European Publications.

Meillassoux, Quentin. 2012. *The Number and the Siren*. Translated by Robin Mackay. New York: Sequence Press.

Plato. 2005. *The Collected Dialogues of Plato*. Edited by Edith Hamilton and Huntington Cairns. Princeton, NJ: Princeton University Press.

Plotnitsky, Arkady. 2000. "Algebra and Allegory: Nonclassical Epistemology, Quantum Theory and the Work of Paul de Man." In *Material Events: Paul de Man and the Afterlife of Theory*, edited by Barbara Cohen, Thomas Cohen, J. Hillis Miller, and Andrzej Warminski, 49–92. Minneapolis, MN: University of Minnesota Press.

—. 2015. "'Wandering Beneath the Unthinkable': Organization and Probability." *International Romantic Review* 26 (3): 301–14.

—. 2016. *The Principles of Quantum Theory, From Planck's Quanta to the Higgs Boson: The Nature of Quantum Reality and the Spirit of Copenhagen*. New York: Springer/Nature.

Rancière, Jacques. 2011. *Mallarmé: The Politics of the Siren*. Translated by Steven Corcoran. New York, NY: Continuum.

Weyl, Hermann. (1918) 1994. *The Continuum: A Critical Examination of the Foundation of Analysis*. Translated by Stephen Pollard and Thomas Bole. New York: Dover.

CHAPTER 21

Modernist Literature and Modernist Mathematics II: Mathematics and Event, with Mallarmé, Gödel, and Badiou

Arkady Plotnitsky

This chapter continues the examination of the relationships between modernist literature and modernist mathematics begun in chapter "Modernist Literature and Modernist Mathematics I: Mathematics and Composition, with Mallarmé, Heisenberg, and Derrida." The previous chapter discussed a movement in literature and mathematics toward independence and self-determination and proposed that the technology of composition can place the ultimate nature of reality beyond a representation or even conception. This chapter continues to address the question of ontology and mathematics in modernism, with a focus on Stéphane Mallarmé's (1842–1898) theoretical ideas and Alain Badiou's (1937–) philosophical work. Mallarmé and Badiou, for whom Mallarmé is a major literary inspiration, share the mathematical groundings of their thinking concerning, respectively, literature, and philosophy. There are, however, significant differences between them in this regard, in the present reading (Badiou reads Mallarmé differently), because Mallarmé's appeal to and use of mathematics are primarily technological while for Badiou, they are ontological. Badiou contends that any rigorous philosophical ontology can only be mathematical, at least as "a thesis … about discourse": "mathematics … pronounces what is expressible about being qua being" (Badiou 2007, p. 8). On the other hand, as discussed in "Modernist Literature and Modernist Mathematics I," Badiou's interest in Mallarmé is equally shaped by Mallarmé's thinking concerning "the power of

A. Plotnitsky (✉)
Purdue University, West Lafayette, IN, USA

© The Author(s) 2021
R. Tubbs et al. (eds.), *The Palgrave Handbook of Literature and Mathematics*, https://doi.org/10.1007/978-3-030-55478-1_21

385

chance," and also concerns Badiou's concept of "event," always an event of *trans*-Being placed beyond ontology (in his sense) and thus mathematics.

Two mathematical theories are especially important for Badiou: set theory and category theory. Set theory, introduced by Cantor in the late-nineteenth century, grounds all of Badiou's work, beginning with his 1988 *Being and Event* (Badiou 2013), where the presence of Mallarmé is massive. The book develops a form of set-theoretical ontology, which, in contradistinction to most previous concepts of ontology, from Plato to Heidegger, is defined by its irreducibly multiple character—by the impossibility of establishing a unity by which the multiple could in principle be subsumed. Badiou speaks of this ontology as that of "the multiple-without-One" (for example, Badiou 2007, p. 35; 2013, p. 29).[1] Badiou's later work, from the early1990s, is marked by its engagement with category theory, developed in the 1940s, and with topos theory (based on category theory), introduced by Alexandre Grothendieck (1928–2014) in the 1960s. In contrast to set theory, topos theory is primarily used by Badiou as logic rather than ontology, although the theory was developed by Grothendieck along ontological lines and was related to logic only subsequently by Lawvere (1937) and others.

Mathematics as ontology, Badiou argues, "functions as a [necessary] condition of philosophy" (Badiou 2007, pp. 43, 54). A necessary, but not sufficient condition! For, while it is also "about identifying what real ontology is," philosophy is the thinking of "event" (Badiou 2007, p. 60). An event is defined by Badiou in relation to what he calls a "situation," as something from within which, but also against which, an event emerges. This eruptive emergence of an event (always a crisis in Badiou) and, thus, the ultimate nature of thought exceeds Being, are "trans-Being," and hence, are beyond ontology, which Badiou defines as mathematics. While, however, the emergence of an event is beyond representation (already regarding set-theoretical ontology), it is, in contrast to the present view, not beyond thought—quite the contrary. This would still make thinking this emergence and Badiou's philosophy of event ontological in the present definition. This is because at stake in Badiou is the thinkable—in the case of the ontology of a situation mathematically, as Being, and in the case of the emergence of an event philosophically, without giving what is thought a name or representation, or almost, because he still speaks of "the being of the event" (Badiou 2007, p. 61). At stake in the present argument is, by contrast, that which perhaps can be named, here as "reality" (without the capital R, because other names are possible, too) but which cannot not be thought of. It is a name without a concept or thought that is or can be associated to this name. This also implies two views of mathematics: axiomatic in Badiou and compositional (not necessarily dependent on axioms) here, and two different readings of Mallarmé, in part in view of the concept of composition, which plays no role in Badiou. The present concept of event, too, applies more broadly, specifically to the emergence of each composition, literary or mathematical.

For the present purposes, it is sufficient to remain within the limits of the so-called naïve set theory, although Badiou uses more rigorous versions of it (for example, axiomatic, constructible, or generic). A set is a collection of certain, usually abstract objects, called elements of the set, such as of numbers between 1 and 10, which is a finite set, or of all natural numbers (1, 2, 3, 4, etc.), which is a countably infinite set. There are greater infinities, such as that of the continuum, represented by the numbers of points in a straight line, which multiplicity already poses difficulty for a naïve concept of set. The ontological multiplicity of sets themselves and thus of infinities is unavailable to unification, to the One, or to naming and representation, and this multiplicity is also inconsistent (in view of several famous paradoxes of set theory), while nonetheless enabling a set-theoretical ontology. Our propositions concerning this inconsistent multiplicity must be consistent, for it would not be mathematically rigorous otherwise. The inexpressible and the multiple become linked in turn, in Badiou without the unthinkable. While "the set has no other essence than to be a multiplicity" and while, with Cantor, we recognize "not only the existence of infinite sets, but also the existence of infinitely many such sets" (and of different magnitudes of infinity), "this infinity itself is absolutely open ended" (Badiou 2007, p. 41; translation modified). In particular, it cannot be contained in a set. There is no "the One" of set theory, because the set of all sets does not exist, or at least cannot be consistently defined.

Crucial further dimensions of this ontology and of the beyond ontology (trans-Being) of events are revealed by Gödel's discovery of the existence of undecidable propositions in mathematics in 1931 and then by the proof, by Paul Cohen (1934–2007) of the undecidability of the continuum hypothesis in 1963. An undecidable proposition is a proposition the truth or falsity of which cannot, in principle, be established by means of the system (defined by a given set of axioms and rules of procedure) in which it is formulated. Gödel's discovery of the existence of such propositions in arithmetic or any system that contains it undermined the thinking of the entire preceding history of mathematics, defined by the idea that every mathematical proposition can, at least in principle, be shown to be either true or false. Gödel proved—rigorously, mathematically—that any system sufficiently rich to contain arithmetic (otherwise the theorem is not true) would contain at least one undecidable proposition. Some among such propositions can in fact be established as true, without, however, it being possible to offer a formal mathematical demonstration of their truth. This is Gödel's "first incompleteness theorem." Gödel made our thinking concerning the grounding of mathematics even more difficult with his "second incompleteness theorem," by proving that the proposition that such a system, say, classical arithmetic, is consistent is undecidable. It follows that the consistency of most of the mathematics we use cannot be proven, although the possibility that this mathematics may be shown to be inconsistent remains open. As explained in "Modernist

388   A. PLOTNITSKY

Literature and Modernist Mathematics I," Derrida relates Mallarmé's logic of literary, compositional mimesis, as a mimesis of technology, to Gödelian undecidability, or a quasi-Gödelian one (Derrida 1981, p. 219). For this undecidability is philosophical rather than mathematical, and it is specifically linked to the undecidable relations between reality and representation, or Being and Appearance. On the other hand, it follows from Gödel's findings that there is an analogous relation between the ultimate nature of mathematical reality and our mathematical-logical proposition concerning it, *if one assumes that there is such a mathematical reality*, as both Gödel (along conventional Platonist lines) and Badiou (in terms of a Platonism of the multiple) do, rather than merely seeing all mathematics as a form of logic. The latter is an Aristotelian view, which is rejected by Badiou, but, while not the same, is in accord with the present technological-compositional view of mathematics.

A dramatic case of undecidability arises from Cantor's continuum hypothesis, which states, roughly, that there is no infinity larger than that of a countably infinite set (such as that of natural numbers: 1, 2, 3, etc.) and smaller than that of the continuum. The latter is intuitively represented by the number of points on the straight line, and defined as the set of all subsets of the set of natural numbers, assumed, by the so-called power set axiom, to contain more elements than the original set. The hypothesis is crucial if one wants to maintain Cantor's hierarchical order of (different) infinities, and hence for the whole edifice of set theory. As noted above, however, the hypothesis was proven undecidable by Cohen. It follows that one can extend classical arithmetic in two ways by considering Cantor's hypothesis as either true or false, that is, by assuming either that there is no such intermediate infinity or that there is. This allows one, by decisions of thought, to extend arithmetic into mutually incompatible systems, ultimately infinitely many such systems, because each such system will contain at least one undecidable proposition—a difficult and, for some, intolerable situation. Badiou, by contrast, finds in it a special appeal:

> As we have known since Paul Cohen's theorem, the Continuum [h]ypothesis is ... undecidable. Many believe Cohen's discovery has driven the set-theoretic project into ruin.... My point of view ... [is] the opposite. [The] undecidability of the Continuum hypothesis ... indicates a set theory's line of flight, the aporia of immanent wandering in which thought experiences itself as an unfounded confrontation with the undecidable. Or, to use Gödel's lexicon: as a continuous recourse to intuition, that is, to decision. (Badiou 2007, p. 99)

This type of situation already accompanied the discovery on non-Euclidean geometry in view of (de facto) the undecidable nature of Euclid's fifth (parallel) postulate. As unfoundable, this confrontation with the undecidable and the emergence of such decisions cannot be handled mathematically. While one's decision on either mathematics is equally legitimate mathematically, something beyond mathematics determines this decision. It is an event, leading to other events. It is, again, difficult to speak of it as either strictly the product of a choice or strictly the result of chance. It arises chaosmically (Joyce 2012, p. 118), as an effect of the reality the ultimate nature of which is, in the present view (but not that of Badiou), beyond thought.

While the wandering through this "garden of forked paths," forked by undecidable propositions, is dramatic by virtue of arising from within mathematics itself, this extra-mathematical logic of an event is general. To still approach this logic mathematically, Badiou appeals to topos theory. The theory is prohibitively difficult in view of its abstractness and mathematical rigor. I shall "solve" this difficulty by only minimally addressing topos theory and by translating it into Mallarmé's theory of literature. It should be mentioned, however, that topos theory is a far-reaching generalization of Descartes's analytic geometry. The latter considered curves and surfaces defined by algebraic (polynomial) equations, objects now known as algebraic varieties, defined in much more abstract terms in modern algebraic geometry, which led Grothendieck to topos theory. Both may be seen as a technological mimesis of Descartes's analytic geometry. We still deal with the connections between geometry and algebra, even if the structure of the compositions and the nature of connections involved is immeasurably more complex. "The great and long period of Descartes," discussed in "Modernist Literature and Modernist Mathematics I," continues, while our mathematical compositions reach new heights.

Badiou sees topos theory as a theory explaining "the plurality of possible logics," rather than as the plurality of possible ontologies, since the ontology (there is only one, however multiple its nature) is provided by set theory (Badiou 2007, p. 166). For Grothendieck, a topos is an algebraic concept of structure or composition, akin to Descartes's algebraic equation of a curve, rather than a *logos*, a new rendering of mathematical logic, as it is for logicians or Badiou. According to Badiou:

> The theory of toposes is descriptive and not really axiomatic. The classical axioms of Set Theory lay out an untotalizable universe of the thought of pure multiple. Say that Set Theory is an ontological decision. Topos theory defines the conditions beneath which it is acceptable to speak of the universe of thought, based on the absolutely impoverished concept of relation-in general. … Set Theory creates … a singular universe in which what there is is thought according to its pure "there is." Topos theory describes the possible universes and their rules of possibility. It is like inspecting the possible universes Leibniz conceived of as in God's Understanding. This is why it is not a mathematics of Being but mathematical logic. Topos theory explains the plurality of possible logics. This point is crucial. Indeed, if Being's local appearing is intransitive to its being, there is no reason why logic, which is the thought of appearing, should be unique. The linkage form of appearing, which is the manifestation of the "there" of Being-there, is itself a [multiplicity]. Topos theory allows us to understand in depth from the mathematicity of possible universes where and how logical variability, which is the contingent variability of appearing as well, is marked in relation to the strict and necessary univocity of Being-multiple. (Badiou 2007, pp. 166–67; also p. 119)

In Badiou, then, topos theory is a kind of mathematical phenomenology: "within ontology, it is the science of appearing, that is, the science of what signifies that every truth of Being is irremediably a local truth. What we can read in it … is that the science of appearing is also and at the same time the science of Being qua Being" (Badiou 2007, p. 167). It is this unity of being and appearing, the single (single-multiple) Being and multiple (multiple-multiple) logics of appearing, that, according to Badiou, should enable us to understand the nature of what he calls an "event," which belongs to trans-Being. Badiou's conception of "the event as trans-Being" establishes the difference between mathematics, as ontology, and philosophy, as that which, while being "all about identifying what real ontology is," is ultimately released from ontology, with topos theory in a decisive role in this task (Badiou 2007, p. 60). Only if the program is accomplished, then understanding "event" could be possible. As he says:

> Only then shall we know why, when a novelty is shown, when the Being beneath our eyes seems to shift its configuration, that this always occurs for want of appearance – in a local collapse of Being's consistency and thus in a provisional termination of any logic. For what then surfaces, what displaces and revokes logic from the place, is Being itself in its redoubtable and creative inconsistency. It is Being in its void, which is the non-place of every place. (Badiou 2007, p. 168)

This collapse is an event: an occurrence in which a given situation and its logic of being and appearance collapse, and a new situation and new logic emerge. Because in the rise a new situation, chance, a new "constellation," is irreducible, Mallarmé's entrance at this juncture is not surprising. The question will, again, be the nature of this chance. As Badiou writes, citing *Coup de dés*:

> This is what I call an "event." All in all, it lies for thought at the inner juncture of mathematics [as ontology] and mathematical logic. The event occurs when the logic of appearing is no longer apt to localize the manifold being of which it is in possession. As Mallarmé would say, at that point one is then in the waters of the wave in which reality as a whole dissolves. Yet one also finds oneself where there is a chance for something to emerge, as far away as where a place might fuse with the beyond, that is, in the advent of another logical place, one both bright and cold, a Constellation [*Coup de dés*, X-XI]. (Badiou 2007, p. 168)

Being, then, comprises, from its two different sides, both the multiple-without-One and the void; and an event is defined by the topos-theoretical logic of the conjunction of Being and Appearing, by a single being-multiple. It reveals a form of existence *beyond mathematical ontology*, as well as beyond univocity, even if the univocity of the multiple, but *not beyond thought*. Badiou's philosophy is ontological in the present definition: it still *thinks* the ultimate nature of reality, even if now beyond mathematics, or almost because an event is still a "set." As he says: "an event is nothing but a set, a manifold. But its emergence and supplementation subtract one of the axioms of the manifold, namely the axiom of foundation.... Taken literally, this means an event is strictly an unfounded manifold. This detection is in fact a pure chance supplement to the manifold-situation for which it is an event" (Badiou 2007, p. 61).

One might, however, conceive of the nature of events differently, on lines of radical modernism, and read Mallarmé's line cited (Mallarmé 2012, p. 236) in terms of the irreducible multiplicity, dissemination, of being-multiples (beings without capital B), as effects of reality beyond thought, a name of what cannot be thought, as against Badiou's thought of without a name. "If Being's local appearing is intransitive to its being, there is no reason why logic, which is the thought of appearing, should be unique," Badiou says, correctly. But there is no reason either why the thought of Being should be unique, as Badiou thinks, or correlatively that the ultimate nature of reality, responsible for multiple effect-ontologies, should be thinkable, or at least some reasons for this alternative. Badiou thinks in the Platonist fashion, even if this Platonism is that of the multiple, the multiple of multiples (sets). While a multiple-without-One, this multiple is still a single Being-multiple, merely governed by different logics, logics that change, by events, from one situation to another. His more recent, category theory oriented works do not change this (Badiou 2009, 2017). In the present reading, Mallarmé's statement, "the wave wherein reality dissolves," says that reality dissolves into a reality beyond thought and, correlatively, all wholeness dissolves into dissemination (Badiou 2007, p. 168). This also means that events change ontologies (in the absence of the ultimate ontology) and logics alike, and differently mix them.

This is allowed by Grothendieck's topos theory and may even be its most radical implication. For Badiou, Mallarmé is "the unconscious contemporary of Cantor," because both discover that the infinite can be an integer (Badiou 2007, p. 124). Whether this (I would argue strained) interpretation of the infinite in *Igitur* is correct or not, I would still see Mallarmé as thinking, philosophically, along the lines of topos theory and not only as the plurality of logics, as Badiou clearly sees him as well, but also and ever primarily as the irreducible plurality of ontologies, worlds, under the conditions of the ultimate reality beyond thought. It is, I argue, this situation that is allegorized in his works. In advancing the project of algebraic geometry, also as against his great predecessor, André Weil (1906–1998), Grothendieck abandoned the idea, central to Weil, of a fixed universal domain, no matter how vast that contains all structures and operations considered. He instead started to think through category theory, in terms of multiple relatable domains or categories, which need not be sets. This is a more radical conception of the multiple, in accord with radical modernism, although Grothendieck himself did not consider this multiple as an effect of a reality beyond thought and would have been unlikely to adopt this view.

392 A. PLOTNITSKY

As a composition—and it is a remarkable composition, just as was Descartes's analytic geometry, which topos theory technologically *imitates*—topos theory does not leave any place to chance. But because it allows for multiple possible logics and ontologies alike (not all of which need to be set-theoretical), our decision concerning which of them we select may involve chance, just as in the case of the garden of the forked paths, brought about by Gödel's undecidable propositions. Here, however, our field of decision is incalculably vast: from the start, the situation, I argue, equally describes the range of our decisions concerning which literature to pursue. While such decisions involved chance, it is more likely that their nature is, again, neither strictly a chance nor strictly a choice, but is chaosmic. But then, again, what kind of chaosmos is it? The answer is in turn decided by another chaosmic decision, which may or may not be subject to the same type of chaosmos. In the (radically modernist) view adopted here, also as an interpretation of Mallarmé, while there might be, and usually are, causal or random surface-level ontological components involved, there is always the irreducible, quantum-like efficacy of this decision, and thus of the rise of a composition, be it literary or mathematical or other. The reality does dissolve "in the indefinite regions of the wave," a quantum-like wave, accompanying these events (Badiou 2007, p. 168). It dissolves in any shape (appearance) we can give it. It dissolves, but does not disappear! It "reveals" itself to be a reality beyond thought, that of the One or the many included, which thus cannot be revealed, while being irreducibly multiple in its manifested effects. Topos theory takes advantage of this ontological multiplicity.

Conclusion: OR

It would take literally a lifetime to address in detail how the chaosmic thinking sketched here works in Mallarmé's literary or theoretical texts (the boundaries between them are, again, fluid and often undecidable). I would like, however, by way of conclusion, to consider the case of "or" (hereafter OR). This is one of Mallarmé's great signifiers, discussed by Derrida in important elaborations in "The Double Session," to which I am indebted, although my focus is different (Derrida 1981, pp. 262–64). Derrida was more interested in showing the irreducibly disseminating (rather than merely polysemic, containable) character of "OR." In contrast, I here focus on "OR" itself as an operator *enacting* dissemination, thus, also indicating another allegorical *dedoublement*: OR is the structure of disseminating signifier, itself included. Derrida also bypasses R as the signifier of the real [*réal*] or reality [*réalité*], which would be closer to *différance*.

"OR" is an essential logical operator, as it is also a conjunction indicating a change in the argument. "OR" joins, is *composed* of, two signifiers O and R, which could be read, for example, as zero, zeRO (the mirror image of OR), nothing, and reality, everything, or, as a reality beyond thought, that which gives rise to everything, or as zero and real numbers, collectively designated as R in mathematics. Mallarmé discusses zero in the context of natural numbers, as increasing, by adding more and more [mORe] zeros, to the infinite, "the improbable," in effect what Kant calls the mathematical sublime (Mallarmé 2013, p. 399; Derrida 1981, p. 262). In its dissemination, in Mallarmé, OR of Mallarmé's branches into O and R, disseminating each as well. "Or" is also the French for gold, and is persistently used by Mallarmé, perhaps more famously in "monstre d'Or" (mOnstRe d'OR) in his poem "Toast funébre," but, as Derrida shows, disseminating across his oeuvre. The English "or" is part of Mallarmé's disseminating play, often taking place between French and English, their *différance* and translation into each other. The English "therefore" (coupled to Mallarmé's famous "igitur," technically meaning "therefore" in Latin, another logical operator) is part of the same play—"for," "or," or "on." Once there is an "or," there is always an "on," also the "on" of ontology, and vice versa. It is tempting to see "OR" as an operator of dissemination. Once it enters, it cannot be stopped, there is more and more of OR, mORe [*majORe*], "encORe" in play in Mallarmé (2013, p. 399). OR is part of Mallarmé's alchemy without the Philosopher's stone, as it creates new gold each time (Mallarmé 2013, p. 400).

The blank space between O and R is not always decidable either, as to whether O and R, "nothing" and "all," are joint or disjoint, in Mallarmé. "Blank/*blanc*" is in turn a key figure for Mallarmé. Every "blank," every actual blank space may be different, each time, physically and conceptually— in a *différance*, along with dissemination of empty space—although certain effects of sameness, which allow us to treat such blank spaces as the same, are produced. The typography of *Coup de dés*, the "geometry" of the signifying algebra, is a famous example. Mallarmé's text is the *différance* of blanks and marks and their folds and unfoldings.

These folds can be read along Derridean (more algebraic) lines, structured by the Gödelian undecidability or along the (more geometrical) lines of Deleuzean/Leibnizean Baroque folding, joining the folds and interfolds or infra-folds of matter and soul, being and appearance, as does Deleuze's *The Fold* (1993).[2] Read along Derridean lines, the movement of writing in Mallarmé's text or (they are one and the same) the movement of Mallarmé's Mime would entail a radical reinterpretation of everything previously thought ontologically, on Platonist, Hegelian, Heideggerian, or other lines, by inserting the undecidability, *or* OR-ness, of being and appearance, truth and falsity, philosophy and literature, mathematics and literature, and so forth. A reconfigurative operator of writing would be attached to everything. There will be writing-speech, or writing-thought, in the first place, writing-philosophy, writing-mathematics, into writing-algebra and writing-geometry, writing-physics, or of course writing-literature, or writing-dancing, as in the movement of Mallarmé's Mime, where all these forms of writing are in play.

394   A. PLOTNITSKY

But one could also, even by the same token, see the situation along the lines of Deleuze and Guattari's thinking in *A Thousand Plateaus* (1980). For one could see ORs between all these forms of writing as the possibility of experimenting within or, while retaining their independence, between different endeavors or forms of thought, fields, models, and so forth, as in the case of different models of the smooth and the striated—textile, maritime, musical, mathematical, physical, aesthetic, and so forth, while keeping possible the relationships and technological mimesis between them in play, here (although perhaps not in Deleuze and Guattari) under the condition of reality beyond thought (Deleuze and Guattari 1987, pp. 474–500). The relationships, the *différance*, between modernist literature and modernist mathematics, in Mallarmé and beyond, is the same type of OR. Given, however, that any *différance, or* any OR, is always a dissemination, this one, too, overflows these borders, into physics, for example, or philosophy, or historically from modernism to Descartes or (a near contemporary) Shakespeare (1564–1616), a great poet of chance and a major presence in Mallarmé.

True events are rare, as each is "a throw of dice launched in eternal circumstances from the depth of a shipwreck [du fond d'un naufrage]" (Mallarmé 2012, p. 226). "Every thought is a throw of dice," but it may well be that it is launching it from the depth of a shipwreck that makes it a composition in literature and mathematics alike, amidst Mallarmé's music of the sea, or that of Claude Debussy (1862–1918), who translated Mallarmé into his music, captured in his *La Mer*. The work is not expressly linked to Mallarmé, but is a Mallarmé-like image of "these indefinite regions of the wave wherein all reality is dissolved/EXCEPT in the height PERHAPS as far away as a place merged with beyond" (Mallarmé 2012, p. 236). It is also the music of chance, and the complex harmonies, mixing chaos and order—chaosmic harmonies. They reemerge as the place of "the smooth and the striated," in Pierré Boulez's *Pli Selon Pli: A Portrait of Mallarmé*, which Deleuze and Guattari linked then to both Riemannian geometry and back to "the very special problem of the sea" (Deleuze and Guattari 1987, pp. 475–85). But before these variations on Mallarmé's themes, these translations of Mallarmé, there are Mallarmé's translations of chaosmic harmonies of the sea in poetry: that of Shakespeare's *Hamlet* (Mallarmé's special preoccupation) and *The Tempest* and of so much else, in literature and mathematics or music (that of Richard Wagner, for example), into French, as the mathematical language of his compositions, and thus never only French. These compositions helped to usher the event of modernism, to throw its dice, an event, a manifold of events, brought about by a reality beyond thought and thus beyond mimesis or translation, literary, mathematical, *or* other. There is always another OR in literature or mathematics, which, too, are better linked by an "or" than by an "and," linked by the gold of the alchemy that incessantly transforms them into each other and changes each in the process.

Notes

1. For convenience, I shall primarily refer to Badiou's 1998 *Briefings on Existence* (Badiou 2007), which, while building on his 1988 *Being and Event*, also discusses category theory and topos theory.
2. I have considered these two ways of reading Mallarmé's folds in Plotnitsky (2003).

Reference List

Badiou, Alain. 2007. *Briefings on Existence*. Translated by Norman Madarasz. Albany, NY: SUNY Press.

———. 2009. *Logics of Worlds*. Translated by Alberto Toscano. New York: Continuum.

———. 2013. *Being and Event*. Translated by Oliver Feltman. London: Bloomsbury.

———. 2017. *Mathematics of the Transcendental*. Translated by A. J. Bartlett and Alex Ling. London: Bloomsbury.

Deleuze, Gilles, and Félix Guattari. 1987. *A Thousand Plateaus*. Translated by Brian Massumi. Minneapolis, MN: University of Minnesota Press.

Derrida, Jacques. 1981. *Dissemination*. Translated by Barbara Johnson. Chicago, IL: University of Chicago Press.

Joyce, James. 2012. *Finnegans Wake*. Oxford: Oxford University Press.

Mallarmé, Stéphane. 2012. "*Coup de dés* [A Throw of Dice]." In *The Number and the Siren*, edited by Quentin Meillassoux, translated by Robin Mackay, 225–71. New York: Sequence Press.

———. 2013. *Oeuvres Completes (Bibliotheque de la Pleiade)*. Paris: French and European Publications.

Plotnitsky, Arkady. 2003. "Algebras, Geometries, and Topologies of the Fold: Deleuze, Derrida, and Quasi-Mathematical Thinking, with Leibniz and Mallarmé." In *Between Deleuze and Derrida*, edited by Paul Patton and John Protevi, 98–119. New York: Continuum.

PART IV

Relations Between Literature and Mathematics

CHAPTER 22

King Lear, Without the Mathematics: From Reading Mathematics to Reading Mathematically

Travis D. Williams

As the essays in this collection and other recent work demonstrate, "literature and mathematics," a field that originated with the study of allusions and overt themes, has evolved to include the examination of mathematics as form and metaphor, as well as historicist excavations of mathematics as a participant in subtle and complex power dynamics operating among persons, cultures, and texts. Shakespeare studies has similarly courted both the overt and phantom presence of dozens of discourses to great profit. Mathematics has not generally appeared in the latter trend, perhaps because the presumed literalness of mathematics' objectivity would seem to require literal presence; certainly because of aversions born of comparatively recent disciplinary distinctions: mathematicians have no feelings, and English professors can't balance their checkbooks (or so I am often told).[1] This chapter proposes to extend the work of literature and mathematics with an argument that deploys readings of mathematics overtly present in literary texts to create a way of reading mathematically when mathematics is only implicit, or not present at all.

Early modern European mathematics was a cultural practice engaged in the paradoxical activity of seeking certainty while constantly creating objects that threatened certainty. Take, for example, the curious objects created by early modern mathematicians as they sought to clarify the relationship of the ancient discipline of geometry to the technical and scribal practices we now call algebra. As I have suggested elsewhere, this work constituted early

---

T. D. Williams (✉)
Department of English, University of Rhode Island, Kingston, RI, USA

© The Author(s) 2021
R. Tubbs et al. (eds.), *The Palgrave Handbook of Literature and Mathematics*, https://doi.org/10.1007/978-3-030-55478-1_22

modern mathematicians' dominant activity, at least from our perspective (Williams 2016). As mathematical thinkers engaged and fought with one another over the merits and demerits of algebra, and whether it was a servant to or distinct from geometry, they created, rediscovered, and/or revived a steady stream of innovative mathematical objects, including negative numbers, irrational numbers, and imaginary numbers. The debates and feuds in European mathematics in the sixteenth and seventeenth centuries considered whether these objects were legitimate or not—if they could have any proper use, especially if they had no geometrical representation. And, if they were legitimate, what could that mean for the canons of classical geometry? In a culture that considered mathematics a divine gift of certitude to a fallen humanity, early modern mathematics constantly upset classical clarity with a steady stream of eminently useful but perversely unmanageable new objects.[2]

Shakespeare's (1564–1616) *King Lear* (1605, rev. 1610) is a play notable for the prominence of mathematical language, and for the imprecision of that mathematical language. *King Lear* presents its textual information in such a way that all experience of it, from inside or outside the play, renders logical coherence impossible.[3] Mathematics in the play stands as a model of reasoning and of theatrical experience that entices characters, readers, and audience with an offer of certainty and stability, and that largely betrays those hopes. Mathematics is therefore not a corrective to interpretational incoherence, but a correlative of it. But *King Lear* also shows that to be betrayed is to betray; every character experiences this irony, and the audience is not exempt. An audience can only grasp eagerly at clarity and objectivity if it makes certain assumptions about the state of affairs that themselves constitute self-betrayal: by refusing to see something that is there, or seeing something that is not there. Nevertheless, such defeat has a critical benefit. In the move from reading mathematics to reading mathematically, there occurs a reunion of mathematical results with mathematical motivations, through an appeal to ethics that might be lost when technical results are the sole object of mathematical attention. Sir Philip Sidney (1554–1586) worries about the mathematician who "might draw forth a straight line with a crooked heart" (Sidney 1965, p. 104). The line is no less straight because the heart is crooked, but the line *must* be understood to carry more than mere straightness now that we know what kind of heart drew it. Sidney's argument is really that the slippage between a perfect line and an imperfect heart shows that mathematics is a "serving science," subservient to the cultivation of personal and public virtue. The line now has ethical encumbrances that the viewer may choose to see and use, or not.[4] Observers now have ethical obligations, placed on them by this opportunity to choose.

Reading mathematically permits the shift from mere exactitude to an exactness that has optionally active ethical supplements. Once we take the option of considering ethics, there is no going back to the merely exact. Of course, the logic of the supplement means that uncomplicated mere exactitude is only seeming; to opt for exactitude is itself an ethically loaded decision, and can never be otherwise. Supplementation, as derived in deconstructive theory, invokes a circumstance that both adds to, and therefore explains, a text, and that subtracts from, and therefore confuses, a text. The supplement, being the thing that is necessary to explain the original text, becomes an inalienable part of the original, demonstrating, in an attempt to repair and explain, the original's essential and eternal inadequacy. Because the supplement both explains and confuses, makes whole and debilitates, it exemplifies the endless regression and deferral of meaning characteristic of all semiotic systems. Traditionally, mathematics stands in opposition to this deferral of certainty: mathematics is, instead, certainty itself. The mathematics of *King Lear* defies this tradition. While imprecise mathematics might call out for correction, any attempt to force objectivity (precision, certainty, ...) only has meaning in the context of the original imprecise state. As a supplement, imprecision is that thing which can never be fully excised or exiled. Having introduced imprecision, the mathematics of the play cannot be rectified without doing violence to the world that produced it. Even ignored, the trace of imprecision is present and potent. I will pursue this line of thought through readings of three moments in *King Lear*. Then, like the old rule of three, I will use what we have learned to extract a lesson about a fourth moment, the scene at Dover "cliff," for which reading mathematically places unexpected and confusing demands on a reader's and viewer's ethics.

Three Hard Lessons

While the opening scene establishes mathematics as a significant theme in *King Lear*, I will defer consideration of it to start with an examination of a later but simpler mathematical moment. The reduction of Lear's knights by his daughters Goneril and Regan from 100 (to 50, to 25, to 10, to 5, to 1) to 0 is so brutally effective because it is no more than a counting game.[5] We can quickly discern the pattern, and so enjoy the moment even more because we can play along.[6] The pleasures of anticipation and participation in the ritualized humiliation of an old man provide ample opportunities for ethical considerations. An audience's pleasurable awareness aligns with Goneril and Regan's ruthless efficiency, and we might consider ourselves to be accessories after the fact. Shakespeare also finds opportunity to link verbal form to the counting exercise. After reductions from 100 to 50, there is a bump in the pattern. "Twenty-five" is not very poetic, nor does it divide elegantly in two. "Five and twenty" is the solution, which allows the pattern to continue to ten and then five (Shakespeare 1997, 2.2.448).[7] Aesthetic admiration for this syntactical fix also perhaps contributes to the ethical dubiety of viewing this scene. After "five," Goneril, Regan, and Shakespeare have made their points. There is no reason not to proceed immediately to the end game: one. But in fact the pattern skips over one and lands on zero, because Regan's "What need one?" (2.2.452) is equivalent to "Why not zero?" As poetic language, these choices are examples of Shakespearean polysemy through numerical inequality: twenty-five is equivalent to twenty; one is equivalent to zero. The same is true of zero when it actually is spoken in the scene, not as a numeral, but as Lear's anguished "O, reason not the need!" which tries too late to teach his daughters some ethics (2.2.453). The letter-word standing obliquely for the numeral draws on the culture's tendency often to think of zero as an "O" (as does the Fool in an earlier scene, considered below).[8] It also makes Lear a participant in his own humiliation. He provides the zero so his daughters don't need to, but this is no more than to be expected, since he is the instrument of his own destruction from the start of the play. Lear, like the audience, has opportunities to notice what he is doing and to pull back. That he does not, even in a moment so starkly drawn with the help of numbers, reinforces the pervasive role of mathematics in the play as a mirage of objectivity and certainty. Until his too-late intervention, Lear is reduced to near speechlessness. He has no power to enter the conversation constructively because he is a zero, as the Fool has told him repeatedly (see below), and because he invests his entire identity in the knights. The knights are the substance of his identity, which he has already given away, according to the Fool: "Thou hast pared thy wit o'both sides and left nothing i'the middle" (1.4.177–79).

Back at the start of the play, Kent and Gloucester initiate the mathematical theme through discussion of Lear's "darker purpose" (1.1.35): "in the division of the kingdom, it appears not which of the dukes he values most; for equalities are so weighed, that curiosity in neither can make choice of either's moiety" (1.1.3–6). Concepts of numerical measurement come thick and fast. Political favor is a literally measurable phenomenon. Though "moiety" can mean either an exact half or just a smaller or greater portion of a whole, Gloucester's point here is that the shares going to the dukes might as well be exactly equal for all that inspection can tell of their differences (hence, "equalities").[9] Also noteworthy from the exchange is that the plan, for all that the earls know, is to divide a kingdom *in two* and to give the shares to *dukes*. In a critically underevaluated shift, the play quickly moves from which of the dukes Lear most affects ("which ... he values most"), to which of the daughters (the dukes' wives) most affect Lear. Two shifts to three, dukes to daughters, love from a sovereign to love for a sovereign, and exact halves turn into infamously inexact thirds.[10] An audience would be forgiven if it thought it had misunderstood what the earls had spoken of once Lear enters the scene and the love test begins.

Lear divides his kingdom into imprecise thirds in exchange for his daughters' expressions of love: "To thee and thine hereditary ever / Remain this *ample* third of our fair kingdom"; "what can you say to draw / A third *more opulent* than your sisters?" (1.1.79–80, 85–86, emphasis added). The modifiers emphasize how easy it is to muddy numerical exactness with unheeded supplements (for example, in the previous paragraph, "exact" and "exactly" are thoroughly unnecessary), and how hard it is to divide anything cleanly beyond the realm of numbers themselves. But Lear is also deliberately courting these inexactitudes, since Shakespeare clearly departs from the arrangements in one of his likely sources. In *The True Chronicle Historie of King Leir* (printed 1605), the adviser Skalliger suggests an unequal division, "To make them eche a Joynter more or lesse, / As is their worth, to them that love professe," but Leir refuses: "No more, no lesse, but even all alike, / My zeale is fixt, all fashiond in one mould: / Wherefore unpartiall shall my censure be, / Both old and young shall have alike for me" (qtd. in Bullough 1957–1975, 7, p. 338).[11] Since division would still be difficult (and technically impossible) despite Leir's insistence, the marked departure from exactitude in Shakespeare's play emphasizes how his characters court imprecision. It is a compulsion that agrees with their sociopathies.

Both Goneril and Regan eagerly fall to the quantification of their love for Lear. Goneril's is "Dearer than eyesight," "Beyond what can be valued, rich or rare," and "makes breath poor" both "more" and "less" (1.1.55–61). Regan's is of "her worth," "a very deed of love" (property rights standing here for monetary value) (1.1.69–76). Goneril begins her glib and oily speech not with money, but with implied weight, "I do love you more than word can wield the matter" (1.1.55), which anticipates Cordelia's studied worry about a love that is so weighty that she cannot lift it to the state of verbal expression: "I am sure my love's / More ponderous than my tongue," "I cannot heave / My heart into my mouth," which terminates with a metaphor of monetary debt: "I love your majesty / According to my bond, no more nor less" (1.1.77–78, 91–93). Amid so many expressions of love correlated to monetary and numerical value, and Regan's insufficiently explained "precious square of sense" (1.1.74), which seems at least to describe, ironically, something like right reason, we should also not forget that all of these actions—weighing, dividing, valuing—are contextually impossible. When mathematics touches love and land, its certainty is shot through with imprecision, and reveals each speaker's ethical state. The apparent political clarity discussed by the earls is now a mess of familial dysfunction, with the pose of certainty especially humiliating because the love test is forced onto women on behalf of their husbands or husband-to-be.[12] Producing the right answer is especially empty because it will in fact have no chance of increasing the amplitude or opulence of the thirds, which have already been determined. The daughters' humiliation is increased by the utter pointlessness of their efforts, as long as they make any effort at all.

Discourse around nothing and zero, my third "moment," is perhaps the play's most sustained mathematical theme, and unites all of the episodes I discuss. Initiated by Cordelia and immediately taken up by Lear, then sustained by Lear and the Fool during the middle scenes,[13] "nothing" comprehends many potential meanings, both destructive and constructive. The Fool calls Lear an "O without a figure; I am better than thou art now. I am a fool, thou art nothing" (1.4.183–85). This simple joke should recall us, though it does not recall Lear, to the full range of a zero's power. Multiplication by zero is destructive. It annihilates. But within the structures of place value,

404   T. D. WILLIAMS

zero also creates. In early modern mathematical terminology, numbers were constructed out of digits (the numerals 1–9, which the Fool calls figures, and zero), articles (a multiple of ten, e.g., 20, 40), and compounds[14] (e.g., 53, 72). In numerical place value, a digit or compound is increased or decreased by an order of ten if a zero is inserted in or removed from a place "before" it.[15] Since this increase (or decrease) is accomplished through multiplication (or division), this is how zero might be understood to have creative capacities. Lear is not a figure, the Fool implies, because he has evacuated himself of all substance, the metaphorical non-zero digit that sits to the left of a zero and is increased by that graphic arrangement. I suggest that zero and nothing have power to explain the play's ethics because the generative potential hinted at by the Fool's riddling speech brings into sharper relief Lear and Cordelia's use of zero as an exclusively destructive object; multiplication by zero achieves zero: nothing. According to Sigurd Burckhardt's still vibrant interpretation, Lear's error is to take words as immediately performative, and without the necessity or possibility of interpretation. Unlike Lear's solely destructive nothing—"nothing will come of nothing. Speak again" (1.1.90)—"the Fool can make use of it.... Lear discovers that his faith in directness was mistaken because it ultimately did rest on an intervening matter of fact: the sovereign's power to make good his words. The power gone, so is the immediacy; power was the integer before the zero" (Burckhardt 1968, p. 246). The full scope of the play permits an equally vibrant inversion of this reading. As Lear deteriorates through the storm, he successively strips himself of his "lendings" (3.4.106) until he is "the thing itself, Unaccommodated man ... a poor, bare, forked animal" (3.4.104–6). Numerically, he is stripping away the zeroes—all the addition and trappings of a king—that inflate his pathetic figure (his "digit" of base human/animal existence), which is all that now remains.[16]

"Nothing" is an amorphous idea that requires interpretation of context; understanding it is the only way we can hope to negotiate the thicket of misunderstanding and cruelty in the play. Depending on how limited one's point of view may be, the meaning of "nothing" will be limited in proportion, and so we return to the beginning of the play:

> *Lear.*   what can you say to draw
>   A third more opulent than your sisters? Speak.
> *Cordelia.* Nothing, my lord.
> *Lear.* Nothing?
> *Cordelia.* Nothing.
> *Lear.* Nothing will come of nothing, speak again. (1.1.85–90)

Both Cordelia and Lear are unable to understand another kind of multiplication with zeroes. When zero is accompanied by a digit it inserts zeroes "before" the multiplicand, creating "something" and preserving existence.[17] The Fool does understand this kind of multiplication. It's clear enough in his joke about Lear's figure: "now thou art an O without a figure. I am better than thou art now, I am a Fool, thou art nothing" (1.4.183–85). The repetitiousness of this speech, *now/thou/art*, may be the Fool's way of mocking Lear and Cordelia's verbal repetition in the first scene.[18] The Fool also introduces a usury theme when he asks Lear, "Can you make no use of nothing, nuncle?" (1.4.128–29). The interest gained through usurious multiplication is one way that "thou shalt have more / Than two tens to a score" (1.4.124–25), but Lear can't see it. His answer to the Fool's question is typically unimaginative, a replication of his answer to Cordelia: "Why no, boy; nothing can be made out of nothing" (1.4.130).

In the Gloucester family "nothing" is also misunderstood, at least by the father:

> *Gloucester*. Why so earnestly seek you to put up that letter?
> *Edmund*. I know no news, my lord.
> *Gloucester*. What paper were you reading?
> *Edmund*. Nothing, my lord.
> *Gloucester*. No? What needed then that terrible dispatch of it into your pocket? The quality of nothing hath not such need to hide itself. Let's see. – Come, if it be nothing, I shall not need spectacles. (1.2.28–36)

However, the reasons for misunderstanding are different, as Burckhardt demonstrates. If Lear's error is to take words as performatively substantial, without the necessity of interpretation, Gloucester's error is to take words as vacuously insubstantial, putting him "at the mercy of report, or hearsay, of signs" (Burckhardt 1968, p. 242). Similarly, while the royal family behaves with narrow stubbornness or narrow selfishness, Gloucester's limited understanding of what "nothing" means is contrasted with two of the most visionary characters in the play. Edmund, though unquestionably wicked, defines himself outside all prevailing standards of ethics or belief in a way that looks forward to behavior that we would otherwise define as skepticism or enlightenment reason. Edgar, on the other hand, possesses powers of persuasion that allow him to produce one of the most famous examples of dramatic *enargeia*, when he convinces his father that they are atop the cliffs at Dover (discussed below). This scene is the culmination of a role in which Edgar evacuates his entire identity—"Edgar I nothing am"—but can still try to survive (unlike his father or Lear) by asserting "in nothing am I changed / But in my garments" (2.2.192, 4.6.9–10).

Before moving to more mathematics and Dover cliff, it is worthwhile to take stock. Shakespeare manages to wring the bitterest sentiments from basic arithmetic and the symbolism of zero. There is nothing, however, so complicated that an attentive audience cannot see what is happening, and its role within it, if they want to, just as the characters are perfectly capable of seeing what they are doing and drawing back. But this is a privileged claim, based on having read the play through and importing its conclusion into the

406   T. D. WILLIAMS

opening. Starting not with the first scene, but rather with the depletion of Lear's knights, emphasizes how quickly both the characters and the audience are put into a state of ethical compromise before they or we are aware of it. We might say we are accessories after the fact, or notice our pleasure after it has occurred, but do we think and notice in the moment? Our slippage into the pleasures of anticipation and wordplay in the later scenes suggest that we don't and that the cruelties of the first scene slip by all too easily, until it's just a bit too late to pause and reflect. Understanding of what is happening is really an understanding of what has *just* happened; it comes rushing into clarify or muddle what seemed complete. It is a supplement.

IMAGINARY NUMBERS AS SUPPLEMENT

Imaginary numbers, as we now call them, appeared in European mathematics in a sustained way in the two centuries that surround the early seventeenth-century moment when Shakespeare composed and then revised *King Lear*.

The square root of a number x is another number y that when multiplied by itself produces x; $y^2 = x$. Thus, 2 is the square root of 4 because $2 \times 2 = 4$. If our number system includes negative numbers, then all positive numbers actually have two square roots. The number 4 also has -2 as a square root, because $-2 \times -2 = 4$. But what about negative numbers? The sign rules of multiplication mean that a positive number multiplied by itself produces a positive number, while a negative number multiplied by itself also produces a positive number. So it would seem that negative numbers cannot have square roots. But early modern mathematics increasingly came up against situations in which it was clearly useful for negative numbers to have square roots.[19] For example, the polynomial equation $x^2 + 1 = 0$ has no solution unless we decide to explore the possibility that a negative number can have a square root, because the equation fundamentally asks what number x, when squared (multiplied by itself), produces -1.

The number system as it existed at the turn of the seventeenth century had nothing in it that would solve the problem of how to state the square root of a negative number. This state of insensibility was consonant with a denial of the relevance of roots for negative numbers, but contradicted the direction mathematics was going with this complicated situation. Early modern mathematics needed a way to express the square root of negative numbers. The eventual solution, though one achieved much later than the period of *King Lear*, was simply the creation of a completely new number. That number was i, defined $i = \sqrt{-1}$. According to Florian Cajori, Leonhard Euler (1707–1783) first used i in 1777. The letter-number was first printed in 1794, and it came into regular use, by Carl Friedrich Gauss (1777–1855), in 1801. It is the notation still used today (Cajori 1928, 2, p. 128). With the innovation of i, every negative number could have two square roots just like every positive number: the square roots of -4 are $2i$ and $-2i$.

In the evolution of the number system, imaginary numbers are a profound leap in abstraction. If negative numbers are a first order abstraction of a practice that existed primarily to count and measure things, then imaginary numbers—the roots of negative numbers—constitute a sort of second-order impossibility. If negative numbers were already a cause of anxiety for early modern mathematical thinkers—since they certainly did not fit nicely with ideas either of number as a construct of multiple units, or of an intuitive counterpart, even equivalent, to the geometric magnitude—the root of a negative number was even more bizarre and unsettling. Girolamo Cardano (1501–1576), the first mathematical writer to seriously treat imaginary numbers, writes simultaneously of their impossibility and of their reality and usefulness (because they provide roots to an equation when there are none at all from the real numbers). In a frequently remarked play on words during the working out of a problem that includes the expression $(5 + \sqrt{-15})(5 - \sqrt{-15})$, Cardano advises his readers, "dimissis incruciationibus." This participial expression has several, simultaneous translations: "the cross-multiples having canceled out," "the imaginary parts being lost," and "putting aside the mental tortures involved" (Cardano 1545, cap. 37; for the English, see Cardano 1968, p. 219). In a prosaic, technical mode, the phrase instructs us to let positive and negative imaginary terms cancel ("cross out") one another: a positive imaginary term and its negation cancel one another (e.g., $2i + (-2i) = 0$). In more poetic, philosophical, and even theological modes, the same phrase also instructs us to ignore the mental tortures ("crucifixions") of mathematical objects that require taking a square root of a negative number. According to Cardano, imaginary numbers are also recondite, "abscondita," something hidden, secret, and concealed (Cardano 1968, n. 6). It was René Descartes (1596–1650) who first labeled these objects "imaginary," a gesture designed to denigrate them and so push them out of the realm of legitimate mathematical thought (see "La Géométrie" in Descartes 1637, p. 380).

Descartes's effort to eliminate imaginary numbers from his mathematical activity by denigrating and then ignoring them is of course an illogical and unrigorous thing to do from our point of view, but it is a corollary of Renaissance habits that Katherine Eggert powerfully explores as "disknowledge": "being acquainted with something and being ignorant of it, both at the same time" (Eggert 2015, p. 3).[20] If disknowledge is a tendency to believe in concepts one knows to be false, imaginary numbers demonstrated the period's proclivity for refusing to believe in or accept ideas that were useful and putatively correct. Mathematicians would prefer not to engage with objects so obviously useful because they radically upset a number system already roiled by negatives and irrationals. If we categorize (perhaps reductively) this behavior as a denial of truth, we can then understand this denialism as a mathematical version of one of the most famous readings of *King Lear*, Stanley Cavell's "The Avoidance of Love" (Cavell 1987, pp. 39–123). Cavell develops "disowning" as the destructive refusal of knowledge, which for Lear means a refusal to be recognized. Cordelia recognizes the nasty game

Lear has forced her into by refusing to play it, and he is thereby enraged. What enrages Lear is not particularly the surface form of Cordelia's refusal to play, her answer of "Nothing," but the recognition of his ethical misbehavior that it reveals. "Nothing" and the play's whole complex of mathematical thought open the ethical economy of the play, which touches and implicates all the characters and the audience.

Today, imaginary numbers make up a numerical field also often called the complex numbers. In current notation, a complex number $a + bi$ has two parts: the real part a and the imaginary part b. The coefficients a and b are real numbers; the number i is the square root of -1. The imaginary part is a supplement of the real part—an extra that solves one problem but creates another. The imaginary portion, bi, exists, as an object, nowhere except in its verbal/graphic (re)presentation. It neither counts nor measures anything. It is a purely human, mental invention, a fantasy or vision in the rhetorical meaning of those terms, orthogonal to the realms of reality. This is true of both the numerical forms and the graphic representations of those forms. In current practice, the standard xy Cartesian axes are adjusted to assign the real term to the x-axis and the imaginary term to the y-axis. We must eliminate one dimension of two-dimensional real space so it can be given over to a part of one-dimensional complex space. The rich dimensionality of real space must be depleted for the imaginary to be accommodated. Hence, in the Derridean sense of the supplement, the imaginary part both explains and disables—it adds and subtracts. We must also realize that, in the fullest consideration of the definition of i, all real numbers x have a phantom imaginary presence, a nihilistic $0i$ term $(x \pm 0i)$, haunting the otherwise stable and confident realms of pure reality. The real number 2 is equal to the complex number $2 \pm 0i$. The realms of reality so confidently asserted and modeled by absolutist mathematics thus suffer predation and a crisis of confidence. Reality now is both compromised and occluded. It becomes fuzzy and unclear in precisely the ways mimicked in the mathematics that, in one standard cultural stereotype, is the equivalent of unarguable objectivity, but in another guise, promoted by the play, is always riddling and riddled, presenting only a patina of certainty while in the same breath insisting on illegibility, irregularity, and imprecision.

"Thus"

And so we come to Dover cliff, a passage that has no overt discourse. The three moments discussed above constitute, respectively, a simple counting game, a contest between exact and inexact, and the contrast of present and absent. Dover cliff gains from mathematical thinking by adding to this list the unsiftable cohabitation of the real and imaginary.

The Dover cliff scene trades on a variety of assumptions that are culturally popular but that the play does not explicitly license, including that it takes place at Dover at all and that the audience knows that Gloucester and Edgar are not at the top of a high cliff.[21] Dover is only presumed because it has been repeated so many times during Gloucester's torture, and because he asks to be led there. Being on the way to Dover and being at Dover are not the same thing. As my analysis (and many others) likewise shows, the audience is not aware that the players are not on top of a high cliff until well into the scene.[22] Therefore, the mental posture of the audience in the early lines of the scene is not one of wise recognition of Gloucester's error, but rather of a contest to determine whether Gloucester or Edgar more reliably describes the situation they are in. The early lines of the scene test and eventually betray the perception of the audience.

In response to Gloucester's opening question, "When shall I come to the top of that same hill?" Edgar's response, "You do climb up it now. Look how we labor," immediately confuses the audience (4.6.1–2). Edgar suggests that Gloucester is climbing and does not realize it, thereby establishing Gloucester's perceptual incapacity: poor balance and spatial perception allied to and consequent of blindness. The synesthesia of "look" for physical exertion is particularly marked. It also calls attention to the audience's ability to parse the scene. Whether the Renaissance stage was flat or slightly raked, the audience sees flatness, just as Gloucester appears to perceive flatness. "Looking" is conspicuously unavailable to Gloucester, so Edgar's imperative is simultaneously awkward, unfeeling, comic, and normal—"looking" as a synonym for perceiving, understanding, noting, was already a dead metaphor. But the audience, however Gloucester interprets the demand to look, is at a double disadvantage. We cannot feel if there is an incline in the fiction of the moment. Nor can we see if we are on an incline. All we might see is if the actors choose to adopt the posture of laborious climbing. Gloucester does not, because if he did he would not ask the question or receive Edgar's answer. Edgar does not because (as we later learn) he is not laboring, and Gloucester is incapable of seeing whether Edgar labors or not. The audience doesn't know what to think, or, more likely, doesn't think about it. Do we accept the facts of our eyes—no laboring, so no hill—or do we let our minds overwrite the literal facts of the stage with an imagined hill, thereby adopting Edgar's perception as truthful and Gloucester's as faulty? Probably, we do the second, since it is all too easy and we've done it many times before.[23] Blindness also creates a bias against Gloucester's perception. Edgar's status as a trickster has so far been applied only to other characters, not to us.[24] The odd logic of stage falsehood expects the audience to assume that it is not subjected to subterfuge or not implicated or implicatable in the ethical event unfolding before it.

Gloucester's reply, "Methinks the ground is even," presents another possibility that is immediately rejected by Edgar, "Horrible steep" (4.6.3). Edgar again characterizes a blind man's correct perception as faulty because he is blind, and by the same trickster logic, we go along with the ruse, without really thinking about it.[25] Here begins Edgar's bravura scene creation. "Hark, do you hear the sea?" he says. "No, truly," Gloucester responds (4.6.4), but we are already disposed to hear the sea, because we trust Edgar and discount

410   T. D. WILLIAMS

Gloucester. We of course hear the sea in our imaginations. Gloucester's hearing now joins the list of incapacitated senses. Our hearing is also inactive, as far as the sea is concerned. Like Gloucester, we have to hear it in our heads, with the "mind's ear." Gloucester's position and the audience's position begin to merge. "No, truly," also alerts Edgar to the fact that he has a blank canvas on which to work. None of Gloucester's real perceptions will get in the way of anything Edgar's chooses to describe, a position ratified by Edgar's suggestion, "Why then, your other senses grow imperfect / By your eyes' anguish," and Gloucester's agreement: "So may it be indeed" (4.6.5–6). Now we come to matters that the audience should be able to verify from its own experience. Gloucester remarks, "Methinks thy voice is altered and thou speak'st / In better phrase and matter than thou didst" (4.6.7–8). Is Edgar's speech different in the ways described by Gloucester? The difficulty is time. Can the audience remember and compare from Edgar's last appearance if his speech has altered in form and content? In fact, the change has begun in the last encounter: "Give me thy arm, / Poor Tom shall lead thee" (4.1.81–82). Whether or not the audience remembers, the point is that Gloucester does, and this memory is subject to the same uncertainties that clouded his claims about steep or flat terrain. Edgar reinforces uncertainty by denying a change: "You're much deceived; in nothing am I changed / But in my garments," a statement belied by its own clarity and elegance (4.6.9–10).

Edgar brushes off Gloucester's last remark of change, "Methinks you're better spoken" (4.6.10), to create one of the greatest descriptive passages in dramatic literature:

> Come on, sir, here's the place. Stand still: how fearful
> And dizzy 'tis to cast one's eyes so low.
> The crows and choughs that wing the midway air
> Show scarce so gross as beetles. Half-way down
> Hangs one that gathers samphire, dreadful trade;
> Methinks he seems no bigger than his head.
> The fishermen that walk upon the beach
> Appear like mice, and yon tall anchoring barque
> Diminished to her cock, he cock a buoy
> Almost too small for sight. The murmuring surge
> That on th'unnumbered idle pebble chafes,
> Cannot be heard so high.   (4.6.11–22)

He prioritizes the extreme height of the cliff. "Crows" and "choughs" bring to mind birds flying high above us, tending to direct the eyes upward from where they have been "cast ... so low" (4.6.13, 12). The birds are in the air, but below us, as "midway" suggests. This is what makes the widely remarked perspectival essence of the moment so famous and effective. Perspective is one thing, but the placement of the viewing origin is unexpected, and unexpectedly shifted in a quite violent way. Of course, it is no more strange for the birds to be half-way up or down the cliff than to be at the base or high up where Edgar and Gloucester supposedly stand, but the *suddenness* of our sense of great height is achieved by violating the cultural convention that birds fly above our heads.

"One that gathers samphire" (4.6.15) subtly contrasts Gloucester's intention—death through falling from a great height—with purposeful and necessarily preservative work. The samphire gleaner risks the height and presumably has death from the danger as far from his thoughts as possible. The gatherer's reduction to "no bigger than his head" (4.6.16) follows the birds' reduction to "show scarce so gross as beetles" (4.6.14), and introduces the bravura sequence of perspectival comparisons that result in the elimination of eyesight: "fisherman ... / Appear like mice," and "yon tall anchoring barque / Diminished to her cock, her cock a buoy / Almost too small for sight" (4.6.17–20). By the time we get to "The murmuring surge / That on th'unnumbered pebble chafes" (4.6.20–21), we are deep into one of the great examples of *enargeitic* (intensely vivid) description. Edgar also here shifts the description from visual to aural, just as he had when Gloucester and the audience were trying to parse the opening of the scene. The sentence about the sound of the surf is strongly enjambed by a line-long relative clause: "that on th'unnumbered pebble chafes." "Murmuring," "surge," and "chafes" all insist on the presence of sound, but the main sentence is "The murmuring surge / ... / Cannot be heard so high" (4.6.20, 22). If as dutiful listeners we have created an image of being atop a high cliff near the sea and then seen this image with our mind's eye, all the mental work we have done to hear the sea with our mind's ear has to be canceled, quickly. We have to hear the murmuring waves in order just as quickly not to hear them. A creation of imagination is degraded to impossibility.

The sound of the surf is *there*, it is just beyond the reach of our senses. But are our senses like those of Edgar or like those of Gloucester? When Edgar "leaves" his father to his suicide, it is reasonable to think that the mental images held by Gloucester and the audience are the same, not necessarily because our senses are debilitated in the same way as Gloucester's, but because we have been carefully maneuvered to accept Edgar's scene creation as unquestioningly as Gloucester has. Gloucester must think it is a mental representation of his reality. The audience must think it is what they would themselves perceive if they were in the world of the fiction. The answer is important, because (jumping over some equally incoherent leave-taking) there soon comes this thunderclap from Edgar: "Why I do trifle thus with his despair / Is done to cure it" (4.6.33–34). Everything collapses for the audience, because now anything that was uncertain or misunderstood becomes, paradoxically, clear. We are now firmly complicit with Edgar, though perhaps against our will, and must now participate in the extremely awkward ethics of this situation: both the dubious ethics of Edgar's clinical treatment and our sense of being duped. Our imaginative forces have been finely played and the result is not the always ephemeral mental image that drama requires, but a deliberate hoax—a false image, not just a fiction. The flat stage is at one with the flat ground on which Gloucester and Edgar have been treading all along, and Gloucester turns out to have been correct and Edgar a liar, and yet we are in an ethical bind because we cannot help but be relieved that Gloucester is not about to destroy himself. Edgar's confident "thus" asserts camaraderie and complicity; it assumes that we are already and have always been party to the subterfuge. Now the confusion is one of trying to decide how we missed it, if we missed it, and when we noticed it, if we noticed it.

Since Edgar seems to assume that his subterfuge is clear and obvious, perhaps we ought to take him at his word. Where, therefore, is the moment when we know him to be bluffing? When do we know that he is not setting up his father for self-destruction, despite all the misleading information adduced above? The only place it could come is from an extra-textual cultural assumption that Edgar would never do such a thing because he is the good son and has already demonstrated as much by thoroughly abasing himself in our eyes in order to accompany and assist his father and the king. And yet if this were sufficient, we could not be unsettled by the parade of pretense that the whole scene presents and which is paradoxically characterized in Edgar's Hamletian statement: "why I do trifle thus with his despair / Is done to cure it."[26] The collusion and knowledge that this line assumes are undermined by the precisely odd filial choices they describe. The result is that the presumption of audience knowledge of Edgar's plan provides a mirage of certainty where none exists; where none *can* exist.

After our perceptual whiplash, we watch helplessly as Gloucester prays to gods that our own recent disillusion suggests are simply not there (4.6.34–41). Our mental image has been discarded as false. We see Gloucester still defining his existence through an image, and its imaginative elaboration, that from our point of view is also false. Our reaction has no clear ethical definition. Agreeing that Gloucester should be allowed his fantasy makes us another Edgar, a liar whose confident assertion of a cure is far from proved. It sets us up for precisely the complacency of Edgar and the rest, shattered when Lear enters howling in the final scene with Cordelia in his arms. To insist that Gloucester's illusion should be shattered seems merely cruel and perhaps abrupt to a degree that would shatter the old man as well. There is no right decision. We simply have to let it play out, impotently, until the awful final moments, trapped in a torturous dramatic device that we ourselves entered with overconfidence and complacency when Kent and Gloucester's "moiety" shifted vaguely from half to third, and everything began to go wrong.

There is much more that might be said about the Dover cliff scene in a similar vein until the key changes with Lear's entrance. I will confine myself to Edgar's description of the cliff "from below," a passage that emphasizes the audience's entrapment in a scheme that now forces it to relive the description from above, in reverse, now with full consciousness of its earlier error and without the possibility of escape. After Gloucester "jumps" and then rises from his "death," he says, "Away, and let me die" (4.6.48), which is a paradox, since his chosen method of death, if successful, would not allow him even this short speech. Subtly, Gloucester returns in a state of apparent insensibility. So does the audience, but now ethically divided from Gloucester, and united with Edgar, uncomfortably. But which Edgar? The one who personated the companion at the top of the cliff? This makes no sense because such a person would not ask "Alive or dead" (4.6.45) and also have seen Gloucester fall to his knees and roll over. Is it a new man, whose underlying "actor" (i.e., Edgar) has recently speculated that Gloucester could indeed have died (from grief), so he must inquire? The audience is returned to a state of insensibility for being unaware of whom they see, but not yet aware that they are unaware.

The cliff from below is the mirror image of the cliff from above: the same, but reversed through the lens of consciousness we did not have before. "Gossamer, feathers" and "egg" recall the birds (4.6.49, 51). Edgar also returns to nautical themes to present a perspective of great height, now from below: "Ten masts at each make not the altitude / Which thou hast perpendicularly fell" (4.6.53–54). "The shrill-gorged lark" flies above our heads, and we must see it and hear its piercing song before suddenly being told that it is "so far" that it "Cannot be seen or heard" (4.6.58–59). We relive the original description again, now with the uncomfortable and unavoidable supplement, acquired so suddenly and perhaps painfully, that what seemed certain was anything but; the audience squirms in recognition that it has been tricked and has been presumed to be in on the trick. Edgar rubs it in. Ultimately, Edgar appeals to a sight that Gloucester cannot exercise, to suggest that a monster was at the top of the cliff, despite the just-established assertion that something at that height would be nearly invisible. Only a construction in the mind will do. Edgar effectively describes himself as the "monster" in his brother Edmund's forged letter (1.2.46–58, 94). It was Gloucester's unheeding acceptance of that mirage that began his own tragedy.

The Dover cliff scene approaches mathematics in the submerged proportions that govern perspective, and in the ten perpendicular masts that measure the impossible height of the cliff. The latter is a purely imaginative construction, as unreal as the perfect halves and imperfect thirds that started the play. It is a thought experiment, urged by Edgar on Gloucester and the audience, whose imaginations have been shown repeatedly to be bad guarantors of stable reality. Edgar's constant, tactless insistence on Gloucester's visual apprehension functions as an imaginary number: something created mentally, purely to serve a need. For Gloucester, specifically, vision can only be mental and imagined. Edgar is the play's early modern mathematician, equally creating, then denigrating, then using Gloucester's and the audience's imagination. But Edgar goes even further than the mathematician, since the ultimate purpose of his work is uncertain to everyone but himself.

The cliff scene ends where it began, with Edgar leading Gloucester by the arm. Did it achieve anything, except to waive the audience's inattention in its face? If a lesson is learned here, it may be that the audience it betrayed, once again, into recognizing that it is always too late to see that what appears

414   T. D. WILLIAMS

certain is not certain, and should not be trusted. The numbers show us the way, but the apparent purity of their example do no better at getting the audience ready for the cliff test. The supplement of deranged certainty is unavoidable. It cancels earlier experience, but even as it looks back, it seems never to look forward to avoid the next disaster.

NOTES

1. On the feelings of mathematicians since the Enlightenment, consider just one data point: Emma Woodhouse's fantasy of a romantic connection between Frank Churchill and Harriet Smith after their encounter with the gypsies in Jane Austen's novel *Emma* (1815): "Could a linguist, could a grammarian, could even a mathematician have seen what she did, have witnessed their appearance together, and heard their history of it, without feeling that circumstances had been at work to make them peculiarly interesting to each other?" (Austen 1923, p. 335).
2. Thomas Hobbes calls geometry "the only science that it hath pleased God hitherto to bestow on mankind" (Hobbes 1996, pt. 1, cap. 4, para. 12).
3. Though there is not space to consider the matter here, "information" presumes a certainty belied by the immensely complicated textual relationship between the Quarto and Folio texts of *King Lear*. The play itself is a set of material objects that participate in an indeterminacy intensified by the mathematics contained within them.
4. Compare Euclid's spare definition in his *Elements*: "A straight line is a line which lies evenly with the points on itself" (Euclid 1925, 1, p. 153).
5. In all I say about the mathematical themes of the play, I wish to record my indebtedness to the late Paula Blank's powerful arguments in *Shakespeare and the Mismeasure of Renaissance Man* (2006). In the reduction of knights, Shakespeare adjusts the dynamics of a source for maximum effect. In a 1574 extension of *The Mirror for Magistrates*, "Leire" starts out with sixty knights. "Gonerell" reduces them by half so Leire goes to "Ragan," with whom he lives peacefully for a year before she, suddenly, reduces his knights to ten and then five. So Leire returns to Gonerell, "but beastly cruell shee, / Bereavde him of his servauntes all save one, / Bad him content him self with that or none." What in the source is diffused by time is concentrated in Shakespeare's play, and made even worse by the double-team of the daughters against which Lear is helpless (Higgins 1574, excerpted in Bullough 1957–1975, 7, pp. 323–32; see specifically lines 127–54).
6. On the numerical dismissal of Lear's knights, see Emrys Jones: "The over-all effect, accelerated here at the climax, is of a countdown: from fifty to naught. The numerical references reinforce the already taut structure of the scene and give it a grimly rigorous inevitability" (1971, pp. 32–33). Jones also discusses the comical counterpart of the episode in *King Lear*: Falstaff's inflation of the number of men he killed at Gad's Hill in *1 Henry IV*, which has its own ethical implications, on stage and off.
7. Act, scene, and line numbers appear parenthetically within the text.
8. Aside from entries in dictionaries and translations from Romance languages, "zero" was not in common use in English before the late seventeenth century.

For zero as the letter O, see Robert Recorde: "marke that there are but x figures, that are used in arithmetike, and of those x one doth sygnifie nothyng, which is lyke an o, and is called privately a cyphar, though all the other sometyme be lykewyse named" (1543, sig. A7v-8r). Note that Recorde's locution "one doth sygnifie nothing" recalls the proverb "one is no number" and contains within it the poetic possibility of $0 = 1$, exploited by Shakespeare.

9. Blank demonstrates that "'equal' does not always mean the 'same' in early modern contexts," and further suggests that the pre-determined divisions include two that are "equal" and a third that is larger (2006, pp. 127, 130). Since Regan gets exactly the same as Goneril, the implication is that Cordelia will get, was always destined to get, the larger share. Even if this is true, and indeed it is plausible, it does not remove the fact of the play's dual assertion of equality and inequality across the board. On the practical problems of equality and equivalence, see also Turner (2006, p. 171).

10. Even without Lear's inexact thirds, the move from two to three is a deterioration in exactitude. Division by two is always easy, but division by three can be messy. While division of a line segment into thirds is possible, the trisection of an angle is a famously impossible problem, if one tries to employ a finite number of compass-and-straightedge constructions. It is possible with an infinite number of constructions. Once decimal notation began to take hold, division of a number by three could result in a messy repeating decimal. Simon Stevin's *De Thiende* ("The Tenth") on decimal numeration was printed in 1585.

11. In the last line, "for" may be a misprint for "from."

12. Goneril and Regan appear to have been married for some time, to husbands chosen for political rather than romantic reasons. The women's subsequent behavior certainly suggests that they have not enjoyed the marriage state.

13. The survival of the "nothing" theme after Cordelia's departure and into the Fool's presence is just one of many ways that the two characters are closely linked, even to the point of possible theatrical doubling, until Lear's near-final, suggestively complex outburst: "And my poor fool is hanged" (5.3.304). On doubling in the play, see Booth (1983) and Gamboa (2018).

14. "The compound is all manner of numbers which are compound or made of the digit and article together" (*An Introduction for to Lerne to Recken with the Pen* 1539, sig. a5r-v); not in *OED*.

15. Because Western numerical place value was borrowed from cultures that often read and write from right to left, lower place values appear "before" (i.e., to the right of) higher place values. This is why the one's place is to the right of the ten's place, even though this violates the left-to-right conventions of European writing systems. European numbers are constructed from right to left, but written and read from left to right. See also Blank's discussion (2006, p. 121) and Ostashevsky (2004).

16. See also Blank: "In the signifying system of *King Lear*, 'addition' stands for the social accommodations of nature—clothes, honors, degrees—but also, it seems, rational and moral 'powers' in excess of our commonalities, our shared feelings, and our mortality. These additions represent that which distinguishes us from other people, the diversity of our (e)qualities. In the social arithmetics of Shakespeare's play, we are 'nothing'—at least, nothing human—without them" (2006, p. 142).

416 T. D. WILLIAMS

17. Recall the Prologue to *Henry V* (1599): "O pardon, since a crooked figure may / Attest in little place a million, / And let us, ciphers to this great account, / On your imaginary forces work" (Shakespeare 1995, Pro.15–18).
18. Stage directions do not require or prevent the Fool's presence in 1.1, though having him there would upset potential doubling (see above).
19. The solution to the problem of square roots for negative numbers was eventually extended and generalized to define nth roots of negative numbers. I will confine my discussion to square roots.
20. Eggert also explains the meaning of "disknowledge" as "not pure ignorance, but rather something more like what Peter Sloterdijk calls 'enlightened false consciousness'" (2015, p. 3), citing Sloterdijk (1987, p. 5). This refinement is important for the present mathematical discussion, because none of the mathematicians concerned is unaware of imaginary numbers; they simply choose to ignore and sidestep them, for a variety of reasons.
21. The audience I invoke in this discussion is a peculiar and particular one. It is an observer of the text of this play (as either a theater audience or a reader), in an unspecified historical moment (including the present), that encounters the textual details of the play either as performed spectacle in a theatrical space, or as an imagined spectacle in an imagined theatrical space. The key factor in this advantageously constructed audience is that it has not encountered the play before and has not already been co-opted by cultural traditions about the play. I justify this constructed audience by appeal to the long rhetorical tradition of critics doing precisely this, a tradition that does not invalidate more hermeneutically or historically precise constructions of an audience or a reader. See Boecker (2015), Farabee (2014), Lopez (2003), and Rodgers (2018).
22. The Dover cliff passage has been the subject of many compelling interpretations, among them Adelman (1978), Armstrong (1995, pp. 51–52), Goldberg (1984), Levin (1959, pp. 97–98), Orgel (1984), Roychoudhury (2018, pp. 110–36), and Turner (2006, pp. 166–69).
23. *Henry V* is an entire play concerned with this habit.
24. This, despite the dominance of "Poor Tom" as a "primary" character (see Adelman 1978, p. 12; Booth 1983, p. 46; Palfrey 2014, pp. 5–24).
25. Of course, Gloucester is literally correct about the actor's situation, but in the situation that Edgar starts to create for the play's world, the ground is steep, and the actor playing Gloucester could be acting the role of a blind man who is incapable of recognizing the truth of his surroundings. In that scenario, the audience is required to imagine a steep hill, since the convention of stage tricksters expects that we will believe Edgar. Shakespeare's *The Winter's Tale* (1609) and Ben Jonson's *Epicoene* (1609–1610) indicate, however, that this convention was increasingly less secure.
26. Compare Hamlet's enigmatic remark: "I must be cruel only to be kind" (Shakespeare 1982, 3.4.180).

References

Adelman, Janet. 1978. "Introduction." In *Twentieth-Century Interpretations of* King Lear: *A Collection of Essays*, edited by Janet Adelman, 1–21. Englewood Cliffs, NJ: Prentice Hall.

Armstrong, Philip. 1995. "Spheres of Influence: Cartography and the Gaze in Shakespearean Tragedy and History." *Shakespeare Studies* 23: 39–70.

Austen, Jane. 1923. *Emma*. Edited by R. W. Chapman. 3rd ed. Oxford: Oxford University Press.

Blank, Paula. 2006. *Shakespeare and the Mismeasure of Renaissance Man*. Ithaca: Cornell University Press.

Boecker, Bettina. 2015. *Imagining Shakespeare's Original Audience, 1660–2000: Groundlings, Gallants, Grocers*. New York: Palgrave Macmillan.

Booth, Stephen. 1983. King Lear, Macbeth, *Indefinition and Tragedy*. New Haven: Yale University Press.

Bullough, Geoffrey, ed. 1957–1975. *Narrative and Dramatic Sources of Shakespeare*. 8 vols. London: Routledge & Kegan Paul.

Burckhardt, Sigurd. 1968. *Shakespearean Meanings*. Princeton: Princeton University Press.

Cajori, Florian. 1928. *A History of Mathematical Notations*. 2 vols. La Salle, IL: Open Court.

Cardano, Girolamo. 1545. *Ars Magna*. Nuremburg.

———. 1968. *The Great Art, or The Rules of Algebra*. Trans. and ed. T. Richard Witmer. Cambridge, MA: MIT Press.

Cavell, Stanley. 1987. *Disowning Knowledge in Six Plays of Shakespeare*. Cambridge: Cambridge University Press.

Descartes, René. 1637. "La Géométrie." In *Discours de La Méthode Pour Bien Conduire Sa Raison, & Chercher La Verité Dans Les Sciences*, 297–413. Leiden: Ian Maire.

Eggert, Katherine. 2015. *Disknowledge: Literature, Alchemy, and the End of Humanism in Renaissance England*. Philadelphia: University of Pennsylvania Press.

Euclid. 1925. *The Thirteen Books of Euclid's Elements*. Edited by Thomas L. Heath. 3 vols. Cambridge: Cambridge University Press.

Farabee, Darlene. 2014. *Shakespeare's Staged Spaces and Playgoers' Perceptions*. New York: Palgrave Macmillan.

Gamboa, Brett. 2018. *Shakespeare's Double Plays: Dramatic Economy on the Early Modern Stage*. Cambridge: Cambridge University Press.

Goldberg, Jonathan. 1984. "Dover Cliff and the Conditions of Representation: *King Lear* 4:6 in Perspective." *Poetics Today* 5: 537–47.

Higgins, John. 1574. *The First Parte of the Mirour for Magistrates*. London: Thomas Marshe.

Hobbes, Thomas. 1996. *Leviathan*. Edited by J. C. A. Gaskin. Oxford World's Classics. Oxford: Oxford University Press.

An Introduction for to Lerne to Recken with the Pen, or with the Counters Accordyng to the Trewe Cast of Algorisme, in Hole Numbers or in Broken, Newly Corrected. 1539. London: Nycolas Bourman.

Jones, Emrys. 1971. *Scenic Form in Shakespeare*. Oxford: Clarendon Press.

Levin, Harry. 1959. "The Heights and the Depths: A Scene from '*King Lear.*'" In *More Talking of Shakespeare*, edited by John Garrett, 87–103. London: Longmans.

Lopez, Jeremy. 2003. *Theatrical Convention and Audience Response in Early Modern Drama*. Cambridge: Cambridge University Press.

Orgel, Stephen. 1984. "Shakespeare Imagines a Theater." *Poetics Today* 5: 549–61.

Ostashevsky, Eugene. 2004. "Crooked Figures: Zero and Hindu-Arabic Notation in Shakespeare's *Henry V*." In *Arts of Calculation: Quantifying Thought in Early*

Modern Europe, edited by David Glimp and Michelle R. Warren, 205–28. New York: Palgrave Macmillan.

Palfrey, Simon. 2014. *Poor Tom: Living* King Lear. Chicago: University of Chicago Press.

Recorde, Robert. 1543. *The Ground of Artes Teachyng the Worke and Practise of Arithmetike, Moch Necessary for All States of Men. After a More Easyer & Exacter Sorte, Then Any Lyke Hath Hytherto Ben Set Forth: With Dyuers Newe Additions, as by the Table Doth Partly Appeare.* London: Reynard Wolfe.

Rodgers, Amy J. 2018. *A Monster with a Thousand Hands: The Discursive Spectator in Early Modern England.* Philadelphia: University of Pennsylvania Press.

Roychoudhury, Suparna. 2018. *Phantasmatic Shakespeare: Imagination in the Age of Early Modern Science.* Ithaca: Cornell University Press.

Shakespeare, William. 1982. *Hamlet.* Edited by Harold Jenkins. Arden (2). London: Routledge.

———. 1995. *Henry V.* Edited by T. W. Craik. Arden (3). London: Routledge.

———. 1997. *King Lear.* Edited by R. A. Foakes. Arden (3). London: Thomas Nelson.

Sidney, Philip. 1965. *An Apology for Poetry or The Defence of Poesy.* Edited by Geoffrey Shepherd. London: Nelson.

Sloterdijk, Peter. 1987. *Critique of Cynical Reason.* Minneapolis: University of Minnesota Press.

Stevin, Simon. 1585. *De Thiende. Leerende door onghehoorde lichticheyt allen rekeningen onder den Menschen noodich vallende, afveerdighen door heele ghetalen sonder ghebrokenen.* Leyden.

Turner, Henry S. 2006. *The English Renaissance Stage: Geometry, Poetics, and the Practical Arts 1580–1630.* Oxford: Oxford University Press.

Williams, Travis D. 2016. "Mathematical *Enargeia*: The Rhetoric of Early Modern Mathematical Notation." *Rhetorica: A Journal of the History of Rhetoric* 34: 163–211.

CHAPTER 23

Newton, Burns, and a Poetics of Figure: Toward a Prehistory of Consilience

Matthew Wickman

Affective differences between humanities and STEM disciplines (science, technology, engineering, and mathematics) are at least as old, and probably much older, than the emergence of these disciplines in their modern forms. And this would be true whether we charted that emergence in the nineteenth century, which is usually cited as the period of the formation of the modern disciplines, or the eighteenth century, when Scottish universities moved from a regent-based to a professorial system (designating a shift from general education, with a single tutor guiding students across multiple subjects, to a more specialized approach resembling modern university practices).[1] These differences achieved philosophical salience in the twentieth century in the work of thinkers as diverse as Martin Heidegger and C. P. Snow. For Heidegger, the most telling symptom was the severing of disciplines from their ontological foundations: mathematics, physics, biology, the human sciences, literature, and theology, he believed, had all been overrun by the ideological paradigm of the empirical sciences, a mode of thinking consumed with facts rather than foundations, or with "collecting results and storing them away in 'manuals'" rather than "inquiring into the ways in which each particular area is basically constituted" (Heidegger 1962, p. 29). Snow's assessment, some two decades after Heidegger, would prove even more definitive, according this division between humanities and science disciplines its most famous formulation: the two cultures, split between "literary intellectuals" and "physical scientists. Between the two a gulf of mutual incomprehension – sometimes … hostility and dislike, but most of all lack of

---

M. Wickman (✉)
Brigham Young University, Provo, UT, USA

© The Author(s) 2021 419
R. Tubbs et al. (eds.), *The Palgrave Handbook of Literature and Mathematics*, https://doi.org/10.1007/978-3-030-55478-1_23

understanding .... Their attitudes are so different that, even on the level of emotion, they can't find much common ground" (Snow 1964, p. 4).

In recent years the duality Snow asserts has been alternately challenged and renewed under the rubric of "consilience." The Harvard biologist Edward O. Wilson revitalized this term, although he attributed it to William Whewell's "1840 synthesis *The Philosophy of the Inductive Sciences*, which was the first to speak of consilience, literally a 'jumping together' of knowledge." As Wilson defined it in 1998, "[t]he greatest enterprise of the mind has always been and always will be the attempted linkage of the sciences and humanities .... Consilience is the key to unification" (Wilson 1998, p. 8). In the two decades following Wilson's pioneering book, consilience has come to designate an "ontological continuity between the human/mental and the non-human/ material, which justifies approaching these two realms of inquiry with a unified explanatory framework" (Slingerland and Collard 2012, p. 11). For Wilson, consilience predicates itself on the vaguely evolutionary idea that life itself inherently conduces to the initial specialization and eventual convergence of knowledge: first we identify differences, then we begin to notice salient patterns. He predicted that "the enterprises of culture will eventually fall out into science, by which I mean the natural sciences, and the humanities, particularly the creative arts. These domains will be the two great branches of learning in the twenty-first century" (Wilson 1998, p. 12).

While this idea has received a fair amount of attention in the popular press, it has not exactly met with a warm embrace among humanities scholars, who have often expressed "indifference or outright hostility" toward the prospect of disciplinary convergence.[2] Geoffrey Galt Harpham, the former director of the National Humanities Center, for example, argues that disciplinary differences "should be preserved and valued not despite their limitations but because of them." Minus these differences, "we would all be poorer – indeed, everything would be terribly confused." This is because "[t]he practices we now call science and the humanities correspond to two ways of knowing the world" that tend in opposite, and complementary, directions: "hard" versus "soft" (or "exact" versus "ambiguous") knowledge, preoccupation with the future versus the past, the production of knowledge versus critical examination of governing assumptions, and so forth (Harpham 2015, p. 223). Efforts to unify these ways of knowing seem motivated, meanwhile, less by the ideals of comprehensive understanding than by administrative efforts to reduce costs by shrinking the number of university departments. And so, while Harpham himself proposes some engaging interdisciplinary projects, two academics whose work ranges across humanities and the sciences, Edward Slingerland and Mark Collard, nevertheless detect residual ambivalence in his thought.[3] And, as they see it, Harpham's case for sustained specialization, as well as his derogatory characterization of the historical call for consilience, effectively functions "as both a plea for dialogue across the sciences/humanities barrier and a paradigmatic example of the sort of mutual incomprehension that allows that barrier to remain standing" (Slingerland and Collard

2012, p. 35). Nature may demand consilience in Wilson's cosmic imaginary, but the politics of union are decidedly more fractious.

The Enlightenment comprises a conflicted, and fascinating, chapter in the history of consilience. In Heidegger's view, the long eighteenth century (roughly 1650–1800) represents an era whose seminal advancements in modern science obscured—or rather, only further obfuscated—human understanding of its own existence, distorting at once the concept of humanity and our relation to the earth.[4] Heidegger expressed particular disdain for mechanical philosophy and the idea that nature could be computed mathematically; in essence, he objected to the "geometrization of space" for which Isaac Newton was celebrated (Koyré 1995, p. 60). But for Wilson, Newton is the hero of consilience, a "stunningly resourceful" mathematician and "inventive experimentalist, one of the first to recognize that the general laws of science might be discovered by manipulating physical processes" (Wilson 1998, p. 29). Such a synthesis provides the basis for a grand reconciliation of all understanding. As Stephen Pinker puts it, "Newton's greatest accomplishment was to subvert the ancient doctrine that there was a fundamental division in the universe between the supralunary sphere of the moon ... and the grubby, chaotic Earth below"; geometrization rendered all phenomena equal, or at least equally legible, by taking diverse qualities and rendering them uniform under the aegis of quantity (Pinker 2012, p. 46). To that extent, Newton's mechanico-mathematical model unified heaven and earth, serving, in effect, as an archetype of modern materialist convictions regarding the continuity of all things, from minerals to organic life, across a continuum of matter.

Of course, Newton's promise, the proto-consilience he presumably achieved, was inherently complex and provisional—as much the product of a capacious and fertile imagination as of the kind of settled synthesis of which Wilson dreams. For starters, Newton's geometrization of space was really the "calculation" of space, the derivation of far-reaching observations about space on the basis of calculus, and more specifically, of the theory of fluxions, Newton's particular ("integral") version of calculus. This theory was itself a complex amalgam of geometry and algebra that ventured into numerical conundrums that would not find formal resolution until the early nineteenth century. Hence, Newton employed a geometry that was not entirely geometric in pursuit of a rational project that would bear no formal resolution. In that way, Newton's calculus illustrates the drama of consilience at the historical point of its mythical conception.

The theory of fluxions thus represents an important chapter, substantively and symbolically, in the history of consilience. More specifically, Newton's theory serves as a self-reflexive meditation on the gap between the disciplines. I call Newton's theory self-reflexive because fluxions—the rate of change of a fluent, a time-varying quantity for measuring motion—purportedly sutured gaps between points along a line. The generation of lines emblematically displayed for Newton the passage of time; and, adds Morris Kline, "[s]ince the results of his mathematical work were physically true"—since he could illustrate them through geometric methods in the tracing of lines—"Newton spent very little time on the logical foundations of the calculus" (Kline 1986, p. 135). Because the mathematical technicalities of this operation had yet to be formalized, its achievement was "a matter of art rather than science," including the kinds of "arts" one associates with the humanities, including linguistic and graphic figuration (Guicciardini 2009, p. 210). Hence, we uncover a paradox at the heart of Newton's work: relative to the ambitions of consilience, the theory of fluxions succeeded precisely because it failed; unable to legitimate its own project on formal grounds, the theory shored up its methods by exemplifying collaboration across the divide separating the arts and sciences. This is the paradox I explore in the present essay. Specifically, I sketch an overview of this paradox as it took root in Newton's theory and analyze, by way of a discussion of mathematics and poetry, a consilient (or, perhaps, pre-consilient) logic of *figure*.

THE CULTURE OF GEOMETRY AFTER NON-GEOMETRIC MATHEMATICS

Eighteenth-century Scotland provides fertile soil for reflection on the prehistory of consilience. For one thing, its universities still subscribed to the tenets of what we would call general education, a broad survey of diverse liberal arts whose conceptual corollary was the common sense philosophy (of overlapping, mutually reinforcing mental faculties and physical senses) persuasively enunciated by Thomas Reid. For another, it implicitly defended a convergence of disciplines in the way it justified instruction in geometry (including advocating for Newton, as I will discuss below). In his pioneering 1961 study *The Democratic Intellect*, George Davie argued that geometry forged links between diverse university subjects, primarily because of its congruity with metaphysics. Metaphysics served as a superintending center of philosophy, a discipline of mind that negotiated the relationship between branches of learning. And geometry, with its traditional emphasis on precision and logical reasoning, was an important language of metaphysics. Hence, geometry functioned "as a cultural subject, not as a technical one, and it was found generally that the best way to maintain the students' interest in the subject was to give courses in which ... elementary mathematics was discussed with special reference to its philosophy and its history." To this extent, geometry was essentially "one of the 'humanities,'" or proto-humanities, in Scottish universities (Davie 1961, p. 109). (While I focus on Scotland, it is worth mentioning that, in the eighteenth century, the same still largely held true in English universities.)

There is a story to tell here about how it was that geometry retained pride of place in Scottish universities and, just as important, what a geometrically inflected "humanities" means relative to the Scottish Enlightenment. This is especially significant given that the most innovative work in mathematics on the continent was generally algebraic. It is a story I tell at some length elsewhere (Wickman 2016). Here, over the next few pages, let me reconstruct a short but pertinent portion of it.

At some point during their careers, many of Scotland's most esteemed intellectuals—Matthew and Dugald Stewart, David Gregory, Adam Ferguson, John Leslie, John Robison, John Playfair, and others—held a chair in mathematics at one of the nation's five universities: St. Andrews, Glasgow, Edinburgh, and King's College and Marischal College (which later merged to form the University of Aberdeen). But Scotland's reputation as a bastion for geometry during the eighteenth century derives principally from the influence of two individuals: Robert Simson, chair of mathematics at Glasgow University from 1711 to 1761, and Colin Maclaurin, appointed chair of mathematics at Marischal College in 1717 and then at Edinburgh University in 1725. Simson played the part of the conservator, Maclaurin the creative defender. In 1756, Simson published a translation of Euclid's *Elements* that became the standard edition in British schools for the next century. In the edition's preface, Simson decried the omissions and errors of Euclid's past editors and announced his intention "to remove such blemishes, and restore the principal Books of the Elements to their original Accuracy" (Simson 1756, n.p., "Preface").[5] More generally, he denounced the habits of mind introduced by algebra. The entry on Simson that appeared in the 1797 edition of the *Encyclopaedia Britannica* described the situation this way:

> Perspicuity and elegance are more attainable, and more discernible, in pure geometry, than in any other parts of the science of measure ... For the same reason [Simson] preferred the ancient method of studying pure geometry, and even felt a dislike to the Cartesian method of substituting symbols for operations of the mind, and still more was he disgusted with the substitutions of symbols for the very objects of discussion, for lines, surfaces, solids, and their affections ... [H]e came at last to consider algebraic analysis as little better than a kind of mechanical knack, in which we proceed without ideas of any kind, and obtain a result without meaning, and without being conscious of any process of reasoning, and therefore without any conviction of its truth. (*Encyclopaedia Britannica* 1797, 17, p. 504)

The affective distinction between geometry and algebra cut across multiple disciplines. Thomas Reid imported it into philosophy, specifically in his reflections on the difference between "natural" and "artificial" language. The former is said "to express human thoughts and sentiments distinctly" through an animated combination of gesture and sound. As inefficient as such language could be—devoid, ostensibly, of alphabetic writing—its refinements, Reid remarked, were not entirely salutary: "Is it not pity that the refinements of a civilized life, instead of supplying the defects of natural language, should root it out and plant in its stead dull and lifeless articulations of unmeaning sounds, or the scrawling of insignificant characters?" (Reid 1872, 1,

p. 118). One can see here a parallel logic to the distinction between geometry and algebra, above. Geometry presents us with distinct objects—"lines, surfaces, [and] solids"—whereas algebra "substitut[es] symbols [literally, alphabetic characters: x, n, and so forth] for operations of the mind." Reid then more overtly draws this comparison: "Artificial signs signify, but they do not express; they speak to the understanding, as algebraical characters may do, but the passions, the affections, and the will, hear them not" (Reid 1872, 1, p. 118). Elsewhere, in his 1783 *Lectures on Rhetoric and Belles Lettres*, Hugh Blair invoked this logic in differentiating figures from tropes. The latter "consist in a word's being employed to signify something that is different from its original and primitive meaning," whereas figures retain "their proper and literal meaning" even if "you vary the words that are used" (Blair 1783, 1, p. 275). Blair also calls this latter use of language "figures of thought," whereas tropes are "figures of words." Metaphors, for example, fall into the category of tropes, whereas exclamations or apostrophes are figures of thought. Tropes are thus algebraic whereas figures are geometric.[6]

I will return to this language of "figures" below. The point I underline here is that for Simson and those, like Reid or Blair, who invoked similar logic, the distinction between geometry and algebra was largely phenomenological, appealing to divergent experiences of the world. While Maclaurin also occasionally gives expression to that view (remarking, for example, that geometry "has been most admired for its evidence," or clarity [Maclaurin 1742, 1, p. 1]), he also engaged readily with algebra. To a large extent, his pursuit of geometry followed from his advocacy of Newton's geometric calculus, or theory of fluxions. Maclaurin, who secured his post at Edinburgh University partly through Newton's endorsement, became Newton's most rigorous and able defender in Europe; indeed, he grasped the theory more thoroughly than did Newton himself. Newton needed the defense Maclaurin was able to provide, both from the continental supporters of the rival, "differential" (or non-geometric) model of calculus developed by Gottfried Wilhelm Leibniz and from detractors closer to home, like George Berkeley, who accused Newton of trafficking in illogical obscurities. For example, Berkeley remarked in 1734, Newton's practice of subdividing quantity until one reached a virtual zero, or limit, was devoid of sense. What are the "evanescent increments," the "infinitesimally small" quantities, one finds in fluxions? "They are neither finite Quantities nor Quantities infinitely small, nor yet nothing. May we not call them the Ghosts of departed Quantities?" (Berkeley 1734, p. 59). In response, Maclaurin observed that fluxions operate according to a different logic. "[A] surface is not considered … as a body of the least sensible magnitude, but as the termination or boundary of a body." Hence, fluxions do not subscribe to a program of number and infinitude, of countable units; instead, "we conceive … quantit[y] to be increased and diminished, or to be wholly generated by motion"—literally, by the tracing of geometric lines and figures (Maclaurin 1742, p. 245).

The use of *figure* is where Maclaurin's defense of Newton intersects with Simson's critique of algebra. Newton set the stage here by negotiating his theory of fluxions through geometry. Not that this would have been intuitive given Newton's early interest in algebra. A eulogy of Newton remarked that Newton had found Euclid "too clear, too simple, too unworthy of taking up his time" (Fontanelle qtd. in Hall 1999, p. 59). Later in life, however, Newton expressed regret at "his mistake at the beginning of his mathematical studies, in applying himself to the work of *Descartes* and other algebraic writers" (qtd. in Hall 1999, p. 79). This regret may have been magnified by his dispute with Leibniz over priority in the invention of calculus, an enormously important mathematical innovation that enabled the calculation of change over time (in, for example, planetary motion, the path of artillery fire, and so on). The differential calculus devised by Leibniz divided curves into a series of minutely spaced points and then measured the difference between these points in order to determine the rate of change. Newton's calculus, by contrast, was based on time rather than space; it calculated the instantaneous velocity of movement along a curved path. It did so through two steps, an "analytical" conversion of curves into algebraic terms (a kind of translation of figure into calculable data) and then a "synthetic" step "deliver[ing] a geometrical demonstration" of the first (a translation of data back into figure) (Guicciardini 2009, pp. 179, 216).

This demonstration was important to the extent that, as Berkeley had complained, Newton's fluxions incorporated "infinitesimals," tiny numbers greater than zero but smaller than any measurable quantity. While Newton employed fluxions to ingenious effect, he was never able to resolve the philosophical webs Berkeley accused him of weaving—webs, for example, asserting that an infinite number of infinitesimally small numbers still did not equate to any measurable quantity. Newton's response to this puzzle, Niccolò Guicciardini explains, was to appeal to "geometrical and kinematic intuition," where to *see* the tracing of a figure would resolve concerns about the logic of its construction. "[T]he continuity observed in physical motions," or in the creation of lines traced out by a moving point, made "it possible to conceive of mathematics as a language applicable to the study of the natural world" (Guicciardini 2009, pp. 222–23). This is why, as I remarked above, Guicciardini considers Newton's theory a matter of art more than science. In the minds of Newton's Scottish defenders like John Keill, however, it was even more a matter of common sense. "[M]any Propositions, which appear conspicuous in [Euclid, are] knotty ... and scarcely intelligible to Learners by [the] Algebraical Way of Demonstration." The reason for this folds into the rationale we already witnessed with Simson, as geometry, Keill continues, displays "Evidence by the Contemplation of Figures" in contrast with the "Symbols, Notes, or obscure Principles" one finds in algebra (Keill 1733, n.p., "Preface"). Clarity and common sense thus compensated, in Newton's fluxions, for the mathematical solutions that would only be found in the early nineteenth century, with Augustin-Louis Cauchy.[7]

In fluxions, figures thus compensated for thought; the tracing of curves effectively materialized, or literalized, a mathematical practice that, as yet, bore no solid ground. Adopting Blair's terminology, the geometric aspect of fluxions was a figure without a corresponding thought, and was thus caught somewhere in the gap—the "infinitesimal" abyss, we might call it—between word and thing, or representation (which fluxions evidently displayed) and operation (which it had yet fully to formalize). While the limits of my argument do not allow me to develop this idea here, it nevertheless seems noteworthy that, more than two centuries later, in 1962, Jacques Derrida would initiate deconstruction by taking issue with Edmund Husserl's 1939 essay "The Origin of Geometry" on similar grounds. Where Husserl appealed to geometry to ascertain the basis of its self-evidence, Derrida sought to expose the gaps—specifically, the implication in language, tropes—on which such self-evidence is predicated (Derrida 1989). In this respect, Derrida was treading ground previously covered by Berkeley, albeit in a different fashion. And in doing so, Derrida also turns our attention to the poetic quality, the tropology, of Newton's theory of fluxions.

The Geometer's Saturday Night

Not to belabor the obvious, but the theory of fluxions and poetry represent distinct domains of Enlightenment thought. And yet, without mapping directly onto each other, they reflect each other and, in places, pursue similar intuitions. This is what makes them relevant to the subject of consilience.

Reflection, of course, is not the same thing as *influence*. Influence would be harder to prove, and for the purposes of this essay it is also beside the point. A simpler task is to show the awareness of Newton's theory by certain poets. These poets include David Mallet, whose long 1728 poem *The Excursion* adopts the language of fluxions in describing the motion of the planets: "with transport I survey / This firmament, and these her rolling worlds, / Their magnitudes and motions" (Mallet, qtd. in Johnson 1969, 14, p. 22). A *magnitude* in this context designates an infinitely little quantity, a fluent, hence the movement of a point in the creation of a curve (Wickman 2016, pp. 164, 245). The poetry of James Thomson presents an even more dramatic case, for it takes up Newtonian subjects (the *Poem Sacred to the Memory of Sir Isaac Newton* addresses astronomy and gravity), meditates on the moral and creative implications of Newton's achievement (by drawing a series of subtle comparisons and distinctions between the poetic and scientific enterprises), and models fluxions in its kinematic survey of topics (most famously in the sweeping catalogues of *The Seasons*, which replicate the effects of a moving gaze).[8]

That last motif, the *survey*, is especially important, for it permeates much eighteenth-century poetry. This is true of descriptive poems of nature (*The Seasons* is the most famous exhibit, but hardly the only one: think of Anna Laetitia Barbauld's "A Summer Evening's Meditation" [1773]), of poems of morals or the human condition (with Alexander Pope's *Essay on Man* [1734] and Samuel Johnson's *Vanity of Human Wishes* [1749] some of the most famous illustrations), of poetico-philosophic expositions (like Mark Akenside's *The Pleasures of Imagination* [1744]), and of poems describing society or class (for example, Oliver Goldsmith's *The Deserted Village* [1770] and Stephen Duck's "The Thresher's Labour" [1736], or the pointed responses they engendered: George Crabbe's *The Village* [1783] and Mary Collier's *The Woman's Labour* [1739]). The conceit of the astronomer's—better said, the geometer's—gaze, poised at some Archimedean point high above the scene and sweeping across broad swathes of observable or psychological territory, is so pervasive that one is hard-pressed to make a case for the "influence" of Newton's theory of fluxions, the period's most sophisticated instrument for measuring such motions. This is true even though fluxions were instrumental to Newtonian thought, which, in turn, informed and even overwhelmed the creative imagination—at least if we believe Joseph Addison's famous observation in *The Spectator* No. 420: "among ... Writers, there are none who more gratifie and enlarge the Imagination, than the Authors of the new Philosophy," or science, whether in scaling up to the measure of planets orbiting on their axes or down to "the smallest Particle of this little World ... capable of being spun out into another Universe" (Addison and Steele 1963, 3, pp. 302–3). This would be a case of consilience *avant la lettre*, for the relationship between poetry and the sciences in Addison's essay (and across the eighteenth century) is, at best, muddled and, at worst, represents more a case of hostile imaginative takeover than of mutual enlightenment.

And yet, as we discussed above, the theory of fluxions is calculus in a less developed, more figurative stage prior to its nineteenth-century elucidation. And in some cases, the logic and operations of fluxions—or, at the very least, its tropology and popular associations—seem to animate eighteenth-century poems and their discursive contexts. Robert Burns's iconic poem "The Cotter's Saturday Night" presents a compelling case study. It is also a curious one, for, unlike self-consciously cosmic poems like Mallet's *The Excursion* or Thomson's *The Seasons*—or the kinds of works Alice Jenkins analyzes in her survey of quasi-"consilient" culture in the nineteenth century, *Space and the "March of Mind."*[9] Burns's poem is by no means overtly mathematical, natural theological, or Newtonian. It purports, with wry modesty, to present "life's sequester'd scene" from the perspective of one who lives it. As William Wordsworth would conceptualize such a premise less than two decades later, Burns effectively chooses "incidents and situations from common life" and "relate[s] or describe[s] them ... in a selection of language really used by men; and, at the same time ... throw[s] over them a certain colouring of imagination" (Wordsworth and Coleridge 2007, p. 59). The poem identifies, and even cites, some important precursors: Robert Fergusson's *The Farmer's Ingle* (to which I will return below), Thomas Gray's *Elegy Written in a Country Churchyard* (whose lines concerning "The short and simple annals of the

428   M. WICKMAN

poor" Burns invokes in his epigraph (Burns 1969, p. 116)), James Beattie's *The Minstrel* (which, like "Cotter's," adapts the Spenserian stanza), and, perhaps most importantly, Thomson (who, himself, had adapted the Spenserian stanza in *The Castle of Indolence* and, in *The Seasons*, charted a similar course between pastoral observation and sententious moral commentary). Burns's seemingly homespun poem thus positions itself as an inheritor of a rich poetic tradition, or body of signifiers, even as it purports to give us a privileged view of life in an eighteenth-century peasant cottage, a pure signified.

Perhaps most relevant for our consideration, Burns fashions his poem after a model of "figure"—specifically, an image of his friend and patron, Robert Aiken, dwelling in an idealized pastoral setting. The opening stanza introduces this figurative exercise: "My lov'd, my honor'd, much respected friend / ... To you I sing, in simple Scottish lays / ... What Aiken in a *cottage* would have been; / Ah! tho' his worth unknown, far happier there I ween!" (Burns 1969, p. 116, ll. 1, 4, 8–9). The apostrophe constitutes what Blair would call a "figure of thought." This is because the idealized setting does not alter the meaning of pastoral life by turning its rural scene into a trope or allegory of something else: there are no symbols in play here, for instance. Rather, it exalts and intensifies its subject matter, ennobling what is overtly and avowedly mundane: Aiken—a successful lawyer, and thus belonging to a part of the social empyrean above Burns's own laboring class—blesses the Scottish peasantry by association with it. The poem thus acknowledges the subjunctive abstraction (the "would have been") in which it participates. This sheds light on what follows, provocatively, as the poet leads us across a series of topoi: the home-bound cotter, his young children (the "*wee-things*" (Burns 1969, p. 117, l. 21)) and "thriftie Wifie" (l. 24), his "*elder bairns*" (l. 28) and love-smitten eldest daughter, her suitor, the meal, and the gathering "round the ingle" (l. 101) to read the Bible. The effect is visual, almost painterly, although as Carol McGuirk remarks, the poem aims for something other than straightforward visual documentation. Unlike "The Farmer's Ingle," for example, which "confines [it]self to what peasants see and eat .... Burns, in an occasionally awkward mixture of high English and vernacular Scots, captures what peasants feel: their mutual affection and their religious faith" (Burns 1993, p. 229). The poem wishes to show, in other words, something that is not outwardly evident.

To take one instance, this is how Burns sets the scene for the family's reading of the Bible—a non-dramatic activity that acquires, here, sentimental gravity:

> The chearfu' supper done, wi' serious face,
>   They, round the ingle, form a circle wide;
> The Sire turns o'er, with patriarchal grace,
>   The big *ha'-Bible*, ance [once] his *Father's* pride:
> His bonnet rev'rently is laid aside,
>   His *lyart haffets* [grizzled locks] wearing thin and bare;

> Those strains that once did sweet in Zion glide,
>   He wales [chooses] a portion with judicious care;
> '*And let us worship God!*' he says with solemn. (Burns 1969, p. 119, ll. 100–8)

The items we see—serious faces, the gathered circle, the Bible, the bonnet—explicate themselves through things we cannot: the supper's "cheer," the father's "grace," the reverence of his motions, the idealized memory of a utopian religious ("Zion") community, and the solemnity of the scene. The poem's form—seen on the page, but registered aurally and conjuring an evocative set of (generic) associations—heightens these impressions. As Nigel Leask notes, "The long Spenserian stanza," featuring in other Scottish poems (by Fergusson, Thomson, and others), "and the pentameter line permit the descriptive depth and extension requisite for such domestic scene-painting" (Leask 2010, p. 222). Just as important, McGuirk would remind us, the lines paint the dwelling's *ethos* as well as its *objects*. The composite effect is less an engagement of anything substantively real than the materialization of atmosphere, the glorified image of a common setting that actually nowhere appears as such.

Burns's cottage, we might say, is to a standard peasant dwelling what a geometric figure is to a shape in nature: cleaner, clearer, devoid of accident. The geometric analogy may seem a little too pat here, but then again, it may not be as far afield as may initially appear. What I mean is that, much as geometry, in the minds of Maclaurin and others, enabled Newton to circumvent some of the philosophical conundrums involving infinitesimals, so the poem's overall effect stylistically purges the poetic image of unresolved conflict. The result is something that looks like (and even purports to be) nature, but is actually something more stylized and hypothetical: "What Aiken in a cottage *would have been*." Consider too that the tenor of much modern scholarship of "The Cotter's Saturday Night," similar to the mathematicians who recognized that Newton's geometric (kinematic) analogy was simply untenable, has been to identify and undercut the poem's idealizing logic—to reintroduce historical contingency into the auratic timelessness of Burns's artifact. The first step in that process has been simply to recognize and reconstruct the poem's own historicity. As Corey E. Andrews reminds us, the poem—and, for that matter, Burns's legacy itself—owe a great deal to James Currie, whose authoritative anthology of Burns's poetry after the poet's death, and whose influential essay "Some Observations on the Character and Condition of the Scottish Peasantry," at once borrowed liberally from Burns's airbrushed portrait of country life and inflated it into an extended, Rousseauian disquisition of enlightened savagery. Hence, Scottish peasants, for Currie, "possess a degree of intelligence generally not found among the same class of men in the other countries of Europe" (Currie qtd. in Andrews 2015, p. 167). And yet, Andrews observes, Currie continually reasserts his own position of comparative privilege, a distinction that collapses his idealizations of the peasantry into fictions (Andrews 2015, pp. 164–70).

430 M. WICKMAN

These were, however, compelling and contagious fictions. Andrew Nash has traced the history of the poem's reception into some important cultural artifacts. This includes David Wilkie's visual rendition of the poem in an 1837 painting that became, Duncan Macmillan argues, "*the* canonical image of Scottish art" (qtd. in Nash 1997, p. 183). But "[t]he most notable text that was to help tie Burns up with visual topography was *The Land of Burns*" (1840) (Nash 1997, p. 184), a series of topographical descriptions and visual prints that portrayed a landscape at once deserted and Edenic, converting Scotland into an iconic location of rural simplicity. *The Land of Burns* "presented to the reading public for the first time a highly influential essay by John Wilson, entitled 'The Genius and Character of Burns.'" This essay, coupled with other writings of Wilson's, "established topics that would reverberate throughout the century: elevating the spiritual status of the poor, representing rural, peasant Scotland as the supposedly essential Scotland, understanding literature through topography, preserving the past through literature[,] and, significantly, understanding Scotland through literature" (Nash 1997, p. 184). This would become most prominently realized in the popular Kailyard (cabbage patch) romances of J. M. Barrie, Ian Maclaren, S. R. Crockett, and others. As Andrew Blaikie recounts, the treacly sentiment and rural imagery promulgated by these romances, "enjoying particular appeal among Scottish emigrants" (Blaikie 2010, p. 106), helped craft the image of Scotland as a land where time stood still. Scotland became "a state of mind, a way of viewing the world" (Herman qtd. in Blaikie 2010, p. 3) against which modern sociology emerged. The Scottish Renaissance movement after World War I would confound this stylized image, with Hugh MacDiarmid perhaps unsurprisingly adopting the poetic artifice of a complex, bookish (indeed, in Thomas Reid's terms, *algebraic*: strictly bookish, non-natural) Scots as a way of drawing attention to the constructed, ideological quality of Scottish—"human"—nature. The tenor here was to formalize the historical processes that a poem like "Cotter's" portrays as self-evident.

As McGuirk indicates, "Cotter's" conduces to a very different set of effects, in this respect, from Fergusson's alternately descriptive and strident portrait of peasant life in "The Farmer's Ingle." Consider the stanza Fergusson devotes to the evening meal awaiting the farmer:

> Weel kens the gudewife that the pleughs require
>> A heartsome meltith [meal], and refreshing synd [drink]
> O' nappy liquor, o'er a bleezing fire:
>> Sair wark and poortith [poverty] douna weel be joined.
> Wi' buttered bannocks [oatcakes] now the girdle reeks,
>> I' the far nook the bowie [keg] briskly reams;
> The readied kail stand by the chimley cheeks,
>> And had the riggin [roof] het [hot] wi' welcome steams,
> Whilk than the daintiest kitchen nicer seems. (MacLachlan 2002, pp. 286–87, ll. 19–27)

The scene is far less sentimental than the correlative passage in "Cotter's," given not to "expectant *wee-things*" but to the correspondence of hard work and hearty fare. Christopher Whyte reads Fergusson's poem as "an idyll of family and labour" (Whyte 2000, p. 55) in the tradition of Virgil's *Georgics*. It also represents a brand of (Scots) straight-talking criticism of the kinds of bourgeois values (and Augustan language) inscribed into "The Cotter's Saturday Night." Take Fergusson's defense of the meal and the nourishment it provides: "Frae this lat gentler gabs a lesson lear [learn]; / Wad they to labouring lend an eidant [busy] hand, / They'd rax [grow] fell strang upo' the simplest fare, / Nor find their stamacks ever at a stand" (MacLachlan 2002, p. 287, ll. 28–31). The poet and the meal dignify the labor that produces it. In "Cotter's," by contrast, the farmer returns home as a shelter from and shunning of his work; indeed, Whyte argues, the poem treats such labor as "an indignity from which the Sabbath offers a brief respite" (Whyte 2000, p. 55).

Curiously, but compellingly, the differences between the poems—one a straightforward defense of (supposed) simplicity, the other a more ambivalent celebration of the same—are manifest in their authors' divergent remarks concerning fluxions. On its surface, this is purely coincidental, but upon reflection it seems more telling. One of Fergusson's earliest poems was a satire of David Gregory, a professor of mathematics at St. Andrews University. Fergusson was a student at St. Andrews in his mid-teens, and Gregory died shortly after young Robert's experience with him. Fergusson's satire captures youthful impatience with obscure ideas:

> He could, by *Euclid* prove lang sine
> A ganging *point* compos'd a line;
> In numbers too he cou'd divine
>   Whan he did read,
> That *three* times *three* just made up nine;
>   But now he's dead. (Fergusson 1954, 2, pp. 1)

The references to Euclid and the "ganging" (or "moving") point indicate that Fergusson was mocking Gregory's instruction concerning fluxions. At once leveled into common sense ("That *three* times *three* just made up nine") and etherealized into virtual hieromancy, the satire voids Gregory's instruction of existential value, rendering it ridiculous, pretentious in the face of death's certainty.[10] (To be sure, Fergusson composed this poem as a very young man, which may partly explain his irreverence, though his early death at age 24 lends his flippancy a poignant edge.) Burns' appeal to fluxions, by contrast, is more complex, labile. We see this in an invocation of deity expressed in a letter Burns wrote to William Nicol in early 1793:

> O thou, wisest among the wise, meridian blaze of Prudence, full-moon of Discretion, & Chief of many Counsellors! – How infinitely is thy

432   M. WICKMAN

> puddle-headed, rattle-headed; wrong-headed, round-headed slave indebted
> to thy goodness, that from the luminous path of thy own right-lined rectitude
> thou lookest benignly down on an erring Wretch, of whom the zig-zag wander-
> ings defy all the powers of Calculation, from the simple copulation of Units up
> to the hidden mystery of Fluxions! (Burns 1985, 2, p. 183)

Where God is the great geometer defined by "right-lined rectitude," Burns is
a "puddle-headed, rattle-headed … slave" given to "zig-zag wanderings" of
an irregular shape that fluxions were designed to measure. Hence, while the
reference to fluxions is casual and self-consciously ridiculous, it seems weirdly
apposite to the confusion Burns describes. If for the youthful Fergusson,
then, Newton's obscure mathematics amounted to an expression of vanity,
it captured for (or, perhaps, indicates in) Burns something of the contradiction of humanity, aspiring as it does after something like Euclidean simplicity ("right-lined rectitude") but consigned instead to psychological, and not
merely mathematical, irrationality and complexity.

This unwonted poetic justice of Burns's appeal to calculus lends greater
poignancy to the Thomsonian strains of "The Cotter's Saturday Night"—
specifically, to those moments when the poet zooms out from the homely
scene and pontificates on the majesty of it all:

> O Scotia! my dear, my native soil!
> > For whom my warmest wish to heaven is sent!
> Long may thy hardy sons of *rustic toil*,
> > Be blest with health, and peace, and sweet content!
> And O may Heaven their simple lives prevent
> > From *luxury's* contagion, weak and vile!
> Then howe'er *crowns* and *coronets* be rent,
> > A *virtuous populace* may rise the while,
> And stand a wall of fire around their much-lov'd Isle. (Burns 1969,
> p. 121, ll. 172–80)

Here, in a move reminiscent of *The Seasons*, the poet takes flight high
above his subject matter and engages in a moralizing master commentary on
it. Having shifted from object to object within his scene of representation,
forging a set of connections between them in a painterly manner (almost
sketching lines between the corners of the cottage and shading them with
sentiment), the poet resorts here to Blair's "figure," an apostrophe that punc-
tuates the image: "O Scotia! …" That apostrophe is significant, and a "figure" in Hugh Blair's sense of the term, because it inflates the rustic scene
inside the cottage into a spiritual image of all of Scotland. However, as Whyte
remarks, "Burns' use of an Augustan English idiom [e.g., 'Scotia'] in 'The
Cotter's Saturday Night' is" problematic, "not … for reasons of nationalist
loyalty, or because he had not mastered it sufficiently, but because it contradicts the poem's explicitly stated stylistic programme" (Whyte 2000, p. 53).
It conflicts, that is, with the language and understanding—the purported

simplicity—of the poem's subjects. Reprising Burns's language in the letter we cite above, it comprises a kind of "zig-zag" that deviates from Fergusson's stylistically purer achievement. The "figure" to which Burns resorts in this stanza is thus an overt abstraction that strives to bring the poem's subject matter into clearer focus, but in doing so it also distracts attention from the gaps and fissures that riddle the poem's depiction of the cotter's family. Whyte outlines several such moral lacunae in the poem: a disdain for the labor of the laboring class, which we mention above; gendered and sexual hierarchies in the poem's commentary on Jenny, the daughter ("With Burns, sexual innocence is exclusively a female issue" [Whyte 2000, p. 59])[11]; and religion as a cover for social injustice. The "figure" of "Scotia" with which the poem concludes, in other words, is less the distillation of pure national truth than an extension of unresolved tensions in the poem. The poem's very form—Spenserian but also, by the 1780s, after "The Farmer's Ingle," Fergussonian—rendered it an ambivalent cultural hybrid. An uncanny, but telling, analogy forms here with Newton's mathematics. There, the geometric, figural aspect of calculus compensated—as "art," Guicciardini remarks: "a matter of art rather than science, a matter of guesswork rather than of algorithmic deduction" (Guicciardini 2009, p. 210) for what Newton's ingenious invention could not render fully consistent on strictly mathematical grounds. This is to say, Newton was never able to resolve how an infinite number of infinitely small numbers might nevertheless amount to less than any measurable number, which is why, Guicciardini remarks, Newton resorted to "geometrical and kinematic intuition" (2009, p. 222), involving the motion of points rather than the strict calculation of infinitesimal differences. Hence, the theory of fluxions itself, like Burns's influential poem, was something of a "zig-zag" accomplishment.

This is what makes the nineteenth-century legacy of "Cotter's" so compelling. Coming of an age in the high era of Newtonian-fueled industrialization, "The Cotter's Saturday Night" fashioned an image of Scotland that would hold its currency for more than a century, until at least after World War I, roughly the time of the Bolshevik Revolution. I mention this era of revolution and the Great War because it brought together, unwittingly, two important intellectual turns. One—evident in physics as well as in broad social unrest—pitted itself against Newtonian science, specifically the latter's impact on the technology and course of industrial progress. This is the intellectual history we associate with, for example, the physics of relativity and the cultural production of the avant-garde: the displacement of Newtonian thought (and of classical linearity) in both the arts and sciences. The other important turn may be seen in the Scottish Renaissance. This revision of Scottish culture and history in no way equals the scale and scope of the Leninist movement (however fascinated Scottish writers like Hugh MacDiarmid may have been by Lenin),[12] but the proponents of new Scottish writing sought in their own way to disrupt the tradition—with Burns in mind, we might call it the Newtonian-Thomsonian legacy—that had exerted such force on the shape

434   M. WICKMAN

of Scottish life. How much the early twentieth century managed to overturn the eighteenth and nineteenth centuries is debatable.[13] But the heavy strain placed on the legacy of Newtonian thought also distends and maybe even severs Wilson's Edenic image of natal consilience—the purported clarity of Newton's synthesis of the arts and sciences—that has made the long eighteenth century seem in some ways more innocent than the long twentieth.

NON-CONSILIENT CONSILIENCE?

Let's step back and review some of the general principles we have discussed. Whatever we make of the ultimate legacy of Newtonian thought, calculus and poetry each reflected it in important ways during the era immediately following Newton, each contributing to what I have called a logic of figure. To that extent, calculus and poetry also factor into the picture of consilience that the latter's proponents identify as an important part of that concept's prehistory. One might expect as much from calculus, a "science" that drew upon "art" before later practitioners formalized it more than a century after its invention. But poetry, an "art," registered the effects of science, and in ways that go far beyond a citation of the "numbers" associated with rhythm and meter.[14] Indeed, the great Newtonian project, the synthesis of theory and experimentation that brought together heaven and earth, exerted a powerful influence on the poetic imagination of James Thomson, as I discussed above. But Newton's influence extended beyond such obvious cases as Thomson's, as I have tried to show in my analysis of Burns's iconic, sentimentalized tour de force on the Scottish peasantry. That influence was not always direct; indeed, it imparted itself at times at a second and third remove (through poets like Thomson, for example). Significantly, though, what it reveals are some of the labyrinths mediating the logic of figure in the eighteenth century, labyrinths where language explained itself through recourse to mathematics (as in Thomas Reid's critique of algebra) and where mathematical operations occasionally took a metaphorical turn. All told, it is a picture that complicates the prehistory of consilience.

Consilience may thus trace its genealogy to the Newtonian Age, but that does not make that era a Golden one for the present-day search for unifying theories across the arts and sciences. Rather, it indicates a fraught origin for a subject that promises to remain controversial, largely because it is likely to become more significant in our era of ever-more-fungible disciplines. Newtonian thought may be no more directly related to modern consilience than it was to Burns's poem "The Cotter's Saturday Night," but I hope my discussion of the Newtonian echoes in that poem indicate the reverberations of Newton's "figural" fluxions in fields that reach far beyond the discipline of mathematics and well into later modernity. Indeed, the connections I have drawn between Burns's poem and Newton's ingenious maneuver reveal the correspondence between, if not always the precise mechanisms of, the fields we struggle to bring into relation.

Notes

1. For an account of the emergence of the modern disciplines within the research university, see (Wellmon 2015); on transformations in the Scottish university system, see Emerson (2003, pp. 18–19).
2. For scores of articles on the practical impact of the humanities in the real world, and for examples of its interface with the sciences, see the Humanities+blog kept at Brigham Young University: humanitiesplus.byu.edu. Accessed 16 June 2017.
3. Harpham's chief example is the study of pleasure, in which literary study would engage with cognitive neuroscience and evolutionary biology (Harpham 2011, pp. 99–122). Slingerland is a professor of Chinese who engages cognitive linguistics and evolutionary psychology. Collard is an evolutionary anthropologist.
4. In his essay "Modern Science, Metaphysics, and Mathematics," for example, Heidegger criticizes Newton, Galileo, and Descartes (Heidegger 1993, pp. 271–305).
5. On errors in editions of Euclid's *Elements*, see Benjamin Wardhaugh's "Rehearsing in the Margins: Mathematical Print and Mathematical Learning in the Early Modern Period" in this volume.
6. For a more extensive analysis of the difference between geometry and algebra at the level of rhetoric, see Wickman (2016, pp. 29–37, 49–52).
7. On the solution of this problem, see Edwards (1979, pp. 308–22).
8. For further discussion of this point, see Wickman (2016, pp. 168–91).
9. Jenkins, for example, undertakes extended discussions of such luminaries as Milton, Coleridge, Wordsworth, and Dickens (Jenkins 2007).
10. On Fergusson's evocation of fluxions, see (MacQueen 1982, pp. 57–58).
11. Consider these lines: "Is there, in human form, that bears a heart – / A Wretch! a Villain! lost to love and truth! / That can, with studied, sly, ensnaring art, / Betray sweet Jenny's unsuspecting youth? [ll. 82–85] As Nigel Leask remarks, 'This is pretty fresh coming from the man who had recently seduced his mother's servant Elizabeth Paton, justifying [the] description of Burns as 'a Country Libertine'" (Leask 2010, p. 224).
12. See "First Hymn to Lenin" and "Second Hymn to Lenin" in MacDiarmid 1985, 1:297–99 and 1:323–28.
13. For a discussion of the uncanny recapitulation of Newtonian thought in the supposed era of its renunciation, see Wickman (2016, pp. 194–222).
14. On the conscientious attention to number and ratio in the long tradition of Scottish poetry, particularly in the early modern period, see Macqueen (2006, pp. 20–26).

References

Addison, Joseph, and Richard Steele. 1963. *The Spectator*. Edited by Gregory Smith. London: Everyman's Library.

Andrews, Corey E. 2015. *The Genius of Scotland: The Cultural Production of Robert Burns, 1785–1834*. Leiden: Brill.

Berkeley, George. 1734. *The Analyst; or, a Discourse Addressed to an Infidel Mathematician*. London: Printed for J. Thomson.

Blaikie, Andrew. 2010. *The Scots Imagination and Modern Memory*. Edinburgh: Edinburgh University Press.

Blair, Hugh. 1783. *Lectures on Rhetoric and Belles Lettres*. London: Printed for W. Strahan.

Burns, Robert. 1969. *Burns: Poems and Songs*. Edited by James Kinsley. Oxford: Oxford University Press.

———. 1985. *The Letters of Robert Burns*. Edited by J. De Lancy Ferguson and G. Ross Roy. Oxford: Oxford University Press.

———. 1993. *Selected Poems*. Edited by Carol McGuirk. London: Penguin.

Davie, George Elder. 1961. *The Democratic Intellect: Scotland and Her Universities in the Nineteenth Century*. Edinburgh: Edinburgh University Press.

Derrida, Jacques. 1989. *Edmund Husserl's Origin of Geometry: An Introduction*. Translated by John P. Leavey, Jr. Lincoln: University of Nebraska Press.

Edwards, C.H. 1979. *The Historical Development of the Calculus*. New York: Springer.

Emerson, Roger. 2003. "The Contexts of the Scottish Enlightenment." In *The Cambridge Companion to the Scottish Enlightenment*, edited by Alexander Broadie, 9–30. Cambridge: Cambridge University Press.

Encyclopaedia Britannica; or, a Dictionary of Arts, Sciences, and Miscellaneous Literature. 1797. 3rd ed. Edinburgh: Printed by A. Bell and C. McFarquhar.

Fergusson, Robert. 1954. *The Poems of Robert Fergusson*. Edited by Matthew P. McDiarmid. Edinburgh: Blackwood.

Guicciardini, Niccolò. 2009. *Isaac Newton on Mathematical Certainty and Method*. Cambridge, MA: MIT Press.

Harpham, Geoffrey Galt. 2011. *The Humanities and the Dream of America*. Chicago: University of Chicago Press.

———. 2015. "Defending Disciplines in an Interdisciplinary Age." *College Literature* 42 (2): 221–40.

Heidegger, Martin. 1962. *Being and Time*. Translated by John Macquarrie and Edward Robinson. San Francisco: Harper. humanitiesplus.byu.edu. Accessed 16 June 2017.

———. 1993. "Modern Science, Metaphysics, and Mathematics." In *Basic Writings*, edited by David Farrell Krell, 271–305. San Francisco: Harper.

Hall, Rupert, ed. 1999. *Isaac Newton: Eighteenth-Century Perspectives*. Oxford: Oxford University Press.

Johnson, Samuel. 1969. *The Works of the English Poets, from Chaucer to Cowper*, 21 vols. 14: 16–24. New York: Greenwood.

Jenkins, Alice. 2007. *Space and the "March of Mind": Literature and the Physical Sciences in Britain, 1815–1850*. Oxford: Oxford University Press.

Keill, John. 1733. "A Preface, Shewing the *Usefulness* and *Excellency* of This Work." In *Euclid's Elements of Geometry, from the Latin Translations of Commandine*, edited by John Keill, n.p. London: Printed for Thomas Woodward.

Kline, Morris. 1986. *Mathematics: The Loss of Certainty*. Oxford: Oxford University Press.

Koyré, Alexander. 1995. "The Significance of the Newtonian Synthesis." In *Newton: Texts, Backgrounds, Commentaries*, edited by I. Bernard Cohen and Richard S. Westfall, 58–72. New York: Norton.

Leask, Nigel. 2010. *Robert Burns and Pastoral: Poetry and Improvement in Late Eighteenth-Century Scotland*. Oxford: Oxford University Press.

MacDiarmid, Hugh. 1985. *The Complete Poems of Hugh MacDiarmid*. Edited by Michael Grieve and W. R. Aitken. Harmondsworth: Penguin.

MacLachlan, Christopher, ed. 2002. *Before Burns: Eighteenth-Century Scottish Poetry*, 286–90. Edinburgh: Canongate.

Maclaurin, Colin. 1742. *A Treatise of Fluxions*. Edinburgh: Ruddimans.

MacQueen, John. 1982. *Progress and Poetry*. Edinburgh: Scottish Academic.

———. 2006. *Complete and Full with Numbers: The Narrative Poetry of Robert Henryson*. Amsterdam: Rodopi.

Nash, Andrew. 1997. "The Cotter's Kailyard." In *Robert Burns and Cultural Authority*, edited by Robert Crawford, 180–97. Iowa City: University of Iowa Press.

Pinker, Steven. 2012. "The Humanities and Human Nature." In *Creating Consilience: Integrating the Sciences and the Humanities*, edited by Edward Slingerland and Mark Collard, 45–55. Oxford: Oxford University Press.

Reid, Thomas. [1872] 2005. *An Inquiry into the Human Mind on the Principles of Common Sense*. Edited by William Hamilton. Reprint, Hoboken, NJ: Elibron.

Simson, Robert. 1756. *The Elements of Euclid*. Glasgow: Printed by Robert and Andrew Foulis.

Slingerland, Edward, and Mark Collard. 2012. "Introduction: Creating Consilience—Toward a Second Wave." In *Creating Consilience: Integrating the Sciences and the Humanities*, edited by Edward Slingerland and Mark Collard, 3–39. Oxford: Oxford University Press.

Snow, C.P. 1964. *The Two Cultures: And a Second Look*. Cambridge: Cambridge University Press.

Wellmon, Chad. 2015. *Organizing Enlightenment: Information Overload and the Invention of the Modern Research University*. Baltimore: Johns Hopkins University Press.

Whyte, Christopher. 2000. "Competing Idylls: Fergusson and Burns." *Scottish Studies Review* 1: 47–62.

Wickman, Matthew. 2016. *Literature After Euclid: The Geometric Imagination in the Long Scottish Enlightenment*. Philadelphia: University of Pennsylvania Press.

Wilson, Edward O. 1998. *Consilience: The Unity of Knowledge*. New York: Knopf.

Wordsworth, William, and Samuel Taylor Coleridge. [1800] 2007. *Lyrical Ballads: Second Edition*. Edited by Michael Mason. Edinburgh: Pearson.

CHAPTER 24

The Mathematics of Associationism in Laurence Sterne's *Tristram Shandy*

Aaron Ottinger

Laurence Sterne's (1713–1768) *The Life and Opinions of Tristram Shandy, Gentleman* (1759–1767) and associationism, the eighteenth-century's dominant psychological theory in Great Britain, have been the subject of numerous studies.[1] Critics have also demonstrated the importance of mathematics for Sterne.[2] This essay combines these conversations to show how the mathematics of associationism played a significant role in conjoining the acts of reading and calculating, ultimately for the sake of eliciting embodied, moral feelings. It confirms that literature served as a mediating device for correcting personal, subjective associations, a practice thought to benefit readers morally. More difficult to appreciate is the way that literary authors like Sterne revised the mathematical approaches associationist philosophers employed.

Sterne's contributions to the mathematics of associationism may have gone unnoticed in light of Wolfgang Iser's representative claim that Tristram's "pre-birth life stories are all marked by the workings of chance" (Iser 1988, p. 3). This outlook neglects the novel's indelicate balance between geometry and propositional logic and probability and associationist logic. The imbalance is embedded in Tristram's birth. Once a month, his father Walter Shandy would wind the family clock—an image of the Newtonian God resetting the mechanical motions of the universe[3]—and in response, Mrs. Shandy "could never hear the said clock wound up – but the thoughts of some other things unavoidably popp'd into her head" (p. 1.4.9).[4] On the occasion Walter forgets the clock, Mrs. Shandy's "strange combination of ideas" never fall into array, and drawing attention to the fact during intercourse, Walter

---

A. Ottinger (✉)
Seattle University, Seattle, WA, USA

© The Author(s) 2021
R. Tubbs et al. (eds.), *The Palgrave Handbook of Literature and Mathematics*, https://doi.org/10.1007/978-3-030-55478-1_24

439

440   A. OTTINGER

ejaculates prematurely, leaving the animal spirits "scattered and dispersed" (p. 1.2.6). Like a planet set into retrograde orbit, the event accounts for Tristram's wayward life, for as he claims, "nine parts in ten of a man's sense or his nonsense depend upon [the animal spirits'] motions and activity":

> and the different tracks and trains you put them into; so that when they are once set a-going, whether right or wrong, 'tis not a halfpenny matter, – away they go cluttering like hey-go-mad; and by treading the same steps over and over again, they presently make a road of it, as plain and as smooth as a garden-walk, which, when they are once used to, the Devil himself sometimes shall not be able to drive them off it. (p. 1.1.5)

Tristram maintains that one's life should follow a straight or regular course of events beginning with a natural cause. This efficient cause ought to be thought of in axiomatic terms, or as a cause with necessary succeeding steps, falling under the sign of geometry. Yet, when the initial cause is accidental, the ensuing events follow a retrograde or irregular order. This cause ought to be thought of in associationist terms, or as a cause with arbitrary succeeding steps, falling under the sign of probability. It is a mistake then to read Tristram's life according to chance and probability alone because it is always being measured against necessity and geometry.

An additional complication emerges from the apparent divide between the chance events in nature (Tristram's birth) and the accidental connections in the mind (Tristram's associations). Yet as Walter explains the post-Newtonian psychology to his brother Toby, associationism relies on just such a bond. Ideas, according to Walter, "follow and succeed one another in our minds at certain distances," just like the planets in orbit; but additionally, such a "train" of ideas ought to appear like "the images in the inside of a lanthorn turned round by the heat of a candle" (p. 3.18.151). Walter embraces a geometrical conformity between world and mind: ideas and their motions on the screen of consciousness should fit objects in a measured and determinate way ("at certain distances"). But upon hearing the word "train" Toby dismantles Walter's purely geometrical framework by incorrectly completing the phrase "train of ideas" with "train of artillery." No doubt, Toby's mind also conforms to external nature. But because his development as a moral subject is altered midway through life due to an injury he suffers to the groin at the Siege of Namur, an external chance accident becomes inseparable from his internal chance habit.[5]

For Tristram and Toby, chance occurrences pry apart the regularity of nature's geometrical order. And the accidents of everyday life are made manifest in the characters' thoughts and behaviors, often without their conscious recognition. It is simultaneously the charm of *Tristram Shandy* and the heart of the novel's problem. For Sterne, it is fair to suffer from chance accidents and to proceed unaware of the accidental associations that ensue. He is more critical of a critic like Walter, who attaches moral judgments to the arrangement of one's thoughts and feelings. Moreover, Sterne departs from the view that representing the world geometrically will resolve the accidental associations of the mind. Pointing out unnecessary associations only triggers stronger defenses. Hence Uncle Toby whistles "Lillibullero" when anyone breaches the subject of his injury and he spends his days reconstructing fortifications from the Siege of Namur.

Yet Sterne remains committed to the mathematics of associationism for the purpose of remedying untoward relations between ideas. Indeed, Sterne's eponymous narrator treats his "book as a *machine*," operating according to mathematical principles which have the power to alter a train of ideas in succession (p. 7.1.385). Accordingly, readers are asked to make calculations for the sake of correcting their own accidental associations regarding events in the narrative, providing them with "half a day to give a probable guess at the grounds of this procedure" (p. 1.10.16). But, as I argue, Sterne's calculations are impossible to solve, and this impossibility reintroduces a brand of certainty into the narrative's associationist progression. Sterne thereby inverts the expectations customarily attributed to geometry and propositional logic and probability and associationist logic: it is the certainty introduced by the latter (probability) and not the former (geometry) that places a limit on thinking, allowing feelings uncoerced by habit to reenter one's personal outlook. On a grander scale, *Tristram Shandy* establishes new causes from within the associationist mind/nature relationship, upsetting the order of cause and effect, the plan for design, and the moral values attributed to reflecting a natural symmetry.

Critics have certainly recognized how the mathematics of associationism played a role in upholding the balance between mind and world. Walter Ong stresses the geometrical conformity between mind and nature, especially in Adam Smith (Ong 1951, p. 17), while Douglas Patey illustrates probability's role in this process, especially for Archibald Alison (Patey 1984, p. 253). More recently, James Chandler combines the geometrical and probabilistic approaches to linking mind and external world, be it nature or the work of art. "There is a long tradition ... of invoking audience credulity or generosity in the realization of some bit of action on the stage or page," writes Chandler, "[but] in the case of the sentimental [mode] these two axes [line of action and line of affect] become curiously hinged to each other" (Chandler 2013, pp. 208, 218–223). Looking especially at Sterne's *A Sentimental Journey* (1768), Chandler claims that the "improbability of the plot is a projection of the anticipated effect of the sentimental text on the collective sensibility of its readers" (Chandler 2013, p. 224). It is the inability of readers to predict the unfolding narrative that elicits an affective response, and in *Tristram Shandy*, it is the impossible calculations involved that universalize this effect.

This line of inquiry intersects with work on reading and embodiment, as explored by cognitive literary studies. Lisa Zunshine sees great value for cognitive scientists and literary critics in turning to texts like *Tristram Shandy*,

442 A. OTTINGER

"to inquire into other, not yet formulated, cognitive regularities underlying our interaction with fictional narrative" (Zunshine 2006, p. 43). Studies of the embodied and extended mind consistently include references to Sterne's novels, no doubt. But there remains a tendency among these inquiries to forego the role of mathematics.[6] Thus, due to the ongoing importance of associationism for contemporary cognitive science, it is incumbent upon us to maintain the earlier emphasis on mathematics in the mind/nature relationship, which reading brings into focus.[7]

Mathematics' uncertain role in criticism may reflect its inconsistent role in associationist thinking, which extends as far back as Thomas Hobbes (1588–1679), and reached a critical state when John Locke (1632–1704) added the chapter "Of the Association of Ideas" to the fourth edition of *An Essay Concerning Human Understanding* (1700).[8] The *Essay* is premised on the idea that life begins axiomatically with a *tabula rasa*, and as simple ideas about the external world combine, forming complex ideas regarding the intellectual realm, the mind becomes vulnerable to two kinds of associations: natural and unnatural. The former is characterized by consistency and regularity, while the latter belong to "another connection of ideas wholly owing to chance or custom" (Locke [1700] 1975, p. 395).[9] In other words, Locke disavows any equation between the propositional thinking of Euclid and the associations of human thoughts and feelings. His reaction is fair because he ultimately fears that unnatural associations that become habit will produce madness. But Locke's concerns also contributed to moral-aesthetic judgments regarding geometry and probability. Chance and custom, which were the province of probability, were regarded as base and low, while geometry and its related terms (regularity, uniformity, symmetry, consistency, and necessity) represented reason and a moral ideal.[10]

Locke's epistemology fueled debates regarding one of the most important questions recurring throughout the period: can moral lives be guided according to the same logic as mathematics?[11] While the question received its first developed mathematical response in Part IV of Jakob Bernoulli's *Ars conjectandi* (1713), if the moral-aesthetic side of humanity was to be improved on a practical level, it would require a medium fitting to the task.[12] Accordingly, many eighteenth-century literary and aesthetic theorists attempted to outline how artworks could be based on fundamental principles, just as Isaac Newton had done for nature. The project is indicative of the traffic between objective and subjective versions of probability in the eighteenth-century, for there was little distinction between probability based on frequency and probability based on opinion, feeling, and authority.[13] Once critics saw that questions of evidence could be measured, it became reasonable to assume that the associationist logic of mind (dealing in morals) could conform to the propositional logic of nature (dealing in causes), or that a probable outcome in one instance could be reduced to geometrical regularity over time.

Accordingly, a critic like Henry Home, Lord Kames (1696–1782) posits in *The Elements of Criticism* (1762) that "we are framed by nature to relish order and connection" (Kames 1762, p. 32). The "sense of order," he explains, "coincides with the order of nature," for which Kames provides a host of examples, from tracing smoke to fire to proceeding "in the order of time" as to "historical facts ... which comes to the same, to proceed along the chain of causes and effects" (Kames 1762, pp. 29–30). Kames privileges a mirror relationship between mind and nature, according to which both sides exhibit necessary connections. His purpose is ultimately a moral one. Without the guidance of necessity and consistency, he claims, "our conduct [in everyday life] would be fluctuating and desultory; and we would be hurried from thought to thought, and from action to action, entirely at the mercy of chance" (Kames 1762, p. 41). Thus things and events in nature have necessary connections; our thoughts ought to mirror these links to develop sound habits; therefore, literary works ought to avoid accidental connections and instead reflect the necessary relations of nature.

This tendency to read nature geometrically began with Galileo Galilei (1564–1642), who first compared nature to a book "written in the language of mathematics" (Galileo 1957, p. 238). His comparison led numerous British philosophers, to "regard mathematics as a language for expressing regularities observed in nature, subject to the conventions of natural and artificial languages" (Sepkoski 2007, p. 4). If the mind truly corresponded with nature, as associationists believed, and nature could be reduced to a kind of language, the ability to predict events in the world, like predicting words in a sentence, confirmed that the mind/nature divide was only an illusion. The emergence of new mathematical methods for prediction, namely probability and calculus, reinforced rather than disrupted this human-centered view: whatever errors emerged from personal associations, the new methods in mathematics could counteract.

Bringing the new mathematics to bear on the mind/nature nexus was one of the major achievements of David Hartley (1705–1757), a significant though often overlooked influence for Sterne (Lamb 1980, p. 285). Due to Hartley and his later popularization through Joseph Priestley (1733–1804), authors could incorporate the various mathematics emerging in the eighteenth century for the purpose of reinforcing epistemological and moral outlooks grounded in the body.[14] Rather than reject this view, Sterne extends its logical consequences to an extreme, inventing new causes that disturb the idea of a natural conformity between mind and nature. Despite this artificial disturbance, Sterne's interventions ultimately appeal to an affective register, eliciting moral feelings undetermined by the habits of association. But to understand how Sterne arrives at these feelings, it is first necessary to appreciate the mathematical methods Hartley rendered legible.

444   A. OTTINGER

HARTLEY'S MATHEMATIZED ASSOCIATIONISM
AND THE BOOK OF NATURE

David Hartley's *Observations on Man, His Frame, His Duty, His Expectations* (1749) resembles the common eighteenth-century account of a developing consciousness, beginning with external stimuli that awaken a discerning subject. To explain this process, Hartley relies on the "doctrine of vibrations"—the theory that small vibrations in the æther fill every pore of bodies and nerves before stimulating the brain—as first posited in Newton's *Principia mathematica* (1725), the *Opticks* (1704), and a letter from Newton to Robert Boyle (1627–1691) (Hartley [1749] 1966, pp. 7–8). By way of the doctrine of vibrations, Hartley provides the first thoroughgoing psychobiological explanation for how internal ideas follow from external things, thus binding occurrences in nature to the succession of ideas (Buckingham and Finger 1997, pp. 21–37). The problem is that human memory affords the mind opportunities to recall old associations and apply them to similar situations, however remote (Hartley [1749] 1966, p. 67–72). Accordingly, Hartley's recourse to mathematics represents an attempt to correct illusory outlooks ensuing from one's personal, subjective history.

Hartley turns to correcting accidental associations in Proposition 87: "To deduce the Rules for the Ascertainment of Truth, and the Advancement of Knowledge, from the mathematical Methods of considering Quantity." The section contains a summary of Abraham de Moivre's (1667–1754) *Doctrine of Chances* (1718), which had presented the mathematics of probability for the first time as an independent discipline (Hacking 1975, p. 166). Hartley then describes the solution to the inverse problem, published posthumously in 1764 but likely obtained through personal correspondence, with either Thomas Bayes (1702–1761) or Nicholas Saunderson (1682–1739) (Stigler 1983, pp. 290–96). Hartley concludes with an outline of Newton's differential or fluxional calculus. For Hartley, all three methods assist in the discovery of unknowns, but the reference to Newton is especially significant because Newton's method required trafficking from geometry to algebra and back to geometry, thereby presenting final results in the form of Euclid's axiomatic method (Wickman 2016, p. 48). Despite unknowns in life, it could still be understood in terms of a certain and necessary order in nature.

Hartley begins Proposition 87 with a court trial—often the model for theorists of epistemological probability[15] where the truthfulness of an event is called into question. Turning to de Moivre, Hartley believes that "the separate Probability of each Evidences must be very great ... to make the proposition [or the story] credible" (Hartley [1749] 1966, p. 335), and the truth of the matter can be calculated where "the Value of each Evidence be $\frac{1}{a}$, and the Number of Evidences be n, then will the resulting Probability be $\frac{1}{a^n}$." (Hartley [1749] 1966, p. 335). So long as the value of each evidence is less than one, the probability will be simultaneously equal for each instance, regardless of the number of evidences or witnesses, thereby diminishing the credibility of the witnesses overall. The formula could thus be used to calculate the credibility of individuals on a case-by-case basis, informing one's decision to assent to or trust each individual encountered.[16]

On a grander scale, Hartley believes the doctrine of chances can "[account] for that Order and Proportion, which we every-where see in the Phenomena of Nature" (p. 338). For every failed event, the doctrine of chances demonstrates the equal probability for that event's occurrence. Thus, if one believes that event x is necessarily the result of a falsely attributed cause, or "various Associations" (p. 337), the doctrine of chances has the power to correct such prejudiced views.

But de Moivre's formula only assists in calculating the probability that a thing or person is behind an effect when it is already known the number of things or persons and their proportion to one another.[17] To predict the probability of an agent when their numbers and proportions to one another are unknown, Hartley draws on Bayes's solution to the inverse problem, that is, a formula which can "determine the Proportions, and, by degrees, the whole Nature, of unknown Causes, by a sufficient Observation of their effects" (p. 339). Bayes's literary executor, Richard Price (1723–1791), elucidates the formula's significance: "tho' in such cases the *Data* are not sufficient to discover the exact probability of an event, yet it is very agreeable to be able to find the limits between which it is reasonable to think it must lie" (Price 1763, p. 418). Price saw the solution as an aid to common sense thinking regarding effects with unknown causes (Daston 1988, p. 257). And although his interpretation would not have satisfied associationists who emphasized necessary causal relations, the emerging probabilists were more comfortable with best guesses. Ultimately the power of Bayes's Theorem for Hartley cannot be overstated: it provided a mathematical formula for tracing present appearances back to unknown causes.

In Hartley's summary of Newton's differential calculus, he makes it explicit that the three methods under consideration are to mathematicians what induction and analogy are to natural philosophers—thereby bridging the gap between ascertaining truths in a purely intellectual realm and a moral-aesthetic one. As Hartley describes it, Newton's method calculates a general law for an unknown curve, where points on the y-axis (ordinates) can be thought of as effects in a series of experiments, and points on the x-axis (abscissae) as the various circumstances of these experiments. Using the general law or equation, one can determine, given certain points on the x-axis, "the Lengths of Ordinates not given" (p. 339). When Hartley compares Newton's method to induction and analogy, his initial purpose is to point out, like Price, that both approaches are subject to uncertainties (Hartley [1749] 1966, p. 340). But when considering the grander implications, Hartley becomes a touch more romantic:

> The analogous Natures of all the Things about us, are a great Assistance in deciphering their Properties, Powers, Laws, &c. inasmuch as what is minute or

446   A. OTTINGER

> obscure in one may be explained and illustrated by the analogous Particular in
> another, where it is large and clear. And thus all Things become Comments on
> each other in an endless Reciprocation. (Hartley [1749] 1966, p. 343)

Hartley describes the world in terms of things referring to each other in a web of relations, united by a single fabric and subject to the general laws of mathematics.

Proposition 87 is motivated by the fear shared among associationist critics that all things and events are products of chance, accidental, or unnatural associations. As Lorraine Daston clarifies, the aforementioned mathematical references were all efforts to counter skeptical philosophers suspicious of design and unconvinced of a proportionate ratio between occurrences and their failures in nature (Daston 1988, pp. 253–57).[18] Indeed, his combination of approaches reinforced the view that the world adhered to regular and necessary patterns. For every failed event in nature, a counter-event occurs. Moreover, if Hartley's methods for ascertaining truth had the power to correct accidental associations and reveal this balance governing the cosmos, the implication was that the mind could in fact be rid of unnatural associations.

Of course, for mathematics to inform associationism, language or the medium of ideas would also need to operate according to mathematical principles. Hence, Hartley claims that, "Language itself may be termed one Species of Algebra; and, conversely, Algebra is nothing more than the Language which is peculiarly fitted to explain Quantity of all Kinds" (Hartley [1749] 1966, p. 280).[19] Based on this reasoning, Hartley imagines "a Dictionary of any language" that would allow everyone "to understand one another perfectly" (Hartley [1749] 1966, p. 285). Hartley reinforces the mathematical underpinnings of his universal language when he predicts at the end of his discussion regarding events in nature that "future Generations should put all Kinds of Evidences and Inquiries into mathematical Forms" (Hartley [1749] 1966, p. 351). Hartley, it appears, combines images of a universal language and Galileo's book of nature. In this book where "all Things become Comments on each other in an endless Reciprocation," things are reduced to words and mathematics provides a means to translate between any domain.

It is by way of Hartley's figure of the dictionary that his image of a mathematically regular and ordered world comes to bear on *Tristram Shandy*. While not identical to it, Hartley's dictionary shares a certain resemblance with Sterne's *Tristrapedia*, the encyclopedia Walter Shandy writes for his youngest son. After losing his first-born Bobby, Tristram writes, "I was my father's last stake … he had lost, by his own computation, full three fourths of me – that is, he had been unfortunate in his three first great casts for me – my geniture, my nose, and name (p. 5.16.298). The three casts of the die refer to probability's deep roots in gambling (Hacking 1975, pp. 2–8), and Walter loses each one of the three casts. Accordingly, the ratio of successes to failures

Hartley speaks of is so far absent in Tristram's life. To test his conclusions, Walter makes "a thousand other observations" of Tristram (p. 1.3.7). Similar to his calculations in the "chapter of chances" regarding the probability of the doctor crushing Tristram's nose ("it will turn out a million to one" [p. 4.9.224]), and Toby wrapping his brother on the shin ("'Twas a hundred to one" [p. 4.9.224]), Walter finally concludes that Tristram will "neither think nor act like any other man's child" (p. 1.3.7). Thus he sets out to right the wrongs of his son's fragmented and backwards life by writing "a Tristrapedia, or system of education ... binding [Walter's thoughts] together, so as to form an INSTITUTE for the government of [Tristram's] childhood and adolescence" (p. 5.16.298). It is a source of knowledge as well as a "design," as the word "institute" signifies, thereby reinforcing Walter's commitment to Hartley's balanced cosmos.

The problem is that any attempt to capture the world in writing will quickly fall short of its purpose. Walter is at work on the *Tristrapedia* for three years, and "by the very delay, the first part of the work, upon which my father had spent the most of his pains, was rendered entirely useless" (p. 5.16.300). The fact of obsolescence poses a practical obstacle to promoting texts as devices for fitting ideas in succession to the unfolding events of nature. Inevitably, the book becomes tethered to a particular time and place: rather than correct accidental associations, it nails them down. What is actually needed is not merely an account of factual references between things, but a way of unbinding the associations already in place. Recourse to mathematics, it would seem, is only good for reinforcing the former, demonstrating that y is in relation to x. But in Tristram's book, the turn to mathematics is ultimately in the service of severing this bond, and thereby freeing the passions.

STERNE'S EMBODIED ASSOCIATIONISM AND THE BOOK OF "INEVITABLE CHANCE"

Because the mathematics of probability did not arise until the decades following 1660, it would be very surprising to see a narrator involve readers in an act of probabilistic conjecture prior to the eighteenth century.[20] Still, Sterne was not the first. Henry Fielding (1707–1754) popularized the address to the reader in *The History of Tom Jones, a Foundling* (1749), and Henry Brooke (1703–1783) made interesting use of it in *The Fool of Quality* (1766–1770) (Folkenflik 2009, pp. 53–54). But Eliza Haywood's (1693–1756) addresses in *The History of Miss Betsy Thoughtless* (1751) actually include invitations to make conjectures regarding the plot's unfolding, "though ten to one but he finds himself deceived" (Haywood [1751] 1986, p. 3). The reader's ability to calculate the probability of an event depends on the associationist principle underlying narrative consistency: if the novel corresponds with the contiguity of events in nature, history, and the other narratives one comes to know, then the reader ought to be able to predict unknowns in the story.[21] But, as

Christina Lupton (2012, p. 29) and Alex Wetmore (2013, p. 3) have recently demonstrated, there were a host of mid-eighteenth-century authors like Haywood that actively aggravated audiences, often through drawing attention to the materiality of the book. It comes as no surprise then when the narrator of *Tristram Shandy* concludes the first volume with an aggressive wager: "if I thought you was able to form the least judgment or probable conjecture to yourself, of what was to come in the next page, – I would tear it out of my book" (p. 1.25.63). Tristram's slight outburst is indicative of a frustrated or confused feeling, as if the impossible invitation he bestows on readers matches something inscrutable rather irritating himself. Indeed, Tristram's irritation coincides with the absolute certainty of the audience's failure which the book guarantees—in this case, by the narrator's own hands. It is my contention that Tristram's impossible wager is the mathematical representation of an "inevitable chance." Walter's euphemism for death (p. 5.3.283). It is a kind of eternal object of irritation because it cannot fit into the associationist's train of ideas without bringing that train to a final terminus. But for the same reason, such inevitabilities are good stopgaps for a mind caught in a cycle of associations and thereby they can liberate a ruminating consciousness.

Accordingly, unlike Hartley's dictionary and Walter's encyclopedia, Tristram's narrative forces readers to make predictions regarding occurrences that cannot be incorporated into the axiomatic of causal relations. It is not that Sterne disavows the mathematical substructure of mind and nature. But through the act of reading, Tristram's narrative intervenes in the book of nature, establishing new and stable causes that inaugurate a different train of moral feelings, written on the body and registered by the heart: "[t]here are trains of certain ideas, which leave prints of themselves about our eyes and eye-brows; and there is a consciousness of it, somewhere about the heart, which serves but to make these etchings the stronger – we see, spell, and put them together without a dictionary" (p. 5.1.278). While these virtual trains of ideas operate irrespective of Hartley's dictionary, for Sterne they are nevertheless natural because they follow from certain causes which the materiality of the book guarantees, namely voids or representations of death. Tristram establishes a new conformity between the reader and a machine that comes with its own certain causes, thereby intervening in the balance between mind and nature.

Laying bare his mathematical approach in volume 1, Sterne illuminates the moral-affective machinery at work in the rest of the novel. For instance, just prior to the book's most famous page, the marbled leaf, Tristram directs an order much like the bet he makes with audiences at the end of the first volume, minus the language of probability:

> Read, read, read, read, my unlearned reader! read, – or by the knowledge of the great saint *Paraleipomenon* – I tell you before-hand, you had better throw down the book at once; for without *much reading*, by which your reverence knows, I mean *much knowledge*, you will no more be able to penetrate the moral

of the next marbled page (motly emblem of my work!) than the world with all its sagacity has been able to unravel the many opinions, transactions and truths which still lie mystically hid under the dark veil of the black one. (p. 3.36.180)

While there is no mention of conjectures or probability, the mathematical approach is embedded in the challenge. Consider, the marbled leaf gives body to Tristram's moral, like print gives body to language.[22] By switching from printed word to marbled page, Tristram ensures that no one can pinpoint the page's meaning, no doubt, through recourse to the method of decipherers, that is, "investigating Words written in unknown Characters," which according to Hartley "is an Algebra of its own Kind" (Hartley [1749] 1966, p. 350). Sterne challenges the assumption that all words and events are decipherable based on contiguous words and events. But he also challenges the moral values one would typically associate with an event that conformed to expectation.

Of course, Sterne does not disavow moral feelings altogether. The marbled leaf only makes explicit that associations leading up to this point do not refer to a necessary moral. The added fact that each copy of the original four thousand books contained a marbled leaf unique to each individual volume prohibited the arrival of a general law based on the many pages across editions (Day 1972, p. 145). In short, the probability that readers would fail to penetrate the moral of this page was equal to one, or absolutely certain. Yet, it is precisely this absolute truth, the meaninglessness and purposelessness associated with the marble page that Tristram's challenge to the reader elicits with certainty. Sterne demonstrates his point. Trains of ideas can ensue from manufactured points of departure, and at the same time they can be grounded in an eternal bulwark against the mind's personal associations. In other words, whatever affect emerges in response to Tristram's challenge is the product of something unnatural but certain, and yet it is also untethered from a subjective personal history.

To clarify the moral-affective import of the marbled leaf, Tristram redirects readers to the "dark veil of the black one." The reference alludes to an early page, covered entirely in black ink, signifying Parson Yorick's death. It also recalls the "inevitable chance" Walter speaks of in reference to his eldest son, Bobby. Plus, these negatives represent an absolute break. For Walter, his son's death points to a final terminus: "the world itself, brother Toby, must – must come to an end" (p. 5.3.284). The various references to death fall under the heading of extinction, as it represents the only idea that can unbind a train of thoughts with absolute certainty.[23] For associationists, while it may be associated with other ideas, extinction signifies nothing beyond itself in the chronological succession of events.[24] Put another way, extinction represents the one failed occurrence in nature without a positive counterpart: the ratio of occurrence is 1:0, no matter what. Accordingly, death and extinction offer a break from the repetition of a ruminating mind, as described by Locke. And it revises the certainty and consistency attached to axiomatic images of life, beginning with a *tabula rasa* at birth. Sterne stresses instead a negative form of certainty, one that necessarily blocks, disrupts, and challenges connections on account of a void, and its supreme example lies at life's end, not its beginning. Of course, within the book, Sterne inserts these voids somewhere in the middle. Their purpose is to draw out the transposition of ideas in succession, from a cognitive to an affective domain. As illustrated in the case of Mrs. Shandy's servant Susannah, blocking the mind's predictive powers liberates the heart. Upon hearing of Bobby's death, Susannah abruptly transitions from imagining a "procession" of Mrs. Shandy's vibrant clothes—countering the grim accoutrements of mourning wear—to "[bursting] into a flood of tears" (p. 5.7.290). The idea of death is enough to put the brakes on associations in a train. Thus Sterne locates the absolute within the machinery of the book, represented here "under the dark veil of the black one."

Bringing audience and book together in these moments of inevitability ultimately introduces a new and perhaps more powerful form of geometrical conformity between mind and text. Tristram even provides geometrical figures of the book's internal structure, thus diagramming the voids. The figures of the chapters Tristram has already written appear retrograde, digressive, and frayed. They reinforce the irregular order of Tristram's thoughts. Still, as if one of the diagrams represented a proposition from Euclid, Tristram assigns letters to significant plot points, marking it with an A, B, and D, plus c, c, c, c, c. He defines the latter as "parentheses and the common *ins* and *outs* incident to the lives of the greatest ministers of state; and when compared with what men have done, – or with my own transgressions at the letters A B D – they vanish into nothing" (p. 6.40.380). Tristram downplays the importance of these nothings, even hinting at another bawdy joke. But his humor is suspicious given Tristram's many attempts to pinpoint causes in his life and suggests on another level that the voids punctuating his diagram are in fact the negatives determining the nonsensical lines of his existence. Lockeans would argue that Tristram is ruminating on something preventing him from seeing the true determining agents operating at these points, be it a word or event. And if they are truly unknown, the only reasonable event these points can represent is death—or at least a partial death. For instance, when a window slams shut on a young Tristram urinating outside, it is "the *murder*" of him (p. 5.18.301). "'Twas nothing," Tristram comments sharply (p. 5.17.301). But his ire is apt. If the mathematics of nature is only another language, and this language is from birth grounded in the body and sense perception, to lose a body part is to introduce a void or nothing into the cosmos. It is an epistemological disaster reinforced by Uncle Toby's attempts to use geometry to pinpoint the exact location of his injury at the Siege of Namur. As Ala Alryyes convincingly demonstrates, Toby's ceaseless calculations irritate the wound, the excluded variable from his science, and thus "*Tristram Shandy* highlights the contrast between the mind's pretensions to certainty and the body's fragility" (Alryyes 2015, pp. 1113–14). But, I would argue, rather than serve as a cautionary tale against certainty, Sterne exploits the certainty

24 THE MATHEMATICS OF ASSOCIATIONISM ...   451

that speculative unknowns ensure, and suggests that the negative of bodily harm can effectuate similar trains of ideas. Moreover, Tristram's irritation at his inability to pinpoint unknowns in his own timeline is made manifest in the many impossible challenges he puts to the reader. No doubt, it is difficult to resist the suspicion that Tristram's unknowns are, in fact, coincident with the challenges to the audience, for these signs without referents designate the points where the book draws on the readers' predictive powers, rendering Tristram's life contingent upon the excluded audience's actions.[25] Conversely, the audience's inability to maintain a consistent and regular line of reading depends on Tristram's inscrutable propositions. If there is a geometrical conformity between text and audience, the moments when this affective link is at its greatest intensity are at the points of the novel's vacancies.

Thus Sterne qualifies his resistance to the geometrical conformity between outside and inside. Geometrical conformity between mind and external stimuli will set off a chain of reactions, regardless if the determining agents are the circumstances surrounding an unplanned birth or a mathematical calculation. But the propositional logic of the latter may not refer to anything beyond itself. While associationists appealed to mathematics to correct accidental associations, they also believed recourse to calculus and probability would uphold an image of design, the whole, and endless reciprocity. Sterne offers a more finite outcome: the same combination of geometrical conformity and probabilistic conjecture may evince a final end. But it is precisely the absolute oneness of death and extinction that Sterne thinks quells the mind's eternal links and provides an outlet for the heart. If there is a moral to be gained from mathematics, it is not through a proposition that leads to a metaphysical moral but a proposition that silences the mind.

CONCLUSION

The mathematics of associationism did not end in the literary realm with *Tristram Shandy*. It is significant that Sterne applied the mathematical methods of the eighteenth century to reading, and thus used the book as a device for altering audience's thoughts and feelings. The same tactic is picked up in a variety of ways by Sterne's successors. Moreover, the absolute void located in Sterne's book is internalized by a number of Romantic-era inheritors as a self-positing act of transcendence—William Wordsworth (1770–1850) being singled out among them (De Man 1969, p. 190). Following this earlier tendency, Arkady Plotnitsky claims that the second generation of Romantic-era writers, represented in England by Percy Bysshe Shelley (1792–1822), can be classified as "non-classical," where the traffic between epistemic and statistical probability becomes less fluid, and connections in the cosmos are finally regarded as probabilistic relations rather than causal links (Plotnitsky 2015, p. 309). Nevertheless, this history is gradual and full of turns, failures, and successes, and is often as clarified as it is distorted by our present, retrospective view. Its picture resembles the

452   A. OTTINGER

wayward lines of Tristram's actual narrative, more so than the straight lines he idealizes ("The emblem of moral rectitude! says *Cicero*" [p. 6.39.380]). Accordingly, if we are familiar with the historical context, the mathematical import of Sterne's challenges to the reader can be detected even, for instance, in the Wordsworth of *Lyrical Ballads* (1800/1802) (see Margaret Kolb's "Romantic Parts and Wholes, Statistical and Literary" in this volume). It thereby falls to literary critics to decode the ways that mathematical approaches entered literature and informed the reading of the mind-nature-text relationship, especially where the tradition of criticism has written mathematics out of the picture.[26]

NOTES

1. Debates regarding associationism and *Tristram Shandy* include a disavowal of Locke's associationist influence (Cash 1955, p. 130), and the reemphasis on Lockean associationism (Anderson 1969, p. 27), before Locke is subordinated to other contemporary philosophers, like David Hume (Bannerjee 1974, p. 698). Another recent comparison to Hume emphasizes the conversation between the novel's characters as the organizing principle behind Tristram's personal psychology (Craig 2007, pp. 145–58).

2. Recent articles on geometry and *Tristram Shandy* explore Uncle Toby's preoccupation with the trajectory of the bullet (Alryyes 2015, p. 1113) and the "entanglement of literature, geometry, military theory, and the media of war in the eighteenth and early nineteenth century" (Engberg-Pedersen 2013, p. 24). One study on probability in the eighteenth century highlights the importance of "chance" as a cause in *Tristram Shandy* and claims that the novel charts incidents rather than accidents in an attempt to eke out some semblance of self (Molesworth 2010, pp. 189–205). Another study touches on *Tristram Shandy* to support the claim that the sentimental mode was conditioned by the emergence of probability (Chandler 2013, pp. 203–28).

3. See Leibniz's fifth response to Samuel Clarke, first printed in 1717 (Alexander 1956, p. 88).

4. The image is shocking, but Blackwell suggests that Elizabeth Shandy as wound-up clock is linked to the supposed speed and efficiency brought about by the male-midwife's technological instruments or obstetrics; and because the man-midwife's instruments backfire, there is a case to be made in favor of Sterne's feminist outlook (Blackwell 2001, p. 83). All further references to *Tristram Shandy* will be included in the body of the essay, starting with the volume and followed by the chapter and page numbers.

5. Carol Stewart also links the obsessive habits of Sterne's characters to madness in Locke. But the "arbitrary association of ideas that drives the narrative" is, according to Stewart, attributable to his Latitudinarianism and his resistance to the dogma of certainty (Stewart 2010, p. 137). As we will see, it is precisely certainty that liberates Sterne's characters.

6. One of the most significant studies on embodied consciousness in the eighteenth and early nineteenth centuries sees it as a literary phenomenon beginning in the romantic period—eliding Sterne—but reinforces the idea that

Hartley (below) marks an important philosophical starting point (Richardson 2001, p. 5, 9ff.). Another recent study treats associationism in *Tristram Shandy* in purely psychological terms (Bullard 2016), and while Kate Tunstall foregrounds embodiment in *Tristram Shandy* (Tunstall 2016, pp. 203–8), neither study includes the role of mathematics.

7. Dawson outlines the historical trajectory from seventeenth-century associationism to the mathematics of an associative memory system in computer simulations (Dawson 2004, pp. 133–69).

8. This view regarding Hobbes and associationism has already been outlined (Kallich 1945, pp. 291–303), but Gert reinforces the main point, explaining that Hobbes's basic principle for motions in the physical domain also applied to the intelligible domain, making for a coherent whole between the two (Gert 1996, p. 158).

9. Throughout the eighteenth century, "chance" referred to a lack of skill or art. But Thomas Bayes identifies chance with mathematical probability by 1763 (Bayes 1763, p. 376).

10. Probability follows from the "low sciences" and "*opinio*." while the high sciences, "such as optics, astronomy, and mechanics, still lusted after demonstration" (Hacking 1975, p. 35).

11. The project of outlining a moral life in geometrical/axiomatic terms follows directly from Locke's claim that from a *tabula rasa* at birth consciousness ensues (Gaukroger 2010, pp. 387–420, at 417).

12. Lorraine Daston explains that Bernoulli established two common lines of argument for moral sciences, at least until the nineteenth century. Moral dilemmas were treated like court cases and their underlying principles were borrowed from the natural/physical realm (Daston 1988, p. 296).

13. On the traffic between epistemic and frequentist or subjective and objective approaches in classical probability see Daston (1988, p. 188).

14. On the far-reaching influence of Hartley see Robertson (1921, pp. 50–117).

15. The law is concerned with evidence, while natural philosophy is concerned with causes (Hacking 1975, pp. 85–91). This distinction coincides with the difference between epistemic probability and statistical probability (ibid., 13–14). Not until the seventeenth century did the mathematics of epistemic probability arise, and according to Hacking, the mathematical approach remained confined to continental Europe (Hacking 1975, p. 129). David Hartley represents an exception to this rule in England.

16. Prior to denoting something measurable in 1662, *probabilitas* was not "evidential support but support from respected people," and as this example from Hartley illustrates, "this sense of 'probability' survived into eighteenth century English -." according to Hacking (1975, p. 23).

17. This explanation is derived from the "urn model of causation" put forth by Condorcet in 1781 (Daston 1988, p. 230).

18. On "design" see (Hacking 1975, pp. 166–75).

19. Hartley's equation of language with algebra follows Berkeley's belief that "arithmetic is simply a formal language, and numbers are created by the mind to serve as marks or signs" (Sepkoski 2007, p. 81).

20. Epistemic probability (and not statistical probability) first denotes something measurable in the Port Royal *Logic* in 1662 (Hacking 1975, p. 73).

454 A. OTTINGER

21. These blank spaces where readers are asked to make conjectures differ from the *Leerstelle* or the "virtual hole in the text" discussed by Roman Ingarden and Wolfgang Iser, and exemplified in Fielding's *Tom Jones*, because Sterne's virtual holes cannot be guessed at (Campe 2012, pp. 290–91).
22. See Fanning on the significance of print technology for making the word material in Sterne (2003, p. 368).
23. Today, extinction is the term used in psychological approaches to healing, such as exposure therapy, to represent the process of unbinding conditioned and unconditioned stimuli. But as has been demonstrated (Bouton 2002), extinction does not necessarily unbind old associations so much as create new ones.
24. On extinction as a non-phenomenological absolute see Brassier (2007, pp. 205–39).
25. While her concern is strictly generic, Patricia Meyer Spacks supports this claim, positing: "Its contemporary readers were already being trained in making sense of narratives that lack obvious coherence and in understanding lack of coherence as a signal to react with feeling. [*Tristram Shandy*] offers incoherence of a new sort, insisting that the disjunction represents a way of thinking – the way most people think – as well as feeling" (Spacks 2006, p. 263).
26. I would like to thank Marshall Brown for his feedback on an earlier draft.

REFERENCES

Alexander, H. G., ed. 1956. *The Clarke-Leibniz Correspondence: Together with Extracts from Newton's* Principia *and* Opticks. Manchester: Manchester University Press.

Alryyes, Ala. 2015. "Uncle Toby and the Bullet's Story in Laurence Sterne's *Tristram Shandy*." *ELH* 82 (4): 1109–1134.

Anderson, Howard. 1969. "Associationism and Wit in *Tristram Shandy*." *Philological Quarterly* 48: 27–41.

Bayes, Thomas. 1763. "An Essay Towards Solving a Problem in the Doctrine of Chances." *Philosophical Transactions of the Royal Society of London* 53: 370–418.

Bannerjee, Chinmoy. 1974. "Tristram Shandy and the Association of Ideas." *Texas Studies in Literature and Language* 15 (4): 693–706.

Blackwell, Bonnie. 2001. "'Tristram Shandy' and the Theater of the Mechanical Mother." *ELH* 68 (1): 81–133.

Bouton, Mark E. 2002. "Context, Ambiguity, and Unlearning: Sources of Relapse after Behavioral Extinction." *Biological Psychiatry* 52 (10): 976–86.

Brassier, Ray. 2007. *Nihil Unbound: Enlightenment and Extinction*. Basingstoke: Palgrave.

Buckingham, Hugh W., and Stanley Finger. 1997. "David Hartley's Psychobiological Associationism and the Legacy of Aristotle." *Journal of the History of the Neurosciences* 6 (1): 21–37.

Bullard, Paddy. 2016. "Eighteenth-Century Minds: From Associationism to Cognitive Psychology." *Oxford Handbooks Online*. Oxford: Oxford University Press. https://doi.org/10.1093/oxfordhb/9780199935338.013.95.

Campe, Rüdiger. 2012. *The Game of Probability: Literature and Calculation from Pascal to Kleist*. Translated by Ellwood H. Wiggins, Jr. Stanford: Stanford University Press.

Cash, Arthur H. 1955. "The Lockean Psychology of *Tristram Shandy*." *ELH* 22 (2): 125–135.

Chandler, James. 2013. *An Archaeology of Sympathy: The Sentimental Mode in Literature and Cinema*. Chicago: Chicago University Press.

Craig, Cairns. 2007. *Associationism and the Literary Imagination: From the Phantasmal Chaos*. Edinburgh: Edinburgh University Press.

Daston, Lorraine. 1988. *Classical Probability in the Enlightenment*. Princeton: Princeton University Press.

Dawson, Michael R.W. 2004. *Minds and Machines: Connectionism and Psychological Modeling*. Malden: Blackwell.

Day, W.G. 1972. Tristram Shandy: The Marbled Leaf. *The Library* 27: 143–5.

De Man, Paul. 1969. "The Rhetoric of Temporality." In *Interpretation: Theory and Practice*, edited by Charles S. Singleton, 173–209. Baltimore: Johns Hopkins University Press.

Engberg-Pedersen, Anders. 2013. "The Refraction of Geometry: *Tristram Shandy* and the Poetics of War, 1700–1800." *Representations* 123 (1): 23–52.

Fanning, Christopher. 2003. "Small Particles of Eloquence: Sterne and the Scriblerian Text." *Modern Philology* 100 (3): 360–92.

Folkenflik, Robert. 2009. "Tristram Shandy and Eighteenth-Century Narrative." In *The Cambridge Companion to Laurence Sterne*, edited by Thomas Keymer, 49–63. Cambridge: Cambridge University Press.

Galilei, Galileo. [1623] 1957. *The Assayer*. Translated by Stillman Drake. In *Discoveries and Opinions of Galileo*, 229–80. New York: Doubleday Anchor.

Gaukroger, Stephen. 2010. *The Collapse of Mechanism and the Rise of Sensibility: Science and the Shaping of Modernity, 1680–1760*. Oxford: Oxford University Press.

Gert, Bernard. 1996. "Hobbes' Psychology." In *The Cambridge Companion to Hobbes*, edited by Tom Sorell, 157–74. Cambridge: Cambridge University Press.

Hacking, Ian. 1975. *The Emergence of Probability: A Philosophical Study of Early Ideas about Probability, Induction, and Statistical Inference*. Cambridge: Cambridge University Press.

Hartley, David. [1749] 1966. *Observations on Man, His Frame, His Duty, and His Expectations (1749)*. Gainesville: Scholars' Facsimiles and Reprints.

Haywood, Eliza. [1751] 1986. *The History of Miss Betsy Thoughtless*. London: Pandora.

Iser, Wolfgang. 1988. *Sterne: Tristram Shandy*. Cambridge: Cambridge University Press.

Kallich, Martin. 1945. "The Association of Ideas and Critical Theory: Hobbes, Locke, and Addison." *ELH* 12 (4): 290–315.

Kames, Henry Home, Lord. 1762. *Elements of Criticism*. 3 vols. Edinburgh: A. Millar.

Lamb, Jonathan. 1980. "Language and Hartleian Associationism in *A Sentimental Journey*." *Eighteenth-Century Studies* 13 (3): 285–312.

Locke, John. [1700] 1975. *An Essay Concerning Human Understanding*. Edited by Peter H. Nidditch. Oxford: Oxford University Press.

Lupton, Christina. 2012. *Knowing Books: The Consciousness of Mediation in Eighteenth-Century Britain*. Philadelphia: University of Pennsylvania Press.

Molesworth, Jesse. 2010. *Chance and the Eighteenth-Century Novel*. Cambridge: Cambridge University Press.

Ong, Walter. 1951. "Psyche and the Geometers: Aspects of Associationist Critical Theory." *Modern Philology* 49 (1): 16–27.

Patey, Douglas. 1984. *Probability and Literary Form: Philosophic Theory and Literary Practice in the Augustan Age*. Cambridge: Cambridge University Press.

Plotnitsky, Arkady. 2015. "Wandering Beneath the Unthinkable": Organization and Probability in Romanticism and the Nineteenth Century." *European Romantic Review* 26 (3): 301–14.

Price, Richard. 1763. Introduction to Thomas Bayes, "An Essay Towards Solving a Problem in the Doctrine of Chances." *Philosophical Transactions of the Royal Society of London* 53: 370–418.

Richardson, Alan. 2001. *British Romanticism and the Science of the Mind*. Cambridge: Cambridge University Press.

Robertson, Warren C. 1921. *A History of the Association Psychology*. New York: Charles Scribner's Sons.

Sepkoski, David. 2007. *Nominalism and Constructivism in Seventeenth-Century Mathematical Philosophy*. New York: Routledge.

Spacks, Patricia Meyer. 2006. *Novel Beginnings: Experiments in Eighteenth-Century Fiction*. New Haven: Yale University Press.

Sterne, Laurence. [1759–1767] 2009. *The Life and Opinions of Tristram Shandy, Gentleman*. Edited by Ian Campbell Ross. Oxford: Oxford University Press.

Stewart, Carol. 2010. *The Eighteenth-Century Novel and the Secularization of Ethics*. London: Routledge.

Stigler, Stephen M. 1983. "Who Discovered Bayes' Theorem?" *The American Statistician* 37: 290–96.

Tunstall, Kate. 2016. *Mind, Body, Motion, Matter: Eighteenth-Century British and French Literary Perspectives*. Toronto: University of Toronto Press.

Wetmore, Alex. 2013. *Men of Feeling in Eighteenth-Century Literature: Touching Fiction*. Basingstoke: Palgrave.

Wickman, Matthew. 2016. *Literature After Euclid: The Geometric Imagination in the Long Scottish Enlightenment*. Philadelphia: University of Pennsylvania Press.

Zunshine, Lisa. 2006. *Why We Read Fiction: Theory of Mind and the Novel*. Columbus: The Ohio State University Press.

CHAPTER 25

Romantic Parts and Wholes, Statistical and Literary

Margaret Kolb

A boom in big data began in British culture at the turn of the nineteenth century. Merchants, reformers, politicians, and bureaucrats began to study extant numerical records, as well as to gather new information and to call for broader data collection. These efforts were manifest most noticeably by the inauguration of a nationwide census in 1801.[1] Such massive projects sparked a host of questions, many of which remain pressing today. How should numbers be aggregated, arranged, and read? What are the limits of numerical representation? Can a part—what we now call a sample—explain a larger whole? In the 1790s, the very novelty of these questions led to haphazard mathematical approaches. Before method was theorized, collection and analysis of data often looked more narrative than quantitative, and were therefore significantly shaped by the tropes and trends of the period's literature.

Literary texts in the Romantic period reflect on related questions. They wonder about the difference between part and whole, how disparate poems might be combined in a single volume, and how numbers and descriptions might mutually constitute narrative.[2] Indeed, it is possible to read texts such as William Wordsworth and Samuel Taylor Coleridge's *Lyrical Ballads* (1798), George Gordon, Lord Byron's *Don Juan* (1819), and Mary Shelley's *The Last Man* (1826) as scaffolded upon such questions. These mutual concerns register how portable mathematical and literary forms were at the turn of the nineteenth century, a point made recently by Matthew Wickman (2015, esp. pp. 13–21). In Wickman's evocative model of exchange, mathematical and literary forms cross-pollinate by necessity, so culturally

M. Kolb (✉)
University of California, Berkeley, Berkeley, CA, USA

© The Author(s) 2021
R. Tubbs et al. (eds.), *The Palgrave Handbook of Literature and Mathematics*, https://doi.org/10.1007/978-3-030-55478-1_25

457

458   M. KOLB

dominant as to absorb one another reflexively. This essay analyzes one such exchange by juxtaposing *Lyrical Ballads* and a contemporaneous big data project: baronet, social reformer, and insurance company founder Frederick Morton Eden's *The State of the Poor: Or, an History of the Labouring Classes in England, from the Conquest to the Present Period* (1797), a three-volume survey of the social and economic factors affecting the rural poor. Self-conscious as *Lyrical Ballads'* presentation is, the volume does not announce itself as a demography, a reflection on the nature of counting, or an account of the ontological and political questions raised by dividing parts and wholes. Yet reading the volume alongside *The State of the Poor* reveals *Lyrical Ballads* to be all of these things.

I read these texts together not to make a historical claim of influence, but rather to uncover resemblances that would otherwise go obscured by subsequent mathematical developments. Recent literary criticism has identified a fixation on numbers in Romantic literature, with fascinating critical studies unpacking early nineteenth-century texts through the varied lenses of Cantor's proof that infinities have different sizes (1874), Dedekind's definition of zero (1888), Russell's paradox (1902), and Gödel's Incompleteness Theorem (1931).[3] These mathematical fireworks notwithstanding, I contend that contemporaneous projects like Eden's offer access to Romantic reflections on number that subsequent nineteenth- and twentieth-century developments cannot afford. Moreover, at a moment when literary historians are themselves turning toward big data projects in search of new configurations of literary history, *The State of the Poor* and *Lyrical Ballads'* representations of number and narrative remind us of the contingency of this relationship, and render *Lyrical Ballads'* insistence both on enumeration, and on the experiences altogether beyond it, newly compelling.

The State of the Poor

The year before Coleridge and Wordsworth began *Lyrical Ballads*, Eden published *The State of the Poor*. Hailed by Karl Marx as "the only disciple of Adam Smith during the eighteenth century that produced any work of importance," Eden grew interested in poor relief after devastating price spikes in food, clothing, and fuel in 1794 and 1795 (Marx 1990, p. 766). He set out to understand "accurate details respecting the present state of the Labouring part of the community, as well as the actual Poor" (Eden 1797, 1, p. i). Noting contemporary arguments over the most effective systems and structures for poor relief, Eden begins *The State of the Poor* by pointing out the absence of pertinent data. "The public mind is once more afloat," he writes, "and, like the dove sent out from the ark, anxiously solicitous to find, if it be possible, amid the surrounding confusion, some spot of permanent tranquility, on which the nation may rest" (Eden 1797, 1, pp. vi–vii). Eden lays out a method of inquiry through which the public mind might achieve stable ground, if not "permanent tranquility." This methodology is at once

The following is a statement of the expence of the usual daily fare of a labourer.

	s.	d.
Breakfast; hasty-pudding and milk	0	1
Dinner; potatoes $\frac{1}{4}$d. butter, or bacon $\frac{1}{2}$d. milk and bread $\frac{1}{2}$d.	0	$1\frac{1}{4}$
Supper; boiled milk, and bread	0	$0\frac{3}{4}$
	0	3

This sum, however, is more, than any poor person expends in a day's provisions.

Fig. 25.1 Frederick Eden Morton, "Cumwhitton," *The State of the Poor*, 2.74 (*Source* Reproduced courtesy of HathiTrust)

The following is a statement of the earnings and expences of a woman, aged 61, and is an instance of Cumberland economy among many others that might be pointed out.

	£.	s.	d.
She spins wool for her neighbours about 15 weeks a year, and earns 4d. a day and victuals,	1	10	0
The remaining 37 weeks, she spins lint at home for a manufacturer, and earns $13\frac{1}{2}$d. a week	2	1	$7\frac{1}{2}$
Total earnings,	£3	11	$7\frac{1}{2}$
Interest of £10.	0	10	0
Total income,	£4	1	$7\frac{1}{2}$

Fig. 25.2 Frederick Eden Morton, "Cumwhitton," *The State of the Poor*, 2.75 (*Source* Reproduced courtesy of HathiTrust)

groundbreaking and symptomatic of contemporary puzzles over how numbers might be aggregated, presented, and interpreted.

The State of the Poor records earnings, rents, occupations, food prices, standard menus, and common land acreage for 186 parishes in England and Wales.[4] The results, organized geographically and arranged in tables, offer unprecedented access to daily economic life in the regions they describe. Two adjoining tables, Figs. 25.1 and 25.2, exemplify the project.

These two tables represent a systemic problem for Eden: scale. Figure 25.1 breaks down the cost of the "usual daily fare" of a worker, while Fig. 25.2 shows the yearly income flows of a particular woman, presented as both an exemplary case of economy and as "an instance … among many others." Alongside this accounting—earnings, interest, and expenditures—Eden narrates the necessities, yearly patterns, and daily rhythms that ground the comparison between one geography and another: the boiled milk and hasty-pudding, the neighbors and regional customs ("Cumberland economy") that make up a life. Though similar in pacing and presentation, the scales—the "usual" daily experience of a regional laborer, and that of a particular woman in a year—are markedly different. This vacillation characterizes Eden's project, which never settles on a single method for depicting a regional population, instead offering diverse descriptions, ranging from the peculiarities of individual lives to the general patterns of groups.

Eden presents these findings as "useful knowledge"—a nascent category described by historian John Brewer as "a type, a category of knowledge, not necessarily of immediate value to those who acquired it, but having the potential to be deployed usefully" (Brewer 1988, p. 228). This unmotivated presentation marks the turn that cultural historian Mary Poovey has identified toward "the modern fact," a rhetorical shift by which tables of information, usually numerical, came to be privileged as objective in ways that narrative accounts could not convincingly be (Poovey 1998, pp. xii–xiii). Eden's framing of his project exemplifies this turn. Comparing his own practices with those of thinkers who "raise their specious systems, without well authenticated facts to support them," Eden claims that he offers only facts, "putting the Public in possession of such facts as were attainable by one individual, to enable them to draw their own conclusions." To that end, Eden writes that he has "purposely, and almost wholly, abstained from drawing conclusions from the facts here presented." Designating the contents of *The State of the Poor* as factual, rather than theoretical, Eden makes a bid at once for objectivity and political efficacy. Unlike "specious systems," concocted without reference to numerical records, Eden's findings could inform a broad range of political questions, including questions that had yet to be asked. His tables could therefore serve as parts of wholes as yet unknown (Eden 1797, 1, pp. xxix, xiii, xiix).

To this end, Eden distinguishes not only fact from system, but also style from substance, boasting that he has not "wasted that time in polishing a sentence, which I thought I could better employ in ascertaining a fact" (Eden 1797, 1, pp. xix–xxx). The tradeoff Eden proposes between polished sentences and authentic facts prophesies a tension between literary and statistical representation that absorbed writers from Charles Dickens to Belgian statistician and astronomer Adolphe Quetelet. In light of this tension, Eden's preface concludes strikingly, with a quotation from Samuel Johnson's own preface to his *Dictionary* in 1755. "I look with pleasure on my Book," Eden quotes Johnson, "however defective, and deliver it to the world, with the spirit of a man, that has endeavoured well" (Eden 1797, 1, pp. xxxi). Aligning Johnson's compilation of the English language with his own assembly of regional populations and economies, Eden proposes statistics as the new language for the nineteenth century. Subsequent statistical writing followed suit, envisioning that statistics would provide factual bases for representations of every kind, including literary representations. Quetelet, for example predicted that "the man of literature" would one day "choose from these materials those which are best suited to the subject of their studies, as the painter borrows from optics the few principles bearing on his art" (Quetelet 1842, p. 98).

Early critics of statistical enquiries noted that facts could not be disentangled from fictions so easily; in other words, style might often *be* substance. "There are innumerable circumstances," essayist and historian Thomas Carlyle observes, "and one circumstance left out may be the vital one on which all turned" (Carlyle 1840, p. 9). A glance at any of Eden's tables suggests how impossible the exclusion of argument must have felt. Eden's parish-by-parish lists of recipients of poor relief, for example, read as causal narratives (Fig. 25.3).

Here we encounter recipients of aid figured as characters, whose professions, genders, and life stories explain their claims on parish resources. Initialed but unnamed, they teeter between particularity and exemplarity. This list, like the tabulations of the lone woman of Cumwhitton's income, suggests how mere inclusion or exclusion—of a number, a person, or a word—could alter readerly conclusions altogether. For mathematician Charles Babbage, compiling a table of numbers relied on judgment of "common character," a point manifestly apparent in Babbage's own proposals for data collection. In 1832, he proposed assembling "Constants of Nature and Art." These constants would include measurements of animal life (including total number of species, with "smallest circumference" for each), alongside the "frequency of occurrence of the various letters of the alphabet in different languages" and the distance "Man, horse, heavy wagon, stage-coach, mail-coach, camel, elephant, steam-carriage" could travel in one day (Babbage 1832, pp. 335–38). The absurdist comedy of Babbage's collection underscores a significant point: choices of subject and group are also stylistic choices, masked only with solidification over time as standard categories.

The following is a lift of the Paupers maintained by the Contractor:

M. Y. a mafon's widow, aged 73.

A. F. a labourer's widow, aged 80.

R. H. a weaver's widow.

M. I. a widow, aged 80.

T. S. a blackfmith's wife, aged 78; rather infane at times; occafionally chargeable.

M. B. a taylor's widow, aged 82.

M. N. receives 10s. annually from this parifh, and 10s. from Cumahitton. The two parifhes, in order to avoid a conteft, agreed to join in maintaining her.

Fig. 25.3 Frederick Eden Morton, "The Parish of Ainstable," *The State of the Poor,* 2.46 (*Source* Reproduced courtesy of HathiTrust)

462   M. KOLB

Though Eden's disavowal of conclusions and polish promises a data set that might speak for itself, Eden also recognizes that his gatherings are incomplete. In *The State of the Poor*, Eden dreams up the kinds of data, not yet collected, that might speak even more powerfully. Repeatedly, Eden laments the facts he could not gather: "what quantity of liquor, and of what sorts, was consumed in each parish," the extent of commons and wastelands, and accurate information concerning the number of friendly societies (grassroots associations for mutual insurance) and their memberships (Eden 1797, 1, pp. xii, xvi, xxiii). Each of these facts would tell a particular story. Poovey has noted similar tendencies in thinkers contemporary to Eden, including Thomas Malthus and Adam Smith.[5] On the other hand, statistical writers and their critics often encountered data so excessive as to be illegible. In one of the first modern compilations of medical statistics, William Black complains that most statistical writers "have obscured their works," to the extent that "the reader must have no small portion of phlegm and resolution to follow them through with attention; they often tax the memory and patience with a superfluity of figures, even to a nuisance" (Black 1781, p. 195). John Sinclair's statistical survey of Scotland, one inspiration for Eden's project, was criticized for being "too voluminous" (qtd. Cullen 1975, p. 11). Carlyle compared statistical tables to cobwebs, "beautifully reticulated, orderly to look on, but which will hold no conclusion" (Carlyle 1840, p. 9). As William Playfair put it, "a man who has carefully investigated a printed table, finds, when done, that he has only a very faint and partial idea of what he has read; and that like a figure imprinted on sand, is soon totally erased and defaced" (Playfair 1786, p. 3).[6]

Numbers should enable readers to "draw their own conclusions," as Eden contended, and must be sufficiently copious to do so. Too often, however, tables prove illegible at worst, inconclusive at best. These constraints, interlocking and opposed, loomed the larger because of the limited possibilities in the period for presenting and aggregating numerical information. Well into the nineteenth century, tables interspersed with sentences introducing them remained the chief way of presenting numerical time series; as a result, understanding population from year to year, annual counts of orphans and the cost of their care, and so on, required scrutiny of these numbers themselves, laid out in tables like Eden's. William Playfair's presentation in 1786 of such time series through innovative graphs—now a prevailing mode of representation—took decades to gain traction.[7] Lacking, too, was methodology that would allow one or two numbers to stand in meaningfully for full tables. Benchmarks like average (popularized in the 1830s), variance (some rudimentary measures took shape in the 1830s), and correlation (defined by Francis Galton in 1886) were not yet theorized for social data. In the absence of these methods, Eden represents diverse numerical information of differing relevance and at varying scales exclusively through tables. We learn in succession the yearly economic life of one single widowed woman, the "usual" daily experience of an undisclosed group of workers, and the particular costs to the parish of maintaining each recipient of relief.

The project, composed of such bits, is itself a representational fragment. Not only does *The State of the Poor* contain only some of Eden's preferred data, but it covers only some of the nation's parishes, focusing on rural parishes and smaller townships. It is therefore a series of case studies, never a totalizing portrait. In absence of a particular method for quantifying the state of England's working poor, *The State of the Poor* nonetheless imagines how such a whole might be constructed—in overlapping, multiply intersecting categories, from varying scales and piecemeal vantage points. Conscious of this limitation, Eden locates himself as a tiny part of a vast group of gatherers of information, constructing "a large experience," that will offer "a multitude of causes" explaining the origins of—and solutions to—poverty. "The edifice of political knowledge cannot be reared without its 'hewers of stone' and 'drawers of water'," Eden writes. He must be "content to work among them." In *The State of the Poor*, building a comprehensive account of the state of the poor remains, as promised, an exercise for the reader (Eden 1797, 1, pp. xviii, xxix).

Eden's attempt to theorize the relation between the disparate scales of *The State of the Poor*—individual, parish, and nation—continued beyond 1797. In 1800, Eden published an *Estimate of the Number of Inhabitants of Great Britain and Ireland*, which relied on the tables compiled in *The State of the Poor* to estimate the population based on small samples (Fig. 25.4).

Eden applies these ratios to tax records and combines this figure with other estimates; after correcting for minor miscalculations, his result comes within one percent of the 1801 census (Scheuren 2003, p. 95). This use of *The State of the Poor* suggests how its fragments might be rendered meaningful mathematically. A part, carefully observed, extrapolates an unobserved whole. Contemporaries, including mathematician Pierre-Simon Laplace, saw that such a result could prove more elegant and accurate than a brute-force enumeration. It was a method John Rickman, who administered the first census, initially proposed as superior to the method employed, a full census—which Rickman saw would be "fraught with trouble and expence," attempting "an accuracy not necessary, or indeed attainable, in a fluctuating subject."[8]

The question of how a part might represent a whole, how "useful knowledge" might best be *used*, were as yet uncharted mathematical territory. Eden's population tally precedes and prophesies the subsequent consolidation of statistical methods as the bedrock of British political discourse. Registering three interlocking puzzles—a rapid accumulation of uncategorized information, dubious legibility of numerical tables, and the difficulty of condensing numbers into a convincing story—Eden resolves these puzzles by allowing numbers to serve as localized narrative accounts—fragmentary descriptions—and as tools for extrapolating a national portrait: a census. In

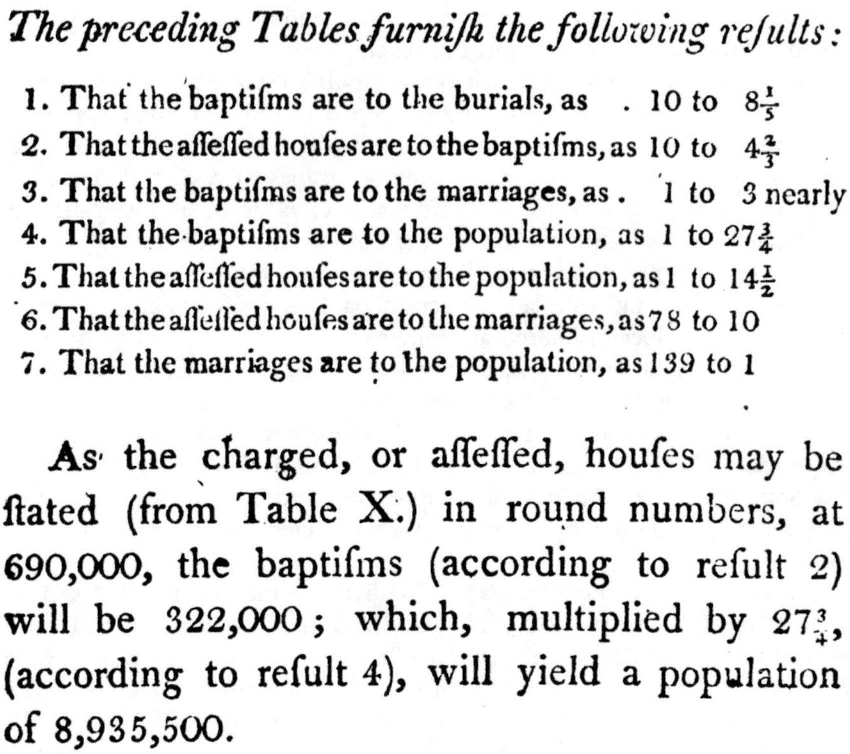

The preceding Tables furnish the following results:

1. That the baptisms are to the burials, as . 10 to $8\frac{1}{5}$
2. That the assessed houses are to the baptisms, as 10 to $4\frac{2}{3}$
3. That the baptisms are to the marriages, as . 1 to 3 nearly
4. That the baptisms are to the population, as 1 to $27\frac{3}{4}$
5. That the assessed houses are to the population, as 1 to $14\frac{1}{2}$
6. That the assessed houses are to the marriages, as 78 to 10
7. That the marriages are to the population, as 139 to 1

As the charged, or assessed, houses may be stated (from Table X.) in round numbers, at 690,000, the baptisms (according to result 2) will be 322,000; which, multiplied by $27\frac{3}{4}$, (according to result 4), will yield a population of 8,935,500.

Fig. 25.4 *Estimate of the Number of Inhabitants of Great Britain and Ireland,* p. 25 (*Source* Reproduced courtesy of HathiTrust)

so doing, he establishes a value both for the extreme specificity of a single woman's daily economic life, and the possibilities of full representation of the nation through the aggregation of such lives—a practice as literary as it was statistical.

LYRICAL BALLADS

William Wordsworth and his collaborator Samuel Taylor Coleridge's rambles looked so much like statistical surveys that one spectator mistook them for spying. In the *Biographia Literaria* (1817), Coleridge recounts the "Spy Nozy" affair, so dubbed for an overheard—and misconstrued—discussion of Spinoza. "Has he not," Coleridge recalls his accuser querying, "been seen wandering on the hills towards the Channel, and along the shore, with books and papers in his hand, taking charts and maps of the country?" Coleridge had assumed such compromising positions while planning a poem, "The Brook." "With my pencil and memorandum-book in my hand," he writes, "I was making studies, as the artists call them, and often moulding my thoughts into verse, with the objects and imagery immediately before my senses." "The Brook" was never finished. But had it been completed and published, Coleridge writes,

> it was my purpose in the heat of the moment to have dedicated it to our then committee of public safety as containing the charts and maps, with which I was to have supplied the French Government in aid of their plans of invasion. And these too for a tract of coast that, from Clevedon to Minehead, scarcely permits the approach of a fishing-boat! (Coleridge [1817] 2009, p. 254)[9]

While Coleridge treats the episode with comic disdain, the incident underscores a homology between the rambles and conversations preceding the publication of *Lyrical Ballads* (1798) and surveys of the same historical moment. *Lyrical Ballads*, consisting of poems composed in the wake of conversations about Spinoza and studies like Coleridge's of the brook, charted and mapped the country in one way; analyses like Eden's, in another.

Lyrical Ballads is a volume consisting of nineteen poems written by William Wordsworth and four poems by Samuel Taylor Coleridge, selected, edited, and ordered collaboratively in the months before its publication in October of 1798. A series of disparate poems by two authors, the volume is a reflection on parts and wholes, often explicitly so. Three poems are fragments themselves, presented without notes or explanation, an editorial choice that critic Marjorie Levinson sees as installing not only a new form that she terms the "Romantic Fragment Poem," but a new readerly practice, substituting "a reading for a writing," and giving rise to an irreconcilable tension between imagined whole, and existent part (Levinson 1986, pp. 23, 26). More recently, Christopher Stratham has proposed that such effects are true not only of poems declared to be fragmented by their authors, but are also revelatory of "the fragmentary imperative" of Romanticism (Stratham 2006, p. 5). And while critics often frame fragments in relation to the Romantic sublime, or as groundbreaking formulations of questions about literary form, William Galperin has pointed out that fragment poems by Coleridge and Byron are often also marked by "banal" or "interminable" turns to the every day (Galperin 2017, pp. 131–33).

Inviting suppositions relating visible part to invisible whole, the fragment invokes a kind of quantitative reasoning. It is surprising, then, that the link between number and fragment in Romantic poetry has gone unexplored. I extend these recent conversations about fragments in Romantic literature to the numerical "facts" presented in *Lyrical Ballads*. The banality of these facts, as well as the readerly responses they invite, resonate powerfully with Galperin and Levinson's arguments. In what follows, I read three numerically oriented poems—which are not themselves fragments—as reflections of the relationship between narrative and number, part and whole, and poetic fragment and poetic form; or, to borrow Stratham's words, as reflections on "the fragmentary imperative." Diverse as they are, these numbers are all bits

466   M. KOLB

spun into metrical wholes. But the complexities of the stories they plot and the methods they invoke undermine any illusion of wholeness that the poems might otherwise conjure.

Lyrical Ballads opens with an "Advertisement," written by Wordsworth, laying claim to the factual nature of the work that follows. "It is the honourable characteristic of Poetry," writes Wordsworth, "that its materials are to be found in every subject which can interest the human mind." Readers should seek "a natural delineation of human passions, human characters, and human incidents" in the volume; in other words, they will find "facts." Even "absolute inventions," like Coleridge's "The Rime of the Ancyent Marinere," are stylistic imitations, or representations of speech patterns—and therefore, Wordsworth claims, also factual. The poems that follow are not only factual on Wordsworth's capacious terms, but on Eden's as well. Wordsworth's poem "Goody Blake and Harry Gill" recounts an imaginary woman, driven to theft by the expense of heating. Heating cost was one of the measurements Eden was particularly eager to obtain for his parishes, and his characters, however real, are at least partially anonymized (Wordsworth and Coleridge 2008, pp. 47, 48). Indeed, Wordsworth's measurements often prove so meticulous as to recall complaints of the overabundance of facts in statistical tables. His poem "The Thorn," which describes an infant's grave, foregrounds an attention to numerical particulars so excessive that it immediately inspired both vehement objections and parodies. Approaching the spot where the child is buried:

> Not five yards from the mountain-path,
> This thorn you on your left espy;
> And to the left, three yards beyond,
> You see a little muddy pond
> Of water, never dry;
> I've measured it from side to side:
> 'Tis three feet long, and two feet wide. (Wordsworth and Coleridge 2008, pp. 103–4)

The description (not revised until 1820, when Wordsworth omitted this oft-ridiculed final measurement) directs us to the pond (*you* espy, *you* see), before recording the narrator's measurement of it (*I've* measured it).[10] These measurements turn out to be literal directions: "I wish that you would go:/ Perhaps when you are at the place/You something of her tale may trace." We are unlikely to trace much of the story, already two decades past when the speaker arrives, and which we are told "no one knows," though "some will say" otherwise. Instead, "The Thorn" encourages us to repeat what Paul Sheats felicitously dubs the narrator's "inane measurement" (Sheats 1991, p. 93). We are to "trace" the tale by checking its coordinates, foot by foot—not only through the speaker's invitation to go ourselves, but also through the experience of measurement the ballad form renders palpable via meter. Rhythmically recalling the experience of step-by-step measurement, the poem knits a flagrantly pointless accumulation of numerical details into a map (Wordsworth and Coleridge 2008, p. 104, ll. 108–10; p. 105, l. 90; p. 109, l. 214).[11]

The parodic specificity of numbers in "The Thorn" echoes through the most numerically obsessed poems in *Lyrical Ballads*. Wordsworth's "The Last of the Flock" is one of these. It is the poem in *Lyrical Ballads* which most clearly enacts a census, or rather, two: one of children, and one of sheep. The census-taker, both shepherd and father, relates these populations causally. The purchase of one "single ewe" leads to more sheep, more sheep to his marriage, marriage to children, and children to the urgent necessity of selling his sheep, one by one. The speaker encounters him in tears, on his way to sell the last of his flock. "And here it lies upon my arm/Alas! and I have none," he mourns. Counting the growing flock inspires other counts: of children, and of feelings. "[D]aily with my growing store/I loved my children more and more," he recalls, as the sheep become the apparent condition for fatherly love. But they also become the condition for poor relief; the man cannot receive aid until he gives up his sheep. Relinquishing them one by one, he finds "I loved my children less" (Wordsworth and Coleridge 2008, p. 113). The peculiarity of the relationship between sheep tallies and fatherly love obscures an otherwise obvious feature of the poem: a story of "years and years" is comprised of numbers. The experiences of adolescence, marriage, and fatherhood are condensed into three body counts—of feelings, of sheep, and of children. Count becomes tale. We can recall the poem through these tallies and the relations that they bear to one another. Yet even as numbers constitute the poem's narrative, they call attention to its absurdity. Numbers structure a scale of value for the poem, but they also explode it.

In Wordsworth's "We are Seven," the most famous example of the volume's attention to numbers, the speaker encounters a child, and immediately asks her "How many may you be?" It is significant that the initial question posed by the speaker to the child is how many (of her) there may be, a question that recalls Eden's impulse to extrapolate from individual to group, and back again. The speaker's polite "may" also hints at an indeterminacy of counting methodology that the speaker is subsequently, and somewhat comically, shocked to discover. The child, who explains that she has two siblings who have died, two away at sea, and two living outside of the family home, adds these partial counts up to answer, "seven," an answer with which the speaker vehemently disagrees. For the speaker, the dead do not count: while the child's legs can "run about," as he brutally reminds her, her sister Jane's and her brother John's cannot. For the child, however, the absence posed by death is structurally identical to her two siblings' absence at sea or at Conway; each pair persists equivalently.[12] Indeed, absence in the poem is strangely double; all the child's siblings are absent, and all are accounted for in pairs. Every member of the seven has a geographic twin, save for the little girl, who is peculiar in that she is present (Wordsworth and Coleridge 2008, pp. 100, 101).

She is also peculiar in that she is numerous. Though nameless, the child can be mapped to a plenitude of numbers: she is both one (she is the only one of her siblings present, a counting mechanism that neither speaker nor child utilizes), and she claims to be (one of) seven, and, as the speaker claims (one of) five. She is also eight, her age. Though she may count unconventionally, the child of "We are Seven" counts often. Numbers justify existence—her dead siblings are there, she insists, because they are "[t]welve steps or more from my mother's door," and though her sister Jane is also named (a privilege in the poem available only to the deceased), she is introduced as "the first that died" (Wordsworth and Coleridge 2008, p. 101). For Frances Ferguson, the poem's numbers point to the orderly array of numbers beginning to map the social world; street and house numbers introduced in London in 1764 located the individual, as population demographics described the nation. Numbering, Ferguson writes, had become "not just a way of saying what there is, but largely a way of making it possible to anticipate one's future." We might follow Ferguson in reading the numerical peculiarities of "We are Seven" as "a more general movement to coordinate meanings among individuals" (Ferguson 1992, pp. 168, 169). We might also, with Aaron Fogel, regard the conversation as a metrical debate, pitting the "4/3 cadence of ballads against" the more rationalized "five-ness" that the speaker proposes (Fogel 2009, p. 37). That is, the poem's content—a conversation about how one should count—restages a question of poetics.

Whether we read the poem as a reflection on form, or as a consideration of efforts to coordinate space through numbers, the poem's defiance of any single count persists. Both the tale and its meter depend on discord. The child is one of both seven *and* five, the remainder once her siblings have been paired in their various absences. She is the part that stands in for the whole, conjuring it up by her singular presence. Yet at the same time, her very presence produces both questions and disagreement about the nature of this whole, and the relation of its constitutive parts. She is a figure composed of disparate counts, methods, and scales, united only by the sense that to discard any of her constitutive bits would be to risk losing "useful knowledge," even as each bit seems potentially excessive. In other words, she embodies the table, in all its contradictions.

As the Romantics conceived it, a poem was reflection on parts and wholes, a form through which, to borrow a fragment of German writer and philosopher Friedrich Schlegel, "every whole can be a part and every part really a whole" (Schlegel 1971, p. 142). This relation was at the heart of the poetic project of *Lyrical Ballads*, and the conversations that preceded its publication. Coleridge's recollections of the Spy Nozy affair in his *Biographia Literaria* recalled him to the questions of parts and wholes that had inspired "The Brook" to begin with. The poem, Coleridge's composition at the time of the Spy Nozy affair, was an attempt to render a scattered geography and economy whole through a single figure. "I sought for a subject," Coleridge writes,

> that should give equal room and freedom for description, incident, and impassioned reflections on men, nature, and society, yet supply in itself a natural connection to the parts, and unity to the whole. Such a subject I conceived myself to have found in a stream, traced from its source in the hills among the

yellow-red moss and conical glass-shaped tufts of bent, to the first break or fall, where its drops become audible, and it begins to form a channel; thence to the peat and turf barn, itself built of the same dark squares as it sheltered; to the sheepfold; to the first cultivated plot of ground; to the lonely cottage and its bleak garden won from the heath; to the hamlet, the villages, the market-town, the manufactories, and the seaport. (Coleridge 2009, p. 254)

Coleridge sought a subject that would unite yellow-red moss and cottage, market-town, and manufactories. In the words of the government agent who overheard Coleridge's conversations with Wordsworth, Coleridge planned to "put Quantock here in print" (Coleridge 2009, p. 254).[13] His chosen muse, a stream, would afford "connection to the parts, and unity to the whole"—a unity not simply aesthetic, but objective, revealing how fluidly interwoven these disparate enclosures, places, and things were. Unsurprisingly, the poem went unfinished. Promising to exemplify the relation between part and whole, market and hamlet, stream drops and seaport, the project outstripped Eden's in ambition and scope. It certainly shared Eden's hopes and his puzzles.

Conclusion

It is no surprise, given the nature of these puzzles, that the numbers of Romantic literature lead contemporary criticism to twentieth-century mathematics. But rather than prophesying subsequent developments, Wordsworth and Coleridge's numbers engage implicitly with studies like Eden's, in which numbers are both fact and story, promising to represent both the particular lived experience of one woman or one child, and to account as if by magic for the nation or the family from a provisional and partial tally. This mathematical wizardry is in fact an important context out of which *Lyrical Ballads* springs: a context in which literature served for statistical writers as a convenient foil to numerical descriptions (fact versus fiction, polished sentence versus careful count), even as statistical and literary writing attempted in strikingly similar ways to map the brooks and cottages that made up the nation, and in so doing, theorize the relation between sample and group, part and whole.

Misquoting Charles Babbage, whose "Collection of Numbers" proposed recording birds and freight trains in the same table, we might call *Lyrical Ballads* a collection of numbers. Much like the velocities Babbage hoped to record, numbers in *Lyrical Ballads* prove excessive, variously legible, and complexly situated. Falling at the crux of early statistical endeavors, theory and practice collided: numbers were becoming individuals, as individuals were numbered; numbers were conceived as narratives in and of themselves, as narratives were increasingly structured by them. Through the act of assembling numerical data, collections like *The State of the Poor* and *Lyrical Ballads* simultaneously map spaces, places, and economies, and reflect on the methodological difficulties of all such projects. In contemplating how we might read

470 M. KOLB

count as chronicle, Wordsworth and Coleridge's numbers plot tales even as they trouble them, stage questions of poetic form, and lay claim to facticity.

The numbers of *Lyrical Ballads*, then, gesture toward the assembly of wholes from parts, even as these parts always threaten to dissemble back into constitutive bits. They recall us to Carlyle's reminder that we are always at risk of missing a crucial circumstance. So full of numbers, the poems are also full of pasts beyond the reaches even of memory, but whose causal power persists nonetheless—the "unremembered pleasure" that the speaker muses may have "no trivial influence" on a life in "Tintern Abbey" (Wordsworth and Coleridge 2008, p. 143). These lost pasts, and the fragmented nature of numerical information, remind us that surveys, from the state of the poor to the state of the brook, are necessarily incomplete, bits to be assembled with methodological (and metrical) ingenuity, but whose appearance of conclusion is always only fictional, and whose identity depends as much on their reading as on their writing. After all, in the rush to find out how many child of "We are Seven" "may" be, Wordsworth realized that he never inquired after her name. Months later, he wrote, he was unable to retrace his steps to ascertain it. Among the most scrupulously tabulated figures of literary history, she remains only a partial story, the embodiment of the forgetting and remembering of measurements, both metrical and statistical.

NOTES

1. This interest was prophesied by political arithmeticians and mathematicians as early as the seventeenth century, including William Petty, John Graunt, and Jacob Bernoulli, all of whom bemoaned a lack of sufficient data, and interest in it. By most accounts, interest in numerical information grew widespread at the end of the eighteenth century (Brewer 1988, pp. 221–50; Innes 2009, pp. 109–79).
2. Fragments are an important formal aspect of Romantic literature, as well as in its criticism. Classics include Levinson (1986); for a more recent example, see Wickman (2013).
3. See for example Levinson (2010, pp. 652–54), Ferguson (1992, pp. 157–79), as well as Wickman (2013).
4. Tables were based on queries to clergyman and surveys conducted by an emissary, as well as on his own tours.
5. Both thinkers, Poovey observes, dreamed of counts that did not—and perhaps could not—exist (Poovey 1998, pp. 239–40). This hope extends back to William Petty, who wistfully imagined mortality data so precise that one might extrapolate optimal counts of the nation's doctors.
6. Alexander Humboldt, a rare disciple of Playfair's graphic techniques, reiterated the point (Humboldt 1811, p. ciii). See also Stigler (1999, p. 368), for an account of the incapacity of government departments for processing the output of official returns.
7. For an account of Playfair's innovation, see Funkouser and Walker (1935).

8. See Rickman (1973, p. 111). See also Wrigley (2009, pp. 711–35), for a description of Rickman's proposed method. Similar sampling methods were proposed by Pierre-Simon Laplace and Quetelet.

9. For a further critical account of the episode, see also Roe (1988, pp. 248–62).

10. For two excellent accounts of the varied reactions to the poem, see Thomas (1983, pp. 237–42), and Sheats (1991, pp. 92–100).

11. The sense that the poems might be maps, as well as poems, was not lost on readers. Later versions of the volume, published with substantive revisions and additional poems by Wordsworth, actually included maps, and the volume quickly became a guidebook of sorts for the Lake District. See Wordsworth and Coleridge (2008, p. 543).

12. One of the difficulties of the census for modern researchers is that it did not include the military: see Wrigley (2011, p. xvi). Indeed, uniformity from census to census was to be lacking, a problem that the poem identifies; see Wrigley (2011, esp. pp. 8–10).

13. "Quantock" refers to the Quantock Hills, where Coleridge lived between 1797 and 1799 and where the Spy Nozy affair transpired.

REFERENCES

Babbage, Charles. 1832. "On the Advantage of a Collection of Numbers to Be entitled the Constants of Nature and Art." *The Edinburgh Journal of Science* 6: 334–40.

Black, William. 1781. *Observations Medical and Political, on the Smallpox*. London: J. Johnson.

Brewer, John. 1988. *The Sinews of Power: War, Money and the English State, 1688-1783*. Cambridge: Harvard University Press.

Carlyle, Thomas. 1840. *Chartism*. London: James Fraser.

Coleridge, Samuel Taylor. 2009. *Biographia Literaria* in *The Major Works*. Edited by H. J. Jackson. Oxford: Oxford University Press.

Cullen, Michael. 1975. *The Statistical Movement in Early Victorian Britain: The Foundations of Empirical Social Research*. Hassocks: Harvester Press; New York: Barnes & Noble.

Eden, Frederick Morton. [1797] 1928. *The State of the Poor: A History of the Labouring Classes in England, with Parochial Reports*. Edited by A. G. L. Rogers. London: Routledge.

Ferguson, Frances. 1992. *Solitude and the Sublime: Romanticism and the Aesthetics of Individuation*. New York: Routledge.

Fogel, Aaron. 2009. "Wordsworth's 'We Are Seven' and Crabbe's 'The Parish Register': Poetry and Anti-census." *Studies in Romanticism* 48 (1): 23–65.

Funkouser, H. Gray, and Helen M. Walker. 1935. "Playfair and His Charts." *Economic History* 3: 103–9.

Galperin, William. 2017. *The History of Missed Opportunities: British Romanticism and the Emergence of the Everyday*. Stanford: Stanford University Press.

Humboldt, Alexander. 1811. *Political Essay on the Kingdom of New Spain: Founded on Astronomical Observations, and Trigonometrical and Barometrical Measurements*. Vol. 1. Translated by John Black. New York: Riley.

Innes, Joanna. 2009. *Inferior Politics: Social Problems and Social Policies in Eighteenth-Century Britain*. Oxford: Oxford University Press.

Levinson, Marjorie. 1986. *The Romantic Fragment Poem*. Chapel Hill: The University of North Carolina Press.

———. 2010. "Of Being Numerous." *Studies in Romanticism* 49 (4): 633–57.

Marx, Karl. 1990. *Capital: Volume 1: A Critique of Political Economy*. Translated by Ben Fowkes. London: Penguin Classics.

Petty, William. 1964. *Political Arithmetick* in *The Economic Writings of Sir William Petty*. Edited by Charles Henry Hull, 233–313. New York: Augustus M. Kelley.

Playfair, William. [1786] 2005. *The Commercial and Political Atlas: Representing, by Means of Stained Copper-Plate Charts, the Progress of the Commerce, Revenues, Expenditure and Debts of England During the Whole of the Eighteenth Century*. Edited by Howard Wainer and Ian Spence. Cambridge: Cambridge University Press.

Poovey, Mary. 1998. *A History of the Modern Fact*. Chicago: University of Chicago Press.

Quetelet, Adolph. 1842. *A Treatise on Man, and the Development of His Faculties*. Edinburgh: William and Robert Chambers.

Rickman, John. 1973. "Thoughts on the Utility and Facility of Ascertaining the Population of England." In *Numbering the People: The Eighteenth-Century Population Controversy and the Development of Census and Vital Statistics in Britain*, David Glass, 106–13. Farnborough: Saxon House.

Roe, Nicholas. 1988. *Wordsworth and Coleridge: The Radical Years*. Oxford: Clarendon Press.

Scheuren, Fritz. 2003. "Looking Back So We Can Look Forward." *The American Statistician* 57 (2): 94–96.

Schlegel, Friedrich. 1971. *Friedrich Schlegel's Lucinde and the Fragments*. Translated by Peter Firchow. Minneapolis, MN: University of Minnesota Press.

Sheats, Paul. 1991. "'Tis Three Feet Long, and Two Feet Wide': Wordsworth's 'Thorn' and the Politics of Bathos." *The Wordsworth Circle* 22 (2): 92–100.

Stigler, Stephen. 1999. *Statistics on the Table: The History of Statistical Concepts and Methods*. Cambridge: Harvard University Press.

Stratham, Christopher. 2006. *Romantic Poetry and the Fragmentary Imperative: Schlegel, Byron, Joyce, Blanchot*. Albany: State University of New York Press.

Thomas, Gordon K. 1983. "'The Thorn' in the Flesh of English Romanticism." *The Wordsworth Circle* 14 (4): 237–42.

Wickman, Matthew. 2013. "Of Tangled Webs and Busted Sets: Tropologies of Number and Shape in the Fiction of John Galt." In *Romantic Numbers*, edited by Maureen McLane, n.p. Romantic Circles. http://www.rc.umd.edu/praxis/numbers/HTML/praxis.2013.wickman. Accessed on 2 February 2018.

———. 2015. *Literature After Euclid: The Geometric Imagination in the Long Scottish Enlightenment*. Philadelphia: University of Pennsylvania Press.

Wordsworth, William, and Samuel Taylor Coleridge. 2008. *Lyrical Ballads*. Edited by Michael Garner and Dahlia Porter. Calgary: Broadview Press.

Wrigley, E. A. 2009. "Rickman Revisited: The Population Growth Rates of English Counties in the Early Modern Period." *Economic History Review* 62: 711–35.

———. 2011. *The Early English Censuses*. New York: Oxford University Press.

CHAPTER 26

"Colours of the Dying Dolphin": Nineteenth-Century Defenses of Literature and Mathematics

Imogen Forbes-Macphail

In his 1869 Presidential Address to the Mathematics and Physics Section of the British Association for the Advancement of Science (BAAS) at Exeter, the mathematician James Joseph Sylvester relates an anecdote about a member of the House of Lords, who declared himself unable to engage in any kind of public oratory unless he had an "adversary before him – somebody to attack or reply to" (Sylvester 1870, p. 106). Finding himself subject to a similar "combative instinct," Sylvester alights on the biologist Thomas Henry Huxley (popularly styled "Darwin's Bulldog") as his own opponent—a somewhat audacious choice, given Huxley's legendary pugnacity, coupled with the fact that he was, at the time, the likely President-elect of the next meeting of the Association. In particular, Sylvester disputes Huxley's claim that mathematics is "almost purely deductive," and that it "knows nothing of observation, nothing of experiment, nothing of induction, nothing of causation" (Huxley, qtd. in Sylvester 1870, pp. 107–8). According to Huxley, Sylvester writes, it would seem that

> the business of the mathematical student is from a limited number of propositions (bottled up and labelled ready for future use) to deduce any required result by a process of the same general nature as a student of language employs in declining and conjugating his nouns and verbs – that to make out a mathematical proposition and to construe or parse a sentence are equivalent or identical mental operations. (Sylvester 1870, p. 107)

I. Forbes-Macphail (✉)
University of California, Berkeley, Berkeley, CA, USA

© The Author(s) 2021   473
R. Tubbs et al. (eds.), *The Palgrave Handbook of Literature and Mathematics*, https://doi.org/10.1007/978-3-030-55478-1_26

474   I. FORBES-MACPHAIL

The article in which Huxley aired these views had been published under the title "Scientific Education – Notes of an After-Dinner Speech," and Sylvester quips that these opinions might have "been couched in more guarded terms by [his] distinguished friend had his speech been made *before* dinner instead of *after*" (Sylvester 1870, p. 107).

Huxley evidently had an unusual talent for causing offense across the entire disciplinary spectrum; he also features, perhaps more famously, as the principal antagonist in Matthew Arnold's 1882 Rede Lecture, "Literature and Science." Arnold's lecture and Sylvester's address are connected through more than just a shared adversary, however. In fact, Arnold directly references Sylvester in his lecture, citing him as a friend who "holds transcendental doctrines as to the virtue of mathematics" (Arnold 1882, p. 224), while Sylvester, upon republishing his address alongside a treatise on the *Laws of Verse* (1870), dedicated the volume to Arnold, "in grateful recognition of much valuable criticism and generous encouragement received at his hands." These entanglements between Arnold's Rede Lecture and Sylvester's BAAS address are emblematic of the intriguing relationship between literature and mathematics during this period, showcasing some of the ways these disciplines reacted and responded to the claims of the sciences, drawing on similar arguments, and sometimes mutual support, to define themselves and to justify their legitimacy and importance in the face of growing competition from more ostensibly practical scientific subjects.

While the defense of poetry is a relatively well-established literary genre, defenses of mathematics have a more problematic status in relation to their discipline. As Sylvester observes, mathematicians are more accustomed "to making mathematics than to talking about them" (Sylvester 1870, p. 102); or, as P. G. Tait would put it a few years later, "the flights of the imagination which occur to the pure mathematician are in general so much better described in his formulae than in words" (Tait 1872, p. 4). Consequently, a defense of mathematics, unlike a defense of literature, is not a typical output of the subject itself.[1] In the latter half of the nineteenth century, however, numerous presidential addresses to the Mathematics and Physics Section of the British Association for the Advancement of Science (as well as general presidential addresses to the society at large, when given by a mathematician) indicate that representatives of the discipline often felt a real and pressing need to justify their subject. Indeed, many of these addresses are framed as "apologies" in a literal (rather than a literary) sense. William Whewell, for instance, notes in 1858 that "the managers of the Association have assigned a small room to this Section," confessing that its members were "very much in the habit ... of treating our subjects in so sublime a manner that we thin the room very speedily," and that many people find mathematics "repulsive" (Whewell 1859, p. 1). Arthur Cayley likewise sympathizes with those audience members who would prefer to have a "different President" deliver a "discourse on a different subject" (Cayley 1884, p. 4). The Earl of Rosse concedes that papers on "so abstruse a subject" as pure mathematics will be of little interest to non-mathematicians (Rosse 1860, p. 1), and Rev. T. R. Robinson, although declaring mathematics the "noblest" discipline, acknowledges that "it can only be attractive to a very few, and the inferior objects of our Section – as Optics, Electricity, Magnetism, Meteorology, and the like – will always be more popular" (Robinson 1858, p. 1).

Notably, many of these addresses register a direct sense of competition from other, more practical, branches of the sciences. As G. Carey Foster laments in his address of 1877:

> long-suffering as British-Association audiences have often shown themselves to be, there is no doubt that before a tenth part could be read of a report on the year's work on the subjects included in this Section, the room would be cleared and most of those who came to hear about Mathematics and Physics would have gone to try whether they could not find in Section E [Geography] or F [Economic Science and Statistics] something appealing more directly to the common sympathies of mankind. (Foster 1878, p. 2)

Arnold also records an anxiety about the encroachment of the physical sciences on the study of literature, claiming in this instance that "the friends of physical science," a "growing and popular body," are not only in competition with, but in "active revolt against" letters:

> To deprive letters of the too great place they had hitherto filled in men's estimation, and to substitute other studies for them, was now the object, I observed, of a sort of crusade with the friends of physical science. (Arnold 1882, p. 216)

Arnold is not alone in perceiving this struggle as a "crusade" or a "revolt." Huxley likewise refers to it as a "battle," and those "ranged round the banner of Physical Science" as a "guerilla force" (Huxley 1895c, p. 136).

Historically, mathematics and classics had held comparably elite positions within the British education system, with classics predominant at Oxford, and mathematics at Cambridge (see Crilly 2011). The claims of the sciences were comparatively new. In 1869, Huxley maintains that even a few years previously he would not have dared to advocate on behalf of general scientific education "without some more or less apologetic introduction" (Huxley 1895b, p. 111); in 1880, he finally considers the debate to be approaching its "crisis" (Huxley 1895c, p. 136). Arnold, writing in 1882, likewise perceives the physical sciences as having emerged into the "meridian radiance" of popular favor over the course of the past ten years (Arnold 1882, p. 216). The supremacy of the dead languages in general education had been, in Lord Rayleigh's words, largely the "result of routine rather than of argument," and advocates of scientific education were beginning to demand a fair share of the curriculum (Rayleigh 1885, p. 22).

One of the critiques frequently leveled at both classics and mathematics is that of elitism. As Huxley complains, for the "great majority of educated

Englishmen," a person exclusively educated in the sciences, no matter how proficient in these, would nevertheless be refused admission into the "cultured caste," for which an education in classical literature was seen to be necessary (Huxley 1895c, pp. 141–42). Huxley takes care to clarify that he does not attribute these opinions to Arnold, citing his familiarity with "the generous catholicity of spirit, the true sympathy with scientific thought, which pervades the writings of our chief apostle of culture," but nevertheless maintains that his works do contain "sentences which lend them some support" (Huxley 1895c, p. 142). Arnold himself acknowledges in *Culture and Anarchy* (1869) that there are those who conceive of a classical education as constituting a mere "smattering of Greek and Latin," valuable "either out of sheer vanity and ignorance, or else as an engine of social and class distinction, separating its holder, like a badge or title, from other people who have not got it" (Arnold 2009, pp. 32–33), while, of course, rejecting such a limited understanding of its value. As Alice Jenkins illustrates in *Space and the 'March of Mind'*, a similar aura of elitism also surrounded mathematics, and in particular classical geometry. The elevated position of mathematics within the universities, and "severe barriers surrounding access," caused mathematics, like classical literature, to acquire a certain "cultural mystique" (Jenkins 2007, pp. 92–93). As Jenkins describes, however, in the "changing educational environment of the first decades of the nineteenth century, geometry ... became a subject of profound contestation," and a strain of critics arose who began to argue that "not only geometry but all higher mathematics were overprivileged in contemporary education and culture" (Jenkins 2007, pp. 166–67). Sylvester, notably, holds some sympathy with this viewpoint as far as it extends to classical Euclidean geometry, proclaiming that he "should rejoice to see ... Euclid honourably shelved or buried 'deeper than e'er plummet sounded' out of the schoolboy's reach" (Sylvester 1870, p. 120). Advocates of literature and of mathematics (especially classical literature and classical mathematics) across this period thus have similar fears about the encroachment of the physical sciences into educational spaces in which they had previously been preeminent, and face accusations that their subjects owe their positions more to entrenched elitism than to genuine educational value.

Knowledge of the sciences, by contrast, was represented as practically useful for those engaged in industry and the workforce. Although Huxley himself claims to be "devoted to more or less abstract and 'unpractical' pursuits" (Huxley 1895b, p. 113)—theoretical, rather than applied, science—he also stresses "the value of a knowledge of physical science as a means of getting on" (Huxley 1895b, p. 114). "There are hardly any of our trades," he writes, "except the merely huckstering ones, in which some knowledge of science may not be directly profitable to the pursuer of that occupation" (Huxley 1895b, p. 114); "the diffusion of thorough scientific education is an absolutely essential condition of industrial progress" (Huxley 1895c, p. 139). His argument in favor of scientific education (at least for those destined for scientific careers), likewise, becomes a largely utilitarian one: "neither the discipline nor the subject-matter of classical education is of such direct value to the student of physical science as to justify the expenditure of valuable time upon either" (Huxley 1895c, p. 141).

Unlike literature, it seems evident that mathematics has an undeniable practical utility in facilitating calculations necessary for many branches of science, engineering, and mechanics, and so ought to be safe from the kind of utilitarian attacks which are still often leveled against the humanities.[2] Indeed (and somewhat ironically) the very ability to make utilitarian judgments is predicated upon mathematics: Jeremy Bentham's proposal to reckon the "value of a lot of pleasure or pain" is dependent upon a series of mathematical calculations (Bentham 1823, pp. 49–54). In reality the situation is not so simple. First, as many mathematicians note, it is almost impossible to pursue mathematics in a straightforwardly utilitarian manner. As mathematician and physicist William Spottiswoode argued, the "utility" of pure mathematics "often crops up at unexpected points," and so can only be pursued indirectly at best (Spottiswoode 1879, p. 25). A. R. Forsyth, likewise, maintained that "in the pursuit of mathematics, the path of practical utility is too narrow and irregular, not always leading far" (Forsyth 1898, p. 546); the Earl of Rosse, that "the mere utilitarian … has been often reminded that discoveries the most important, the most fruitful in practical results, have frequently in the beginning been apparently the most barren" (Rosse 1860, p. 2). Additionally, many mathematicians, including Sylvester, object to the perception of their subject as merely "the torch-bearer leading the way, or the handmaiden holding up the train of Physical Science" (Sylvester 1870, p. 122). This sentiment is reiterated by Forsyth, who observes that while practical men often regard mathematics "as a machine which is to provide [them] with tables" (Forsyth 1898, p. 543), certain natural philosophers seek to render it "not indeed the drudge, but the handmaid of the sciences" (Forsyth 1898, p. 544). Rather than attributing value to the subject purely on the basis of its utilitarian or practical byproducts, many mathematicians prefer to consider the study of mathematics as justified for its own sake. As Cayley, in his address to the BAAS, says:

> I am not making before you a defense of mathematics, but if I were I should desire to do it – in such a manner as in the "Republic" Socrates was required to defend justice – quite irrespectively of the worldly advantages which may accompany a life of virtue and justice, and to show that, independently of all these, justice was a thing desirable in itself and for its own sake – *not* by speaking to you of the utility of mathematics in any of the questions of common life or of physical science. Still less would I speak of this utility before, I trust, a friendly audience, interested or willing to appreciate an interest in mathematics in itself and for its own sake. (Cayley 1884, pp. 4–5)

In 1890 J. W. L. Glaisher acknowledged that to the outsider, pure mathematics might seem "to be stretching out into limitless symbolic wastes, without

478   I. FORBES-MACPHAIL

producing any results at all commensurate with its expansion"; nevertheless, he maintains that "the search after abstract truth for its own sake, without the smallest thought of practical application or return in any form, and the yearning desire to explore the unknown, are signs of the vitality of a people, which are among the first to disappear when decay begins" (Glaisher 1891, pp. 722, 725).

However, even if mathematics is studied without regard to its applications in industry or the physical sciences, it may nevertheless yield incidental benefits of other sorts. For instance, mathematicians frequently argue that their subject has a unique capacity to inculcate mental discipline. In Tony Crilly's words, even if "the knowledge of mathematics was not claimed to be useful in itself ... it was believed that the study of mathematics would develop and strengthen the faculties of the mind, and *after* the completion of this study one could go on to other fields and be more effective in them." (Crilly 2011, p. 19) As Sylvester writes, "there is no study in the world which brings into more harmonious action all the faculties of the mind," raising its followers to "higher and higher states of conscious intellectual being" (Sylvester 1870, pp. 120–21). These effects of mental discipline also had ramifications for mental (and indeed physical) health. Jenkins' article, "Mathematics and Mental Health in Early Nineteenth-Century England," details some of the problematic and conflicting conceptions surrounding mathematics and its effect on the mind in the earlier part of the nineteenth century. Mathematics, as Jenkins writes, was sometimes felt to have "unbalancing effects" (Jenkins 2010, p. 93), especially on those already pre-disposed to mental illness, an effect aggravated by the demanding and competitive nature of the Cambridge Mathematical Tripos; on the other hand, and somewhat paradoxically, mathematical study could also be prescribed as a "therapy for mental illness," as long as it was studied as part of a balanced education, and not to the exclusion of other types of learning (Jenkins 2010, p. 98). William Wordsworth is cited as one notable example of a person "self-medicating" with mathematics (Jenkins 2010, p. 102), reporting, in *The Prelude* (1805), to find in geometry a "charm" for a troubled mind (Wordsworth 2000, VI.178). Poetry is often considered to have a similar consolatory effect; just as Wordsworth sought solace in the *Elements* of Euclid, John Stuart Mill, in the midst of his own emotional breakdown, found relief in the poetry of Wordsworth, which turned out to be the "precise thing for [his] mental wants at that particular juncture," a "medicine for [his] state of mind" (Mill 1989, pp. 120, 121). Arnold, in "Literature and Science," goes so far as to claim that a knowledge of Greek might even carry something like an evolutionary advantage. Rather ironically, given his somewhat dismissive view of the natural and particularly biological sciences, Arnold concludes his lecture with a defense of the study of Greek literature in oddly Darwinian terms: "if the instinct for beauty is served by Greek literature as it is served by no other literature, we may trust to the instinct of self-preservation in humanity for keeping Greek as a part of our culture" (Arnold 1882, pp. 229–30). Sylvester, in turn, was a firm believer in the health benefits of his own subject, arguing that the study of mathematics was directly responsible for the "extraordinary longevity" of a number of famous mathematicians (Sylvester 1870, p. 121).

In addition to potentially healthful effects on the mind, mathematics and poetry also have the capacity to provide pleasure. However, the question of how to evaluate pleasure, especially in the context of intellectual pursuits, is a vexed one. As the numerous apologies scattered throughout the Presidential Addresses to the Mathematics and Physics Section of the BAAS attest, mathematics did not necessarily provide a great deal of pleasure to the ordinary person. In *Utilitarianism* (1861), however, Mill argues that some types of pleasure, which appeal to the higher faculties, are "more desirable and more valuable than others" (Mill 2004, p. 279), an argument crafted, in part, in response to Bentham's assertion that "the game of push-pin is of equal value with the arts and sciences of music and poetry" if it yields an equivalent amount of pleasure (Bentham 1825, p. 206). Sylvester reports being confronted with similar objections in relation to mathematics, citing a recent magazine article which questioned whether mathematics was, "in itself, a more serious pursuit, or more worthy of interesting an intellectual human being, than the study of chess problems or Chinese puzzles" (Sylvester 1870, p. 122). Mathematician G. H. Hardy would later address the comparison of mathematics with chess in *A Mathematician's Apology* (1940), likening chess problems to the "hymn-tunes of mathematics," in that they have broad popular appeal, but lack the higher aesthetic qualities which make for truly great mathematics (Hardy 1992, p. 87). As Hardy writes, "The beauty of a mathematical theorem *depends* a great deal on its seriousness, as even in poetry the beauty of a line may depend to some extent on the significance of the ideas which it contains" (Hardy 1992, p. 90). He argues that great mathematics must be "*serious* as well as beautiful"—chess problems "are *unimportant*" (Hardy 1992, pp. 88–89).

Sylvester and Arnold appear to feel that the claims made by Huxley and other detractors fundamentally misrepresent the nature of their subjects, trivializing literature, on the one hand, as a belletristic "smattering of Greek and Latin," "slight and ineffectual" (Arnold 1882, p. 219); and, on the other, reducing mathematics to a merely deductive science, little more intellectually engaging than chess, or useful only to the extent that it can be of service to more practical fields. In short, they refute attempts to characterize their subjects as unserious or unimportant for their own sakes. At the core of both Sylvester's and Arnold's defenses is a concern with the means by which individual facts or ideas come to be brought together into something of greater value, and then connected to a sense of the beautiful or the meaningful. Arnold maintains that most people have, firstly, an impulse to "combine the pieces of our knowledge together, to bring them under general rules, to relate them to principles" (Arnold 1882, p. 223), and then secondly (and more importantly) to relate what they have learnt "to the sense which we have in us for conduct, to the sense which we have in us for beauty"

(Arnold 1882, p. 223). There is "weariness and dissatisfaction," he argues, when this "desire is balked" (Arnold 1882, p. 223). While the sciences are certainly capable of performing the first part of this operation, amalgamating individual facts into larger and more complex systems, Arnold believes they are incapable of performing the subsequent step of relating them to a sense of beauty and conduct; this "the men of science will not do for us, and will hardly, even, profess to do" (Arnold 1882, p. 225). Arnold cites Darwin's evolutionary hypothesis, the claim that "our ancestor was a hairy quadruped furnished with a tail and pointed ears, probably arboreal in his habits," as an example of the type of proposition which is offered as a culmination of the accumulation of scientific facts (Arnold 1882, p. 224). "Interesting" though these sorts of propositions might be, Arnold insists, they still leave us in the "sphere of intellect and knowledge," rather than beauty and conduct, and so after a period of time become "unsatisfying" and "wearying" (Arnold 1882, pp. 224–25). Arnold argues, however, that poetry has the capacity to act as a medium which can "relate the results of modern science to our need for conduct, our need for beauty" (Arnold 1882, p. 227). As George Levine puts it, for Arnold, scientific knowledge brings "light without sweetness" (Levine 1988, p. 146). The sciences must therefore look outside themselves to humane letters if they wish to be elevated to the sublime heights to which, as Arnold believes, all forms of knowledge aspire. Arnold, however, is overly dismissive of the sciences; as Levine remarks, "for the scientific education against which Arnold contended, facts are nothing unless they can be generalized and seen as part of a system revealing the laws of nature" (Levine 1988, p. 146), and, after all, whether or not Darwin was "*attempting* to appeal to [the senses of conduct and beauty] *The Descent of Man* is very much about whether we have them, and if so, how we got them" (Levine 1988, p. 152). Darwinian theory, furthermore, did have an "enormous impact on the culture's sense both of conduct and of beauty," and helped to "naturalize morality and aesthetics": "We might prefer Milton's creation myth to Darwin's; but Darwin's also had the force of myth." As Levine's article goes on to show, moreover, contemporary scientists, including Huxley, Tyndall, Clifford, and Lewes, make strong arguments for the capacity of the sciences to appeal to the imagination, to the sense of the beautiful, to conduct, and indeed "the full range of human needs" (Levine 1988, pp. 153–54).

Helen Small, in *The Value of the Humanities*, explores how the Huxley/Arnold dispute on the respective merits of literature and science developed into the more "aggravated antagonism" of the Snow/Leavis "two cultures" debate in the twentieth century (Small 2013, p. 45). Mathematics today is often grouped with the sciences as part of this dichotomy. As Ian Stewart argues, however, in many respects mathematics ought to be viewed not as a branch of the sciences but rather as a "third culture," which "overlaps art and science but is contained in neither, not even in their union" (Stewart 2013). Mathematics also holds a liminal position within Arnold's scheme. Arnold discusses mathematics within the context of what he terms "instrument-knowledges": knowledges which "cannot be directly related to the sense for beauty, to the sense for conduct," but which "lead on to other knowledge, which can" (Arnold 1882, p. 224). As examples, Arnold cites both formal logic and the study of Greek accents, ironically doing exactly that which Sylvester had earlier objected to in Huxley, and comparing mathematics with grammar. It is in the context of this discussion of instrument-knowledges that Arnold references Sylvester:

> My friend Professor Sylvester, who holds transcendental doctrines as to the virtue of mathematics, is far away in America; and therefore, if in the Cambridge Senate House one may say such a thing without profaneness, I will hazard the opinion that for the majority of mankind a little of mathematics, also, goes a long way. Of course this is quite consistent with their being of immense importance as an instrument to something else; but it is the few who have the aptitude for thus using them, not the bulk of mankind. (Arnold 1882, p. 224)

If Arnold will not allow that mathematics could ever hold the same *popular* appeal as poetry, he seems to acknowledge that for Sylvester, at least, it has a similar *kind* of appeal. Referring to formal logic as a mere "instrument-knowledge" may seem disparaging, but Arnold at least concedes that these instrument-knowledges have the potential to lead to something more for the select few who have the "aptitude" to use them in this way. Whereas, for Arnold, the sciences cannot be linked to the desire for beauty or the development of moral conduct without the assistance of the humanities, instrument-knowledges explicitly have the potential to develop into kinds of knowledge which do have this capacity.

However inaccessible mathematics may be to the "bulk of mankind," Sylvester, in any case, was clearly one of those who *did* have the ability to perceive it as beautiful, and to find pleasure in it. As he remarks rather charmingly at one point in his address, "I really love my subject" (Sylvester 1870, p. 102). Sylvester had been satirized some years earlier by *Punch* for his inordinate love of mathematics, with the magazine feigning to doubt his capacity to be really passionately attached, or "spooney," over the subject (*Punch* 1855, p. 149). If he were indeed a kind of "Mathematical *Romeo*, with one of EUCLID's figures for his *Juliet*," *Punch* maintains, he would surely be "jealous of every other clever individual who pays attention to the alleged object of his affections," and "writing odes to Problem 1, or sighing over the *Pons Asinorum*." The article concludes by stating its conviction that Sylvester merely has a "sensible regard" for the subject. While clearly exaggerated for the sake of amusement, this article indicates the apparent difficulty many felt in imagining anyone genuinely possessing the kind of passion for mathematics that Sylvester manifestly felt.

Sylvester does concede that individual mathematical facts—the fact, say, that "the three angles of a triangle are equal to two right angles"—may often seem "dull, stale, flat and unprofitable" in isolation (Sylvester 1870, pp.

122–23). However, he argues that this is "like judging of architecture from being shown some of the brick and mortar ... or of painting from the colours mixed on the palette" (Sylvester 1870, p. 123). When mathematics is properly appreciated as a system, on the other hand:

> The world of ideas which it discloses or illuminates, the contemplation of divine beauty and order which it induces, the harmonious connection of its parts, the infinite hierarchy and absolute evidence of the truths with which it is concerned, these, and such like, are the surest grounds of the title of mathematics to human regard, and would remain unimpeached and unimpaired were the plan of the universe unrolled like a map at our feet, and the mind of man qualified to take in the whole scheme of creation at a glance. (Sylvester 1870, pp. 123–24)

For Sylvester, therefore, mathematics fulfills precisely the same kinds of yearnings that Arnold would later identify in "Literature and Science." It draws one's thoughts beyond the mere facts themselves and connects them to a greater purpose, a contemplation of "divine beauty and order." Sylvester's address was made thirteen years earlier than Arnold's, and it seems likely that Arnold would have been familiar with it, given its later inclusion in a volume dedicated to him. As such, these passages afford interesting speculation on the extent to which Sylvester's ideas may have shaped Arnold's own.

The passages cited above clearly indicate that Sylvester shared Arnold's interest in the way in which *systems* of knowledge are constructed, with individual facts combined into larger networks organized by "connection" and "hierarchy." The ability to perceive mathematics as a connected system of knowledge is integral to the sense of beauty, interest, and pleasure yielded by its study. Sylvester considers this a relatively new development in the history of mathematics. In the past, "all parts of the subject were dissevered," and "algebra, geometry, and arithmetic either lived apart or kept up cold relations of acquaintance confined to occasional calls upon one another" (Sylvester 1870, p. 124). At the time of his address, however, he believed these different branches of mathematics were "constantly becoming more and more intimately related and connected by a thousand fresh ties." In this context, the importance of Sylvester's objections to Huxley's claim that mathematics is a "purely deductive" science becomes clearer. Huxley's understanding of mathematical inquiry is that "the mathematician starts with a few simple propositions, the proof of which is so obvious that they are called self-evident, and the rest of his work consists of subtle deductions from them" (Huxley 1895b, p. 126). He contrasts this elsewhere with the work of biologists: "while the Mathematician is busy with deductions *from* general propositions, the Biologist is more especially occupied with observation, comparison, and those processes which lead *to* general propositions" (Huxley 1895a, p. 57). This model of mathematics, in which a profusion of deductions can be made from an initially limited number of propositions, is antithetical to the quality Sylvester values in the discipline, its ability to amalgamate and connect numerous diverse fields of inquiry. According to Sylvester, in fact, induction is one of the "principal weapons" of mathematics:

> Induction and analogy are the special characteristics of modern mathematics, in which theorems have given place to theories, and no truth is regarded otherwise than as a link in an infinite chain. "Omne exit in infinitum" is their favourite motto and accepted axiom. No mathematician now-a-days sets any store on the discovery of isolated theorems, except as affording hints of an unsuspected new sphere of thought. (Sylvester 1870, p. 118n)

For Sylvester, the strength of mathematics is its ability to connect ideas to each other. In order to accomplish this vision, mathematics cannot be considered as merely a deductive science. As Joan L. Richards writes, Sylvester "mounted his defense of mathematics within the same progressive camp which supported Huxley's attack," and "welcomed Huxley's suggestion that the spirit of open-ended, inductive sciences be made central" to mathematical education (Richards 1988, pp. 134–35).

In addition to his interest in connections *within* the sphere of mathematics, Sylvester also seeks to combine the discipline of mathematics with that of poetry through what he calls the "great law of Continuity" (pp. 14–15), ostensibly his reason for republishing the BAAS address alongside his *Laws of Verse*. For Sylvester, the whole of mathematics revolves around the concept of continuity "as contained in our notions of space" (Sylvester 1870, p. 125); in relation to poetry, "continuity of sound and continuity of mental impression" is likewise his "guiding star" (Sylvester 1870, p. 15). This emphasis on continuity is connected to Sylvester's interest in projective geometry, a comparatively new development in mathematics which, as Richards observes, was represented as having more in common with the inductive methods of the natural sciences than traditional Euclidean geometry (Richards 1988, pp. 136–38).

In the spirit of the "continuity" he believes to exist between these two subjects, Sylvester imports a number of mathematical terms (most notably "synectic" and "syzygy") into his theory of versification. His explanation of their application to poetry, unfortunately, is often somewhat obscure, leading his biographer Karen Hunger Parshall to describe the resulting discussion as "largely incomprehensible" and "virtually unintelligible" (Parshall 2006, p. 214). Daniel Brown's *The Poetry of Victorian Scientists*, which discusses Sylvester's poetical writings at length, helps to clarify much of Sylvester's "esoteric" mathematical idiom in greater detail than is possible here (Brown 2013, p. 207). Sylvester himself, however, does provide some guidance for the non-mathematical reader, explaining that the word "Synectique" was first used by Cauchy in his theory of functions, "the true and very insufficiently acknowledged foundation and origin of Rieman's [*sic*] great doctrine of Continuity" (Sylvester 1870, p. 65n). In relation to poetry, he uses it to refer to the "continuous," rather than the "discontinuous," aspect of poetic art (Sylvester 1870, p. 10). The synectic element of poetry and mathematics, therefore, relating to principles of continuity *within* both subjects, thus also helps to forge a continuity between them, "lead[ing] down from the Alps of Cauchy and Riemann to the flowery plains of Milton and Byron" (Sylvester 1870, p. 64n).

The continuous or synectic element of poetry, however, is only one part of a larger scheme. For Sylvester, poetry has three main divisions: the Pneumatic, the Linguistic, and the Rhythmic (Sylvester 1870, p. 9). His treatise is concerned with the last of these, which itself has "three principal branches," the Metric, Chromatic and Synectic (Sylvester 1870, p. 10). "The Metric," he writes, "is concerned with Accent, Quantity and Suspensions" (the latter of which "includ[es] the theories of Pauses, Rests and Synthesis or Syllabic Groupings"), while the Synectic likewise has three "channels," Anastomosis, Symptosis, and Phonetic Syzygy (Sylvester 1870, pp. 10–11). Importantly, each time Sylvester's scheme yields new branches, it divides into precisely three new areas. Sylvester claims he was not "on the look-out for any such an arithmetical law," but found that these topics "grouped themselves as it were spontaneously, according to this law of trichotomy, each joint of the arborescence ... throwing off from itself three branches" (Sylvester 1870, p. 12n). Sylvester suggests that this mathematical form may not be purely arbitrary:

> It may be the case that a similar ternary law of development would apply to music, painting, sculpture, in a word, to the elucidation of the higher principles of all the fine arts. Following out this clue it is conceivable that we might succeed in laying the foundations of a science of Comparative Aesthetic, or at least of a general Aesthetico-technic. (Sylvester 1870, p. 12n)

According to this hypothesis, even subjects outside the sphere of poetry may be governed by a similar mathematical structure. If the structure of these subjects did follow such a "ternary law of development," then the congruence of these mathematical patterns would enable one to connect poetry, music, painting, and sculpture into one "Comparative Aesthetic," in which they would be unified and brought together under what Arnold would call "general rules."

Sylvester also believes mathematical laws are capable of being applied to the analysis of individual poems. He makes a "profession of faith" in the principles of Edgar Allan Poe's "Rationale of Versification" (Sylvester 1870, p. 64), a work which maintains that "nine tenths" of the subject of versification "appertain to the mathematics" (Poe 2008, p. 81), and he himself discusses the possible groupings of syllables in a line of poetry with reference to permutations and the theory of partitions (Sylvester 1870, pp. 35–36). Additionally, *The Laws of Verse* provides several examples of mathematical forms in the context of larger poetic structures such as the stanza. The first poem discussed, one of Horace's odes, is described by Sylvester as having a "Perdualistic or

Dichotomous Plan," through which the poem is divided up into a tree-like structure, "bearing two stems, four branches, eight branchlets, and sixteen twigs or terminals" (Sylvester 1870, p. 25):

[…]

And in like manner each of the eight duads above may be rendered into separate stanzas containing a single and distinct subject or image.

Sylvester does not supply an illustration or diagram of this himself, but his scheme is easily translated into one:

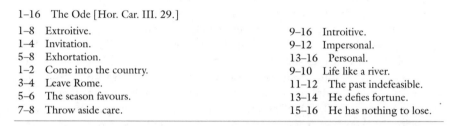

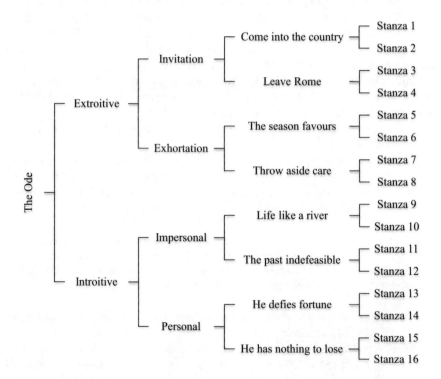

The divisions within this poem follow a similar kind of law to that which Sylvester describes as governing poetics as a whole: each individual section branches out into an equal number of subdivisions, apparently governed by

486   I. FORBES-MACPHAIL

an arithmetical rule, except that in this case each node yields two, rather than three, new branches. This bifurcation, Sylvester argues, springs from "one fundamental root-idea, carried out ... upon a principle of strict arithmetical precision" (Sylvester 1870, p. 37). This mathematical form thus allows individual stanzas, and the ideas they contain, to be brought together into an ordered hierarchy, so that the perfect precision of Horace's form becomes intelligible, in turn contributing to the reader's apprehension of the poem's beauty.

Further on in this analysis, Sylvester represents the metrical scheme of the Alcaic stanza, the verse form used in this ode, as a "square matrix," and, highlighting recurring patterns within this structure, argues that the matrix has a "pure algebraical or tactical deep-seated harmony of its own" (Sylvester 1870, p. 47). He next observes that if each unique *pair* of metrical feet in the stanza is considered as a unit, the matrix would contain the "complete combination system" of these duads, and claims that the above schemes are evidence of the "strong mathematical bias of Horace's mind, wherein perhaps is to be sought the secret of the peculiar incisive power and diamond-like glitter of his verse" (Sylvester 1870, p. 48). As Brown writes, "Sylvester's analysis of verse here is like his work in geometry, dedicated to finding deeper algebraic principles in its phenomena and representations" (Brown 2013, p. 226). Sylvester thus implies that a similar type of beauty can be found within both poetry and mathematics, and moreover, that we may require some degree of mathematical understanding in order to apprehend and appreciate the beauty of certain poetic forms. Whereas Arnold had defended literature on the grounds that it could relate the natural sciences to our need for conduct and beauty, in Sylvester's example it is instead *mathematics* which relates this particular verse form to our sense of the beautiful.

In addition to using mathematical ideas to help elucidate the beauty of poetry, Sylvester is also highly sensible of the rhetorical power of poetry to render his discussion of mathematics more appealing. He cites numerous poets throughout his BAAS address, and at one point, he says,

> I could tell a story of almost romantic interest about my own latest researches in a field where Geometry, Algebra, and the Theory of Numbers melt in a surprising manner into one another, like sunset tints or the colours of the dying dolphin, "the last still loveliest". (Sylvester 1870, p. 117)

This rather unusual allusion derives from Byron's *Childe Harold's Pilgrimage*, Canto IV (1818), and refers to the fact that the dolphinfish changes color when removed from the water (Byron 2008, IV, 258–61). A similarly evocative passage is used in *The Laws of Verse*, in a section discussing a type of phonetic beauty "which depends on gradation of tones, on the agreeable succession of allied sounds" (and which Sylvester claims, interestingly, to be especially prevalent in Byron): "the continuity is like that of the colours and tints in the solar spectrum, and the pleasure like that we feel in watching a sunset or seeing a peacock unfold its tail, or a rainbow glow and fade away in the sky" (Sylvester 1870, p. 45n). His aesthetic response to this particular poetic feature, and his use of sunset imagery, is similar to that which he uses to describe his latest mathematical researches; in both cases, the beauty is derived from a variety of elements blending together in harmony.

In 1878, nine years after Sylvester's address, Spottiswoode likewise used his Presidential Address to the BAAS to advance a defense of mathematical studies. Like Sylvester's address, which he commends for its able handling of the observational, experimental, inductive and imaginative aspects of mathematics (Spottiswoode 1879, p. 27), this speech also emphasizes the importance of continuity and connection between mathematical ideas. While Spottiswoode believes that at its highest level every science is "pursued alone for its own sake, and without reference to its connection with, or its application to, any other subject," his speech chooses to focus on the "other side of the question," seeking "points of contact" between "mathematics and the outer world" (Spottiswoode 1879, p. 11). This speech, however, is even more ambitious than a general justification of the discipline; a core part of the address is devoted to defending some of the most avant-garde and controversial developments at the forefront of contemporary mathematics, including imaginary quantities, higher dimensional spaces, and non-Euclidean geometry—developments which were sufficiently controversial that they often needed to be defended even to a mathematical or scientific audience. In these methods, as Spottiswoode writes, mathematicians are "often thought to have exceeded all reasonable limits of speculation" (Spottiswoode 1879, p. 17). Fascinatingly, a significant portion of his defense relies on a claim that it is possible to find precedents for these mathematical ideas in art and in literature, using the fact that art and literature have "unconsciously employed methods similar in principle" to argue that mathematics has not "outstepped its own legitimate range" (Spottiswoode 1879, p. 17). Spottiswoode inverts Sylvester's approach in the *Laws of Verse*; rather than applying mathematical terminology to describe literary features, he uses literary analogy to explain mathematical concepts.

At the beginning of the nineteenth century, mathematical methods involving imaginary numbers had excited considerable controversy—even the use of negative numbers was viewed with suspicion by some more traditional mathematicians such as William Frend (see Pycior 1981, 1982). In 1830, George Peacock's *Treatise on Algebra* argued that while "arithmetical" algebra might confine itself to positive, real numbers, "symbolical" algebra could admit negative and imaginary quantities, an idea which, according to De Morgan, at first seemed "something like symbols bewitched, and running about the world in search of a meaning" (cited in Pycior 1982, p. 395). Spottiswoode expresses a similar sense of wonder at William Rowan Hamilton's 1843 discovery of quaternions, numbers which consist of one real and three imaginary parts; he remembers, in the "early days," gingerly handling quaternions "as a magician's page might try to wield his master's wand" (Spottiswoode 1879,

488   I. FORBES-MACPHAIL

p. 17). According to Spottiswoode, these methods continued to meet with a degree of skepticism even decades later:

> It is objected that, abandoning the more cautious methods of ancient mathematicians, we have admitted into our formulae quantities which by our own showing, and even in our own nomenclature, are imaginary or impossible; nay, more, that out of them we have formed a variety of new algebras to which there is no counterpart whatever in reality; but from which we claim to arrive at possible and certain results. (Spottiswoode 1879, p. 17)

Nevertheless, these "imaginaries," Spottiswoode insists, are "called up by legitimate processes of our science," and not only "[suggest] ideas," but "[conduct] us to practical conclusions" (Spottiswoode 1879, p. 19). He then resorts to literary analogy, citing the fact that writers are frequently driven "to imagery, to an analogy, or even to a paradox, in order to give utterance to that of which there is no direct counterpart in recognised speech"; and yet, we can still find a "meaning for these literary figures, an inward response to imaginative poetry, to social fiction, or even to ... tales of giant and fairyland" (Spottiswoode 1879, p. 20). Moreover, in order to make sense of this kind of imaginative literature, to "reanimate these things with a meaning beyond that of the mere words," we must behave "like the mathematician," "enlarge the ideas with which we started," and go back to first principles in order to find something in common between "the idea conveyed" and "the subject actually described" (Spottiswoode 1879, p. 20).

The second example Spottiswoode gives concerns higher dimensional or "manifold" space. Manifold space, Spottiswoode writes, "is not seriously regarded as a reality in the same sense as ordinary space; it is a mode of representation, or a method which, having served its purpose, vanishes from the scene" (Spottiswoode 1879, pp. 22–23).[3] Turning again to literary precedent, he asks: "when in legendary tales, or in works of fiction, things past and future are pictured as present, has not the poetic fancy correlated time with the three dimensions of space, and brought all alike to a common focus?" (Spottiswoode 1879, p. 23). As a further instance, "when space already filled with material substances is mentally peopled with immaterial beings, may not the imagination be regarded as having added a new element to the capacity of space, a fourth dimension of which there is no evidence in experimental fact?" (Spottiswoode 1879, p. 23). Six years later, in 1884, Edwin A. Abbott sought to illustrate the concept of spatial dimensions through literature in his novella *Flatland: A Romance of Many Dimensions*, which was in fact recommended by Sylvester as a helpful aid in obtaining "a general notion of the doctrine of space of *n* dimensions" (Sylvester 1998, p. 255). In his BAAS address, a full fifteen years before *Flatland* was published, Sylvester himself had attempted to explain *n*-dimensional space by inviting his audience to imagine an "infinitely attenuated bookworm" who can conceive of space in only two dimensions (Sylvester 1870, p. 113). Thus literature, both directly, as in Abbott's work, and by analogy, as in Spottiswoode's examples, turns out to be of great service in clarifying and explaining this somewhat difficult-to-conceptualize mathematical idea.

Spottiswoode's address is fascinating, therefore, for its claim that literature can be used to defend and explain, not merely mathematics in general, but even the most advanced and theoretical branches of the subject, by rendering its underlying principles intelligible to a skeptical audience. Toward the conclusion of his speech, Spottiswoode directly inserts himself into the Huxley/Sylvester/Arnold debate concerning the competition between the arts and the sciences. He expresses a wish that his views will act as a "solvent of that rigid aversion which both Literature and Art are too often inclined to maintain towards Science of all kinds" (Spottiswoode 1879, p. 30), blaming this on the "severance" which exists between these two subjects in early education. The sciences, Spottiswoode writes, may appear in a "severe garb ... little attractive" in comparison with the "light companionship" of the arts, but he hopes for a day when the "outcomings of Science, which at one time have been deemed to be but stumbling-blocks scattered in the way, may ultimately prove stepping stones which have been carefully laid to form a pathway over difficult places for the children of 'sweetness and light'" (Spottiswoode 1879, pp. 30–31), quoting Arnold's famous phrase from *Culture and Anarchy* (Arnold 2009, passim). Spottiswoode's address, in attempting to explicate difficult mathematical concepts by means of more familiar literary ideas, strives to render the trappings of mathematics a little less outwardly severe, and so assist this rapprochement.

Literature and mathematics thus became somewhat unlikely allies in nineteenth-century debates over the educational curriculum. In part, they were thrust together by circumstance, and by the new institutional challenges presented by the growth of the natural, physical and mechanical sciences. However, as these speeches by Arnold, Sylvester, and Spottiswoode demonstrate, there are also a number of inherent similarities in the way advocates of these subjects approach their disciplines—a resistance to straightforward utilitarian aims, a valuing of knowledge for its own sake, and an appreciation of beauty. One of the most important threads running through these defenses is the capacity for both mathematics and literature to forge connections not only among previously disparate ideas *within* their own disciplines, but also outside and between them. For Arnold, literature has a unique ability to connect science to our sense for beauty and conduct, something which he believes science alone lacks the capacity to do, while for Sylvester, mathematics discloses a world of "divine beauty and order," "harmonious connection" and "infinite hierarchy," and is united with the study of poetry through the principle of continuity. If poetry has the power to connect the sciences to our instinct for beauty, mathematics, too, as Sylvester's analysis of Horace's ode seeks to demonstrate, has the ability to reveal new connections, structures, and harmonies within poetic forms which similarly increase our appreciation of the beauty of poetry. As Sylvester suggests, there is a kind of continuity

490   I. FORBES-MACPHAIL

present not just within but also between mathematics and poetry, the two subjects "melt[ing] in a surprising manner into one another, like sunset tints or the colours of the dying dolphin, 'the last still loveliest'."

NOTES

1. G. H. Hardy, in *A Mathematician's Apology* (1940)—possibly the best-known defense of the subject—is particularly severe on this point: "It is a melancholy experience for a professional mathematician to find himself writing about mathematics. The function of a mathematician is to do something, to prove new theorems, to add to mathematics, and not to talk about what he or other mathematicians have done .... If then I find myself writing, not mathematics but 'about' mathematics, it is a confession of weakness, for which I may rightly be scorned or pitied by younger and more vigorous mathematicians" (Hardy 1992, pp. 61–63).
2. As Helen Small notes, it is in the Victorian period that we "[see] emerging the now familiar pressure to justify expenditure on educating students in the humanities in the face of resistance from many political economists" (Small 2013, p. 7).
3. Joan L. Richards, discussing the views of other significant mathematicians just prior to this period, writes that "Neither Salmon nor Cayley would have considered the possibility that spaces of higher dimension might be real in the same sense that Euclid's was, and there were few who would seriously have disagreed with them" (Richards 1988, p. 55); although, James Joseph Sylvester was "somewhat more daring" in his BAAS address, advocating for such generalized space to be considered a legitimate part of geometry (Richards 1988, pp. 55–56).

REFERENCES

[Anon.]. 1855. "The Romance of Euclid." *Punch* 29 (October 13): 149.

Arnold, Matthew. 1882. "Literature and Science." *The Nineteenth Century* XII (July–December): 216–30.

———. 2009. *Culture and Anarchy*, edited by Jane Garnett. Oxford: Oxford University Press.

Bentham, Jeremy. 1823. *An Introduction to the Principles of Morals and Legislation*. A new ed., corrected by the author. Vol. 1. London: W. Pickering.

———. 1825. *The Rationale of Reward*. London: J. and H. L. Hunt.

Brown, Daniel. 2013. *The Poetry of Victorian Scientists: Style, Science and Nonsense*. Cambridge: Cambridge University Press.

Byron, George Gordon Noel. 2008. "Childe Harold's Pilgrimage." In *Lord Byron: The Major Works*, edited by Jerome J. McGann, 19–206. Oxford: Oxford University Press.

Cayley, Arthur. 1884. "President's Address." In *Report of the Fifty-Third Meeting of the British Association for the Advancement of Science; Held at Southport in September 1883*, 1–37. London: John Murray.

Crilly, Tony. 2011. "Cambridge: The Rise and Fall of the Mathematical Tripos." In *Mathematics in Victorian Britain*, edited by Raymond Flood, Adrian Rice, and Robin Wilson, 17–32. Oxford: Oxford University Press.

26 "COLOURS OF THE DYING DOLPHIN": NINETEENTH-CENTURY DEFENSES ... 491

Forsyth, A. R. 1898. "Address" to the Mathematical and Physical Science Section, Under "Transactions of the Sections." In *Report of the Sixty-Seventh Meeting of the British Association for the Advancement of Science; Held at Toronto in August 1897*, 541–49. London: John Murray.

Foster, G. Carey. 1878. "Address" to the Mathematics and Physics Section, Under "Notices and Abstracts of Miscellaneous Communications to the Sections." In *Report of the Forty-Seventh Meeting of the British Association for the Advancement of Science; Held at Plymouth in August 1877*, 1–8. London: John Murray.

Glaisher, J. W. L. 1891. "Address" to the Mathematical and Physical Science Section, under "Transactions of the Sections." In *Report of the Sixtieth Meeting of the British Association for the Advancement of Science Held at Leeds in September 1890*, 719–27. London: John Murray.

Hardy, G. H. 1992. *A Mathematician's Apology*. Cambridge: Cambridge University Press.

Huxley, Thomas H. 1895a. "On the Educational Value of the Natural History Sciences [1854]." In *Science and Education: Essays*, 38–65. New York: D. Appleton and Company.

———. 1895b. "Scientific Education: Notes of an After-Dinner Speech [1869]." In *Science and Education: Essays*, 111–33. New York: D. Appleton and Company.

———. 1895c. "Science and Culture [1880]." In *Science and Education: Essays*, 134–59. New York: D. Appleton and Company.

Jenkins, Alice. 2007. *Space and the 'March of the Mind': Literature and the Physical Sciences in Britain 1815–50*. Oxford: Oxford University Press.

———. 2010. "Mathematics and Mental Health in Early Nineteenth-Century England." *BSHM Bulletin: Journal of the British Society for the History of Mathematics* 25: 92–103.

Levine, George. 1988. "Matthew Arnold's Science of Religion: The Uses of Imprecision." *Victorian Poetry* 26: 143–62.

Mill, John Stuart. 1989. *Autobiography*, edited by John M. Robson. London: Penguin Classics.

———. 2004. *Utilitarianism* in *John Stuart Mill and Jeremy Bentham: Utilitarianism and Other Essays*. Edited by Alan Ryan, 272–338. London: Penguin.

Parshall, Karen Hunger. 2006. *James Joseph Sylvester: Jewish Mathematician in a Victorian World*. Baltimore: John Hopkins University Press.

Parsons, William, Third Earl of Rosse. 1860. "Introductory Remarks" to the Mathematics and Physics Section, under "Notices and Abstracts of Miscellaneous Communications to the Sections." In *Report of the Twenty-Ninth Meeting of the British Association for the Advancement of Science; Held at Aberdeen in September 1859*, 1–3. London: John Murray.

Poe, Edgar Allan. 2008. "The Rationale of Verse." In *Poe's Critical Theory: The Major Documents*, edited by Stuart Levine and Susan F. Levine, 77–144. Champaign: University of Illinois Press.

Pycior, Helena M. 1981. "George Peacock and the British Origins of Symbolical Algebra." *Historia Mathematica* 8: 23–45.

———. 1982. "Early Criticism of the Symbolical Approach to Algebra." *Historia Mathematica* 9: 392–412.

Richards, Joan L. 1988. *Mathematical Visions: The Pursuit of Geometry in Victorian England*. San Diego: Academic Press.

Robinson, Rev. T. R. 1858. "Opening Address" to the Mathematics and Physics Section, Under "Notices and Abstracts of Miscellaneous Communications to the Sections." In *Report of the Twenty-Seventh Meeting of the British Association for the Advancement of Science; Held at Dublin in August and September 1857*, 1–2. London: John Murray.

Small, Helen. 2013. *The Value of the Humanities*. Oxford: Oxford University Press.

Spottiswoode, William. 1879. "Address by the President." In *Report of the Forty-Eighth Meeting of the British Association for the Advancement of Science; Held at Dublin in August 1878*, 1–32. London: John Murray.

Stewart, Ian. 2013. "The Third Culture: The Power and Glory of Mathematics." *New Statesman*, May 21. https://www.newstatesman.com/sci-tech/2013/05/third-culture-power-and-glory-mathematics.

Strutt, John William, Third Baron Rayleigh. 1885. "Address by the President." In *Report of the Fifty-Fourth Meeting of the British Association for the Advancement of Science; Held at Montreal in August and September 1884*, 1–23. London: John Murray.

Sylvester, James Joseph. 1870. *The Laws of Verse or Principles of Versification Exemplified in Metrical Translations: Together with an Annotated Reprint of the Inaugural Presidential Address to the Mathematical and Physical Section of the British Association at Exeter*. London: Longmans, Green.

———. 1998. Letter to Arthur Cayley, 2 November 1884. In *James Joseph Sylvester: Life and Work in Letters*, edited by Karen Hunger Parshall. Oxford: Oxford University Press.

Tait, P. G. 1872. "Address" to the Mathematics and Physics Section, Under "Notices and Abstracts of Miscellaneous Communications to the Sections." In *Report of the Forty-First Meeting of the British Association for the Advancement of Science; Held at Edinburgh in August 1871*, 1–8. London: John Murray.

Whewell, William. 1859. "Address" to the Mathematics and Physics Section, Under "Notices and Abstracts of Miscellaneous Communications to the Sections." In *Report of the Twenty-Eighth Meeting of the British Association for the Advancement of Science; Held at Leeds in September 1858*, 1–2. London: John Murray.

Wordsworth, William. 2000. *The Prelude*, in *The Major Works*, edited by Stephen Gill. Oxford: Oxford University Press.

CHAPTER 27

Combinatorial Characters

Andrea Henderson

In 1866 a reviewer of Charles Dickens's *Our Mutual Friend* remarked that "[t]he closer we look at Mr. Dickens's characters, the more we detect the trickery of an artificer. ... his art is *extra naturam*" (*Westminster Review* 1866, p. 582). This complaint, although founded on assumptions twenty-first-century critics do not necessarily share, speaks to a feature of this and other of Dickens's late novels that critics continue, in various ways, to notice: the characterization, although rich in concrete particulars, feels oddly abstract in conception. As literary critic Ayse Celikkol argues, in *Our Mutual Friend* Dickens pointedly undermines the Romantic ideal of organic form: "individuals are interchangeable" and are associated with "automation and mechanization" (Celikkol 2016, pp. 1, 11). What Alex Woloch would call the character system of the novel—"the arrangement of multiple and differentiated character-spaces ... into a unified narrative structure" (Woloch 2009, p. 14)—calls more attention to itself than it properly should. The plot is driven less by its characters' intrinsic natures than by their relations to one another; the system trumps its individual constituents. Put another way, characters in these novels are not the fundamental units of social meaning that earlier nineteenth-century novelistic protocols would lead us to expect. They are instead like atoms in a compound or letters in a word; they form variable combinations with others, and it is the combinations that matter. I will argue that this focus on clusters of people rather than singular individuals has its historical grounding in the logic of late Victorian capitalism, an abstract logic of ramifying networks in which meaning and value inhere less in particular people or things than in their links to other people and things. As we will see, that logic is epitomized in a branch of mathematics of particular interest

A. Henderson (✉)
University of California, Irvine, Irvine, CA, USA

© The Author(s) 2021
R. Tubbs et al. (eds.), *The Palgrave Handbook of Literature and Mathematics*, https://doi.org/10.1007/978-3-030-55478-1_27

to Victorian mathematicians: combinatorics. If Victorian literature sometimes dramatizes developments in mathematics, as for instance in the many Victorian novels and stories that make use of the idea of a fourth dimension, here mathematics brings into focus a development in literature. This essay will explore the power of combinatorics to illuminate the structure of late Victorian characterization, focusing particularly on Dickens's *Our Mutual Friend*. It will show that there is a good historical reason why, in the words of the reviewer cited above, the characters of this novel could be described as "a number of automatons moving about," "tattooed with various characteristics" (*Westminster Review* 1866, p. 583): nodes in a social network, they are the literary embodiment of a conception of personhood as mobile, extrinsically defined, and combinatory.

Arranging Schoolgirls

We begin with two typically Victorian mathematical puzzles. The first is "The Icosian Game," developed by mathematician William Rowan Hamilton, which requires players to map a route between points without visiting any one of them more than once—in mathematical terms, the route must describe a "Hamiltonian cycle" (see Fig. 27.1). The two-dimensional figure on the game board is a schematic representation of a dodecahedron, the twelve-sided Platonic solid that first piqued Hamilton's interest in cycles.

Some of that three-dimensionality is restored in a later iteration of the game, "The Traveller's Dodecahedron, or A Voyage Round the World"

Fig. 27.1 Hamilton's "Icosian game" (Photo courtesy of the Puzzle Museum. Copyright 2017 Hordern-Dalgety Collection. https://puzzlemuseum.org)

(see Fig. 27.2). In this game, the player connects lettered pegs with string, thereby travelling to all the cities signified by those initial letters.

The second puzzle was popularly known as the "Kirkman's Schoolgirls" problem. This is how it reads as "versified by a lady" in *The Educational Times*:

A governess of great renown
Young ladies had fifteen,
Who promenaded near the town,
Along the meadows green.

But as they walked
They tattled and talked,
In chosen ranks of three,
So fast and so loud,
That the governess vowed
It should no longer be.

So she changed them about,
For a week throughout,
In threes, in such a way

Fig. 27.2 Traveller's Dodecahedron, or A Voyage Round the World (Photo courtesy of the Puzzle Museum. Copyright 2017 Hordern-Dalgety Collection. https://puzzlemuseum.org)

That never a pair
Should take the air
Abreast on a second day;
And how did the governess manage it, pray? (Miller 1870, p. 79)

Thomas Kirkman first posed the problem in 1850 in the pages of another magazine, *The Lady's and Gentlemen's Diary*, in the next issue he provided an answer (see Fig. 27.3).[1]

Both of these puzzles are instances of problems in combinatorics. As the name suggests, this is a branch of mathematics that concerns itself with combinations or permutations. Such combinations can take many forms: some of the more famous problems in combinatorics include the Königsberg Bridge Problem (is it possible to walk through Königsberg crossing each of its seven bridges only once?); the "three friends or three strangers" problem (what is the smallest number of people who must be present at a gathering to be certain that among them three are mutual friends or three are mutual strangers?); the four-color problem (can the regions on any given geographical map be colored with four or fewer colors without adjacent regions ever being of the same color?); and the travelling salesman problem (how can a salesman minimize the distance he travels to visit a certain set of cities?). As these examples make clear, practical problems in combinatorics often involve social or spatial networks.

While such problems have been called "a very Victorian recreation," they also engaged professional mathematicians, who made combinatorics one of the most productive subfields within Victorian mathematics (Wilson 2011, 377). Among laypeople combinatorial problems were typically narrativized

---

VI. QUERY; *by the* Rev. THOS. P. KIRKMAN, *Croft, near Warrington*.
Fifteen young ladies in a school walk out three abreast for seven days in succession: it is required to arrange them daily, so that no two shall walk twice abreast.

Answered by the Rev. Mr. KIRKMAN, *the Prop oser*.

Denoting the ladies by $a_1, a_2, a_3; b_1, b_2, b_3; c_1, c_2, c_3; d_1, d_2, d_3; e_1, e_2, e_3$, the following arrangement will be found to answer the question :

$$
\begin{array}{ccc|ccc|ccc|ccc|ccc|ccc|ccc}
a_1 a_2 a_3 & a_1 b_1 c_1 & a_1 d_1 e_1 & a_1 b_2 d_2 & a_1 c_2 e_2 & a_1 b_3 e_3 & a_1 c_3 d_3 \\
b_1 b_2 b_3 & a_2 b_2 c_2 & a_2 d_2 e_2 & a_2 b_3 d_3 & a_2 c_3 e_3 & a_2 b_1 e_1 & a_2 c_1 d_1 \\
c_1 c_2 c_3 & a_3 d_3 e_3 & a_3 b_3 c_3 & a_3 c_1 e_1 & a_3 b_1 d_1 & a_3 c_2 d_2 & a_3 b_2 e_2 \\
d_1 d_2 d_3 & b_3 d_1 e_2 & d_3 b_1 c_2 & b_1 c_3 e_2 & c_1 b_3 d_2 & b_2 c_3 d_1 & c_2 b_3 e_1 \\
e_1 e_2 e_3 & c_3 d_2 e_1 & e_3 b_2 c_1 & d_1 c_2 e_3 & e_1 b_2 d_3 & e_2 c_1 d_3 & d_2 b_1 e_3 \\
\end{array}
$$

This is the symmetrical and only possible solution. All others differ from this only in disturbing the alphabetical order, or that of the three subindices in certain triplets of the first column, or in both these together.

Fig. 27.3 Kirkman's solution to "Kirkman's Schoolgirls" problem (From *The Lady's and Gentleman's Diary* 148 [London: J. Greenhill, 1851], p. 48)

and had an administrative tenor—"The same six persons filled a six-seated railway carriage every week-day for twenty-weeks. How did they arrange themselves so that no three of the six persons ever twice occupied the same three seats?" (*Lady's and Gentleman's Diary* 1860, p. 72); professional mathematicians abstracted and generalized these problems. Asserting priority over Kirkman, James Joseph Sylvester claimed that "in connection with my researches in combinatorial aggregation … I had fallen upon the question of forming a heptatic aggregate of triadic synthemes comprising all duads to the base 15, which has since become so well known, and fluttered so many a gentle bosom, under the title of the fifteen school-girls' problem" (Sylvester 1861, p. 371). Variations of Kirkman's problem engaged not just Sylvester but also his friend and colleague Arthur Cayley, and in 1852 William Spottiswoode generalized the problem to consider $2^{2n}-1$ young ladies.

Although Kirkman did not use one, problems in combinatorics can be solved using a special kind of schematic graph that looks very like the Icosian Game. Neither a Cartesian graph (see Fig. 27.4) nor a descriptive map (see Fig. 27.5, a map of the bridges of Königsberg), a graph of this kind is a collection of points (vertices) and lines (edges), where vertices are joined by an edge if they are related.

Thus, for instance, in a graph of the Königsberg Bridge Problem landmasses are represented as vertices and bridges as edges (see Fig. 27.6). The presence or absence of edges is the only pertinent information for the problem and in the graph itself—the length and shape of the lines is irrelevant. Thus simplified, it becomes apparent that a circuit that traces every edge only once can only be formed where every vertex has an even number of edges.[2]

Fig. 27.4 Cartesian coordinate graph. 345Kai at the English language Wikipedia (Public domain, GFDL [http://www.gnu.org/copyleft/fdl.html] or CC-BY-SA-3.0 [http://creativecommons.org/licenses/by-sa/3.0/]), via Wikimedia Commons

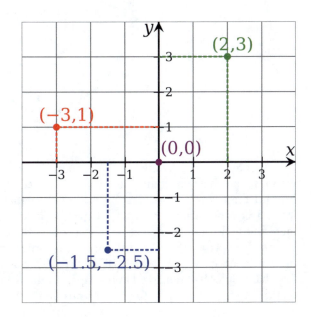

498 A. HENDERSON

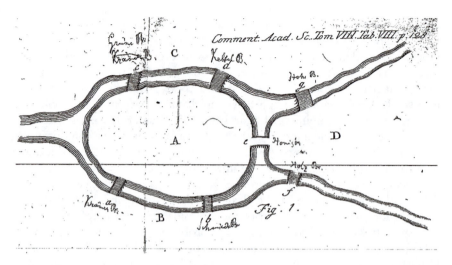

Fig. 27.5 Map of the bridges of Königsberg. Leonhard Euler, "Solution of a Problem in the Geometry of Position," *Commentarii Academiae Scientarum Imperialis Petropolitanae* 8 ([1736] 1741), Plate VIII

Fig. 27.6 Graph of the Königsberg Bridge Problem

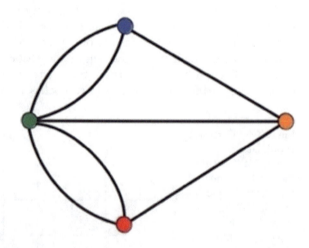

It was Sylvester, drawing on the terminology of contemporary chemistry, who dubbed these diagrams "graphs."[3] Sylvester not only imported chemist Edward Frankland's term "graph" into mathematics but also helped to clarify its abstractly formal character: "Chemical graphs … are to be regarded as mere translations into geometrical forms of trains of priorities and sequences having their proper *habitat* in the sphere of order and existing quite outside the world of space" (Sylvester 1878b, p. 79). Such problems belonged, as Leonhard Euler had remarked in his paper on the Königsberg Bridge Problem, to the much-neglected "geometry of position" rather than the geometry of space (Euler [1736] 1741, p. 128). As Sylvester explains, the shape forming a linkage is unimportant; "the question is one purely of colligation or linkage in the abstract." The graph, that is, images order itself. It is for this reason that Sylvester considers the new chemistry fundamentally mathematical—and therefore fundamentally modern: "By the new Atomic Theory I mean that sublime invention of Kekulé which ... bases its laws on pure relations of form." Indeed, Sylvester regards chemistry as a practical application of the purely conceptual, mathematical study of combinations: "chemistry is the counterpart of a province of algebra as probably the whole universe of fact is, or must be, of the universe of thought" (Sylvester 1878b, pp. 71, 64, 83). By these lights, combinatorics, as the study of order generally, should be applicable to most anything.

It is little surprise, then, that Sylvester and his colleague William Clifford saw in graphs an opportunity to study permutations of mathematical invariants: feeling that there was an

> affinity if not identity or object between the inquiry into compound radicals and the search for "Grundformen" or irreducible invariants, I was agreeably surprised to find, of a sudden, distinctly pictured on my mental retina a chemico-graphical image serving to embody and illustrate the relations of these derived algebraical forms to their primitives and to each other. (Sylvester 1878b, p. 64)

As historian of mathematics Norman Biggs notes, the analogy between chemical permutations and permutations of quantics (homogeneous polynomials of two or more variables) was "superficial" and ultimately proved unproductive, but however misguided, Sylvester's enthusiasm for the enterprise foregrounds the extent of his faith in the general significance of colligation and the power of combinatorics to elucidate it (Biggs et al. 1976, p. 67). Even as it served as a "Victorian recreation" then, combinatorics also spoke to a more encompassing contemporary fascination with linkage per se.

A VOYAGE ROUND THE WORLD

If on the one hand the combinatorial graph is notable for its abstracting character, on the other it is prone to telling metaphoric literalizations. The terms Victorian mathematicians used to describe the combinatorial graph reflect contemporary habits of mind: thus, for instance, Cayley coined the term "tree" to describe connected acyclic graphs, a naturalizing metaphor and one that might seem to have its source in the genetic preoccupations of Victorian England (Cayley 1857). Yet Cayley and his peers focused precisely on the variability of these combinations—their potentially inorganic, adventitious character. As fitting as the tree metaphor is as a description of acyclic graphs, it certainly did not function as a conceptual constraint: Cayley solved problems with "rooted trees" by removing the roots.[4] Sylvester, meanwhile, remarked that while a connected graph could be described as a family, "families" could be conceived any number of ways: "social individuals differ as egregiously as Isomers in their capacity for forming multifarious attachments" (Sylvester 1875, p. 214). Indeed, Sylvester understood graphic trees as capable of infinite ramification and interconnection:

> I have found it a profitable exercise of the imagination, from a philosophical point of view, to build up the conception of an *infinite* arborescence and to dwell on the relations of time and causality which such a concept embodies. An example of the good to be gained by these limitless mental constructions (new tracts and highways, so to say, opened out in the all-embracing "grand continuum" which we call space) is afforded by ... [Crofton's] new idea of an infinite reticulation (warp and woof), every finite portion of which contains an infinite number of meshes, being formed by the crossings of two sets of parallel lines all infinitely extended in both directions. (Sylvester 1878b, p. 90)

This arborescence is more akin to the extension of a rail network than the growth of a tree, as Sylvester's description of graphs as having "nodes, branches and *terminals*" implies (Sylvester 1878b, p. 90; my emphasis). Moreover, and as the example of the loom suggests, the combinatorial graph resonates not just with the contemporary expansion of transportation networks but also with the massive expansion of the British textile industry with the mechanization of spinning and weaving. Sylvester's fantasy of the network as all-embracing and perpetually growing is the mathematical expression of the economic imperatives of late Victorian England.

Sylvester sought, moreover, to produce a general "theory of *linkages*"; as he explains, "underlying" "the new mechanism, the new chemistry, and the new algebra" is "the theory of pure colligation" (Sylvester 1878a, p. 284). That Sylvester returns so insistently to the domain of "pure" theory when discussing linkage is itself a symptom of the historicity of his thought, for as important as linkages were to transportation and communication networks, they were at least as important to the more abstract bonds of finance capitalism—links that bound together bankers, merchants, and shareholders across the globe. For all their practical consequences, these bonds often appeared to contemporaries to be insubstantial and even fictional. And yet they, like the other linkages described here, seemed to generate value, leveraging capital just as mechanical linkages leveraged and redirected force. As the character named Optimist explains in "Review of Some Extraordinary Operations":

> I was at home ... looking over that crowded sheet, called the railway shares list, when my eyes lighted on the Lombardo-Venetian Stock. ... I meditate a trip to the metropolis, and at the same time a venture in Lombard-Venetian shares. ... I went to a London banker [...] he could introduce me to a broker. ... Clever bank manager – link the first: I sought the broker – link the second: here was I on the high road to gambling. (Evans [1861] 1968, pp. 22–24)

Banker and broker facilitate connections between London and Venice, home and the high road. Venice and Lombard Street are collapsed into a hyphenated compound adjective, as if an Italian city and England's banking and finance hub could be yoked together and used to define a purchasable asset. The links formed here are not only tenuous but also fortuitous: one link happens to lead to another, and even the initial decision to purchase this particular stock is made only because Optimist's eyes happen to "light on" it. The combinatorial graph effectively dramatizes this species of network building: it is, as it were, the fully schematized version of this already schematic narrative of connection. The graph, after all, captures just two features of the set it describes: the number of members that comprise it and the connections that obtain between them. It thereby highlights interchangeability on the one hand and the possibilities of combination on the other. This particular practice of mathematical abstraction thus formalizes the requirements of late Victorian capital: in its emphasis on the construction of reconfigurable networks it recalls the rapid growth and increasing complication of investment, production, and distribution networks. It reflects the extension of the fundamental capitalist principle of exchange, which relies upon the abstraction of commodities, to the persons involved in the exchange process, who become themselves "exchangeable." Just as exchange strips commodities of their sensuous particularity, reducing them to a number value, the networks of late Victorian capitalism, to be maximally flexible and capable of growth, required that their members be reduced to the fact of their numerical "valence," their capacity, like chemical elements, to bond. In any given supply chain, economic actors can be swapped out, middlemen added or eliminated, investor rolls distended, new markets added. I can walk with these schoolgirls today, and another pair tomorrow, and another pair the next day, and so on. Little wonder then that for Sylvester linkage serves as a master metaphor for describing the machine of the cosmos and our place within it: "the whole universe may constitute one great linkage" (Sylvester 1875, p. 214).

Mutual Friends and Mutual Others

Even regarded as a theoretical rather than a technical practice, combinatorics would seem to have little to do with characterization in the Victorian novel, especially given the Romantic presumption that great novels present us with characters whose individuality is defined precisely in opposition to the social networks within which they must act. But in its prioritization of the network over the unit, combinatorics forcibly reminds us that in the latter part of the nineteenth century the tendency toward combination was understood as a fundamental law of matter. Under such a dispensation, as Walter Pater makes clear, the very notion of the "individual" becomes nothing more than a fiction:

> Like the elements of which we are composed, the action of ... forces extends beyond us: it rusts iron and ripens corn. Far out on every side of us those elements are broadcast, driven in many currents; and birth and gesture and death and the springing of violets from the grave are but a few out of ten thousand resultant combinations. That clear, perpetual outline of face and limb is but an image of ours, under which we group them – a design in a web, the actual threads of which pass out beyond it. (Pater 1893, pp. 186–87)

If Pater conceives personhood as a local design in a much larger web, Dickens, in his late novels, explores what it means for a person to be a mere node in a larger social "linkwork." As Alex Woloch remarks, "Dickens's novels increasingly revolve around the structure of social interconnection, and the three late omniscient masterpieces all build interconnected plot edifices and overtly thematize this interconnection" (Woloch 2009, p. 221). One of these late novels, *Our Mutual Friend*, is a particularly instructive instance of this combinatorial logic: its characters draw their meaning from their piecemeal embodiment on the one hand and their place in an abstract social network on the other. These are not the "deep" characters of Romantic aesthetics or the "rounded" characters of modernism.[5] Indeed, Henry James complains that the characters of *Our Mutual Friend* are merely "mechanical": they are neither fully realized individuals nor recognizable "types." For James, this lack at the level of the individual character undermines the effectiveness of the novel as a whole: he argues, for instance, that for the opposition of Bradley Headstone and Eugene Wrayburn to be artistically effective, "it is requisite that they should *be* characters: Mr. Dickens, according to his usual plan, has made them simply figures" (James 1865, p. 787).

Anna Gibson rightly links this characterological flatness to the book's account of relations *between* characters: each "installment indeed brings together people and things in new combinations that change the direction of narrative this way and that to form a dynamic web of nomadic characters who change as a result of forces that make them collide, combine, and break apart." Gibson argues that "the most explicit use of the language of networks to turn static structures into plastic forms was found in Victorian physiology," and she develops a compelling account of the way agency in the novel plays out in terms of quasi-Darwinian interactions rather than conscious individual decisions (Gibson 2015, pp. 65, 67). But during the late nineteenth century the logic of the network was most fully articulated in mathematics. The web of interconnections in *Our Mutual Friend* is indeed reminiscent of Darwinian ecosystems, just as it is reminiscent of the language of chemistry. But in a more fundamental way characterization in this novel speaks to the primacy of the "theory of linkage" itself. James observed that the characters of *Our Mutual Friend* "have nothing in common" with "mankind at large" (James 1865, p. 787). But if they do not share some nominal humanity they do share something else, as the title suggests: a friend, a "node" to link them.

Of course, social bonds had always been crucial to the English novel, where family connections in particular had long served to organize both characterization and plot. Dickens himself organized some of his best-loved novels around the revelation of such lineal entanglements, what Franco Moretti calls "Dickens' notorious family romances" (Moretti 1998, p. 129). But *Our Mutual Friend* is, in some respects, the inverse of a novel like *Bleak House*: rather than organize its plot around the revelation of the meaningfulness of the characters' bonds, it deliberately explores contingent and adventitious social connections, connections that are founded not on familial ties or genuine intimacy but chance encounters, the exchange of capital, and the "infinite aborescence" of mutual friendships. This contingency characterizes even marriage bonds in the novel.[6] Sophronia and Alfred Lammle, both adventurers, each learn from their "mutual friend" Veneering that the other is rich, and they marry for that reason, only to discover that Veneering knew nothing of their characters or their fortunes. Even more strikingly, Bella Wilfer and John Harmon/John Rokesmith/Julius Handford are brought together by the terms of a will deliberately perverse, in that Harmon's father selected for him a bride whom he saw only once, as a toddler manifesting ill-temper while on a walk with her father. The Harmons' "courtship" takes the form of an elaborate charade designed to infuse meaning into an otherwise arbitrary connection.

The book's exploration of persons as points in a network begins, in fact, with its first paragraph:

> In these times of ours, though concerning the exact year there is no need to be precise, a boat of dirty and disreputable appearance, with two figures in it, floated on the Thames, between Southwark bridge which is of iron, and London Bridge which is of stone, as an autumn evening was closing in. (Dickens [1865] 1997, p. 13)

Rather than characters in a setting we are given two figures in a schematic map: with the exception of the boat's "disreputable" appearance, the facts we are given—one bridge is of iron and one is of stone—are strikingly concrete, but not obviously meaningful. They position the characters relative to a set of coordinates but little more. The next paragraph seeks, as it were, to fill in these Jamesian "figures," to make this diagram look more like a group portrait: we are told something of the man's appearance, and that the girl looks enough like him to be his daughter. We then proceed to a lengthy discussion of the man's occupation—another traditional "characterizing" move. But we learn about this occupation only negatively through a process of elimination, one that focuses on the *absence* of objects on the boat. This process, like the initial description of the bridges, is not attuned to human feelings or values; it appears, on the face of it, to be deductive, but the repeated phrase "*and he could not be*" calls attention to its unmotivated quality. Our narrator, in

504 A. HENDERSON

other words, is not only schematic in his characterization of the boatman, as if it involved nothing but a relative positioning of objects, but "types" him through a random process of eliminative induction: it is as if he were putting an "x" through all the positions the boatman cannot occupy. Just as the boatman tries to "make something" of the flotsam and jetsam floating on the river, we as readers struggle to piece these schemata together to make meaning. It is primarily his occupation that particularizes this man, but even that category is as slippery as the "ooze" at the bottom of the Thames to which he is likened; like Riderhood, his former partner, Hexam is simply a "waterside character," amphibious and ill-defined.

The second chapter might be expected to develop these characters further, but it turns instead to a scene from high life—one that will make clear that the schematic and incomplete rendering of Hexam and his daughter was not simply an expression of their peripheral class status. If the Hexams are mobile points between two bridges, floating along the current of the Thames and pursuing mysterious ends, the Veneerings are ostensibly fixed points: because they are wealthy, we expect them to be literally landed and metaphorically grounded. Their name, however, links them not to real estate but its portable accoutrements: like cheaply made modern furniture, their "surface smelt a little too much of the workshop." The Veneerings, "if they had set up a great-grandfather," would have received him from "the Pantechnicon, without a scratch upon him, French polished to the crown of his head" (Dickens [1865] 1997, p. 17). Unacknowledged by genuine forebears, they would instead collect ersatz heirlooms, the commodified signs of family ties. This conceit—of people as indistinguishable from furnishings—runs throughout the novel: Bella Wilfer complains that she could not like John Harmon, having been "left to him in a will, like a dozen of spoons," and when Harmon asks to be secretary to Mr. Boffin, Boffin misunderstands him to be asking to be a secretary table (Dickens [1865] 1997, p. 45). These people do not derive their meaning from their place in a lineal order, an order signified by a family name indissolubly bound to a family estate. Instead, they function as the exchangeable signs of such order; like furniture, they are commodities amenable to arrangement and rearrangement. These characters are, as James said, "mechanical" figures. Indeed, they are quite specifically like cogs in a machine: they derive their meaning and value from their relations to other people and things—and those people and things, in turn, derive their meaning and value from them. We never see the Veneerings "set up a great-grandfather," but they do "set up," or introduce, many of the novel's characters to each other.

Because meaning inheres less in persons themselves than in their arrangement, it is the function of these human furnishings to constellate those around them into new sets of "friends":

> There was an innocent piece of dinner-furniture that went upon easy castors and was kept over a livery stable-yard in Duke Street, Saint James's, when not

27 COMBINATORIAL CHARACTERS 505

> in use …. The name of this article was Twemlow. Being first cousin to Lord Snigsworth, he was in frequent requisition, and at many houses might be said to represent the dining-table in its normal state. Mr. and Mrs. Veneering, for example, arranging a dinner, habitually started with Twemlow, and then put leaves in him, or added guests to him. Sometimes …. Twemlow was pulled out to his utmost extent of twenty leaves. (Dickens [1865] 1997, p. 17)

Like an atom with a valence of twenty or a vertex from which as many as twenty edges might extend, Twemlow is less a man in his own right than a node in a mobile system of connections. His is a society comprised entirely of bonds forged indiscriminately by means of soi-disant mutual friends: "Twemlow had first known Veneering at his club, where Veneering then knew nobody but the man who made them known to one another, who seemed to be the most intimate friend he had in the world, and whom he had known two days." The nodes that form such networks can ultimately be fined down to mere offices or even objects: "At the man's were a Member, an Engineer, a Payer-off of the National Debt, a Poem on Shakespeare, a Grievance, and a Public Office, who all seemed to be strangers to Veneering." Each man can be summed up in a single social fact, for it is that fact through which he mediates "friendships" between others (Dickens [1865] 1997, p. 18).

The novel's very title, of course, foregrounds this dynamic. In the most explicit reference to the title, Boffin resists calling Rokesmith by his name: "I may call him Our Mutual Friend.…. What sort of a fellow *is* Our Mutual Friend, now? … I'm not particularly well acquainted with Our Mutual Friend" (Dickens [1865] 1997, p. 115). That these intimacies are no intimacies at all is a joke to which the novel repeatedly returns. At the Lammle wedding, where Reverend Blank Blank and Reverend Dash Dash preside and bridesmaids who do not know the bride "come by railroad from various parts of the country" like items in a trousseau, Lady Tippins reckons up "the rest of the characters" right down to the "Attendant unknowns": "Bride; five-and-forty if a day, thirty shillings a yard, veil fifteen pound, pocket-handkerchief a present. Bridesmaids; kept down for fear of outshining bride, consequently not girls, twelve and sixpence a yard" (Dickens [1865] 1997, pp. 121, 123). Even more striking examples of faux intimacy center around that expandable table, Twemlow. In being given, along with Podsnap, an "unceremonious invitation like an old, old friend" to plan the Lammle's wedding, Twemlow initially believes he and Podsnap must be like family to the Veneerings: "then there are only two of us, and he's the other." As Twemlow learns of more and more people who were invited to participate in the planning, he keeps adding to the number of "old, old friends": "then there are three of us, and *she's* the other. [T]hen there are four of us, and *he's* the other," and so on (Dickens [1865] 1997, pp. 119, 120). At each new addition the newcomer becomes "the other," a placeholder and little more. Like Sylvester's infinite arborescence, this addition of nodes can be endlessly repeated, with each new addition functioning simply as another "other." In an even more striking example of the meaningless accumulation of new nodes of connection, Riderhood, who calls Lightwood "Governor," refers to Lightwood's friend Wrayburn as "T'other Governor." When Riderhood chances to meet Headstone outside Lightwood's lodgings, he dubs him "T'otherest Governor" (Dickens [1865] 1997, p. 170). Headstone is in every way unlike Wrayburn, whom he longs to kill; that the two should form a cluster of "others" linked by Lightwood is, however, par for the course in this novel.

When not mutual "friends" or "others," characters in the novel form pairs or clusters, but as with the application of the word "other" to entire sets of people, this clustering serves primarily to mark the interchangeability of the persons involved. Thus Veneering "begs to make his dear Twemlow known to his two friends, Mr. Boots and Mr. Brewer – and clearly has no distinct idea which is which." Twemlow, too, "noted the fusion of Boots in Brewer and Brewer in Boots" (Dickens [1865] 1997, pp. 19, 20). Like atoms forming molecules, this pair, in turn, is subsumed into a larger cluster of four elements: "Boots and Brewer, and two other stuffed Buffers interposed between the rest of the company and possible accidents" (Dickens [1865] 1997, p. 21). In this role, all four men function identically and so need not be distinguished by us as readers or even by other characters:

> The four Buffers, taking heart of grace all four at once, say:
> "Deeply interested!"
> "Quite excited!"
> "Dramatic!"
> "Man from Nowhere, perhaps!"

The Buffers always function in this manner: "the four Buffers, again mysteriously moved all four at once, exclaim, 'You can't resist!'" (Dickens [1865] 1997, p. 23). When one of the Buffers "astound[s] the other three, by detaching himself, and asserting individuality," that individuality is short-lived: "herein perishes a melancholy example; being regarded by the other three Buffers with a stony stare, and attracting no further attention from any mortal" (Dickens [1865] 1997, pp. 25, 26). The Buffers' very name captures their purely relational significance: even as a set they matter only as intermediaries within the larger set of dinner guests. And while the word "buffer" was used in the nineteenth century to describe a neutral state lying between two hostile ones, it is fitting that its primary use was as a name for a mechanical shock absorber, especially those attached to the front and back of trains and to train terminal walls.[7] Buffers ensure that vertices are connected while being kept at a safe distance from one another: like Twemlow, they facilitate connections by mediating them, functioning as "mutual friends." Adventitious clusters like the Buffers appear throughout the novel; their arbitrary quality is captured in the change of a single letter in a series of names: Mortimer's office boy, Blight, "rings alphabetical changes" on imaginary clients, as "Mr. Aggs, Mr. Baggs, Mr. Caggs," or "Mr. Alley, Mr. Bally, Mr. Cally" and so on. Nor is clustering simply a sign of the loss of individuality under the pressure of social expectations; Betty Higden names two of her beloved young charges "Poddles" and "Toddles," and they seem always to move as a unit, hand in hand (Dickens [1865] 1997, pp. 92, 199).

The economics of Dickensian characterization have been explored most explicitly by Woloch, who, focusing on the distinction between minor and principal characters, brilliantly demonstrates the ways "minor characters are simultaneously torn away (from their own full selves) and bound up (to other social fragments)"—they conjure metonymic chains of associations rather than the metaphoric symbols that reveal psychological depth. For Woloch, this fragmentation of the minor character is a reflection of his proletarian function: "The functional automaton and the deviant eccentric, are both the results of the division of labor" (Woloch 2009, pp. 201, 162). Minor characters are necessarily incomplete, even as their labor produces "whole" gentlemen. This is a compelling argument, and it effectively describes characterization in Woloch's primary source text, *Great Expectations*. But the logic of fragmentation and combination pervades the entire social fabric of *Our Mutual Friend*, as if characterological "minorness" had become ubiquitous. And in a sense, it has. If the proletarian has been rendered a mere node in a system by his labor, so too have middle-class middlemen and even the aristocratic speculator. Finance capitalism, after all, is the economy of the mutual friend in its most abstract form; it is also inimical to a traditional notion of character, which, as J. G. A. Pocock argues, was founded on an ideal of landed as opposed to mobile property (Pocock 1985). The narrator of *Our Mutual Friend* cynically remarks: "As is well known to the wise in their generation, traffic in Shares is the one thing to have to do with in this world. Have no antecedents, no established character, no cultivation, no ideas, no manners; have Shares." The possession and multiplication of bonds is all that matters: "oscillate on mysterious business between London and Paris, and be great. Where does he come from? Shares. Where is he going to? Shares. What are his tastes? Shares. Has he any principles? Shares" (Dickens [1865] 1997, p. 118). Nor is the reduction of character to contingent bonds limited to the domain of finance capital. At the lowest level of society the traffic in bonds takes the literal form of snatching bodies out of the flowing river and binding them to one's boat; the bodies themselves are interchangeable, and, indeed, often indistinguishable. In his home Hexam displays handbills with descriptions of recovered bodies, differentiating them only "by the places on the wall": they are known by their relative position, like the two figures between two bridges of the novel's first paragraph (Dickens [1865] 1997, p. 31).

The plot of the novel, fittingly enough, tracks the ramifications of bond formation. Franco Moretti maps the characters' meetings and movements in *Our Mutual Friend* to make the point that the novel's middle-class characters provide crucial socio-geographic links between its East End proletariat and its West End aristocracy (see Fig. 27.7). But the appropriateness of this mapping technique to this particular novel arises also from a more general truth about the novel's construction: Moretti's schematic map speaks to the book's own (mathematically) graphic nature, its connect-the-dots social logic. And while the formation of new bonds within London is the most common form of social networking, it is nevertheless the case that the characters of *Our Mutual Friend* aspire to global linkages, from Veneering, who began as a "traveler or commission agent," to pawnshop owner Pleasant Riderhood, who "may have had some vaporous visions of far-off islands in the southern seas or elsewhere ... waiting for ships to be wafted from the hollow ports of civilization" (Dickens [1865] 1997, pp. 41, 346).

Just as Sylvester found it a "profitable exercise of the imagination ... to build up the conception of an infinite aborescence," so too do the men and women of *Our Mutual Friend*. Not only does Bella Wilfer profit by her impersonal financial connection with the "man from Somewhere" but her fantasies of wealth and happiness all take a global maritime form: she imagines that her father is enriched by bringing home opium from China, or that her husband is an immensely wealthy merchant who owns every boat on

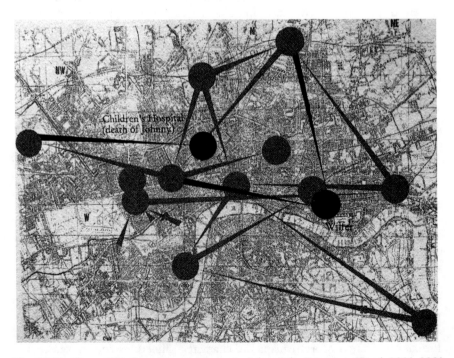

Fig. 27.7 Figure 58h from Franco Moretti, *Atlas of the European Novel, 1800–1900* (London: Verso, 1998)

the water, or that he is an Indian prince who was a "Something-or-Other" (Dickens [1865] 1997, p. 316). The flatness of the characterization here is a reflection both of the abstractions of finance at home and the necessary anonymity of connections abroad: "Where does he come from? Shares. Where is he going to? Shares."

"That Popular Character, Called Another"

Ultimately, then, although it is not explicitly concerned with mathematics, *Our Mutual Friend* is governed by a mathematical logic. One need not know anything about combinatorics to see that the novel's characters are "flat," that the Veneerings seek to advance themselves by pretending to intimacies they have not earned, or that Dickens is satirizing the ascendency of finance capital. But familiarity with combinatorics enables us to see that all these features spring from a single common principle: the novel is, as much as any essay of Sylvester's, a theory of linkages.

For novelist and mathematician alike, the nodes in a system are significant less in themselves than in the bonds they form and their capacity to change or multiply those bonds. For Sylvester, the priorities of this formal combinatorial logic pose no intellectual, ethical, or professional problems; indeed, he embraces them and explores all their ramifications. For Dickens, however, this logic necessarily elicits deep ambivalence. This is true first of all for ethical reasons: if a person's moral nature is most obviously manifest in his or her relationships with others, the reduction of those others to go-betweens is itself a register of social and ethical impoverishment—*mutual* friends are a poor replacement for mutual *friends*. The plot is designed to convert the former into the latter: thus the original "mutual friend," John Harmon, is discovered to be not the shiftless "man from somewhere" Boffin initially fears but the grown version of the boy he and his wife used to love. This revelation and the moral development of a few key characters notwithstanding, it is telling that the concluding chapter, "The Voice of Society," returns us to the domain of heartless abstraction, which remains normative, countered only by Lightwood's pleasure at Twemlow's defense of Wrayburn's decision to marry the proletarian Lizzie Hexam: "The feelings of a gentleman I hold sacred, and I confess I am not comfortable when they are made the subject of sport or general discussion" (Dickens [1865] 1997, p. 819). As the narrator had cuttingly remarked of Veneering, who sought to be M. P. for a borough he had never visited, it is entirely reasonable for a man like him to serve as a "representative man," for he *is* representative of the strictly formal bonds that structure contemporary English society. Dickens's own "representative men" are, like Twemlow, living in a society where they function like expandable pieces of furniture. Twemlow's final plea is Dickens's own; Dickens, too, longs for the return of "character," both in the form of personal integrity and in the form of the organic novelistic construct.

Ultimately then, just as problems in combinatorics gave rise to a new mode of mathematical representation—the modeling that serves as the basis of what is now termed graph theory—so the combinatorial logic of social life in the 1860s transformed novelistic representation. Jonathan Grossman argues that early in his career Dickens used the serial form of the novel to highlight the networked character of modern public transit—that is, he presented the network as a means to social synthesis, a way of bringing pieces, places, and people together (Grossman 2012, p. 53). But synthesis at the level of the network can easily suggest the insignificance of its individual elements. In this late work, Dickens signals the threat this abstraction poses to the novel as a genre by introducing the dominant narrative of Harmon's inheritance and its attendant network of relationships as mere formula. The characters of this tale, abstractly and relationally conceived, are referred to with monikers like "a man from somewhere," and "that popular character ... called Another" (Dickens [1865] 1997, p. 24). Dickens will flesh out these abstractions in his own development of the tale, but that rendering "real" will require extensive artifice, not only extradiagetically but also within the diegesis, on the part of Harmon and the Boffinses.[8] Dickens recognizes that this is the way we do realism now: the second chapter, in which Lightwood tells his story of the man from somewhere over dinner at the Veneerings, includes an extended rendering of the dining room as reflected in a large looking glass. This mirror brings the characters' relative positions into focus even as it produces a morbidly distorted image of the individuals involved: it reflects

> Lady Tippins on Veneerings' right, with [a] ... face, like a face in a tablespoon ... pleased to patronize Mrs. Veneering opposite, who is pleased to be patronized. Reflects a certain "Mortimer," ... who sits disconsolate on Mrs. Veneering's left .... Reflects Eugene, friend of Mortimer; buried alive in the back of his chair, behind a shoulder. (Dickens [1865] 1997, p. 21)

These guests, along with their "filmy" host and hostess, are, like Mortimer, people in scare quotes, people who exist not in themselves but in relation to each other. Mimesis "in these times of ours" has become inhuman—a mirroring performed by a piece of furniture, and one that is schematic rather than "natural." We might amend Sylvester's claim that "the theory of pure colligation" "[underlies] the new mechanism, the new chemistry, and the new algebra," to include the new novel (Sylvester 1878a, p. 284).

NOTES

1. Four other readers of the *Lady's and Gentleman's Diary* offered a solution in which the young ladies are labeled not in sets of three but as numbers 1 through 15 (1851, p. 48).
2. Leonhard Euler, who originally solved the problem, did not use a graph, but instead foregrounded the relevant information by assigning land masses letters and counting the number of bridges that connected with each.

3. Biggs, Lloyd, and Wilson claim that Sylvester was the first to use the word "graph" in the way that it would be used in mathematical graph theory (Biggs et al. 1976, p. 65).
4. Essentially, Cayley simplified the problem by eliminating the vertices with one edge (those representing the hydrogen atoms) altogether.
5. See, for instance, Deidre Lynch's account of "the psychological depth of the 'new style of novel'" so dear to the hearts of Romantic-era readers and critics (Lynch 1998, p. 126); E. M. Forster used the term "round character" in his *Aspects of the Novel* (Forster [1927] 2005, p. 73).
6. Lizzie and Eugene meet only because Eugene is the friend of the lawyer who handles the estate of a man believed to be the same as a dead man Lizzie's father finds in the river.
7. Fittingly enough, the word "buffer" would also come to be used as a term for a chemical compound that neutralizes the acidity or alkalinity of a solution.
8. As many critics have noted, Mr. Boffin in particular, in supporting Harmon's ruse, has to play the part of a miser for some weeks, making it appear that his personality has entirely changed.

References

Biggs, Norman, Keith Lloyd, and Robin Wilson, eds. 1976. *Graph Theory*. Oxford: Clarendon Press.

Cayley, Arthur. 1857. "On the Theory of the Analytical Forms Called Trees." *Philosophical Magazine* 13, 4th series (January–June): 172–76.

Celikkol, Ayse. 2016. "The Inorganic Aesthetic in Dickens's *Our Mutual Friend.*" *Partial Answers* 14 (1): 1–20.

Dickens, Charles. [1865] 1997. *Our Mutual Friend*. Edited by Adrian Poole. London: Penguin.

Euler, Leonhard. [1736] 1741. *Commentarii academiae scientiarum Petropolitanae* 8: 128–40.

Evans, D. Morier. [1861] 1968. "Review of Some Extraordinary Operations." In *Speculative Notes and Notes on Speculation, Ideal and Real*, 22–24. New York: Burt Franklin.

Forster, E. M. [1927] 2005. *Aspects of the Novel*. Edited by Oliver Stallybrass. London: Penguin.

Gibson, Anna. 2015. "*Our Mutual Friend* and Network Form." *Novel* 48 (1): 63–84.

Grossman, Jonathan. 2012. *Charles Dickens's Networks: Public Transport and the Novel*. Oxford: Oxford University Press.

James, Henry. 1865. "Review of Charles Dickens' *Our Mutual Friend.*" *The Nation* (21 December): 786–87.

The Lady's and Gentleman's Diary 148. 1851. London: J. Greenhill.

The Lady's and Gentleman's Diary 157. 1860. London: J. Greenhill.

Lynch, Deidre. 1998. *The Economy of Character*. Chicago: University of Chicago Press.

Miller, W. J. C. 1870. *Mathematical Questions with Their Solutions, from "The Educational Times."* London: C. F. Hodgson and Son.

Moretti, Franco. 1998. *Atlas of the European Novel, 1800–1900*. London: Verso.

Pater, Walter. [1893] 1980. *The Renaissance: Studies in Art and Poetry*. Edited by Donald L. Hill. Berkeley: University of California Press.

512 A. HENDERSON

Pocock, J. G. A. 1985. "The Mobility of Property and the Rise of Eighteenth-Century Sociology." In *Virtue, Commerce, and History*, 103–24. Cambridge: Cambridge University Press.

Sylvester, James Joseph. 1861. "Note on the Historical Origin of the Unsymmetrical Six-Valued Function of Six Letters." *Philosophical Magazine* XXI (141): 369–77.

———. 1875. "History of the Plagiograph." *Nature* 12 (July 15): 214–16.

———. 1878a. "Chemistry and Algebra." *Nature* 17 (February 7): 284.

———. 1878b. "On an Application of the New Atomic Theory to the Graphical Representation of the Invariants and Covariants of Binary Quantics." *American Journal of Mathematics* 1 (1): 64–104.

Westminster and Foreign Quarterly Review. 1866 (April 1).

Wilson, Robin. 2011. "Combinatorics: A Very Victorian Recreation." In *Mathematics in Victorian Britain*, edited by Raymond Flood, Adrian Rice, and Robin Wilson, 377–95. Oxford: Oxford University Press.

Woloch, Alex. 2009. *The One vs. the Many.* Princeton: Princeton University Press.

CHAPTER 28

Datelines

Steven Connor

TELLING THE TIME

Where can we live, Philip Larkin wonders, but days? (Larkin 1968, p. 67). Days are the way in which the force and necessity of number seem to intersect with what we call "lived time." We have clocks to mark the passing of the hours, but days seem to keep the beat on their own account. It is easy to appreciate the arbitrariness of divisions like hours and minutes and seconds, since we could get along perfectly well with hours and minutes and seconds of different durations; but days seem like a spontaneous self-numbering of nature. "They come, they wake us," writes Larkin (1968, p. 67). Days are like numbers in that they seem to be naturally given to us as exactly equivalent units. We need a clock or some other system of measurement to tell us when an hour has passed, but days themselves tell us of their coming and going. If there is variation in the length of days over the course of the year, there is no variation in that pattern of variation, and no let-up in the endless succession of days that, like numbers, are each one the same, and yet each one different, but only in the fact of simply being the next in sequence. The givenness of days as units is indicated by the fact that they provide the principal reference point or gearing for all human reckonings of historical time, a week, month or year being reckoned as a certain number of days, not a certain number of hours or minutes.

And yet, number-like as they are, days are not, of course, self-numbering. They have the cardinal dimension of number, but not the ordinal. The word "date" is from "datum," that which has been given, but really dates are just the opposite of data, since they are never merely given, but carefully made and maintained. If the apparent capacity for self-timing of days can give the

S. Connor (✉)
Cambridge University, Cambridge, UK

© The Author(s) 2021

513

R. Tubbs et al. (eds.), *The Palgrave Handbook of Literature and Mathematics*, https://doi.org/10.1007/978-3-030-55478-1_28

sense that days are indeed doling themselves out and ticking themselves off, dating is precisely how we pluck and peak the empty, open, isotonic succession of days into shape by numbering, allowing both for the rhythm and scansion of individual lives and, most importantly, for the coordination of time lived and experienced in common. Dates are both arbitrary and empty and yet also charged with significance, particularly in relation to birthdays and anniversaries, the secular shadows of the sanctified days of the religious calendar, which represent the intersection of temporality and timeless. Indeed if the word "sacred" indicates that which has been set apart from a sequence, "secular," which came to mean a period of a hundred years, is from *saeculum*, a generation, or lifetime, sometimes thought of as thirty-three years or a third of a century. It is suggested that *saeculum* derives from Greek σάω, to sift, preserve, or observe, cognate with Latin *sero*, to sow, from which we derive the word *series*. These serialities give a kind of magical power to recurrences that have reference only to our dating systems, which is made to keep the cosmic beat only by a kind of nip-and-tuck adjustment that is a kind of equal temperament.

The historian Benedict Anderson famously compares the workings of a novel to that of a newspaper, in its function of forming an imagined community. Both the novel and the newspaper, he suggests, depend upon the "meanwhile," upon the idea of lives convened and made comparable by the arbitrary fact of taking place simultaneously, synchronized by the date (Anderson 2006, pp. 24–25). Indeed, a newspaper is sometimes used as a kind of timepiece, as when hostages are photographed holding a newspaper as proof that they were still alive on a particular day.

For the most part, our calendrical systems serve as a way of giving ourselves to the givenness of time: as Virginia Woolf's Mrs Dalloway thinks to herself, "that one day should follow another; Wednesday, Thursday, Friday, Saturday; that one should wake up in the morning; see the sky; walk in the park; meet Hugh Whitbread; then suddenly in came Peter; then these roses; it was enough" (Woolf [1925] 2000, p. 104). And yet that steady succession is mined with menace. "After that," Mrs Dalloway immediately thinks, "how unbelievable death was! – that it must end" (p. 104). I may experience, as folklore suggests, a sudden shiver when somebody "walks over my grave." But, especially for those of us without access to family vaults, we need not necessarily know or have set foot in the place on the earth on which we have had our fitful residence where our long home will turn out to be. Yet, even though, as Samuel Beckett puts it, "Death has not required us to keep a day free" (Beckett 1965, p. 17), the existence of dates and dating means that I will in fact have met my deathday unawares and lived it through as many times through as my birthday (plus or minus one). Days are where we live, but one day in particular will be, and will always have been, the day I was to die, the day that can happen only once but, once I have survived my first year, can never happen for the first time. Any day in a series of days may be, arbitrarily, the day that will bring the series to an end; and it cannot not be one of the days in the repeated series. There is indeed, a kind of fatality in dates, a kind of enunciative force, fatality being from the past participle of Latin *fari*, to speak or utter: so the fatal is that which has always been fore-said, or *predicted*. Time, it seems, must be told, implying telling of as well as telling off: as one washerwoman says to another in James Joyce's *Finnegans Wake*, "every telling has a taling" (Joyce [1939] 2012, p. 2013), a telling entailing a tailing off, an accounting and an en-taling.

The relations between time and timing became newly demanding in the nineteenth century, as the increasing speed of travel and communications meant that the world needed to be ever more synchronized. This produced greater consistency and commensurability in temporal experience—there could no longer be Reading time and Cheltenham time when you could travel from one to the other quickly enough to register the difference—but also produced paradox, especially when it came to taking the measure of global time. Lewis Carroll was much taken up with the problem of the date line, and included it in a series of mathematical puzzles that he set in the *Monthly Packet* magazine during the 1880s, and published as *A Tangled Tale* in 1885. One of the problems in this book concerns the question of when, or, what puzzlingly comes to the same thing, where precisely the date changes.

> "Well, now, suppose it's midnight here in Chelsea. Then it's Wednesday *west* of Chelsea (say in Ireland or America) where midnight hasn't arrived yet: and it's Thursday *east* of Chelsea (say in Germany or Russia) where midnight has just passed by?"
>
> "Surely," Balbus said again. Even Lambert nodded this time.
>
> "But it isn't midnight, anywhere else; so it can't be changing from one day to another anywhere else. And yet, if Ireland and America and so on call it Wednesday, and Germany and Russia and so on call it Thursday, there *must* be some place – not Chelsea – that has different days on the two sides of it. And the worst of it is, the people *there* get their days in the wrong order: they've got Wednesday *east* of them, and Thursday *west* – just as if their day had changed from Thursday to Wednesday!" (Carroll 1973, p. 918)

The problem is not very clearly posed, but Carroll seems to mean that, some-where on the opposite side of the world to London, there must be a place where Thursday reverts to being Wednesday, since, if one were to follow (at a speed greater than that of the earth's rotation) the Thursday that has begun just east of Chelsea all the way round the world, there would come a point at which one would have to find oneself back in Wednesday. But when (or where) is that point?

That question was answered by *fiat* with the establishment of the inter-national date line. The International Meridian Conference took place in

516   S. CONNOR

October 1884 in Washington DC, and agreed that the Greenwich Meridian should be universally agreed to be 0° longitude. Everything was coordinated in relation to time, space therefore being subordinated to time rather than time to space, but it was only possible for this to happen if time itself was localized, so that one knew where to count from. There were imperial reasons for London to have been chosen for the Prime Meridian, but also certain practical considerations. For it was necessary to decide, not just what the time was, but where the date changed from one day to another. One of the advantages of choosing Greenwich for the Prime Meridian was that the antimeridian, 180° degrees away from it on the opposite side of the globe, would pass mostly through uninhabited ocean, where the disruption of the date-change would have less effect. Rapid travel across large landmasses like the USA posed a similar difficulty of coordinating space and time. One solution proposed for the San Francisco to New York railway was for the train to keep San Francisco time, in a sort of travelling timelock, until the point at which the passengers alighted in New York and changed their watches. The train became, as aeroplanes are today, a kind of chronopher, or time-carrier.

The precise course of the international date line was never in fact formally agreed: rather it was from the beginning a matter of changeable convention and consent. As a result, the line has dodged and woven unpredictably. But there were other ways of putting oneself at the centre of time. Paris succeeded in making itself the centre of world time by establishing a system for transmitting the time through wireless signals in the International Conference on Time of 1912. If London was the place from which time was measured, Paris, and, more specifically the Eiffel tower, would be the place from which time would be transmitted (Bartky 2007, pp. 138–48).

Carroll published solutions to all his problems in the *Monthly Packet*, often teasing or upbraiding his contributors. But the date line problem is the only one for which he could not offer a solution, declaring instead: "I must postpone, *sine die*, the geographical problem … because I am myself so entirely puzzled by it; and when an examiner is himself dimly hovering between a second class and a third how is he to decide the position of others?" (Carroll 1973, p. 965). The sly little joke of the Latin tag *sine die*, which, when applied to a postponement, simply means "indefinitely," but literally means "without a day," along with the blurring between degree classes and railway travel classes, points to the very condition of temporal deterritorialization that puzzles Carroll. The day seems to be a period that is given to us externally, rather than an arbitrary division imposed on time, because of the very place where we live, meaning that the day is spatially "there" in a way that the minute, the hour, the week or the month are not. But if there is no precise "there" at which one day changes to another, if one is always dimly hovering between one temporal position or another, where after all, is the day?

Naming the Day

During the nineteenth century, novels came closer to the condition of news-papers than ever before because they were more often themselves produced periodically. Charles Dickens lived his writing life subject to the relentless tick of the hours and the days, keyed in as they were with columns of type and printed pages. His novels were not only written by numbers, but the monthly parts were known as "numbers" themselves, just as turns in the music hall would later be. We know that Dickens began to develop a kind of internal time-sense in his writing, so that his sentences and chapters were not so much subject to the clock and the calendar as themselves meting them out. But his internal word-clock could sometimes go a little haywire (a word that entered English in the sense of tangled or confused only as late as 1929), overpro-ducing or, even more alarmingly, undershooting, so that he found himself on one occasion, having underwritten a number of *Dombey and Son* (1848) by two pages, needing to travel from Paris to London to make up the shortfall (Forster 1876, 1.523, 2.42). The reading of novels was also subject to this rhythm of anticipation, publication and distribution, forming a kind of pros-ody of the commodity.

And yet, if novels are in this sense held in lockstep with collective and commercial time, they are also hostile or allergic to it. The fact that there are so many Sundays and Thursdays, Mays, Novembers and Januarys in Dickens's writing is an apparent proof of the novels' worldly secularity, but fiction tends to recoil from actual, historical dates, because they seem to belong too much to the deathly, entropic one thing after another order of *chronos* rather than the order of *chairos*, sacred or significant time. The growing association of novels with time spent travelling, notably on the railways, associated them with the time-out-of-time of that experience of idle speed, at rest in haste.

Just as there are two kinds of figurings of time, the singular and the figu-rative (things that happened once and things that would often happen) there are two kinds of date, which we may call the calendrical and the categorical. A calendrical day is singular and unrepeatable. An actual, that is to say, unique date, is the equivalent of a proper name. Dates confront literary texts with the order of the irreversible. The point about the "[i]mplacable November weather" (Dickens [1853] 1971, p. 49) of the beginning of *Bleak House* is that it is not any particular November that is being specified, it is November as a quality rather than a place in a sequence. November is in fact the month that is mentioned most commonly in Dickens's novels, because of being the richest in associations of blear and dreariness, which was always a source of paradoxically energetic comic protest in Dickens. In *American Notes* we meet "four morsels of girls (of whom one was blind) [who] sang a little song, about the merry month of May, which I thought (being extremely dismal) would have suited an English November better" (Dickens [1842] 2000, p. 58). In *Bleak House*, much of the action of which seems to occur in an arrested November, the smoky and explosive associations of the month for

518   S. CONNOR

English readers are also drawn on, for example in narrating the arrival of the guy-like figure of Grandfather Smallweed in "a group at first sight scarcely reconcilable with any day in the year but the fifth of November" (Dickens [1853] 1971, p. 425). Of course, the Fifth of November is, like the Fourth of July, not a calendrical value but a topos, and a poetic one at that, which helps to smooth its passage from the alien order of number into the order of language, as in David Copperfield's efforts to compose a note to Agnes: "I even tried poetry. I began one note, in a six-syllable line, 'Oh, do not remember' – but that associated itself with the fifth of November, and became an absurdity" (Dickens [1850] 1999, p. 411).

When a fiction incorporates an actual date, it can act as a kind of authentication, suggesting the historicity of the events in question, but is also a kind of foreign body, which must somehow be digested to the parallel order of symbolic or represented time that novels always substitute for calendrical time, even, and perhaps especially, if they are representing what purport to be historical events. There are quite a few particular dates in Dickens's *American Notes*, and *Pictures from Italy*, but these are the mark of a kind of writing that advertizes itself as chronicle rather than fiction. Even here, Dickens can take a sort of comic revenge on the order of number through a kind of quantitative hyperbole, in the opening words of *American Notes*:

> I shall never forget the one-fourth serious and three-fourths comical astonishment, with which, on the morning of the third of January eighteen-hundred-and-forty-two, I opened the door of, and put my head into, a "state-room" on board the Britannia steam-packet, twelve hundred tons burden per register, bound for Halifax and Boston, and carrying Her Majesty's mails. (Dickens [1842] 2000, p. 9)

When actual historical dates make their appearance in Dickens's fiction, they are often subject to this kind of assimilation to the symbolic order of the novel, as public time gives way to private fantasy. This friction tends to be a feature of novels dealing with historically dateable events, like *A Tale of Two Cities*:

> The scene was Mr. Cruncher's private lodging in Hangingp. sword-alley, Whitefriars: the time, half-past seven of the clock on a windy March morning, Anno Domini seventeen hundred and eighty. (Mr. Cruncher himself always spoke of the year of our Lord as Anna Dominoes: apparently under the impression that the Christian era dated from the invention of a popular game, by a lady who had bestowed her name upon it.) (Dickens [1859] 1994b, p. 55)

Much more typical is the recoil from particular dates we find at the beginning of *Oliver Twist*, in the interest of creating a kind of mythic or chairotic time:

28 DATELINES 519

> Among other public buildings in a certain town, which for many reasons it will be prudent to refrain from mentioning, and to which I will assign no fictitious name, there is one anciently common to most towns, great or small: to wit, a workhouse; and in this workhouse was born; on a day and date which I need not trouble myself to repeat, inasmuch as it can be of no possible consequence to the reader, in this stage of the business at all events; the item of mortality whose name is prefixed to the head of this chapter. (Dickens [1838] 1994a, p. 3)

There are quite a few moments of coyness about dates in Dickens. "[O]ne November morning," writes *Bleak House*'s Esther Summerson, "I received this letter. I omit the date" (Dickens [1853] 1971, p. 74). Why does Dickens not simply omit mention of the omitted date, rather than informing us of the omission? The tell-tale locket in *Oliver Twist* is described by the work-house matron: "'It has the word 'Agnes' engraved on the inside,' said the woman. 'There is a blank left for the surname; and then follows the date; which is within a year before the child was born'" (Dickens [1838] 1994a, p. 274). There are two kinds of blankness here, the blank left for the surname and the date which is present but blanked out by the narrative. Though Pip can remember the "character and turn" (Dickens [1861] 2003, p. 3) of the inscriptions on his parents' tombstones, no mention is made of the dates of their decease. It is almost as though fiction maintains a deliberate blindness with respect to dates, equivalent to the principle that most people are una-ble to read writing that may present itself in dreams or, even more intrigu-ingly, that light switches never work in dreams (Hearne 1981, p. 98). Just as switching on the light is a kind of *felo-de-se* for the dreamwork, so the letting into fiction of the toxic force of purely serial succession might signify a kind of dissolution of its power to give shape to time.

Dates are, for the most part, external to literary works, precisely because dates belong to the order of death. This is highlighted when dates are pro-moted into titles, when there is a certain tendency to spell the dates out as words rather than numerals, as in Victor Hugo's *Quatrevingt-treize* (1874), George Orwell's *Nineteen Eighty-Four* (1949) or Adam Thorpe's *Nineteen Twenty-One* (2001), as though to neutralize the alienness of number by digesting it into the order of words. Philip Larkin's poem "MCMXIV" goes even further by spelling the date out in Roman numerals, which mimic those on memorials. Dickens does something similar at the beginning of *Pickwick Papers*, where we read:

> That punctual servant of all work, the sun, had just risen, and begun to strike a light on the morning of the thirteenth of May, one thousand eight hundred and twenty-seven, when Mr Samuel Pickwick burst like another sun from his slum-bers, threw open his chamber window, and looked out upon the world beneath. (Dickens [1837] 1972, pp. 72–73)

520   S. CONNOR

The trial of Pickwick makes much play with the specificity of dates, which stands as one of the characteristics of legal discourse, coming into comic collision with fictional discourse:

> Let me read the first: – "Garraway's, twelve o'clock. Dear Mrs. B. – Chops and Tomata sauce. Yours, PICKWICK." Gentlemen, what does this mean? Chops and Tomata sauce! Yours, Pickwick! Chops! Gracious heavens! and Tomata sauce! Gentlemen, is the happiness of a sensitive and confiding female to be trifled away, by such shallow artifices as these? The next has no date whatever, which in itself is suspicious. "Dear Mrs. B., I shall not be at home till to-morrow. Slow coach." And then follows this very remarkable expression. "Don't trouble yourself about the warming-pan." The warming-pan! Why, gentlemen, who *does* trouble himself about a warming-pan? (Dickens [1837] 1972, p. 563)

The most significant date in the trial is the unspecified but "particular morning in July last" on which Pickwick is held to have proposed to Mrs Bardell. Naming the day is put into play in different ways in the testimony of Susannah Sanders:

> Susannah Sanders was then called, and examined by Serjeant Buzfuz, and cross-examined by Serjeant Snubbin. Had always said and believed that Pickwick would marry Mrs. Bardell; knew that Mrs. Bardell's being engaged to Pickwick was the current topic of conversation in the neighbourhood, after the fainting in July …. Thought Mrs. Bardell fainted away on the morning in July, because Pickwick asked her to name the day; knew that she (witness) fainted away stone dead when Mr. Sanders asked *her* to name the day, and believed that everybody as called herself a lady would do the same, under similar circumstances. (Dickens [1837] 1972, p. 571)

The point here once again seems to be to mobilize in order to neutralize the friction between number and word (unless perhaps it is the other way round). Dates belong to the order of the law rather than of fiction and, especially in the matter of wills, dates are caught up in calculation, contracting and transaction; but the play with the date here is what makes it clear that this kind of legal discourse is in fact itself extravagantly fictional in method and intent.

If dates offer a way of keeping track of one's life, they can often dizzy and disorientate too. Phil's method of calculating his own age in *Bleak House* involves a labour that is at once numerical and number-blind:

> "I was just eight," says Phil, "agreeable to the parish calculation, when I went with the tinker …. That was April Fool Day. I was able to count up to ten; and when April Fool Day come round again, I says to myself, 'Now, old chap, you're one and a eight in it.' April Fool Day after that, I says, 'Now, old chap, you're two and a eight in it.' In course of time, I come to ten and a eight in it; two tens and a eight in it. When it got so high, it got the upper hand of me; but this is how I always know there's a eight in it." (Dickens [1853] 1971, p. 421)

Mr Dick's uncertain grasp on time and memory in *David Copperfield* is expressed through his simultaneous fixation upon and distraction by one special date, that of King Charles's execution:

> "The first time he came," said Mr. Dick, "was – let me see – sixteen hundred and forty-nine was the date of King Charles's execution. I think you said sixteen hundred and forty-nine?"
>
> "Yes, sir."
>
> "I don't know how it can be," said Mr. Dick, sorely puzzled and shaking his head. "I don't think I am as old as that."
>
> "Was it in that year that the man appeared, sir?" I asked.
>
> "Why, really" said Mr. Dick, "I don't see how it can have been in that year, Trotwood. Did you get that date out of history?" (Dickens [1850] 1999, p. 282)

But, precisely because of the deathly nature of their stubbornly neutral seriality, dates can actually be taken up into projective fantasies of time, as in John Chivery's funereal *rêverie* in *Little Dorrit* about his happy-ever-after afterlife with Amy Dorrit:

> they would glide down the stream of time, in pastoral domestic happiness. Young John drew tears from his eyes by finishing the picture with a tombstone in the adjoining churchyard, close against the prison wall, bearing the following touching inscription: "Sacred to the Memory Of JOHN CHIVERY, Sixty years Turnkey, and fifty years Head Turnkey, Of the neighbouring Marshalsea, Who departed this life, universally respected, on the thirty-first of December, One thousand eight hundred and eighty-six, Aged eighty-three years. Also of his truly beloved and truly loving wife, AMY, whose maiden name was DORRIT, Who survived his loss not quite forty-eight hours, And who breathed her last in the Marshalsea aforesaid. There she was born, There she lived, There she died." (Dickens [1857] 1967, p. 256)

Another way of turning the fatality of dates to account is to give them the force of a kind of necessity, as we commonly do in celebrating anniversaries. Sometimes this tallies, or at least dallies, deliciously with the deathliness of date-recurrence, turning blank arbitrariness into poetic potency, as in the murder story told by Solomon Daisy in *Barnaby Rudge*:

> The crime was committed this day two-and-twenty years – on the nineteenth of March, one thousand seven hundred and fifty-three. On the nineteenth of March in some year – no matter when – I know it, I am sure of it, for we have always, in some strange way or other, been brought back to the subject on that day ever since – on the nineteenth of March in some year, sooner or later, that man will be discovered. (Dickens [1841] 1973, p. 58)

522   S. CONNOR

If there is a fondness in Dickens for things that are "out of date and out of purpose," like the Clennams' house in *Little Dorrit* (Dickens [1857] 1967, p. 85), characters can also find a kind of stability and purpose through the periodicity of dates, which, far from exposing them to time, can seem to immunize them from it, as in the careful accountancy of Tim Linkinwater in *Nicholas Nickleby*:

> It was a sight to behold Tim Linkinwater slowly bring out a massive ledger and day book, and, after turning them over and over, and affectionately dusting their backs and sides, open the leaves here and there, and cast his eyes, half mournfully, half proudly, upon the fair and unblotted entries.
>
> "Four-and-forty year, next May!" said Tim. "Many new ledgers since then. Four-and-forty year!" (Dickens [1839] 2008, p. 471)

In fact, a certain kind of date-fetishism can become a way of resisting the erosive arbitrariness of seriality, nowhere more so than in the superstitious determination of Mrs Badger in *Bleak House* never to allow variation in her personal calendar of attachments:

> "I was barely twenty," said Mrs Badger, "when I married Captain Swosser of the Royal Navy. I was in the Mediterranean with him; I am quite a Sailor. On the twelfth anniversary of my wedding-day, I became the wife of Professor Dingo."
>
> ("Of European reputation," added Mr Badger in an undertone.)
>
> "And when Mr. Badger and myself were married," pursued Mrs Badger, "we were married on the same day of the year. I had become attached to the day."
>
> "So that Mrs. Badger has been married to three husbands – two of them highly distinguished men," said Mr. Badger, summing up the facts; "and each time upon the twenty-first of March at Eleven in the forenoon!"
>
> We all expressed our admiration. (Dickens [1853] 1971, p. 225)

The skirmishing which takes place in Dickens's writing with regard to the question of dates is part of a much larger and tenser story, of the tense interpenetration of language and number that became ever more a feature both of writing and of social-symbolic life during the nineteenth century. The phenomenology of number humanizes the alien and inhumanly indifferent order of pure number, so that Mrs Badger can become attached to the day on which she is accustomed to get attached. The novel both accommodates us to living in and by numbers and also offers the fantasy that there might be some other mode of living in time, as hinted at with sad ridiculousness by Mr Dick: "'I suppose history never lies, does it?' said Mr. Dick, with a gleam of hope" (Dickens [1853] 1971, p. 282). Keeping track of the date is a task that is always divided between Mr Dick and Mr Dickens.

Dates are an important part of the cultural phenomenology of novel reading not just because novels have sometimes been produced serially, but because of the contingent fact that novels cannot usually be read in one sitting, meaning that the coordination of the patterns of reading and resumption with the sequences of day and night, work and leisure, week and weekend, is always a factor of novel-time. There is always a relation of "contre-temps" between real and represented time (Connor 1996), as the tempo of the narrative is lined up and counted off against the external time-signature of the reading experience. And there is a social dimension to this too. Dates are an increasingly important part of the way in which we measure literary history and our relationship to it. Reciprocally, the recurrence of dates in anniversaries, centenaries and other markers of cyclical recurrence is perhaps one of the literary features of our timekeeping, in the way it imparts a kind of rhyme-scheme to time, which allows for sameness and difference to coexist.

MODULATIONS

The kind of mathematics that makes sense of this structure of temporal recurrence is known as modular arithmetic, first formalized by Carl Friedrich Gauss (1777–1855) in his *Disquisitiones Arithmeticae* (1801), though its principles have been familiar to casters of horoscopes, makers of orreries and readers of analogue clocks for centuries. Knowing that it will be 1 o'clock three hours after 10 o'clock means being able to perform a modulo 12 calculation. Working out what day Christmas will fall on involves a polyrhythmic coordination of modulo 7 and modulo 28, 29, 30 and 31 cycles. In fact there is no system of arithmetic that does not make use of some such cycle of recurrence. Decimal arithmetic is modulo 10 arithmetic and binary arithmetic is modulo 2 arithmetic. Decimal arithmetic shares a recurrence property with modulo 10 arithmetic, and binary arithmetic with modulo 2 arithmetic: if you successively add 1 to a decimal integer, its ones digit will increase up to 9 and then become 0, so change modulo 10. The digits of a number expressed in binary form will change modulo 2. In the "wrap-around" arithmetic of modulo calculations, what matters are not equalities but congruences, numbers that seem to rhyme with each other, as 3 and 15 do in a 24-hour clock when converted to a 12-hour system.

The most elementary periodicity of any modular system is binary, which enacts the oscillation between event and structure that governs both temporality and representation. In this two-stroke rhythmic engine, each moment breaks out from the preexisting continuum of time, but is then immediately—one, two, buckle my shoe—absorbed back into that continuum. Every number is a new and absolutely unique appearance in the number series, but it can only be so because every number has exactly the same value as every other number; so in order to be a number at all, it must be another number 1. That primary number itself cannot be understood as a number except by reference to another number with exactly the same value, so that 1 cannot be 1 until it is 1 no longer, because another 1 has come about and there has been 2.

In the business of poetic rhythm, literary numbers are much taken up with the difference in value between the odd and the even. We think of completed cycles as even, or self-identical: we round numbers off, rather than odding them out. Even numbers offer the satisfaction of the completed sequence, like the diastolic relaxation following the systolic contraction. The even seems to signify the known, the decided, the rounded-off, and seems to promise the ideal alignment of the orders of the real and the rational: you need double-entry book-keeping to balance accounts. And yet there is a kind of terror in this abolition of all error or remainder, for it amounts to a desire for the end of all difference, a big crunch in which all the unresolved adventure of things coils back into itself. And yet the number 2 may be regarded as an odd number, since it represents the unresolved exteriorization of the One into the multitude, the entry of the catastrophe of number itself and what Christian theology calls *economy*, or the pluralising of the deity as three divine persons, into the immaculate self-sufficiency of the Godhead. Once there was the number 2, the universe must always have had to choose between the left-hand and the right-hand branch.

This is in fact the reason for the mystic powers of the number 3, the number that aims to invert the unbalancing effects of the number 2, to advance through time while leaving nothing behind, and thus to subtract the addition to the One made by 2 and to bring the One home to itself, as though it had never gone out. If the number 1 is occurrence, and the number 2 is recurrence, then we would need the word *decurrence* (not far from the word *discourse*) for the number 3. Perhaps all obsessive-compulsive routines are an attempt to master time through this logic, a logic one must equally struggle to avoid being mastered by. And so, perhaps all human enactments of this desire to leave, not only no stone unturned, but no stone not returned to its original state, share in the surrogate death-drive of the millenarian zealot, which "lies in bringing everything down to the number one, which tolerates no one and nothing beside itself" (Sloterdijk 2009, p. 96), and replays vicariously the pang of regret the Deity must instantly have felt at the first moment of Creation.

In the one thing after another of pure succession, no mode of counting would be possible—or rather, there would be counting with no numerical succession or accumulation, the "one and one and one and one and one and one and one and one and one and one" with which Alice is confronted by the White Queen in *Through the Looking Glass* (1871) (Carroll 2009, p. 226). All counting involves a tangled tale of *reculer* and *sauter*, in which, as soon as there is 2 there can have been 1 (you do not have your first birthday until a year after you have been born, which only then is confirmed as your birthday). The paper-chaining of the number line, in which each number is no sooner completed than it already implies the next in the sequence, is already a kind of music, which makes for the patterns of tension and resolution that constitute prosody, or "numbers" as it was still sometimes known in the nineteenth century.

There is something intolerable, because absurd, either surplus or residue, about an uncompleted cycle, a surd being an irrational quantity that cannot be rendered as a finite decimal. Because it can never end, or at least never be stated, the cube root of 9 can never in fact have taken place in time, being an eruption of the infinite into finite series. Latin *surdus*, deaf, was used to render Greek ἄλογος, *alogos*, irrational and therefore unspeakable, through the interposition of Arabic *açamm*, deaf, in *jaðr açamm*, surd root. A surd can never be spoken or spelled out, it is a void or stammer in representation. And yet, there is a matching intolerability in anything that has come definitively to an end. "What are the odds?" we ask of an event that is yet to come. Hearing of the death of Antony, Cleopatra laments

> The soldier's pole is fallen; young boys and girls
> Are level now with men; the odds is gone,
> And there is nothing left remarkable
> Beneath the visiting moon. (Shakespeare [1623] 2006, pp. 268–69)

The oddity of the phrase *the odds*, which was employed simply to refer to any disparity between two quantities, an understanding preserved in the expression "it makes no odds," is that it should itself seem to hesitate between the condition of singular and plural, the tick and the tock of the counting clock.

Frequency

Alfred North Whitehead argues that "[w]e are comparing objects in events whenever we can say, 'There it is again.' Objects are the elements in nature which can 'be again'" (Whitehead 1920, pp. 142–43). There can only be finite and countable numbers because we understand each number to be exactly congruent to every other number, and therefore an instance of this recognition principle.

This rhyming recurrence is known to theorists of information as redundancy, defined as the total amount of information transmitted in a message minus the smallest amount that is strictly necessary for its transmission. In ordinary usage the redundant is the superfluous or the inessential, though until the eighteenth century it tended to mean positive abundance rather than useless excess (Connor 2011). But for information theorists redundancy is in fact a superfluity that is essential to any system of relations, which, in order to be recognized and function as a system, must have repeatable or non-unique elements. One may say that redundancy, which literally means the turning back of a wave, or re-undulation, involves not just the recognition or recurrence of elements but also the system's own reference, or turning back to, its own systematic character. Repetition always tends towards the recursive, turning in on the system of which it is a part, and which it thereby brings to notice. This is why it can be so uncomfortable to be asked to repeat your name, if it has been misheard, or perform your signature again if your first attempt does not seem recognizable or convincing. Repeating something which is usually (that is to say, repeatedly) an unwilled or spontaneous performance enjoins a kind of awareness of or attention to its way of being, rather than the raw fact of its being, turning an event into an intention.

In constituting this kind of self-reference, periodic recurrence in any system, whether numerical or verbal, pushes it inwards towards autopoeisis, rather than outward towards the contingency of the external world. Redundancy allows for checking and self-reinforcement in any communication which is subject to noise or randomness and so is equivalent to predictability. Cultures, languages, dialects, lyric poems, hurricanes and physiognomies all depend on redundancy to remain in being.

Redundancy cannot however be identified with order as such. Absolute redundancy—the absorbed self-equivalence without remainder of God before the Creation for example—would allow for no information at all, just as, at the other end of the scale, the absolute maximum of noise, in which no patterns could be discerned or predicted, would also reduce information to zero. Information occurs as a ratio of redundancy to noise: birthdays and deathdays seem significant, for example, largely because the actual and possible days are in a ratio of 1:366. For me to celebrate my birthday every time Thursday came round, or to go in for Humpty Dumpty's idea of "un-birthday presents" (Carroll 2009, p. 189), would make the whole event seem much more, well, workaday.

We may say that the implicit ratio of rarity to predictability is an important part of the heightened expectation of pattern or self-resembling orderliness that is a feature of the works that tend to be regarded as literary. Literary texts are iterative or periodic texts, texts that keep the beat in relation to themselves, turning external temporalities into internalized tempi. You can only keep time if there is something that keeps threatening to escape from it (an escape which is actually time itself), so keeping time is really keeping time from happening, even if it also discloses the background fact of that happening. For Michel Serres, all narrative, from the metanarrative of Genesis, to the narrative of any beginning or emergence, depends on a ratio between event and repetition, or noise and redundancy, rupture and restoration of equilibirum (Serres 2011, p. 110). Perhaps, as I have suggested in my *Living by Numbers*, the most inclusive definition of literature is language that has a high iterability index, or probability of being reread (Connor 2016, p. 170). This is why the mere act of subjecting something to repeated rereadings, for purposes other than consultation, can be enough to focus attention on its way of being in self-resemblance and so make it seem like literature.

The fact that literature is systemic language, the means whereby language realizes, in both senses, its own systematicity, makes it congruent with the patterned self-reference of mathematics. This is a convergence or co-agitation of congruences, the self-resemblance of numbers rhyming with the self-resemblance of words. The fact that words and numbers are external to each other's system, so that neither can be simply reduced to the other (and indeed each would in fact be the death of the other) is precisely what makes their congruence significant, and able to seem so repeatedly new. The divergence between congruence and identity is both what time lets in and what allows for time. Perhaps we may think of literature and mathematics as the different frequencies on which time transmits, to itself.

REFERENCES

Anderson, Benedict. 2006. *Imagined Communities: Reflections on the Origin and Spread of Nationalism*. London: Verso.

Bartky, Ian R. 2007. *One Time Fits All: The Campaigns for Global Uniformity*. Stanford: Stanford University Press.

Beckett, Samuel. 1965. *Proust and Three Dialogues with Georges Duthuit*. London: Calder and Boyars.

Carroll, Lewis. 1973. *Complete Works*. London: Nonesuch Press.

———. (1865, 1871) 2009. *Alice's Adventures in Wonderland and Through the Looking-Glass: And What Alice Found There*. Edited by Peter Hunt. Oxford: Oxford University Press.

Connor, Steven. 1996. "Reading: The Contretemps." *Yearbook of English Studies* 26: 232–48.

———. 2011. "The Poorest Things Superfluous: On Redundancy." http://steven-connor.com/redundancy.html. Accessed 17 December 2017.

———. 2016. *Living by Numbers: In Defence of Quantity*. London: Reaktion.

Dickens, Charles. (1857) 1967. *Little Dorrit*. Edited by John Holloway. Harmondsworth: Penguin.

———. (1853) 1971. *Bleak House*. Edited by Norman Page. Harmondsworth: Penguin.

———. (1837) 1972. *The Posthumous Papers of the Pickwick Club*. Edited by Robert L. Patten. Harmondsworth: Penguin.

———. (1841) 1973. *Barnaby Rudge*. Edited by Gordon Spence. Harmondsworth: Penguin.

———. (1838) 1994a. *Oliver Twist*. Edited by Steven Connor. London: J.M. Dent.

———. (1859) 1994b. *A Tale of Two Cities*. Edited by Norman Page. London: J.M. Dent.

———. (1850) 1999. *David Copperfield*. Edited by Paul Bailey. Oxford: Oxford University Press.

———. (1842) 2000. *American Notes: For General Circulation*. Edited by Patricia Ingham. London: Penguin.

———. (1861) 2003. *Great Expectations*. Edited by Charlotte Mitchell. London: Penguin.

———. (1839) 2008. *Nicholas Nickleby*. Edited by Paul Schlicke. Oxford: Oxford University Press.

Forster, John. 1876. *The Life of Charles Dickens*. 2 Vols. London: Chapman and Hall.

Hearne, Keith. 1981. "A 'Light-Switch Phenomenon' in Lucid Dreams." *Journal of Mental Imagery* 5: 97–100.

Joyce, James. (1939) 2012. *Finnegans Wake*. Edited by Robbert-Jan Henkes, Erik Bindervoet, and Finn Fordham. Oxford: Oxford University Press.

Larkin, Philip. 1968. *Collected Poems*. Edited by Anthony Thwaite. London: Marvell Press/Faber and Faber.

Serres, Michel. 2011. *Musique*. Paris: Le Pommier.

Shakespeare, William. (1623) 2006. *Antony and Cleopatra*. Edited by John Wilders. London: Arden Shakespeare.

Sloterdijk, Peter. 2009. *God's Zeal: The Battle of the Three Monotheisms*. Translated by Wieland Hoban. Cambridge: Polity.

Whitehead, Alfred North. 1920. *The Concept of Nature*. Cambridge: Cambridge University Press.

Woolf, Virginia. (1925) 2000. *Mrs Dalloway*. Edited by David Bradshaw. Oxford: Oxford University Press.

CHAPTER 29

The Metaphor as an Equation: Ezra Pound and the Similitudes of Representation

Jocelyn Rodal

In "Vorticism," Ezra Pound tells us that "In a Station of the Metro" first came to him as an equation: "I found, suddenly, the expression. I do not mean that I found words, but there came an equation" (Pound [1914d] 1974, p. 87). Elsewhere, he writes that "[p]oetry is a sort of inspired mathematics, which gives us equations" (Pound 1910, p. 5). He describes equation as the seed for Imagism, his self-proclaimed movement, declaring that "[b]y the 'image' I mean such an equation" and explaining that, in defining Imagism, he aims particularly "to formulate more clearly my own thoughts as to the nature of some mystery or equation" ([1914d] 1974, p. 92; 1915b, p. 349). Pound is explicit that he uses this term in its mathematical sense— citing $(x-a)^2 + (y-b)^2 = r^2$ as an example—but he also makes it clear that the ramifications are literary: "Great works of art contain this... sort of equation. They cause form to come into being" ([1914d] 1974, pp. 91, 92). Equation, for Pound, offered an understanding of how things that look different— indeed, things that are different—might nonetheless disclose sameness. Abstract art and experimental poetry can seem foreign to common human experience. For Pound, however, they held hidden samenesses. As such, for Pound, equation offered to mediate poetic abstraction and concrete reality.

This essay examines, mathematically as well as literarily, what Pound's equations entail. My argument will turn to modern mathematical definitions of equality (which reach well beyond simple equalities between numbers, able to link far more various and complicated ideas) in order to analyze Pound's use of similitude and sameness. Modern mathematical definitions of equality

J. Rodal (✉)
Princeton University, Princeton, NJ, USA

© The Author(s) 2021
R. Tubbs et al. (eds.), *The Palgrave Handbook of Literature and Mathematics*, https://doi.org/10.1007/978-3-030-55478-1_29

529

can offer us an understanding of sameness's form—that is, of the patterns attendant on its application. In characterizing the patterns born of similarity and commonality, equation can offer an understanding of metaphor in terms of both mimesis and abstraction. Equality, for Pound, promised to model mimetic representation and at the same time to explain how the most abstract art could designate meaning. As such, equality subtended the potentialities of representation itself.

WRITING EQUATION

"Equal, adj.," according to the *Oxford English Dictionary*:

> 1.a. Of magnitudes or numbers: Identical in amount; ... Of things: Having the same measure; identical in magnitude, number, value, intensity, etc.
> 2.a. Possessing a like degree of a (specified or implied) quality or attribute; on the same level in rank, dignity, power, ability, achievement, or excellence; having the same rights or privileges.

The *OED* separates the more mathematical usage from the more sociocultural one, but the concept remains the same. For two things to be equal, they do not need to be identical in all ways, but they do need to be the same in some aspect (or aspects). That is true irrespective of whether those aspects be number or dignity, magnitude or human rights.

When Pound describes "In a Station of the Metro" as an equation he declares, literally, that it states an equality, and as such he directs attention to some sameness or commonality that it locates. "In a Station," that odd exemplar of modernist brevity, is exactly two lines long:

> The apparition of these faces in the crowd;
> Petals on a wet, black bough. (Pound [1916] 1990, p. 111)[1]

The faces in the crowd seem to resemble petals on a wet, black bough. Pound identifies the faces with petals. He represents the faces as petals. In a sense, this is the definition of metaphor: two differing things enter into analogy as one is named as the other. In the *Poetics*, Aristotle wrote that "the successful use of metaphor is a matter of perceiving similarities" ([circa 335 BCE] 1996, p. 37), a passage that we know Pound attended to with interest.[2] "In a Station" offers us a "successful use of metaphor" insofar as it offers us an especially surprising similarity. We do not expect faces in a crowded Metro station to share very much in common with petals on a wet, black bough, but they do.

At the same time, "In a Station" avoids stating the relationship between the faces and petals outright. Pound omits the connecting words that would form the pivot point of the entire poem, and the central semicolon wordlessly hinges the two phrases together, setting them on parallel planes without explained connection. A more typical metaphor would pin the phrases together with "are," or any number of other relating verbs or prepositions. Instead, the semicolon leaves readers with a relation that is at once ostentatious and unspoken, at once known and undisclosed. This substitution is never named as such, opening a crevice in which "In a Station" stages and investigates the relationship between representation and likeness.

In "Vorticism," when Pound describes the metaphor central to "In a Station" as an equation, he first recalls that, in that Metro station, he saw a repeating commonality:

> Three years ago in Paris I got out of a "metro" train at La Concorde, and saw suddenly a beautiful face, and then another and another, and then a beautiful child's face, and then another beautiful woman, and I tried all day to find words for what this had meant to me, and I could not find any words that seemed to me worthy, or as lovely as that sudden emotion. And that evening, as I went home along the Rue Raynouard, I was still trying, and I found, suddenly, the expression. I do not mean that I found words, but there came an equation ([1914d] 1974, pp. 86–87).

Pound emphasizes a sense of the same thing happening again and again. He renders the faces of several distinct people as, in his perception, iterations of one common quality of beauty. He describes struggling to represent that correspondence, finding himself unable to locate words that describe the pattern of "another and another." Here Pound seeks not only a description, but words that are "as lovely as," and "worthy" of, the emotion evoked by the original experience. "Satisfaction lay not in preserving the vision, but in devising with mental effort an abstract equivalent for it" (Kenner 1971, p. 184).[3] Pound perceived a series of repeating beauties, and then sought a poetic beauty that could sit alongside the others, that was equal to them. He settled, finally, on an expression that he explicitly tells us did not itself—at first—consist of words. Instead, he found "an equation." Pound saw faces in the crowd; he deemed them equal to petals on a wet, black bough.

If we read this equality as not only an account of Pound's writing process but also an illustration of Pound's theory of representation, we can see a whole network of equations. Pound's original experience in the Metro station revolved around the recognition of a series of equal beauties. Then he attempted to set his experience equal to written lines, to the poem itself. In turn, he frames the poem's central figures as equal to one another. The poem's two lines become the two sides of one equation. With his description of "In a Station" Pound emphasized not only that the commonality staged by this poem is mathematical, but that the mathematical understanding of commonality informs poetic composition, readerly interpretation, and representation generally. Representation, after all, poses one thing as another, and here Pound frames that act of representation as one of the equation.

532   J. RODAL

When he declared that "In a Station" first came to him as an equation, Pound also described the poem's relational and syntactic process. The semicolon grammatically renders these phrases parallel to map "faces" onto "petals," and, indeed, a semicolon is a kind of grammatical equals sign: it joins two sentences in a homology that requires no further explanation (or, it places listed phrases in grammatically equal positions). The two spare lines of "In a Station" also visually mimic an equals sign, floating parallel just above and below each other in a kind of textual illustration of "=." The semicolon itself does something similar, because with two marks lying one on top of the other, the two signs are typographically related (; to =). In fact, an earlier version of the poem offers even more visual equivalence. As it was first published in 1913, Pound uses a colon, set off by extra spacing:

> The apparition   of these faces   in the crowd :
> Petals   on a wet, black   bough . (Pound 1913b, p. 12)

The colon offers more visual equality, with identical marks floating one above the other. The colon also offers a more direct link, since grammatically what follows a colon typically offers a specification or enumeration of whatever precedes it. Yet the semicolon that Pound later settled on establishes more complete syntactical equality, since the two independent clauses before and after a semicolon carry equal weight, reversible in their identity. And, as Daniel Albright has pointed out, Pound's images characteristically insist on this kind of reversibility (Albright 1997, pp. 140–41).

Pound's theories of representation place great emphasis on visual, pictorial writing, and that is also intrinsic here. The sign of equality ($=$) originated with a succinct pictorial equality in mind. As Robert Recorde, inventor of the sign, wrote in 1557: "to avoide the tedious repetition of these woordes: is equalle to: I will sette ... a paire of paralleles, or Gemowe [twin] lines of one lengthe, thus: $=$, because noe .2. thynges, can be moare equalle" (Recorde 1557, p. 238). By design, the equals sign not only signifies equality but visually illustrates and exemplifies equality. The equals sign is, in fact, an ideogram—the directly signifying mark that so fascinated Pound. As Pound defined it, an ideogram (unlike written English) "does not try to be the picture of a sound, but it is still the picture of a thing; of a thing in a given position or relation, or of a combination of things. It means the thing or the action or situation, or quality germane to the several things that it pictures" (Pound [1934] 1960, p. 21). Mathematical notation, in fact, is often ideogrammic. Mathematical and logical marks such as $=, >, <, \approx$, or $\rightarrow$ are all designed to directly, pictorially embody their meaning, creating a firm and immediate link between signifier and signified that is very rare elsewhere in written language.[4]

For Pound, the relations between the parts of an ideogram promise to visually mimic the relations between the parts of that ideogram's meaning. That is, it is important that an ideogram is not just a holistic picture of its meaning, but that it can also provide a visual understanding of the meaning's various components, and of how those components relate to each other. In the case of "=", the relation between the parts of the ideogram is itself the entire meaning of the ideogram, because "=" *is* a relation. Here two little lines stand parallel, neither intersecting nor coinciding but coexisting, side by side, as two things distinct but equal. Such illustrating of how components connect, disconnect, and jostle among each other allows ideograms to make their grammars visible. Pound writes that ideograms particularly highlight the "relation" or "combination" among things (Pound [1934] 1960, p. 21). Meanwhile, mathematical ideograms such as $=, >, <, \approx$, or $\rightarrow$ particularly tend to address modes of relation—unlike non-ideogrammic marks such as 6 or x, which indicate things rather than relations between things. Relations, distilled from the things they relate, can seem terribly abstract, but for Pound ideograms promised to render them visible. And while equality is an abstract mathematical relation, in "In a Station" Pound gave it visual and human immediacy.

T. S. Eliot famously argued that "[t]he only way of expressing emotion in the form of art is by finding an 'objective correlative'; in other words, a set of objects, a situation, a chain of events which shall be the formula of that *particular* emotion" (Eliot [1920] 2009, p. 67). The objective correlative is a "formula" that works via equality between symbol and affect, and Eliot describes it in terms of equivalence four times over. He describes it as an "exact equivalence." He writes that the objective correlative should offer an "adequate equivalent," an "objective equivalent." It is supposed to correlate two things with "immediate," "inevitab[le]" reliability, and equality promises that absolute correlation (Eliot [1920] 2009, pp. 67, 68, 67). The objective correlative offers up its meaning with all the directness of an ideogram, and it does so, Eliot tells us, by becoming "equivalent" to that meaning.

Eliot speaks particularly of equivalence, while Pound refers most often to equation. There are differences here. With "equivalence" Eliot seems to highlight the substitutability and interchangeability of the objective correlative and its attendant emotion, while with "equation" Pound draws further attention to the larger phrase or relational system which articulates an equality between two terms. Yet nonetheless we have accumulated, to this point, a collection of synonyms and near-synonyms—equivalence, equality, equation, similarity, sameness, substitutability—and this list could continue. What needs to be examined is the commonality among these commonalities. To do that, we will need a broader and deeper understanding of mathematical equality: one that reaches beyond and outside of number, because "In a Station of the Metro" is not so straightforward, nor so finalistic, as the grade-school arithmetic problems that "=" most readily calls to mind. Faces and petals are not numbers, and "In a Station" is not $2+2=4$. Mathematicians of the modernist era were developing exactly these ideas, studying equality as a further-reaching articulation of commonality.

534   J. RODAL

EQUALITY AND THE MODERNIST EQUIVALENCE RELATION

Gottlob Frege's "On Sense and Reference," the 1892 essay about meaning and denotation that became foundational to the philosophy of language, begins not with language but, apparently, with mathematics. Frege's opening words are: "Equality gives rise to challenging questions which are not altogether easy to answer. Is it a relation?" (Frege [1892] 1997, p. 151). To analyze meaning Frege begins by interrogating what it means for something to be equal to something else. But, emphatically, here he does not refer to equality between numbers or values, but far more broadly to equality between any manner of things. "I use this word [equality] in the sense of identity and understand '$a = b$' to have the sense of 'a is the same as b'" (Frege [1892] 1997, p. 151). He examines equality to better characterize the relation by which words attach to their referents.

Words are not the same as what they describe. As theorists of language, we know that the relation between word and meaning is fraught and inexhaustibly complex. Yet as practitioners of language, we routinely treat it as simple. Asked to explain what a table is, we might point to a wooden surface standing on four legs. Asked, in turn, to explain what that four-legged surface is, we might simply call it a table. That process, whereby word and thing are reliably and immediately correlated one with the other, exemplifies Eliot's objective correlative in more banal surroundings: not Hamlet wrestling with his feelings about this mother, just a person naming or describing a thing. This process, whereby one thing directly stands for or stands in for something else, works through substitution and equality, insofar as we so often intuitively use a word as if it is the same as its meaning, forgetting so many complexities of signification that stand between the two.

Frege defines equality as identity, the state wherein two things are identical. Equality is most commonly imagined between numbers, while identity can exist between all manner of things, but equality and identity share something fundamental in common, because both assert commonality, both locate some sameness. When we say that two triangles are congruent—that they have exactly the same size and shape—that relation bears an intuitive unity with the equality that exists between numbers. For example, looking at the three triangles below, it feels natural to say that A and B are somehow equal to each other in a way that neither is to C (Fig. 29.1).

We could cut out A, rotate it slightly, and we would find that it fits exactly on top of B. That is, although A and B might differ in position and orientation, they are identical in size and shape. In geometry, we would thus say that A and B are congruent, that $A \equiv B$. Congruence is a relation of sameness (sameness of shape and size), and equality, too, is a relation of sameness (in arithmetic, sameness of numeric value). But for the vast majority of mathematical history, from the ancient Greeks up until the 1890s, mathematicians had to define these relations separately.[5] They had no well-defined way to speak about both equality and congruence at once. Whatever $x = y$ had in

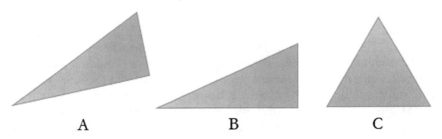

Fig. 29.1 Three triangles, two of which are congruent

common with A ≡ B, there was no precise way to talk about that commonality, no single definition that described the pattern of what those assertions of commonality shared in common.

It was around the turn of the twentieth century that modernist mathematicians built that kind of further-reaching definition of identity relations.[6] In 1884 Frege noted that "[t]he relationship of equality [*Gleichheit*] does not hold only amongst numbers. From this it seems to follow that it ought not to be defined specially for this case" (Frege [1884] 1997, p. 110). In 1889 Giuseppe Peano defined equality by a means that can also describe sameness more generally, naming the broader properties of reflexivity, symmetry, and transitivity as the only criteria for equality (Peano [1889] 1973, p. 109). A decade later David Hilbert published a definition of congruence that was similarly far-reaching, capable of also describing much of equality and identity (Hilbert [1899] 1971, pp. 10–11). Hilbert made a point of defining points, lines, and planes only as "distinct sets of objects" with "certain mutual relations" determined by his stated axioms, allowing his definition of congruence to apply to many mathematical systems beyond traditional geometry (Hilbert [1899] 1971, p. 3). In 1903 Bertrand Russell wrote about how Peano's properties offered to characterize what "[a]ll kinds of equality have in common" ([1903] 1996, p. 159). In 1926 Helmut Hasse gave a name to all these different kinds of equality: equivalence relation (in German, *Äquivalenzrelation*), which continues in general use today. Hasse stated directly that his definition of the equivalence relation was utterly nonspecific about the particular relations that might exemplify its properties, and that it was "indifferent to which meaning the signs possess"[7] ("*gleichgültig welche Bedeutung dabei den Zeichen ... zukommt*" [Hasse 1926, p. 17]), indicating instead more general form, structure, and behavior. Rather than defining any particular linkage such as congruence or numerical equality, Hasse used the new concept of an equivalence relation to articulate what a whole class of linkages shared in common, to specify their single form independent of their many meanings.[8]

What equality, equivalence, identity, and sameness have in common is, mathematically, a single form; that is, the way in which they relate things,

536   J. RODAL

the pattern of their use. Equality's form does not include its meaning—sameness—but does include the logical syntax that all kinds of sameness share in common. Equality's form in fact characterizes much broader kinds of similarity: the state of sharing something—anything—in common defines an equivalence relation. For example, the property of sharing the same surname with someone else is an equivalence relation, even though that property is not mathematical. Equivalence relations pinpoint the commonalities of identity, equality, congruence, similarity, and many other related relations, all under one single, global form, and they were only defined for the first time in the modernist era. Here, in fact, the mathematicians have brought us back to the *OED* definition of equality which I offered at the opening of this paper: equality describes the state of being alike *in some aspect*. This definition had existed well before mathematicians defined equivalence relations in their general form, and has been in use for hundreds of years in fields as disparate as mathematics and politics. Thus, in an important sense, when mathematicians defined equivalence relations at the turn of the last century they were only catching up to an older sociocultural idea. However, in giving that broader cultural understanding of equality a logical definition, mathematicians were the first to discover what it technically entailed. They were articulating the precise contours of something that, previously, had only been known intuitively.

It turns out that sameness, commonality, and equality, in their many varieties, have a unique syntax that is entirely their own. An equivalence relation can be completely defined by three properties laid out by Peano.[9] Imagine that "~" is some unknown relation between two things. Then "~" is an equivalence relation if and only if it possesses all three of the following features:

1. Reflexivity: $x \sim x$
2. Symmetry: If $x \sim y$, then $y \sim x$
3. Transitivity: If $x \sim y$ and $y \sim z$, then $x \sim z$

If we know that "~" has these three properties, then we know that "~" is an equivalence relation, no matter what x, y, or z are. And, if we know that "~" is an equivalence relation, then we always know that "~" is reflexive, symmetric, and transitive.

To take a less technical example, let's say that x, y, and z are three people named Mona, José, and Blair, and let's say that "~" is the property of sharing a surname. That property is reflexive (because Mona must have the same surname as herself), symmetric (because if Mona and José share a surname, then José and Mona share a surname), and transitive (because if Mona and José share a surname and José and Blair share a surname, then Mona and Blair must share a surname). Hence, it is an equivalence relation. Conversely, if we knew that "~" was an equivalence relation but didn't know what it described, we would know that it must be reflexive, symmetric, and transitive, no matter whether x, y, or z are a triangle, the number 2, or a kitchen table.

In the *Principia Mathematica*, Russell and Alfred North Whitehead lay out these three properties to define identity, and then assert that "no new definition of the sign of equality is required in mathematics. All mathematical equations in which the sign of equality is used in the ordinary way express some identity, and thus use the sign of equality in the above sense" (Whitehead and Russell [1910] 2009, pp. 22–23). Notice that Whitehead and Russell speak here not of "equivalence relations" but of "equality"; the former term had not yet been coined, and in its absence Whitehead and Russell were comfortable referring to this whole class of relations as equality, which, for them, marks simply the "express[ion of] some identity" ([1910] 2009, p. 23). Terms get used interchangeably here not because Whitehead, Russell, and others did not take care to be precise, but rather because the intersection among these terms is exactly what is being discussed. After the interventions of modernist mathematics, mathematical equality does not imply only numerical sameness; it can characterize commonality much more inclusively.

Equality promises a way to contemplate similarity without denying difference, but also without depending on difference. It is tempting to conceive of sameness simply as the absence of difference, and of difference simply as the state wherein things are not the same, but in fact each can be accounted for without any reference to the other. The mathematical definition of equivalence relations—a rigorous account of the properties of sameness—never makes any reference to either difference or inequality. (Difference, it turns out, is symmetric but not reflexive, and not necessarily transitive.) Sameness and difference are relations with their own syntaxes, their own unique attendant patterns, and although they are complementary, they are also self-sufficient in their own right.

Reflexive Tautology and Pound's "Is"

The move from numerical equality to the broader notion of an equivalence relation, from specific identity to general commonality, constituted a remarkable expansion in scope and a characteristically modernist shift in perspective. Modernist mathematicians and logicians such as Frege, Peano, Russell, and Whitehead were the first to abstract from particular equalities to define not only equality's semantic content (sameness) but also its syntactic form (reflexivity, symmetry, transitivity). Yet, in another sense, the understanding of equality's structure is much older than this; identity is too basic a concept for its characterization to have first come about with the modernists. As early as 300 bce, Euclid's first Common Notion stated that "[t]hings which are equal to the same thing are also equal to one another" (Euclid [circa 300 BCE] 2013, p. 2). That property (often called the Euclidean property) involves a

538   J. RODAL

combination of transitivity and symmetry, and, mathematically, if a relation is both Euclidean and reflexive then it must also be an equivalence relation in the modern sense. All that was really missing from Euclid's definition is reflexivity, the declaration that anything must always equal itself. Euclid might have simply deemed this fact too obvious to need pointing out. On the other hand, he may have considered equality a relation which one could only want to state between two things. To point out that one thing is equal to itself would have been, perhaps, not only a tautology but an absurdity, senselessly simple-headed.

The careful investigation of that which had previously seemed too basic to warrant serious attention was in fact crucial to modernism across fields. James Joyce rewrote the *Odyssey* as one mundane day in the life of an advertising agent while Virginia Woolf declared that we must "not take it for granted that life exists more fully in what is commonly thought big than in what is commonly thought small" (Woolf [1925] 1967, p. 107). That kind of foundational analysis—which great thinkers such as Euclid had skipped over for hundreds or thousands of years, knowing tacitly without caring to precisely analyze and clarify—was typical of modernism, which grew in large part from a commitment to the complexity of the simple. In some ways, my argument here is also a deliberate turn back to the complexity of the obvious.

Metaphor pivots on some measure of commonality shared by the two things being brought into comparison. As I. A. Richards writes, metaphors

> work through some direct resemblance between the two things, the tenor and vehicle, ... [or else] through some common attitude which we may (often through accidental and extraneous reasons) take up towards them both. The division is not final or irreducible, of course. *That we like them both* is, in one sense, a common property that two things share, though we may, at the same time, be willing to admit that they are utterly different. (Richards 1936, p. 118)

Metaphor locates some sameness residing in two distinct things. It springs from some commonality shared by tenor and vehicle. Richards traces how this commonality is sometimes straightforward, easy to locate, while at other times it may be intrinsically vague, difficult if not impossible to name. Nonetheless, it is there.

This commonality indicated by metaphor has the same syntactic form as $a = b$, because *any* commonality has the syntactic form of $a = b$. Let's examine a deliberately flat-footed example. If I say that "she is a lion," I might imply that she is fierce like a lion. (That interpretation is susceptible to argument: perhaps I imply instead that she is brave like a lion; perhaps I simply voice some nebulous shared feeling toward her and lions. Interpret whatsoever shared commonality you will. Regardless, the following analysis can apply to that commonality.) Shared fierceness is symmetric (if she and the lion are both fierce, then the lion and she are both fierce) and transitive (if she and the lion are both fierce, and if the lion and the tiger are both fierce, then she and the tiger are both fierce). This leaves reflexivity, the final defining property of equivalence relations and the one that mathematicians missed until the modernist era. Reflexivity does pertain: the lion should be exactly as fierce as the lion is fierce, and she should be neither more nor less fierce than she is fierce. Yet this last formulation seems perhaps even less natural to our common understanding of metaphor than do those above.

The reflexivity of identity—the very simple fact that a is a—marks some of the most haunting and characteristic lines of modernist poetry. It asserts itself aggressively and paradoxically in Gertrude Stein's "Rose is a rose is a rose is a rose" (Stein [1913] 1998, p. 395) and it girds the religious stabilities of much of Eliot's late poetry: "Because I know that time is always time / And place is always and only place" ([1930] 1968, p. 85); "the roses / Had the look of flowers that are looked at" ([1943] 1968, p. 176). While "a rose is a rose is a rose" begins as a senselessly simple-headed fact, with every repetition of "rose" the word sounds and looks weirder, leaving us increasingly estranged from the rose that we had previously taken for granted, as though the fact that the rose is a rose reveals that it is so much more and so much else. In "Ash Wednesday," Eliot underscores the fact that time is a complicated, paradoxical thing, and yet when he reminds us "time is always time" he grounds that poem with the universalities that we are nonetheless capable of knowing, and which somehow enclose complexities that seem unenclosable.

Pound's early poetry turns repeatedly to reflexivity as a good that merits reflection in and of itself, underlying and enabling both his deployments of metaphor and his characterizations of being. In Pound's essays, reflexivity is fundamental to his theories of representation. It supports his rhetoric of insistence, his characteristic emphasis and reemphasis of assertions.[10] The understanding and insistence that x is always x offers confidence in some universal truth while at the same time it makes space for the multiplicity of particulars. Pound argues for the explanatory value of a "presentative method," which "calls a calf a calf," because "[t]he presentative method is equity" (Pound 1913a, p. 662). He identifies a thing and then presents that thing, a pattern of reassertion that provides definite emphasis, but that also calls peculiar attention to the self-commonality of the thing itself, as if we must pause to affirm that the thing is indeed itself. In "Salutation the Third":

> The taste of my boot?
>   Here is the taste of my boot (Pound [1914a] 1990, pp. 75–76)

> Let us deride the smugness of "The Times":
> GUFFAW! (Pound [1914a] 1990, p. 75)

Lest we think Pound is being merely suggestive with the imperative to deride the smugness of *The Times*, he provides the actual manifestation of that derision, putting forward the act that he had already described. The "GUFFAW!" does match the derision promised, and Pound makes sure readers will be able

540   J. RODAL

to verify that fact. x does equal x, as is standard, as is always expected, yet he takes care to announce and verify that which we already knew to be true.

Elsewhere in Pound's poems, the complexity of the object seems only to further underline the importance of its reflexive identity:

> Your mind and you are our Sargasso Sea
>
> ...
> No! there is nothing! In the whole and all,
> Nothing that's quite your own.
>   Yet this is you. (Pound [1912b] 1990, pp. 57–58)

"Your mind" and "you" are not quite the same. But they are utterly intertangled, each defined by the other, and to combine them additively under a single "are" is to insist on a kind of tautology which "is"—when it identifies two things as one and the same—actually always entails and implies.[11] Pound emphasizes this inherent repetition when he follows "are" immediately with "our," as the homophone for "are" marks the repetition inherent in that term's meaning. The woman Pound writes about may be many things, but first and foremost she is herself. Outside of that, "there is nothing," "nothing that's quite your own. / Yet this is you." If Pound's deictic "this" does indicate the lady before him, that final line cannot possibly impart new information. Of course she is she. But he returns to where he started—"your mind and you," "this is you"—because the reflexivity of her selfhood is hauntingly important and maddeningly complicated even as it remains obvious. That applies, insistently, everywhere and always, "in the whole and all"—a quantifier for ubiquity that itself insists on its own reflexivity, in meaning as well as sound.

Such repetitions might seem even more characteristic of Stein's writing, for example, "Rose is a rose is a rose is a rose" (Stein 1913, p. 395). However, Stein's reiterations emphasize metonymic associations—the slippery mental contiguities by which things melt into things other than themselves ("Rose" calls to mind at once a flower and a woman's name, as well as a vibrant red and Shakespeare's "a rose / By any other name" [1597, 2.2.46–47]). Pound's reflexive repetitions definitely reveal semantic multiplicities, but they tend not to tolerate metonymic slippages such as Stein's poetry invites. If we try to perform similar readings of "In a Station" we tend to come up short, because that poem violates metonymic contiguities with its starkly surprising juxtaposition. The urban crowds evoked by the title and first line jar against the pastoral landscape of the second line. The technological modernity of the title calls to mind sooty locomotives and great grinding metal wheels: harsh indeed against fluttering petals after spring rain.[12] Moreover, the clipped precision of "In a Station" provides nearly no context by which metonymic associations can proliferate. Pound, in fact, showed no love for metonymy in his early work: criticizing the French Symbolists, he wrote that they "dealt in 'association' …. They degraded the symbol to the status of a word. They made it a form of metonymy" (Pound [1914d] 1974, p. 84).[13] Pound's repetitions insist instead on some stability: whereas Stein's reiterations so often estrange terms from their original referents, Pound's repetitions foreground constancy, drawing out an understanding of solidity communicated by terminological consistency. Across texts, Pound repeats the same words with striking uniformity, as when he writes so many times of his "watchword" of precision (see, for example: Pound 1913d, p. 126; 1915a, p.277; 1996, p.731; 1914d, p.89; and the entirety of 1913c). He does so with an insistence that forestalls prevarications about the term itself.

Often, Pound's tautologies depend on grammar and logic instead of semantic or oral repetition, for example in "Salvationists":

Let us apply it in all its opprobrium
To those to whom it applies. (Pound [1914b] 1914b, p. 100)

Perhaps the speaker could attempt to apply "it" to those to whom it did not apply, except that by definition then it would apply, since he had applied it. Pound is stating an obvious kind of logical fact which cannot be escaped any more than the truth of a tautology can be escaped. But the truth of a tautology is not its problem. No one denies the truth of tautologies. People deny, instead, their usefulness. Yet that which is true everywhere is everywhere pertinent, and Pound values the reflexive repetition intrinsic to tautology because it provides—very usefully—certainty, constancy, and the capacity to characterize the nature of the identities that objects, ideas, and even people intrinsically possess.

At the same time, Pound often seizes on these identities only to leave them prominently unspoken. The poems of *Lustra*, in particular, repeatedly interrogate forms of similitude, but they very often do so without verbs of identity or similitude. Frequently, as in "In a Station," a colon or semicolon displaces "is" or "is like." One way or another, the central "is" generally goes unspoken.[14] In "Vorticism," Pound explains this process:

"The footsteps of the cat upon the snow:
(are like) plum-blossoms."
The words "are like" would not occur in the original, but I add them for clarity.
(Pound [1914d] 1974, p. 89)

Pound adds the "are like" that he explicitly tells us would not be voiced, using parentheses as though to mark the unnecessary intrusion of that statement of semblance even as he also informs us that it is, indeed, the correct interpretation. "For clarity," then, Pound implies that we could fairly read such additions into "In a Station": the faces in the crowd (are like) petals on a wet, black bough. In explication he has translated the metaphor into a simile, as though to indicate that the similitude of simile is already embedded in metaphor: a characterization of metaphor with which theorists as disparate as Richards and Aristotle might concur.

542   J. RODAL

In "Vorticism," Pound continues: "In a poem of this sort one is trying to record the precise instant when a thing outward and objective transforms itself, or darts into a thing inward and subjective" (Pound [1914d] 1974, p. 89). He remembers the faces in the crowd while he imagines the petals on a bough: the former is outward, objective, the latter inward, subjective.[15] The relation between the two, between objective and subjective, between tenor and vehicle, is the heart of the poem itself. Pound's theory of metaphor reveals itself most fully here in his claim that the outward, objective thing "transforms itself." The tenor does not transform into the vehicle, it transforms itself into the vehicle. That reflexive "itself" implies that, in order for the metaphor to work, we must already believe in a kind of preexisting "is," because to talk about the identity of the faces with the petals is to presuppose the possibility of identity, which is to presuppose the identity of the faces with themselves. This reflexivity both precedes and prefigures the metaphor itself.

Pound, with his insistence on reflexivity, imagines a sign that is an identity, a metaphor that is an equation, a symbol that is its own referent. "In a Station" never speaks the "is," but that only works to underline how much it stands upon and also interrogates that absent, unspoken verb. This is not tautology so much as an examination of the general relational forces that render tautologies true. With his interrogation of the very possibility of locating sameness in "In a Station," Pound locates and instantiates an intrinsic form embedded in any metaphor. "Form," used in this sense, is profoundly abstract. It does not describe visible or audible aspects of the poem, nor of what the poem describes. It refers instead to the patterns attendant on the poem, and to the patterns attendant on that which it describes.[16] Such patterns have ramifications in our world, some of them accessible to the senses, but the patterns themselves can never be directly pointed at except via thought and imagination. I can reason out the fact that x must always equal x, but I can never directly see or hear reflexivity.

Daniel Albright writes that "In a Station" "summarizes the luminous faces on the subway platform and the petals on the bough, glistening with wetness, into a single, highly contrasted linear arrangement. Pound's term *non-representative* is somewhat misleading: the poem is *bi-representative*, abstract because doubly concrete" (Albright 1997, p. 139). The concreteness of Pound's sensuous first and second lines only drives home the fundamental abstraction of the relation between them. Here Pound's equality is abstract in the same way that mathematics is abstract. Mathematics has a remarkable relationship with reality, because it is both a maddeningly ethereal realm of thought and form, and also an undeniably descriptive tool that reliably predicts physical phenomena. It usefully describes the physical world without residing in that world. The semicolon at the center of "In a Station" cannot be found in nature in the way that petals and faces can be. Nonetheless, it characterizes that nature.

IDENTITY AND ABSTRACTION

Pound's attention to equation as descriptive abstraction offers an understanding of how art can designate meaning via identity. In "Vorticism," Pound describes equality as a basis for the fundamental signifying link:

> [T]he equation $(x - a)^2 + (y - b)^2 = r^2$ governs the circle. It is the circle. It is not a particular circle, it is any circle and all circles. It is nothing that is not a circle. It is the circle free of space and time limits .... It is in this way that art handles life. (Pound [1914d] 1974, p. 91)

Pound describes two equalities here: the equation between $(x - a)^2 + (y - b)^2$ and r^2 is one, but more astonishing and far-reaching is the equality between that equation and that which it describes, because that equation does not just signify circles, it exists in identity with circles. In Cartesian geometry, $(x - a)^2 + (y - b)^2 = r^2$ is the algebraic form of any and all circles, while circles are only the plotted geometric form of that equation. Form and referent become, here, symmetrical, inseparable, and one and the same.

The relationship between a circle and its Cartesian equation is immutable in a way that the relationship between meaning and language—especially literary language—typically is not. Pound is attending partly, here, to the way that words can *seem* to be one with what they describe, to the way that, as casual speakers of language, we so easily forget that our words do not definitively designate our intended meanings. The ambiguities and complexities of words as they get drawn out in poetry can work to obscure the remarkable effectiveness of words in day-to-day practice, the way they function most of the time without us noticing the gaps between words and their meanings. In that sense, Pound seeks to reunite poetry with the effectiveness that we more naturally attribute to everyday speech.

From this perspective, we can think of equality as Pound's standard for mimesis in its flattest sense: a kind of representation that so directly, obviously, "realistically" mirrors its subject as to make its meaning impossible to misunderstand. Many centuries of philosophy and art criticism, from Plato to Erich Auerbach to the present day, have established that mimesis is not actually so simple as that—indeed, that mimesis in this flat sense may not actually exist at all. But if the ideal of transparent mimesis is a fiction, it is nonetheless a useful fiction: useful for modeling how language can work so efficiently and so automatically in spite of its inexhaustible complexity; useful as a shorthand for how a work of art can seem so alike its subject even as it remains materially foreign to it. Pound seized on mathematical equality to exemplify this useful fiction.

Pound himself would not have used the word "mimesis." He wrote in 1912 that "[i]n every art I can think of we are dammed and clogged by the mimetic"—but in the same paragraph he also declared that "[w]orks of art attract by a resembling unlikeness" (1912a, p. 370). As Joseph Kronick

544   J. RODAL

writes, here "Pound attacks mimetic art as a slavish effort to reproduce the likeness of the original, but his concern is with stoppage or blockage, not with resemblance" (1993, p. 225). Pound's "resembling unlikeness" in fact proffers a surprisingly classic description of metaphor as well as an apt characterization of "In a Station." Kronick concludes that the good mimesis, for Pound, is "translation, the archive of non-sensuous or linguistic correspondences" (1993, p. 225), because while the things translated may be sensuous and concrete, their mode of correspondence is abstract.

Pound sought a kind of poetry that did not slavishly mimic its subject, did not even *re*-present its subject, but instead *was* its subject. In some literal sense, that could always only be a dream: "In a Station" does not actually embody faces in a Metro station, nor petals on a wet, black bough. Yet, if we shift our understanding of what it is that is being described, this union of symbol and symbolized becomes more plausible. In an important sense, Pound is right to declare that $(x - a)^2 + (y - b)^2 = r^2$ does not just describe the circle, but *is* the circle—if by "circle" we understand the mathematical definition thereof, the properties which make circles circles. $(x - a)^2 + (y - b)^2 = r^2$ is not the only such way of defining a circle, but it is a complete definition nonetheless.[17] Pound is defining something abstract, something that is capable of translation.

Roman Jakobson used equality to characterize definition itself, that is, meaning abstracted from context:

> In the theory of language, since the Middle Ages, it has repeatedly been asserted that the word out of context has no meaning. The validity of this statement is, however, confined to aphasia or, more exactly, to one type of aphasia. In the pathological cases under discussion, an isolated word means actually nothing but "blab." ... When repeatedly asked what a bachelor was, the patient did not answer and was "apparently in distress." A reply like "a bachelor is an unmarried man" or "an unmarried man is a bachelor" would present an equational predication and thus a projection of a substitution set from the lexical code of the English language into the context of the given message. The equivalent terms become two correlated parts of the sentence and consequently are tied by contiguity. The patient was able to select the appropriate term *bachelor* when it was supported by the context of a customary conversation about "bachelor apartments" but was incapable of utilizing the substitution set *bachelor = unmarried man* as the topic of a sentence because the ability for autonomous selection and substitution had been affected. The equational sentence vainly demanded from the patient carries as its sole information: "*bachelor* means an unmarried man" or "an unmarried man is called a *bachelor*". (Jakobson [1956] 1995, p. 122)

Jakobson argued that there are two fundamental processes of language: combination and selection. The former process both creates and depends on context, the stitching between words that get said together. The aphasic patients described above used that linguistic process perfectly well, but they lacked the selection process, in which "[a] selection between alternatives implies the possibility of substituting one for the other, equivalent in one respect and different in another" (Jakobson [1956] 1995, p. 119). Jakobson argued that metonymy depends on the former process, metaphor on the latter, because "[s]imilarity connects a metaphoric term with the term for which it is substituted" ([1956] 1995, p. 132). He begins by characterizing definition as an assertion of sameness (because to define *a* as *b* is to assert that *a* and *b* have the same meaning) and then he extrapolates to understand metaphor as an outgrowth of definition, as a kind of creative *re*-definition that attends to similitudes not normally noticed between two things.

Mimesis, as I described it above, pairs symbol and subject via (an illusion of) obvious sameness. Metaphor, as Jakobson describes it, pairs tenor and vehicle by observing hidden sameness. Both are equality relations. They differ simply in hiding or highlighting novelty. Pound brought them together when he used equation as a standard both for immutable, absolute meaning and for abstract, novel meaning. He felt that the mathematics of equality was capable of explaining how these seemingly opposite representational poles can come together, can turn out to have been, reflexively, one and the same all along:

> The poet's true and lasting relation to literature and life is that of the abstract mathematician to science and life. As the little world of abstract mathematicians is set a-quiver by some young Frenchman's deductions on the functions of imaginary values – worthless to applied science of the day – so is the smaller world of serious poets set a-quiver by some new subtlety of cadence. Why?
>
> A certain man named Plarr and another man whose name I have forgotten, some years since, developed the functions of a certain obscure sort of equation, for no cause save their own pleasure in the work. The applied science of their day had no use for the deductions, a few sheets of paper covered with arbitrary symbols – without which we should have no wireless telegraph. (Pound [1912c] 1973, pp. 361–62)

It is a timeworn fact of mathematical history that work created in abstraction so often finds applications in science later, applications undreamt of by the mathematician whose work gets applied. This happened with non-Euclidean geometry and Einsteinian relativity, with Hilbert spaces and quantum mechanics, with number theory and cryptography: it is so common a story as to become a tired tale. For Pound, this process was not just an analogy for how abstract art describes human life, but a ubiquitous and necessary process, intrinsic to art itself. It offered to explain how similitudes can go hidden for so long and yet seem obvious and immutable once noticed. It offered to model how art can seem so abstract, so foreign, and yet have expansive, irrefutable effects on human life.

Originally, in that station of the metro, Pound tells us that he saw "a beautiful face, and then another and another, and then a beautiful child's face, and then another beautiful woman" (Pound [1914d] 1974, pp. 86–87). At first there were no petals: it was the relation between faces and faces that he sought to describe. Here similitude, distilled to its essence, is not only how one thing resembles another, but, more fundamentally, how anything resembles itself, because that self-resemblance seeds the very possibility of resemblance. Equality, for Pound, offered to characterize how things which seem nothing alike can be alike, while at the same time characterizing the possibility of being alike in the first place. But here, for Pound, likeness everywhere slips into meaning, so that equality shows itself capable of characterizing both representation that resembles its meaning (mimesis) and representation that resembles nothing in nature (abstraction). In this sense, it seemed to him to offer something so broad as a unified theory of art.

Notes

1. "In a Station" was first published in *Poetry* in 1913. Over the following three years Pound republished it several times with varying punctuation and spacing. The version I cite here is Pound's final version, as it appeared in *Lustra* in 1916. Years later, Pound reprinted "In a Station" in its 1916 form in *Personæ*. See Chilton and Gilbertson (1990) for a complete treatment of Pound's revisions to "In a Station."

2. Pound quotes from this sentence, in his own translation, in a note on Ernest Fenollosa's *The Chinese Written Character* (Fenollosa and Pound [1919] 2008, p. 54).

3. See also Bell (2012, p. 141) and Witemeyer (1969, pp. 34–37).

4. Brian Rotman has argued that an even broader range of mathematical symbols are ideogrammic, speaking of "ideograms, such as '+,' '*X*,' '*I*,' '2,' '3,' '=,' '>,' '...,' 'sin *s*,' 'log *s*,' and so on" (Rotman 2000, p. 55). However, for Rotman an ideogram is any symbol that directly designates an idea, unmediated by either alphabetic or diagrammatic systems, whereas for Pound an ideogram must be substantially pictorial. Pound's use of 'ideogram' is closer to Rotman's use of the word "diagram." I argue here that mathematical marks such as $=, >, <, \approx$, or are ideogrammic in Pound's sense of the word.

5. Regarding how far historical definitions of identity were and were not capable of generalization, from Euclid up until the modernist era, see Asghari (2009). Arguably, in the seventeenth century René Descartes provided another way of understanding equality and congruence together, but that method depended on a kind of translation from geometry into arithmetic, rather than the more direct, universal generalization that nineteenth- and twentieth-century modernist mathematicians developed.

6. On modernism as a wide-ranging movement in mathematics, see Mehrtens (1990), Gray (2008), and Alexander (2010).

7. My translation.

8. For more on this history, see Mancosu (2016, pp. 12–22), and Weyl ([1927] 1927, pp. 9–10).

9. Peano was the first to define equality as the possession of these three properties, but he was not the first to observe that these properties attach to equivalence relations. As Mancosu points out, Carl Friedrich Gauss got very close to doing so as early as 1801 (Mancosu 2016, pp. 29–30).

10. Reiteration was critical to Pound's poetics, and he quoted and requoted his own writing astoundingly often. For example, in 1910, in *The Spirit of Romance*, Pound published an analysis of how Cartesian geometry could inform literary form and meaning; then in 1912, in "The Wisdom of Poetry," he repeated the claim that the Cartesian formula for a circle could model the relationship between literary form and meaning; in 1914, in "Vorticism," he published exactly the same explanation, this time in more expanded form; finally, in 1916, he republished "Vorticism" in its entirety in his *Memoir of Gaudier-Brzeska*, explaining that "I reprint that article entire because it shows our grounds for agreement ... we wished a designation that would be equally applicable to a certain basis for all the arts." It is no coincidence that he repeats this essay, because repetition is part of the point (Pound 1910, p. 127, [1912c] 1973, p. 362, 1914c, pp. 461–71, [1914d] 1974, p. 81).

11. In his *Tractatus Logico-Philosophicus*, Ludwig Wittgenstein distinguished three different kinds of "is" (Wittgenstein [1921] 2001, 3.323). There is "is" as in "to exist" ("there is an author of this poem"); "is" expressing the possession of a property, usually attributing an adjective to a noun ("the poem is short"); and "is" as a marker of identity, usually equating two nouns ("he is Ezra Pound"). It is the final category that I indicate here.

12. See Kenner (1971, pp. 186–87).

13. Importantly, this viewpoint shifted later in his career. The *Cantos* accumulate metonymy richly and everywhere. See Perloff ([1981] 1999, p. 172).

14. See Perloff ([1981] 1999, pp. 183–84).

15. See Witemeyer (1969, pp. 34–35), Childs (1986, p. 37), and Kronick (1993, pp. 224–25).

16. Levine 2015 has written about this kind of form at length, but this understanding of the term preceded her. See also Williams ([1976] 1983, pp. 137–40) and Russell (1919, pp. 199–200), as well as Pound himself on form as "structure" and "arrangement" (Pound 1917, p. 104, [1912c] 1973, p. 360).

17. We could, for example, use polar coordinates, in which case $(x - a)^2 + (y - b)^2 = r^2$ becomes simply $r = R$ (centered at the origin). Polar coordinates offer a different mathematical language for converting geometric forms into algebraic equations, and they are every bit as valid as the Cartesian system that Pound uses to describe the circle.

References

Albright, Daniel. 1997. *Quantum Poetics: Yeats, Pound, Eliot, and the Science of Modernism*. Cambridge: Cambridge University Press.

Alexander, Amir. 2010. *Duel at Dawn: Heroes, Martyrs, and the Rise of Modern Mathematics*. Cambridge: Harvard University Press.

Aristotle. [ca. 335 BCE] 1996. *Poetics*. Translated by Malcom Heath. New York: Penguin.

Asghari, Amir. 2009. "Experiencing Equivalence but Organizing Order." *Educational Studies in Mathematics* 71 (3): 219–34.

Bell, Ian F. A. 2012. "Ezra Pound and the Materiality of the Fourth Dimension." In *Science in Modern Poetry: New Directions*, edited by John Holmes, 130–48. Liverpool: Liverpool University Press.

548   J. RODAL

Childs, John Steven. 1986. *Modernist Form: Pound's Style in the Early Cantos.* Selinsgrove: Susquehanna University Press.

Chilton, Randolph, and Carol Gilbertson. 1990. "Pound's '"Metro" Hokku': The Evolution of an Image." *Twentieth Century Literature* 36 (2): 225–36.

Eliot, T. S. [1920] 2009. "Hamlet and His Problems." In *The Sacred Wood*, 64–69. Dodo Press.

———. [1930] 1968. "Ash Wednesday." In *Collected Poems: 1909–1962*, 83–95. New York: Harcourt Brace & Company.

———. [1943] 1968. *Four Quartets.* In *Collected Poems*, 173–209. New York: Harcourt Brace.

"equal, adj. and n." 2018. *Oxford English Dictionary Online.* Oxford University Press. http://www.oed.com.ezaccess.libraries.psu.edu/view/Entry/63695?rskey=wSQSWT&result=1&isAdvanced=false. Accessed March 2, 2018.

Euclid. [ca. 300 BCE] 2013. *Euclid's Elements.* Translated by Thomas L. Heath. Edited by Dana Densmore. Santa Fe: Green Lion Press.

Fenollosa, Ernest, and Ezra Pound. [1919] 2008. *The Chinese Written Character as a Medium for Poetry.* Edited by Haun Saussy, Jonathan Stalling, and Lucas Klein. New York: Fordham University Press.

Frege, Gottlob. [1884] 1997. *The Foundations of Arithmetic: A Logico-Mathematical Investigation into the Concept of Number.* Translated by Michael Beaney. In *The Frege Reader*, edited by Michael Beaney, 84–129. Malden: Blackwell.

———. [1892] 1997. "On Sinn and Bedeutung" ["On Sense and Reference"]. Translated by Max Black. In *The Frege Reader*, edited by Michael Beaney, 151–71. Malden: Blackwell.

Gray, Jeremy. 2008. *Plato's Ghost: The Modernist Transformation of Mathematics.* Princeton: Princeton University Press.

Hasse, Helmut. 1926. *Höhere Algebra.* Vol. I. Berlin: Walter de Gruyter & Co.

Hilbert, David. [1899] 1971. *Foundations of Geometry.* Translated by Leo Unger. La Salle: Open Court.

Jakobson, Roman. [1956] 1995. "Two Aspects of Language and Two Types of Aphasic Disturbances." In *On Language*, edited by Linda Waugh and Monique Monville-Burston, 115–33. Cambridge: Harvard University Press.

Kenner, Hugh. 1971. *The Pound Era.* Berkeley: University of California Press.

Kronick, Joseph. 1993. "Resembling Pound: Mimesis, Translation, Ideology." *Criticism* 35 (2): 219–36.

Levine, Caroline. 2015. *Forms: Whole, Rhythm, Hierarchy, Network.* Princeton: Princeton University Press.

Mancosu, Paolo. 2016. *Abstraction and Infinity.* New York: Oxford University Press.

Mehrtens, Herbert. 1990. *Moderne, Sprache, Mathematik: Eine Geschichte des Streits um die Grundlagen der Disziplin und des Subjekts formaler Systeme.* Frankfurt: Suhrkamp.

Peano, Giuseppe. [1889] 1973. "The Principles of Arithmetic, Presented by a New Method." In *Selected Works of Giuseppe Peano.* Edited and Translated by Hubert C. Kennedy, 101–36. Toronto: University of Toronto Press.

Perloff, Marjorie. [1981] 1999. *The Poetics of Indeterminacy.* Evanston: Northwestern University Press.

Pound, Ezra. 1910. *The Spirit of Romance.* London: J. M. Dent & Sons.

—. 1912a. "I Gather the Limbs of Osiris, XI." *The New Age* X (16) (15 February): 369–70.

—. [1912b] 1990. "Portrait d'une Femme." In *Personæ*, 57–58. New York: New Directions.

—. [1912c] 1973. "The Wisdom of Poetry." In *Selected Prose: 1909-1965*. Edited by William Cookson, 359-62. New York: New Directions.

—. 1913a. "The Approach to Paris." *The New Age* XIII (23) (October 2): 662–64.

—. 1913b. "In a Station of the Metro." *Poetry* 2 (1) (April 1913): 12.

—. [1913c] 1968. "The Serious Artist." In *Literary Essays of Ezra Pound*, edited by T. S. Eliot, 41–57. New York: New Directions.

—. 1913d. "Status Rerum." *Poetry* 1 (4): 123–27.

—. [1914a] 1990. "Salutation the Third." In *Personae*, 75–76. New York: New Directions.

—. [1914b] 1990. "Salvationists." In *Personæ*, 100. New York: New Directions.

—. 1914c. "Vorticism." *Fortnightly Review* 96 (573): 461–71.

—. [1914d] 1974. "Vorticism." In *A Memoir of Gaudier-Brzeska*, 81–94. New York: New Directions.

—. 1915a. "Affirmations, II: Vorticism." *The New Age* XVI (11): 277–78.

—. 1915b. "Affirmations, IV: As for Imagisme." *The New Age* XVI (13): 349–50.

—. [1916] 1990. "In a Station of the Metro." In *Personæ*, 111. New York: New Directions.

—. 1917. "Arnold Dolmetsch." *The Egoist* IV (7) (August 1917): 104–105.

—. [1934] 1960. *ABC of Reading*. New York: New Directions.

—. 1996. *The Cantos*. New York: New Directions.

Recorde, Robert. 1557. *The Whetstone of Witte*. Internet Archive. http://archive.org/details/TheWhetstoneOfWitte. Accessed March 2, 2018.

Richards, I. A. 1936. *The Philosophy of Rhetoric*. New York: Oxford University Press.

Rotman, Brian. 2000. *Mathematics as Sign: Writing, Imagining, Counting*. Stanford: Stanford University Press.

Russell, Bertrand. [1903] 1996. *The Principles of Mathematics*. New York: W. W. Norton.

—. 1919. *Introduction to Mathematical Philosophy*. London: George Allen and Unwin.

Shakespeare, William. [1597] 1960. *Romeo and Juliet*. Baltimore: Penguin.

Stein, Gertrude. [1913] 1998. "Sacred Emily." In *Stein: Writings, 1903–1932*, vol. 1, 387–96. New York: Library of America.

Weyl, Hermann. [1927] 1949. *Philosophy of Mathematics and Natural Science*. Translated by Olaf Helmer. Princeton: Princeton University Press.

Whitehead, Alfred North, and Bertrand Russell. [1910] 2009. *Principia Mathematica*. Vol. 1. Merchant Books.

Williams, Raymond. [1976] 1983. *Keywords*. New York: Oxford University Press.

Witemeyer, Hugh. 1969. *The Poetry of Ezra Pound: Forms and Renewal, 1908–1920*. Berkeley: University of California Press.

Wittgenstein, Ludwig. [1921] 2001. *Tractatus Logico-Philosophicus*. Translated by D. F. Pears and B. F. McGuinness. New York: Routledge.

Woolf, Virginia. [1925] 1967. "Modern Fiction." In *Collected Essays*, vol. II, 103–10. New York: Harcourt, Brace & World.

PART V

Mathematics as Literature

CHAPTER 30

Rehearsing in the Margins: Mathematical Print and Mathematical Learning in the Early Modern Period

Benjamin Wardhaugh

Arithmetical computation and proportional reasoning, algebraic reasoning and geometrical proof: these were all experienced in the early modern period—as they still are today—as special kinds of performance. They were transmitted by processes involving demonstration by a teacher, private rehearsal, and specimen performance by the student. Printed and manuscript texts recorded the traces of successful (or unsuccessful) performances, and learners used slates, waste paper, and the margins of printed texts as rehearsal space in which to perfect their own performances. The result was a pedagogy and a mathematical culture that used the written word—printed or manuscript—in unique, highly distinctive ways. This chapter examines the evidence for that culture and the influences on it of ancient and early modern examples.

THE FRAGMENTED TEXT

Early modern readers experienced mathematical text as fragmented and non-linear, as negotiable and malleable, and as a model for imitation and from which to assimilate praxis (Raphael 2013, 2015, 2016, 2017; Oosterhoff 2015). Consider the famous story related by John Aubrey about Thomas Hobbes:

B. Wardhaugh (✉)
Oxford University, Oxford, UK

© The Author(s) 2021
R. Tubbs et al. (eds.), *The Palgrave Handbook of Literature and Mathematics*, https://doi.org/10.1007/978-3-030-55478-1_30

554   B. WARDHAUGH

He was (vide his life) 40 yeares old before he looked on geometry; which happened accidentally. Being in a gentleman's library in ..., Euclid's Elements lay
open, and'twas the 47 El. libri I. He read the proposition. "By G—," sayd he,
"this is impossible!" So he reads the demonstration of it, which referred him
back to such a proposition; which proposition he read. that referred him back to
another, which he also read. Et sic deinceps, that at last he was demonstratively
convinced of that trueth. This made him in love with geometry. (Aubrey 1898,
p. 332)

Broadly similar stories were told for instance about Isaac Newton: for example, D. T. Whiteside cites an account given by Abraham de Moivre (Whiteside
1967, p. 6). The story is plausible in general terms, and part of its plausibility
is the familiarity to early modern ears of the model of engagement with the
Euclidean text it narrates. Hobbes did not begin at the beginning of the text,
with the meticulous logical basis of axioms, postulates, and common notions.
Nor did he begin at the beginning of the propositions in Book 1—simple
constructions and proofs to do with circles and triangles—and work up
from there. He began *in medias res* with the forty-seventh proposition in the
Elements (the Pythagorean Theorem) and having formed his initially skeptical
response he continued by working through the text selectively and in reverse
order, guided presumably by the printed cross-references that adorned the
main text or the margins of most early modern editions, jumping from proposition to proposition until he arrived, finally, at the most basic material, and
the definitions and axioms. Compared with our usual images and assumptions
about the reading of a printed book, this is a rather extraordinary way to proceed; but nothing Aubrey says suggests that it was unusual. As we will see,
the evidence we have indicates that for mathematical books it was the norm.

The nonlinear experience of the text produced by readerly rearrangement
and selection of course responds directly to the textual conventions which
governed mathematical print in this period; Hobbes would hardly have been
able to work as he did had not the printed text been reasonably dense with
cross-references. The division of mathematical knowledge into discrete theorems, for which a serial order is necessarily somewhat unhelpful, and its
transformation into a meaningful network of logical relationships by printed
(or manuscript) cross-references, surely invited selective, nonlinear styles of
reading, and this even at the most elementary level of arithmetic primers and
practical manuals.

That many readers were thus guided by the cross-references printed in
mathematical books is clearly indicated by the fact that so many of them
added supplementary cross-references or corrected those that were printed
when they found them to point to the wrong place.[1] That they read selectively is also indicated by the frequency with which they adorned the text
with marks of attention applied to selected parts: selected theorems, selected
chapters or sections, selected exercises or examples. The handwritten, marginal cross indicating "I studied this" or "I will study this" or "this is worthy

of attention"—and implicitly that everything not so marked was less worthy of attention—is an absolutely ubiquitous device in early modern mathematical books as marked by their early modern readers (Wardhaugh 2020a).[2] Some readers of mathematical books showed a concern to make the text as complete as possible within the sections they studied, remedying omissions in the operational instructions, however slight; recopying diagrams to bring them into better proximity with the text to which they related; supplementing the proofs with the definitions, or the definitions with examples of their use. Some simply wrote "done" by the exercises they had completed. Many readers went further in their restructuring of the text, numbering the sections to be studied, marking up the contents page or index or supplying a manuscript contents page or index to direct attention exactly where they wanted it.[3] For example, late seventeenth-century users of a copy of the *Elements* of Euclid in the popular teaching edition by Christoph Clavius, at the Queen's College, Oxford, went through the contents list, picking out and numbering twenty-three of the propositions from Books 1–4 of the text. Users of another copy of the same text at Trinity College, Cambridge did the same, marking a total of forty-three propositions in Books 1, 3, 4 and 6.[4]

For many readers, like Hobbes in his apparently private setting, these activities seem to have been autonomous. Out of about a hundred annotated early modern copies of the *Elements*, I have yet to see two in which the selection of theorems picked out for attention seems to be substantially the same, suggesting that such decisions were frequently made locally if not individually (see below for one example of the transmission of Euclidean annotations, however). On the other hand, many readers' attention will certainly have been directed by a teacher. Individual ownership of textbooks did not become the norm in British schools until the second half of the eighteenth century, and up to—and in many places beyond—that point the typical practice was for the teacher to copy into the pupils' exercise books selected parts of a printed text (Walkingame 1751, a2^r–a2^v; Denniss 2012; Ellerton and Clements 2012, 2014). Something similar can be seen happening in university teaching, where the teaching notebooks of tutors consist—on the mathematical side—almost invariably of *excerpta* from classical mathematical texts: William Poole discusses in particular the manuscript volumes in the Queen's College, Oxford, MSS 425–32 (Poole 2018), and Ann Blair the autonomous and the teacher-directed modes of note-taking (Blair 2010). And in university lecturing too, the lectures that survive approach the classics of mathematics not as wholes to be worked through page by page but as sources from which to select, rearrange and reassemble in a new order, thought suitable for the particular audience at hand (see Wardhaugh 2020b and John Wallis's manuscript lectures on Euclid's *Elements*).[5] Henry Savile at Oxford, for instance, chose a fairly small selection of Euclid's propositions for discussion in his *Prælectiones* (1621) but gave rather fuller attention to his preliminary definitions, as well as to general questions about philology and geometrical method.

The impulse to rearrange certainly influenced the production of printed mathematics too; from the seventeenth century onwards there began to appear with increasing frequency textbooks whose titles invoked Euclid or Apollonius or Archimedes but which consisted in fact of *excerpta* rearranged according to the fancy of the editor. Giovanni Alfonso Borelli's (1658) *Euclides restitutus* is typical of the type; its seven books contain the propositions of *Elements* 1–6 and 11–13, rearranged and partly rewritten: compare (Pardies 1671) and (Mohr 1673), which both also appeared in English. The evidence provided by readers' annotations shows that these books too were received by readers not as wholes but as sources from which to select and rearrange.

Nonlinear reading easily crossed the boundaries between one text and another. Almost as soon as the texts of the university curriculum were put into print their margins were used by students to record something of what their teachers had said about those texts (Oosterhoff 2018; Groote 2013; Grafton and Leu 2013; Grafton 1981). And only slightly later it became quite normal for the attentive student to customize a printed book not only with internal cross-references and internal excerpting and rearrangement, but also to supplement it with sections copied or even cut out from other books. A 1638 copy of John Wells's *Sciographia: or, The Art of Shadowes* was supplemented a generation later by an anonymous reader who copied in summaries of the principles of geometry as well as diagrams cut from other books.[6] A 1659 copy of Norwood's *Epitome* was customized in the mid-eighteenth century by owners Roger Sherman and Elijah Porter, who added extracts from the 1707 *Thesaurium Mathematicæ* of John Taylor and William Alingham and from John Love's 1688 *Geodaesia*.[7] Another similar example is a copy of John Seller's 1677 *Pocket Book*, with dense annotations indicating that despite its author's practical intentions it was studied by university students.[8] Printed mathematical books were thus customized into compendia, and nonlinear reading took on a new dimension emphasizing still further that the one thing no reader seems to have done with printed mathematics was to read it in the order in which it was printed.

The Negotiable Text

Mathematics presented special difficulties for the early modern print shop. Geometry involved a wealth of diagrams, which were expensive to commission and required either wastefully wide margins for their insertion or the complication of inserting them within the text itself. When copperplate engraving became available the new technology was soon adopted for the printing of geometrical diagrams, but this raised other problems. Diagrams were now relegated to a plates section at the back of the book, distant from the portion of text to which they related. One expedient was to print the plates on wide fold-outs, sufficiently wide that they could be folded out and remain visible while the main text block was turned back to the theorem in question: see for example the note "To the Bookbinder" (Tacquet 1703, P8ᵛ). Another solution was for the reader to cut up the plates and paste the separate diagrams into the book where they were needed.[9]

The points in these geometrical diagrams were conventionally labeled using upper case letters starting from the beginning of the alphabet, and the diagrams were referred to from the main text using these labels: often over and over again. It is occasionally clear that the print shop ran out of the required upper case letters and had to substitute letters from another font. And geometry is repetitious in other ways, too. When almost every paragraph began with one of two or three fixed phrases, the supply of decorated initials could also run low: Achates' edition of Ratdolt (1491) uses in twenty locations a decorated S which is faulty (it is reversed left-right), and d'Étaples (1516) has at least two uses of a similarly faulty decorated N. Algebra in its sixteenth- and seventeenth-century forms required some altogether new symbols or the adaptation of old ones, and it was easy to make mistakes that resulted in incomprehensibility.

More frequent and more serious, however, were the perfectly usual errors of the press in which a single character was omitted, inverted, or substituted for another: in prose this would be easily ignored, but in mathematics it could turn a comprehensible statement into nonsense, whether the wrong symbol was a number, an algebraic letter, or a geometrical label. Such errors were easy to make and relatively easy to spot for a reader who was working closely through the mathematical content of a particular passage: but they were hard for proofreaders working rapidly to detect compared with typographical blunders of similar magnitude in printed prose or poetry. A wrong character in the midst of a geometrical proof does not leap off the page in the same way as a wrong character in a line of verse. Indeed, authors, editors, and printers routinely acknowledged that it was especially difficult to correct the press thoroughly for mathematical books. John Ward's *Young Mathematician's Guide* (1707) contained the following characteristic lament:

> If the Reader were but Sensible of the Great Care and Difficulty that unavoidably attends Correcting the Press to Books of this Nature; he would the more readily Excuse and Amend the following Errata's. (Ward 1707, Mmm2ᵛ)

Later in the eighteenth century, it would become the norm for mathematical authors to correct the press for their own and one another's books.[10] We do not know for certain what the usual practice was in the sixteenth or the seventeenth century (to my knowledge no correctors' sheets for mathematical books are known to survive, nor have I seen a publisher's "correction copy" with interleaved blanks, although editors certainly did make an effort to collect errata in mathematical books). But, certainly, not a few errata slips were introduced by laments like Ward's, admitting at least implicitly that print shops printing mathematics did not feel they had the resources required to correct it adequately. Jacques Lefèvre d'Étaples, printing his great edition of Euclid's *Elements* in 1516, ended the errata leaf typically by stating that he had noted the errors he judged worthy of correction but left it to the reader to correct any others while reading (d'Étaples 1516, errata leaf).

At the other end of the scale, John Tapp introduced his *Sea-Man's Kalendar*—a small compendium of practical mathematics—in 1672: "intreating the courteous Readers to do me that favour, as to correct what they shall find amiss, either in the Printer's over-sight or mine own errour," and promised to "endeavour the mending of them in the next Impression" (Tapp 1674, A2^r). A long printed errata list implicitly made the same point, and some mathematical books came with very long lists indeed.

Thus, readers of early modern mathematical print routinely found themselves faced with a direct invitation to correct faults in the text and to find others for themselves, and they were faced with a book which in order to be effectively used needed to be so corrected. They were also faced with a situation in which the very mathematical skills the book aimed to transmit must be deployed in order to determine whether the book itself was correct at a given point: whether the calculation was accurate or the algebra correct. One of the functions of the printer, therefore, was to provide the reader with an adequate writing surface on which to make corrections, and one of the underexplored functions of white space in mathematical layouts is just that.

Many printers did provide adequate surfaces for this purpose, and very many readers did indeed get involved with such negotiations about what the text should say. Correction is not unique to mathematical text—indeed, a "culture of correction" (Grafton 2011) has rightly been identified with respect to printed and manuscript texts quite generally in the sixteenth century—but that culture persisted, and extended socially, much longer and wider for mathematical texts than for any other genre. Something like eighty percent of surviving early modern mathematical books bears readers' corrections, making this one of the most distinctive features of mathematical reading in this period (Wardhaugh 2020a).

A number of studies of early modern annotation (such as Jackson 2001, p. 164) have remarked on the frequency of adversarial annotation, the kind that says "nonsense!" or "the author is a fool." This type proves to be very rare among readers' responses to early modern mathematical print (for some exceptions see Goulding 2005); but detailed, usually autonomous correction effectively took its place. Furthermore, those who corrected autonomously outnumbered by far those who carried out the corrections directed in the printed errata (Wardhaugh 2020a). Here, as in the selection of material, readerly autonomy seems to have been the norm.

Corrections could of course take a variety of forms, and could go well beyond cleaning up single-character errors within mathematical working. A significant dimension of the "mathematical classicism" that informed the response to ancient mathematical texts, particularly in the British Isles, took the form of grammatical and linguistic improvement—a particle here, a spelling or a Byzantinism there—or of philological collation with another printed or manuscript version of the same text. Many are the copies of Euclid's *Elements*, in particular, that bear evidence of this kind of correction—even on every page—as well as or instead of correction of the mathematical content.[11]

Furthermore, and as some authors admitted, it was all too easy for mathematics to be wrong in grosser ways. Ward admitted in his *Young Mathematician's Guide* that "at Question 5. Page 93. The Operation or Work is all Mistaken; for the Answer should be only 13 oz" (Ward 1707, Mmm3^v). And Isaac Barrow, in the 1659 impression of his translation of the *Elements*, included ten dense pages of "annotations," in which errors and obscurities in the printed text were emended and whole sections replaced (Barrow 1659, pp. 343–52). Just as lists of errata prompted readers to find small errors, these large-scale authorial second thoughts invited them to get involved on a similar scale, and there were those who took up the challenge, substituting on what appears to have been their own initiative alternative methods of proof or even entirely different sequences of theorems. Such work could result in customized copies, such as the interleaved volumes occasionally prepared by zealous teachers, sometimes for presentation to pupils, in which a printed text was supplemented *in extenso* with alternative, supplementary or explanatory material.[12]

A culture of correction is self-perpetuating. Early modern mathematical readers were served by a brisk trade in second-hand books, and it was undoubtedly the case—sometimes the sale catalogues tell us so—that many of the books sold bore the corrections of previous readers. While "clean" could in some cases be a selling point, for mathematical books it was also on occasion noted as a positive feature that a book for sale had been "corrected." Thus, a substantial fraction—very possibly the majority—of readers encountered mathematical books that contained not only printed invitations to correct but the corrections of one or more previous readers. Some, inevitably, corrected the corrections, or supplemented them, perhaps choosing different parts of the text to work on. Corrections, then, could become sociable in the hands of subsequent readers, and occasionally they were copied from book to book: from teacher to student, or on occasion from scholar to scholar, say in the context of projects to re-edit a text. John Chambers, for instance, studied Euclid with Savile at Oxford, and his own copy of the book bears annotations which are in many cases copied from Savile, while others report or paraphrase Savile's discoveries, making it clear that Chambers had both Savile and Savile's annotated copy of the text with him as he studied.[13] Annotations and emendations by Savile would subsequently be used by both Edward Bernard and David Gregory in their projects to re-edit the Latin and Greek texts of the *Elements*; Bernard incorporated material from other Oxford annotators including a Dr Pain of Christ Church, while Gregory made some use of Bernard's own notes after the failure of the latter's editorial and publication project.[14]

The Imitable Text

Donating a copy of the *Elements* to Corpus Christi College, Oxford in the early sixteenth century, John Claymond—president of the college—indicated that he intended it for the use of students and, specifically, so that they could copy out Euclidean theorems from it.[15] We have already heard that in a school context copying was a usual practice, carried out until the mid-eighteenth century, usually by the teacher. It is also the case, of course, that a culture of commonplacing governed much of both the learned and the student response to printed and manuscript texts in this period: so we certainly cannot say that copying out sections in itself made for a distinctive culture of mathematical reading. Commonplacing as such did not really occur in mathematical contexts, as far as our evidence goes—indeed, such a principle of reorganization might have sat rather uneasily with what I have argued was the normal experience of mathematical text as essentially fragmented and nonlinear. But, as we saw above, it was common to construct volumes of *excerpta*, particularly for teaching purposes.

The students at Corpus Christi College certainly did copy out parts from Claymond's gift; it is visible on the pages of the volume that sheets have been laid over certain of the diagrams and the diagrams traced, leaving scratches on the printed page itself (Wardhaugh 2020b). Around Oxford, other college-owned copies of similar texts bear similar marks as well as the tell-tale prick of a compass point through the centers of diagrams where copies have been made.[16] Many readers seem to have copied onto loose sheets and to have been none too careful about letting the ink dry before closing the book onto the loose sheet, leaving in some cases clearly legible impressions of diagrams or even of texts by ink transfer.[17] Actual copies made onto separate sheets very rarely survive; the few exceptions seem to show they were often reused as waste paper: such as Gerard Langbaine's manuscript catalogue of the Savilian Library,[18] written over large geometrical diagrams that suggest group teaching, or a catalogue of coins in the Bodleian Library written over a set of small geometrical diagrams that might suggest individual study or teaching.[19] But copies into notebooks or exercise books are common, and wholesale recopying of an entire mathematical textbook remained a reasonably common practice into the second half of the eighteenth century: for example, regulations at the Royal Military Academy at Woolwich in force in the 1780s (see Jones 1895) envisaged each cadet making a fair copy of Thomas Simpson's *Algebra* (Simpson 1745 or a subsequent edition). Robert Sandham's letters from the RMA in the 1750s also mention such activities: "I have written all *Mr Muller's Artillery*, which is forty octavo pages" (Hogg 1963, p. 7).

Such practices raise the question of what the copies were for, and take us deeper into what was distinctive about mathematical reading and the uses of mathematical text. The evidence shows that mathematics—arithmetic, proportional reasoning, algebra, and geometry—was thought of and treated as a set of practices, and that texts were received as models not just to be copied in the obvious sense but from which to learn, to assimilate those practices (see Jardine and Grafton 1990). This mode of use can be referred right back to the earliest evidence for mathematical texts and their construction in the Greek tradition, when they appear to have been the by-products of deduction or of *viva voce* teaching, the mathematician constructing a diagram, talking through the construction, and leaving a diagram and eventually a written description of its construction and meaning as the traces of that process (Netz 1999, p. 167 and *passim*).

Thus, every mathematical text in this tradition is in a sense a double one. On the one hand, there is the description of operations performed, whether arithmetical, geometrical, or of some other kind. On the other hand, there are the traces those operations leave: arithmetical or algebraic working, or a diagram. Mathematical books interweaved the two modes section by section, or presented them side by side: diagrams beside proofs, or calculations interspersed with the description of methods. In the era of print, and certainly from the seventeenth century onwards, it became quite common to adopt a two-column arrangement for the two parts of the mathematical text: such as for example John Hill's *Arithmetick* (1713), which frequently uses a two-column layout to place a verbal description of an arithmetical operation side by side with the "working."

Nevertheless, a printed or manuscript text cannot achieve all of the same things as *viva voce* teaching. Neither of the two parts of the text—description or traces—is really meant to be copied as such. In live teaching, a diagram or written calculation is the outcome of a specimen performance that the learner(s) have observed, the imitation of which is the expected next step. In a manuscript or printed book, a diagram is the supposed outcome of a specimen performance that the learners have *not* seen, and for which the accompanying text now functions as a narrative description. The process which leads to the traces was supposed to be learned as a practice, a *habitus*. And it was to be learned by repetitive private practice whose traces were essentially inferior versions of the traces that appeared in printed books. So, almost as frequently as they corrected errors in the printed text, readers of mathematical books used the margins and other white space as places in which to re-create for themselves the diagrams, calculations or proofs as they read, using the descriptive text as a set of instructions and the printed diagrams or calculations as a model not for copying but for comparison with what they had done themselves.

Thus we find time and again the margins of printed mathematics books used as a kind of rehearsal space, bearing more or less hesitant and inadequate versions of geometrical diagrams, faulty or correct rehearsals of arithmetical working, and faulty or, less often, correct rehearsals of algebraic working. These are not mere copies of what is in the printed text; usually they differ in their detailed approach, and often they also differ in being not quite correct.

562   B. WARDHAUGH

It is to be presumed that a vast amount of such rehearsal also took place on surfaces that are not now available to the historian: slates for school arithmetic, waste paper for higher level students; occasionally readers even took a sharp point and pricked or scored geometrical diagrams into the leather covers of their books. Blackboards and even sand boxes are evidenced as teaching tools in the Middle Ages and Renaissance, and may well have been used by at least some students for these purposes of private rehearsal.[20]

The outcome of the rehearsal was proficiency at a practice, and it was shown to the teacher in the form of a demonstration performance whose successful enactment permitted the student to move on to the next level of material. In schools this was a continuous process, the students being called out one by one or in small groups to perform what they had just been practicing to see if they were fluent at it (Ellerton and Clements 2012). W. D. Jones shows the inspector of the Royal Military Academy in 1772 commending "the daily practice of calling up to their desks their respective pupils" (1895, p. 17). In advanced schools, such as the military academies, the public examinations took exactly the same form, with students performing set pieces from Euclid or their teachers' textbooks before a public— sometimes a distinguished—audience (Jones described public examinations at the Royal Military Academy in 1779; see 1895, p. 26). It is unfortunate that we do not have evidence to tell us if the same kinds of mathematical demonstration performances took place in university contexts, although, judging from the evidence of marginal annotations, the same kinds of private rehearsal undoubtedly did.

Some readers, moreover, rehearsed their mathematical skills in a somewhat different way: not by writing out the material in the form in which they had read it or been shown it, but specifically by translating it into a different mathematical language: words into algebra, algebra into words; words into diagrams, or diagrams into algebraic equivalents; algebra into numerical examples, or sometimes vice versa (Wardhaugh 2020a). An eighteenth-century reader of Ward's *Young Mathematician's Guide*, for example, translated verbal rules into algebra throughout the text; a century earlier a reader of Barrow's edition of the *Elements* systematically turned the diagrams into examples by adding numbers to them.[21] The practice was so common that it seems it must have been taught or at least encouraged in schools and universities, and Renee Jennifer Raphael persuasively suggests that transformation of this kind was an important part of scholarly and pedagogical practice (Raphael 2015, pp. 6–21). Indeed, we find it at all levels from the elementary textbook to such distinguished cases as Newton reading Euclid: he worked through much of Book 10 putting the theorems about commensurability into algebraic terms.[22] Like other modes of mathematical reading, this tendency to translation also found its way into the printed texts themselves. The provision of numerical examples (for which there were medieval precedents) became a feature first of editions of Euclid's *Elements* from the later sixteenth century onwards, and subsequently of textbooks on other parts of mathematics. For instance, the editions of Christoph Clavius (1574 and subsequently) added numerical examples in the preamble to Book 5; Henry Billingsley (1570) added them to Book 2. The act of moving from the particular case to the general one would become an important part of mathematical pedagogy in the long run, as well as an important part of the early modern debate about the proper ways to do mathematics and to conceive the relationship between—say—algebra and geometry.

CONCLUSION

Throughout the early modern period, changes in the physical characteristics of print proceeded hand-in-hand with changes in the ways readers used texts, both printed and manuscript. The production and uses of mathematical print are part of this process, but they are a highly distinctive part. Other genres of early modern print, and other kinds of text more generally, were sometimes approached nonlinearly and in fragmented ways; they were corrected, sometimes compulsively, and their contents reorganized or renegotiated; and they were sometimes used as models for more or less literal "copying" as well as for sources of imitation at the higher level of the skill acquisition they embodied. Yet no other genre of early modern print was received in these ways all the time, and in no other genre was the reception, reading, and marginal response so dominated by these modes. Thus we can conclude that there was a distinctive culture of mathematical reading in the early modern period, and that mathematical print and mathematical texts were received in a distinctive way.

Mathematical knowledge was conceived primarily in terms of special kinds of performance: hybrid performances with both mental and manual aspects, producing a diagram and/or a text that served as sign of a successful performance. If educational theory chiefly valued the mental side and the notion that mathematical study would improve the mind, pedagogical practice—as witnessed by readers' marks in mathematical books—appears to have valued the manual side at least as much.

Some of the processes discussed here can be evidenced in readers' use of mathematical manuscripts before the era of print, and it seems certain that some of what was distinctive about mathematical reading carried over from older, even ancient assumptions about what mathematics was and what mathematical text was.[23] On the other hand, early modern ways of reading mathematical text and transmitting mathematical knowledge had consequences for the longer term cultural profile of mathematics. They bear on questions such as how and why thinkers mainly associated with other disciplines attached to mathematics a particular kind of importance, and how mathematics came to be used as a regular source of culturally transformative ideas from the early modern period into the modern.

Notes

1. Copies of Ward 1719 and 1734 in University of Michigan Library with shelf-marks QA35.W259 1719 and QA35.W259y 1734, with annotations including supplementary cross-references. Copy of Briggs 1620 in All Souls College, University of Oxford, with shelfmark 4:SR.59.c.23, with dense annotations including many added cross-references.

2. Among many examples a copy of Fisher (1731) in the British Library with shelfmark 8506.bb.11, with marginal exclamation marks; a copy of Fisher (1767) in the National Library of Scotland with shelfmark YY.8/2, with marginal Xs; and a copy of Ward 1771 in Regent's Park College, Oxford, with shelfmark 24.g.33, with pencil ticks, crosses, and the note "done" throughout.

3. Copy of Ward 1758 in the British Library with shelfmark 08535.a.70, with index marked up (Qqq1^v–2^r, 3^r–4^r) by Charles Moore and/or William Canter RN, showing sections studied; copy of Ward 1771 in the University of Michigan with shelfmark QA35.W259 1771, with supplements to contents list on A4^v and partial indexing on the rear end-paper.

4. Copy of Clavius (1607) in Trinity College, University of Cambridge with shelfmark T.38.13–14; copy of Clavius (1607) in the Queen's College, University of Oxford with shelfmark 40a.A.18–19, donated for the use of the "taberdars" of that college.

5. Bodleian Library, University of Oxford, MS Don. d. 45.

6. Copy of Wells (1635) in the Huntington Library with shelfmark RB 271,828, described in Sherman (2008), p. 9, n. 29.

7. Copy of Norwood (1659) in the Smithsonian Library with shelfmark QA33. N67 1659, described in Bedini (2001); information from Smithsonian online catalogue entry and from Yelda Nasifoglu (pers. comm.).

8. Copy of Seller (1677) in the London Science Museum with shelfmark O.B. SEL SELLER (1), described in Tracey (2020).

9. Copy of Whiston (1765) in the National Library of Ireland with shelfmark Dublin 1765(23).

10. For example, Cambridge University Library, RGO 14/5, p. 344 (Board of Longitude minutes, 6 March 1779), with a mention of Charles Hutton correcting the press for a volume of tables by Bernoulli.

11. For example copies of Grynäus (1533) in Merton College, University of Oxford with shelfmark 40.J.15, marked probably by Henry Savile, and in Balliol College, University of Oxford with shelfmark St Cross 0625 d 06, apparently marked at that college.

12. For example a copy of Brown 1753 with interleaving by the teacher Samuel Davis, in a private collection: see Wardhaugh (2020a). For other examples of interleaved and extended books see Woolf (2000), p. 87 with note 19 and Grenby (2011), p. 246.

13. A copy of d'Étaples (1516) in Oxford, Bodleian Library with shelfmark Savile W 12, annotated by John Chambers; information from Renee Raphael, pers. comm.

14. See the heavily annotated printed copies of the *Elements* in the Bodleian Library with shelfmarks Auct. S 1.12–15, 8° B 16 Linc. and 8° C 134 Linc., and a similarly annotated 1625 copy of the *Data* with shelfmark Auct. S 2.22; also Beeley (2020b).

15. Copy of d'Étaples (1516) in Corpus Christi College, University of Oxford, Rare Books Collection, with shelfmark Δ.10.1, inscription on title page: "in usum discipulorum + ut inde exscriberent theoremata euclidis."
16. Copy of Rudd (1651) in Trinity College, University of Cambridge with shelfmark NQ.7.82, with prick-marks indicating copying using a compass (very likely by Isaac Newton) throughout pp. 1–40.
17. Copy of Fine (1536) in Trinity College, University of Cambridge with shelfmark S.10.72, p. 17: a probable ink transfer of a diagram on a loose leaf.
18. Oxford, Bodleian Library, MS Savile 107.
19. Oxford, Bodleian Library, MS Lister 39.
20. Mordechai Feingold, presentation at Research workshop on "Teaching Mathematics in the Early Modern World," Oxford, December 2016.
21. Copy of Ward 1752 in a private collection; copy of Barrow 1659 in University College, London with shelfmark Strong Room Euclid Octavo 1659 (2).
22. Copy of Barrow (1655) in Trinity College, University of Cambridge with shelfmark NQ.16.201[1], with algebraic explanations added in books 2, 5 and 10; see also Warwick (2003), p. 31.
23. Erfurt, Biblioteca Amploniana F. 377, f. 23^r has an instance of a reader adding a translation of the text into diagrammatic form, in the context of Alfonsine astronomical material of the fourteenth century.

References

Aubrey, John. 1898. *"Brief Lives," Chiefly of Contemporaries*. Edited by Andrew Clarke. Oxford: Clarendon Press.
Barrow, Isaac. 1655. *Euclidis Elementorum Libri XV. Breviter Demonstrati*. Cambridge: Cambridge University Press.
———. 1659. *Euclidis Elementorum Libri XV. Breviter Demonstrati*. London: Roger Daniel for William Nealand.
Bedini, Silvio A. 2001. "History Corner: Roger Sherman's Field Survey Book." *Professional Surveyor Magazine* 21: 70.
Beeley, Philip, et al. 2020a. *Reading Mathematics in the Early Modern World*. London: Routledge.
———. 2020b. "'A Designe Inchoate': The Curious Story of Edward Bernard's Edition of Euclid's *Elements*." In Beeley et al. 2020a.
Billingsley, Henry. 1570. *The Elements of Geometrie of the Most Auncient Philosopher Euclide of Megara*. London: John Daye.
Blair, Ann. 2010. "The Rise of Note-Taking in Early Modern Europe." *Intellectual History Review* 20: 303–16.
Borelli, Giovanni Alfonso. 1658. *Euclides Restitutus*. Pisa: Francisco Honophri.
Briggs, Henry. 1620. *Εὐκλείδου Στοιχειῶν Βιβλία ῖγ*. London: William Jones.
Brown, W. 1753. *Problems in Practical Geometry*. Birmingham: T. Aris.
Clavius, Christoph. 1574. *Euclidis Elementorum Libri XV*. Rome: Vincenzo Accolti.
———. 1607, 1654. *Euclidis Elementorum Libri XV*. Frankfurt: Nikolaus Hoffman and Jonas Rhodius.
d'Étaples, Jacques Lefèvre. 1516. *Euclidis Megarensis Geometricorum Eleme[n]torum libri XV*. Paris: Henri Estienne.

566   B. WARDHAUGH

Denniss, John. 2012. *Figuring It Out: Children's Arithmetical Manuscripts, 1680–1880*. Oxford: Hexley Press.

Ellerton, Nerida, and M.A. Clements. 2012. *Rewriting the History of School Mathematics in North America 1607–1861: The Central Role of Cyphering Books*. Dordrecht: Springer.

———. 2014. *Abraham Lincoln's Cyphering Book and Ten other Extraordinary Cyphering Books*. Dordrecht: Springer.

Fine, Oronce. 1536. *In Sex Priores Libros Geometricorum Elementorum Euclidis Megarensis Demonstrationes*. Paris: Simon de Colines.

Fisher, George. 1731. *Cocker's Arithmetic*. London: Edward Midwinter.

———. 1767. *The Instructor*. Edinburgh: A. Donaldson.

Goulding, Robert. 2005. "Polemic in the Margin: Henry Savile Against Joseph Scaliger's Quadrature of the Circle." In *Scientia in Margine: études sur les Marginalia dans les Manuscrits scientifiques du Moyen Age à la Renaissance*, edited by Danielle Jacquart and Charles Burnett, 241–59. Geneva: Droz.

Grafton, Anthony, and Urs Leu. 2013. *Henricus Glareanus's (1488–1563) Chronologia of the Ancient World: A Facsimile Edition of a Heavily Annotated Copy Held in Princeton University Library*. Leiden: Brill.

Grafton, Anthony T. 1981. "Teacher, Text and Pupil in the Renaissance Class-Room: A Case Study from a Parisian College." *History of Universities* 1: 37–70.

———. 2011. *The Culture of Correction in Renaissance Europe*. London: British Library.

Grenby, Matthew O. 2011. *The Child Reader 1700–1840*. Cambridge: Cambridge University Press.

Groote, Inga Mai. 2013. "Studying Music and Arithmetic with Glarean: Contextualizing the Epitomes and Annotationes among the Sources for Glarean's Teaching." In *Heinrich Glarean's Books: The Intellectual World of a Sixteenth-Century Musical Humanist*, edited by Iain Fenlon and Inga Mai Groote, 195–222. Cambridge: Cambridge University Press.

Grynäus, Simon. 1533. *Εὐκλείδου Στοιχείων βιβλ. ιε*. Basel: Johann Herwagen.

Hill, John. 1713. *Arithmetick, Both in the Theory and Practice*. London: J. M. for D. Midwinter.

Hogg, O. F. G. 1963. *The Royal Arsenal: Its Background, Origin, and Subsequent History*. London: Oxford University Press.

Jackson, H. J. 2001. *Marginalia: Readers Writing in Books*. New Haven, CT: Yale University Press.

Jardine, Lisa and A. T. Grafton. 1990. "'Studied for Action': How Gabriel Harvey Read his Livy." *Past and Present* 129: 3–51.

Jones, W. D. 1895. *Records of the Royal Military Academy*, 2nd ed. Woolwich: [Printed at the Royal Artillery Institution].

Mohr, Georg. 1673. *Compendium Euclidis Curiosi*. Amsterdam: Joannes Janssonius van Waesberge.

Netz, Reviel. 1999. *The Shaping of Deduction in Greek Mathematics*. Cambridge: Cambridge University Press.

Norwood, Richard. 1659. *Norwood's Epitomie*. London: R. and W. Leybourn for G. Hurlock.

Oosterhoff, Richard J. 2015. "A Book, a Pen, and the Sphere: Reading Sacrobosco in the Renaissance." *History of Universities* 28: 1–54.

———. 2018. *Making Mathematical Culture: University and Print in the Circle of Lefèvre d'Étaples*. Oxford: Oxford University Press.

Pardies, Ignace Gaston. 1671. *Elemens de geometrie*. Paris: Sebastien Mabre-Cramoisy.

Poole, William. 2018. "A Royalist Mathematical Practitioner in Interregnum Oxford: the Exploits of Richard Rawlinson (1616–1668)." *The Seventeenth Century*, 33 (5): 557–86.

Raphael, Renee. 2013. "Teaching through Diagrams: Galileo's *Dialogo* and *Discorsi* and his Pisan Readers." *Early Science and Medicine* 18: 201–30.

———. 2015. "Reading Galileo's *Discorsi* in the Early Modern University." *Renaissance Quarterly* 68: 558–96.

———. 2016. "Galileo's *Two New Sciences* as a Model of Reading Practices." *Journal of the History of Ideas* 77: 539–65.

———. 2017. *Reading Galileo: Scribal Technologies and the Two New Sciences*. Baltimore, MD: Johns Hopkins University Press.

Ratdolt, Erhard. 1491. *Preclarissimus liber elementorum Euclidis perspicacissimi*. Vicenza: Leonardus Achates de Basilea and Gulielmus de Papia.

Rudd, Thomas. 1651. *Euclides Elements of Geometry*. London: Robert and William Leybourn for Richard Tomlins and Robert Boydell.

Savile, Henry. 1621. *Praelectiones ... in Principium Elementorum Euclidis*. Oxford: John Lichfield and James Short.

Seller, John. 1677. *A Pocket Book, Containing Several Choice Collections*. London: John Seller.

Sherman, William. 2008. *Used Books: Marking Readers in Renaissance England*. Philadelphia, PA: University of Pennsylvania Press.

Simpson, Thomas. 1745. *A Treatise of Algebra*. London: John Nourse.

Tacquet, Andreas. 1703. *Elementa Geometriæ*. Translated by William Whiston. Cambridge: Corn. Crownfield for the University Press.

Tapp, John. 1674. *The Sea-Man's Kalendar*. London: W. Godbid, for E. Hurlock.

Tracey, Kevin. 2020. "'Several choice collections' in Geometry, Astronomy, and Chronology: Using and Collecting Mathematics in John Seller's *A Pocket Book* (1677)." In Beeley et al. 2020a.

Walkingame, Francis. 1751. *The Tutor's Assistant*. London: Dan Browne for the author.

Ward, John. 1707, 1719, 1734, 1752, 1758, 1771. *Young Mathematician's Guide*. London: Edw. Midwinter, for John Taylor.

Wardhaugh, Benjamin. 2020a. "'The Admonitions of a Good-Natured Reader': How Georgians Read Mathematics." In Beeley et al. 2020a.

———. 2020b. "Defacing Euclid: Reading and Annotating the *Elements of Geometry* in Early Modern Britain". In *Early Modern Universities: Networks of Higher Learning*, edited by Anja Silvia Goeing et al. Leiden: Brill.

Warwick, Andrew. 2003. *Masters of Theory: Cambridge and the Rise of Mathematical Physics*. Chicago: University of Chicago Press.

Wells, John. 1635. *Sciographia: or, The Art of Shadowes*. London: Thomas Harper for Andrew Hebb.

Whiston, William. 1765. *The Elements of Euclid*. Dublin: Isaac Jackson.

Whiteside, D. T. 1967. *The Mathematical Papers of Isaac Newton*. Cambridge: Cambridge University Press.

Woolf, D. R. 2000. *Reading History in Early Modern England*. Cambridge: Cambridge University Press.

CHAPTER 31

Mathematics, Narrative, and Temporality

Marcus Tomalin

This chapter explores a form of mathematical criticism that is similar in spirit to that advocated by the philosopher Imre Lakatos over forty years ago. In essence, it contends that mathematical texts (especially proofs) possess literary qualities which render them amenable to literary critical scrutiny. The discussion focuses on several classic proofs from the eighteenth and nineteenth centuries that were propounded by influential mathematicians such as Leonhard Euler and Carl Friedrich Gauss. Various aspects of narrative structure manifest in these texts are examined, particularly the artful shifts in tense that occur at pivotal moments. Although it is often assumed that mathematical proofs exist only in a disembodied form of the present tense, the examples explored in this chapter show that temporal transitions frequently reveal underlying conceptual limitations and methodological uncertainties.

Lakatos's *Proofs and Refutations* (1976) is written in the form of a Platonic dialogue in which a group of students discuss the relation between the number of vertices (V), edges (E), and faces (F) in three-dimensional polyhedra. Taking Leonard Euler's celebrated equation $V - E + F = 2$ as a starting point, they modify that conjecture in response to increasingly troublesome counterexamples, such as a cube with a four-sided hole drilled through it. As the conversation progresses, one of the students, Gamma, becomes frustrated by the "cheap, *trivial*" modifications that are introduced merely to accommodate dubious anomalous cases, and warns against needlessly devising "complicated, pretentious formulas" (Lakatos 1993, p. 97). These concerns lead him to draw an analogy between the cultural practices of mathematics and those of literature: "[w]hy not have mathematical critics just as you have literary critics, to develop mathematical taste by public criticism?" (Lakatos 1993,

---

M. Tomalin (✉)
Trinity Hall, Cambridge University, Cambridge, UK

© The Author(s) 2021 569
R. Tubbs et al. (eds.), *The Palgrave Handbook of Literature and Mathematics*, https://doi.org/10.1007/978-3-030-55478-1_31

p. 98). Choosing not to elaborate upon this, he assumes the comparison is clear. But what exactly would these "mathematical critics" do? Presumably, they need not necessarily write mathematical treatises themselves, since (in the analogical domain) literary critics are not always, or even usually, poets, novelists, or playwrights. Instead, their task would be to reveal the strengths, weaknesses, beauties, and infelicities inherent in existing mathematical texts. According to Gamma, the ultimate purpose of this kind of critical scrutiny would be to improve "mathematical taste", with the aim of ameliorating the quality of mathematical research in general.

Whether this seems an odd methodology or not may depend on your understanding of what constitutes mathematics proper in the first place. Those indoctrinated into the ideology of Formalism (especially its more hard-line varieties) are unlikely to consider the processes of discovery and the acquisition of taste as constituting part of mathematics at all. Either a conjecture has a complete and correct formal proof, in which case it is a theorem, or it does not—and the body of accepted proofs constitutes the totality of mathematics. Proof-theorists such as Gaisi Takeuti are unflinching: "[m]athematics is a collection of proofs. This is true no matter what standpoint one assumes about mathematics – platonism, anti-platonism, intuitionism, formalism, nominalism, etc. Therefore, in investigating 'mathematics', a fruitful method is to formalize the proofs of mathematics and investigate the structure of these proofs" (Takeuti 1987, p. 1). Seemingly, there can be no gainsaying: this is the only way of proceeding.

Yet Lakatos argued that these dominant attitudes were detrimental. Although they have their origins in the axiomatic-deductive method influentially exemplified in Euclid's *Elements*, circa 300 BCE, they gathered considerable ideological force during the great debates about the foundations of mathematics in the early twentieth century. Lakatos's dissatisfaction with this pervasive creed resurfaces later in *Proofs and Refutations* when he returns to the contrast between mathematical and literary criticism:

> It was the infallibilist philosophical background of Euclidean method that bred the authoritarian traditional patterns in mathematics, that prevented publication and discussion of conjectures, that made impossible the rise of mathematical criticism. Literary criticism can exist because we can appreciate a poem without considering it to be perfect; mathematical or scientific criticism cannot exist while we only appreciate a mathematical or scientific result if yields perfect truth. (Lakatos 1993, p. 139)

There is much here that requires unpacking. The central idea is that the prevailing obsession with formal proofs suppresses a more generous mathematical culture in which informal or partial arguments can be discussed and assessed. But even before considering the alleged deleterious influence of "infallibilist" Euclideanism, there is (once again) the small matter of the analogy. Since Lakatos's notion of "mathematical criticism" is constructed via direct comparison with literary criticism, it presupposes a degree of internal stability and coherence in the latter. In other words, it assumes some kind of consensus about what literary criticism *is*, and what it is *for*. Yet, for many literary critics, these remain moot points. Starting in the mid-1950s, the literary critic René Wellek began to delineate a history of the practice from the eighteenth century onwards; he characterised it as involving "the study of concrete works of literature with emphasis on their evaluation" (Wellek 1963, p. 35). In subsequent decades, this has become (for some) a conventional way of distinguishing literary criticism from its conjoined twin, literary theory. The literary critic Patricia Waugh, for example, defines the difference thus: "[w] hereas literary criticism tends to emphasize the experience of close reading and evaluation and explication of individual works, literary theory insists that assumptions underlying reading practices must be made explicit, and that no reading is ever innocent or objective or purely descriptive" (Waugh 2006, p. 2). Even if such contrasts are accepted as being plausibly sustainable, the difficulties of pluralism remain. While Gary Day identifies two broad and dominant subtypes of literary criticism—"rhetorical criticism" which involves the belief that "literature can mould behaviour", and "grammatical criticism" which is "editing texts and establishing their authenticity"—there are, in truth, roughly as many ways of practising literary criticism as there are critics who practise it (Day 2008, p. 1). Any attempt to identify a shared underlying, or overarching, methodology is usually doomed to failure, unless a few high-level platitudes about a small number of loosely identifiable schools are deemed sufficient. Lakatos's chosen analogical domain is, therefore, protean at best—a fact that necessarily complicates his subsequent remarks concerning "mathematical criticism".[1]

Nonetheless, despite these definitional intricacies, if the study of mathematics and the study of literature are to be compared, then it is reasonable to expect that someone familiar with both might usefully elucidate the discussion, and certainly the nineteenth-century mathematician and poet William Rowan Hamilton wrote insightfully about the two disciplines (see Tomalin 2009). While still in his teens, Hamilton thought deeply about the related tasks of constructing and presenting axiomatic-deductive systems. In a dialogue entitled *Waking Dream* (1822), Pappus (Hamilton) interrogates Euclid concerning the *Elements*, probing differences between processes of heuristic discovery and the formal presentation of the final proofs. At one point, he compliments Euclid for having been "so successful ... in disguising the Analysis which you pursued, that to this day even the learned are doubtful whether your discoveries were made by a gradual process ... or whether they were imparted as an immediate gift from Him who constructed for the Bee its wondrous habitation" (Graves 1882, p. 662). As the dialogue continues, Pappus considers apparent fissures between discovery and presentation, and suggests that all mathematical treatises deceive. The structures of formal proofs conceal the particular lines of reasoning that prompted their construction in the first place, just as a published poem masks its own redraftings and

572   M. TOMALIN

revisions. Once the components of a proof have been successfully determined and elaborated, they are then reordered, reconfigured, and re-expressed to ensure that the finished version is as concise, captivating, and convincing as possible. In Hamilton's dialogue, Euclid responds to these suggestions with admirable candour:

> It was not unintentionally that I adopted, as the medium of communicating to my contemporaries those results at which I arrived, a Synthesis, which presented them under a form the best adapted to excite astonishment, and to disguise the process of discovery. To exoterics the science appeared more interesting as it was more mysterious. ... The inventor of a curious piece of mechanism does not expose his artifice to the vulgar eye; nor does an architect, when he has erected a magnificent edifice, leave the scaffolding behind. (Graves 1882, p. 662)

Hamilton's Euclid readily acknowledges that dissimulation is central to proof presentation. The veiling of the discovery process engenders admiration and astonishment, so the best proofs (like the best poems) do far more than merely conform to formal structural constraints. These inherent tendencies may be conspicuous and important, yet for many theorists they fall outside the domain of mathematics proper (Curtin 1982, p. 26ff.). In his well-known textbook *How To Prove It* (1994), the mathematician Daniel J. Velleman acknowledges that "there's a difference between the reasoning you use when you are figuring out a proof and the steps you write down when you write the final version of the proof". But he asserts bluntly that "[t]he first is psychology; the second mathematics" (Velleman 1994, pp. 87–88). Hamilton's Pappus views the matter quite differently. He remains preoccupied with the actual discovery procedures Euclid used, and considers the genesis of the definitions, postulates, and axioms to be the main work of the mathematician. He is eager to know "by what intuition you selected *à priori* all that could be necessary or useful and nothing besides?". Euclid's response is characteristically telling:

> You are not to suppose that they were received at once, or as you have expressed it *à priori*, that form in which they now appear. The Definitions arose, some out of necessity of making my own ideas precise, and of communicating them to others; some I introduced that I might from the statement of a simple property deduce by geometrical reasoning properties less obvious and remote; some were suggested by analogy, and others invented afterwards, to present under a more systematic form the introduction to the science. In a word, no part of the *Elements* has received more alterations as I proceeded than the collection of Definitions with which they commence. ... It was not at once that I perceived the smallest number of data that were sufficient to resolve all geometrical problems, and effect all geometrical constructions. (Graves 1882, p. 663)

This all sounds very Lakatosian: mathematics as an experimental science; the construction of an axiomatic-deductive system as an heuristic art form, a dynamic process of iterative redefinition and adaptation. In summary, the sequential ordering of the mathematical argument in the *Elements* differs greatly from the order in which the various "discoveries" were made—and Hamilton's awareness of this enables him to recognise the beauty, artistry, and structural magnificence of certain mathematical arguments. And many other practitioners of the art have shared this appreciation. G. H. Hardy claimed that a mathematician "like a painter or poet, is a maker of patterns", before famously declaring that "there is no permanent place in the world for ugly mathematics", while Paul Erdős often spoke wryly of "The Book" in which God had written down all the most elegant proofs (Hardy 1967, pp. 84–85; Aigner and Ziegler 2014, p. v).

Lakatos's notion of "mathematical criticism", and Hamilton's appreciation of the aesthetic potency of formal arguments, encourage a more attentive consideration of the structural properties of mathematical texts. And far from merely confirming or questioning validity or soundness, such assessments could also identify characteristic stylistic features. An undertaking of this kind would certainly provide opportunities for something akin to literary critical scrutiny. It is important to remember that, despite the strictures of Formalism, mathematical arguments have taken many different forms over the centuries, and have been expressed with varying degrees of punctiliousness—from speculative hypothesising in private letters, to partially complete expositions in pedagogical textbooks, to exhaustive proofs published in journal articles and monographs. Ever since the appearance of the first periodicals specifically devoted to the mathematical sciences (for example *Journal des sçavans* (1665), *Giornale de'letterati* (1668), *Acta Eruditorum* (1682)), such publications have become the primary forum for the sharing of new mathematical research. Yet stylistic variation abounds even in scholarly articles that have been stringently peer-reviewed. Most mathematical texts (including proofs themselves) are usually written in some kind of narrative form: along with sequences of formal mathematical symbols, they offer sentences in natural language. Hence the conventional use of the phrase "narrative proofs" when the different subtypes of proof are identified (Wohlgemuth 1990, p. iii; Healy and Hoyles 2000, pp. 415–25).

To illustrate by way of an example, I will briefly discuss one of the most famous mathematical texts from the late twentieth century—namely, Andrew Wiles's "Modular Elliptic Curves and Fermat's Last Theorem" (1995). This monumental article finally provided a proof for Fermat's notorious 358-year-old conjecture that there are no nonzero integers a, b, c, such that $a^n + b^n = c^n$, with $n > 2$. Published in the prestigious bi-monthly journal *Annals of Mathematics*, the text extends to more than one hundred pages, consisting of an introduction followed by five chapters, an appendix, and references. Although the whole article is frequently referred to as Wiles's *proof*, it is helpful to be more precise than that (Cornell et al. 1997, p. 2). The introduction, for example, which is written in a discursive narrative mode, summarises the various stages in the informal discovery process that enabled Wiles to piece his argument together. It contains statements such as:

574   M. TOMALIN

> I began working on these problems in the late summer of 1986 immediately on learning of [the mathematician Ken] Ribet's result [which demonstrated a conjectured connection between modular forms, elliptic curves, and Fermat's Last Theorem]. For several years I had been working on the Iwasawa conjecture for totally real fields and some applications of it. In the process, I had been using and developing results on ℓ-adic representations associated to Hilbert modular forms. (Wiles 1995, p. 449)

These passages provide historical context, and they are predominantly in the style of a conventional intellectual autobiography. Although the introduction also offers an informal overview of the article's most important technical contributions, the core of the mathematical argument appears in the five main chapters. Therefore these (presumably) compose Wiles's proof. Yet, once again, it is possible to be more specific than that. The chapters are divided into numbered subsections, and in addition to presenting expository summaries of the central mathematical ideas, they also contain sub-subsections explicitly classified as "*Proof*" and "*Remark*". Indeed, the chapters contain thirty distinct "*Proof*" sub-subsections in total, with a further two appearing in the appendix. These all take the familiar form of deductive narrative proofs overtly associated with stated propositions, theorems, or lemmas. For instance, Lemma 1.10 case (i) is expressed as follows

> *If* im ρ_0 *has order divisible by p then*:

> I. *It contains an element* γ_0 *of order* $m \geq 3$ *with* $(m, p)=1$ *and* γ_0 *trivial on any abelian quotient of* im ρ_0. (Wiles 1995, p. 475)

And here is the corresponding proof for this lemma:

Proof. (i) Let $G=$ im ρ_0 and let Z denote the centre of G. Then we have a surjection $G' \to (G/Z)'$ where the ' denotes the derived group. By Dickson's classification of the subgroups of $GL_2(k)$ containing an element of order p, (G/Z) is isomorphic to $PGL_2(k')$ or $PSL_2(k')$ for some finite field k' of characteristic p or possibly to A_5 when $p=3$ ... In each case we can find, and then lift to G', an element of order m with $(m, p)=1$ and $m \geq 3$, except possibly in the case $p=3$ and $PSL_2(\mathbf{F}_3) \simeq A_4$ or $PGL_2(\mathbf{F}_3) \simeq S_4$. However, in these cases $(G/Z)'$ has order divisible by 4 so the 2-Sylow subgroup of G' has order greater than 2. Since it has at most one element of exact order 2 (the eigenvalues would both be -1 since it is in the kernel of the determinant and hence the element would be -I) it must have an element of order 4 (Wiles 1995, p. 476).

The technical details of this proof, including the specifics of the formalism, need not concern us here. Considering it from the perspective of narrative stylistics, though, the use of the *let*-imperative, the temporal conjunctions ("then"), the modal auxiliaries ("we can find"), the present tense ("it has at most one element"), and so on, all characterise the text and serve to delineate its temporal structure. Linguistic forms such as these are typical of many proofs in modern mathematical literature, and some of them will be considered at greater length in this chapter. Therefore, Wiles's celebrated text certainly demonstrates that the word *proof* is glaringly polysemic. A text classified as a "proof" may contain informal expositions, autobiographical passages, and illustrative asides, as well as formal arguments that establish the validity of theorems, lemmas, corollaries, and the like. The terminology is remarkably slippery. Consequently, for the purposes of the present article, a distinction will be made between a *proof text* and its constituent *proofs*. The former is the entire text in its full stylistic heterogeneity (for example Wiles's whole journal article), while the latter are those particular subsections of a given proof text that contain narratives specifically designed to establish the validity of specific mathematical statements (for example Wiles's proof of Lemma 1.10 case (i)). In other words, while the former might contain autobiographical elements, descriptions of the discovery process, speculative remarks about domains of future research, the latter by contrast consist only of rigorous axiomatic-deductive arguments that establish particular propositions, theorems, or lemmas.

Even with this clarifying distinction, the denotation of "proof" still encompasses a forbiddingly wide range of structural and stylistic possibilities due to the variety of available proof techniques. The most common include direct proofs (for example, if P then Q: assume P, deduce Q), contrapositives (for example, if P then Q: assume $\neg Q$, deduce $\neg P$), proofs by contradiction (for example, if P then Q: assume P and $\neg Q$; deduce $R \wedge \neg R$; therefore the assumption that P does not imply Q is false P), as well as existence and uniqueness proofs, proof by cases, and so on. Just as certain novelists and poets fashion a distinctive literary style that is immediately recognisable, so too do the proofs created by some mathematicians manifest identifiable stylistic characteristics. A desire to understand more deeply this aspect of proof texts has prompted a small number of literary critics, classicists, mathematicians, philosophers, and cultural historians to explore mathematical writings from a range of literary critical perspectives. The mathematician John Allen Paulos's *Once Upon A Number* (1998), for instance, considers various interconnections between stories and statistics; and, influenced by Paulos's work, R.S.D. Thomas summarised core similarities and differences between narratives and proofs, suggesting that lemma were like "flash-backs" since "they allow us to prove things out of order" (Thomas 2002, p. 45). Thomas's analysis does not engage especially closely with literary critical theorising about narratives; he uses "story" and "narrative" virtually interchangeably at times, which obfuscates his argument somewhat. However, more recent studies have been more attentive to such matters. Reviel Netz has written at length about the narrative structures that characterise Hellenistic mathematical writings, focusing on the artful deployment of surprise and profusions of detail; while Alan J. Cain has sought to explain how unexpectedness and inevitability can (despite appearances) be compatible aesthetic characteristics in formal proofs (Netz 2009; Cain 2010). More recently still, Apostolos Doxiadis and

576   M. TOMALIN

Barry Mazur's volume *Circles Disturbed: The Interplay of Mathematics and Narrative* (2012) has offered numerous insights into such diverse topics as the role(s) of storytelling, dreams, biography, and narrative structures, both in mathematical writings themselves and in the formation of mathematical intuitions. As Doxiadis and Mazur put it: "[a] close reading of certain mathematical treatises with a view to their characteristics as narratives reveals the troubled self-questioning of their authors, the drama, and the false moves that accompany the actual process of research" (Doxiadis and Mazur 2012, p. viii). These are all aspects of mathematical texts that have deservedly begun to receive more extensive literary critical attention during the past decade.

This chapter seeks to contribute to this inchoate field by probing more deeply certain narrative-related aspects of mathematical works. In essence, the approach will be to consider proofs as texts that can be subjected to forms of literary critical analysis which elucidate their underlying forms and strategies. Such a broad topic could be approached from a variety of different theoretical perspectives, of course. But the specific methodology adopted here will be essentially the one developed by Gérard Genette concentrating primarily on narrative discourse and temporality. This is partly because Genette's work has been especially influential in several distinct academic domains over the last few decades, but also because it has helpfully foregrounded the centrality of linguistic structures. Influenced by the Bulgarian philosopher and literary critic Tzvetan Todorov, Genette elaborated an analytical framework for narrative discourse (that is, *discours du récit*) that was founded on traditional grammatical categories associated with verbs (for example tense, aspect, mood) (Genette 1983, p. 29). The system he fashioned explored, among other things, the temporal orderings and disorderings of stories and narratives that are manifest as discernible anachronies, such as analepsis (essentially, a flashback) and prolepsis (essentially, a flash-forward) (Genette 1983, pp. 35–36). Given the reliance of narrative on temporal sequences, this provides a reasonable starting point, since, from Genette's perspective, all narrative is "the expansion of a verb" (Genette 1983, p. 30). And mathematicians have started to adopt a similar approach to proof texts, with Michael Harris recently offering comparable advice: "[t]o detect a narrative structure in a mathematical text, first look at the verbs" (Harris 2012, p. 139). This is sensible enough, but it is important to remember that narrative temporality is conveyed by far more than merely the verb complex alone. The tense and aspect of "I am travelling to France" may (in certain contexts) imply presently ongoing activity, but the addition of an adverbial phrase, "I am travelling to France *on Wednesday*", undeniably locates the action in the future. As Mark Currie has reminded us, "[p]hilosophers and linguists broadly accept that temporal reference is not determined by tense alone, that any single tense, be it past, present or future, is capable of expressing past time, present time and future time" (Currie 2007, p. 34). Consequently, if we seek to read proofs (or any other texts, for that matter) with a particular regard for their manipulation of temporality, we must be prepared to consider the full complexity of the temporal references they contain.

Before assessing such phenomena more extensively, though, it is necessary first to justify such an undertaking. Whether proofs can be classified as narratives at all depends in part on what constitutes a narrative (that is, *récit* in Genette's critical lexicon), and how this type of discourse is distinguished from its associated story (that is, *histoire*). For the philosopher Jerome Bruner, there are two entirely distinct modes of thought: the paradigmatic mode which "attempts to fulfil the ideal of a formal, mathematical system of description and explanation", and the narrative mode which "deals in human or human-like intention and action and the vicissitudes and consequences that mark their course" (Bruner 1986, p. 13). Dichotomising analyses of this kind have encouraged the notion that certain characteristic features can be associated with narratives but not with proofs (and vice versa). Thinking in this vein, the philosopher of mathematics Mark Colyvan has stated that "[p]roofs, whatever they are, are not narratives about causal histories"—which at least leaves open the possibility of them being narratives about something else (Colyvan 2008, p. 64). Yet all such considerations depend on the denotation of contentious words such as *action* or *event*. In the Genettian framework, which was devised initially to elucidate Proustian narrative discourse, the noun "narrative" itself is acknowledged to be inherently ambiguous, but Genette chose to prioritise one particular interpretation: "the oral or written discourse that undertakes to tell of an event [*un événement*] or a series of events [*une séquence d'événements*]" (Genette 1983, p. 25). In the latter case, this does not clarify whether the events must be causally related or not, or whether they must involve humans, or at least human-like entities. Predictably, literary critical debates about such matters have rarely considered the case of proofs specifically. Yet given the many uncertainties surrounding the definitions and denotations, there are many opportunities for identifying *events* in mathematical narratives too. For instance, if a formal proof begins with a statement such as "let $x=2$", then an event of some kind has presumably occurred. At the very least, a variable has been assigned a particular value. If a different event/assignment had taken place, then x might equal 3 or 47 or $\sqrt{5}$ instead. Therefore even something as trivial as the establishing of an equivalence, or the application of an operation (for example, addition, multiplication), or the performing of a computation (for example, differentiation, integration) can plausibly be classified as an event. This is why recent studies of mathematical pedagogy have recognised that it is possible to elaborate "a narrative-like causal representation of a proof" (Fletcher et al. 2010, p. 196). These causal representations, which possess the properties of a narrative, are constructed around the sequences of events that take place during the course of the proof.

At this point it is worth recalling the distinction between heuristic discovery procedures and formal proofs, since these are generally associated with distinct respective temporalities. Adapting the English version of Genette's terminology slightly, the first of these will be referred to here as the *discovery story*. This denotes the steps or stages in the actual process of mathematical discovery itself, so this story (or "histoire") consists of a sequence of events ordered in historical time. This is precisely the kind of sequence that Wiles detailed in the Introduction to his 1995 proof text, and it constitutes a narrative in a fairly unambiguous and uncontroversial sense: sentences such as "I began working on these problems in the late summer of 1986" refer to specific occurrences and establish temporal relationships between them (Wiles 1995, p. 449). Every proof necessarily has an associated discovery story, since it has already been constructed before being presented in its final form. However, proofs do not always allude overtly to their corresponding discovery stories, though conventions about this have certainly changed over the centuries. Proof texts from the seventeenth and eighteenth century are far more likely to refer to the discovery story in the midst of the formal mathematical argument, while more recent proof texts will generally confine these components to different sections of the discourse (as Wiles does). Consequently, the second temporality at play is manifest in what will be referred to here as the *proof narrative*. This is the narrative discourse elaborated by the (formal) proof itself. In other words, to use the terms defined above, a *proof text* often contains subcomponents that are *proofs*, and the arguments presented in those proofs constitute *proof narratives*. As mentioned above, the proof narrative may conceal its own discovery story entirely, ensuring that the two temporalities remain distinct, or it may allude to it at strategic moments, thereby introducing external anachronies. Either way, it generally seeks to present a sequence of mathematical events inferentially to create a persuasive axiomatic-deductive argument (for example Wiles's proof for Lemma 1.10 case (i)). While the discovery story is rooted in (past) time and establishes a diachronic chronology (for example a specific breakthrough occurred in a specific year), the proof narrative is often expressed in a logical order. Drawing upon Claude Lévi-Strauss's claim that myths are "instruments for the obliteration of time" Harris has recently argued that mathematical proofs serve a similar purpose: "[m]athematics, whose proofs admit the impersonal imperative and the conditional but whose conclusions are stated in the eternal present tense, is another such instrument" (Harris 2017, p. 255). It is unclear, though, what is meant here by "the eternal present tense". As noted above, temporal reference involves much more than merely the tense classification of the verb complex, and proofs frequently manifest far more temporal variation than is commonly acknowledged. Indeed, as we shall see, much of this variation occurs when the proof narrative is punctuated by references to its associated discovery story. The latter is necessarily temporally prior to the former, and therefore any such references are instances of external analepsis—that is, they extend back in time beyond the start of the proof narrative itself.

At this point, it is worth considering a few illustrative examples, and Leonhard Euler's celebrated treatise *De fractionibus continuis dissertatio* (wrt. 1737; pub. 1744) provides a convenient focus. Euler was one of the greatest mathematicians of the eighteenth century, and this important work was his first published account of the theory of continued fractions. In it he seeks to develop an expansion of e, the irrational constant that is the base of the natural logarithms. An analytical assessment of the verb complexes present in Euler's proof text soon reveals that, unsurprisingly, he utilises several different kinds of temporal reference. The following extract is fairly representative:

> Thus, a alone will be less than the value of the continued fraction [*habebitur valor minor vero*], ignoring the entire fractional part. However, $a+\alpha/b$ will have a value larger than the true value [*valor habebitur maior vero*], since b is smaller than the fractional denominator. But if it is assumed that [*Sin autem sumatur*]
>
> $$a + \cfrac{\alpha}{b + \frac{\beta}{c}}$$
>
> then its value will be smaller than the true value [*habebitur iterum valor iusto minor*] since the fraction β/c makes the denominator $b+\beta/c$ too large. And so in this manner successive truncations of the continued fraction will produce alternately greater and lesser values [*alternatiue valores iusto maiores et minors prodibunt*]. Therefore it will be possible to approach the true value of the continued fraction as closely as desired.
>
> The following sequence of expressions will be therefore obtained:
>
> $$a, a + \frac{\alpha}{b}, a + \cfrac{\alpha}{b + \frac{\beta}{c}}, a + \cfrac{\alpha}{b + \frac{\beta}{c + \frac{\gamma}{d}}}, etc.$$
>
> of which the odd ones in the order (first, third, fifth, etc.) are less than the value of the continued fraction [*prima, tertia, quinta, etc. minors sunt vero fractionis continuae valore*], while the even ones in the order will be greater than that value [*pares autem erunt maiores eodem*]. (Euler 1744, p. 102)

The future passive (for example "*habebitur*") is dominant initially, yet the verb complex does not introduce an overt prolepsis (either internal or external) since there is no attempt to establish any kind of specific forward temporal reference, either to a later part of the proof narrative itself or to any other posterior time locus. The future is a hypothetical one—just as the conditional clause signalled by the conjunction "*Sin*" explicitly creates a hypothetical present, which in turn leads to a reintroduction of the future ("*prodibunt*"). Crucially, though, the predominant future tense actually insinuates a *past* activity. Euler is able to expound his argument only because the successive truncations of the continued fractions *have already been* calculated. They are certainly not being computed for the first time in the narrator's present or future (though the *reader* may not have performed the calculations previously). In essence, therefore, the future reference is essentially a forward projection of a past sequence of events. The main algebraic expansion that provides the focus for the discussion is anticipated by the use of the same future passive as previously ("*habebitur*"), and the subsequent brief excursion to the present tense ("*sunt*") is swiftly followed by an abrupt return to the future ("*erunt*"). If these convoluted changes from one tense (and voice) to another, with their associated explicit and implicit temporal references, must all be classified merely as manifestations of the eternal present, then clearly the analytical vacuity of the latter phrase is a cause of concern.

The most prominent of these temporal shifts is of especial interest since it is prompted primarily by conventions of form. It occurs as a result of the relationship between the preliminary assertion (that is, the statement of a theorem, lemma, or corollary) and the associated proof that follows it. As Wiles's statement of Lemma 1.10 case (i) demonstrates, the preliminary assertion of the lemma is not strictly a component of the proof itself, yet its status as a lemma depends entirely upon the existence of said proof. Indeed, "Fermat's Last Theorem" was famously misnamed for 358 years. In the absence of a proof, it was in truth nothing more than a *conjecture* until 1995, at which point it finally became a *theorem*. And the affiliation between a preliminary assertion and its proof is an intriguing one, not least because it displays many of the qualities that characterise what Jacques Derrida called *supplements*, being an indispensable non-essential element (Derrida 2016, Chapter 2). An author's preface to a novel is not strictly part of the novel itself; but neither is it not part of the novel since the latter would be incomplete without it. Similarly, a preliminary assertion is both outside the scope of the proof proper, and yet simultaneously within that scope too since the one depends upon the other. Despite this quasi-paradoxical relationship, the preliminary assertions are unambiguously part of the associated *proof narratives*. And the same is true of "□", the Q.E.D. symbol conventionally used to denote the completion of the formal argument. The proof is still (technically) complete without this symbol since it only appears *after* the proof has been expounded—which is why, in practice, it is not always deployed. Schematically this all suggests a structure such as the following:

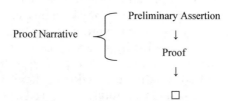

If this captures the nature of the relationship between the preliminary assertion and the proof narrative sufficiently accurately, then it is self-evident that the latter always begins with a proleptic excursion. Proof narratives start by anticipating that which will not be accomplished until the very end. They commence with their own conclusions. While this pattern is conventional in mathematical literature, it is less common in fiction, though certain novels occasionally adopt comparable strategies. Famously, in *Barchester Towers* (1857) the novelist Anthony Trollope created a dramatic scenario in which Eleanor Bold seems likely to marry either the oleaginous Obadiah Slope or the feckless Ethelbert Stanhope. Nonetheless, with thirty-eight chapters of the novel remaining, Trollope decided to reassure his readers:

> But let the gentle-hearted reader be under no apprehension whatsoever. It is not destined that Eleanor shall marry Mr. Slope or Bertie Stanhope. And here perhaps it may be allowed to the novelist to explain his views on a very important point in the art of telling tales. He ventures to reprobate that system which goes so far to violate all proper confidence between the author and his readers by maintaining nearly to the end of the third volume a mystery as to the fate of their favourite personage. ... And what can be the worth of that solicitude which a peep into the third volume can utterly dissipate? What the value of those literary charms which are absolutely destroyed by their enjoyment? (Trollope [1857] 1950, pp. 342–43)

In this celebrated example, Trollope questions the established temporal conventions of Victorian narrative fiction by means of a wry internal prolepsis. In the context of proof narratives, the particular temporal (dis)ordering introduced by the initial temporal excursion is manifest overtly in the use of "□" or its corresponding abbreviation "Q.E.D.". This familiar acronym began to be affixed at the end of proofs during the Renaissance. It is derived from the Latin phrase *quod erat demonstrandum* (literally "that which was to be demonstrated"), and the use of the imperfect active suggests that a temporal transition of some kind has taken place, that an event has occurred. The Latin phrase was in turn derived ultimately from the Greek phrase ὅπερ ἔδει δεῖξαι deployed by Euclid, Archimedes, and other early Greek mathematicians. This phrase means "which was to be shown", and, once again, the main verb ("ἔδει") is conjugated for the active imperfect (Alsina and Nelsen 2010, p. xxii). This distinctive characteristic of proof narratives to commence with a flash-forward potentially raises several troublesome theoretical problems. Recent literary critical discussions of Genette's notion of prolepsis (that is, a flash-forward) have queried whether it is possible for a narrative to *begin* with a proleptic reference (Currie 2007, pp. 35–36). In the case of a proof narrative, for instance, if the preliminary assertion establishes the "first narrative" (or *récit premier*), then arguably it determines a temporality in relation to which the whole of the ensuing proof is an extended internal analepsis (that is, a flashback). The difficulty of distinguishing between these cases in practice has always been a weak point in Genettian approaches to narrative discourse, and proof narratives just happen to highlight the problem in a particularly illuminating manner.

Further light may be shed on these issues by other proof narrative analepses—and especially those that do not arise primarily from conventions of

582   M. TOMALIN

form. Many of these particular anachronies occur when a preliminary assertion established earlier in the same proof text is referred to in a subsequent proof. Since axiomatic-deductive treatises are necessarily accumulative, internal analepses of this kind are routine. However, there are less common, and therefore more intriguing, flashbacks in many proof texts, and Euler's (1744) treatise contains several pertinent examples. Like so many of his contemporaries, Euler included external analepses at strategic moments, and these back references often take the form of allusions to the heuristic methods he deployed while forming his intuitions and developing his formal arguments. For some, his willingness to incorporate elements from discovery stories into his proof narratives is indicative of his broader tendency to be satisfied with partial or pre-formal proofs, an approach the author Julian Havil has described as evincing a "characteristic disregard for rigour" (Havil 2003, p. 38). Yet such critiques are historiographically naïve, since they anachronistically back-project standards of methodological strictness that did not emerge until at least the nineteenth century. Be that as it may, even Euler's more rigorous partial or pre-formal proofs contain frequent analepses which allude to calculations that are not presented within the proof narrative itself. Near the start of *De fractionibus continuis dissertatio*, for instance, he states that

> Cum igitur iam pridem in his fracionibus continuis examinandis laborauerim, atque plura cum ad earum vsum tum inuentionem pertinentia non parui momenti obseruauerim, ea hic exponere constitui, quo aliis viam easdem tractandi planiorem efficerem. (Euler 1744, p. 99)

> (Since I have now been studying continued fractions for a long time, and I have observed many important facts relating both to their use and their derivation, I have decided to discuss them here.)

The temporal adverbs "*iam*" and "*pridem*" combine with the perfect subjunctive of the main verb ("*laborauerim*") to denote an activity that has already occurred over a long period of time, and which certainly was taking place before the commencement of the time locus of the pre-formal proof narrative itself. And both the tense (perfect) and mood (subjunctive) are sustained in the ensuing verbal forms (for example "*obseruauerim*"). These instances of analepsis enable the proof narrative to reach outside of itself and refer to its associated discovery story, thereby alluding to a prior temporal period. Comparable examples emerge later when Euler finally starts to consider the task of creating a continued fraction expansion of e. Having briefly summarised his practical method of calculation, he comments on it as follows:

> Hoc igitur logarithmis aliisque expressionibus transcendentibus tentans deprehendi in eiusmodi fractiones continuas deduci, si numerus cuius loganthmus hyperbolicus est vnitas, eiusque potestates quaeque confiderentur. Posito igitur hoc numero $=e$, erit $e=2, 71\ 828182845904$. (Euler 1744, p. 120)

(Therefore, testing this method by means of logarithms and other transcendental quantities, I have observed that the number whose natural logarithm is 1, and its powers, lead to continued fractions of this kind. Therefore I set this number$=e$, so that $e=2.71828182845904$.)

Once again the perfect tense (*"deprehendi"*) provides the past temporal context for the actions undertaken, deviating from present tense of the main mathematical argument, and thereby creating an external analepsis. The various calculations alluded to (those involving logarithms and other transcendental quantities) have clearly taken place prior to the time locus of the proof narrative itself. Nonetheless, the consequences of those calculations have immediate significance in the present, since the events that occur during the proof (for example the setting of e to the value 2.71828182845904) depend for their justification entirely upon the events in the discovery story that has preceded them.

In external analepses such as these, the discovery story becomes explicitly manifest within the proof narrative, and connections between these two distinct, but related, temporalities are established. By adopting this methodology, Euler overtly acknowledges that the more formal arguments of his proof narrative depend in some way upon his prior informal heuristic explorations. The resulting anachronies can serve a wide range of purposes, some of which will be discussed in the remaining paragraphs of this chapter. In certain cases, though, back references to the discovery story can be used to justify the purposeful omission of important steps within the proof, and there is sometimes a sociocultural dimension to such things. The acceptability of such excursions may depend in part on the reputation of the mathematician concerned, since such manoeuvres presuppose the existence of a relationship of trust between author and reader. If a highly-regarded mathematician declares that he or she has obtained certain results, then that declaration alone may carry the force of a formal proof. In his 1744 treatise, for instance, Euler describes what Christian Goldbach had dubbed "the law of progression" (that is, *"lex progressionis"* [Goldbach 1728, p. 164]; when each term is determined by a number of its predecessors), but then notes that

Lex quidem haec ex ipsa inspection harum fractionum, si vlterius continentur, facile obseruatur; sed eadem etiam ex ipsa fractionum continuarum natura deduci potest: quam demonstrationem autem hic apponere superfluum iudico. (Euler 1744, p. 104)

(This law is easily observed from the inspection of these fractions, if they are continued further, but it is possible to deduce their nature from the continued fractions themselves; however, in my opinion there is no point in including the proof here.)

This is a tease; a remnant of seventeenth-century-style mathematics, when the task of devising the proof was left as an exercise for the reader, to stimulate

584   M. TOMALIN

engagement with the problem. The same technique is, of course, widely deployed in modern pedagogical mathematics textbooks. In this case, the law is capable of being demonstrated (*"deduci potest"*), which suggests that the determining of it has already taken place, prior to the temporal period of the proof narrative itself. Yet, according to Euler, this stage in the argument can be accomplished so easily (*"facile"*), that it is unnecessary for it to be included explicitly. This is a classic example of a technique sometimes wryly referred to as "proof by obviousness", and such methods frequently involve the proof narrative being punctuated by external analeptic references to their associated discovery stories (Gaither and Cavazos-Gaither 2003, p. 274). This is necessary since the allusions to the earlier time locus indicate that an important part of the axiomatic-deductive reasoning has already been accomplished.

The foregoing examples illustrate some of the forms of anachrony that are regularly encountered in proof narratives, and especially those from earlier centuries. But the task of determining their narratological function and purpose is often a complex and subtle one. For instance, quick transitions from prolepsis to analepsis (or vice versa) can signal a crucial moment of axiomatic-deductive fragility, especially in the pre-formal texts of the long eighteenth and early nineteenth centuries. Such transitions regularly reveal either a lurking conceptual weakness in the argument being developed, or at least a localised anxiety concerning the strength of the underlying reasoning. Discovery stories are frequently referred to at these junctures to bolster dubious inferences. Carl Friedrich Gauss's celebrated (1799) proof of the so-called Fundamental Theorem of Algebra offers a classic example. Put simply, this theorem states that every polynomial equation of degree n, with complex coefficients, has n, possibly complex, roots. In the 1790s, Gauss took a slightly different version of this as his starting point, and he offered four different proofs over a fifty-year period. In his 1799 attempt, he sought to demonstrate that an nth degree polynomial with real coefficients of the form

$$P(X) = x^n + Ax^{n-1} + Bx^{n-2} + \cdots\cdots + Lx + M$$

has n roots. As his proof proceeds, he realises that his argument depends upon the assumption that, if an algebraic curve enters a delimited space, then it must of necessity eventually leave that space. Although confident that this hypothesis was beyond reasonable doubt, Gauss clearly felt some trepidation, which he confided in a footnote:

Satis bene certe demonstratum esse videtur, curuam algebraicam neque alicubi subito abrumpi posse (vti e.g. euenit in curua transscendente, cuius aequatio $y = 1/log\ x$), neque post spiras infinitas in aliquo puncto se quasi perdere (vt spiralis logarithmica), quantumque scio nemo dubium contra rem mouit. Attamen si quis postulat, demonstrationem nullis dubiis obnoxiam alia occasione tradere suscipiam. (Gauss 1799, p. 36)

(It seems to have been proved with sufficient certainty that an algebraic curve can neither suddenly break off anywhere (as happens, for example, with the transcendental curve whose equation is $y = 1/\log x$) nor lose itself, so to speak, in some point after infinitely many coils (like the logarithmic spiral). As far as I know, no one has raised any doubts about this. However, should anyone demand it, then I will take it upon myself to provide a proof that is not subject to any doubt, on some other occasion)

The passive infinitive (*"demonstratum esse"*) implies that the proving of the properties of algebraic curves has already occurred prior to the time locus of the proof narrative, and therefore the idea is securely established in the present with sufficient rigour (*"[s]atis bene certe"*). This is yet another example of the discovery story being incorporated analeptically into a proof narrative, to strengthen a frail argument. But things soon become more intricate than that. Anticipating doubters, Gauss immediately transitions to the future tense (*"suscipiam"*) and offers a promissory note of eventual action: if the supplementary proof is required, then it will be provided on another occasion. This is a classic instance of what Currie dubs "rhetorical prolepsis"—that is, "the anticipation of an objection to an argument" (Currie 2007, p. 29). Gauss is here seeking to remove his readers' *future* doubts.

To achieve this, two further temporalities, distinct from that of the discovery story, are incorporated into the proof narrative; and they are both located in the future. First, there is the time locus of the reader, which is (of course) posterior to that of the narrator. Put simply, while promulgating his proof, Gauss is anticipating impending objections. In addition, though, he also refers to another time locus that is chronologically posterior to that of the reader—namely, the imagined future in which he himself will provide the promised proof. Consequently, in rapid succession, an external analeptic reference to the discovery story is followed by two external proleptic references to two distinct future time loci—and it is these shifts, signalled initially by tense-based transitions, that reveal the presence of the "immense gap" that lies at the core of Gauss's proof (Smale 1981, p. 4; also see Basu and Velleman 2017). In fact, despite his promise, Gauss *never* provided a rigorous demonstration of these properties of algebraic curves. Indeed, no mathematician succeeded in furnishing one until Alexander Ostrowski valiantly accomplished the task in 1920 (Ostrowski 1920). And the fact that Gauss went on to publish three more attempted proofs of the same theorem, suggests that he was well-aware of the flaws in his 1799 effort. From a strict ideological perspective, a flawless proof (that is, one in which the steps in the argument can be justified by means of strict inference alone) need not include *any* external anachronies at all. Temporal references of that kind should not be essential to the construction of a formal argument. However, the fact that proofs frequently do involve something more than mere axiomatic-deductive inference is itself revealing, and at such times the lurking anachronies become most clearly manifest. These temporal excursions can take a wide range of forms, and can serve a variety of discourse purposes, but, despite their multifarious nature, they usually disclose conceptual instabilities in the proof narrative in which they are embedded.

This chapter has only considered a few aspects of the complex relationship between mathematics, narrative, and temporality. Since the study of proofs as narratives is a relatively new domain of academic enquiry, the main emphasis in the foregoing paragraphs has necessarily been partly on justifying the view that such texts are amenable to some kind of narratological scrutiny at all. This undertaking requires the established analytical lexis of narrative discourse (e.g., "event" /événement, "story" /histoire, prolepsis, analepsis) to be reassessed in the specific context of mathematical arguments—and the specific sub-type of "proof narratives" has provided the primary focus here. In some respects, proof narratives clearly share common properties with their more purely literary counterparts (for example novels, short stories, poems). Yet they also evince several characteristics that accord them a distinctive form and function as narratives. Fictional stories do not usually start with their conclusions, but proof narratives generally commence in that way—and the narrative-related purposes of these traits have only recently begun to receive the sustained (literary) critical attention they obviously deserve. Inevitably, many specifics remain uncertain and opaque at this stage, and the analyses offered in this chapter are speculative at best. Nonetheless, if the preliminary discussion presented here has at least succeeded in demonstrating that the temporal structure of mathematical proof narratives is a rich and intricate subject that merits serious consideration, then it will have more than served its purpose.

NOTE

1. For a discussion of literary criticism and pluralism, see Metzidakis (2012).

REFERENCES

Aigner, Martin, and Günter M. Ziegler (eds.). 2014. *Proofs from The Book*, 4th ed. Berlin, Heidelberg, New York: Springer-Verlag.

Alsina, Claudi and Roger B. Nelsen. 2010. *Charming Proofs: A Journey into Elegant Mathematics*. Washington DC: The Mathematical Association of America.

Basu, Soham, and Daniel J. Velleman. 2017. On Gauss' First Proof of the Fundamental Theorem of Algebra. *American Mathematical Monthly* 124 (8): 688–94.

Bruner, Jerome. 1986. *Actual Minds, Possible Worlds*. Cambridge, MA: Harvard University Press.

Cain, Alan J. Cain. 2010. "Deus Ex Machina and the Aesthetic of Proof." *Mathematical Intelligencer* 32 (3): 7–11.

Colyvan, Mark. 2008. "Mark Colyvan." In *Philosophy of Mathematics: 5 Questions*, edited by Vincent F. Hendricks and Hannes Leitgeb, 75–85. Pennsylvania: Automatic Press.

Cornell, Gary, Joseph H. Silverman, and Glenn Stevens (eds.). 1997. *Modular Forms and Fermat's Last Theorem*. New York: Springer-Verlag.

Currie, Mark. 2007. *About Time: Narrative, Fiction, and the Philosophy of Time*. Edinburgh: Edinburgh University Press.

Curtin, Deane W. Curtin, ed. 1982. *The Aesthetic Dimension of Science*. New York: Philosophical Library.

Day, Gary. 2008. *Literary Criticism: A New History*. Edinburgh: Edinburgh University Press.

Derrida, Jacque. 2016. *Of Grammatology*. Translated by Gayatri Chakravorty Spivak, 40th anniversary edition. Baltimore: John Hopkins University Press.

Doxiadis, Apostolos, and Barry Mazur (eds.). 2012. *Circles Disturbed: The Interplay of Mathematics and Narrative*. Princeton and Oxford: Princeton University Press.

Euler, Leonhard. 1744. De fractionibus Continuis Dissertation. In *Commentarii Academiae Scientiarum Petropolitanae* 9: 98–137.

Fletcher, Charles R., Sarah Lucas, and Corinne M. Baron. 2010. "Comprehension of Mathematical Proofs." In *Narrative Comprehension, Causality, and Coherence: Essays in Honor of Tom Trabasso*, edited by Susan R. Goldman, Arthur C. Graesser and Paul Van Den Broek, 195–208. London and New Jersey: Lawrence Erlbaum Associates.

Gaither, Carl C., and Alma E. Cavazos-Gaither (eds.). 2003. *Astronomically Speaking: A Dictionary of Quotations on Astronomy and Physics*. Bristol and Philadelphia: Institute of Physics Publishing.

Gauss, Carl Friedrich. 1799. *Demonstratio nova theorematis omnem functionem algebraicam rationalem integram unius variabilis in factores reales primi vel secundi gradus resolvi posse*. Helmstedt.

Genette, Gérard. 1983. *Narrative Discourse: An Essay in Method*. Translated by Jane E. Lewin. Ithaca, New York: Cornell University Press [based on 'Discours du récit', a portion of *Figures III* (1972)].

Goldbach, Christian. 1728. *De terminis generalibus serierum* In *Commentarii academiae scientiarum imperialis Petropolitanae* 3: 164–73.

Graves, Robert Percival. 1882. *Life of Sir William Rowan Hamilton*, vol. 1. Dublin: Hodges Figgis.

Hardy, G.H. 1967. *A Mathematician's Apology*. Cambridge: Cambridge University Press.

Harris, Michael. 2012. "Do Androids Prove Theorems In Their Sleep?" In *Circles Disturbed: The Interplay of Mathematics and Narrative*, edited by Apostolos Doxiadis and Barry Mazur, 130–82. Princeton and Oxford: Princeton University Press.

———. 2017. *Mathematics Without Apologies: Portrait of a Problematic Vocation*. Princeton, NJ: Princeton University Press.

Havil, Julian. 2003. *Gamma: Exploring Euler's Constant*. Princeton, NJ: Princeton University Press.

Healy, Lulu and Celia Hoyles. 2000. "A Study of Proof Conceptions in Algebra." *Journal for Research in Mathematical Education* 31 (4): 396–428.

588 M. TOMALIN

Lakatos, Imre. 1993. *Proofs and Refutations: The Logic of Mathematical Discovery*, reprint. Cambridge: Cambridge University Press.

Metzidakis, Stamos. 2012. *Difference Unbound: The Rise of Pluralism in Literature and Criticism*. Amsterdam: Rodopi.

Netz, Reviel. 2009. *Ludic Proof: Greek Mathematics and the Alexandrian Aesthetic*. Cambridge: Cambridge University Press.

Ostrowski Alexander. 1920. "Über den ersten und vierten Gaussschen Beweis des Fundamentalsatzes der Algebra." In *Nachrichten der Gesellschaft der Wissenschaften Göttingen* 2: 1–18. Reprinted in Carl Friedrich Gauss. 2011. *Werke*, vol. 10, part 2, various editors, 299–316. Cambridge: Cambridge University Press.

Paulos, John Allen. 1998. *Once Upon a Number: The Hidden Mathematical Logic of Stories*. New York: Basic Books.

Smale, Stephen. 1981. "The Fundamental Theorem of Algebra and Complexity Theory." *Bulletin of the American Mathematical Society* 4: 1–36.

Takeuti, Gaisi. 1987. *Proof Theory*, 2nd ed. Amsterdam: North Holland.

Thomas, R.S.D. 2002. "Mathematics and Narrative." *Mathematical Intelligencer* 24 (3): 43–46.

Tomalin, Marcus. 2009. "William Rowan Hamilton and the Poetry of Science." *Romanticism and Victorianism On the Net* 54. http://www.erudit.org/revue/ravon/2009/v/n54/038763ar.

Trollope, Anthony. 1950. *Barchester Towers*. New York: Random House.

Velleman, Daniel J. 1994. *How to Prove it: A Structured Approach*. Cambridge: Cambridge University Press.

Waugh, Patricia (ed.). 2006. *Literary Theory and Criticism: An Oxford Guide*. Oxford: Oxford University Press.

Wellek, René. 1963. *Concepts of Criticism*. New Haven and London: Yale University Press.

Wiles, Andrew. 1995. "Modular Elliptic Curves and Fermat Last Theorem." *Annals of Mathematics* 114: 443–551.

Wohlgemuth, Andrew. 1990. *Introduction to Proof in Abstract Mathematics*. Philadelphia: Saunders College Publishing.

CHAPTER 32

A Cognitive and Quantitative Approach to Mathematical Concretization

Marc Alexander

> We have suggested that ... we tend to structure the less concrete and inherently vaguer concepts (like those for the emotions) in terms of more concrete concepts, which are more clearly delineated in our experience.
>
> (Lakoff and Johnson 1980, p. 112)

> as soon as one gets away from concrete physical experience and starts talking about abstractions or emotions, metaphorical understanding is the norm.
>
> (Lakoff 1993, p. 205)

Mathematical texts for non-expert readers, whether they are presented as literature or as popular science, need to engage with the fundamental problem of how to describe and represent abstract mathematical entities using a natural language (one developed and evolved through usage) such as English. Human beings, as embodied minds perceiving and construing the world around them, generally linguistically encode abstractions in terms of either concrete entities or events readily perceived by the physical senses. This process—which is variously called metaphor, analogy, reification, or concretization—is a key one for books popularising mathematics as a means of using natural language to describe what would otherwise be encoded in challenging notation. However, such a process has to draw carefully on its concrete 'sources' in order to render on a reader an accurate and coherent impression of a given mathematical concept.

M. Alexander (✉)
University of Glasgow, Glasgow, UK

© The Author(s) 2021
R. Tubbs et al. (eds.), *The Palgrave Handbook of Literature and Mathematics*, https://doi.org/10.1007/978-3-030-55478-1_32

589

590   M. ALEXANDER

This chapter analyses at close range a rich example of this sort of concretization, and does so using both quantitative and cognitive methods. I am particularly interested in identifying how, within a large text, rich analogies can be reliably found, and also in describing the cognitive process of constructing one such analogy. To do this, there are two approaches which will be used in turn: firstly, quantitative methods are necessary in order to examine long texts for areas of particularly significant or dense concretizations for more detailed analysis, and secondly, a linguistic approach known as *conceptual blending* will be used to describe the cognitive 'construction' of such a dense concretization in the mind of the reader.

In doing this, in addition to undertaking a close reading of a mathematical analogy, I hope to show the potential use of a new method of finding analogical material in electronic collections of text, and apply for the first time a conceptual blending analysis to a piece of popular science focusing on mathematics, and demonstrate where its utility may be found. I also propose some new methods and refinements to existing methods for stylistic analysis.

LITERATURE AND QUANTITATIVE ANALYSIS

In order to find, with minimal human intervention, an area of concretization-dense popular mathematics, the first part of this chapter uses some new computational methods. Such a quantitative approach to literature is 'founded on principles of maximal attainable objectivity, procedures that are as replicable as possible' (Sinclair 2007, p. 6). Such an approach traces its origins to pre-computational quantitative work, of which T. C. Mendenhall's article in 1887 is the most widely cited. For Mendenhall's studies (1887, on the frequency of different word lengths in major authors, and 1901 on the Shakespeare authorship question), two women undertaking 'very exhausting' work for several months took the place of what would be done on a modern computer in an instant (Mendenhall 1901, p. 102). With the increasing availability of electronic texts, the field of linguistic stylistics was prominent in embracing the additional toolset that quantitative approaches to language provided to literary analysis. For example, Leech and Short in 1981 stated:

> The more a critic wishes to substantiate what he says about style, the more he will need to point to the linguistic evidence of texts; and linguistic evidence, to be firm, must be couched in terms of numerical frequency .... If challenged, I ought to be, and can be, in a position to support my claim with quantitative evidence. (Leech and Short 1981, pp. 46–47)

This approach has become standard within stylistics—including the study of authorship—and after a long period of having little to no impact on the study of literature elsewhere, has cross-fertilised in recent years with the well-established field of broad-scale literary history (as outlined by Underwood 2017) and other experimental and quantitative approaches to literature which are often placed under the umbrella of digital humanities (although note that this umbrella also includes regular and somewhat tiresome debates about its own definition, approach, coverage, and aims; more of this can be found in Terras et al. 2013). Despite the growing use of quantitative methods, John Burrows' observation that in the humanities '[c]omputer-based evidence, especially when it incorporates statistical analysis, is too often regarded with special deference or special scepticism' (1992, p. 91) remains valid today.

Deference aside, for the present study computational and quantitative methods provide something which in the words of Scott Selisker (2012) can 'humbly supplement – usually by just a little, and perhaps one day by a great deal – what we know about literature' (Selisker 2012, n.p.). It is most likely that the most significant impact on literature of such quantitative methods is that they can be used to provide, as they do in this chapter, a starting-point for the close reading which is so essential to the work of the humanities.

METHODOLOGY

The quantitative methodology developed for the work in this chapter aims to join a detailed stylistic analysis of a dense mathematical analogy to a desire to find—as objectively as is reasonable—sufficiently dense areas of mathematical concretization to warrant analysis. The reason for this is that there can be a danger when constructing an argument about texts that one relies on incomplete evidence, often as an unintentional result of selective attention. This problem is particularly acute in linguistics (and linguistic stylistics), where any given native speaker's intuitions about the language they speak are often neither consistent nor accurate (Gibbs 2006, p. 4). It is not always easy to find evidence which is not gained through selective attention—the process of a person reading a book is not a purely mechanical one to be made automatic—and so I aim in this section to develop and employ some quantitative methods to find, from a bottom-up perspective driven by the data itself, an area to which an analyst could usefully pay most attention. The methodology is new, and tailored to this particular genre and research question, but improvements can likely be made, and an adapted version of it would be usable for any study of metaphor or analogy.

I apply this methodology to a popular science text which is particularly full of concrete metaphors for mathematical concepts: Marcus du Sautoy's *The Music of the Primes* (2003; hereafter *MP*). Of the increasing number of popular mathematics books, it fulfils some technical criteria (substantial length, highly ranked in book review sites, available as an electronic text, and written by a professional mathematician, not a journalist) as well as being a well-regarded and popular book. Even the title of the book indicates its enthusiastic engagement with metaphor and analogy, in contrast to the generally more analogy-free material found in textbooks. *MP* focuses on the highly abstract field of prime number theory, and so du Sautoy's analogies are likely to have a strong contrast between abstract and concrete. It is approximately 125,000 words long, which allows, as suggested above, for a computational 'distant-reading' approach to the content designed to identify a text's highly analogical sections. I will also briefly analyse a technical mathematical textbook, which is very low in analogical and metaphorical context, as a comparison to the popular style of *MP*—this means that if the work I describe below to identify analogical concretizations finds some in *MP* and none in the comparison text, we can be more confident that the process is actually finding analogies in the text, rather than producing haphazard results. The comparison text is W. A. Coppel's (2009) *Number Theory* textbook.

With these two books in hand, the general outline of what we need to do with them is firstly to manipulate the texts into a way in which the *meaning* of the words in them can be analysed (that is, to allow us to work with the words and their content), and then to find where these meanings are analogical (that is, where the text is discussing something not to do with mathematics in order to illustrate mathematics). Finally, I then want to identify areas particularly dense in that analogical material. My assertion is that the 'real' content of both books is mathematical, with some biographical and historical background, and that any other text in *MP* could potentially be there for the purposes of analogy.

For the first part, manipulating the texts ready for us to look at their meaning, the work was accomplished by a series of custom Python programs I produced using a series of routines known as the Natural Language Toolkit (NLTK; Bird et al. 2009). The programs firstly separate each text from one long-running document into 400-word files which represent smaller 'chunks' of each book. Within each file, the words in that text were then reduced to their lower-case lemmas (sometimes known as a word's dictionary form, so that *run, Runs, ran, running* are all reduced to *run*, while *Cat* and *cats* are reduced to *cat*; see also Alexander and Dallachy 2019). Next, frequently occurring function words were removed from the files on the basis of a standard pre-existing 'stop list' of these words (thus eliminating, for example, *the, of, and, to, a, in, is,* and so on). Thus far, these are all straightforward digital humanities approaches to counting words. Finally, the number of times each remaining lemma occurs in each 400-word chunk of text is then counted and summed in a database.

In the second stage, all these words (and their frequencies) need to be connected to the concepts they encode. To connect words to meanings, the database of *The Historical Thesaurus of English* (Kay et al. 2019; hereafter *HT*) is ideal. The *HT* is a meaning-based rearrangement of the complete contents of the first edition of the *Oxford English Dictionary*, together with the later supplements and the 1990s' three-volume *Additions* series (being approximately equivalent to the second edition of the *OED* in 1989 with the later *Additions*). It therefore lists every meaning recorded in English along with every word which has been used to realise that meaning. Despite the word *Historical* in its title, it should be noted that when filtered to the present day, the *HT* forms the largest thesaurus of contemporary English currently available. In a database program, I took the lists of frequencies (for each 400-word chunk of text) of both books and connected the words to their corresponding *HT* categories. The data now consists, for each chunk of each text, of the words, their frequencies, and each *HT* category which is 'activated' by each word. Note that no disambiguation is performed—unlike in Alexander et al. (2015)—because a simple statistic in the next stage will deal with the issue of multiple meanings for one word (such as when *cosh*, as an abbreviation for the hyperbolic cosine, overlaps with identical word forms of *cosh* meaning a hovel and a bludgeon).

This simple statistic is known as log-likelihood (Dunning 1993), which lets us look at where frequencies of words or meanings are statistically significant in a text. For example, if a book uses the words *garden* and *flowers* and *lawn* so often that, by comparison with other books, all those words are statistically significant, it probably means the book is 'about' gardening, or at least has a decent amount of its content about gardens. Of course, even though the word meanings of a text may be 'about' gardening, the text itself may actually be saying something about personal growth, or loss, or environmental sustainability; the computer can tell the first, and the reader the second. For our purpose, we want to find where in *MP* there are particularly significant occurrences of meanings 'about' something other than mathematics, as in the next section I want to analyse how they are used to help a reader understand prime number theory. The log-likelihood statistic itself identifies those meanings in our two texts which are activated unusually frequently in comparison with the meanings in a reference corpus, which gives a norm against which our mathematical texts are compared. The reference corpus used in this essay is a set of one million random sentences taken from Wikipedia, as a source of a very large amount of expository text. The log-likelihood figure for each of the *HT* categories activated in the texts indicates whether that meaning category occurs significantly more often than one might expect given the reference corpus baseline. For example, if the *Number Theory* textbook used as the comparison text has 5858 references to *HT* category *01.06.04.07 Algebra*,[1] the first step in looking at log-likelihood is to provisionally assume that both *Number Theory* and the reference corpus should have identical proportions of words activating that category (this is known as the null hypothesis; it is the default position one might expect should the category *not* be key to the text, and so exists to be disproved by the statistical test). The *expected* frequency of that category in the text is calculated using the reference corpus, which in this case gives a prediction of 911 occurrences in *Number Theory*, much smaller than the actual occurrence of 5858. The log-likelihood formula then takes the difference between the prediction and the actuality and returns a score of significance; as McIntyre and Walker put it, 'the higher the log-likelihood value the more key or statistically significant an item is or the higher the likelihood that the unusually high or low occurrence of an item is not due to chance' (2010, p. 517). In the case of this example, the log-likelihood value is 12,111, which is an extraordinarily high figure. Rayson et al. (2004) suggest figures above 15.13 to be significant, while figures above 37.3 mean there is a probability of only one in one billion that the difference is not statistically significant. The log-likelihood value in this example suggests very strongly that the semantic category of *01.06.04.07 Algebra* is significant in *Number Theory*, which should be no surprise given the topic of the textbook. The log-likelihood test is an established one (for a range of users see, *inter alia*, Baayen 2008; Baker 2010; Larose 2006; Mendoza-Denton et al. 2003), and in the case of *Number Theory* gives us just a very large list of meanings—as represented by *HT* categories—all to do with mathematics (with one exception, discussed in the following section). This means *Number Theory* does its job well as a comparison text; there are no analogies here, no meanings on our list about anything other than numbers and mathematics. *MP*, however, has a much wider range of entries—some about mathematics, and some not. Thus far, the method is providing what it was intended to produce.

Reading the frequency lists of *MP*, there are four categories of entry: relevant material (about mathematics), supporting material, analogical material, and accidental 'noise'. The next task is to remove everything but analogical material from our lists. Firstly, 'noise' are those semantic categories which are present due to the interference of homonymous or polysemous terms. As an example, in the *Number Theory* comparison text the irrelevant meaning category *01.02.08.01.22.06.06 Preparation of bread* comes out as supposedly statistically significant. This can be discounted because the category is entirely activated by one polysemous word (*proof*, with its multiple meanings of leaving bread to rise and also its mathematical sense), and so the category's 'lexical spread' (how many words are used to represent that meaning) is low. By comparison, a category of *Gardening* in a gardening text would have a huge range of words used to represent meanings and concepts about gardening. Therefore, an entire semantic domain—such as *Clothing*, *Law*, or *Music*—which is potentially key but consists only of one polysemous word used over and over again can be discounted for this analysis as irrelevant 'noise'. Fortunately, the log-likelihood process also removes much other 'noise' by focusing only on the significant concepts.

Putting all of the *MP* chunks back together to see the text as a whole, those hundred meaning categories with the highest log-likelihood score (the 'key categories') after removing the 'noise' as above, give us a list of categories either about mathematics (and related concepts) or potentially analogical. Our final step is to contrast those categories which might be analogical with those which are consistent with the *literal* topic of the text. This judgement is a decision which needs to be made by the analyst, and so for this essay I followed some basic guidelines to categorise the key categories. Firstly, as *MP* is about mathematics, I categorised as relevant material any category within the *HT* structure of mathematics, number, and computer science (which the *HT*, with its historical perspective, currently has as an adjunct of mathematics). Such categories came from *HT* sections including *01.06.04.x Mathematics* and *02.01.x Mental capacity* (describing processes of thought, understanding, and knowledge), as well as the more specific domains *Infiniteness* and *Logical syllogism*. Secondly, areas of supporting material which are of importance to the context of the text were also considered 'relevant' to the literal topic; for example, as the text is academically related, the high log-likelihood of *03.06.06.02 College/university* is to be expected, as is *01.01.06.10.01 United States* and *01.02.07.08.04 Person*, where the author describes the academics involved in research and their locations. Elsewhere, categories of *Determination, Persistence,* and *City* (describing where mathematicians interviewed or discussed came from or worked) were also relevant. In terms of noise, the polysemy of *prime* was the source of most of the domains removed for a low lexical spread (including senses in fencing, preparing a boiler, and the 6 am canonical hour). All these relevant and noisy domains were therefore set aside.

We now have a list of semantic categories which are statistically 'key' to the text when compared to general expository discourse, and which are not immediately relevant to the literal topic of the text, and which are not irrelevant 'noise'. It is here that our sought-after concretized analogies are to be found.

THE MUSIC OF THE PRIMES

Taking these *HT* categories which represent potential analogies, there are two with a very high log-likelihood score: *01.05.07.01.01 Distance/farness* and *01.05.07.06 Direction*. Figure 32.1 below shows the dispersion of these categories throughout the text, indicating that they are not clustered in only some parts of the book but rather appear often and repeatedly across the whole book.

Fig. 32.1 A dispersion plot of key analogical categories in *MP*; the plot represents the book from its start on the left to its end on the right, and each small vertical line indicates a point in *MP* where either *HT* category is used

596 M. ALEXANDER

```
blem As mathematicians navigate their way across the mathematical terrain it is
0 bc they had evolved a very physical way of understanding what it is amongst a
thor Not only do they finally see the way to the peak but also they understand
new mathematics encountered along the way The problems allow for exploration of
ling of exhilaration at discovering a way to reach the summit of some distant p
ily rigid geometry There was only one way that the landscape could be expanded
r case mathematicians finally found a way to avoid having to cross the summit o
dy and Littlewood were fighting their way across Riemann s strange landscape so
 height of the Riemann landscape This way he might be able to find a point at s
important breakthrough He had found a way to show that points at sea level in a
othesis for primes but he had found a way to show that points at sea level in r
 be seen until one had travelled very far north in the landscape As Littlewood
al test By the time you had gone this far north if the graph still did not cros
 primes never run dry Or perhaps this far off peak is just a mathematical mirag
further you count Gauss had seen this far off mountain peak but it was left to
os in his imaginary map Remarkably so far his analysis had worked without him h
s finally powerful enough to navigate far enough north in Riemann s zeta landsc
ician Hugh Montgomery about the zeros far up Riemann s ley line Odlyzko recogni
ng the largest percentage of zeros so far to be found on Riemann s line Fry emb
up in a straight line He couldn t see far enough across his landscape to tell w
```

Fig. 32.2 Twenty random concordance lines of *way* and *far* from MP, output using the author's programs. Concordances such as these are read by running one's eye down the central column, in bold, and reading just enough context as is necessary to establish the sense of each word. Only a set amount of characters on either side of the central word are reproduced in a concordance, and concordances should not be read like normal sentences

Looking back at the database which stores not just the *HT* categories used, but also the words from the text which activate those categories, there are words such as *way, far, line, level,* and *point* all used to encode the semantic categories *Distance/farness* and *Direction.* It is clear that these words are used often as analogies in *MP.* A random set of concordance lines (in Fig. 32.2) shows an extended analogy of a landscape or topography.

While some uses of *way* here are not analogical, there are clear indications that references to a landscape—*terrain, a very physical way of understanding, peak, encountering on the way, summit, landscape, sea level, imaginary map, his landscape,* and so on—are used to explicate abstractions around prime number theory. A brief examination of *line* (used both in the entrenched metaphor of *number line,* see Mazur (2004), and in lines on a landscape) and *level* (used primarily in the phrase *sea level*) can support that point.

Using these indicative words, it is possible to plot where in *MP* they occur (or, more accurately, in how many of *MP*'s 327 chunked-up files of 400 words each), in order to find dense analogical areas for analysis. A graph of this is given in Fig. 32.3. There is a high spike representing the word *line* in file 224, and two clusters, centring around files 97 and 277. From the graph, it is apparent that the cluster at 97 is more mixed with different words than the one at 277, as the latter consists mainly of the lexeme *level,* whereas the 97 cluster has a wider range of analogical items. As I aimed to identify a complex analogy, I chose the material centring around file 97 to analyse, as it is both a dense cluster of analogy and consists of the most mixed terms.

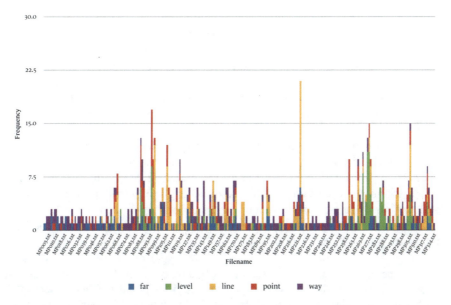

Fig. 32.3 Frequency of key terms in *MP*, arranged by the 400-word 'chunks' the book was split into for analysis

Combining Concepts

The second half of this chapter will engage in a close reading of this cluster of concretized analogy. Many different approaches could be taken, but as a linguist I am primarily concerned with how the language of a passage gives rise to meaning in the mind of a reader. The approach below comes from the framework of conceptual blending theory,[2] a relatively new linguistic approach to the construction of meaning in the same general approach as George Lakoff and Mark Johnson's classic *Metaphors We Live By* (1980) and subsequent work.[3] Blending is 'a theoretical framework for exploring human information integration' (Coulson and Oakley 2000, p. 176), where analysis involves looking at the 'processes in which the imaginative capacities of meaning construction are invoked to produce emergent structure' (Coulson 2001, p. 115). In short, blending is a basic mental operation where two sets of information held in the mind are integrated in order to gain a 'compressed' and insightful meaning. It is fundamental to the construction of meaning, especially of meanings that we have to 'simulate' as readers or hearers, and has even been suggested as the most crucial element of linguistic meaning.

Blending works by taking more than one mental 'input' and blending them to create a resulting concept: a sentence like *University Avenue runs through our campus*, an example of what linguists call fictive motion, blends the concept of firstly an absolutely static paved avenue, secondly that avenue's static context of a campus, and thirdly the active concept of 'running', consisting of a mobile element travelling along a path. The resulting blended idea combines all of these elements, applying a motion verb to an immobile object and giving a 'compressed', graspable idea which simulates motion where no motion is placed and takes advantage of the bias of human perception towards motion, which is more easily understood than a stative concept. For our purposes with regards to popular mathematics, the analogous concretizations looked at in this chapter take advantage of our easy comprehension of physical situations as opposed to the challenge of understanding abstract mathematical concepts. Blending can therefore describe, in an empirically supported way, our ability as readers and hearers to combine two different sorts of information to arrive at an integrated understanding. It is ideally suited for an analysis of analogy. More detail is given below about the theory and its aims, before I apply it to a passage from *MP*.

Blending's leading theorists, Gilles Fauconnier and Mark Turner, describe the wider implications of the theory:

> [Conceptual blending] serves a variety of cognitive purposes. It is dynamic, supple, and active in the moment of thinking. It yields products that frequently become entrenched in conceptual structure and grammar, and it often performs new work on its previously entrenched products as inputs. Blending is easy to detect in spectacular cases but it is for the most part a routine, workaday process that escapes detection except on technical analysis. It is not reserved for special purposes, and is not costly.[4]

> In blending, structure from input mental spaces is projected to a separate, 'blended' mental space. The projection is selective. Through completion and elaboration, the blend develops structure not provided by the inputs. Inferences, arguments, and ideas developed in the blend can have effect in cognition, leading us to modify the initial inputs and to change our view of the corresponding situations.
>
> (Fauconnier and Turner 1998, p. 133)

The theory has developed from its origins in the 1990s into an extensive cognitive approach to the construction of meaning in natural language. Crucially, while blending is used to describe the integration of different concepts into a unified whole (for the purposes of this chapter, the blending of mathematical concepts with concrete analogical material), it also describes a process of *compression* of complex concepts towards what is called human scale. Ungerer and Schmid believe compression towards human scale to be 'the ultimate goal of the whole blending process' (2006, p. 260). They continue:

> The crucial effect of compression is that the conceptual complexity of the inputs from several sources is reduced considerably. A newly integrated and unified conceptual structure emerges that is cognitively manageable and thus has ... 'human scale'. (2006, p. 260)

Human scale can be characterised as a situation which is easily processed, being based on simple, familiar, and embodied concepts. Fauconnier and Turner describe this as follows:

> The most obvious human-scale situations have direct perception and action in familiar frames that are easily apprehended by human beings: An object falls, someone lifts an object, two people converse, one person goes somewhere. They typically have very few participants, direct intentionality, and immediate bodily effect and are immediately apprehended as coherent. (2002, p. 312)

They further state that human scale 'is the level at which is it natural for us to have the impression that we have direct, reliable, and comprehensive understanding' (2002, p. 323). The obvious consequence of this for this essay is that the reduction of complex situations to ones which are at human scale, and thus more easily apprehended, is a clear example of the sort of explication and concretization which characterises popular writing on mathematics.

An analysis of an instance of conceptual blending—such as the one below from *MP*—is an attempt to represent a process going on in the mind when a reader or hearer attempts to comprehend the meaning of a text or utterance. *Blending* is both the name for the theory and for the overall process of comprehension, which takes 'mental space' inputs (knowledge structures in the mind which contain the various information we have about a thing or concept) and when these input spaces are selectively blended together we have another mental space, the *blend*, which should be at human scale, and compressed in the process of blending into something we can easily grasp. This work goes on in parallel (including alternative interpretations), and takes only milliseconds.

For example, the metaphorical blend *to be in the dark about something* contains two 'input spaces' (that of darkness, and that of comprehension—the first from the words used and the second from the context), and they have in common the goal of perception. From the mental space of darkness we take knowledge about the difficulty of perceiving objects in the dark and the potential danger of doing so, and from the mental space of comprehension we take information about comprehension being a desired goal of many activities, a necessary part of experience, and a goal which gives rise to frustration if being blocked. The entrenched COMPREHENSION IS TOUCHING (*I've grasped your idea, hold onto that thought*, etc.) and UNDERSTANDING IS LIGHT (*she enlightened me, a bright idea, seen in this light I don't agree*, etc.) metaphors may also form an additional input space here, as they are standard (and even default) manners of thinking about perception. All this information in the mind is activated by the words used in the expression, and the mind generates the final blend, which combines a goal of comprehension with a difficulty (or impossibility) of perception in the present environment, the danger of attempting to continue without that knowledge, and the frustration of

600   M. ALEXANDER

being in that environment. The technical process of the mind blending all of this—which is too complex to discuss here, but is gone into in detail in either Fauconnier and Turner (2002) or Coulson (2001)—involves taking all of this information and producing the most relevant understanding, which is the output blend. Finally, such a blend can have *emergent structure* by the juxtaposition of elements; darkness is not in itself a negative thing (and can have pleasant connotations, for example during sleep or relaxation) but when put alongside the comprehension mental space and the touching/light entrenched metaphors, the mind can connect a goal-frustration scenario to darkness and so convey the frustration intended to be communicated by the *in the dark* expression.[5]

To analyse the long 'chunk' of *MP* which has been identified as highly analogical, at the level of detail of the above example, would be both tedious and unreasonably lengthy. Instead, an analysis is best formed by examining the mental spaces evoked by the passage (by the words used in the text) and describing the output blends created by the process of blending the mind goes through when encountering these mental spaces. Mental spaces are traditionally very hard to describe, but here we can name them according to their *HT* category name for that concept, principally through the headword nouns which occur in the passage, and this is an advantage of this chapter's new methodology. As the *HT* is a meaning-based organisation of every recorded English concept which can be expressed by a word or phrase, it is ideal to be used as a supplementary aid to describing conceptual blending.

The large analogical passage from *MP* (pages 84−86 in the original text, at the start of the fourth chapter) is too long to reproduce here but key quotations are given below. It comes as part of a description of work done in the distribution of the prime numbers. The ability to predict the distribution of primes is a highly sought-after goal in mathematics, and in 1859 Bernhard Riemann hypothesised that primes can be partly understood by looking at where the graph of a particular function (known as the Riemann zeta function) crosses the x-axis of the graph (that is, where it equals zero). Du Sautoy's primary challenge in the passage under analysis is that 'To illustrate his graph, Riemann needed to work in four dimensions' (*MP* p. 85; all following quotes are pp. 84−86). Each input is a complex number $a+bi$ and each output is a complex number $c+di$, so that its graph exists in two-dimensional complex space, which can be viewed as a four-dimensional real space.

The passage can be analysed in stages, taking each input space in turn as it occurs throughout the extract in order to show the creation of this complex analogical blend. The passage's goal is to generate an understanding of this four-dimensional space for non-mathematicians (who do not naturally relate to higher-dimensional mathematics easily, as any reader of popular science from Edwin A. Abbott's *Flatland* (1884) onwards can understand from the energy many authors have put into its explication). The last stage of the quantitative analysis above gave us an indication that the extract in question focuses on the language of a landscape to discuss the prime numbers (which du Sautoy does by analogizing a landscape to the four-dimensional graph mentioned above). From the passage the idea of dimensional reduction (a 4-dimension graph to a 3-dimension landscape) is formed from a number of intersecting blends of knowledge about graphs, shadows, and landscapes. There is a total of twenty-seven input spaces in the passage, in four major sections. I work through each in turn below.

Mathematics (4 inputs)

The extract begins at the start of a chapter, and opens by reintroducing Riemann and his work combining the mathematics of prime and imaginary numbers in the Riemann zeta function. (Input spaces: mathematics, prime numbers, imaginary numbers, and the zeta function).

Alchemy (4 inputs)

Du Sautoy then compares Riemann's work to discovering, 'like some mathematical alchemist', a mixture of elements like a sought-after treasure. These four input spaces (treasure, alchemy, admixture, and elements) are not used again, and form an illustrative comparison which does not link to the key landscape analogy, but does connect the zeta function to the idea of treasure. The idea of treasure does not reoccur in the extract, but rather is a brief digression to emphasise the important nature of Riemann's work. With this done, the opening discussion then moves to the key analogy.

Landscape (14 inputs)

In the main part of the extract, the zeta function is compared to a vista, and then a large number of additional input spaces are brought in. A reader is taken by du Sautoy from the Reimann zeta function (the same mental space we have already encountered) to new elements, introducing mathematical functions in general, then equations, then graphs of equations. After discussing cosmological analogies for four-dimensional mathematics (where time is the fourth dimension) as well as an economic analogy, du Sautoy describes three-dimensional economic graphs as a 'landscape' with peaks and valleys; this description brings the reader to the idea of a dimension-as-landscape, which will be du Sautoy's major analogy for the Riemann zeta function.

Shadow Side-Analogy (5 inputs)

At this point, du Sautoy pauses his landscape analogy to argue that a reduction of a complex idea to a simpler one can still help you understand that

602   M. ALEXANDER

complex idea. He argues this point through an analogy between reducing a three-dimensional face to a two-dimensional shadow or silhouette, and reducing Riemann's four dimensions to a 'three-dimensional shadow' of them:

> Looking at shadows is one of the best ways to understand them. Our shadow is a two-dimensional picture of our three-dimensional body. From some perspectives the shadow provides little information, but from side-on, for example, a silhouette can give us enough information about the person in three dimensions for us to recognise their face. In a similar way, we can construct a three-dimensional shadow of the four-dimensional landscape that Riemann built using the zeta function which retains enough information for us to understand Riemann's ideas. (*MP*, p. 85)

This creates the idea of reduction of four dimensions to three dimensions so that du Sautoy can then complete and return to his analogy of a landscape; as a landscape exists in three dimensions, the analogy can't begin until we are introduced to the idea that dimensions can be reduced in a representation like a shadow.

Landscape Continued

There are only a few remaining input spaces. The landscape analogy is returned to with a section which combines the language of a graph with the language of a map:

> Gauss's two-dimensional map of imaginary numbers[6] charts the numbers that we shall feed into the zeta function. The north-south axis keeps track of how many steps we take in the imaginary direction, whilst the east-west axis charts the real numbers. We can lay this map out flat on a table. What we want to do is to create a physical landscape situated in the space above this map. The shadow of the zeta function will then turn into a physical object whose peaks and valleys we can explore. (*MP*, p. 85)

The vocabulary of landscape is broadened to an immersive three-dimensional world, with peaks, hills, and sea level:

> As Riemann began to explore this landscape, he came across several key features. Standing in the landscape and looking towards the east, the zeta landscape levelled out to a smooth plane 1 unit high above sea level. If Riemann turned round and started walking west, he saw a ridge of undulating hills running from north to south. The peaks of these hills were all located above the line that crossed the east-west axis through the number 1. Above this intersection at the number 1 there was a towering peak which climbed into the heavens. It was, in fact, infinitely high. As Euler had learned, feeding the number 1 into the zeta function gives an output which spirals off to infinity. Heading north or south from this infinite peak, Riemann encountered other peaks. None of these peaks, however, were infinitely high. (*MP*, p. 86)

This is where the extract identified by the computational analysis above ends, although the analogy continues for some time. The chapter continues, still using figurative material (as can be seen by the dispersion plot in Fig. 32.1), but in a less concentrated manner.

A Map of the Riemann Landscape

This network of input spaces in this passage from du Sautoy's book, with their textual connections to each other shown, is illustrated in Fig. 32.4. The subscript numbers indicate the order in which the input spaces appear in the text (starting from bottom left), and the spaces have had to be arranged in an order which tries as much as possible to correspond to the order of the spaces in the text while also allowing for their display. Each of the twenty-seven input spaces is listed, and the lines between them indicate which spaces are connected to each other in the text (so when du Sautoy discusses a shadow as a two-dimensional picture of a three-dimensional object, the connective verbs and prepositions connect *shadow* to *picture* and *object*, which is what the lines between spaces in the top right of Fig. 32.4 represent).

Our final step in this analysis is to describe the blends created through the complex interaction of these spaces.

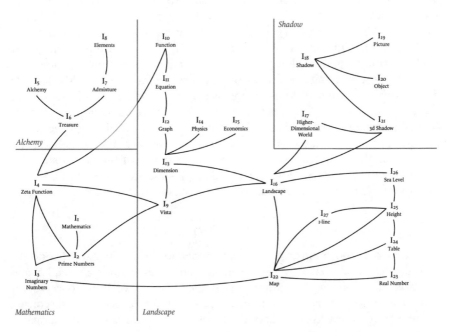

Fig. 32.4 A 'map' of the input spaces in the analysed extract from *MP*

The Blends

There are four blends which are created in the course of the extract, coming together to form the final fifth blend, which itself represents the full set of concretized human-scale knowledge created by the passage.

As discussed above, the first blend straightforwardly uses the alchemy/treasure comparison to make explicit that the zeta function is a valuable result. The other three blends are ones which redefine mathematical concepts introduced and refined through analogy. Blend two concludes the 'shadow' textual group by integrating information from that group into a single conceptual blend, which profiles the nature of a three-dimensional shadow of a four-dimensional entity; the third blend builds on this and integrates it with the landscape analogy to redefine the landscape mental space into one which takes account of the higher-order dimensionality it describes (the four-dimensional space of which the three-dimensional landscape is a shadow); and the fourth blend integrates the measures of the map's axes with the concept of height above the map.

These combine with each other into the full arch-blend, which is the final Riemann landscape. It explicitly combines the second blend above (a 'shadow' of a four-dimensional object, which gives a perspective on the object it is a shadow of, but loses some information) with the third blend (a three-dimensional thing can be thought of as a landscape, with hills, valleys, a sea level, and so on) and the fourth blend (the 'height' of an item, as with du Sautoy's infinitely high peak, which relates to the output of the Riemann zeta function) to come up with the final blended concept of the Riemann landscape. This landscape has hills, troughs, directions, sea level, and a particular line marking the number one (all brought in from input spaces). It tracks three-dimensional elements of interest which correspond to these hills and troughs and infinite peaks, all of which represent axes of a 'shadow' or dimensional reduction of four-dimensional space, which is the blended input space of the valuable zeta function. It also has emergent structure (in conceptual blending terms) which is the nature of the core analogy of the whole passage; that this landscape of the zeta function may have three axes but as we travel through this space, the human-scale experience of the varied shape of the land, with its valleys and infinite peaks, gives a reader a sense of comprehension of the more complex four-dimensional zeta function itself.

Conclusion

As the analysis above illustrates, the concretization of mathematical concepts in a non-technical text (with significant narrative elements) is a complex and recursive situation. When authors make their choices of how to represent mathematics for readers other than experts in a shared common discourse, the natural option is to take established, concrete physical concepts—a map, a shadow, or a landscape—in order to relate the technical material to graspable ideas at 'human scale'.

Such mappings, realised here in the cognitive linguistics paradigm as input spaces, are not entirely unrestricted—in the same way that we can say a theory has *strong foundations* and is *well buttressed* and is *a towering approach* but can't quite say the theory has *good French windows* or *a typical wooden theoretical staircase*, our conceptual structure does not permit us to concretize popular mathematics with any old analogy (the restrictions here are known as mapping scope; see Lakoff 1993). The approach described here, when applied to a sufficiently large corpus of mathematical texts, could in a large-scale way identify the *Historical Thesaurus* categories which are statistically prevalent in the collection of texts and so describe the areas of our conceptual 'inventory' which are available as analogies for popular or literary mathematics. These sorts of big data projects, enabled for the *Historical Thesaurus* by other computational advances such as those described in Piao et al. (2017), will still need to be matched with a close reading—whether of the type above or in another style entirely—of the texts themselves in order to determine not just what can and cannot be analogically mapped to mathematics, but why those mapping scope restrictions exist. Similarly, I hope the conceptual blending analysis presented here—although much abbreviated, especially when considering the richness of blending theory—illustrates the potential power of blending for a close analysis of the construction of meaning in dense and tightly packed passages of analogy. In parallel with my aim of selecting a passage for analysis as objectively as is reasonably possible, the blending approach requires acknowledging every mental space activated in a section of text and accounting for each space's links both to its neighbours and to the human scale blends to which they contribute.

The study of literature and mathematics—taking literature in its broadest sense—can gain to an extent from an engagement with cognitive and quantitative approaches, even if the details of such approaches may appear rather technical and overly narrow for some purposes. Quantitative approaches cannot and should not replace an analyst's reading of a text, but they can supplement our existing methods for finding areas worth studying. Similarly, cognitive approaches should not be used for their own sake, but can focus our attention on the fundamental methods by which texts give rise to meaning in the mind—from the shortest passage to the broadest genre—and so enact the fundamental processes of language.

Notes

1. The long number is the *HT* hierarchical reference code for this concept.
2. The theory has also been known as *conceptual integration theory* (which is treated as synonymous with *conceptual blending* in the literature).
3. Recently, Woźny (2018) has applied conceptual blending theory to fundamental mathematics (not popularised or literary representations), and the theory is

608   M. ALEXANDER

congruent with Lakoff and Nuñez's (2000) approach to mathematics (see also Fauconnier and Lakoff [2010]).
4. 'Costly' here appears to be used in the sense of requiring extra mental processing time. [My note].
5. For many more examples and worked demonstrations, including psychological justifications for conceptual blending's approach, see Fauconnier and Turner (2002) and Coulson (2001).
6. Correctly speaking, Gauss refers to a two-dimensional map of *complex* numbers. Complex numbers are expressed in the form $a+bi$. The real part a is indicated on the horizontal number line (stretching west-east in du Sautoy's terms) and the imaginary part indicated on the perpendicular number line (running north-south). See further Rozenfeld (1988). [My note].

WORKS CITED

Abbott, Edwin A. 1884. *Flatland: A Romance in Many Dimensions*. London: Seely and Co.

Alexander, Marc and Fraser Dallachy. 2020. "Lexis". In *The Routledge Handbook of English Language and Digital Humanities*, edited by Svenja Adolphs and Dawn Knight. London: Routledge. 164-84.

Alexander, Marc, Fraser Dallachy, Scott Piao, Alistair Baron and Paul Rayson. 2015. "Metaphor, Popular Science, and Semantic Tagging: Distant Reading with the *Historical Thesaurus of English*." *Digital Scholarship in the Humanities* 30 (s1): i16–i27.

Baayen, R. Harald. 2008. *Analyzing Linguistic Data. A Practical Introduction to Statistics Using R*. Cambridge: Cambridge University Press.

Baker, Paul. 2010. *Sociolinguistics and Corpus Linguistics*. Edinburgh: Edinburgh University Press.

Bird, Steven, Ewan Klein, and Edward Loper. 2009. *Natural Language Processing with Python*. Sebastopol, CA: O'Reilly.

Burrows, John F. 1992. "Not Unless You Ask Nicely: The Interpretative Nexus between Analysis and Information". *Literary and Linguistic Computing* 7: 91–109.

Coppel, William A. 2009. *Number Theory: An Introduction to Mathematics*, 2nd ed. London: Springer.

Coulson, Seana. 2001. *Semantic Leaps: Frame-shifting and Conceptual Blending in Meaning Construction*. New York: Cambridge University Press.

Coulson, Seana and Todd Oakley. 2000. "Blending Basics". *Cognitive Linguistics* 11 (3–4): 175–96.

Dunning, Ted. 1993. "Accurate Methods for the Statistics of Surprise and Coincidence". *Computational Linguistics* 19: 61–74.

du Sautoy, Marcus. 2003. *The Music of the Primes: Why an Unsolved Problem in Mathematics Matters*. London: Harper Perennial.

Fauconnier, Gilles and George Lakoff. 2010. *On Metaphor and Blending*. http://www.cogsci.ucsd.edu/~coulson/spaces/GG-final-1.pdf (1 August 2019).

Fauconnier, Gilles and Mark Turner. 1998. "Conceptual Integration Networks". *Cognitive Science* 22: 133–87.

———. 2002. *The Way We Think: Conceptual Blending and the Mind's Hidden Complexities*. New York: Basic Books.

Gibbs, Raymond W. 2006. "Introspection and Cognitive Linguistics: Should We Trust Our Own Intuitions?" *Annual Review of Cognitive Linguistics* 4: 135–51.

Kay, Christian, Marc Alexander, Fraser Dallachy, Jane Roberts, Michael Samuels and Irené Wotherspoon (eds.). 2019. *The Historical Thesaurus of English*, version 4.21. https://ht.ac.uk/.

Lakoff, George. 1993. "The Contemporary Theory of Metaphor". In *Metaphor and Thought*, edited by Andrew Ortony, 2nd ed. Cambridge: Cambridge University Press. 202–51.

Lakoff, George, and Mark Johnson. 1980. *Metaphors We Live By*. Chicago: University of Chicago Press.

Lakoff, George, and Rafael Nuñez. 2000. *Where Mathematics Comes From*. New York: Basic Books.

Larose, Daniel. 2006. *Data Mining Methods and Models*. Hoboken, NJ: Wiley.

Leech, Geoffrey, and Mick Short. 1981. *Style in Fiction: A Linguistic Introduction to English Fictional Prose*. London: Longman.

McIntyre, Dan and Brian Walker. 2010. "How Can Corpora be Used to Explore the Language of Poetry and Drama?" In *The Routledge Handbook of Corpus Linguistics*, edited by Anne O'Keeffe and Michael McCarthy. London: Routledge. 516–30.

Mazur, Barry. 2004. *Imagining Numbers: Particularly the Square Root of Minus Fifteen*. Harmondsworth: Penguin.

Mendenhall, T. C. 1887. "The Characteristic Curves of Composition". *Science* IX (214). 237-248.

———. 1901. "A Mechanical Solution of a Literary Problem". *Popular Science Monthly* LX (7): 97–105.

Mendoza-Denton, Norma, Jennifer Hay, and Stefanie Jannedy. 2003. "Probabilistic Sociolinguistics: Beyond Variable Rules". In *Probabilistic Linguistics*, edited by Rens Bod, Jennifer Hay, and Stefanie Jannedy, 97–138. Cambridge, MA: MIT Press.

Piao, Scott, Fraser Dallachy, Alistair Baron, Jane Demmen, Stephen Wattam, Philip Durkin, James McCracken, Paul Rayson and Marc Alexander. 2017. "A Time-sensitive Historical Thesaurus-based Semantic Tagger for Deep Semantic Annotation." *Computer Speech and Language* 46: 113–35.

Rayson, Paul, Damon Berridge, and Brian Francis. 2004. "Extending the Cochran Rule for the Comparison of Word Frequencies between Corpora". *7th International Conference on Statistical Analysis of Textual Data*. http://eprints.comp.lancs.ac.uk/893/ (5 August 2019).

Rozenfeld, Boris Abramovich. 1988. *A History of Non-Euclidean Geometry: Evolution of the Concept of a Geometric Space*. London: Springer.

Selisker, Scott. 2012. "The Digital Inhumanities?". *Los Angeles Review of Books*, 5 November 2012. https://lareviewofbooks.org/essay/in-defense-of-data-responses-to-stephen-marches-literature-is-not-data.

Sinclair, John M. 2007. "The Exploitation of Meaning: Literary Text and Local Grammars". In *Challenging the Boundaries*, edited by Isil Bas & Donald C. Freeman, 1–36. Amsterdam: Rodopi.

Terras, Melissa, Julianne Nyhan, and Edward Vanhoutte. 2013. *Defining Digital Humanities: A Reader*. Farnham: Ashgate.

Underwood, Ted. 2017. "A Genealogy of Distant Reading". *Digital Humanities Quarterly* (*online*) 11 (2).

Ungerer, Friedrich, and Hans-Jorg Schmid. 2006. *An Introduction To Cognitive Linguistics*, 2nd ed. Harlow: Longman.

Woźny, Jacek. 2018. *How We Understand Mathematics: Conceptual Integration in the Language of Mathematical Description*. New York: Springer.

Correction to: Introduction: Relationships and Connections Between Literature and Mathematics

Nina Engelhardt and Robert Tubbs

Correction to:
Chapter 1 in: R. Tubbs et al. (eds.),
The Palgrave Handbook of Literature and Mathematics,
https://doi.org/10.1007/978-3-030-55478-1_1

Chapter 1 was previously published non-open access. It has now been changed to open access under a CC BY 4.0 license and the copyright holder updated to 'The Author(s)'. The book has also been updated with this change.

The updated version of this chapter can be found at
https://doi.org/10.1007/978-3-030-55478-1_1

© The Author(s) 2022
R. Tubbs et al. (eds.), *The Palgrave Handbook of Literature and Mathematics,*
https://doi.org/10.1007/978-3-030-55478-1_33

INDEX

A

Abbott, Edwin A., 286, 488, 489, 600
   Flatland: A Romance of Many Dimensions, 488, 600
Abstraction, 5, 8, 53, 62, 125, 145, 189, 201–203, 248, 257, 258, 300, 341, 355, 407, 428, 433, 501, 509, 510, 529, 530, 542, 543, 545, 546, 589, 596
Accounting, 5, 42, 46, 51, 90, 92, 101, 459, 515, 605
Addison, Joseph, 191, 427
Aeronautics, 151
Agnesi, Maria, 72
Albee, Edward
   Who's Afraid of Virginia Woolf, 247
Albert the Great, 170–172, 177, 184
Albright, Daniel, 532, 542
Alcuin of York, 27
Aleatory, 163, 208, 209, 211, 213, 216, 218, 221, 234, 239, 380
Alexander of Aphrodisias, 171
Algebra, 6, 42, 64, 83, 98, 99, 106, 176, 180, 182, 287, 290, 324, 330, 364, 370–372, 377, 379, 389, 393, 399, 400, 421, 423–425, 434, 444, 446, 449, 482, 486, 499, 500, 510, 557, 558, 560, 562
Algebraic variety, 389

Algorithms, 161, 208, 217, 219, 220, 245, 247, 250–253, 258, 272
Alighieri, Dante, 169, 170, 173, 174, 183
   Convivio, 172
   Divine Comedy, 170–172, 182
Anachronies, 576, 578, 582, 583, 585
Anagrams, 163, 234
Analepsis, 576, 578, 581–584
Analogy, 11, 24, 31, 67, 71, 75, 86, 145, 164, 210, 212, 221, 294, 305, 310, 323, 330, 349, 369, 373, 429, 433, 445, 483, 487–489, 499, 530, 545, 569, 570, 572, 589–592, 596–598, 601, 602, 604
Analytic geometry, 31, 32, 370, 371, 389, 392
Anderson, Benedict, 514
Angle trisection, 415
Anti-Platonist, 350
Aphasia, 544
Archimedes, 132, 174–176, 178, 179, 181, 556, 581
Aristotle, 29, 33, 44, 45, 171, 175, 194, 328, 530, 541
   Categories, 172, 174
   Nicomachean Ethics, 44
   Physics, 24, 32
   Posterior Analytics, 171
   Sophistical Refutations, 171

© The Editor(s) (if applicable) and The Author(s), under exclusive
license to Springer Nature Switzerland AG 2021
R. Tubbs et al. (eds.), *The Palgrave Handbook of Literature
and Mathematics*, https://doi.org/10.1007/978-3-030-55478-1

610 INDEX

Arithmetic, 6, 23, 26, 30, 46, 48, 52, 82, 107, 116, 120, 164, 227, 308, 330, 340, 343, 354, 387, 388, 405, 482, 523, 533, 554, 560, 562
Arithmetical proportion, 44
Arnold, Matthew, 9, 474–476, 478–482, 489
  Culture and Anarchy, 476, 489
  "Literature and Science", 478, 482
  Rede Lecture, 474
Arrival, 276
Ashton, Jennifer, 339, 340, 342, 344
Associationism, 439–442, 444, 446, 447, 451
Astrolabe, 23, 30
Atelier de Littérature Assistée par la Mathématique et les Ordinateurs (ALAMO), 220
Aubrey, John, 553, 554
Audience, 10, 34, 42, 44, 47, 49, 51–54, 58, 83, 87, 132, 133, 140, 145, 222, 246, 249, 252–254, 400–402, 405, 406, 408–414, 416, 441, 448, 450, 451, 474, 475, 477, 487, 488, 562
Audin, Michèle, 217, 228, 255
Aufklärung, 300, 301
Augé, Marc, 237
Austen, Jane, 273–275, 277
  Emma, 414
  Pride and Prejudice, 274
Average, 49, 251, 324, 462
The axiomatic-deductive method, 570
Axioms, 87, 110, 164, 221, 249, 323, 386, 387, 389, 391, 535, 554, 572
Ayckbourn, Alan, 246
  Intimate Exchanges, 246

B
Babbage, Charles, 68, 72, 78, 196, 197, 461, 469
  "Constants of Nature and Art", 461
Badiou, Alain, 364, 370, 372, 375, 376, 380, 385–391
Baker, Humfrey, 47
Barad, Karen, 348, 349, 355
Barron, Brainerd, 246
Bayes's theorem, 445
Bayes, Thomas, 368, 380, 444, 445

Beardsley, Monroe C., 202, 203
  "The Concept of Meter: An Exercise in Abstraction", 202
Beauty of poetry, 486, 489
Beckett, Samuel, 2, 7, 244, 248, 249, 319–322, 324, 325, 327, 329, 330, 332, 334, 335, 514
  Endgame, 327
  How It Is, 320, 331
  Quad, 244, 247–249, 324
  Texts for Nothing, 335
  Waiting for Godot, 331
  Watt, 322
Bee-keeping, 91
Beer, Gillian, 11–13, 69, 339
  Alice in Space, 69
Being and appearing, 390
Bell, Clive, 292
Beller, Greg, 251
  Emotional-Based Music Synthesizer, 251
Bell, Vanessa, 292
Bense, Max, 210, 211, 216
Bens, Jacques, 212, 216–218, 231
Bentham, Jeremy, 477, 479
Berge, Claude, 207, 217–219, 228, 235
Berkeley, George, 141, 320, 333, 334, 424–426
  The Analyst, 333, 334
Bernoulli, Jakob, 442
Bernstein, Charles, 3
Beum, Robert, 189, 192, 201–203
Black, William, 462
Blake, William, 272
Boccaccio, Giovanni, 32, 34
  Criseida, 34
  Il Filostrato, 32
Boethius, 35, 174
Bohr, Niels, 340, 348–351, 355, 369, 370
  indeterminacy principle, 348, 350
Bolyai, János, 4
Book of nature, 3, 377, 444, 446, 448
Boolean logic, 245, 247
Boole, George, 106, 107, 110
Borelli, Giovanni Alfonso
  Euclides restitutus, 556
Boulez, Pierre, 212, 213, 394
Bourbaki, Nicolas, 221

B

Bourbaki, 232
Bradwardine, Thomas, 28, 30, 31, 33, 34
Braffort, Paul, 218–220, 228
Brecht, Bertolt, 257, 283
British Association for the Advancement
   of Science (BAAS), 68, 69, 71, 72,
   77, 83, 473, 474, 477, 479, 483,
   486–488
Broch, Hermann, 282, 295, 296, 301,
   311–313
   The Death of Virgil, 312
   The Guiltless, 295
   The Sleepwalkers, 312
   The Unknown Quantity, 312
Brome, Richard, 53
Brouwer, L.E.J., 282, 286, 296, 312
Bruner, Jerome, 577
Bruno, Giordano, 301
Bryson, 170–172, 178
Buridan, Jean, 30
Burns, Robert, 9, 71, 427–429, 432, 433
Butler, Octavia, 270, 272
   Parable of the Sower, 270
Byron, George Gordon Noel, 457, 465,
   484, 486
   Childe Harold's Pilgrimage, 486

C

Cage, John, 208, 211–214
Calculus, 9, 64, 77, 78, 99, 116–118,
   133, 150, 151, 153, 300, 303–305,
   307, 309, 311, 320–322, 325–327,
   332–334, 421, 424, 425, 432, 434,
   443, 444, 451
   of infinitesimals, 321
Calvino, Italo, 219, 229
Cambridge, 75, 475
Cambridge Mathematical Tripos, 68, 478
Cantor, Georg, 7, 120, 158, 161, 291,
   300, 305, 340, 342, 345, 376,
   386–388, 391, 458
   set theory, 158
Capital, 55, 58, 235, 327, 386, 391,
   500, 503, 509
Capitalism, 10, 41, 43, 74, 131, 133,
   277, 363, 493, 500, 507
Cardano, Gerolamo, 366, 407
Carlyle, Thomas, 461, 462, 470

Carroll, Lewis, 62, 63, 69, 70, 77, 106,
   157, 515, 516, 524, 526
   Alice in Wonderland, 106
   The Hunting of the Snark, 69, 70
   A Tangled Tale, 515, 524
   Through the Looking Glass, 524
Cartesian axes, 408
Category theory, 386, 391
Cauchy, Augustin-Louis, 77, 78, 334,
   425, 483, 484
Cavendish, Margaret, 169, 170, 175,
   177, 178, 183
   "The Circle of the Brain cannot be
   Squared", 177
   "A Circle Squared in Prose", 177
   Poems and Fancies, 176, 177
Cayley, Arthur, 70, 474, 477, 497, 499
Census, 9, 457, 463, 467
Chance, 5, 25, 26, 207, 209–212, 214,
   216, 217, 219, 221, 222, 246, 332,
   362–364, 366–369, 375, 379, 380,
   386, 392, 440, 446, 449, 503
Chance procedures, 208, 222
Chaosmos, 376, 380, 392
Chaos theory, 4, 10, 129–135, 139–142,
   144–146, 253, 380
Chaotic system, 129, 130, 133, 134,
   140, 141, 244, 254
Characterization, 2, 10, 232, 420, 493,
   494, 501–504, 507, 509, 537, 539,
   541, 544
Chaucer, Geoffrey, 23–36
   The Book of the Duchess, 36
   "The Nun's Priest's Tale", 28
   "The Pardoner's Tale", 25, 26
   "Summoner's Tale", 26–28, 30
   Treatise on the Astrolabe, 30
   Troilus and Criseyde, 23, 31–33, 35
Chemistry, 306, 498–500, 502, 510
Chiang, Ted
   "The Story of Your Life", 276
Cicero, 212
Circles, 86, 87, 109, 115, 118, 122, 162,
   170, 179, 208, 238, 254, 282, 344,
   370, 543, 544, 554
Circular causality, 264, 269
Clark, John
   the Eureka, 196
Classical education, 62, 64, 68, 476, 477

612 INDEX

Classical geometry, 69, 138, 144, 476
Clavius, Christoph, 555, 563
Clifford, William Kingdon, 104, 499
Clinamen, 150, 153–155, 157, 159–164, 219
Closed system, 267
Cognitive literary studies, 441
Cohen, Mark
    The Fractal Murders, 134
Cohen, Paul, 387, 388
Coke, Sir Edward, 55
Coleridge, Samuel Taylor, 63–65, 67, 81, 464, 465, 469
    Biographia Literaria, 464, 468
    "The Eolian Harp", 67
    "Kubla Khan", 65, 78
    Lyrical Ballads, 457, 458
    "A Mathematical Problem", 4, 63, 70
Colligation, 499, 510
Combinations, 26, 27, 117, 198, 207, 216, 219, 246, 247, 249, 304, 493, 496, 499, 502
Combinatorial school, 308
Combinatorics, 10, 25, 207, 208, 217, 219, 228, 244, 245, 247, 248, 250, 251, 494, 496, 497, 499, 501, 509, 510
Commodity, 55, 501, 517
Compass-and-straightedge constructions, 415
Complex numbers, 14, 64, 152, 312, 408, 600
Complex space, 408, 600
Complicite, 243, 244, 253–257
Conceptual blending theory, 597
Concretization, 589–592, 598, 599, 604
Congruent, 525, 526, 534
Connections *within* the sphere of mathematics, 483
Conrad, Joseph, 3
    The Inheritors, 3
Consilience, 9, 420–422, 426, 427, 434
Constructible number, 179, 181
Continuity, 31, 77, 78, 152, 303, 327, 330, 376, 420, 421, 425, 483, 486, 487
Continuum, 28, 32, 34, 179, 291, 303, 328, 331, 335, 387, 421, 523
    divisibility of, 30

Continuum hypothesis, 387, 388
Correction, 401, 558, 559
Correlation, 462, 533
Countably infinite set, 387, 388
Craig, John, 332
Creative process, 5, 216, 263, 267, 269, 271, 273
Crilly, Tony, 478
Crisis in foundations, 299
Currie, Mark, 576, 581, 585
Cusanus, Nicholas, 301–305, 309, 315
    On Learned Ignorance (*De Docta Ignorantia*), 302
Cycloid, 161

D

Dada, 210, 212, 213, 215
d'Alembert, Jean le Rond, 227
Darcy v Allein, 55
Darwin, Charles, 12, 92, 324, 480, 502
Day, Gary, 571
Decimal notation, 199, 415
Decimal numeration, 415
Decomposition, 256
Dee, John, 44, 48
    Golden Rule, 48
    Rule of Bartering, 48
    Rules for Exchange, 48
    Rules of Fellowship, 48
Defense
    of mathematics, 474, 477
    of poetry, 474
Defense of the study of Greek literature, 478
De Fermat, Pierre, 25, 371, 573
Dekker, Thomas, 50, 51
Deleuze, Gilles, 248, 376, 380, 393, 394
DeLillo, Don, 150, 151, 157–160, 164
    Ratner's Star, 150, 157, 161
Democritus, 322, 366
De Moivre, Abraham, 444, 445, 554
De Morgan, Augustus, 6, 87, 93, 98, 99, 101, 102, 106, 107, 487
Derrida, Jacques, 365, 366, 370, 373, 374, 376–379, 385, 388, 392, 393, 408, 426, 580
De Saussure, Ferdinand, 7, 290

Descartes, René, 176, 178, 179, 227, 330, 331, 335, 355, 364, 370–372, 389, 394, 407, 546
  Discours de la méthode, 330
  imaginary number, 291
  La Géométrie, 3, 330, 407
  theory of matter, 304
  "universal mathematics", 304
Determinants, 70
De Vetula, 25
Diagrams, 10, 70, 208, 213, 248, 266, 273, 308, 450, 485, 498, 503, 546, 555–557, 560–563
Dice, 25, 26, 209, 211, 212, 221, 366, 375, 376, 394
Dickens, Charles, 9, 84, 460, 502–510, 517–520, 522
  American Notes, 517, 518
  Barnaby Rudge, 521
  Bleak House, 503, 517, 519
  David Copperfield, 518
  Dombey and Son, 517
  Little Dorrit, 521, 522
  Nicholas Nickleby, 522
  Oliver Twist, 518, 519
  Our Mutual Friend, 493
  Pickwick Papers, 519
  Pictures from Italy, 518
  A Tale of Two Cities, 518
Difference demands distinction, 33
Differential and integral calculus, 64
Digital humanities, 17, 591, 592
Digits, 199, 404, 523
Dilthey, Wilhelm, 300
Dimensions, 3, 13, 58, 64, 118, 123, 145, 208, 286, 287, 290, 330, 370, 488, 558
Dinu, Mihai, 247
Discovery procedures, 572, 577
Discovery story, 578, 582, 583, 585
Distributive justice, 44, 55
Dodecahedron, 494
Dodgson, Charles, 69, 70, 82, 84
  Euclid and his Modern Rivals, 69
Döhl, Reinhard, 210, 211
Donne, John, 43, 175
  "Upon the Translation of the Psalms by Sir Phillip Sydney, and the Countess of Pembroke, his Sister", 175

Dorsen, Annie, 247, 249, 251, 252
  A Piece of Work, 244, 250, 251
Doxiadis, Apostolos, 16, 17, 575, 576
du Sautoy, Marcus, 11, 255, 591, 600–602, 604

E
Earl of Rosse, 474, 477
Economics, 9, 41, 42, 45, 46, 507
Eco, Umberto, 208, 229
Eden, Frederick Morton, 458–463, 467
  Estimate of the Number of Inhabitants of Great Britain and Ireland, 463
  The State of the Poor: Or, an History of the Labouring Classes in England, from the Conquest to the Present Period, 458–460, 462, 463
Einstein, Albert, 116, 121, 287, 365, 366, 371
Elements of the Analysis of the Infinite, 305
Eliot, George, 4, 97
  Adam Bede, 99, 100, 102
  Daniel Deronda, 103, 104
  Felix Holt, 103, 107
  Middlemarch, 97, 103, 106, 107
  The Mill on the Floss, 88, 89, 93, 99, 101
  Romola, 103
Eliot's mathematics, 97, 98
Eliot, T.S., 289, 323, 533
Embodiment, 153, 244, 253, 254, 257, 258, 341, 441, 470, 494, 502
Enard, Jean-Pierre, 246
Enargeia, 405
Enargeitic, 411
Encyclopedia, 149, 150, 162, 304, 307, 446, 448
Encyclopedic novel, 149–151, 158, 162–164
Engelhardt, Nina, 17, 152, 153
English Hexameter Translations, 75
Enlightenment, 91, 150, 160, 281, 285, 296, 304, 305, 321, 405, 421, 422, 426, 427
Entropy, 151, 152

614   INDEX

Equality, 9, 44, 45, 181, 194, 529–537, 542, 544, 546
Equation, 3, 9, 17, 70, 73, 74, 115–117, 137, 145, 152–157, 179, 190, 254, 268, 294, 326, 330, 344, 371, 377, 406, 529, 531, 532, 542, 543, 569
Equivalence relation, 9, 535–537, 539
Erdős, Paul, 573
Errata, 557–559
Ethics, 33, 45, 49, 75, 329, 400–406, 408, 409, 411–414
Euclid, 44, 62, 132, 290, 442, 537, 570
    "Euclid's *Elements*", 12, 34, 62, 63, 66, 68, 69, 175, 179, 181, 192, 414, 554
    Prælectiones, 555
Euclid's axioms, 97, 98, 100, 104
Eudoxus (of Cnidus), 150
Euler, Leonhard, 10, 300, 334, 406, 498, 569, 579
Event, 16, 29, 55, 119, 209, 213, 264, 332, 367, 368, 372, 375, 376, 380, 381, 386, 394, 586
Examinations, 69, 90, 307, 562
Exponentiation, 116, 117, 309

F
Fermat's Last Theorem, 10, 573, 574, 580
Fichte, Johann Gottlieb, 307
Fluent, 421, 426, 562
Fluxions, 64, 78, 334, 421, 424, 425, 434
    theory of, 421
Fluxus group, 212
Ford, Ford Madox, 3
    The Inheritors, 3
Formalism, 6, 7, 282, 292, 296, 570, 573, 574
Forsyth, A.R., 477
Foster, G. Carey, 475
Foundational crisis, 7
Four-dimensional geometry, 286, 287, 290
Fournel, Paul, 220, 246
Fractal geometry, 4, 10, 13, 129–135, 137–140, 143–146, 253, 254
Fragment, 310, 372, 375, 463, 465, 468

Frankland, Edward, 498
Franklin, Benjamin, 41, 58
Free indirect discourse, 275
Frege, Gottlob, 282, 288, 289, 300, 323, 340, 342, 344, 345, 351, 534, 537
Freytag, Gustav, 245
Fundamental Theorem of Algebra, 584
Fungibility, 252

G
Galilei, Galileo, 364, 443
Galton, Francis, 462
Gambling, 25, 26, 55, 354, 446, 500
Game, 77, 163, 212, 219, 230–232, 235, 245, 264, 276, 282, 288, 401, 407, 408, 479, 494, 518
Gauss, Carl Friedrich, 10, 77, 158, 300, 305, 313–315, 406, 523, 546, 569, 584, 585, 602
Genette, Gérard, 576, 577, 581
Geometrical optics, 163
Geometrical proportion, 44, 330
Geometrization of space, 421
Geometry, 4, 12, 23, 64, 65, 67, 81–83, 85, 86, 88, 90, 91, 93, 99, 113, 121, 123, 132, 137, 138, 151, 234, 295, 389, 399, 400, 440, 441, 450, 482, 556, 557
German Enlightenment, 300, 303
German idealists, 307
German mystics, 302
Gibson, Anna, 502
Gödel, Kurt, 150, 163, 164, 300, 324, 351, 365, 376, 387, 388, 458
    incompleteness theorems, 164, 323
God is the "Absolute Maximum", 302
Godwin, William, 268, 269
    Caleb Williams, or Things as They Are, 268
Goethe, Johann Wolfgang, 245, 306
    Theory of Colors, 306
Golden section, 66
Grahn, Judy, 343
Grammatical loop, 265
Graph, 220, 228, 232, 235, 331, 462, 497–499, 501, 596, 600, 602
Gray, Jeremy, 7, 282, 364, 366, 369

Gresham College, 47
Grothendieck, Alexandre, 386, 389, 391
Group theory, 247
Gyroscope, 156

H
Haller, Albrecht, 304
    "Uncompleted Poem on Eternity",
        304
Halley, Edmond, 333
Hamilton, William Rowan, 14, 64, 152,
        290, 494, 571
    Hamiltonian cycle, 494
    "The Icosian Game", 494
    "The Tetractys", 64
Hamletian, 412
Handmaiden, 477
Hardy, G.H., 243, 255, 256, 479, 573
    beauty of a mathematical theorem, 479
    A Mathematician's Apology, 243, 256,
        479
Harmony, 5, 174, 190, 192, 195, 304,
        335, 487
    of spheres, 66–68
Harris, Michael, 576, 578
Hartley, David, 443–446, 449
    Observations on Man, His Frame, His
        Duty, His Expectations, 444
Hasse, Helmut, 535
Haughton, William, 50, 51
Haydon, Benjamin, 62, 63
Haywood, Eliza, 447, 448
Heidegger, Martin, 300, 364, 419, 421
Heisenberg, Werner, 324, 348, 362, 365,
        376, 377, 385
    uncertainty principle, 348
Hellenga, Robert
    The Fall of a Sparrow, 139
Henslowe, Philip, 42
Herschel, Sir John, 64
Herzman, Ronald, 173, 174
Hexameter craze, 75
Heytesbury, William, 29, 32
Higher dimensional spaces, 487
Hilbert, David, 151, 164, 282, 286, 292,
        294, 366, 535, 545, 574
    Foundations of Geometry, 220
Hobbes, Thomas, 414, 442, 553–555

Homer, 72, 183
    Iliad, 70
Hood, Thomas, 47
Hopkins, Gerard Manley, 61, 63, 68,
        74, 76
    "Let me be to Thee", 68
    "On the Origin of Beauty: A
        Platonic Dialogue", 61
Horace, 75, 486, 489
    Odes, 68, 69, 484
Human scale, 598, 599
Huxley, Thomas Henry, 473, 475, 476,
        480, 482, 489
    claim that mathematics is "Purely
        deductive" science, 482
Hylles, Thomas, 47–49, 51
Hypatia, 72

I
I Ching, 212–214
Idealism, 67, 103, 374, 375, 378
Ideogram, 532, 533
Imaginary numbers, 2, 3, 10, 64, 118,
        119, 123, 284, 291, 400, 406–408,
        416, 487, 601, 602
Imagism, 529
Incarnation, 4, 45, 173
Indeterminacy, 150, 155, 208, 212, 213,
        218, 293, 340, 344, 348, 349, 351,
        352, 367, 368
Indivisibility, 28, 30, 328
Induction, 255, 445, 473, 483, 504
Inductive method, 483
Infinite, 30, 65, 104, 114, 124, 151,
        176, 180, 208, 300–305, 309, 310,
        312, 327, 604
Infinitesimally small, 424, 425
Infinitesimals, 65, 303, 307, 322, 327,
        425, 429
Instantaneous change, 24, 33, 34
Integration, 153, 237, 307, 321, 325,
        326, 577, 597, 598
Interdependence, 264
International date line, 515, 516
Intuitionism, 7, 81, 107, 141, 255, 282,
        283, 285, 286, 288, 296, 306, 309,
        341, 353, 369, 425, 570, 582
Invariants, 499

616 INDEX

Inverse problem, 444, 445
Inversion, 151, 162–164, 404
Irrational numbers, 2, 179, 199, 284, 322, 326, 332, 400, 407, 525
Irreducible, 362, 364–366, 369, 380, 391, 538
Iser, Wolfgang, 439

J

Jakobson, Roman, 114, 544, 545
James, Henry, 502
James, William, 339, 375
Johnson, Samuel, 426, 427
  Dictionary, 460
Jonson, Ben
  The Alchemist, 51
  Bartholomew Fair, 50, 57
  The Devil is an Ass, 3, 43
  Epicoene, 416
  Volpone, 51
Jouet, Jacques, 5, 217, 236–238
Joyce, James, 161, 169, 181–183, 282, 293–295, 363, 376, 538
  Finnegans Wake, 378, 515
  Ulysses, 180–182, 293, 295; *Circe*, 181; *Ithaca*, 180
Jurassic Park, 131, 132, 134

K

Kafka, Franz, 215, 222, 363
  The Castle, 211, 314
Kames, Henry Home, Lord
  The Elements of Criticism, 443
Kandinsky, Wassily, 248, 362
  Point and Line to Plane, 248
Kant, Immanuel, 81, 300, 306, 307, 314, 320, 373, 393
  The Critique of Pure Reason, 306
Kay, Christian
  Historical Thesaurus of English, 592
Keats, John, 61–63
  Lamia, 61
  "Negative Capability", 63
Kehlmann, Daniel, 301, 313–315
  Measuring the World, 313
Kenner, Hugh, 181, 215, 319, 531
Kepler, Johannes, 150, 158

Kharms, Daniil, 113, 114, 119, 120, 123, 125
Khlebnikov, Velimir, 3, 113–116, 118, 121, 282, 291
  "Tables of Fate", 116
Kirkman, Thomas, 496, 497
  Kirkman's Schoolgirls, 495
Knight, Christopher, 346, 347
Knight's Tour, 235
Knowles, Alison, 214, 215
  The House of Dust, 215
Koch Snowflake, 142
Königsberg Bridge Problem, 496–498
Kovalevskaya, Sofia, 254, 255, 257
Kronick, Joseph, 543, 544
Kuhn, Thomas, 13, 15
Kyd, Thomas, 45, 46

L

Lacoue-Labarthe, Philippe, 208
Ladies' College, 99
Lagrange, Joseph-Louis, 334
Lakatos, Imre, 15, 569, 570, 573
  Proofs and Refutations, 15
Lakoff, George, 16, 354, 589, 597, 605
Lamb, Charles, 62, 63
Landscape, 65, 83, 137, 140, 146, 208, 230, 266, 271, 313, 430, 540, 596, 601, 602, 604, 605
Lanier, Sydney, 76, 77
Laplace, Pierre-Simon, 463
Larkin, Philip, 513, 519
Latour, Bruno, 13, 355
Lautréamont, Comte de, 227, 255
  Chants de Maldoror, 227, 255
Law of small numbers, 154
Lawrence, D.H., 282, 284, 285, 290
  Lady Chatterley's Lover, 284, 285
  The Rainbow, 284, 290
Leibniz, Gottfried Wilhelm, 7, 64, 150, 153, 178, 192, 196, 207, 300, 301, 303–305, 315, 319–321, 323, 325, 329, 334, 335, 424
  Monadology, 323, 332
  petites perceptions, 321
Le Lionnais, François, 216, 218, 220, 228, 229, 233, 245, 246
Leslie, John, 83, 423

INDEX 617

Levine, Richard, 174
Lévi-Strauss, Claude, 578
Lewes, George, 99, 104, 480
Linearization, 155
Linear sentences, 265
Linguistics, 202, 591
Linkage, 49, 389, 420, 499–501, 509, 535
Lipogram, 217
Literary form, 5, 6, 207, 222, 230, 231, 282, 294, 457, 465
Liu, Cixim
    The Three-Body Problem, 140, 146
Lobachevsky, Nicholai Ivanovich, 4, 113, 121, 123
Locke, John, 141, 442, 449
    An Essay Concerning Human Understanding, 442
Logical linguistics, 24
Logicism, 7, 282, 288, 289, 292, 296, 323, 340
Logic of *figure*, 422
Log-likelihood, 593–595
Lovelace, Ada, 72
Lucretius, 150, 163, 320
Lutz, Theo, 211, 212, 222

M
Maclaurin, Colin, 86, 87, 90, 93, 423, 424, 429
Mac Low, Jackson, 213–216
Maimon, Salomon, 307
Mallarmé, Stéphane, 207, 209, 213, 221, 222, 283, 362, 363, 365, 368–376, 378, 379, 385, 386, 391–394
    Un coup de dés jamais n'abolira le hasard, 209
Malthus, Thomas, 462
Mandelbrot, Benoît, 4, 13, 132
Mandelbrot set, 135–137, 142–144, 265
Manwaring, Edward, 192–194, 197
    Stichology: Or, a Recovery of the Latin, Greek, and Hebrew Numbers, 192
Marcus, Solomon, 246, 247
Margins, 216, 236, 254, 553, 554, 556, 561
Markov chains, 245, 252
Markov process, 212
Marvell, Andrew, 175

"Upon Appleton House", 175
Marx, Karl, 458
Massinger, Philip, 53, 54
Masterson, Thomas, 47
Mathematical classicism, 558
Mathematical form, 6, 11, 65, 66, 115, 218, 245, 301, 304, 484, 486
Mathematical laws, 190, 315, 484
Mathematical logic, 106, 164, 210, 221, 323, 373, 389, 390, 509
"The Mathematical Man", 299
Mathematical Platonism, 365, 368, 374
Mathematics and classics, 475
Mathematics and physics section, 71, 473, 474, 479
Mathews, Harry, 219, 221
Matrix, 70, 212, 246, 377, 486
Maxwell, James Clerk, 62, 64, 67–70, 72, 78, 84
    "In Memory of Edward Wilson, Who Repented of what was in his Mind to Write after Section", 71, 72
    "A Problem in Dynamics", 70, 71
    "A Student's Evening Hymn", 67
    "To the Committee of the Cayley Portrait Fund", 70
    "A Vision Of a Wrangler, of a University, of Pedantry, and of Philosophy", 71
Mazur, Barry, 16, 17, 576, 596
McBurney, Simon, 254, 255, 257
    A Disappearing Number, 244
McGovern, Iggy, 64
Mean Speed Theorem, 30, 32
Mechanical Turk, 251
Mehrtens, Herbert, 8, 283
Melissus, 170
Merchants, 41, 47, 48, 500
Mersenne, Marin, 176
Metaphor, 2, 4, 5, 9, 10, 12, 13, 16, 32, 75, 105, 116, 146, 160, 175, 178, 244, 253, 257, 258, 268, 291, 323, 326, 341, 353, 399, 424, 499, 507, 530, 531, 538, 541, 542, 545, 589, 591, 599
Metaphysics, 64, 76, 303, 306, 307, 321, 323, 422
Meter, 189, 190, 194, 195, 199–203, 228, 268, 468
"Methodically Constructed", 295

618 INDEX

Method of exhaustion, 328
Metonymy, 540, 545
Metro poems, 5, 217, 236, 237
Meyer, Steven, 339, 341, 342
Mill, John Stuart, 85, 86, 97, 100, 478, 479
  Utilitarianism, 479
Milton, John, 72, 78, 100, 191, 480, 484
Mimesis, 13, 373–375, 377–379, 388, 394, 530, 543, 546
The Mirror for Magistrates, 414
Modernism, 2, 6, 41, 115, 209, 222, 281, 284, 285, 293, 301, 310, 339, 340, 344, 345, 361–365, 369, 370, 372, 379, 434, 502, 538, 540
Modernist literature, 8, 281–285, 288, 290, 296, 361, 368, 369, 394
Modernist mathematics, 8, 283, 361, 366, 369, 385, 394, 537
Monad, 320–323, 328, 332
Money, 27, 41, 45, 57, 312
Monopoly, 41, 55, 65
Moretti, Franco, 503, 507, 508
Musil, Robert, 283, 299, 301, 311, 312
  The Confusions of Young Törless, 311
  The Man Without Qualities, 311
Mysteries, 62, 63, 173, 305, 315, 333, 368

N
Naden, Constance, 62, 74, 78
  "Scientific Wooing", 74
Nancy, Jean-Luc, 208
Narrative, 10, 12, 17, 34, 49, 73, 78, 84, 114, 119, 150, 153, 155, 157, 162, 220, 268, 269, 286, 314, 369, 441, 447, 457, 465, 493, 519, 526, 573, 577
Narrative discourse, 576, 577, 581, 586
Narrative proofs, 573, 574
Natural philosophy, 11, 35, 67, 281
Naturphilosophie, 306
Negative numbers, 120, 400, 406, 407, 416, 487
Networks, 482, 493, 500, 501
Neufeld, Victoria, 99, 246
New Prosody, 195, 196, 200

Newton, Isaac, 9, 11, 62, 63, 86, 87, 93, 150, 178, 325, 327, 333, 334, 421, 424, 425, 434, 442, 554
  "Method of Fluxions", 64
  Philosophiae Naturalis Principia Mathematica, 11, 281, 323
Nicolson, Marjorie Hope, 12
Non-Euclidean geometry, 13, 63, 69, 77, 94, 98, 104, 110, 121, 161, 182, 300, 487, 545
*Non*linear, 155
Nonlinearity, 5, 263–267, 272, 275
Nonlinear logics of everyday language, 263
Nonlinear pedagogy, 271
Nonlinear play, 265
Nonlinear relationship, 264
Nothing, 35, 43, 131, 145, 192, 266, 402–405, 408, 415
Nothyng, 415
Novalis, 301, 306–310
  Heinrich von Ofterdingen, 307, 310
  Hymns to the Night, 307
  magical theory of language, 308
  Spiritual Songs, 307
Novel, 3, 9, 97, 99, 102, 104, 105, 109, 114, 117, 122, 125, 131, 135, 140, 141, 150, 153, 155–160, 162–164, 221, 235, 269, 275, 277, 285, 287, 312, 335, 439, 451, 493, 494, 502, 503, 506, 508, 510, 517, 518, 580, 586
Numbers, 51, 57, 113
Number theory, 140, 218, 228, 244, 253, 314, 545
Numeracy, 43, 46, 190
Nuñez, Rafael, 354

O
Objective correlative, 78, 533, 534
Objectivity, 15, 16, 102, 282, 284, 289, 351, 353, 399–402, 460, 590
Odd and even value, 524
Olivi, Peter, 25
Ontology, 8, 164, 332, 361, 362, 364–367, 369, 371–373, 380, 385–387, 390, 393
Oresme, Nicole, 24, 31

INDEX 619

Oulipo, 5, 208, 216, 217, 228, 231, 234, 239, 244, 245, 283
Oxford, 23, 47, 61, 475
"Oxford Calculators", 28. *See also* Bradwardine, Thomas; Heytesbury, William; Swineshead, Richard

P

Parallel lines, 75, 314, 500
Par cas, 26
Pardies, Ignace Gaston, 180, 181, 556
 Short, but yet Plain Elements of Geometry, 180
Parmenides, 170, 329, 373
Partial differential equation, 3, 155, 156
Partition, 256
Parts and wholes, 452, 458, 465, 468
Pascal, Blaise, 25, 87, 93, 227, 321, 366
Pater, Walter, 501, 502
Patmore, Coventry, 76, 195, 196, 200
 "Essay on English Metrical Law", 76, 195
Peacock, George
 Treatise on Algebra, 487
Peacock, Thomas Love, 81
Peano, Giuseppe, 535, 537
Perec, Georges, 217–219, 233, 235
Performance, 10, 42, 143, 192, 202, 213, 244, 247, 250, 253, 258, 526, 553, 562, 563
Perloff, Marjorie, 214, 340
Permutations, 209, 211, 215, 219, 222, 249, 322, 484, 496, 499
Peyret, Jean-François, 244, 250, 253–255, 257
 Le Cas de Sophie K., 244
Phonetic Syzygy, 77, 484
Physics, 31, 99, 210, 313, 325, 329, 330, 348, 349, 361, 363, 364, 367, 369, 371, 376, 394
Pierpont, James, 243, 257
Place value, 403, 404, 415
Plato, 29, 84–86, 103, 368, 373, 543
 The Republic, 378, 477
Platonism, 341, 373, 375, 388, 570
Platonist philosophy, 354
Play, 237
Playfair, William, 462

Plenitude, 158, 378, 467
Plotinus, 328
Plotnitsky, Arkady, 8, 15, 16, 365, 368, 451
Poe, Edgar Allan, 76, 77, 194, 196, 484
 "Notes upon English Verse", 194
 "The Rationale of Verse", 76
 "Rationale of Versification", 484
Poetry, rhymed, 268
Poincaré, Henri, 6, 139, 244, 255, 257, 290
Poisson's equation, 154
Poisson, Siméon Denis, 154
Polti, Georges, 245
Poovey, Mary, 460, 462
Pope, Alexander, 191, 192, 427
 An Essay on Criticism, 191
Popular science, 589
Postmodernism, 164, 340, 344, 345
Pound, Ezra, 9, 282, 287–289, 294, 529–532, 539–541, 543
 "Dogmatic Statement on the Game and Play of Chess: Themes for a Series of Pictures", 287
 Vorticism, 287, 288, 294, 529, 531, 541–543
Power set axiom, 388
Practical mathematics, 48, 98, 558
Precession, 151, 153, 156, 164
Prediction, 101, 133–135, 140, 141, 257, 272, 368, 375, 443, 445, 593, 600
Prime number theory, 591, 593, 596
Principle of continuity, 77, 489
Prins, Yopie, 75, 196
Probability, 10, 25, 212, 215, 313, 332, 366, 367, 439–442, 444, 446, 451, 594
Probability theory, 25, 315, 366, 367, 379
Profit, 2, 41, 43, 45, 48, 52, 55, 399, 508
Profiteering, 43, 45, 53, 54
Project, 14, 16, 41, 57, 196, 210, 233, 272, 308, 368, 372, 388, 434, 460, 468, 582
Projector, 43, 50, 53, 54, 56, 57
Prolepsis, 576, 579, 581, 584–586

620   INDEX

Proof, 3, 4, 10, 15, 70, 93, 179, 198, 310, 345, 346, 458, 553, 557, 569, 570, 572, 575, 578, 585
Proof narrative, 578–584, 586
Proof text, 575, 576, 578, 579, 582
Proportion, 31, 46, 52, 73, 190–192, 194, 195, 198, 200, 264, 413, 445, 593
Propositional logic, 10, 439, 441, 442, 451
Propositiones, 27
Prosody, 189
Ptolemy, Claudius, 30, 83, 150
Pynchon, Thomas, 3, 149–154, 156, 157, 160
   Against the Day, 152
   Gravity's Rainbow, 150, 152, 157
Pythagoras, 62, 78, 190, 199, 302, 325
Pythagorean heritage of mathematics, 301
The Pythagorean Theorem, 30, 34, 133, 554

Q

Q.E.D., 74, 104, 340, 341, 345, 346, 355, 581
Quadrature of the circle, 1, 169, 170, 173, 177, 181, 183
Quantics, 499
Quantitative method, 352, 590, 591
Quantum mechanics, 151, 362, 363, 366, 368, 371, 376, 377, 545
Quaternions, 14, 64, 152, 290, 487
Queneau, Raymond, 5, 216, 218, 228–231, 235, 245
   "Les Fondements de la littérature d'après David Hilbert", 221
   Odile, 218
   One Hundred Thousand Billion Poems (*Cent mille milliards de poèmes*), 218
Quetelet, Adolphe, 460, 461
Queval, Jean, 212

R

Rabaté, Jean-Michel, 320, 334
Rabelais, François, 30
Ramanujan, Srinivasa, 255, 256

Randomness, 198, 208, 211, 212, 214, 364, 366, 367, 379, 526
Rankine, W.J.M., 62, 73, 74, 78
   "The Mathematician in Love", 73
Rayleigh, Lord, 475
Realism, 51, 114, 163, 266, 324, 365, 369, 510
Real numbers, 118, 342, 393, 407, 408, 487, 602
Real space, 408, 600
Recorde, Robert, 47, 532
Rectificatory justice, 44
Rectify its circumference, 181
Reduction, 2, 116, 211, 244, 246, 247, 256, 401, 507, 599, 601, 604
Redundancy, 525, 526
Reflection, 12, 43, 92, 105, 118, 119, 151, 158–160, 164, 248, 273, 350, 423, 458, 468, 509, 539
Reflexivity, 535–540, 542
Reid, Thomas, 91, 423, 424, 434
   Algebraic, 430
Relativity, 121, 285, 325, 363, 364, 369, 433, 545
Rhyme, 5, 66, 70, 75, 78, 119, 238, 265, 523
Rhythm, 5, 67, 119, 194, 200, 201, 215, 221, 238, 434, 523, 524
Richards, I.A., 200–202, 294, 538
   Practical Criticism, 200
Rickman, John, 463
Riemann, Bernhard, 77, 78, 300, 484, 601, 602
Riemannian geometry, 369, 394
Riemann zeta function, 600, 604
Rigg, Joshua, 107, 108
"Rimbaudelaire" poems, 220
Rimini Protokoll, 249
   100% City, 249
Robinson, Rev. T.R., 475
Romanska, Magda, 244, 245
Romanticism, 116, 253, 300, 306, 465
Rosen, Leonard, 130, 134
   All Cry Chaos, 130
Rosenstiehl, Pierre, 228, 238
Rosetus, Roger, 24
Rotman, Brian, 16, 354, 355, 546
Roubaud, Jacques, 217, 220, 221, 228, 231–233, 245

INDEX **621**

Royal Prussian Academy of Sciences, Berlin, 300
Royal Society, 13, 68
Rucker, Rudy, 141–143
  Mathematicians in Love, 141
"Ruelle-Takens domains", 130, 132
Russell, Bertrand, 7, 281, 282, 289, 322–324, 335, 340, 342, 345, 535, 537
  A Critical Exposition of the Philosophy of Leibniz, 322
  Principia Mathematica, 281, 537
Russell's paradox, 322–325, 458

S
Savile, Henry, 555, 559
Scale, 13, 52, 144, 196, 274, 445, 459, 462, 467, 559, 598, 599, 604, 605
Schelling, Friedrich, 307
Schlegel, Friedrich, 1, 306, 310
Scholastic, 24, 25, 30, 31, 33, 36, 330
Schubert, Hermann, 182
  Mathematical Essays and Recreations, 182
Science Wars, 15, 16
Scott, Greg
  Strange Attractors, 134
Scripture, Edward Wheeler, 199, 200
Self-evident truths, 100, 110
Series, 2, 63, 68, 75, 109, 115, 181, 195, 197, 200, 203, 211, 256, 273, 296, 308, 322, 426, 462, 514, 525, 592
Serres, Michel, 355, 526
Sestina, 218, 228, 233, 238
Set theory, 131, 158, 300, 340, 342, 344, 345, 370, 386, 389
Shakespeare, William, 10, 42, 57, 251, 394, 401, 403, 414, 525, 590
  Antony and Cleopatra, 525
  The Comedy of Errors, 246
  Hamlet, 250, 416
  Henry V, 416
  1 Henry IV, 414
  King Lear, 400, 401, 406, 414, 415
  The True Chronicle Historie of King Leir, 403
  The Winter's Tale, 416
Shapely sentences, 273, 276

Shapiro, Karl, 189, 192, 201–203
Shelley, Mary, 87, 457
Shelley, Percy Bysshe, 67
  Queen Mab, 67
Shepherd-Barr, Kirsten, 243, 244, 253
Sidney, Sir Philip, 190, 400
Sierpinski Gasket, 161
Sierpiński, Wacław Franciszek, 161
Sign rules, 406
Simson, Robert, 423–425
Sinclair, John, 462, 590
Singleton, Charles, 173, 174
Smith, Adam, 458
Snell, Ada, 199, 200
Snow, C.P., 149, 419
Snow and Leavis
  "two cultures", 480
Somerville, Mary, 72
Sonnets, 5, 57, 72, 218, 229, 371
Soule, Charles, 135
  Strange Attractors, 131, 134
Souriau, Etienne, 245
Spencer, Herbert, 100
Spiritual, 3, 190, 285–287, 295, 333, 432
Spottiswoode, William, 9, 477, 487–489, 497
  "utility" of pure mathematics, 477
Square root, 1, 406, 407, 416
Stan's Cafe, 247, 249
  Of All the People in All the World, 249
  Simple Maths, 249, 250
St. Aquinas, Thomas, 170
Stein, Gertrude, 7, 339–350, 352–355, 539, 540
  "Are There Arithmetics", 342
  The Autobiography of Alice B. Toklas, 341
  The Geographical History of America, 353
  "How Writing is Written", 341
  "Patriarchal Poetry", 352
  "Roastbeef", 348, 350
  Tender Buttons, 346, 351
  To Do; A Book of Alphabets and Birthdays, 343
Sterne, Laurence, 10, 439, 441, 443, 448, 450–452
  The Life and Opinions of Tristram Shandy, Gentleman, 439

622 INDEX

Stevens, Brett, 248, 249, 324
Stochastic literature, 208, 211
Stoppard, Tom, 10, 133, 137, 138, 140, 144, 146, 253, 254
    Arcadia, 10, 132, 137, 138, 140, 144, 146, 244, 253
Strode, Ralph, 23, 31, 35
Structuralist theory, 267
Stuttgart school, 208, 210, 216
Supplement, 11, 304, 391, 400–402, 406, 408, 413, 414, 605
Surd, 325, 326, 525
Surrealism, 13, 218, 227, 232
Surveys, 129, 314, 464, 465, 470
Swerve, 150, 153, 156, 157, 160, 164, 320
Swineshead, Richard, 32
Sylvester, James Joseph, 69, 473, 489, 497
    divisions, poetry; the Linguistic, 484; the Pneumatic, 484; the Rhythmic, 484
    Fliegende Blätter; Supplement to the Laws of Verse, 77
    "great law of Continuity", 483
    Laws of Verse, 77, 474
    "three principal branches"; Chromatic, 484; Metric, 484; Synectic, 484
    "To Rosalind", 78
Symbols, 2, 3, 66, 123, 155, 256, 290, 292, 353, 423, 428, 557, 573
Symmetry, 120, 163, 234, 257, 441, 442, 535, 536, 538
Systems novel, 158
Systems of knowledge, 164, 482

T
Tait, P.G., 69, 474
Takeuti, Gaisi, 570
Tapp, John, 47, 558
Tatham, Emma, 68
Temporality, 576–578, 581, 586
Tenney, James, 215
Tennyson, Alfred, 62
    "Locksley Hall", 62

Tense, 330, 346, 347, 522, 569, 574, 576, 578, 582, 585
Terminator, 269
Textbooks, 69, 83, 154, 201, 555, 556, 562, 563, 573, 584, 591
Textile industry, 500
Tf2, 244, 254, 255, 258
Theory of partitions, 484
Things As They Are, 345
Thomson, William, 69
    A Treatise on Natural Philosophy, 69
Time, 2, 5, 17, 29, 31, 42, 51, 56, 64, 73, 78, 93, 113, 116, 129, 134, 155, 178, 195, 233, 244, 253, 255, 269, 288, 306, 314, 346, 367, 461, 514, 516, 523, 543, 601
Todhunter, Isaac, 66, 75, 76
Topology, 142, 159, 160, 217, 228, 247, 364
Topos theory, 386, 389, 391, 392
Törless, 311, 312
Towsley, Gary, 173, 174
Transitivity, 535, 536, 538
Trees, 36, 100, 103, 355, 500
Trollope, Anthony, 581
Tubbs, Robert, 4, 17, 227, 283
Turing, Alan, 198, 255
Twentieth-century modernism, 7, 300
Tzara, Tristan, 213

U
Unconscious, 119, 210, 218, 219, 321, 335, 487
Undecidable proposition, 373, 376, 387, 388, 392
Universal language, 4, 114, 121, 446
Universe is spatially infinite, 301
University of Göttingen, 300
University of Leipzig, 307
University of Vienna, 312
Untaught geometer, 82, 84–86, 89, 92, 93
Useful knowledge, 460, 463, 468
Usury, 25, 45, 405

V

Valéry, Paul, 227, 283
Variance, 35, 462
Velleman, Daniel J., 572, 585
Venn, John, 110
Vernacular mathematics, 46
Versification, 76, 189, 190, 192, 194, 199, 484
Vibrations, 199, 444
Victorian earnestness, 82
Virgili, Polidoro, 190
Voltaire, 98, 334
Von Helmholtz, Hermann, 199
   "The Physiological Causes of Harmony in Music", 199
Von Humboldt, Alexander, 313, 315
Von Lindemann, Ferdinand, 181, 182
Vortex theory, 11
V-2 rocket, 3, 153, 154, 156, 160

W

Wallace, David Foster, 150, 161–164
   Everything and More: a Compact History of Infinity, 161
   Infinite Jest, 150, 160; in inversion, 160
Wallis, John, 179, 180, 555
Waugh, Patricia, 571
Weber, Max, 41, 43, 281, 285
Weil, André, 391
Wellek, René, 571
Wells, H.G., 3
Westminster Review, 100
Weyl, Hermann, 369
Whewell, William, 62, 68, 69, 72, 73, 75, 78, 85, 86, 93, 100, 420, 474
Whitehead, Alfred North, 281, 322, 339, 356, 525, 537
   Principia Mathematica, 281, 322, 537
Wickman, Matthew, 9, 423, 426, 457
Wiegand Brothers, Dometa, 65
Wiles, Andrew, 10, 573–575, 578, 580
Wilson, Edward O., 9, 420, 421
Wimsatt, W.K., 202, 203

"The Concept of Meter: An Exercise in Abstraction", 202
Wittgenstein, Ludwig, 14, 322, 323, 547
   Tractatus Logico-Philosophicus, 324
Wolff, Christian, 304
Woloch, Alex, 493, 502, 507
Woolf, Virginia, 273, 282, 514, 538
   Mrs Dalloway, 286, 514
   Night and Day, 287
Wordsworth, William, 4, 61, 63, 64, 66, 78, 427, 451, 457, 465, 466, 470, 478
   Arab's Dream, 65
   "Goody Blake and Harry Gill", 466
   "The Last of the Flock", 467
   Lyrical Ballads, 452, 457, 458; "Advertisement", 466
   The Prelude, 65, 478
   "The Thorn", 466, 467
   "We are Seven", 467
Writing and language, 5, 263
Wundt, Wilhelm, 199
Wyclif, John, 28, 34, 35

X

Xenakis, Iannis, 213

Y

Young, La Monte, 213, 214
   An Anthology of Chance Operations, 213

Z

Zamyatin, Evgenij, 113, 114, 117–119, 122, 123, 125, 285
Zamyatin, Yevgeny, 282
   We, 283
Zeno's paradox, 327
Zero, 120, 328, 401–405, 414, 415
Zola, Emile, 269
   "The Experimental Novel", 269
Zunshine, Lisa, 275, 441, 442

Printed in the United States
by Baker & Taylor Publisher Services